Microbial Electrochemical Technologies

Editors

Sonia M. Tiquia-Arashiro

Department of Natural Sciences, University of Michigan
Dearborn, MI, USA

Deepak Pant

Separation & Conversion Technology
Flemish Institute for Technological Research (VITO)
Boeretang, Mol, Belgium

CRC Press
Taylor & Francis Group
Boca Raton London New York

CRC Press is an imprint of the
Taylor & Francis Group, an **informa** business

A SCIENCE PUBLISHERS BOOK

Cover credit:
Top image: General scheme of different bioelectrochemical systems including anodic oxidation of wastewater coupled to oxygen reduction at cathode and microbial electrosynthesis at a biocathode coupled to a oxygen evolution anode.
Credits: Shiv Singh

Bottom image: Two identical microbial electrolysis cells (MEC) designed for pig slurry valorization. Each cell has a total volume of 16L (bicameral configuration)
Credits: Adrian Escapa

Left image: Plant-Microbial Fuel Cell installed in a rice paddy field in Indonesia
Credits: Emilius Sudirjo (Government of Landak Regency, West Kalimantan, Indonesia & Wageningen University & Research, The Netherlands).

CRC Press
Taylor & Francis Group
6000 Broken Sound Parkway NW, Suite 300
Boca Raton, FL 33487-2742

First issued in paperback 2021

© 2020 by Taylor & Francis Group, LLC
CRC Press is an imprint of Taylor & Francis Group, an Informa business

No claim to original U.S. Government works

Version Date: 20191003

ISBN-13: 978-1-138-59711-2 (hbk)
ISBN-13: 978-1-03-208208-0 (pbk)

Library of Congress Cataloging-in-Publication Data

Names: Tiquia-Arashiro, Sonia M., editor. | Pant, Deepak, editor.
Title: Microbial electrochemical technologies / editors, Sonia
   Tiquia-Arashiro, Deepak Pant.
Description: Boca Raton : CRC Press, [2020] | Includes bibliographical
   references and index.
Identifiers: LCCN 2019041192 | ISBN 9781138597112 (hardcover ; alk. paper)
Subjects: MESH: Electrochemical Techniques | Microbiological Techniques |
   Water Purification--methods | Biosensing Techniques | Electron
   Transport--physiology
Classification: LCC QD553 | NLM QD 553 | DDC 541/.37--dc23
LC record available at https://lccn.loc.gov/2019041192

Visit the Taylor & Francis Web site at
http://www.taylorandfrancis.com

and the CRC Press Web site at
http://www.crcpress.com

Preface

Microbial electrochemical technology (MET) is a rapidly growing environmental technology, bringing together the disciplines of microbiology, electrochemistry, materials science and engineering. For a long time, microbial electrochemistry has been the interest of mainly fundamental researchers. This has considerably changed during the last decade and this field has gained interest from applied researchers and engineers. These researchers have taken the microbial fuel cell (MFC) from a concept to technology. First prototypes (e.g. microbial fuel cells and microbial electrolysis cells) have been installed and commercialization seems possible. At the same time, the detailed understanding and knowledge-driven engineering of the MET components, especially their complex interplay, are differently elaborated. Since then, a plethora of derivative technologies such as microbial electrolysis cells, microbial desalination cells, photomicrobial fuel cells, microbial electrosynthesis, biocomputing and bioelectronics have been developed. The growing number of systems is often referred to in the literature as the bioelectrochemical system (BES), electrobiotechnology or microbial electrochemical technology. Today, research on METs is highly represented by a young and dynamic scientific society that organized itself among others like the International Society for Microbial Electrochemistry and Technology (ISMET). ISMET (https://www.is-met.org/) is pooling researchers from various areas of science and engineering, spanning from microbiology and electrochemistry to chemical engineering and material science.

Microorganisms are the key players in microbial electrochemical technologies. The technology basically relies on the metabolic activity of electroactive microorganisms for oxidizing or reducing certain compounds that could lead to the synthesis of chemicals, bioremediation of polluted matrices, the treatment of contaminants of interest as well as the recovery of energy. Keeping these possibilities in mind, there is a growing interest in the use of METs for wastewater treatment with simultaneous power generation, and the possibility of merging this technology with constructed wetlands for intensified wetland system with high-performance and lower footprint. Despite the increasing research efforts, not many processes and devices based on microbial electron transfer are yet available on the market. Regardless of their final application, this is mostly because of their low conversion efficiency, limited reliability and complex scalability. While our knowledge of model organisms like *Geobacter* and *Shewanella* provided details of the interactions between microorganisms and electrodes on the cellular as well as subcellular level, very little is known about electroactive microbiomes comprising biofilms as well as planktonic cells that are enriched from complex inocula (e.g. wastewater or sludge). Knowledge of the individual microbe-electrode interaction, the microbial interactions within the biofilm and the interactions between the biofilm and the surrounding planktonic cells are important to proactively steer microbiome-based METs. These are some of the main scientific and technical challenges addressed in this book.

Furthermore, the book puts together the most recent research and practical updates on microbial electrochemical technologies in all its dimensions. It provides a holistic view of METs detailing their functional mechanisms, operational configurations, influencing factors governing the reaction progress and integration strategies. The book will not only provide a historical perspective of the technology and its evolution over the years but also give the most recent examples of up-scaling and near future commercialization.

The book is organized into six parts. Part I underpins the fundamentals of microbial electrochemistry, including novel microbial pathways and electron transfer mechanisms. Otero et al. reported the current state of the characterization of intracellular components of extracellular electron transport pathways for *G. sulfurreducens* and *S. oneidensis* and the challenges in the detailed characterization of extracellular electron transport pathways. The work reported in this study expanded our understanding of electron transfer between outer membrane cytochromes and their soluble and solid redox partners, suggesting that the production of extracellular electron shuttles by this microorganism is crucial to sustaining enhanced electron transfer rates. Shrestha et al. addressed the recent knowledge gaps related to the use of electrochemical impedance spectroscopy (EIS). The readers will learn to use the EIS data for understanding the extracellular electron transfer mechanisms and resistance and capacitive effects of both the biofilms and electrical double layer.

Part II focuses on optimizing microbial electrochemical systems. Lai and Krömer demonstrated an anode-driven process with *Pseudomonas putida*. *P. putida* is the first obligate aerobe confirmed to be capable of using an electrode as electron sink via a mediator-based route. The chapter summarizes the research efforts in the past few years regarding this topic and addresses the benefits and the remaining bottlenecks. *P. putida* was able to perform anode-dependent anoxic glucose catabolism and produce a high-yield carbon product. This work summarizes our current knowledge on electron transport processes and uses a theoretical approach to predict the impact of different modes of transfer on energy metabolism. It adds an important piece of fundamental understanding of microbial electron transport possibilities to the research community and will help to optimize and advance microbial electrochemical techniques.

Parts III and IV explore the applications of microbial electrochemical technologies in wastewater treatment, bioenergy, biosensors, electrosynthesis and bioremediation. The articles demonstrated high diversity and versatility of the technology, which could be a promising sustainable solution incorporating renewable energy, waste treatment and remediation. METs are increasingly being recognized as highly flexible and versatile systems, offering several opportunities in the wastewater treatment sector. Some of these are highlighted in this section of the book, which describes bioelectrochemical system configuration capable of simultaneously oxidizing organic matter, removing and recovering of ammonium-nitrogen and reducing CO_2 to methane, a valuable gaseous biofuel. Bioremediation has emerged as another important application of METs, with a remarkable potential for exploitation. The use of METs for bioremediation purposes is herein addressed in three chapters.

Part V reports the materials for METs. Tremlay et al. described how graphene has been used in recent years in the fabrication of high-performance bioelectrochemical system electrodes. In this chapter, readers will learn that graphene alone or as part of composite electrodes microbial electrochemical systems or even as biosensor systems performed much better compared to similar reactors equipped with electrodes made of other metallic or carbonaceous materials. In certain cases, graphene-based electrodes had comparable performance to Pt/C electrodes, which are significantly more expensive.

Finally, Part VI is devoted to scale-up issues which represent one of the most challenging aspects of METs aiming at energy saving and/or recovery during wastewater treatment. Alonso et al. analyzed some of these difficulties and discuss the ways researchers have dealt with them in their endeavor to develop technically and commercially feasible METs for energy valorization of waste streams. The number of scale-up developments aimed at exploring this potentiality has proliferated significantly during the past decade, giving the impression that commercial application might be within reach. The recent advances in the scale of ~100 L allow for some optimism in this regard. MFCs operating on real wastewater have proved to be capable of producing enough energy to power ancillary equipment (pumps, control systems, etc.) and MECs can produce enough gas fuel (either hydrogen or methane) to produce a positive energy balance. Researchers and engineers have come up with a plethora of novel reactor architectures and flow configurations, always with the aim of minimizing internal losses (overpotentials) and developing technically and economically competitive designs. Other relevant practical issues such as obtaining usable power or developing suitable control and optimization strategies have also been addressed, all of which is helping to pave the way toward market development.

With these key features, we hope that the book will provide an added edge to the current state-of-the-art and generate interest among a wide range of people active in the field of renewable energy generation and sustainable environmental research.

Sonia M. Tiquia-Arashiro
Deepak Pant

Contents

PART-IV: Applications of Microbial Electrochemical Systems in Bioremediation

PART-V: Materials for Microbial Electrochemical Technologies

PART-VI: Design of Microbial Electrochemical Systems: Toward Scale-up, Modeling and Optimization

Part-I

Fundamentals of Microbial Electrochemistry: Novel Microbial Pathways and Electron Transfer Mechanisms

Extracellular Electron Transport in *Geobacter* and *Shewanella*: A Comparative Description

Fernanda Jiménez Otero[1*], Matthew D. Yates[2] and Leonard M. Tender[2]

[1] George Mason University, Fairfax, Virginia 22030, USA
[2] Center for Bio/Molecular Science and Engineering, Naval Research Laboratory, Washington, DC, USA

1. Introduction

Extracellular electron transport (EET) is the process by which certain microorganisms are able to transport electrons to or from insoluble electron acceptors, or donors as part of their metabolism. Two of the most studied EET-capable organisms are *Shewanella oneidensis* and *Geobacter sulfurreducens*. These cells are able to oxidize organics intracellularly, deriving energy for growth and cell maintenance, and transport the liberated electrons via electron transport chains across the cell envelope to an insoluble extracellular terminal electron acceptor, such as an electrode maintained at a sufficiently oxidizing potential (Bond et al. 2002; Marsili et al. 2008). *S. oneidensis* was isolated in the 1980s from the sediment of Lake Oneida, NY, by Ken Nealson's group when they were studying manganese oxide reduction (Myers and Nealson 1988; Venkateswaran et al. 1999). *G. sulfurreducens* was similarly isolated from the sediment of a hydrocarbon-contaminated ditch in Norman, OK, by Derek Lovley and his coworkers in 1994 (Caccavo et al. 1994). These two studies laid the groundwork for a new paradigm in microbiology: the extracellular reduction of insoluble metal oxide terminal electron acceptors by cells that occupy environmental niches, where soluble electron acceptors, such as oxygen, nitrate and sulfate, are limited but insoluble metals, such as iron, are abundant. Although the electron transport chains that *Shewanella* and *Geobacter* species utilize to move electrons from the inner membrane to the extracellular space share common features (i.e., both involve membrane-associated multi-heme cytochromes) (Figure 1), the specifics of their electron transport pathways are remarkably different. The focus of this chapter is to present an up-to-date description of the biochemical pathways that allow *S. oneidensis* and *G. sulfurreducens* to respire extracellular substrates.

2. *Shewanella* and *Geobacter*

To begin the discussion of the differences between the EET pathways of species in the *Geobacter* and *Shewanella* genera, it is useful to first point out some of the differences in their physiology. While both genera include several species known to reduce insoluble metal oxides, *Geobacter* species are members of the δ-proteobacteria whereas *Shewanella* species belong to the γ-proteobacteria. *Shewanella* species are facultative anaerobes that thrive in oxic environments but can use a wide variety of soluble and/or

*Corresponding author: Fernanda.jimenezotero.ctr.mx@nrl.navy.mil

extracellular terminal electron acceptors when oxygen is depleted, while organisms of the *Geobacteraceae* are anaerobes with varying degrees of oxygen sensitivity and thrive in metal-rich environments.

As most environmental microbes, *S. oneidensis* and *G. sulfurreducens* are biofilm-forming organisms. Both organisms can form biofilms on glass and polystyrene surfaces (Lies et al. 2005; Rollefson et al. 2011) but interact only transiently with iron oxide particles (as found in sediments and soils) without forming robust biofilms (Harris et al. 2010; Levar et al. 2012; Chan et al. 2017). This effect is due to both the toxicity of Fe(II)—produced during Fe(III)-reduction to cells—and the limited capacity for a given particle to accept electrons before characteristics of the metal, such as reduction potential and surface area, change and its capacity to serve as terminal electron acceptor is exhausted. However, when these organisms are grown on a solid surface that is able to mimic the redox potential of natural terminal electron acceptors but as an unchanging and unlimited electron acceptor (i.e., an anode – an electrode maintained at a sufficiently oxidizing potential), both organisms form robust biofilms on the electrode surface. On a poised electrode, *G. sulfurreducens* forms multi-cell thick biofilms that can routinely exceed 100 μm in thickness (Bond et al. 2002; Bond and Lovley 2003; Torres et al. 2008; Rollefson et al. 2011; Snider et al. 2012; Bonanni et al. 2013; Stephen et al. 2014; Tejedor Sanz et al. 2018; Yates et al. 2018c) and have even been shown to reach ~1 mm thickness when grown under certain conditions (Renslow et al. 2013; Li et al. 2016). This ability of *G. sulfurreducens* to form thick biofilms makes their EET capabilities all the more remarkable, since they not only have to employ a strategy to move electrons from the inner membrane to the extracellular space (referred to a membrane-associated EET) across the cell/electrode interface (referred to as heterogeneous EET following the electrochemical nomenclature in which one reactant is an electron transport reaction in an electrode) (Richter et al. 2009; Strycharz et al. 2011; Snider et al. 2012; Strycharz-Glaven et al. 2014; Yates et al. 2015; Phan et al. 2016; Yates et al. 2016; Zhang et al. 2017; Yates et al. 2018c; Yates et al. 2018b; Yates et al. 2018a), but the cells must also then have a strategy to form an electron transport network in the biofilm itself to carry electrons over multiple cell lengths (referred to as multi-cell-length EET or long-distance EET) from cells, not in direct contact with the electrode surface (Yates et al. 2018c). *S. oneidensis*, in contrast, only forms very thin biofilms, often a monolayer or submonolayer of cells (Baron et al. 2009; Coursolle et al. 2010), on electrode surfaces that, nonetheless, are capable of long-distance EET laterally across glass surfaces between electrodes (Xu et al. 2018) as also observed for *G. sulfurreducens* (Snider et al. 2012). The reasons behind these differences are unknown and raise interesting questions about how each organism has adapted to perform vertical and lateral long-distance EET.

Of great importance is the motility of these model organisms. *G. sulfurreducens* PCA (ATCC 51573) is used in pure culture studies and is a non-motile variant with a transposon insertion repressing flagellar transcriptional regulation (Caccavo et al. 1994), although other closely related species (e.g., *Geobacter metallireducens*) have retained motility (Lovley et al. 1993). *S. oneidensis* MR-1, often used in *Shewanella*-focused studies, has its motility intact and demonstrates a swarming motility phenotype in the presence of insoluble iron oxide (Harris et al. 2010). We point this out because, while flagella are involved in early biofilm establishment, they can also affect mature biofilm morphology. In other model organisms such as *Pseudomonas*, mutants lacking flagella outcompete wild type within biofilms (Klausen et al. 2003) and non-flagellated isolates are often found in biofilms of chronically infected patients (Luzar et al. 1985; Mahenthiralingam et al. 1994). It is possible therefore that non-motile *G. sulfurreducens* cells form advantageous non-dynamic biofilms, with cells rarely detaching.

3. Long-Distance Extracellular Electron Transport

As noted above, *G. sulfurreducens* has adapted to be able to transport electrons over multiple cell lengths to support thick biofilms that are able to respire a surface for their metabolic benefit. Different components of the extracellular matrix have been proposed to be important in electron transport through the biofilm between the cells and the surface. Much literature favors redox conductivity as the prevailing mechanism of long-distance EET through anode-grown *G. sulfurreducens* biofilms, in which electron self-exchange reactions are proposed to occur among multi-heme *c*-type cytochromes on the outer membrane of cells or in the extracellular matrix or both (Richter et al. 2009; Strycharz-Glaven et al. 2011; Bond et al. 2012; Snider et al. 2012; Robuschi et al. 2013; Lebedev et al. 2014a; Lebedev et al. 2014b; Boyd et al. 2015). Evidence for the specific role of *c*-type cytochromes acting as the charge carriers in the biofilm is primarily provided by the spectral signatures of anode-grown *G. sulfurreducens* biofilms (Liu et al. 2011; Virdis et al. 2012; Liu and Bond 2012; Robuschi et al. 2013; Ly et al. 2013; Lebedev et al. 2014b; Schmidt et al. 2017; Golden et al. 2018) and the abundance of homologous protein sequences encoded in the *G. sulfurreducens* genome (Caccavo et al. 1994). Other literature, however, favors the proposition that long-distance extracellular electron transport occurs along pili proposed to possess metallic-like electron transport abilities due to the proposed alignment of aromatic residues in a manner analogous to that of organic semiconductors (Vargas et al. 2013; Malvankar et al. 2011; Malvankar et al. 2015). A complicating factor in distinguishing the roles of cytochromes and pili in long-distance EET is the roles that pili play in cell attachment and biofilm cohesion, both of which may indirectly affect long-distance EET and the role of pili play in localization of cytochromes that may directly affect long-distance EET (Richter et al. 2012; Richter et al. 2017; Steidl et al. 2016). It is important to note that as this is being written, a number of presentations have been recently made by a key proponent of the aromatic amino acid pili model in which the filaments that were thought to be pili may, in fact, be linearly polymerized outer membrane cytochrome S (OmcS) (Yalcin et al. 2018; Malvankar et al. 2018). Moreover, supramolecular assemblies of proteins have been shown to exhibit conduction attributes similar to organic semiconductors; although metallic conductivity was not invoked and the authors did not definitively provide a mechanism for electron transport (Ing et al. 2017). It is possible that other unidentified components of the extracellular matrix are involved in long-distance EET. Despite *G. sulfurreducens* being one of the most well-characterized electroactive organisms, there are still significant gaps in our understanding of long-distance electron transport through *G. sulfurreducens* biofilms. The major outstanding question is that we still do not know how the extracellular electron transport mediators are spatially organized in the biofilm.

S. oneidensis is also capable of long-distance electron transport, but uses multiple strategies to transport charge over multiple cell lengths. One such mechanism that *S. oneidensis* cells use is a production of soluble flavin redox mediators, to use them as physically diffusing extracellular electron carriers between cells and terminal electron acceptors at a distance (Marsili et al. 2008; Coursolle et al. 2010). Soluble redox mediators as means to dispose of electrons extracellularly are emerging as a widely distributed strategy in both organisms with Gram-negative, i.e., *Pseudomonas aeruginosa* (Glasser et al. 2017) and Gram-positive physiology, i.e., *Listeria monocytogenes* (Light et al. 2018). In *S. oneidensis* this strategy for long-distance EET was first observed when voltammetry performed on the cell-free supernatant of a reactor that was used to grow *S. oneidensis* (Marsili et al. 2008). The voltammetry of the filtered medium exhibited a signal comparable to what was observed before the cells were removed that indicated something in the medium was responsible for the observed current. It was later demonstrated that the deletion of the *bfe* gene encoding a flavin exporter resulted in significantly less current production in *S. oneidensis* (Kotloski and Gralnick 2013). A cytochrome-bound role has also been proposed for flavins, accelerating extracellular electron transport between *c*-type cytochromes and an electrode (Xu et al. 2016). In addition to producing soluble redox shuttles for long-distance extracellular electron transport, *S. oneidensis* is able to produce outer membrane extensions under anoxic conditions that seem capable of long-distance EET via redox conduction involving outer membrane cytochromes (El-Naggar et al. 2010; Pirbadian et al. 2014; Xu et al. 2018).

4. Extracellular Electron Transport Pathways

Microbial metal reduction presents a rich source of possibilities for biotechnological applications, as it represents a mechanism for transport of electrons stored within nonreactive organic compounds to inorganic redox-active compounds. However, two main characteristics make metal respiration distinct from soluble-substrate respiration strategies that have been successfully exploited for biotechnological applications. The first hurdle of metal oxide respiration is that metal oxides exist as large and insoluble particles in most soil and sediment environments, making them extracellular substrates (they cannot enter cells). Both *S. oneidensis* and *G. sulfurreducens* cells have Gram-negative type physiology that means that electrons from intracellular oxidation must cross the two membranes and a periplasmic space before reaching an extracellular terminal electron acceptor. The second challenge for the cells using metal oxides as respiration substrates is that they exist in soils and sediments as very diverse crystalline metal species and their characteristics constantly change (Majzlan 2013). A cellular strategy is, therefore, necessary to be able to respire metal oxides, regardless of the shifts in their surface or redox characteristics. Particular solutions to these two challenges have evolved in the EET pathways of both *Shewanella* and *Geobacter* genera that are tailored to their specific lifestyles.

5. *Shewanella oneidensis*

As mentioned above, the *Shewanella* species have a remarkable metabolic versatility and are able to use a diverse soluble and insoluble terminal electron acceptors including oxygen, fumarate, succinate, arsenate, selenite, glycine, iodate, nitrate, nitrite, sulfite, thiosulfate, elemental sulfur, dimethyl sulfoxide (DMSO), trimethylamine-N-oxide (TMAO), Fe(III), Mn(IV), Cr(VI), U(VI), among others (Nealson and Scott 2006). *Shewanella* species have been isolated from environments that routinely fluctuate between oxic and anoxic, such as fish guts, aquatic environments and sediments (Hau and Gralnick 2007). Therefore, the extracellular respiration of metals and other soluble but toxic compounds, such as DMSO are only some of the strategies *S. oneidensis* can use. To enable this versatility *Shewanellae* have evolved a 'centralized' extracellular electron transport pathway (see Figure 1) with a single inner membrane pathway for the electron transport between the menaquinone pool and the periplasm: the tetraheme cytochrome Cym A (Myers and Myers 1997; Myers and Myers 2000; Schwalb et al. 2003; Myers et al. 2004; Li et al. 2014) that is a NapC/NirT homolog. Once electrons enter the periplasm, this cellular compartment is large enough, ~25 nm, (Asmar et al. 2017) that the electron transport between inner and outer membrane-anchored proteins is necessarily mediated by additional soluble periplasmic cytochromes. *S. oneidensis* has two such periplasmic cytochromes that have been shown to interact with outer membrane cytochromes and assist electron transport across the periplasm (Alves et al. 2015): FccA (Schuetz et al. 2009) and CctA (Coursolle and Gralnick 2010), both of which are highly abundant.

The last cellular barrier of *S. oneidensis* that electrons must cross is the outer membrane, and it is there that the extracellular electron transport pathway of *S. oneidensis* diverges in a substrate-dependent manner. The model of outer membrane electron conduits was, in fact, first described in this organism (Hartshorne et al. 2009). These are outer membrane complexes consisting of an integral outer membrane protein with multiple transmembrane domains that aligns two *c*-type cytochromes from either side of the membrane close enough to facilitate electron transport from one to another (Clarke et al. 2011; Richardson et al. 2012). In this way, electrons can come in from the periplasm and be transported across the outer membrane into the extracellular space. *S. oneidensis* cells use the electron conduit MtrCAB consisting of the integral outer membrane protein MtrB and the periplasmic and extracellular *c*-type cytochromes MtrA and MtrC correspondingly. This outer membrane complex is involved in electron transport through the outer membrane to Fe(III)-oxide, Fe(III) citrate, Mn(IV)-oxide and electrode reduction (Bretschger et al. 2007; Coursolle et al. 2010). There is a homolog of MtrCAB encoded in the *S. oneidensis* genome, MtrDEF, for which a function has not been found yet, but components of this complex can substitute their analogous partner in MtrCAB (Coursolle and Gralnick 2010). During DMSO reduction, *S. oneidensis* cells use the DsmEFA electron conduit that contains homologs to the periplasmic and anchor proteins MtrAB (DmsEF) but a DMSO reductase (DmsA) as an extracellular component (Gralnick et al. 2006).

Beyond the cell, the distance becomes much more significant variable for cells using an extracellular electron acceptor. *S. oneidensis* solves this problem by secreting flavins into the extracellular space. These small, redox-active riboflavin derivatives serve as the mediators between outer membrane cytochromes and long-distance terminal electron acceptors (Marsili et al. 2008; Coursolle et al. 2010) through physical

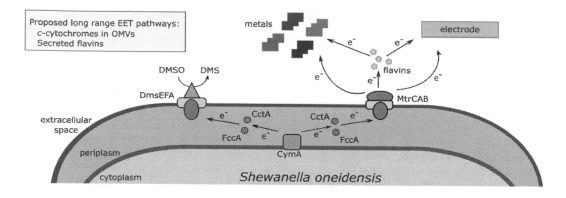

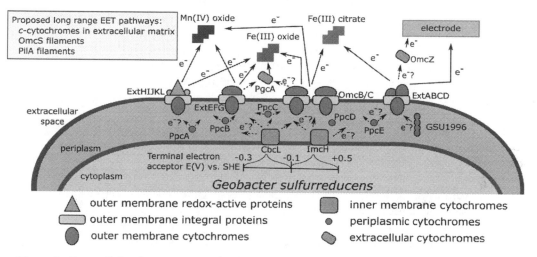

Figure 1: Extracellular electron transport pathways of *S. oneidensis* and *G. sulfurreducens*. Schematic representation of known (solid arrows) biochemical pathways for transferring electrons into the extracellular space as well as those that are yet to be determined (dashed arrows).

Abbreviations: EET- Extracellular Electron Transport, EM- Extracellular Matrix, OMVs- Outer Membrane Vesicles, OmcB/C refers to OmbB-OmaB-OmcB and OmbC-OmaC-OmcC.

diffusion. It has also been proposed that flavins bound to cytochromes can accelerate the rate of electron transport between cytochromes and extracellular electron acceptors, such as electrodes (Xu et al. 2016).

As a final comment on *S. oneidensis* metabolism, it is worth noting that the respiration of extracellular substrates is not the preferred respiration strategy for this organism. This survival strategy gives *S. oneidensis* cells a competitive advantage when oxygen is low but is only expressed under specific conditions (Saffarini et al. 2003; Sundararajan et al. 2011). Biofilm formation is also difficult to obtain under completely anaerobic conditions, with high regulation of this respiration strategy possibly preventing commitment to a surface while conditions are not the preferred aerobic alternative. *S. oneidensis* is, however, as simple as any aerobe to work with in the laboratory, making it an ideal model organism for the study of complicated processes such as extracellular electron transport.

6. *Geobacter sulfurreducens*

The extracellular electron transport pathway of *G. sulfurreducens* cells is characterized by its intricacy (Figure 1). Every step from the inner membrane to the extracellular space contains more than two alternative pathways none of which are homologous to each other (with the exception being a plethora of homologous periplasmic *c*-type cytochromes). *Geobacter* species tend to dominate electrode-enriched communities (Bond et al. 2002; Tender et al. 2002; Holmes et al. 2004; Tejedor Sanz et al. 2018; Yates et al. 2018a) and have been isolated from sediments across the biosphere (Lovley et al. 1993; Caccavo et al. 1994; Straub and Buchholz-Cleven 2001; Nevin et al. 2005; Sung et al. 2006; Shelobolina et al. 2007; Shelobolina et al. 2008; Zhou et al. 2014) that indicates this complex extracellular electron transport pathway is advantageous in their environment. As mentioned above, rather than using the electrodes, metal oxides are the natural substrate for extracellular respiration. In the case of the model organism *G. sulfurreducens*, the complexity of the respiration pathway matches that of the substrate.

At the inner membrane, yield-differing pathways are used in a redox potential-dependent manner (Levar et al. 2017). The tetraheme *c*-cytochrome, ImcH, from the NapC/NirT family is used to transport electrons from the quinone pool to the periplasm when the extracellular substrate has a reduction potential of -0.1 V vs. the standard hydrogen electrode (SHE) or higher (Levar et al. 2014). Alternatively, when the extracellular substrate has a redox potential below -0.1 V vs. SHE, the *c*- and *b*-type cytochrome CbcL is used instead (Zacharoff et al. 2016). The burden of constitutively expressing two inner membrane electron transport pathways is alleviated by the different growth rates and the yields supported by each with cells that can only use ImcH (ΔcbcL) displaying a higher yield (Levar et al. 2017) and a faster growth rate (Zacharoff et al. 2016) than a wild type.

The electron transport step across the periplasm in *G. sulfurreducens* is far from resolved. There are five homologous periplasmic triheme *c*-type cytochromes (PpcA-E) and several more with the Ppc triheme domain repeated in tandem, producing cytochromes resembling four or more Ppc's aligned in a row (Pokkuluri et al. 2011; Alves et al. 2016). This high level of redundancy results in difficulty identifying the individual roles for each periplasmic cytochrome. The transcript levels are highest for *ppcA*, and a Δ*ppcA* strain is significantly impaired in the reduction of extracellular substrates (Lloyd et al. 2003; Ueki et al. 2017), but it is so far unclear as to why four more homologs as well as augmented-copies of Ppc's are conserved. These are, however, due to their small size and solubility, the only cytochromes from *G. sulfurreducens* that have a solved crystal structure, which has enabled detailed thermodynamic characterization of each periplasmic cytochrome (Dantas et al. 2015) and *in vitro* interactions among them (Fernandes et al. 2017).

All electron transport pathways across the outer membrane of Gram-negative organisms known to date follow the electron conduit model. In *G. sulfurreducens*, however, there are at least five different gene clusters predicted to encode outer membrane electron conduits. In contrast to the electron transport across the inner membrane, the outer membrane electron transport pathways are substrate-dependent in a redox-independent manner (Jiménez Otero et al. 2018). The OmcB-OmcC electron conduit duplication is involved in Fe(III) reduction regardless of the redox potential or solubility of the Fe(III). Mn(IV)-oxide reduction is not impaired by the absence of any one of the electron conduits on its own but removing all five eliminates Mn(IV) reduction. Electrode reduction, on the other hand, decreases by 80-90% only when the electron conduit ExtABCD is absent regardless of the redox potential of the poised electrode. This substrate dependency, at the last cellular electron transport step, is concurrent with the location of these proteins where they may interact with either the extracellular proteins that then interact with the terminal electron acceptor or directly with the terminal electron acceptor itself.

In the extracellular space, *G. sulfurreducens* cells surround themselves in a conductive extracellular matrix. This matrix is composed of secreted cytochromes (Mehta et al. 2005; Nevin et al. 2009; Leang et al. 2010; Inoue et al. 2011; Qian et al. 2011; Peng and Zhang 2017; Zacharoff et al. 2017), polysaccharides (Rollefson et al. 2011) and pili (Reguera et al. 2005; Reguera et al. 2007; Cologgi et al. 2011; Adhikari et al. 2016; Liu et al. 2018). All of these components are thought to be important for the cell attachment and/ or the long-range electron transport, as mentioned above, but the degree to which each contributes to the latter is still highly debated. Of the cytochromes secreted by *G. sulfurreducens*, evidence shows substrate specificity for PgcA and OmcZ during Fe(III) and electrode reduction, respectively (Zacharoff et al. 2017; Peng and Zhang 2017; Inoue et al. 2011). Moreover, the outer membrane cytochromes may contribute to the long-distance EET by lateral EET across cell surfaces as proposed to occur for *S. oneidensis* outer membrane-periplasmic extensions (El-Naggar et al. 2010; Pirbadian et al. 2014).

7. Conclusions and Future Perspectives

Here, we present a snapshot in time of the molecular-level understanding of EET pathways utilized by *S. oneidensis* and *G. sulfurreducens*. The extent to which the specifics endure is, of course, unknown as new discoveries will most certainly enhance understanding of some details while completely displacing others. Yet, the recent advances describing extracellular respiration pathways in detail have allowed the field of extracellular respiration to accomplish one of the hallmark experiments that have been possible with other respiration strategies for decades, i.e., to transport the ability to respire a substrate to a heterologous organism. The Ajo-Franklin group has shown that it is possible to express functional MtrCAB complexes in *Escherichia coli*, allowing this heterologous host to produce anodic current and reduce Fe(III) oxides (Jensen et al. 2010; TerAvest et al. 2014; Jensen et al. 2016). As more pathways for extracellular electron transport are discovered (Light et al. 2018), including those that can uptake electrons into the cell (Jiao and Newman 2007; Bose et al. 2014), the possibilities for the applications of the respiration of extracellular substrates to biotechnology are significant.

While most heterologous host expression and synthetic biology efforts currently use the CymA/Ccta/MtrCAB pathway of *S. oneidensis*, further characterization of the different EET pathways in *G. sulfurreducens* will enable substrate specific pathway engineering. Defining which, if any, specific periplasmic partner interacts with ImcH and CbcL as well as which outer membrane electron conduit can take electrons from each periplasmic component are the last steps needed to outline the complete intracellular component of EET pathways (Figure 1). On the extracellular space, however, it is still unclear how the electrons reached the secreted cytochromes PgcA and OmcZ each of which is only important during Fe(III) oxide and electrode reduction, respectively. Additionally, both OmcS polymers and pili have been proposed as the long-range EET pathways but their physiological role is still debated. Creating substrate-tailored EET pathways would enable the biological systems primed for the metal reduction of single substrates and the reduction of anodes only at desired redox potential that is not possible with present setups.

References

Adhikari, RY, Malvankar, NS, Tuominen, MT, Lovley, DR (2016) Conductivity of individual *Geobacter* pili. RSC Adv 6: 8354–8357.

Alves, MN, Fernandes, AP, Salgueiro, CA, Paquete, CM (2016) Unraveling the electron transport processes of a nanowire protein from *Geobacter sulfurreducens*. Biochim Biophys Acta BBA - Bioenerg 1857: 7–13.

Alves, MN, Neto, SE, Alves, AS, Fonseca, BM, Carrêlo, A, Pacheco, I, Paquete, CM, Soares, CM, Louro, RO (2015) Characterization of the periplasmic redox network that sustains the versatile anaerobic metabolism of *Shewanella oneidensis* MR-1. Front Microbiol 6: 665.

Asmar, AT, Ferreira, JL, Cohen, EJ, Cho, S-H, Beeby, M, Hughes, KT, Collet, JF (2017) Communication across the bacterial cell envelope depends on the size of the periplasm. PLoS Biol 15: e2004303.

Baron, D, LaBelle, E, Coursolle, D, Gralnick, JA, Bond, DR (2009) Electrochemical measurement of electron transport kinetics by *Shewanella oneidensis* MR-1. J Biol Chem 284: 28865–28873.

Bonanni, PS, Bradley, DF, Schrott, GD, Busalmen, JP (2013) Limitations for current production in *Geobacter sulfurreducens* biofilms. ChemSusChem 6: 711–720.

Bond, DR, Holmes, DE, Tender, LM, Lovley, DR (2002) Electrode-reducing microorganisms that harvest energy from marine sediments. Science 295: 483–485.

Bond, DR, Lovley, DR (2003) Electricity production by *Geobacter sulfurreducens* attached to electrodes. Appl Environ Microbiol 69: 1548–1555.

Bond, DR, Strycharz-Glaven, SM, Tender, LM, Torres, CI (2012) On electron transport through *Geobacter* biofilms. ChemSusChem 5: 1099–1105.

Bose, A, Gardel, EJ, Vidoudez, C, Parra, EA, Girguis, PR (2014) Electron uptake by iron-oxidizing phototrophic bacteria. Nat Commun 5: 3391.

Boyd, DA, Snider, RM, Erickson, JS, Roy, JN, Strycharz-Glaven, SM, Tender, LM (2015) Theory of redox conduction and the measurement of electron transport rates through electrochemically active biofilms. In Biofilms in Bioelectrochemical Systems. John Wiley & Sons, Ltd, pp. 177–210.

Bretschger, O, Obraztsova, A, Sturm, CA, Chang, IS, Gorby, YA, Reed, SB, Culley, DE, Rearson, CL, Barua, S. Romine, MF, Zhou, J, Beliaev, AS, Bouhenni, R, Saffarini, D, Mansfeld, F, Kim B-H, Fredrickson, JK, Nealson, KH (2007) Current production and metal oxide reduction by *Shewanella oneidensis* MR-1 wild type and mutants. Appl Environ Microbiol 73: 7003–7012.

Caccavo, F, Lonergan, DJ, Lovley, DR, Davis, M, Stolz, JF, McInerney, MJ (1994) *Geobacter sulfurreducens* sp. nov., hydrogen-and acetate-oxidizing dissimilatory metal-reducing microorganism. Appl Environ Microbiol 60: 3752–3759.

Chan, CH, Levar, CE, Jiménez Otero, F, Bond, DR (2017) Genome-scale mutational analysis of *Geobacter sulfurreducens* reveals distinct molecular mechanisms for respiration and sensing of poised electrodes vs. Fe(III) oxides. J Bacteriol 119: e00340-17.

Clarke, TA, Edwards, MJ, Gates, AJ, Hall, A, White, GF, Bradley, J, Reardon, CL, Shi, L, Beliaev, AS, Marshall, MJ, Wang, Z, Watmough, NJ, Fredrickson, JK, Zachara, JM, Butt, JN, Richardson, DJ (2011) Structure of a bacterial cell surface decaheme electron conduit. Proc Natl Acad Sci 108: 9384–9389.

Cologgi, DL, Lampa-Pastirk, S, Speers, AM, Kelly, SD, Reguera, G (2011) Extracellular reduction of uranium via Geobacter conductive pili as a protective cellular mechanism. Proc Natl Acad Sci 108: 15248–15252.

Coursolle, D, Baron, DB, Bond, DR, Gralnick, JA (2010) The Mtr respiratory pathway is essential for reducing flavins and electrodes in *Shewanella oneidensis*. J Bacteriol 192: 467–474.

Coursolle, D, Gralnick, JA (2010) Modularity of the Mtr respiratory pathway of *Shewanella oneidensis* strain MR-1. Mol Microbiol 77: 995–1008.

Dantas, JM, Morgado, L, Aklujkar, M, Bruix, M, Londer, YY, Schiffer, M, Pokkuluri, PR, Salgueiro, CA (2015) Rational engineering of *Geobacter sulfurreducens* electron transport components: a foundation for building improved *Geobacter*-based bioelectrochemical technologies. Front Microbiol 6: 752.

El-Naggar, MY, Wanger, G, Leung, KM, Yuzvinsky, TD, Southam, G, Yang, J, Lau, WM, Nealson, KH, Gorby, YA (2010) Electrical transport along bacterial nanowires from Shewanella oneidensis MR-1. Proc Natl Acad Sci 107: 18127–18131.

Fernandes, AP, Nunes, TC, Paquete, CM, Salgueiro, CA (2017) Interactions studies between periplasmic cytochromes provide insights on extracellular electron transport pathways of *Geobacter sulfurreducens*. Biochem J 474: 797-808.

Glasser, NR, Saunders, SH, Newman, DK (2017) The colorful world of extracellular electron shuttles. Annu Rev Microbiol 71: 731-751.

Golden, J, Yates, MD, Halsted, M, Tender, L (2018) Application of electrochemical surface plasmon resonance (ESPR) to the study of electroactive microbial biofilms. Phys Chem Chem Phys *PCCP* 20: 25648–25656.

Gralnick, JA, Vali, H, Lies, DP, Newman, DK (2006) Extracellular respiration of dimethyl sulfoxide by *Shewanella oneidensis* strain MR-1. Proc Natl Acad Sci 103: 4669–4674.

Harris, HW, El-Naggar, MY, Bretschger, O, Ward, MJ, Romine, MF, Obraztsova, AY, Nealson, KH (2010) Electrokinesis is a microbial behavior that requires extracellular electron transport. Proc Natl Acad Sci 107: 326–331.

Hartshorne, RS, Reardon, CL, Ross, D, Nuester, J, Clarke, TA, Gates, AJ, Mills, PC, Zachara, JM, Shi, L, Beliaev, AS, Marshall, MJ, Tien, M, Brantley, S, Butt, JN, Richarson, DJ (2009) Characterization of an electron conduit between bacteria and the extracellular environment. Proc Natl Acad Sci 106: 22169–22174.

Hau, HH, Gralnick, JA (2007) Ecology and biotechnology of the genus *Shewanella*. Annu Rev Microbiol 61: 237–258.

Holmes, DE, Bond, DR, O'Neil, RA, Reimers, CE, Tender, LR, Lovley, DR (2004) Microbial communities associated with electrodes harvesting electricity from a variety of aquatic sediments. Microb Ecol 48: 178–190.

Ing, NL, Nusca, TD, Hochbaum, AI (2017) *Geobacter sulfurreducens* pili support ohmic electronic conduction in aqueous solution. Phys Chem Chem Phys PCCP 19: 21791–21799.

Inoue, K, Leang, C, Franks, AE, Woodard, TL, Nevin, KP, Lovley, DR (2011) Specific localization of the *c*-type cytochrome OmcZ at the anode surface in current-producing biofilms of *Geobacter sulfurreducens*. Environ Microbiol Rep 3: 211–217.

Jensen, HM, Albers, AE, Malley, KR, Londer, YY, Cohen, BE, Helms, BA, Weigele, P, Groves, JT, Ajo-Franklin, CM (2010) Engineering of a synthetic electron conduit in living cells. Proc Natl Acad Sci 107: 19213–19218.

Jensen, HM, TerAvest, MA, Kokish, MG, Ajo-Franklin, CM (2016) CymA and exogenous flavins improve extracellular electron transport and couple it to cell growth in Mtr-expressing *Escherichia coli*. ACS Synth Biol 5: 679–688.

Jiao, Y, Newman, DK (2007) The *pio* operon is essential for phototrophic Fe(II) oxidation in *Rhodopseudomonas palustris* TIE-1. J Bacteriol 189: 1765–1773.

Jiménez Otero, F, Chan, CH, Bond, DR (2018) Identification of different putative outer membrane electron conduits necessary for Fe(III) citrate, Fe(III) oxide, Mn(IV) oxide, or electrode reduction by *Geobacter sulfurreducens*. J Bacteriol 200: e00347-18.

Klausen, M, Heydorn, A, Ragas, P, Lambertsen, L, Aaes-Jørgensen, A, Molin, S, Tolker-Nielsen, T (2003) Biofilm formation by *Pseudomonas aeruginosa* wild type, flagella and type IV pili mutants. Mol Microbiol 48: 1511–1524.

Kotloski, NJ, Gralnick, JA (2013) Flavin Electron Shuttles Dominate Extracellular Electron transport by *Shewanella oneidensis*. mBio 4: e00553-12.

Leang, C, Qian, X, Mester, T, Lovley, DR (2010) Alignment of the *c*-type cytochrome OmcS along pili of *Geobacter sulfurreducens*. Appl Environ Microbiol 76: 4080–4084.

Lebedev, N, Strycharz-Glaven, SM, Tender, LM (2014a) High-resolution AFM and single-cell resonance Raman spectroscopy of *Geobacter sulfurreducens* biofilms early in growth. Front Energy Res 2: 34.

Lebedev, N, Strycharz-Glaven, SM, Tender, LM (2014b) Spatially resolved confocal resonant Raman microscopic analysis of anode-grown *Geobacter sulfurreducens* biofilms. ChemPhysChem 15: 320–327.

Levar, CE, Chan, CH, Mehta-Kolte, MG, Bond, DR (2014) An inner membrane cytochrome required only for reduction of high redox potential extracellular electron acceptors. mBio 5: e02034-14.

Levar, CE, Hoffman, CL, Dunshee, AJ, Toner, BM, Bond, DR (2017) Redox potential as a master variable controlling pathways of metal reduction by *Geobacter sulfurreducens*. ISME J 11: 741–752.

Levar, CE, Rollefson, JB, Bond, DR (2012) Energetic and molecular constraints on the mechanism of environmental Fe(III) reduction by *Geobacter*. In Microbial Metal Respiration: From Geochemistry to Potential Applications. Gescher, J., and Kappler, A. (eds). Springer Berlin Heidelberg, Berlin, Heidelberg. pp. 29–48.

Li, C, Lesnik, KL, Fan, Y, Liu, H (2016) Millimeter scale electron conduction through exoelectrogenic mixed species biofilms. FEMS Microbiol Lett 363: fnw153.

Li, D-B, Cheng, Y-Y, Wu, C, Li, W-W, Li, N, Yang, Z-C, Tong, Z-H, Yu, H-Q (2014) Selenite reduction by *Shewanella oneidensis* MR-1 is mediated by fumarate reductase in periplasm. Sci Rep 4: 3735.

Lies, DP, Hernandez, ME, Kappler, A, Mielke, RE, Gralnick, JA, Newman, DK (2005) *Shewanella oneidensis* MR-1 uses overlapping pathways for iron reduction at a distance and by direct contact under conditions relevant for biofilms. Appl Env Microbiol 71: 4414–4426.

Light, SH, Su, L, Rivera-Lugo, R, Cornejo, JA, Louie A, Iavarone, AT, Ajo-Franklin, CM, Portnoy, DA (2018) A flavin-based extracellular electron transfer mechanism in diverse Gram-positive bacteria. Nature 562: 140–144.

Liu, X, Zhuo, S, Rensing, C, Zhou, S (2018) Syntrophic growth with direct interspecies electron transport between pili-free *Geobacter* species. ISME J 12: 2142.

Liu, Y, Bond, DR (2012) Long-distance electron transport by *Geobacter sulfurreducens* biofilms results in accumulation of reduced *c*-type cytochromes. ChemSusChem 5: 1047–1053.

Liu, Y, Kim, H, Franklin, RR, Bond, DR (2011) Linking spectral and electrochemical analysis to monitor *c*-type cytochrome redox status in living *Geobacter sulfurreducens* biofilms. ChemPhysChem 12: 2235–2241.

Lloyd, JR, Leang, C, Hodges Myerson, AL, Coppi, MV, Cuifo, S, Methe, B, Sandler, SJ, Lovley, DR (2003) Biochemical and genetic characterization of PpcA, a periplasmic *c*-type cytochrome in *Geobacter sulfurreducens*. Biochem J 369: 153–161.

Lovley, DR, Giovannoni, SJ, White, DC, Champine, JE, Phillips, EJP, Gorby, YA, Goodwin, S (1993) *Geobacter metallireducens* gen. nov. sp. nov., a microorganism capable of coupling the complete oxidation of organic compounds to the reduction of iron and other metals. Arch Microbiol 159: 336–344.

Luzar, MA, Thomassen, MJ, Montie, TC (1985) Flagella and motility alterations in *Pseudomonas aeruginosa* strains from patients with cystic fibrosis: relationship to patient clinical condition. Infect Immun 50: 577–582.

Ly, HK, Harnisch, F, Hong, S-F, Schröder, U, Hildebrandt, P, Millo, D (2013) Unraveling the interfacial electron transport dynamics of electroactive microbial biofilms using surface-enhanced Raman spectroscopy. ChemSusChem 6: 487–492.

Mahenthiralingam, E, Campbell, ME, Speert, DP (1994) Nonmotility and phagocytic resistance of *Pseudomonas aeruginosa* isolates from chronically colonized patients with cystic fibrosis. Infect Immun 62: 596–605.

Majzlan, J (2013) Minerals and aqueous species of iron and manganese as reactants and products of microbial metal respiration. In Microbial Metal Respiration. Gescher, J., and Kappler, A. (eds). Springer Berlin Heidelberg, pp. 1–28.

Malvankar, NS, Vargas, M, Nevin, K, Tremblay, P-L, Evans-Lutterodt, K, Nykypanchuk, D, Martz, E, Tuominen, MT, Lovley, DR (2015) Structural basis for metallic-like conductivity in microbial nanowires. mBio 6: e00084-15.

Malvankar, NS, Vargas, M, Nevin, KP, Franks, AE, Leang, C, Kim, B-C, Inoue, K, Mester, T, Covalla, SF, Johnson, JP, Rotello, VM, Tuominen, MT, Lovley, DR (2011) Tunable metallic-like conductivity in microbial nanowire networks. Nat Nanotechnol 6: 573–579.

Malvankar, NS, Yi, SM, Gu, Y, Neu, J, O'Brian, JP, Vu, D, Huynh, W, Yalcin, SE, Varga, T, Jiang, H, Xia, Q, Ao, G, Zheng, M, Batista, V, Schmuttenmaier, C, Malvankar, N (2018) Mechanism of metal-like transport in bacterial pili protein nanofilaments.

Marsili, E, Baron, DB, Shikhare, ID, Coursolle, D, Gralnick, JA, Bond, DR (2008) *Shewanella* secretes flavins that mediate extracellular electron transport. Proc Natl Acad Sci 105: 3968–3973.

Mehta, T, Coppi, MV, Childers, SE, Lovley, DR (2005) Outer Membrane *c*-type cytochromes required for Fe(III) and Mn(IV) oxide reduction in *Geobacter sulfurreducens*. Appl Environ Microbiol 71: 8634–8641.

Myers, CR, Myers, JM (1997) Cloning and sequence of *cymA*, a gene encoding a tetraheme cytochrome *c* required for reduction of iron(III), fumarate, and nitrate by *Shewanella putrefaciens* MR-1. J Bacteriol 179: 1143–1152.

Myers, CR, Nealson, KH (1988) Bacterial manganese reduction and growth with manganese oxide as the sole electron acceptor. Science 240: 1319–1321.

Myers, JM, Antholine, WE, Myers, CR (2004) Vanadium(V) reduction by *Shewanella oneidensis* MR-1 requires menaquinone and cytochromes from the cytoplasmic and outer membranes. Appl Environ Microbiol 70: 1405–1412.

Myers, JM, Myers, CR (2000) Role of the tetraheme cytochrome CymA in anaerobic electron transport in cells of *Shewanella putrefaciens* MR-1 with normal levels of menaquinone. J Bacteriol 182: 67–75.

Nealson, KH, Scott, J (2006) Ecophysiology of the Genus *Shewanella*. In The Prokaryotes: Volume 6: Proteobacteria: Gamma Subclass. Dworkin, M, Falkow, S, Rosenberg, E, Schleifer, K-H, and Stackebrandt, E (eds). Springer New York, New York, NY. pp. 1133–1151.

Nevin, KP, Holmes, DE, Woodard, TL, Hinlein, ES, Ostendorf, DW, Lovley, DR (2005) *Geobacter bemidjiensis* sp. nov. and *Geobacter psychrophilus* sp. nov., two novel Fe(III)-reducing subsurface isolates. Int J Syst Evol Microbiol 55: 1667–1674.

Nevin, KP, Kim, BC, Glaven, RH, Johnson, JP, Woodard, TL, Methé, BA, DiDonato, RJ, Covalla, SF, Franks, AE, Liu, A, Lovley, DR (2009) Anode biofilm transcriptomics reveals outer surface components essential for high density current production in *Geobacter sulfurreducens* fuel cells. PLOS ONE 4: e5628.

Phan, H, Yates, MD, Kirchhofer, ND, Bazan, GC, Tender, LM, Nguyen, T-Q (2016) Biofilm as a redox conductor: a systematic study of the moisture and temperature dependence of its electrical properties. Phys Chem Chem Phys PCCP 18: 17815–17821.

Peng, L, Zhang, Y (2017) Cytochrome OmcZ is essential for the current generation by *Geobacter sulfurreducens* under low electrode potential. Electrochem Acta 228: 447-452.

Pirbadian, S, Barchinger, SE, Leung, KM, Byun, HS, Jangir, Y, Bouhenni, RA, Reed, SB, Romine, MF, Saffarini, DA, Shi, L, Gorby, YA, Golbeck, JH, El-Naggar, MY (2014) *Shewanella oneidensis* MR-1 nanowires are outer membrane and periplasmic extensions of the extracellular electron transport components. Proc Natl Acad Sci 111: 12883–12888.

Pokkuluri, PR, Londer, YY, Duke, NEC, Pessanha, M, Yang, X, Orshonsky, V, Orshonski, L, Erickson, J, Zagyanskiy, Y, Salgueiro, CA, Schiffer, M (2011) Structure of a novel dodecaheme cytochrome *c* from *Geobacter sulfurreducens* reveals an extended 12nm protein with interacting hemes. J Struct Biol 174: 223–233.

Qian, X, Mester, T, Morgado, L, Arakawa, T, Sharma, ML, Inoue, K, Joseph, C, Salgueiro, CA, Maroney, MJ, Lovley, DR (2011) Biochemical characterization of purified OmcS, a *c*-type cytochrome required for insoluble Fe(III) reduction in *Geobacter sulfurreducens*. Biochim Biophys Acta BBA - Bioenerg 1807: 404–412.

Reguera, G, McCarthy, KD, Mehta, T, Nicoll, JS, Tuominen, MT, Lovley, DR (2005) Extracellular electron transport via microbial nanowires. Nature 435: 1098–1101.

Reguera, G, Pollina, RB, Nicoll, JS, Lovley, DR (2007) Possible nonconductive role of *Geobacter sulfurreducens* pilus nanowires in biofilm formation. J Bacteriol 189: 2125–2127.

Renslow, RS, Babauta, JT, Dohnalkova, DC, Boyanov, MI, Kemner, KM, Majors, PD, Fredrickson, Beyenal, H (2013) Metabolic spatial variability in electrode-respiring Geobacter sulfurreducens biofilms. Energy Environ Sc 6: 1827.

Richardson, DJ, Butt, JN, Fredrickson, JK, Zachara, JM, Shi, L, Edwards, MJ, White, G, Baiden, N, Gates, AJ, Marritt, SJ, Clarke, TA (2012) The "porin–cytochrome" model for microbe-to-mineral electron transport. Mol Microbiol 85: 201–212.

Richter, H, Nevin, KP, Jia, H, Lowy, DA, Lovley, DR, Tender, LM (2009) Cyclic voltammetry of biofilms of wild type and mutant *Geobacter sulfurreducens* on fuel cell anodes indicates possible roles of OmcB, OmcZ, type IV pili, and protons in extracellular electron transport. Energy Environ Sci 2: 506–516.

Richter, LV, Franks, AE, Weis, RM, Sandler, SJ (2017) Significance of a posttranslational modification of the PilA protein of *Geobacter sulfurreducens* for surface attachment, biofilm formation, and growth on insoluble extracellular electron acceptors. J Bacteriol 199: e00716-16.

Richter, LV, Sandler, SJ, Weis, RM (2012) Two isoforms of *Geobacter sulfurreducens* PilA have distinct roles in pilus biogenesis, cytochrome localization, extracellular electron transport, and biofilm formation. J Bacteriol 194: 2551–2563.

Robuschi, L, Tomba, JP, Schrott, GD, Bonanni, PS, Desimone, PM, Busalmen, JP (2013) Spectroscopic slicing to reveal internal redox gradients in electricity-producing biofilms. Angew Chem Int Ed 52: 925–928.

Rollefson, JB, Stephen, CS, Tien, M, Bond, DR (2011) Identification of an extracellular polysaccharide network essential for cytochrome anchoring and biofilm formation in *Geobacter sulfurreducens*. J Bacteriol 193: 1023–1033.

Saffarini, DA, Schultz, R, Beliaev, A (2003) Involvement of cyclic AMP (cAMP) and cAMP receptor protein in anaerobic respiration of *Shewanella oneidensis*. J Bacteriol 185: 3668–3671.

Schmidt, I, Pieper, A, Wichmann, H, Bunk, B, Huber, K, Overmann, J, Walla, PJ, Schröder, U (2017) *In situ* autofluorescence spectroelectrochemistry for the study of microbial extracellular electron transport. ChemElectroChem 4: 2515–2519.

Schuetz, B, Schicklberger, M, Kuermann, J, Spormann, AM, Gescher, J (2009) Periplasmic electron transport via the *c*-type cytochromes MtrA and FccA of *Shewanella oneidensis* MR-1. Appl Environ Microbiol 75: 7789–7796.

Schwalb, C, Chapman, SK, Reid, GA (2003) The tetraheme cytochrome CymA is required for anaerobic respiration with dimethyl sulfoxide and nitrite in *Shewanella oneidensis*. Biochemistry (Mosc) 42: 9491–9497.

Shelobolina, ES, Nevin, KP, Blakeney-Hayward, JD, Johnsen, CV, Plaia, TW, Krader, P, Woodard, T, Holmes, DE, Vanpraagh, CG, Lovley, DR (2007) *Geobacter pickeringii* sp. nov., *Geobacter argillaceus* sp. nov. and *Pelosinus fermentans* gen. nov., sp. nov., isolated from subsurface kaolin lenses. Int J Syst Evol Microbiol 57: 126–135.

Shelobolina, ES, Vrionis, HA, Findlay, RH, Lovley, DR (2008) *Geobacter uraniireducens* sp. nov., isolated from subsurface sediment undergoing uranium bioremediation. Int J Syst Evol Microbiol 58: 1075–1078.

Snider, RM, Strycharz-Glaven, SM, Tsoi, SD, Erickson, JS, Tender, LM (2012) Long-range electron transport in *Geobacter sulfurreducens* biofilms is redox gradient-driven. Proc Natl Acad Sci 109: 15467–15472.

Steidl, RJ, Lampa-Pastirk, S, Reguera, G (2016) Mechanistic stratification in electroactive biofilms of *Geobacter sulfurreducens* mediated by pilus nanowires. Nat Commun 7: 12217.

Stephen, CS, LaBelle, EV, Brantley, SL, Bond, DR (2014) Abundance of the multiheme *c*-type cytochrome OmcB increases in outer biofilm layers of electrode-grown *Geobacter sulfurreducens*. PLOS ONE 9: e104336.

Straub, KL, Buchholz-Cleven, BE (2001) *Geobacter bremensis* sp. nov. and *Geobacter pelophilus* sp. nov., two dissimilatory ferric-iron-reducing bacteria. Int J Syst Evol Microbiol 51: 1805–1808.

Strycharz, SM, Malanoski, AP, Snider, RM, Yi, H, Lovley, DR, Tender, LM (2011) Application of cyclic voltammetry to investigate enhanced catalytic current generation by biofilm-modified anodes of *Geobacter sulfurreducens* strain DL1 vs. variant strain KN400. Energy Environ Sci 4: 896–913.

Strycharz-Glaven, SM, Roy, J, Boyd, D, Snider, R, Erickson, JS, Tender, LM (2014) Electron transport through early exponential-phase anode-grown *Geobacter sulfurreducens* biofilms. ChemElectroChem 1: 1957–1965.

Strycharz-Glaven, SM, Snider, RM, Guiseppi-Elie, A, Tender, LM (2011) On the electrical conductivity of microbial nanowires and biofilms. Energy Environ Sci 4: 4366–4379.

Sundararajan, A, Kurowski, J, Yan, T, Klingeman, DM, Joachimiak, MP, Zhou, J, Naranjo, B, Gralnick, JA, Fields, MW (2011) *Shewanella oneidensis* MR-1 sensory box protein involved in aerobic and anoxic growth. Appl Environ Microbiol 77: 4647–4656.

Sung, Y, Fletcher, KE, Ritalahti, KM, Apkarian, RP, Ramos-Hernández, N, Sanford, RA, Mesbah, NM, Löffler, FE (2006) *Geobacter lovleyi* sp. nov. strain SZ, a novel metal-reducing and tetrachloroethene-dechlorinating bacterium. Appl Env Microbiol 72: 2775–2782.

Tejedor Sanz, S, Fernández Labrador, P, Hart, S, Torres, CI, Esteve-Núñez, A (2018) *Geobacter* dominates the inner layers of a stratified biofilm on a fluidized anode during brewery wastewater treatment. Front Microbiol 9: 387.

Tender, LM, Reimers, CE, Stecher, HA, Holmes, D.E., Bond, D.R., Lowy, D.A., Pilobello, K, Fertig, SJ, Lovley, DR (2002) Harnessing microbially generated power on the seafloor. Nat Biotechnol 20: 821–825.

TerAvest, M.A., Zajdel, T.J., Ajo-Franklin, C.M. (2014) The Mtr pathway of *Shewanella oneidensis* MR-1 couples substrate utilization to current production in *Escherichia coli*. ChemElectroChem 1: 1874–1879.

Torres, C.I., Kato Marcus, A., Rittmann, B.E. (2008) Proton transport inside the biofilm limits electrical current generation by anode-respiring bacteria. Biotechnol Bioeng 100: 872–881.

Ueki, T., DiDonato, L.N., Lovley, D.R. (2017) Toward establishing minimum requirements for extracellular electron transport in *Geobacter sulfurreducens*. FEMS Microbiol Lett 364: fnx093.

Vargas, M, Malvankar, NS, Tremblay, P-L, Leang, C, Smith, JA, Patel, P, Synoeyenbos-West, O, Nevin, KP, Lovley, DR (2013) Aromatic amino acids required for pili conductivity and long-range extracellular electron transport in *Geobacter sulfurreducens*. mBio 4: e00105-13.

Virdis, B, Harnisch, F, Batstone, DJ, Rabaey, K, Donose, BC (2012) Non-invasive characterization of electrochemically active microbial biofilms using confocal Raman microscopy. Energy Environ Sci 5: 7017–7024.

Xu, S, Barrozo, A, Tender, LM, Krylov, AI, El-Naggar, MY (2018) Multiheme cytochrome mediated redox conduction through *Shewanella oneidensis* MR-1 cells. J Am Chem Soc 140: 10085–10089.

Xu, S, Jangir, Y, El-Naggar, MY (2016) Disentangling the roles of free and cytochrome-bound flavins in extracellular electron transport from *Shewanella oneidensis* MR-1. Electrochimica Acta 198: 49–55.

Yalcin, SE, O'Brian, JP, Acharya, A, Gu, Y, Huynh, W, Yi, SM, Chaudhuri, S, Batista, V, Malvankar, N (2018) Large-scale conformational changes induce tunable electronic and mechanical functionality in proteins.

Yates, MD, Barr Engel, S, Eddie, BJ, Lebedev, N, Malanoski, AP, Tender, LM (2018a) Redox-gradient driven electron transport in a mixed community anodic biofilm. FEMS Microbiol Ecol 94: fiy081.

Yates, MD, Eddie, BJ, Lebedev, N, Kotloski, NJ, Strycharz-Glaven, SM, Tender, LM (2018b) On the relationship between long-distance and heterogeneous electron transport in electrode-grown *Geobacter sulfurreducens* biofilms. Bioelectrochemistry Amst Neth 119: 111–118.

Yates, MD, Golden, JP, Roy, J, Strycharz-Glaven, SM, Tsoi, S, Erickson, JS, El-Naggar, MY, Calabrese Barton, S, Tender, LM (2015) Thermally activated long-range electron transport in living biofilms. Phys Chem Chem Phys PCCP 17: 32564–32570.

Yates, MD, Strycharz-Glaven, S, Golden, J, Roy, J, Tsoi, S, Erickson, J, El-Naggar, MY, Calabrese Barton, S, Tender, LM (2018c) Characterizing electron transport through living biofilms. J Vis Exp 136: e54671.

Yates, MD, Strycharz-Glaven, SM, Golden, JP, Roy, J, Tsoi, S, Erickson, JS, El-Naggar, MY, Calabrese Barton, S, Tender, LM (2016) Measuring conductivity of living *Geobacter sulfurreducens* biofilms. Nat Nanotechnol 11: 910–913.

Zacharoff, LA, Morrone, D, Bond, DR (2017) *Geobacter sulfurreducens* extracellular multiheme cytochrome PgcA facilitates respiration to Fe(III) oxides but not electrodes. Front Microbiol 8: 2481.

Zacharoff, LA, Chan, CH, Bond, DR (2016) Reduction of low potential electron acceptors requires the CbcL inner membrane cytochrome of *Geobacter sulfurreducens*. Bioelectrochemistry 107: 7–13.

Zhang, X, Philips, J, Roume, H, Guo, K, Rabaey, K, Prévoteau, A (2017) Rapid and quantitative assessment of redox conduction across electroactive biofilms by using double potential step chronoamperometry. Chem Electro Chem 4: 1026.

Zhou, S, Yang, G, Lu, Q, Wu, M (2014) *Geobacter soli* sp. nov., a dissimilatory Fe(III)-reducing bacterium isolated from forest soil. Int J Syst Evol Microbiol 64: 3786–3791.

Electrochemical Impedance Spectroscopy: A Noninvasive Tool for Performance Assessment of Bioelectrochemical Systems

Namita Shrestha[1], Govind Chilkoor[2], Bhuvan Vemuri[2] and Venkataramana Gadhamshetty[2,3*]

[1] Civil and Environmental Engineering, University of Wisconsin-Platteville, 139 Ottensman Hall, 1 University Plaza Platteville, WI 53818, USA

[2] Civil and Environmental Engineering, South Dakota School of Mines and Technology, 501 E. St. Joseph Street, Rapid City, SD 57701, USA

[3] Surface Engineering Research Center, South Dakota School of Mines and Technology, 501 E. St. Joseph Street, Rapid City, SD 57701, USA

1. Introduction

Electrochemical impedance spectroscopy (EIS) provides noninvasive methods for studying temporal and spatial dynamics of charge transfer resistances and intrinsic losses (activation losses and concentration polarization) in model BESs. EIS is routinely used to quantify ohmic losses in electrical circuits, electrodes, electrolytes and membranes, at both working and counter electrodes. For a typical electrode controlled by a potentiostat, a small alternate current (AC) signal (1 to 10 mV), is adequate for perturbing interfacial bioelectrochemical reactions and measuring their steady state responses (typically the cell current) at a given excitation frequency.

EIS qualitatively deciphers the underlying extracellular electron transfer mechanisms of previously unexplored exoelectrogens, especially in response to changes in physiological conditions (e.g., redox potentials, electrolyte chemistry, nutrients and metabolic preferences) (Sharma et al. 2014), mass transport parameters and material properties of electrodes and membranes (Dominguez-Benetton et al. 2012). Exoelectrogens refer to a class of microorganisms that use extracellular electron transfer (EET) mechanisms to transfer the electrons from the cytoplasm to insoluble terminal electron acceptors (electrodes). They are also referred to as electrochemically active bacteria, anode-respiring bacteria and electricigens.

The intrinsic losses obtained from the EIS results can be correlated with biofilm growth parameters in BESs (Ramasamy et al. 2009). The EIS data can be used to quantify the exchange current density of participating charge transfer reactions (Ramasamy et al. 2009) and identify the major sources of impedance to the flow of electrons to electrode and external circuits. Thus, the EIS provides preliminary knowledge for optimizing performance parameters, including electrochemical power output, coulombic efficiency and chemical oxygen demand removal efficiency of BESs.

The EIS methods have been successfully used to optimize performances of microbial fuel cells (MFCs) (Dominguez-Benetton et al. 2012), microbial desalination cells (MDCs) (Jacobson et al. 2011),

*Corresponding author: Venkata.Gadhamshetty@sdsmt.edu

microbial capacitive deionization cells (MCDCs) (Forrestal et al. 2015; Forrestal et al. 2012; Stoll et al. 2015), microbial solar cells (Call and Logan 2008; Cusick et al. 2012; Strik et al. 2011) and microbial electrolysis cells (Wang et al. 2015). The EIS has also been used to identify factors that impede the Faradaic and non-Faradaic reactions associated with bioelectrochemical oxidation of biomass substrates, including algal residues (Gadhamshetty et al. 2013), pomace (Shrestha et al. 2016), food scraps (Chandrasekhar et al. 2015), sludge (Ge et al. 2013), rice husks (Wang et al. 2013), leaves (Song et al. 2014), marine sediments (Haque et al. 2015), wheat straw (Song et al. 2014) and high molecular weight petroleum compounds of super-saline flowback water (Shrestha et al. 2018).

To gain a comprehensive understanding of BESs, EET mechanisms and EIS, the readers are encouraged to review relevant literature (Beyenal and Babauta 2015) (Dominguez-Benetton et al. 2012) (Sekar and Ramasamy 2013) (See Table 1 for recommended articles). This chapter will briefly introduce the concepts of EIS, electrical equivalent circuits and EET mechanisms. The following four topics related to the use of EIS will be discussed in detail: (1) frequency-dependent EIS spectra for noninvasive detection of redox mediators, (2) impedance analysis of oxygen reduction reactions, (3) impedance behavior of foulants and metal contaminants on electrode surfaces and (4) impedance behavior of microbially driven capacitive deionization electrodes fed with super-saline wastewater.

2. Electrochemical Impedance Spectroscopy

Impedance spectroscopy was first introduced for biological systems in 1920 to analyze the resistance and capacitance of cells of vegetables and the dielectric response of blood suspensions (Orazem and Tribollet 2011). The first use of complex plane analysis in aqueous electrochemistry was proposed by Sluyters and Oomen in 1960 (Sluyters and Oomen 1960). Other remarkable contributions in the early 19th century included the EIS analysis for chemical applications (Nernst 1894), distribution of relaxation time constants (Cole and Cole 1941; Davidson and Cole 1951), impedance characteristics of mass transfer reactions (Warburg 1899) and electrical circuitry (Dolin and Ershler 1940). The details of these EIS studies have been documented elsewhere (Lasia 2002; Lvovich 2012).

EIS is a steady state technique for studying relaxation phenomena with relaxation times varying over several orders of magnitude. In electrochemical systems that feature typical resistive-capacitive behavior, the relaxation phenomena can be viewed as the 'energy/charge storage' and 'relaxation of the stored energy'. This ability of EIS to capture the steady state characteristics enables the signal averaging from a wide range of frequencies (10^6-10^{-4} Hz) and with high precision (Dominguez-Benetton et al. 2012). The studies shown in Table 1 provide an overview of EIS methods for studying BESs.

The working electrode in a BES is perturbed using a small alternating signal (5 to 20 mV), and the steady state electrochemical response is monitored to study the charge transfer reactions occurring on electrode surfaces. The Ohm's law defines the electrical resistance (R) as the ratio of input voltage (V) to current output (I).

$$R = \frac{V}{I} \tag{1}$$

Impedance is the ratio of the voltage to the current, both of which are a function of phase shift and amplitude. Impedance is a complex resistance that occurs when current flows through a circuit composed of resistors, capacitors and inductors. It defines the ability of a circuit to resist the flow of electrons (real impedance) and store electrical energy (imaginary impedance). It is measured using an AC voltage signal (V) with a small amplitude (V_A, Volts) applied at the desired frequency (f, Hz), generating a current response with a small amplitude (I_A, Amperes). The complex impedance can then be defined in terms of input voltage $V(t)$ and output current I(t) as follows:

$$Z^* = \frac{V(t)}{I(t)} = \frac{VA \sin(\omega t)}{IA \sin(\omega t + \phi)} = ZA \frac{\sin(\omega t)}{\sin(\omega t + \phi)} \tag{2}$$

Table 1: Summary of existing EIS review studies on Bioelectrochemical systems

Review Paper	BES	Conclusion	Reference
EIS for better electrochemical measurements	None	An overview of EIS for studying the intrinsic losses in BES	(Park and Yoo 2003)
Exploring the use of EIS in MFC studies	MFC	EIS as an efficient and convenient tool to analyze bioelectrochemical processes in MFCs	(He and Mansfeld 2009)
Techniques for the study and development of microbial fuel cells: an electrochemical perspective	MFC	Electroanalytical techniques used in recent MFC studies. Discusses the principles, experimental procedures, data processing requirements, capabilities and weaknesses of these techniques	(Zhao et al. 2009)
The accurate use of impedance analysis for the study of microbial electrochemical systems	All BESs	Apply EIS for studying microbial electrochemical systems to improve the fundamental and practical knowledge thereof	(Dominguez-Benetton et al. 2012)
Electrochemically active biofilms: facts and fiction	All BESs	Use the electrochemical techniques for studying EET in electrochemically active biofilms	(Babauta et al. 2012)
EIS for MFC characterization	MFC	Correlate the EIS parameters to measurable physical quantities and discuss the need for standardization of procedures	(Sekar and Ramasamy 2013)
Application of EIS in biofuel cell characterization: a review	Biofuel cells	Harness EIS to get information about the electrode and electrolyte resistance, charge transfer resistance and capacitive current in biofuel cells	(Kashyap et al. 2014)
Biofilms in BES: from laboratory practice to data interpretation	Three electrode electrochemical cell	Examples from real biofilm research to illustrate the EIS techniques used for electrochemically active biofilms	(Beyenal and Babauta 2015)
The study of thin films by EIS	Biofuel cell/ electrode surface	A variety of studies on thin films by EIS	(Cesiulis et al. 2016)

As shown, the impedance is defined in terms of both the magnitude ($Z_A = |Z|$) and phase shift (φ). EIS data are typically represented in form of a Nyquist plot (the real part of the impedance on the x-axis and the imaginary part on the y-axis) and a Bode plot (the logarithmic frequency plotted on the x-axis and the absolute value of impedance or the phase angle on the y-axis).

2.1 Electrical Equivalent Circuits: Impedance Aspects of the Working Electrode (Anode)

EIS can probe relaxation phenomena over a frequency range of 1 mHz to 1 MHz. It can also be used to study impedance to Faradaic and non-Faradaic processes occurring at bioelectrochemical interfaces. The electrical equivalent circuits (EECs) shown in Table 2 are used to develop mechanistic models based on EIS data from typical BESs (Sekar and Ramasamy 2013). The EECs are used to propose and test hypotheses involving charge transfer reaction sequences, mass transfer reactions and factors related to biokinetics, potential and current, mass transfer, surface coverage, hydrodynamics and heterogeneities of electroactive biofilms (Dominguez-Benetton et al. 2012).

Table 2a and 2b provide a summary of the typical Nyquist curves and the representative EECs used to fit the EIS results. Table 3 summarizes the typical EECs used to fit the EIS data, the representative impedance values in typical BESs and the factors influencing EIS measurements. The EECs are comprised of three circuit elements including resistors, capacitors and inductors. The EECs are designed to fit the Helmholtz model (Park and Yoo 2003) or variations of the Randles circuit. EIS can characterize both the interfaces as well as the electrolytes, separating the Faradaic currents from the non-Faradaic signals, and enabling the development of meaningful EECs. The sequential Faradaic processes are typically modeled as circuit elements in a series configuration and simultaneously occurring processes in the parallel configurations.

While modeling diffusion processes the electrochemistry experiments should be designed to eliminate uncertainties related to mass transport (e.g., minimize stagnant conditions), enabling the use of EIS analysis to study less understood electrode biokinetics. Simple resistance-capacitance (RC) elements can model diffusion processes that occur in homogeneous materials or physical properties (e.g., charge mobilities) that remain identical on a spatial scale. A constant phase element (CPE) will be required to simulate the distributed nature of microscopic material properties. For instance, the CPE can simulate the diffusion processes when the microscopic properties of electrode/electrolyte interfaces are not smooth and uniform and when it is characterized by surface defects, kinks, jags, ledges, two-phase and three-phase regions and adsorbed species. The R-C contributions of the reaction on an electrode change spatially over a certain range around a mean but only the average effects over the entire surface can be observed. The macroscopic impedance, which depends on the reaction rate distribution across an interface, is measured as an average over the entire electrode. Consequently, in the Nyquist plot, the semicircle becomes flattened. To account for these effects, CPE can be used to replace the double layer capacitance and other typical capacitances in the circuits (Macdonald and Barsoukov 2005).

2.2 Electric Equivalent Circuits: Impedance Aspects of the Counter Electrode (Cathode)

The slow kinetics of electrochemical reactions on the surfaces of counter electrodes, e.g., oxygen reduction reactions (ORR), in BES represents a dominant bottleneck to the commercialization of BES technologies. The rotating disk electrode (RDE) and rotating ring-disk electrode (RRDE) are typically used to evaluate the performance of the ORR catalysts (Géniès et al. 2003). The DC analysis is disadvantageous as it offers summative manifestations of multiple polarization effects on the cathode surface and fails to differentiate them. Conversely, EIS can distinguish the charge transfer resistance from mass transfer impedances (kinetic, ohmic and diffusion) to ORR as they elicit different EIS spectra at different frequency ranges. EIS can characterize a range of mass and charge transfer processes associated with the ORR (Raistrick 1990; Singh et al. 2015; Springer et al. 1996). A classic model that serially combines flooded-agglomerate electrode model with a thin electrolyte film model is commonly used to study the ORR activity on porous electrodes (Raistrick 1990; Springer et al. 1996). At higher frequencies, the current density increases with the increase in the steady state concentration of participating reactants of ORRs. At lower frequencies, the current density decreases because of the mass transport limitations. A study by Springer et al. demonstrates the use of EIS for characterizing the effects of mass transfer limitations and humidification on the performance of air breathing cathodes (Springer et al. 1996). Their EIS study revealed two unique features of the air breathing cathode: (i) the interfacial charge transfer resistance and the catalyst layer properties appeared in the higher frequency range and (ii) the gas-phase transport limitations appeared in the lower frequency range. The EIS studies can also be used to understand the impacts of electrolyte crossover on the ORR activity (Perez et al. 1998; Piela et al. 2006). Table 2b shows a typical EIS behavior of the air breathing cathodes that sustain ORRs.

2.3 Validation of the EIS Data

When compared with the classic transient techniques, EIS techniques are beneficial because the EIS data can be readily validated using the Kramers-Kronig transforms. To further corroborate the physiological

Table 2a: Equivalent electrical circuits commonly used for representing single electrode interfaces (anode and cathode) and full cell BES systems

Electrode	Circuit	Nyquist shape	Time constant	Summary of circuit
Single electrode (anode or cathode)			1	The Nyquist plot shows a typical arc shape at the high frequency region but a typical beeline at the low frequency region that suggests that the oxidation in the anodic/cathodic compartment is governed by both activation and diffusion.
			1	The Nyquist diagram showed an arc shape in the entire frequency range, indicating that the anode/cathode reaction was controlled by activation.
			2	Microorganisms attached to the anode/cathode surface, enhancing the electron transfer kinetics but also determined the occurrence of a film that the electrons and/or the compounds must diffuse through. First RC circuit represented the electrical double layer and second RC circuit represented the diffusion boundary layer.
			2	When the electrical double layer and diffusion boundary layer are two successive reactions, such that when the current flowing through the circuit elements are different but the potential drop is the same.
			3	Electron is between the substrate and terminal electron acceptor, involving electron shuttles or mediators. The three processes identified are bioelectrochemical substrate oxidation charge transfer impedance, electron shuttle redox processes and polarization impedance at the anode.

(Contd.)

Table 2a: (*Contd.*)

Electrode	Circuit	Nyquist shape	Time constant	Summary of circuit
			3	The cathode processes, identified as mediator charge transfer process, polarization impedance at the cathode and charge transfer impedance corresponding to catholyte reduction.
Full Cell			2	R_{ohm} and R_{mem} represented electrolyte resistance, including resistance from the membrane, anolyte and catholyte. R_p represented charge transfer resistance of anode and cathode, related to the activation energy of the anode and cathode. Q_{dl} represented the constant phase element (CPE) of anode and cathode that suggested a rough electrode surface. A capacitor is formed at the interface between the electrode and its surrounding electrolyte (shown as a double layer) when charges in the electrode are separated from those in the electrolyte.
			2	R_{ohm} and R_{mem} represented electrolyte resistance, including resistance from the membrane, anolyte and catholyte. R_p represented charge transfer resistance of anode and cathode, related to the activation energy of the anode and cathode. Q_{dl} represented the constant phase element (CPE) of anode and cathode, that suggested a rough electrode surface. A capacitor is formed at the interface between the electrode and its surrounding electrolyte (shown as a double layer) when charges in the electrode are separated from those in the electrolyte. Additional Q_{dl} is the double layer capacitor caused by a diffusion layer at the cathode.

Abbreviations: R_{ohm}, ohmic resistance; Q_{dl}, CPE associated to the double layer at the electrode; R_p, polarization resistance at the respective electrode; Z_d, Warburg impedance associated to diffusional limitations; Q_{film}, constant phase element associated to a film developed at the electrode; R_{film}, resistance at the respective film developed at the electrode; R_{mem}, membrane resistance. R_A, electron transfer from substrate; R_{ES}, electron transfer from electron shuttle; and R_C, electron transfer to oxidant reduction at the cathode. R_{other}: resistance due to electrode polarization

Table 2b: Typical EIS behavior of the air breathing cathodes that sustain ORRs

System	Material	Circuit	Nyquist shape	Summary	Reference
Direct methanol fuel cell	Pt black			This study shows the effects on ORR activity of the cathode with methanol crossover in a direct methanol fuel cell	(Piela et al. 2006)
Thin-film rotating disk electrode	Pt black			This study investigates the ORR kinetics at mass transfer region (725-675mV).	(Singh et al. 2015)
Thin-film rotating disk electrode	Pt/C (20 wt%) catalyst layer with low ionomer content			This study investigates the ORR kinetics at mass transfer region (725-675mV).	(Singh et al. 2015)

Abbreviations: R_S, ohmic resistance; $Q_{dl}/Q_1/Q_2$, constant phase element associated to the double layer at the electrode; R_p, polarization resistance; $R_{ct}/R_1/R_2/R_L$, charge transfer resistance; C_{dl}, capacitive double layer; L, inductive loop

Table 3: Resistance of BES calculated by EIS Adapted

Type of cell	Electrode	Circuit parameter	Influencing factors	Resistance	Reference
Dual Chamber	Graphite felt	Ohmic resistance	Electrolyte	2 Ω	(Ramasamy et al. 2009)
		Charge transfer resistance	Substrate oxidation	291 Ω	
		Charge transfer resistance	Mediator oxidation	39 Ω	
		Polarization resistance	Other redox processes	4 Ω	
Single chamber	Carbon felt	Charge transfer resistance	Cathode reduction	18 Ω	(Martin et al. 2011)
		Ohmic resistance	Electrolyte	2 Ω	
		Polarization resistance	Other redox processes	25 Ω	
				4.4 Ω	
Single chamber	Platinum coated on carbon cloth	Ohmic resistance	Electrolyte	30 Ω	(He et al. 2008)
		Polarization resistance	Electrode polarization resistance	1.7 Ω	
Dual chamber	Electroplated platinum on bare graphite electrode	Ohmic resistance	Electrolyte	1.5 Ω (anode) and 5.5 Ω (cathode) 8×10^{-6} Ω (anode) and 8×10^{-3} Ω (cathode)	(Manohar et al. 2008)
	Anode: bare graphite electrode	Polarization resistance	Electrode polarization resistance		
Submersible	Carbon paper	Ohmic resistance	Electrolyte	7 Ω	(Min et al. 2012)
		Charge transfer resistance	N/A	2 Ω	
Single chamber sediment MFC	Anode: carbon cloth	Charge transfer resistance	Electron transfer rate from the biofilm to the anode in the presence of higher dissolved oxygen	28-65 Ω	(He et al. 2007)
	Cathode: platinum coated cathode	Ohmic resistance	Electrolyte	40-50 Ω	
Microfluidic MFC	Graphite electrode	Ohmic resistance	Electrolyte and membrane	45.5 Ω	(Ye et al. 2013)

(Contd.)

		Charge transfer resistance	Other redox processes	2.3 kΩ	
		Diffusion resistance	Electron transfer resistance from the biofilm to the anode	0.00033 Ω	
Floating MFC	Anode: granular graphite	Ohmic resistance	Electrolyte and membrane	1.5 Ω	(Huang et al. 2012)
	Cathode: platinum wire	Polarization resistance	Cathode polarization and reduction resistance	20 Ω	
Implantable biofuel cell	Anode: Au/ Au NPs/ enzyme	Ohmic resistance	Electrolyte and membrane	5.5 Ω	(Andoralov et al. 2013)
	Cathode: Au/Au NPs/ Bilirubin oxidase	Charge transfer resistance	Electrical parameters along the electrodes and the depth of the porous layer	N/A	
Enzymatic biofuel cell	Platinum cathode	Ohmic resistance	Electrolyte	N/A	(Li et al. 2012)
		Charge transfer resistance	Diffusion limited electron transfer process on the electrode surface	873 Ω	

Source: International Journal of Hydrogen Energy 39(35): 20159-20170

models that are developed using the EIS data, they should be validated with complementary results from the DC methods (chronoamperometry and polarography), system biology tools, spectroscopic methods or surface characterization tests. For instance, EIS can qualitatively confirm the presence of endogenous redox active mediators and their role in mediated electron transfer (MET) mechanisms. Cyclic voltammetry and differential pulse voltammetry techniques can be used to identify the redox active species. High-resolution mass spectrometry techniques and redox mediator assays could be used to further confirm the physical and chemical characteristics of the redox active species.

3. EIS for Performance Assessment of BESs

3.1 EIS for Studying the Extracellular Transfer Electron Mechanisms

To establish the electrical communication with electrodes, the exoelectrogens respire the electrodes using the direct electron transfer (DET) or the MET mechanisms. The DET mechanisms orient the enzymes to overlap the electronic state of its active center (or other conductive structures) with that of the electrode surface (Marsili et al. 2008; Logan 2009; Torres et al. 2009). The MET mechanisms are based on the use of endogenously secreted redox active species that ferry the electrons from within the cell, through the cell wall and finally to the electrode.

3.1.1 EIS for Studying the MET Mechanisms

Biological species facilitate slow and heterogeneous electron transfer at the electrodes and consequently, they exhibit irreversible electrochemical behavior. Such behavior is attributed to poisoning of the electrode surface by the adsorbed biocomponents or insulation of the electroactive centers in the molecule by the surrounding protein matrix (Fultz and Durst 1982). Exoelectrogens circumvent this problem by endogenously secreting redox mediators that facilitate redox coupling between the substrate oxidation and the electrode reduction. Mediators can also be supplemented in the electrolyte to enhance the performance of biofuel cells. Examples of mediators include thionine, methyl viologen, 2-hydroxy-1, 4-naphtoquinone, neutral red, anthraquinone2,6-disulfonate and humic acids (Kim et al. 2002; Liu et al. 2004). The DET mechanisms become less dominant when exoelectrogens fail to adhere to the anode surface, and in such cases, the microbial communities use soluble mediators to facilitate the electron transfer (Bond and Lovley 2003; Clauwaert et al. 2007; Gregory et al. 2004; Rabaey et al. 2004; Blake et al. 1994; Clauwaert et al. 2008; Park and Zeikus 2000).

EIS can elucidate the details of participating charge transfer processes in the bioenergetic pathway for the diffusible redox mediators. Here, we provide an example of the EIS methods for detecting riboflavin compounds secreted by *Shewanella oneidensis* DSP10 in two-compartment MFCs that is configured with ferricyanide catholyte system (Ramasamy et al. 2009). Figure 1 depicts a hypothetical overview of the flow of the electrical current (flow of charges) in the MFC. The charges, which originate from the bioelectrochemical oxidation of the lactate (electron donor), flow toward the surface of graphite felt anode under the electric potential provided by the MFC. At the anode, *Shewanella* oxidizes lactate to electrons that are transferred to the endogenously produced electron shuttle. The impedance to this charge transfer reaction is denoted as R_A. The electron shuttle then transfers electrons to the graphite felt anode, overcoming the second charge transfer impedance reaction that is denoted as R_{ES}. The electrons are then transferred via the external circuit to the cathode, where it reduces the terminal electron acceptor after overcoming the cathodic charge transfer impedance, R_C.

Electron transfer from *Shewanella oneidensis* to the graphite felt anode was hypothesized to be facilitated by the MET pathways as depicted in Figure 1. The EIS tools were used to prove this MET hypothesis (Ramasamy et al. 2009). The Bode phase angle plots shown in Figure 2 depict more than one electrochemical reaction for both anode and cathode. These plots depict the unique electrochemical reactions that are characterized by distinct time constants in a specific frequency range. The slope of the curve provides qualitative information about the variation of charge transfer resistance as a function of frequency. For instance, the lower charge transfer resistance will shift the phase angle maxima in Bode plots to a higher frequency range, or vice versa.

The anode exhibited three-time constants, as indicated by the three bell shaped curves in different frequency regions (Figure 2a), and these time constants represent different Faradaic processes (in this case a charge transfer reaction). The Nyquist arc, in the medium frequency region (1 to 100 Hz), attributed to the charge transfer impedance of the electron shuttle redox processes on the anode (Figure 2a). The high frequency process attributed to the activity of metal salts in the electrolyte and ferricyanide, which could have seeped into the anode compartment. The Nyquist arc, within the low frequency domain, attributed to the charge transfer impedance to the lactate oxidation (Figure 2a).

3.1.2 Other EIS Studies for Detection of Redox Mediators in Impedance Biosensors

Table 4 provides an overview of the EIS techniques for studying charge transfer resistance associated with the mediators, which are typically observed in the low-to-mid frequency regions. The literature on biosensors provides an overview of the EIS methods for detecting and tracking the mediators. The EIS techniques are beneficial for the detection of redox mediator when compared with the cyclic voltammetry techniques as the latter uses either lower or higher potential, both of which are ineffective for detecting the redox active species. At the lower potential, the interfacial electron transfer kinetics and diffusion kinetics

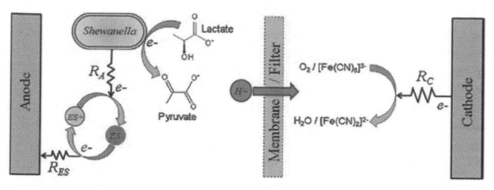

Figure 1: Sequence of MFC electrochemical reactions and their corresponding charge transfer resistances: R_A, electron transfer from the substrate; R_{ES}, electron transfer from electron shuttle and R_C, electron transfer to oxidant reduction at the cathode. Source (Ramasamy et al. 2009). Copyright © Biotechnology and Bioengineering (John Wiley and Sons). Reproduced with permission.

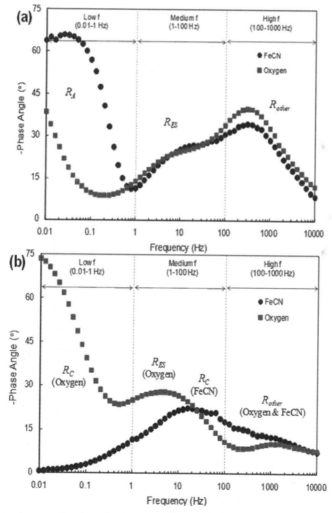

Figure 2: Bode phase angle plots for ferricyanide (circles) and oxygen (squares) based MFCs for: (a) anode and (b) cathode. Source (Ramasamy et al. 2009). Copyright © Biotechnology and Bioengineering (John Wiley and Sons). Reproduced with permission.

Table 4: Mediators for use in electrochemical studies of BES

Mediator	V NHE	Result	EIS application	Circuit used	Reference
Anthraquinone-2,6-disulfonate	-0.184	Output power and current at maximum power was improved significantly. They observed higher charge transfer resistance in unmodified electrode surfaces ($R_{sol} = 0.81\ \Omega$ and Rct = 170 Ω). They found rate of the anodic bioelectrochemical reaction was considerably improved by using conductive film modified electrodes ($R_{sol} = 0.76\ \Omega$ and Rct = 7.9 Ω).	One-time constant circuit model containing ohmic resistance, charge transfer resistance and Warbug's diffusion element. High frequency response was attributed to ohmic resistance. Low and mid frequency response as the semicircle curve was attributed to charge transfer resistance.		(Feng et al. 2010)
Humic acid	0.778 to -1.24	Increase in power by 24%. The cathode was not the limiting factor. The anode gave higher impedance due to the lower ionic strength.	EIS was used to compare internal resistance determined from the polarization curves.		(Thygesen et al. 2009)
Riboflavins	-0.21	Phase angle shift and curvature occurrence in high frequency in the abiotic anode. In bioanode low frequency impedance charge transfer was reduced from 531 to 162 Ω. No significant response was observed in high frequency reaction.	Models two sequential charge transfer reactions in the bioanode. The circuit element consists of ohmic resistance and charge transfer reaction (one in high frequency and other one associated with a low frequency).		(Jung et al. 2012)
Thionine	0.08 to -0.25	Decrease in total internal resistance by 1 to 1.5 folds along with increased power density (36 to 149 mW/m²).	One-time constant circuit model containing ohmic resistance and charge transfer resistance.	The real part of impedance value in the Nyquist plot was used to obtain charge transfer resistant.	(Chen et al. 2015)
Methyl viologen	-0.449	Semicircle part at a higher frequency range and a linear part at a lower frequency range. Charge transfer resistance decreased (13 Ω to 3Ω) and 1.5 folds increased in the power density (9 W/m² to 14 W/m²).	Three-electrode system with anode as the working electrode, a Pt sheet as counter electrode and silver/silver chloride electrode as a reference electrode.	The real part of impedance value in the Nyquist plot was used to obtain charge transfer resistant.	(Liu et al. 2016)

(Contd.)

Pyocyanin	-0.25 to -0.3	3.64-folds increase in maximum power density accompanied by a decrease in the internal resistance of MFC (from 500 Ω to 100 Ω).	Not applicable.	(Wu et al. 2014)
Phenazine-1-carboxamide	-0.116	Well-defined semicircle can be observed at high frequencies, followed by a straight line at low frequencies. Charge transfer resistance decreased (6.7 Ω to 4.2 Ω) that reflected the enhancement of the external electron transfer efficiency and the electrical power output.	One-time constant model was used which consisted of ohmic resistance, charge transfer resistance and the diffusion resistance.	(Feng et al. 2018)
Methyl orange	0.995	Decreases the anodic charge transfer resistance and increases in power density by 2.4 folds each.	Two-electrode configuration was used. Counter and reference electrodes of the impedance measuring apparatus were connected to the anode and working electrode was electrically fixed to the cathode of the fuel cell.	(Hosseini and Ahadzadeh 2013)

occur rapidly; whereas at the higher potential the interfacial electron transfer, enzyme or redox diffusion processes are characterized by faster kinetics. An article written by Daniels and Pourmand provides an overview of EIS methods for detecting biological molecules in biosensors (Daniels and Pourmand 2007). These biosensors are specifically designed to track impedance responses of the captured probe layers (immobilized probes) that capture the target mediators. The charge transfer resistance in the impedance biosensors represents a manifestation of the effects, both due to the energy overpotential associated with the oxidation or reduction events at the electrode and the contributions of the electrostatic repulsion or steric hindrance to the energy barrier of the redox species reaching the electrode (Daniels and Pourmand 2007). A study by Jeuken et al. describes cyanide inhibition strategies to confirm the presence or absence of redox mediators during the EIS analysis (Jeuken et al. 2008).

3.2 EIS Studies on Membrane Fouling in BESs

EIS can reveal the temporal dynamics of ohmic losses associated with the fouling of nonconducting membranes in BESs. The fouling refers to deposition of organic matter, living microorganisms, cell debris and inorganic substances on the surfaces or inside the pores of a membrane (Xia et al. 2019). This fouling phenomena cause operational problems and reduce the life of membranes. The fouled membranes in BESs can increase the overall impedance by nearly an eight-fold in a short duration (Miskan et al. 2016). The EIS can enable early detection of the fouling, *in situ* monitoring of the ohmic losses as a function of biofilm growth and the development of membrane regeneration strategies. A study by Cen et al. used the EIS data to correlate the water flux as a function of fouled reverse osmosis (RO) membranes (Cen et al. 2013). The irreversible fouling components were observed in the lower frequency region (0.01–0.1 Hz) and the reversible components in the mid frequency range (0.1–100 Hz).

An easy method to quantify the impedance of the membrane fouling is to measure the ohmic losses of BES under the following two different configurations: (1) the BES with the membrane and (2) the identical BES without the membrane. The first configuration (with the membrane) represents the ohmic losses due to electrolyte, membrane and all the other electrical connections. The second configuration (without the membrane) represents the ohmic losses due to electrolyte and electrical connections. The difference between the two configurations can be used to quantify the impedance losses associated with the membranes (Bakonyi et al. 2018). This EIS technique can assess the long term performances of yet to be discovered membranes for BES applications (Kim et al. 2009).

EIS analysis can also be used to study the underlying fouling mechanisms of the conductive membranes that are often referred to as reactive electrochemical membranes (REMs). Jing and Chaplin used the EIS techniques to assess the membrane fouling aspects of the REMs. They used EIS methods to quantify the impedances associated with the following mechanisms: (1) fouling at either the active or support layers, (2) fouling of the outer membrane surface, (3) intermediate pore blockage, (4) pore constriction and (5) monolayer adsorption (Jing and Chaplin 2016). They used EIS for studying the interfacial processes and the surface geometry aspects of the fouling and regeneration processes (Jing and Chaplin 2016). Each of the five mechanisms exhibited distinct EIS spectra, and the corresponding impedances were quantified using the transmission line model (Jing and Chaplin 2016). These EIS approaches could be potentially applied for studying the fouling mechanisms on the nonconducting polymeric membranes, including cation and anion exchange membranes, by adding a conducting layer on the membrane surfaces. For instance, the surface of the cation exchange membranes could be sputtered with the conducting layers of platinum or graphene.

The fouled membranes induce mass transfer limitations, such as by retarding the flow of protons from the anode to cathode to increase the pH gradients and reducing the membrane performance. This pH-gradient can be modeled as a resistor element in the EEC. Thus, EIS can model the pH gradients and study their impacts on membrane resistance as a function of the electrical double layer resistance and diffusion boundary layer resistance, both of which dominate at neutral pH conditions and lower electrolyte concentrations. Harnisch et al. used a numerical model to simulate the charge balancing ion transfer across the monopolar ion exchange membranes (Harnisch et al. 2009). Their study highlighted that a decrease in the proton conductivity of the membrane is due to an increase in the membrane resistance. A research by Sevda et al. compared the performances of Zirfon and Fumasep separators to study the underlying mechanisms that govern the proton transport from anode to cathode chambers (Sevda et al. 2013). Their results demonstrated that the internal resistance increased from a four to fivefold due to the low proton transport capabilities of Fumasep (Sevda et al. 2013) (see Figure 3).

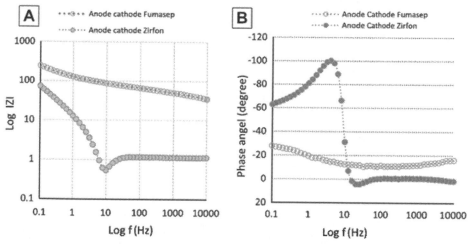

Figure 3: Impedance spectra due to membrane fouling. Source (Sevda et al. 2013) Copyright © Chemical Engineering Journal (Elsevier). Reproduced with permission.

The EIS data can be used to quantify the impedance contributions from the separators, especially as a function of the type and the concentration of the supporting electrolyte (Harnisch et al. 2008). The EIS has been effectively used to quantify the impedance contributions of a range of membranes, including salt-bridge (Sevda and Sreekrishnan 2012), anion exchange membranes (Harnisch et al. 2009; Kim et al. 2007; Torres et al. 2008), cation exchange membranes (He et al. 2005; Kim et al. 2007), bipolar membrane (Ter Heijne et al. 2006) and ultrafiltration membranes (Zuo et al. 2007). Li et al. provides an overview of the characterization aspects of these membranes and compares their advantages and disadvantages (Li et al. 2011). In the coming few years, the EIS methods will play a pivotal role in optimizing the currently available membranes and developing the next generation separators designed to dramatically improve the BES performances.

3.3 EIS for Analysis of Microbial Desalination and Microbial Capacitive Deionization Cells

Considering that the EIS methods can characterize the biofilm capacitance, the electron transfer mechanisms, the reactor design and electrode materials aspects of BES, they can be effectively used to characterize new BES. For instance, a series of earlier EIS studies have been used successfully on MDCs (Forrestal et al. 2015; Forrestal et al. 2012), that consist of an anion exchange membrane next to the anode and a cation exchange membrane next to the cathode, yielding an additional compartment (middle chamber between the membranes) to facilitate the desalination process (Forrestal et al. 2012). For instance, the EIS was used to track the progress of the desalination, marked by the increasing ohmic resistance on a temporal scale (Cao et al. 2009). The EIS tools were successfully used to quantify the ohmic resistance, electrochemical charge transfer resistance, double layer capacitance and diffusional impedance of the MDC system. The drawback of the MDC is that it removes the salt from the middle chamber and transfers it to the treated water (in the anode chamber) or catholyte.

A microbial capacitive desalination cell (MCDC) represents a variant of the BES that addresses challenges of salt migration and pH fluctuation in MDCs (Forrestal et al. 2012). A recent study by our group discusses the use of EIS tools for characterizing the MCDC that is used to desalinate the produced water from a hydraulically fractured oil field. The MCDC system consists of three chambers: anode, cathode and middle chamber (capacitive deionization unit or CDI) (Figure 4e). MCDCs are driven by microbes to simultaneously treat wastewater, desalinate water and produce electricity. Recent developments in ion selective membranes and porous carbon electrode designs can potentially develop novel MCDC designs that outperform classic forms of desalination technologies. In order to advance the

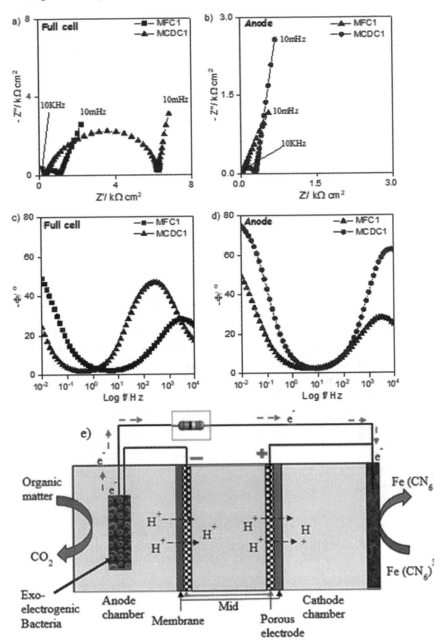

Figure 4: Nyquist plots for MFC1 and MCDC1 between frequencies 10 kHz and 10 mHz: a) full cell configuration; b) anode configuration, Bode phase angle plots for MFC1 and MCDC1 between frequencies 10 kHz and 10 mHz; c) full cell configuration; d) anode configuration; e) operating principle of microbial capacitive deionization. Source (Shrestha et al. 2018). Copyright © Bioelectrochemistry (Elsevier). Reproduced with permission.

MCDC technologies, it is essential to delineate and characterize the underlying electrochemical processes that affect the electrode stability, energy efficiency, ion removal and MCDC performance. Thus, our group carried out a study with the primary objective to evaluate the impedance spectra of MCDCs in detail, describing the kinetics in double layer charging/discharging processes.

We compared the performance of two-compartment MFCs with three-compartment MCDCs under batch-fed mode using identical conditions related to the mixed microbial consortia, ferricyanide catholyte and anolyte-based on produced water in both anode and CDI units. The MCDC registered a very large impedance (6600 Ω cm^2) compared with the MFCs (870 Ω cm^2) (Table 5). The CDI unit contributed to more than 50% of the overall impedance of MCDC (Figure 4a-d). However, the MCDC achieved two-fold higher removal of the total dissolved solids. The EIS results suggest that electrodes and membranes in both the MFCs and the MCDCs suffered from the fouling issues (Shrestha et al. 2018). Both the non-Faradaic (ion electrosorption) and the Faradaic (oxidation-reduction) reactions can impede the performance of the CDI units, specifically by reducing the charge transport ability of porous carbon electrodes (He et al. 2016; Porada et al. 2013). EIS can be used to track the dynamics of ohmic resistance and impedance within the intraparticle pores during both salt adsorption and desorption cycles (Yoo et al. 2014).

Table 5: Results for equivalent circuit fitting of MFC and MCDC test reactors (Shrestha et al. 2018)

	MFC	**MCDC**
Test duration (d)	53	53
Total Impedance (Ω cm^2)	870	6600
Total Impedance (Ω)	22	933
Anode Impedance (Ω cm^2)	280	1150
Anode Impedance (Ω)	7	163
Cathode Impedance (Ω cm^2)	980	1490
Cathode Impedance (Ω)	25	211
Ohmic resistance (Ω cm^2)	180	110
Ohmic resistance (Ω)	5	16

Note: Chi squared (χ^2) values (goodness of fit) for the impedance data was in range of 0.00003 to 0.0002.
Source: Bioelectrochemistry 121: 56-64

3.4 Inductive Loops in Bioelectrochemical Systems

EIS can investigate the protein adsorption, generation of redox species and charge transfer resistances during different stages of biofilm growth on the electrode surfaces in BESs (Duuren et al. 2017). EIS analysis can reveal the biofilm interfacial processes, often observed as inductive loops in the low frequency (LFIL) regions of Nyquist plots. These 'inductive loops' refers to the impedance plot in the fourth quadrant. The following discussion will help the reader interpret the LFIL loops and predict the biofilm interfacial phenomena.

Considering that the metal electrodes based on stainless steel and copper are being promoted in BESs (Baudler et al. 2015), we can use the EIS tools to detect the inductance behavior of metal electrodes exposed to foulants (salt and metal contaminants) (Chilkoor et al. 2018a; Huang et al. 2014). The EIS can also be used for early detection of corroding metal electrodes, especially based on dominant LFIL spectra that signifies effects of passivation (due to oxide film formation), localized corrosion pits (Chilkoor et al. 2018b) and the formation of extracellular polymeric substance (EPS) and metal complexes. All of these effects are commonly observed in the microbial corrosion process. We provide a few examples from our ongoing research efforts focusing on the microbial corrosion behavior of metals coated with few layers of 2D materials, including graphene and hexagonal boron nitride (Chilkoor et al. 2018b). We will use this microbial corrosion platform to depict the use of EIS tools for characterizing the inductance effects (Chilkoor et al. 2018b). We observed the LFIL loop occurs due to the metal dissolution and consequent pit formation, both of which are associated with the microbial respiration processes or attack by the biogenic corrosive compounds (H^+, HS^-) (Chilkoor et al. 2017; Karn et al. 2017). In another unpublished study related to the microbial corrosion of copper electrodes coated with multi-layered graphene (MLG-Cu), we observed the LFIL that typically occurred due to the adsorption of atomic hydrogen and its subsequent formation to hydrogen sulfide (Reza 2008).

The formation of EPS is a key step during the biofilm development in BESs. The presence of LFIL in EIS data can be used to predict the formation of EPS layer on the electrode surface. Jin et al. ascribed the appearance of inductive semicircles in a low frequency region to the formation of EPS film and an adsorbed EPS-iron complex on cast iron surface (Jin et al. 2014). The LFIL can be used to study the material degradation and formation of complexes due to the formation of intermediates and side reactions. Breur et al. obtained two inductive loops in the formation of biological vivianite [$Fe_3(PO_4)_2.8H_2O$]. The first LFIL was attributed to adsorption of iron with EPS complex while the second loop was attributed to the formation of iron phosphate (Breur et al. 2002). In an MFC unit, undesirable reactions and their products can foul the electrodes and membrane. EIS can reveal the occurrence of such poisoning effects in an MFC (Roy et al. 2007). For instance, the sulfide, methanol and chloride contaminants in wastewater poison the exposed ORR catalysts of platinum-based cathodes (Yuan et al. 2016). We can also expect the formation of an oxide layer on the ORR cathodes. Roy et al. noticed inductive loops in an ORR mechanism due to the formation of peroxide intermediate species that reacted with a platinum electrode to form an oxide layer (Roy et al. 2007). Several authors have reported inductive loops due to poisoning of the anode by carbon monoxide in PEM fuel cells (Jiang et al. 2005; Zamel and Li 2008).

Inductance in an electrical circuit is mathematically represented as (Cesiulis et al. 2016)

$$E(t) = L\frac{dI(t)}{dt} \tag{3}$$

where $E(t)$ is sinusoidal potential that induces a sinusoidal current $I(t)$ at time t. L is the inductance.

The impedance (Z) of an inductor is given as:

$$Z = j\omega L \tag{4}$$

where j is a complex number that is ω is the radial frequency.

The impedance of inductance is proportional to the frequency, and the current through it lags the voltage by a phase angle of $90°$ (Cesiulis et al. 2016).

Readers are cautioned to minimize the misinterpretation of the LFIL that occur due to the artifacts related to interfacial phenomena. For instance, the low-conductivity electrolytes generate artifacts (inductive loops) due to the parasitic current flow and the stray capacitance in an electrochemical cell and electrical connections (Scully et al. 1993). The LFIL can appear in an EIS analysis when the system does not satisfy the stability conditions (i.e. linear, stable and casual). Figure 5 shows an example of the unstable system generating nonstationary artifacts in a low frequency inductive loop for copper coated with single layer hexagonal boron nitride (SL-hBN-Cu). The formation of LFIL artifacts caused by nonstationary conditions (unsteady state) can be validated with Kramers-Kronig relations (Chilkoor et al. 2018a). Figure 5 clearly shows that the impedance data do not conform to the Kramers-Kronig fit.

4. Conclusions

This chapter reviews the use of EIS for detecting and tracking redox active mediators, fouling and metal contaminants on the surfaces of electrodes and membrane. The EIS can also be used to analyze, monitor and characterize the resistance and capacitance aspects of biofilms. Future studies could potentially use the low frequency inductance behavior phenomena of the adsorbed proteins to study the different stages of biofilm development. We discussed the need to complement the EIS data with tests based on cyclic voltammetry, differential pulse voltammetry, scanning electrochemical microscopy, HPLC, UV-vis spectroscopy and LC-MS. Finally, owing to the complexity of the underlying mathematics (complex number theory and differential equations) and the skills required to analyze the EIS data, researchers

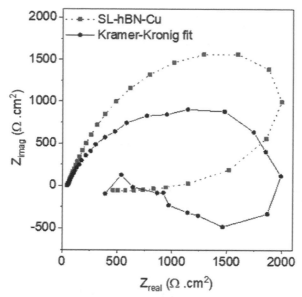

Figure 5: Nyquist plot showing nonstationary artifacts low frequency inductive loops for copper coated with single layer hexagonal boron nitride exposed to lactate medium containing sulfate reducing bacteria in a three-electrode electrochemical cell.

have often restricted the use of the EIS to simple equivalent electrical circuits. The power of EIS for deriving the mechanistic and kinetic information of the interfacial bioelectrochemical processes is yet to be unleashed.

5. Future Perspectives

Researchers have successfully used EIS technique for qualitative and quantitative analysis of biological phenomena in BESs, including biofilm growth, electron transfer mechanisms, biofouling, especially under the influence of electrochemical parameters (poised potential and choice of electron donors and electron acceptors), and material properties (electrodes, coating and catalysts). This fundamental knowledge will provide new strategies to optimize the BES reactor designs that will minimize the electron transfer resistances. Comprehensive EIS studies are required to address the complex questions concerning EET analysis and verify if this fundamental knowledge holds true in pilot-scale BES studies. The use of EIS techniques in a scaled-up BESs could face significant challenges due to complex issues related to the requirements for higher sensitivity, uniquely customized electrical circuit models, control experiments and ability to negate false-positive results. BES literature will benefit from a well-groomed approach to describe the complex EIS data interpretation. Advanced EIS protocols are required to assess the performance of next generation electrodes and membranes based on nanostructured materials, including carbon nanotubes, graphene and other inorganic nanomaterials. The EIS's high sensitivity and its ability to separate various events occurring on the electrode surface make it a versatile tool for developing bioelectrochemical technologies.

Acknowledgment

Gadhamshetty acknowledges the funding support from National Science Foundation CAREER Award (#1454102); NASA (NNX16AQ98A) and the South Dakota Board of Regents under the auspices of the Surface Engineering Research Center (SERC). Financial support was also provided by the National Science Foundation in the form of BuG ReMeDEE initiative (Award # 1736255).

References

Andoralov V, Falk M, Suyatin DB, Granmo M, Sotres J, Ludwig R, Popov VO, Schouenborg J, Blum Z, Shleev S (2013) Biofuel cell based on microscale nanostructured electrodes with inductive coupling to rat brain neurons. Sci. Rep. 3: 3270.

Babauta J, Renslow R, Lewandowski Z, Beyenal H (2012) Electrochemically active biofilms: facts and fiction. A review. Biofouling 28(8): 789-812.

Bakonyi P, Koók L, Kumar G, Tóth G, Rózsenberszki T, Nguyen DD, Chang SW, Zhen G, Bélafi-Bakó K, Nemestóthy N (2018) Architectural engineering of bioelectrochemical systems from the perspective of polymeric membrane separators: A comprehensive update on recent progress and future prospects. J. Membr. Sci. 564: 508-522.

Baudler A, Schmidt I, Langner M, Greiner A, Schröder U (2015) Does it have to be carbon? Metal anodes in microbial fuel cells and related bioelectrochemical systems. Energ. Environ. Sci. 8(7): 2048-2055.

Beyenal H, Babauta JT (2015) Biofilms in Bioelectrochemical Systems: From Laboratory Practice to Data Interpretation. John Wiley and Sons, Hoboken, NJ.

Blake RC, Howard GT, McGinness S (1994) Enhanced yields of iron-oxidizing bacteria by *in situ* electrochemical reduction of soluble iron in the growth medium. Appl. Environ. Microb. 60(8): 2704-2710.

Bond DR, Lovley DR (2003) Electricity production by Geobacter sulfurreducens attached to electrodes. Appl. Environ. Microb. 69(3): 1548-1555.

Breur H, de Wit J, Van Turnhout J, Ferrari G (2002) Electrochemical impedance study on the formation of biological iron phosphate layers. Electrochim. Acta 47(13-14): 2289-2295.

Call D, Logan BE (2008) Hydrogen production in a single chamber microbial electrolysis cell lacking a membrane. Envir. Sci. Tech. 42(9): 3401-3406.

Cao X, Huang X, Liang P, Xiao K, Zhou Y, Zhang X, Logan BE (2009) A new method for water desalination using microbial desalination cells. Envir. Sci. Tech. 43(18): 7148-7152.

Cen J, Kavanagh J, Coster H, Barton G (2013) Fouling of reverse osmosis membranes by cane molasses fermentation wastewater: detection by electrical impedance spectroscopy techniques. Desalin. Water Treat. 51(4-6): 969-975.

Cesiulis H, Tsyntsaru N, Ramanavicius A, Ragoisha G (2016) The study of thin films by electrochemical impedance spectroscopy. *In*: Tiginyanu, I., Topala, P., Ursaki, V. (eds). Nanostructures and Thin Films for Multifunctional Applications. Springer, Cham, Switzerland, pp. 3-42.

Chandrasekhar K, Amulya K, Mohan SV (2015) Solid phase bio-electrofermentation of food waste to harvest value-added products associated with waste remediation. Waste Manage. 45: 57-65.

Chen B-Y, Xu B, Yueh P-L, Han K, Qin L-J, Hsueh C-C (2015) Deciphering electron-shuttling characteristics of thionine-based textile dyes in microbial fuel cells. J. Taiwan Inst. Chem. E. 51: 63-70.

Chilkoor G, Upadhyayula VK, Gadhamshetty V, Koratkar N, Tysklind M (2017) Sustainability of renewable fuel infrastructure: a screening LCA case study of anticorrosive graphene oxide epoxy liners in steel tanks for the storage of biodiesel and its blends. Environ. Sci.: Processes Impacts 19(2): 141-153.

Chilkoor G, Shrestha N, Soeder D, Gadhamshetty V (2018a) Corrosion and environmental impacts during the flowback water disposal associated with the Bakken shale. Corros. Sci. 133: 48-60.

Chilkoor G, Karanam SP, Star S, Shrestha N, Sani RK, Upadhyayula VK, Ghoshal D, Koratkar NA, Meyyappan M, Gadhamshetty V (2018b) Hexagonal boron nitride: the thinnest insulating barrier to microbial corrosion. ACS Nano 12(3): 2242-2252.

Clauwaert P, Van der Ha D, Boon N, Verbeken K, Verhaege M, Rabaey K, Verstraete W (2007) Open air biocathode enables effective electricity generation with microbial fuel cells. Envir. Sci. Tech. 41(21): 7564-7569.

Clauwaert P, Aelterman P, De Schamphelaire L, Carballa M, Rabaey K, Verstraete W (2008) Minimizing losses in bio-electrochemical systems: the road to applications. Appl. Microbiol. Biot. 79(6): 901-913.

Cole KS, Cole RH (1941) Dispersion and absorption in dielectrics I. Alternating current characteristics. J. Chem. Phys. 9(4): 341-351.

Cusick RD, Kim Y, Logan BE (2012) Energy capture from thermolytic solutions in microbial reverse-electrodialysis cells. Science 335(6075): 1474-1477.

Daniels JS, Pourmand N (2007) Label-free impedance biosensors: Opportunities and challenges. Electroanal. 19(12): 1239-1257.

Davidson DW, Cole RH (1951) Dielectric relaxation in glycerol, propylene glycol, and n-propanol. J. Chem. Phys. 19(12): 1484-1490.

Dolin P, Ershler B (1940) The kinetics of discharge and ionization of hydrogen adsorbed at Pt-electrode. Acta Physicochim URSS 13: 747.

Dominguez-Benetton X, Sevda S, Vanbroekhoven K, Pant D (2012) The accurate use of impedance analysis for the study of microbial electrochemical systems. Chem. Soc. Rev. 41(21): 7228-7246.

Duuren JB, Müsken, M, Karge, B, Tomasch J, Wittmann C, Häussler S, Brönstrup M (2017) Use of Single-Frequency Impedance Spectroscopy to Characterize the Growth Dynamics of Biofilm Formation in Pseudomonas aeruginosa. Sci. Rep. 7(1): 5223.

Feng C, Li F, Liu H, Lang X, Fan S (2010) A dual-chamber microbial fuel cell with conductive film-modified anode and cathode and its application for the neutral electro-Fenton process. Electrochim. Acta 55(6): 2048-2054.

Feng J, Qian Y, Wang Z, Wang X, Xu S, Chen K, Ouyang P (2018) Enhancing the performance of Escherichia coli-inoculated microbial fuel cells by introduction of the phenazine-1-carboxylic acid pathway. J. Biotechnol. 275: 1-6.

Forrestal C, Xu P, Ren Z (2012) Sustainable desalination using a microbial capacitive desalination cell. Energ. Environ. Sci. 5(5): 7161-7167.

Forrestal C, Stoll Z, Xu P, Ren ZJ (2015) Microbial capacitive desalination for integrated organic matter and salt removal and energy production from unconventional natural gas produced water. Environ. Sci.-Wat. Res. 1(1): 47-55.

Forrestal C, Haeger A, Dankovich IV L, Cath TY, Ren ZJ (2016) A liter-scale microbial capacitive deionization system for the treatment of shale gas wastewater. Environ. Sci.-Wat. Res. 2(2): 353-361.

Fultz ML, Durst RA (1982) Mediator compounds for the electrochemical study of biological redox systems: a compilation. Anal. Chim. Acta. 140(1): 1-18.

Gadhamshetty V, Belanger D, Gardiner C-J, Cummings A, Hynes A (2013) Evaluation of Laminaria-based microbial fuel cells (LbMs) for electricity production. Bioresource Technol. 127: 378-385.

Ge Z, Zhang F, Grimaud J, Hurst J, He Z (2013) Long-term investigation of microbial fuel cells treating primary sludge or digested sludge. Bioresource Technol. 136: 509-514.

Géniès L, Bultel Y, Faure R, Durand R (2003) Impedance study of the oxygen reduction reaction on platinum nanoparticles in alkaline media. Electrochim. Acta 48(25-26): 3879-3890.

Gregory KB, Bond DR, Lovley DR (2004) Graphite electrodes as electron donors for anaerobic respiration. Environ. Microbiol. 6(6): 596-604.

Harnisch F, Schröder U, Scholz F (2008) The suitability of monopolar and bipolar ion exchange membranes as separators for biological fuel cells. Envir. Sci. Tech. 42(5): 1740-1746.

Harnisch F, Warmbier R, Schneider R, Schröder U (2009) Modeling the ion transfer and polarization of ion exchange membranes in bioelectrochemical systems. Bioelectrochemistry 75(2): 136-141.

Haque N, Cho D, Kwon S (2015) Characteristics of electricity production by metallic and nonmetallic anodes immersed in mud sediment using sediment microbial fuel cell. IOP Conference Series: Mater. Sci. Eng. IOP Publishing. pp. 012072.

He D, Wong CE, Tang W, Kovalsky P, Waite TD (2016) Faradaic reactions in water desalination by batch-mode capacitive deionization. Environ. Sci. Tech. Let. 3(5): 222-226.

He Z, Minteer SD, Angenent LT (2005) Electricity generation from artificial wastewater using an upflow microbial fuel cell. Envir. Sci. Tech. 39(14): 5262-5267.

He Z, Shao H, Angenent LT (2007) Increased power production from a sediment microbial fuel cell with a rotating cathode. Biosens. Bioelectron. 22(12): 3252-3255.

He Z, Huang Y, Manohar AK, Mansfeld F (2008) Effect of electrolyte pH on the rate of the anodic and cathodic reactions in an air-cathode microbial fuel cell. Bioelectrochemistry 74(1): 78-82.

He Z, Mansfeld F (2009) Exploring the use of electrochemical impedance spectroscopy (EIS) in microbial fuel cell studies. Energ. Environ. Sci. 2(2): 215-219.

Hosseini MG, Ahadzadeh I (2013) Electrochemical impedance study on methyl orange and methyl red as power enhancing electron mediators in glucose fed microbial fuel cell. J. Taiwan Inst. Chem. E. 44(4): 617-621.

Huang B, Jia N, Chen L, Tan L, Yao S (2014) Electrochemical impedance spectroscopy study on polymerization of L-lysine on electrode surface and its application for immobilization and detection of suspension cells. Anal. Chem. 86(14): 6940-6947.

Huang Y-X, Liu X-W, Xie J-F, Sheng G-P, Wang G-Y, Zhang, Y-Y, Xu A-W, Yu H-Q (2011) Graphene oxide nanoribbons greatly enhance extracellular electron transfer in bio-electrochemical systems. Chem. Commun. 47(20): 5795-5797.

Huang Y, He Z, Kan J, Manohar AK, Nealson KH, Mansfeld F (2012) Electricity generation from a floating microbial fuel cell. Bioresource Technol. 114: 308-313.

Jacobson KS, Drew DM, He Z (2011) Efficient salt removal in a continuously operated upflow microbial desalination cell with an air cathode. Bioresource Technol. 102(1): 376-380.

Jeuken LJ, Weiss SA, Henderson PJ, Evans SD, Bushby RJ (2008) Impedance spectroscopy of bacterial membranes: Coenzyme-Q diffusion in a finite diffusion layer. Anal. Chem. 80(23): 9084-9090.

Jiang R, Kunz HR, Fenton JM (2005) Electrochemical oxidation of H_2 and H_2/CO mixtures in higher temperature (T cell 100° C) proton exchange membrane fuel cells: Electrochemical impedance spectroscopy. J. Electrochem. Soc. 152(7): A1329-A1340.

Jin J, Wu G, Zhang Z, Guan Y (2014) Effect of extracellular polymeric substances on corrosion of cast iron in the reclaimed wastewater. Bioresource Technol. 165: 162-165.

Jing Y, Chaplin BP (2016) Electrochemical impedance spectroscopy study of membrane fouling characterization at a conductive sub-stoichiometric TiO_2 reactive electrochemical membrane: Transmission line model development. J. Membrane Sci. 511: 238-249.

Jung S-H, Ahn Y-H, Oh S-E, Lee J-H, Cho K-T, Kim Y-J, Kim M-W, Shim, J.-M, Kang, M-S (2012) Impedance and thermodynamic analysis of bioanode, abiotic anode, and riboflavin-amended anode in microbial fuel cells. Bull. Korean Chem. Soc. 33(10): 3349-3354.

Karn SK, Fang G, Duan J (2017) Bacillus sp. Acting as Dual Role for Corrosion Induction and Corrosion Inhibition with Carbon Steel (CS). Front. Microbiol. 8: 2038.

Kashyap D, Dwivedi PK, Pandey JK, Kim YH, Kim GM, Sharma A, Goel S (2014) Application of electrochemical impedance spectroscopy in bio-fuel cell characterization: A review. Int. J. Hydrogen Energ. 39(35): 20159-20170.

Kim HJ, Park HS, Hyun MS, Chang IS, Kim M, Kim BH (2002) A mediator-less microbial fuel cell using a metal reducing bacterium, Shewanella putrefaciens. Enzyme Microb. Tech. 30(2): 145-152.

Kim JR, Cheng S, Oh S-E, Logan BE (2007) Power generation using different cation, anion, and ultrafiltration membranes in microbial fuel cells. Envir. Sci. Tech. 41(3): 1004-1009.

Kim JR, Premier GC, Hawkes FR, Dinsdale RM, Guwy AJ (2009) Development of a tubular microbial fuel cell (MFC) employing a membrane electrode assembly cathode. J. Power Sources 187(2): 393-399.

Lasia A (2002) Electrochemical impedance spectroscopy and its applications. In: Conway, B.E., Bockris, J.O., White, R.E. (eds). Modern Aspects of Electrochemistry. Springer, Boston, MA, pp. 143-248.

Li Y, Chen S-M, Wu T-Y, Ku S-H, Ali MA, AlHemaid FM (2012) Immobilization of laccase into poly (3, 4-ethylenedioxythiophene) assisted biocathode for biofuel cell applications. Int. J. Electrochem. Sc. 7: 11400-11413.

Li W-W, Sheng G-P, Liu X-W, Yu H-Q (2011) Recent advances in the separators for microbial fuel cells. Bioresource Technol. 102(1): 244-252.

Liu H, Ramnarayanan R, Logan BE (2004) Production of electricity during wastewater treatment using a single chamber microbial fuel cell. Envir. Sci. Tech. 38(7): 2281-2285.

Liu X, Li Z, Yang Y, Liu P, Zhang P (2016) Electricity generation from a refuelable glucose alkaline fuel cell with a methyl viologen-immobilized activated carbon anode. Electrochim. Acta 222: 1430-1437.

Logan BE, Hamelers B, Rozendal R, Schröder U, Keller J, Freguia S, Aelterman P, Verstraete W, Rabaey K (2006) Microbial fuel cells: Methodology and technology. Envir. Sci. Tech. 40(17): 5181-5192.

Logan BE (2009) Exoelectrogenic bacteria that power microbial fuel cells. Nat. Rev. Microbiol. 7(5): 375.

Lvovich VF (2012) Impedance Spectroscopy: Applications to Electrochemical and Dielectric Phenomena. John Wiley and Sons. New York, NY.

Macdonald JR, Barsoukov E (2005) Impedance spectroscopy: Theory, experiment, and applications. History 1(8): 1-13.

Manohar AK, Bretschger O, Nealson KH, Mansfeld F (2008) The use of electrochemical impedance spectroscopy (EIS) in the evaluation of the electrochemical properties of a microbial fuel cell. Bioelectrochemistry 72(2): 149-154.

Marsili E, Rollefson JB, Baron DB, Hozalski RM, Bond DR (2008) Microbial biofilm voltammetry: Direct electrochemical characterization of catalytic electrode-attached biofilms. Appl. Environ. Microb. 74(23): 7329-7337.

Martin E, Tartakovsky B, Savadogo O (2011) Cathode materials evaluation in microbial fuel cells: A comparison of carbon, Mn2O3, Fe2O3 and platinum materials. Electrochim. Acta 58: 58-66.

Meng F, Jiang J, Zhao Q, Wang K, Zhang G, Fan Q, Wei L, Ding J, Zheng Z (2014) Bioelectrochemical desalination and electricity generation in microbial desalination cell with dewatered sludge as fuel. Bioresource Technol. 157: 120-126.

Min B, Poulsen FW, Thygesen A, Angelidaki I (2012) Electric power generation by a submersible microbial fuel cell equipped with a membrane electrode assembly. Bioresource Technol. 118: 412-417.

Miskan M, Ismail M, Ghasemi M, Jahim JM, Nordin D, Bakar MHA (2016) Characterization of membrane biofouling and its effect on the performance of microbial fuel cell. Int. J. Hydrogen Energ. 41(1): 543-552.

Nernst W (1894) Methode zur bestimmung von dielektrizitätskonstanten. Zeitschrift für Physikalische Chemie 14(1): 622-663.

Orazem ME, Tribollet B (2011) Electrochemical Impedance Spectroscopy. John Wiley and Sons. Hoboken, NJ.

Park DH, Zeikus JG (2000) Electricity generation in microbial fuel cells using neutral red as an electronophore. Appl. Environ. Microb. 66(4): 1292-1297.

Park S, Yoo J (2003) With impedance data, a complete description of an electrochemical system is possible. Anal. Chem. 75(21): 455-461.

Perez J, Gonzalez ER, Ticianelli EA (1998) Impedance studies of the oxygen reduction on thin porous coating rotating platinum electrodes. J. Electrochem. Soc. 145(7): 2307-2313.

Piela P, Fields R, Zelenay P (2006) Electrochemical impedance spectroscopy for direct methanol fuel cell diagnostics. J. Electrochem. Soc. 153(10): A1902-A1913.

Porada S, Zhao R, Van Der Wal A, Presser V, Biesheuvel P (2013) Review on the science and technology of water desalination by capacitive deionization. Prog. Mater. Sci. 58(8): 1388-1442.

Rabaey K, Boon N, Siciliano SD, Verhaege M, Verstraete W (2004) Biofuel cells select for microbial consortia that self-mediate electron transfer. Appl. Environ. Microb. 70(9): 5373-5382.

Raistrick I (1990) Impedance studies of porous electrodes. Electrochim. Acta 35(10): 1579-1586.

Ramasamy RP, Gadhamshetty V, Nadeau LJ, Johnson GR (2009) Impedance spectroscopy as a tool for non-intrusive detection of extracellular mediators in microbial fuel cells. Biotechnol. Bioeng. 104(5): 882-891.

Reza J (2008) Microbiologically Influenced Corrosion an Engineering Insight. Springer-Verlag London Limited.

Roy SK, Orazem ME, Tribollet B (2007) Interpretation of low frequency inductive loops in PEM fuel cells. J. Electrochem. Soc. 154(12): B1378-B1388.

Scully JR, Silverman DC, Kendig MW (1993) Electrochemical Impedance: Analysis and Interpretation. ASTM Philadelphia.

Sekar N, Ramasamy RP (2013) Electrochemical impedance spectroscopy for microbial fuel cell characterization. J. Microb. Biochem. Technol. S6(2).

Sevda S, Sreekrishnan T (2012) Effect of salt concentration and mediators in salt bridge microbial fuel cell for electricity generation from synthetic wastewater. J. Environ. Sci. Heal., Part A 47(6): 878-886.

Sevda S, Dominguez-Benetton X., Vanbroekhoven K, Sreekrishnan T, Pant D (2013) Characterization and comparison of the performance of two different separator types in air–cathode microbial fuel cell treating synthetic wastewater. Chem. Eng. J. 228: 1-11.

Sharma M, Bajracharya S, Gildemyn S, Patil SA, Alvarez-Gallego Y, Pant D, Rabaey K, Dominguez-Benetton X (2014) A critical revisit of the key parameters used to describe microbial electrochemical systems. Electrochim. Acta 140: 191-208.

Shrestha N, Fogg A, Wilder J, Franco D, Komisar S, Gadhamshetty V (2016) Electricity generation from defective tomatoes. Bioelectrochemistry 112: 67-76.

Shrestha N, Chilkoor G, Wilder J, Ren ZJ, Gadhamshetty V (2018) Comparative performances of microbial capacitive deionization cell and microbial fuel cell fed with produced water from the Bakken shale. Bioelectrochemistry 121: 56-64.

Singh RK, Devivaraprasad, R, Kar T, Chakraborty A, Neergat M (2015) Electrochemical impedance spectroscopy of oxygen reduction reaction (ORR) in a rotating disk electrode configuration: effect of ionomer content and carbon-support. J. Electrochem. Soc. 162(6): F489-F498.

Sluyters J, Oomen J (1960) On the impedance of galvanic cells: II. Experimental verification. Recueil des Travaux Chimiques des Pays-Bas 79(10): 1101-1110.

Song T-S, Wang D-B, Han S, Wu X-Y, Zhou CC (2014) Influence of biomass addition on electricity harvesting from solid phase microbial fuel cells. Int. J. Hydrogen Energ. 39(2): 1056-1062.

Springer T, Zawodzinski T, Wilson M, Gottesfeld S (1996) Characterization of polymer electrolyte fuel cells using AC impedance spectroscopy. J. Electrochem. Soc. 143(2): 587-599.

Stoll ZA, Forrestal C, Ren ZJ, Xu P (2015) Shale gas produced water treatment using innovative microbial capacitive desalination cell. J. Hazard. Mater. 283: 847-855.

Strik DP, Timmers RA, Helder M, Steinbusch KJ, Hamelers HV, Buisman CJ (2011) Microbial solar cells: applying photosynthetic and electrochemically active organisms. Trends Biotechnol. 29(1): 41-49.

Ter Heijne A, Hamelers HV, De Wilde V, Rozendal RA, Buisman CJ (2006) A bipolar membrane combined with ferric iron reduction as an efficient cathode system in microbial fuel cells. Envir. Sci. Tech. 40(17): 5200-5205.

Thygesen A, Poulsen FW, Min B, Angelidaki I, Thomsen AB (2009) The effect of different substrates and humic acid on power generation in microbial fuel cell operation. Bioresource Technol. 100(3): 1186-1191.

Torres CI, Kato Marcus A, Rittmann BE (2008) Proton transport inside the biofilm limits electrical current generation by anode-respiring bacteria. Biotechnol. Bioeng. 100(5): 872-881.

Torres CI, Krajmalnik-Brown R, Parameswaran P, Marcus AK, Wanger G, Gorby YA, Rittmann BE (2009) Selecting anode-respiring bacteria based on anode potential: Phylogenetic, electrochemical, and microscopic characterization. Envir. Sci. Tech. 43(24): 9519-9524.

Wang C-T, Liao F-Y, Liu K-S (2013) Electrical analysis of compost solid phase microbial fuel cell. Int. J. Hydrogen Energ. 38(25): 11124-11130.

Wang Q, Cen Z, Zhao J (2015) The survival mechanisms of thermophiles at high temperatures: An angle of omics. Physiology 30(2): 97-106.

Warburg E (1899) Ueber das Verhalten sogenannter unpolarisirbarer Elektroden gegen Wechselstrom. Annalen der Physik 303(3): 493-499.

Wu C-H, Yet-Pole I, Chiu Y-H, Lin C-W (2014) Enhancement of power generation by toluene biodegradation in a microbial fuel cell in the presence of pyocyanin. J. Taiwan Inst. Chem. E. 45(5): 2319-2324.

Xia L, Vemuri B, Saptoka S, Shrestha N, Chilkoor G, Kilduff J, Gadhamshetty V (2019) Antifouling membranes for bioelectrochemistry applications. *In*: Mohan, S.V., Varjani, S., Pandey, A. (eds). Microbial Electrochemical Technology, Elsevier, pp. 195-224.

Yoo HD, Jang JH, Ryu JH, Park Y, Oh SM (2014) Impedance analysis of porous carbon electrodes to predict rate capability of electric double-layer capacitors. J. Power Sources 267: 411-420.

Yuan H, Hou Y, Abu-Reesh IM, Chen J, He Z (2016) Oxygen reduction reaction catalysts used in microbial fuel cells for energy-efficient wastewater treatment: A review. Mater. Horiz. 3(5): 382-401.

Zamel N, Li X (2008) Transient analysis of carbon monoxide poisoning and oxygen bleeding in a PEM fuel cell anode catalyst layer. Int. J. Hydrogen Energ. 33(4): 1335-1344.

Zhao F, Slade RC, Varcoe JR (2009) Techniques for the study and development of microbial fuel cells: an electrochemical perspective. Chem. Soc. Rev. 38(7): 1926-1939.

Zuo Y, Cheng S, Call D, Logan BE (2007) Tubular membrane cathodes for scalable power generation in microbial fuel cells. Envir. Sci. Tech. 41(9): 3347-3353.

Understanding Bioelectrochemical Limitations via Impedance Spectroscopy

Abhijeet P. Borole*

Chemical and Biomolecular Engineering, The University of Tennessee, Knoxville
Bredesen Center for Interdisciplinary Research and Education, The University of Tennessee,
Knoxville 37996, USA

1. Introduction

Electrochemical impedance spectroscopy (EIS) has been applied to a number of electrochemical systems to gain insights into factors limiting the processes occurring at the interface of electrodes and electrolytes, or chemicals used for the production of electricity or use of electrons for generation of other products, such as hydrogen. Some of the applications include fuel cells, batteries, electrolysis cells, corrosion cells forming at reactive oxidative surfaces and other electrochemical systems. It is a unique and powerful technique to delineate the impedances and a method to represent the various components of the systems via resistive, inductive and capacitive elements within the electrochemical circuit. The challenge is to construct an accurate electrochemical representation of the electrical as well as nonelectrical components of the circuit. This is typically done via the use of equivalent circuit models (ECMs). Developing an accurate ECM is one of the most difficult tasks in the application of EIS, and describing the electrochemical phenomenon via representative models is the key to understanding the complexity in the real systems and the factors limiting the performance of these systems.

Electron transfer in biological systems is an upcoming area where EIS has found applications in recent years. Examples include protein bioelectrochemistry, biosensors, biological corrosion and bioelectrochemical systems (BESs). The interest in the use of renewable and sustainable resources for providing energy and materials for the modern society has led to the development of many applications using the BES concept that require the exploration of the interface between biology and electrochemistry. In this chapter, the focus is on the application of impedance spectroscopy to understand BESs. Some of the early work in this area was conducted by two groups, led by F. Mansfeld and D. Bond, who examined microbial fuel cells (MFCs) via EIS (Manohar et al. 2008; Marsili et al. 2008). An early review of the application of the technique to MFCs, reported by He and Mansfeld, described the basics of EIS for the application to MFCs (He and Mansfeld 2009). The use of EIS in improving the understanding of MFCs has occurred steadily over the last decade and has been a learning experience for researchers in the bioelectrochemical field since most of these researchers are not electrochemists by training. However, much advancement has been made with several key reviews documenting the improvements in understanding and discussing the mechanistic as well as application aspects of BESs

---

*Corresponding author: aborole@utk.edu

(Dominguez-Benetton et al. 2012; Sekar and Ramasamy 2013; Strik et al. 2008). Strik et al. reported on measurement modes and data validation for the use of EIS for microbial electrolysis cells (MECs) (Strik et al. 2008). The review by Sekar and Ramasamy focused on MFC studies conducted during the period 2009 to 2013 while providing a deep account of the methods and critical limitations impacting the progress of the MFC technology (Sekar and Ramasamy 2013). Dominguez-Benetton et al. (2012) described the use of graphical techniques at length to improve the application of EIS and discussed the use of nonideal components, such as constant phase elements for enabling a better understanding of the complex phenomena in BESs.

This chapter describes the EIS method in detail, followed by the equivalent circuit models and potential correlation of model parameters to physical, chemical, biological and electrochemical processes. A review of the current understanding of BESs using EIS is included with a focus on two primary BESs: MFCs and MECs. Reports identifying kinetic, mass and charge transfer characteristics of these systems via EIS analysis are discussed. Finally, the cell impedance is tied to further economic development of the BESs for commercializing MFCs and MECs.

2. Method

EIS is a nonintrusive or nondestructive technique that is operated in a two-electrode or three-electrode mode. In the three-electrode mode, the anode or cathode forms the working electrode, while the other serves as the counter electrode. This mode allows the examination of an individual electrode (the working electrode). A reference electrode (e.g., Ag/AgCl) is used as the third electrode that allows control of the potential applied to the working electrode. There are two primary modes of inquiry: potentiostatic and galvanostatic EIS based on voltage and current control respectively. In a two-electrode mode, the counter electrode serves as the reference electrode, in addition to serving as the electron receiver/provider that has some advantages for probing cells in which access to the reference electrode becomes difficult. A hybrid EIS method is another mode of application that is particularly used with the two-electrode mode. This mode allows examination of the whole cell, although the deconvolution of the EIS spectra becomes more complicated due to the inclusion of processes occurring at both electrodes.

The EIS technique uses a transient electrical signal, applied to the working electrode, which passes through the cell, and the response is being then received and collected at the counter electrode. The changes in the signal provide a blueprint of the impedances between the working and the counter electrode. The signal is an alternating current with amplitude of a few mV and frequencies in the range of 100 kHz to 1 mHz. Applied voltage for MFCs has ranged from 1 to 10 mV. The applied voltage is kept small enough so as not to affect the performance of the MFC but allow diagnosis of the impedances of the electrode/cell. In a simple electrical circuit where the current is flowing through a wire, Ohm's law describes the resistance to flow of electrons as:

$$R = \frac{V}{I} \tag{1}$$

The complex electrochemical phenomenon, however, requires the inclusion of additional impedance elements, including the imaginary component described by:

$$Z(\omega) = Z_0 \left(\cos\varphi + j\sin\varphi \right) \tag{2}$$

The data obtained from the impedance spectroscopy is commonly analyzed via two different graphical representations: the Bode plot and the Nyquist plot. The Nyquist or the Cole-Cole plot includes the real part of the impedance on the x-axis and the imaginary part of the impedance on the y-axis. Each frequency generates one point on this plot but is not represented on the plot itself. The impedance and phase angle can be represented as follows:

$$|Z| = \sqrt{Z_r^2 + Z_j^2} \tag{3}$$

$$\varphi = \tan^{-1}\left[\frac{Z_r}{Z_f}\right] \tag{4}$$

The Bode plot shows the impedance as well as phase angle vs. the frequency. Examples of the two plots are shown in Figure 1. The value of the phase angle is indicative of the resistive vs. capacitive components of the impedance. EIS is a very powerful technique because the output includes changes in the magnitude as well as the phase of the signal and the response is obtained at a steady state. However, experiments have to be done over a relatively short time frame (e.g., 20-30 minutes) over which the system does not change significantly. For biological systems, the dynamic response of the system corresponding to changes in process parameters can be obtained since microbial growth is slower. It should, however, be noted that expression changes can occur in the biological systems during this time frame and proper judgment should be exercised in attributing cause-effect relationships. The validation of the data for assessing should be performed via attention to the following parameters: linearity, causality, stability and finiteness as reported by Strik et al. (Strik et al. 2008).

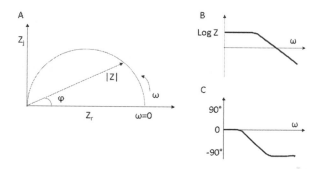

Figure 1: (A). Nyquist plot, (B) and (C) Bode plots showing amplitude and phase angle vs. frequency.

The data shown on the plots is fitted with ECM-based fitting to obtain parameters describing the processes occurring in the system. The circuit typically consists of resistors, capacitors and inductors. The impedance relationships for these elements in their ideal form can be given as follows:

$$R = Z_r = \frac{V}{I} \tag{5}$$

$$Z_L = j\omega L \tag{6}$$

$$Z_C = \frac{1}{j\omega C} \tag{7}$$

The various elements can be arranged in series or parallel to obtain a representative model of the BES. The mathematical treatment of the circuits has been explained in general terms by several authors and by commercial application package vendors associated with electrochemical equipment providers (Beasley 2015; Sekar and Ramasamy 2013).

2.1 Electrochemical Impedance Elements

Key elements of the ECM are described in this section.

2.1.1 Electrolyte Resistance

The space between the anode and cathode is usually filled with a solid or liquid electrolyte to allow a continuous path for ions to move. In BESs it is typically an aqueous solution, with or without a membrane partition. The resistance of the medium between electrodes is an essential element to be included in the ECM. The reader is referred to Beasley (2015) for mathematical descriptions of the electrochemical elements. The resistance can be determined by fitting the data to a model as described in Section 2.2.2.

2.1.2 Double Layer Capacitance

Any electrode surface in contact with liquid electrolyte can result in accumulation of ions upon polarization. The charges result in attracting counter ions to form a double layer. There is a small gap at the interface between the electrode and surrounding electrolyte, on the order or angstroms that acts as an insulating layer. This gives rise to double layer capacitance. The roughness of typical BES electrodes affects the capacitance and can result in a nonideal behavior. For such materials, a modified capacitance element is used called constant phase element (CPE).

2.1.3 Polarization Resistance

When an electrode is poised at a potential different from its potential under open circuit conditions, it is said to be polarized. Due to this polarization, it can cause current to flow through electrochemical reactions that can occur at the electrode surface. The current is determined by the kinetics of the reactions occurring at the electrode and the diffusion of reactants and products. The resistance associated with this polarization is called polarization resistance. It is typically used in corrosion studies to describe the corrosion current and resistance. There are two reactions occurring during corrosion that result in the anodic and cathodic current. However, the concept is only applicable when the anodic or cathodic reaction is occurring that may be the case in BESs.

2.1.4 Charge Transfer Resistance

Current can also be generated by a single electrochemically controlled reaction at equilibrium with its surroundings. An example from conventional electrochemistry is a metal dissolution reaction. In BESs, it can be a biochemical reaction occurring at the electrode that is producing or consuming electrons. The reaction essentially results in a charge transfer and has a resistance associated with it that is defined as charge transfer resistance (CTR). CTR can be described by the following equation, where i_0 is the exchange current density:

$$R_{CT} = \frac{RT}{nFi_0} \tag{8}$$

The exchange current density can be calculated from R_{CT} that can be determined via EIS modeling.

2.1.5 Diffusion

In BESs, the biochemical reactions occurring at the anode/cathode can encounter diffusion resistance due to concentration gradients between bulk liquid and surface of the electrode. The diffusion can be electrochemically described via a Warburg Impedance. The diffusion frequency and diffusion resistance can be calculated from the impedance data to derive the boundary layer thickness, δ, and the diffusion resistance (Beasley 2015; Borole and Lewis 2017; Ter Heijne et al. 2011).

2.2 Analysis

2.2.1 Framework Representing the Flow of Charges in BES

Understanding the processes occurring in a BES requires an accurate representation of the flow of charges associated with the electrochemical changes accompanying physical, chemical and biological processes. Therefore, a good characterization of the physical boundaries and the processes that can occur in the system is essential. In BESs, electrons can be generated via a microbial/enzymatic reaction from an organic electron donor or substrate. Abiotic anodes can be used as well with biotic cathodes. Figure 2 shows a series of steps through which charges are carried over from the electron donor to the terminal electron acceptor and across the electrodes. The first step is the substrate delivery to the reactive surface (Step A1 in Figure 2) that can be the electrode surface or the catalyst surface. The reactive surface does not necessarily need to be in direct contact with the electrode at the time of the reaction. Due to the possibility of mediated electron transfer (MET), the reaction producing electrons can take place in the bulk liquid, away from the electrode surface. The reaction generating electrons is the second step in the charge transfer process (A2) where the electrons go from the substrate to the product of the anodic half-reaction. The product can be an electron itself or an electron carrier. The following step is the transfer of electrons from the reactive surface to the electrode (A3a) that can be a mediated or direct electron transfer (DET). The electrochemical reaction also generates a counter ion that is typically a proton. The next step, thus, is proton transfer or charge transfer (A3b) that occurs simultaneously with the electron transfer. This may involve multiple components if the anode matrix is complex, such as a three-dimensional porous electrode. The initial step of charge/proton transfer may be its movement from the catalytic surface to the bulk liquid. The follow-up step will be its movement from the bulk liquid in anode to the bulk liquid in cathode across the membrane (step 3b). If an anion exchange membrane is used between the anode and the cathode, the anions have to transfer from the cathode to the anode to balance the charge. In the absence of a membrane or separator, this step still exists but may only consist of transfer of the charge between bulk liquid near the anode to bulk liquid near the cathode. Alternately, the reverse flow from the cathode to the anode is also possible, both of which may take place via diffusion or convection. The final step on the anode side is the product transfer from the reactive surface to the bulk liquid (A5) that may further consist of one or more mass transfer steps, depending on the complexity of the electrode structure.

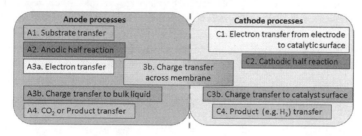

Figure 2: Anode and cathode processes occurring in BES involving electron, proton or charge transfer.

On the cathode side, electrons and protons have to be available at the reactive surface for the cathodic half-reaction to proceed. Thus, electron transfer from the cathode to the catalytic site is the first step (C1), followed by the reaction (C2). In addition to charge transfer across membrane, further transport of the charge to or from the cathode catalytic surface to the bulk (C3) may also exist, similar to that in the anode. The charge transfer process is essential to the overall process since without this step the rate of the catholic half-reaction and, consequently, the overall reaction rate comes close to zero. The last step is product transfer from the catalytic surface to the bulk cathode fluid (C4) that is subsequently removed from the reactor and collected outside the cell.

Figure 3 depicts each of these process and reaction steps and localizing them within a BES to allow visualization of the sequence of these steps. Here, we use the example of an MEC with a porous anode electrode and a nonporous material at the cathode. Different configurations are used at the anode and the cathode to allow a discussion of relevant processes associated with each of these electrode structures. The process includes all the steps shown in Figure 2. The steps include convection and/or diffusion through the chamber/porous electrode and reaction/s as well as charge transfer influenced by potential or mass transfer gradients. The products of the reaction are CO_2, protons and electrons; each will need to be transferred out of the microbial reaction compartments to the external environment and then into the bulk. More sophisticated models may take into account of the individual transfer steps for the three products; however, typically, only substrate, electron and proton transfer are considered.

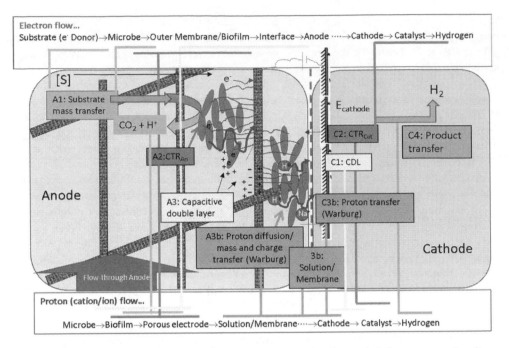

Figure 3: Sub-processes involved in bioelectrochemical reactors and their representation for model development. CTR: Charge Transfer Reaction, CDL, capacitance due to double layer.

A novel feature of BESs is the capacitive behavior of microbial biofilms. This is due to the presence of cytochromes in the membrane and extracellular appendages present in the biofilm. While this is internal to the cell, it can significantly impact the electron transfer process and the overall impedance. The transfer of protons from the reaction site to the bulk liquid can be represented by Warburg impedance. The flow of protons from the substrate in the anode to the product generated in the cathode is shown in Figure 3 as well. The liquids used in BESs are usually based in water that serves as a medium or electrolyte for the proton flow. The protons originate from the substrate that is first transported from the bulk anode liquid to the porous biofilm, then to the microbe and later enters into the microbial cell where the reaction takes place and the proton is released from the substrate. The proton then has to migrate outside the cell, travel through the biofilm to the bulk liquid, then to the membrane separating the anode from the cathode and from there, cross the membrane and reach the cathode electrode where it reacts with the electron and any other substrate at the catalytic surface to produce the product. Each of these steps results in impedance that is shown via a colored line in Figure 3.

In BESs, using the anion exchange membrane, protons are prevented from moving from the anode to the cathode, however, the charge transfer has to take place for the reaction to proceed. In this case, the anions present or generated in the cathode have to move from the cathode to the anode and combine with the protons or cations in the anode to reach charge neutrality. Thus, protons, cations or anions have to move from one electrode to another for the reaction to proceed.

2.2.2 ECMs and Circuits Used in BES

A typical ECM used to describe BES is shown in Figure 2. It includes a resistor and a capacitor for the anode and the cathode each and a resistor representing the membrane/electrolyte between the electrodes. A resistor and a capacitor in parallel with resistance, as shown for the membrane, comprise of a Randles cell that is a very common circuit in EIS analysis. The anode and the cathode can be assessed separately to reduce complexity of the resulting data that is a common practice. This is modeled by removing the other electrode components, but the solution resistance is common to both half-cell ECMs.

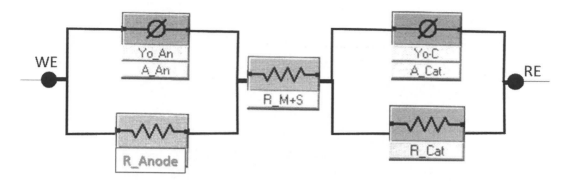

Figure 4: A typical equivalent circuit diagram for a BES. It includes a resistor and a non-ideal capacitor (constant phase element) in parallel for each of the two electrodes, anode and cathode and a resistor for the membrane/solution resistance.

The components of a BES, however, can be quite varied and as the field of BES is growing, many new types of BESs are being developed that results in the need to study new processes. As such, there is a need to expand the model used to describe BES to allow a deeper understanding of the individual steps occurring in the system. Impedance spectroscopy has the potential to provide clues to every individual step in the process if an ECM can be constructed and modeled. Alternatively, other methods using graphical techniques can be used (Dominguez-Benetton et al. 2012).

Figure 5 shows a more expanded version of the BES equivalent circuit model incorporating steps to describe the electron, proton and substrate transfer steps. While it is not a comprehensive diagram for all the components, it includes many of the common steps that may be encountered in a BES. The example shown is hypothetical and should not be used directly for modeling a BES. Rather, the elements shown should be used as a guide to developing models representative of the system being studied. It is important to understand the significance of various components in the BES, so Figure 5 can be used as a knowledge base to construct a suitable circuit diagram.

To begin with, the substrate has to be transferred from the bulk to the microbial surface for the reaction to take place. This is represented by a Warburg impedance. The substrate may encounter a double layer if one is present on the microbial surface that is represented by a CPE. A nonideal capacitance is applicable when a three-dimensional electrode is used or when a biofilm is present. However, this should be ascertained using the actual data (Strik et al. 2008; Zhao et al. 2009). This part of the circuit is followed by another component containing R_An_CTR that is the resistance due to charge transfer reaction. This represents the main reaction producing the electrons and protons. A CPE is included here in parallel to represent the double layer that may be encountered by the charged species during their transfer from the reaction site to the electrode. A Warburg impedance is included to reflect mass transfer of species, such as protons or other products. Figure 5 shows the electron as well as proton/ion transfer pathways in BESs but additionally points to the parallel nature of these two transfer trains. It is important to note that either can be a limiting step in the overall process, and thus both have to be considered equally in the analysis of these systems. Although the two pathways cannot be isolated in terms of their influence on the individual circuit elements, distinguishing between the two may be possible via variation of process, design or reaction parameters. This is where the design of experiments can become important. This can lead to the determination of the individual steps that are limiting. While ECMs for BES reported in the literature is limited in variety, many different models with a wide range of complexity have been developed for other electrochemical systems. Studies of corrosion, fuel cells and other systems have been conducted since many decades and may provide additional insights for readers via analogy to these systems (Orazem and Tribollet 2006; Springer et al. 1996).

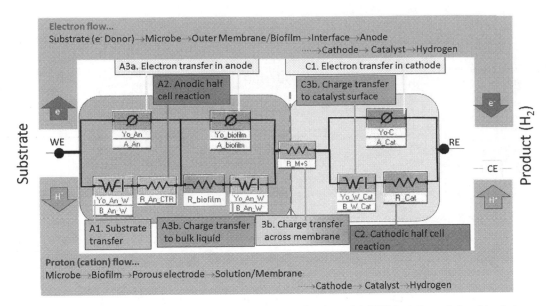

Figure 5: An example of hypothetical, complex equivalent circuit model (ECM). A few relationships between ECM impedance parameters and electron transfer or kinetic and mass/charge transfer processes are shown. However, these are typically not 1:1, and individual processes can influence multiple model parameters, therefore, experimental data should be carefully evaluated using appropriate models that take into account the BES design and process conditions being studied.

The more complex the circuit, the more the number of model parameters that have to be fitted. Thus, the experiments needed to confidently delineate individual model parameters multiply and require a high level of control to keep other components of the circuit relatively stable so as to isolate the effects of the parameters being studied. The rich nature of EIS data with changes in amplitude and phase angle over frequencies ranging 5-6 orders of magnitude can enable teasing out the effects even without large sets of data; however, the results may remain ambiguous. The key is the accuracy of ECM and data collection under stable operating conditions (Strik et al. 2008; Zhao et al. 2009). Identifying the correct configuration of expected elements in the circuit is the hardest task in the analysis of EIS data.

3. Understanding MFCs via EIS

3.1 Effect of MFC Design and Biofilm Growth

The study of MFCs initially focused on examining the influence of microbially-assisted electron transfer on impedance via application of the EIS to the MFC anodes. An early study by Mansfeld's group reported the effect of the growth of *Shewanella oneidensis* on anode impedance (Manohar et al. 2008). A 100-fold reduction in polarization resistance was reported with the growth of *S. oneidensis* MR-1. The lowest resistance obtained was 10.2 kΩ. The ECM they used included a simple version of the circuit, similar to that shown in Figure 4, except that they used half cell measurements and a simple capacitance rather than CPE. Several other researchers have also reported the effect of biofilm on anode impedance using other organisms or microbial consortia. Wang et al. reported a reduction in whole cell impedance under two different anode operation conditions to study the effect of a positively poised potential (Wang et al. 2009). This was attributed to differences in anode biofilm growth; however, half cell analysis on the anode was not conducted to confirm the observation. Ramasamy et al. reported changes in bioanode impedance during the initial biofilm growth (Ramasamy et al. 2008). A 70% reduction in the anode resistance was reported with a 120% increase in power density. The ECM they used included a Warburg resistance in series with

CTR and a CPE in parallel with those two elements. In addition, a solution resistance was also used in the series. The anode CTR was reported to drop to 0.48 kΩ cm^2 after three-weeks of microbial growth.

A study of bioanode performance reported by ter Heijne investigated the effect of substrate concentration on the anode CTR (ter Heijne et al. 2015). The lowest CTR reported was 0.144 kΩ cm^2 that reached after 24 days of growth. Microbial growth in the anode can continue for a long time, however, it is not likely that the impedance can drop forever. Borole et al. showed this by studying microbial biofilms in anode over the long term (Figure 6A) (Borole et al. 2010). After a period of about four months, the impedance of the anode was shown to have reached a steady state of about 0.017 kΩ cm^2 (1.31 Ω). Measurements taken in the fifth and sixth months indicated a similar anode and total impedance (Figure 6C). The impedance reached after the fourth month of biofilm growth (0.017 kΩ cm^2) was about 28-fold lower than that reached after three weeks of biofilm growth reported by Ramasamy et al. (Ramasamy et al. 2008). Figure 6B shows results from modeling of anode half cell using a CPE, CTR and solution resistance. The inset in Figure 6B shows the results in a range similar to that used in Figure 6A. The resemblance indicates the major role played by the anode during the initial period when the biofilm is growing and impacting anode CTR. These studies have been reported for the so-called mediatorless MFCs where conditions are optimized to reduce the influence of mediators on current production via their removal or use of flow through systems (Borole et al. 2010; ter Heijne et al. 2015). In these systems, electron transfer is typically maximized via direct electron transfer facilitated by conductive appendages, such as nanowires growing from electroactive microbes to electrode surfaces (Beegle and Borole 2017; Reguera et al. 2005).

The electron transfer within the bioanode can be accomplished by microbes capable of producing mobile mediators besides direct electron transfer. It has been reported that the mediated electron transfer may be slower than the direct electron transfer process (Torres et al. 2010) proposed by Reguera et al. and others (Gorby et al. 2006; Reguera et al. 2005). Ramasamy et al. conducted a detailed impedance analysis of mediated electron transfer in MFCs for determining the resistance of the electron transport using mediators (Ramasamy et al. 2009). They used *S. oneidensis* DSP-10 as the anode biocatalyst with oxygen and ferricyanide-based cathode to determine the impedances. They included three separate impedance components in their circuit for representing the microbial reaction, electron transfer and other electrochemical reactions. This is a good example of deciphering the complexity of a bioanode in an MFC. Furthermore, they modeled different catholytes used in their MFC using different ECMs to distinguish between the limiting steps in the overall system. They reported the CTR for air-cathode was much higher (1154 Ω) than that for ferricyanide cathode (110 Ω) (Ramasamy et al. 2009). They reported that in the bioanode they studied the resistance of the electron transfer step ranged from 16-39 Ω while for the charge transfer reaction, involving substrate conversion by the microbe, it was 291-417 Ω that corresponded to 5.8 kΩ cm^2 for their system. This is several orders of magnitude higher than that reported for biofilm-based anodes discussed earlier. Thus, these results also suggest that direct electron transfer can be faster than mediated electrons transfer; however, further research is needed with similar MFC designs and microbial catalysts to confirm the findings. A study focused on *Geobacter sulfurreducens* biofilms, capable of direct electron transfer, investigated the impedance of the biofilm itself. This is presented by the term R_biofilm in Figure 5 (Malvankar et al. 2012). They reported the determination of the biofilm resistance using EIS data as well as conductivity measurements and compared the magnitude of this resistance vs. charge transfer resistance of the anode. The biofilm resistance for one of the strains studied *G. sulfurreducens*, DL-1, was 1,208 Ω determined via EIS that compared well with 874 Ω determined via conductivity-based measurements. The anode CTR was reported to be 356 Ω indicating that the rate of electron generation was faster than the rate of electron transfer. Malvankar et al. also reported a correlation between the biofilm conductivity and current density suggesting a direct relationship; however, the data was not consistent for all strains they reported, particularly the highest performing strain KN400 (Malvankar et al. 2012). This was attributed to additional factors impacting the current production including proton transfer. This is certainly possible; however, research in this area using KN400 has not been reported.

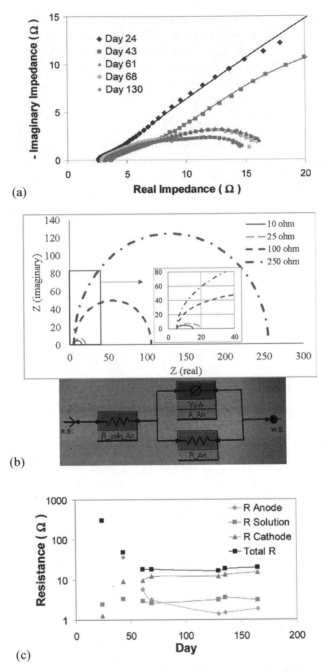

Figure 6: (A) Nyquist diagram showing effect of long-term microbial growth on anode impedance. (B) Results from modeling using a simple Randles cell shown in Figure, (C) Changes in impedance of anode, cathode and whole cell resistance as a function of time. The anode resistance obtained after 4 months was about 1.31 Ω (equivalent to 0.017 kΩ cm^2). (*Source:* Borole et al. 2010) Copyright: ACS Publications, reproduced with permission.

A study conducted by Fan et al. in 2008 used an alternate methodology for determining the internal resistance of a MFC (Fan et al. 2008). It is worth mentioning this study, even though they did not use EIS due to the interesting findings they reported. The study used a polarization curve and the intercepts on the curve to determine the internal resistance of the cell using a least squares method. A comparative analysis of the components of MFC revealed that the anode was not the limiting factor which had allowed more than one year for microbial growth in the anode for enabling the establishment of mature biofilms. The anode resistance they reported was 0.032 kΩ cm^2 (Fan et al. 2008) that was relatively low, indicating a well-developed bioanode. The cathode resistance, on the other hand, was reported to be 0.284 kΩ cm^2, suggesting that the overall power production was limited by the cathode. They also compared single and two-chamber MFCs that indicated the membrane to contribute a significant resistance on the order of 2.2 kΩ cm^2. Removal of the membrane and use of cathode with 14 times larger surface area was reported to reduce the cathode resistance to below that of the anode (Fan et al. 2008).

3.2 Effect of Process Conditions

The biological and electrochemical processes occurring in the anode and cathode obviously depend on the conditions present in the cell, so it is important to understand their effect. The MFCs use an aqueous electrolyte that is typically a nutrient medium for microbial growth but also serves as a buffer. Several studies have reported the effect of ionic strength on MFC performance but not in terms of impedance analysis. Aaron et al. quantified the effect via impedance spectroscopy (Aaron et al. 2010) through varying the ionic strength by an order of magnitude from 0.37 M to 0.037 M. Figure 7 shows the Nyquist plot from their experiment, which illustrates the effect on solution resistance, identified as the intercept of the impedance curve on the x-axis. The electrolyte properties also affect the CTR that is indicated by the change in the diameter of the arc seen in the Nyquist plot. A simple ECM consisting of the solution resistance in series with a parallel RC circuit (consisting one resistor in parallel with a capacitor) was used to illustrate the relationship (Figure 7). The solution resistance (R_soln_An) was varied from 2.5 to 4.5 ohm in this example using the ECM shown in Figure 6B. The ECM analysis reported by Aaron et al. was, however, for the whole cell that showed additional changes in the Nyquist curve. This was analyzed further by Aaron et al. via cathode impedance modeling and determined to be a large change in the cathode resistance, decreasing it from 16 Ω to 9 Ω that changed with ionic strength. It is interesting to note that although the nutrient medium was changed in the anode, the cathode resistance was found to change. This is explained by the proton diffusion phenomenon that depends on proton transfer from anode to cathode; thus, anode conditions influence the cathode impedance. Aaron et al. also reported the effect of fluid flow rate in anode on performance and impedance (Aaron et al. 2010). When the flow rate was increased from zero to 25 mL/min, the power density increased from around 500 mW/m^2 to 900 mW/m^2, and the cathode resistance decreased from 12 Ω to 8 Ω. This was suggested to be due to the increased charge transfer from the anode to cathode to support the electron generation at the anode.

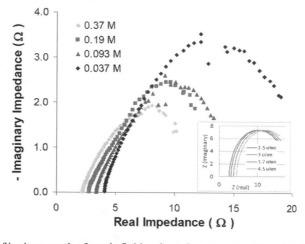

Figure 7: Effect of ionic strength of anode fluid on impedance spectra. The plot shows the change in solution resistance and electrode impedance. Inset shows modeling results using ECM in Figure 6B. (Aaron et al. 2010)

Proton transfer is an important step in the MFC and besides the effect of ionic/buffer strength, pH can be an important determinant of charge transfer in the MFC. He et al. reported the effect of pH on impedance in an MFC that they identified as polarization resistance (He et al. 2008). They reported minima in anode resistance around neutral pH, while cathode resistance decreased with increasing pH (Figure 8). It is important to note the order of magnitude difference between the anode and the cathode resistance. This was a single chamber system, so the same pH exists at the anode and the cathode that might explain the continued decrease in the cathode resistance at high pH. The slight drop in the cathode resistance at higher pH could be due to the anode reaction becoming limiting at high pH; however, this hypothesis needs to be examined further.

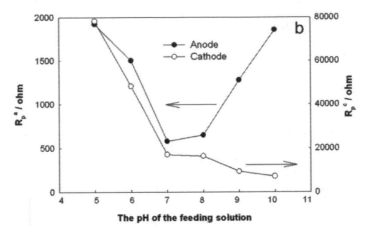

Figure 8: Effect of pH on anode and cathode impedance. (*Source*: He et al. 2008), Copyright: Elsevier, reproduced with permission.

Jung et al. reported a detailed examination of the effect of pH on anode impedance (Jung et al. 2011). They modeled two steps of the electron transfer process; however, the authors did not relate the model structure to physical processes in the anode. Due to the complexity of the microbial biofilms in anode, the authors preferred not to provide an exact correlation of processes, such as substrate conversion by the microbes, electron transfer, proton transfer, etc., to the impedance parameters. The ECM included the two components as shown in Figure 9a.

This model includes a pore resistance and associated CPE in parallel, together in series with the anode CTR. This model results in two arcs for the Nyquist curve (Figure 9b, top section), the one toward the left (high frequency) corresponding to the anode CTR and the large one on right (medium to lower frequency) corresponding to the pore resistance. The effect of pH and the anode half saturation voltage were evaluated. The authors reported that increasing pH from 6 to 8 favored anode performance and decreasing the resistance, at least until the anode potential was slightly above E_{ka}, the half-saturation voltage for the anode. The experimental results were nicely modeled by their ECM showing the effects in the high frequency region (HFR) and the medium frequency region; however, the results did not fit well in the low frequency range (LFR). The pH can have multiple effects on the processes occurring in the anode. This includes the biochemical reaction, double layer formation as well as the transfer of protons out of the biofilm. Their model did not specifically address the proton diffusion out of the biofilm. This was attributed to a lack of a Warburg diffusion element in the model, which will typically result in 45° line after the second arc that begins to appear as shown in Figure 9b.

The kinetic and capacitive effects of pH were, however, clearly illustrated, that primarily affected the pore resistance localized in the middle frequency region. The capacitance C2 showed a reverse behavior compared to resistance that increased with pH, up to the E_{ka}, and then decreased thereafter. The report identifies important EIS signatures related to overpotentials that are generally due to the electron and proton transfer processes but are not correlated to specific steps by the authors. The steps can be grossly linked to the physical or electrochemical processes as discussed in the text above. Correlating EIS data to the ECM model parameters is complicated, and this study shows how one parameter (pH) can affect multiple components of the model. The study also shows how EIS can be powerful in understanding the bioanode processes. Incorporating additional features into the ECM can lead to deeper deconvolution of the complex EIS data resulting in newer insights into optimization of BESs.

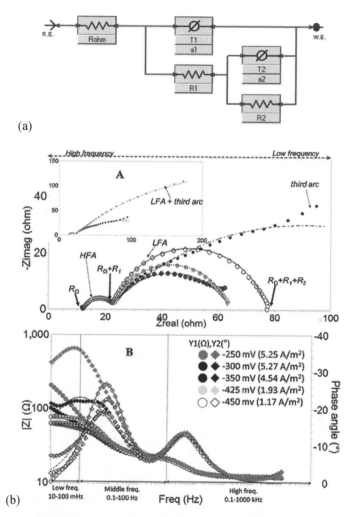

Figure 9: (A) ECM for a bioanode modeling effect of pH and Eh, reported by Jung et al. The circuit is representative of anode biochemical kinetics (R1) and pore resistance R2 and capacitive elements associated with it, (B) Nyquist and Bode plots for bioanode at various anode voltage (E_{an}). (*Source*: Jung et al. 2011), Copyright: ACS Publications, reproduced with permission.

3.3 MFC Impedance and Economic Feasibility

The cell impedance can be an important parameter with respect to economic feasibility of the MFCs as well. Sleutels et al. reported an analysis of the internal cell resistance and economics of MFCs at various loading rates and current densities (Figure 10). The resistance of 25 kΩ cm^2, represented by the dashed line, is necessary for economic feasibility. Assuming that the overall cell impedance is a sum of the anode, cathode and membrane/solution resistance, a cell impedance of 0.21 kΩ cm^2 was achieved by Borole et al., however, the maximum current density reached was only 6 A/m^2 (Borole et al. 2010). Based on Figure 10, a current density above 25 A/m² is required in combination with the low cell resistance. Thus, the performances reported to date indicate that economic feasibility has not been reached. Further improvements are, therefore, needed to commercialize the MFC technology. Reactor design as well as process conditions have to be identified to improve current density while maintaining the low resistance of the cell.

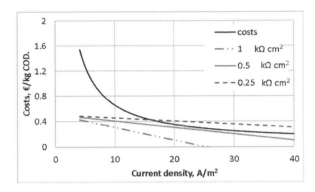

Figure 10: Cost and earnings for MFCs expressed per kg of COD removed for different internal resistances as a function of current density. The economically feasible region is the area above the solid line. For MFCs, an internal resistance 0.4 kΩ cm² has to be reached in combination with a current density of 25 A/m² for economic feasibility. Data courtesy of Dr. Tom Sleutels (Sleutels et al. 2012).

4. Understanding MECs via EIS

The ability to apply a potential difference across anode and cathode introduces a driving force for electrons to be transferred from the anode to the cathode. This changes the dynamics in the MEC vs. A MFC where the potential difference is determined by the natural conditions in the anode and cathode. The method to determine cell impedance is similar; however, the impedances are likely to be different, even for the anode, for the two systems. While a number of reports on determining the internal resistance of MECs exist, a detailed analysis of the MEC impedance via EIS is rare. In this section, data on impedance and overpotential analysis reported via other methods are included to give insights into the impedance of these systems with reference to the few studies that do report the use of EIS for understanding MECs.

A report recently used mature biofilms to evaluate limiting factors in achieving high current densities in MECs to enhance hydrogen production (Borole and Lewis 2017). EIS was used to determine the impedance of the anode, cathode and solution/membrane components. Figure 11 shows results at various conditions. The total cell impedance, based on EIS analysis that is the sum of all resistances reported, including the equilibrium resistance of 0.6 kΩ cm² (Sleutels et al. 2013), was 0.8 kΩ cm² while that based on current-voltage data was 1 kΩ cm². The discrepancy may be due to the existence of additional resistance not modeled in the ECM. The authors also reported a comparison of the overall cell impedance with other studies reporting a high performance in MECs (An and Lee 2013; Ki et al. 2016; Sleutels et al. 2013; Tartakousky et al. 2011; Zhang et al. 2010). It was found that the total cell impedance for these studies reported optimized MECs ranged from 0.8 to 1.9 kΩ cm². The total cell impedance is not particularly useful in optimizing MECs, so it is worthwhile to look deeper into the individual contributions from anode, cathode, etc.

Ki et al. recently studied the effect of pH and cathode materials in MECs and reported the overpotentials at the anode, cathode and membrane/solution that are indicative of impedance. The results were illustrated as voltage vs. current plots that provide an insight into the effect of the process conditions on impedance at various current densities. Figure 12 shows the overpotentials identifying the contributions of ohmic, anode, cathode as well as the pH gradient. The largest contribution was that from the cathode, especially at high current densities, which was investigated in the presence and absence of carbon dioxide bubbled into the cathode as a means to reduce bulk pH. They determined that the operational cell voltage could be reduced from 0.99 V to 0.89 V at a current density of 10 A/m^2 with addition of CO_2. This represented a total cell impedance of around 0.9 to 1 kΩ cm^2. At this current density, the anode impedance was about 0.21 kΩ cm^2 while the cathode contribution was 0.51 kΩ cm^2 with inclusion of CO_2 in the anode. The data plotted in Figure 15 shows that in high current densities the cathode as well as the anode impedances rise further while diminishing the beneficial effect of the CO_2.

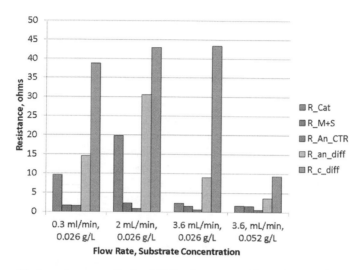

Figure 11: Impedance results for MEC using mature bioanode at various anode flow rates and substrate concentrations. R_Cat: cathode CTR, R_M+S: membrane+ solution resistance, R_An_CTR: anode CTR, R_an_diff and R_c_diff: diffusion resistance in anode and cathode, respectively. (*Source*: Borole and Lewis 2017). Copyright: Royal Society of Chemistry, reproduced with permission.

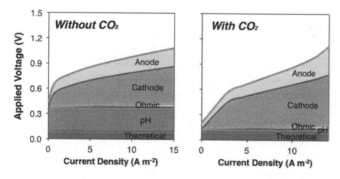

Figure 12: Cell resistance map for MEC investigating effect of pH and addition of CO_2 in cathode to control pH, (*Source*: Ki et al. 2016), Copyright: Elsevier, reproduced with permission.

A study by Sleutels et al. reported the effect of applied voltage on the impedances too using the overpotential method. At an applied potential of 1 V, the cell impedance they reported was similar to that reported by Ki et al. that was equal to 0.98 kΩ cm^2 (Sleutels et al. 2013). This included the anode and cathode resistance in the range of 0.1 to 0.15 kΩ cm^2. It should be noted that the contribution of the pH gradient between anode and cathode was included in equilibrium overpotential by Selutels et al., however this was shown separately by Ki et al. in Figure 12. In the study reported by Borole et al., the contribution to the overall resistance from cathode diffusion was the highest in the order of 0.1 kΩ cm^2 (Borole and

Lewis 2017). The studies using the overpotential method cannot distinguish between the contributions of kinetics vs. diffusion to the cathode resistance but EIS can. Ki et al. have discussed the high cathode overpotential due to proton limitations indicating the presence of diffusion issues (Ki et al. 2016). The current understanding of MEC impedances definitely points to the cathode as the limitation, but additional studies using EIS will help obtain greater insight into the kinetic vs. mass transfer limitations.

4.1 MEC Impedance and Economic Analysis

The performance of MECs is critically dependent on total cell impedance. An understanding of the individual impedance contributions is the key to optimizing MEC performance, however, reducing total cell impedance is key to improving performance since the impedances add up to affect the cell performance. Commercialization of the MEC technology requires minimization of total cell impedance. Understanding the relationship between economic feasibility and total cell impedance is, therefore, important. Sleutels et al. have reported an assessment of economic feasibility as a function of current density (Figure 13) (Sleutels et al. 2012). At a cell impedance of ~ 0.8 kΩ cm^2, the MEC requires a current density of < 20 A/m^2 to be economically feasible. As discussed earlier, several researchers have shown the ability to reach the cell impedance around 1 kΩ cm^2. The maximum current densities reached were in the range of 10-14 A/m^2 (An and Lee 2013; Borole and Lewis 2017; Ki et al. 2016; Sleutels et al. 2013; Zhang et al. 2010). Thus, at least at the laboratory scale, the performance of MECs has reached that required for economic feasibility. Further work should focus on scale-up of this technology while maintaining the low cell impedance and high performance that can lead to commercialization in the near future.

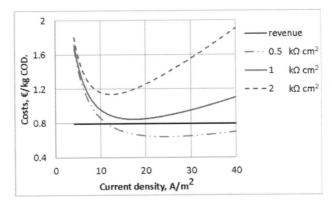

Figure 13: Relationship between current density and MEC costs at various cell impedances. Data courtesy of Dr. Tom Sleutels (Sleutels et al. 2012).

5. Conclusions

Bioelectrochemical systems are a novel technology with the potential for generation of clean energy from waste feedstocks. Many issues plague the commercialization of this technology; however, the complexity of the system prevents a deeper understanding of the limitations. Electrochemical impedance spectroscopy is a tool, which has a significant potential, to be applied in order to understand these issues. MFCs have been investigated in detail over the past decade, enabling identification of many factors affecting performance. Several of the known issues have been overcome via better design and electroactive biofilm development. This has helped faster development of MECs that have an anode similar to MFCs. Further work on investigating MECs using EIS can lead to targeted development of the technology and commercialization in the near future.

6. Future Perspectives

MFC and MEC are important new technologies for integrating waste management and clean energy production. Their continued development will lead to significant environmental benefits. The economic feasibility of MFCs, however, requires current densities above 25 A/m^2 that have not been reached except under special circumstances. Microbial electrolysis cells, on the other hand, have shown performance reaching overall cell impedance below 1 kΩ cm^2 and current densities in the 10-14 A/m^2 range. Charge and mass transfer issues in the cathode of MECs need further investigations to allow additional optimization. It is important to undertake scale-up studies using designs and process conditions that are identified optimal for MECs. If current densities above 15 A/m^2 can be reached at larger scales while maintaining cell impedance below 1 kΩ cm^2, MECs are likely to receive commercial consideration in the coming years.

Acknowledgment

Funding for this work from the Oak Ridge National Laboratory Seed Money Program, via the University of Tennessee, Knoxville, is acknowledged. The author would also like to acknowledge Dr. Tom Sleutels from the Wetsus Institute, Netherlands, for providing data for two figures (Figures 10 and 13).

References

Aaron D, Tsouris C, Hamilton CY, Borole AP. (2010) Assessment of the effects of flow rate and ionic strength on the performance of an air-cathode microbial fuel cell using electrochemical impedance spectroscopy. Energies 3(4): 592-606.

An J, Lee H-S. (2013) Implication of endogenous decay current and quantification of soluble microbial products (SMP) in microbial electrolysis cells. Rsc Advances 3(33): 14021-14028.

Beasley C. (2015) Basics of electrochemical impedance spectroscopy: Application note, http://www.gamry.com/application-notes/EIS/basics-of-electrochemical-impedance-spectroscopy/ Warminster, PA.

Beegle J, Borole AP. (2017) Exoelectrogens for MFC's. *In*: Progress and Recent Trends in Microbial Fuel Cells. Vol. 1st Edition, (Eds.) P Kundu, K Dutta, Elsevier, pp. 193-230.

Borole AP, Aaron DS, Tsouris C, Hamilton CY. (2010) Understanding long term changes in microbial fuel cells using electrochemical impedance spectroscopy. Environ. Sci. Technol. 44(7): 2740-2745.

Borole AP, Lewis AJ. (2017) Proton transfer in microbial electrolysis cells. Sustainable Energy & Fuels 1: 725.

Dominguez-Benetton X, Sevda S, Vanbroekhoven K, Pant D. (2012) The accurate use of impedance analysis for the study of microbial electrochemical systems. Chemical Society Reviews 41(21): 7228-7246.

Fan Y, Sharbrough E, Liu H. (2008) Quantification of the internal resistance distribution of microbial fuel cells. Environmental Science & Technology 42(21): 8101-8107.

Gorby YA, Yanina S, McLean JS, Rosso KM, Moyles D, Dohnalkova A, Beveridge TJ, Chang IS, Kim BH, Kim KS, Culley DE, Reed SB, Romine MF, Saffarini DA, Hill EA, Shi L, Elias DA, Kennedy DW, Pinchuk G, Watanabe K, Ishii S, Logan B, Nealson KH, Fredrickson JK. (2006) Electrically conductive bacterial nanowires produced by Shewanella oneidensis strain MR-1 and other microorganisms. Proc. Natl. Acad. Sci. USA 103(30): 11358-11363.

He Z, Huang Y, Manohar AK, Mansfeld F. (2008) Effect of electrolyte pH on the rate of the anodic and cathodic reactions in an air-cathode microbial fuel cell. Bioelectrochemistry 74(1): 78-82.

He Z, Mansfeld F. (2009) Exploring the use of electrochemical impedance spectroscopy (EIS) in microbial fuel cell studies. Energy Environ. Sci. 2(2): 215-219.

Jung S, Mench MM, Regan JM. (2011) Impedance characteristics and polarization behavior of a microbial fuel cell in response to short-term changes in medium pH. Environmental science & technology 45(20): 9069-9074.

Ki D, Popat SC, Torres CI. (2016) Reduced overpotentials in microbial electrolysis cells through improved design, operation, and electrochemical characterization. Chemical Engineering Journal 287: 181-188.

Malvankar NS, Tuominen MT, Lovley DR. (2012) Biofilm conductivity is a decisive variable for high-current-density Geobacter sulfurreducens microbial fuel cells. Energy & Environmental Science 5(2): 5790-5797.

Manohar AK, Bretschger O, Nealson KH, Mansfeld F. (2008) The use of electrochemical impedance spectroscopy (EIS) in the evaluation of the electrochemical properties of a microbial fuel cell. Bioelectrochem. 72(2): 149-154.

Marsili E, Rollefson JB, Baron DB, Hozalski RM, Bond DR. (2008) Microbial Biofilm Voltammetry: Direct Electrochemical Characterization of Catalytic Electrode-Attached Biofilms. Appl. Environ. Microbiol. 74(23): 7329-7337.

Orazem M, Tribollet B. (2006) *Electrochemical impedance spectroscopy.* Wiley InterScience., New York.

Ramasamy RP, Gadhamshetty V, Nadeau LJ, Johnson GR. (2009) Impedance spectroscopy as a tool for non-intrusive detection of extracellular mediators in microbial fuel cells. Biotechnology and bioengineering 104(5): 882-891.

Ramasamy RP, Ren ZY, Mench MM, Regan JM. (2008) Impact of initial biofilm growth on the anode impedance of microbial fuel cells. Biotechnology and Bioengineering 101(1): 101-108.

Reguera G, McCarthy KD, Mehta T, Nicoll JS, Tuominen MT, Lovley DR. (2005) Extracellular electron transfer via microbial nanowires. Nature 435(7045): 1098-1101.

Sekar N, Ramasamy RP. (2013) Electrochemical Impedance Spectroscopy for Microbial Fuel Cell Characterization. J. Microbial & Biochemical Technology S6(004).

Sleutels T, Ter Heijne A, Buisman CJN, Hamelers HVM. (2012) Bioelectrochemical Systems: An Outlook for Practical Applications. Chemsuschem 5(6): 1012-1019.

Sleutels TH, Ter Heijne A, Buisman CJ, Hamelers HV. (2013) Steady-state performance and chemical efficiency of Microbial Electrolysis Cells. International Journal of Hydrogen Energy 38(18): 7201-7208.

Springer TE, Zawodzinski TA, Wilson MS, Gottesfeld S. (1996) Characterization of polymer electrolyte fuel cells using AC impedance spectroscopy. J. Electrochem. Soc. 143(2): 587-599.

Strik DP, Ter Heijne A, Hamelers HV, Saakes M, Buisman C. (2008) Feasibility study on electrochemical impedance spectroscopy for microbial fuel cells: measurement modes & data validation. ECS Transactions 13(21): 27-41.

Tartakousky B, Mehta P, Santoyo G, Guiot SR. (2011) Maximizing hydrogen production in a microbial electrolysis cell by real-time optimization of applied voltage. International Journal of Hydrogen Energy 36(17): 10557-10564.

Ter Heijne A, Schaetzle O, Gimenez S, Fabregat-Santiago F, Bisquert J, Strik D, Barriere F, Buisman CJN, Hamelers HVM. (2011) Identifying charge and mass transfer resistances of an oxygen reducing biocathode. Energy & Environmental Science 4(12): 5035-5043.

ter Heijne A, Schaetzle O, Gimenez S, Navarro L, Hamelers B, Fabregat-Santiago F. (2015) Analysis of bio-anode performance through electrochemical impedance spectroscopy. Bioelectrochemistry 106: 64-72.

Torres CI, Marcus AK, Lee HS, Parameswaran P, Krajmalnik-Brown R, Rittmann BE. (2010) A kinetic perspective on extracellular electron transfer by anode-respiring bacteria. FEMS Microbiology Reviews 34(1): 3-17.

Wang X, Feng Y, Ren N, Wang H, Lee H, Li N, Zhao Q. (2009) Accelerated start-up of two-chambered microbial fuel cells: effect of anodic positive poised potential. Electrochimica Acta 54(3): 1109-1114.

Zhang Y, Merrill MD, Logan BE. (2010) The use and optimization of stainless steel mesh cathodes in microbial electrolysis cells. International Journal of Hydrogen Energy 35(21): 12020-12028.

Zhao F, Slade RCT, Varcoe JR. (2009) Techniques for the study and development of microbial fuel cells: an electrochemical perspective. Chemical Society Reviews 38(7): 1926-1939.

PART-II

Optimizing Microbial Electrochemical Systems: Microbial Ecology of Biofilms and Electroactive Microorganisms Strain Improvement

Steering Redox Metabolism in *Pseudomonas putida* with Microbial Electrochemical Technologies

Bin Lai* and Jens Krömer*

Systems Biotechnology Group/Department of Solar Materials, Helmholtz Centre for Environmental Research (UFZ), Permoserstraße 15, 04318 Leipzig, Germany

1. Introduction

The world's population was recently predicted to be 9.8 billion by 2050 (United Nations 2017), 400 million more compared to estimates a decade ago (Lewis and Nocera 2006). The growing population raises serious concerns about future demand for energy and living resources. So far, fossil fuels still dominate the energy sector but new renewable energies, e.g., solar, wind, biofuel, etc., have steadily increased their share (IEA 2014). In the materials sector, however, petrochemicals and natural gas still dominate the chemical industry by a large margin with renewable processes still only playing a minor role. This has tremendous implications for basically every aspect of our everyday life.

Industrial biotechnology is considered as a potential solution for chemical production independent of fossil feedstock. The research in this field has been significantly expanded over the last decades, and it is today possible to produce more chemicals sustainably from renewable resources than ever before (Erickson et al. 2012; Tiquia-Arashiro and Mormile 2013). Despite the fast-growing biofeedstock market, only a few bio-produced compounds are cost competitive. There are many limitations in bioprocess scale-up that need to be overcome, but from the microbial catalysts point of view, limitations based on redox balance, carbon yields or product toxicity are most pressing.

Recently, a class of microbes that was recognized as a promising new platform for the production of harsh chemicals, which are often toxic even to the microbial production strain, were the pseudomonads, especially the model species *Pseudomonas putida* (Nikel et al. 2014). Like in many pseudomonads, the catabolism of *P. putida* is efficient in the supply of redox power (Blank et al. 2008) while exhibiting a low cellular energy demand needed for cell maintenance; in other words, it has a high net NAD(P)H generation for enzymatic reactions (Ebert et al. 2011). This feature has endowed *P. putida* for efficient production of chemicals with high NAD(P)H demand as well as a high solvent resistance that also requires high regeneration rate of cellular reducing power (Singh et al. 2007). However, *P. putida* solely relies on oxygen as the terminal electron acceptor and does not exhibit fermentative phenotypes (Escapa et al. 2012; Nikel et al. 2014). This strictly aerobic metabolism of *P. putida* could also lead to higher costs when it comes to industrial application, for instance, high capital cost (e.g., expensive reactor design coming

---

*Corresponding author: bin.lai@ufz.de, jens.kroemer@ufz.de

from gas transfer limits) and high substrate costs due to carbon loss (e.g., full oxidation of substrate to CO_2). Therefore, it would be beneficial to obtain a system where *P. putida* can produce products under anaerobic conditions.

One solution could be to introduce fermentative pathways of other microbes into *P. putida*, even though this was attempted with limited success (Nikel and de Lorenzo 2013; Steen et al. 2013), this would also lead to loss of carbon in the form of by-products that need to be made for redox balancing. Instead, one could also provide the cells with an alternative to O_2 as the final electron acceptor that does not involve carbon as an electron acceptor. The prime solution for such an endeavor is the use of bioelectrochemical methods employing electrodes as solid electron acceptors. This concept was only recently demonstrated. This chapter will provide the first review regarding the research progress in this field and also address their future perspectives. It will discuss the central carbon metabolism of *P. putida* and then introduce the use of an anode for oxygen free chemical production in this obligate aerobic organism, and, finally, we will provide an outlook on the remaining bottlenecks to establish an electrode-driven biochemical production process using *P. putida*.

2. *Pseudomonas putida*

Pseudomonas putida is a Gram-negative rod that is motile through polar flagella (Harwood et al. 1989). It can widely be found in diverse nutritional environments ranging from plant surfaces, soils, water and insects to the human body. The flagella enable a highly motile behavior toward nutrients in the natural environment (Harwood et al. 1984), and as a consequence *P. putida* biofilm formation is hardly observed either in the lab or in nature. One of the features that make this species very interesting for biotechnological application (for both biosynthesis and bioremediation) is its versatile metabolism, including xenobiotics as carbon sources (Jimenez et al. 2002; Nelson et al. 2002; Silby et al. 2011; Wu et al. 2011; Belda et al. 2016).

The carbohydrate central metabolism provides precursor metabolites for the biomass formation as well as the fuel for synthesis, growth and maintenance in the form of ATP (Noor et al. 2010). It lays the basis for the robust metabolic capacity of *P. putida*. Compared to other industrial microorganisms, *P. putida* performs a special cyclic glycolysis pathway that is a combination of the Entner-Doudoroff pathway (ED), the Embden-Meyerhof-Parnas pathway (EMP) and the Pentose phosphate pathway (PP) (Figure 1) (Sudarsan et al. 2014; Nikel et al. 2015).

The exogenous glucose is imported into the periplasm via an outer membrane porin channel (Saravolac et al. 1991; Wylie and Worobec 1995). Once the glucose is imported, there are several pathways available for further converting the glucose to a core intermediate gluconate-6-phosphate (6PGNT). At one hand, the glucose could be oxidized to be gluconic acid (GA) and/or 2-ketogluconic acid (2KGA) with PQQ/PQQH2 or FAD/FADH2 as the redox cofactors. The GA and 2KGA can then be transported into the cytoplasm and be phosphorylated. On the other hand, the glucose could also be directly imported into the cytoplasm via ABC transporter and be phosphorylated. The key difference in these routes is the NADPH pool. One NADPH would be generated from glucose to 6PGNT, whereas one NADPH would be consumed from 2KGA to 6PGNT and no NADPH would be involved from GA to 6PGNT. Based on the carbon flux, the NADPH pool would be dynamically controlled. Another NADPH regulation is the cyclic loop of ED-EMP pathway. This happens while the oxidative stress increases in the environment (Nikel et al. 2015) to enable a high yield of NADPH from glycolysis. The dynamically-controlled high regeneration of NADPH provides the cells essential reducing power for the resilience against oxidative stress and for the biosynthesis of secondary metabolites (Singh et al. 2007; Chavarría et al. 2013; Ng et al. 2015; Spaans et al. 2015).

In addition to the high regeneration of NADPH for oxidative stress endurance, *P. putida* is also equipped with multiple intrinsic efflux pumps that significantly improve its solvent tolerance (Ramos et al. 2002; Fernandez et al. 2012; Udaondo et al. 2012; Hosseini et al. 2017). One of the species in this genus that was isolated from a soil sample in the late 1980s could grow up to 50% (v/v) toluene (Inoue and Horikoshi 1989), a highly toxic chemical that could kill most microorganisms at a concentration as low as 0.1% (v/v). In addition, *P. putida* could also adapt to a high alcohol stress leading to a high alcohol production (Nielsen et al. 2009; Rühl et al. 2009). Those efflux pumps-based processes are energy dependent and the inhibition of respiratory energy regeneration system could significantly affect their activity (Isken and de Bont 1996; Ramos et al. 1997; Ramos et al. 2002). Moreover, an oversupply of glucose will not lead to a typical overflow metabolism in *P. putida* (Blank et al. 2008), which is commonly observed in other microbial hosts, e.g., aerobic acetate production of *Escherichia coli* or the Crabtree effect in yeast. Instead, the oxidative tricarboxylic acid cycle (TCA cycle) and NAD(P) H regeneration rate were significantly improved under the high glucose conditions. This feature will i) benefit downstream processing due to a lower complexity of culture broth and ii) meet the high demand of reducing power during production of chemicals exhibiting solvent toxicity (Nijkamp et al. 2005; Wierckx et al. 2005; Nijkamp et al. 2007; Verhoef et al. 2009; Verhoef et al. 2010; Meijnen et al. 2011; Lang et al. 2014). Ultimately, this underscores the position of *P. putida* as an appealing platform for biochemical production (Meijnen et al. 2011; Nikel et al. 2014).

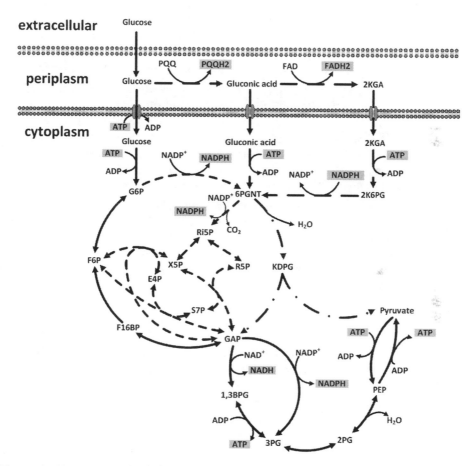

Figure 1: Glucose central metabolism of *Pseudomonas putida*. The long dash indicates the peripheral oxidative pathway; short dash refers to Pentose Phosphate pathway; dash-dot refers to Entner-Doudoroff (ED) pathway; the solid line refers to Embden-Meyerhof-Parnas (EMP) pathway. Abbreviations: 2KGA, 2-ketogluconic acid; G6P, glucose-6-phosphate; 6PGNT, gluconate-6-phosphate; 2K6PG, 2-ketogluconate-6-phosphate; KDPG, 2-dehydro-3-deoxy-phosphogluconate; GAP, glyceraldehyde-3-phosphate; F16BP, fructose-1,6-biphosphate; F6P, fructose-6-phosphate; 1,3BPG, 1,3-biphosphoglycerate; 3PG, 3-phosphoglycerate; 2PG, 2-phosphoglycerate; PEP, phosphoenolpyruvate; OAA, oxaloacetic acid.

3. Constraints for Applying *Pseudomonas putida* in Industry

As described above, the metabolism of *P. putida* positions this organism as a prime candidate for bioprocess development, but there are also some constraints for its industrial application. Apart from the optimization of yields, rates and titres as commonly required in biotechnological processes (Vickers et al. 2010), the dependency of *P. putida* metabolism on oxygen also limits the reactor scale and dictates operating and substrate costs during the industrial application. The pros and cons of aerobic vs. anaerobic processes are addressed below.

3.1 Bioenergetics

Under the presence of oxygen, glucose can be fully oxidized to CO_2 and the ATP yield can reach about 29.85 mol ATP per mol glucose (Rich 2003), while the anaerobic process can only reach a yield of about 2.8-3.2 mol ATP per mol glucose (Muir et al. 1985). Cells are compensating this partially by increasing the glucose uptake rate under anaerobic conditions to meet the energy demand of cell growth and maintenance (Varma and Palsson 1994; Chen et al. 2011), but overall, the specific ATP regeneration rate is higher in an aerobic process. It is commonly observed that the aerobic culture has a higher growth rate that on the one hand leads to higher productivity but on the other hand also results in a higher carbon loss in terms of CO_2 from the TCA cycle (Noor et al. 2010). Moreover, a higher by-product formation can be observed in an anaerobic fermentation process for redox and energy balancing purpose, while the intracellular and extracellular pH value is generally lower for cells grown under anaerobic conditions compared to aerobic conditions (Diaz-Ricci et al. 1990; Chen et al. 2011).

3.2 Thermodynamics

In addition to the higher energy regeneration rate and metabolic turnover of aerobic cultures, the complete catabolism of carbohydrates with oxygen as an electron acceptor, thermodynamically, will also release more free energy mostly in terms of heat. The Gibbs free energy of glucose oxidation with oxygen is about -2870 KJ/mol$_{glucose}$ which is far exceeding than that for the glucose fermentation without O_2 (about -218 KJ/mol$_{glucose}$) (Tran and Unden 1998). This is consistently observed regardless of the fermenter size: heat generation is generally 10 times higher in aerobic fermenters compared to anaerobic fermenters at identical reactor scale (Hannon et al. 2007). Since most biological transformations operate in narrow temperature optima, this means that more process energy will be required to remove the produced heat from the bioreactor under aerobic conditions.

3.3 Gas Transfer

Another significant energy input for the aerobic reactor comes from aeration. The solubility of oxygen in water is very low (only about 37 ppm at 30 °C) (Kolev 2012), and to maintain sufficient oxygen transfer, high stirring speeds and high gas flow are required. Not only the energy cost but also oxygen transfer poses a major limitation for the scale-up of the aerobic bioprocess. The k_La is the volumetric mass transfer coefficient determining the oxygen transfer rate in aerobic reactors (Tribe et al. 1995) and, therefore, is the key parameter during the scale-up of an aerated/gassed fermenter. Figure 2 presents a basic requirement for the k_La value to maintain a relative dissolved oxygen concentration in the water phase. While it is possible to achieve a k_La value of over 0.208 s^{-1} in a well-stirred laboratory size bioreactor, it will be impractical to achieve a similar value or even just values above 0.138 s^{-1} at identical conditions (temperature, pH, etc.) in large scale (Nielsen et al. 2003). A limited supply of oxygen will restrict the growth of aerobic microbe, and what is even worse is that it will significantly change the glucose metabolism toward different products in facultative anaerobes (Ferreira et al. 2013). Running a bioreactor at high pressure could improve the dissolved oxygen concentration in the liquid phase, but it has been observed that the growth of bacteria and yeast are inhibited by elevated oxygen concentration (oxygen partial pressure above 1 atm) (Baez and Shiloach 2014).

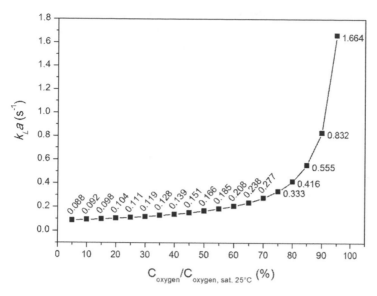

Figure 2: Relative k_La value required to maintain different dissolved oxygen concentrations in water. Data were calculated based on the condition 25 °C and 1 atm, according to the equations (e.g., 10.2) given by Nielsen et al. (Nielsen et al. 2003).

Due to the abovementioned reasons, both the largest and average sizes of aerobic fermenters are much smaller than those of anaerobic fermenters (Hannon et al. 2007) which have been a major bottleneck for aerobic industrial biotechnology.

4. Anoxic Phenotype of *Pseudomonas putida* Driven by Electrode

There has been a desire to genetically engineer an anaerobic mutant for *P. putida* for long due to the constraints of aerobic bioprocess as discussed above. Two papers were published regarding this topic by two groups almost at the same time in 2013. Since the anaerobic physiology of *P. putida* was not accessible through traditional microbiology procedures because no anaerobic cultivations were possible then, both papers started by addressing the missing pathways normally observed for the anaerobes or facultative anaerobes according to the genome data and conducted strain engineering accordingly. Nikel et al. (2013) engineered pyruvate decarboxylase, alcohol dehydrogenase and acetate kinase into *P. putida* KT2440 while aiming to create a fermentative pathway for the strain. The recombinant strain showed improved survival without oxygen compared to the parental strain and could catalyze the decolorization process. The work from Steen et al. (Steen et al. 2013) used another angle to address this issue. By engineering a nitrate/nitrite enzymatic system into *P. putida* KT2440, the authors created a recombinant strain that could do anaerobic respiration using the nitrate/nitrite couple rather than producing a carbon product as the electron sink. However, similar to the work of Nikel et al. (2013), only an improved survival was observed, and no anaerobic growth could be detected that indicated the cells were still strictly metabolically constrained. To further engineer the phenotype of *P. putida*, more fundamental understanding of the metabolic constraints under oxygen-free conditions is needed. As mentioned above, obtaining this understanding is currently hampered by the inability to study cells under anaerobic conditions in the first place.

Using carbon metabolites as electron sinks would intrinsically limit the product spectrum of the future process, and applying nitrate as the terminal electron acceptor would require continuous nitrate dosing ultimately leading to nitrite toxicity (Bollag and Henninger 1978) to the cells as well as demand more expensive nitrogen removal from the process waste streams. Instead, our approach uses an electrode for electron balance. Using a bioelectrochemical system (BES) as an electron sink (i.e., replacing the function of oxygen in cell respiration), the limit to carbon product spectrum would be overcome and redox power could be regenerated sustainably by electricity that could be then used by the *P. putida* cells to balance the cellular redox equivalents and gain energy to support its anaerobic metabolism or even growth.

4.1 Anode as Electron Sink for *Pseudomonas putida*

While direct electron transfer has been confirmed for many organisms during the past decades, we could not demonstrate it for *P. putida* (at least for the strains F1 and KT2440). Many trials with different working electrode potentials, ranging up to 1 V vs. the standard hydrogen electrode (SHE), were conducted and no significant current could be detected over the course of one week. However, it was previously shown that ferricyanide could serve as an electron acceptor during the oxidation of nicotinic acid by *Pseudomonas fluorescens* (Ikeda et al. 1996). In addition, the purified glucose dehydrogenase and gluconate dehydrogenase could also use a range of redox chemicals as artificial electron acceptors *in vitro* (Matsushita et al. 1979; Matsushita et al. 1980).

In fact, it was confirmed by our group that when the glucose-based mineral medium (DM9) was supplemented with 1 mM potassium ferricyanide as electron acceptor, *P. putida* F1 (Lai et al. 2016) anaerobically reduced ferricyanide (oxidized form, $[Fe(CN)_6]^{3-}$) to ferrocyanide (reduced form, $[Fe(CN)_6]^{4-}$) within 55–90 hours. This was also visually detected by a change in color of the solution from yellow-green ($[Fe(CN)_6]^{3-}$) to colorless ($[Fe(CN)_6]^{4-}$). This is a clear indication that ferricyanide can also be an electron acceptor for *Pseudomonas putida in vivo*.

The ability to continuously utilize ferricyanide as an electron acceptor was further tested in a BES, where electrochemical oxidation of $[Fe(CN)_6]^{4-}$ at the anode allowed continuous regeneration of $[Fe(CN)_6]^{3-}$ for microbial metabolism. In the absence of a mediator, *P. putida* was not able to transfer electrons to the anode since no catalytic current was detected when no mediator was added. Several mediators with different formal redox potentials were tested as shown in Figure 3. Surprisingly, neutral red and riboflavin, which were reported to be effective mediators in the former studies for *E. coli* and *Shewanella* (Park and Zeikus 2000; Marsili et al. 2008), were both not showing any function of shuttling electron transfer. Furthermore, among the seven tested redox mediators, only those with redox potentials of higher than 0.2 V (vs. SHE) were able to withdraw electrons from *P. putida* cells. This suggested that the redox chemicals could possibly interact with the cytochromes in the periplasm and could get reduced. The cytoplasmic membrane is strongly hydrophobic, and the soluble redox chemicals are charged; it is, thus, quite unlikely that passive diffusion into the cell could occur. In this case, the cytochrome c reductase on the cytoplasmic membrane could possibly be the functional site for interaction with the external redox chemicals according to the redox potential sorting.

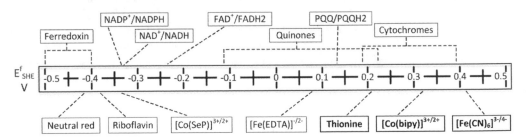

Figure 3: Formal redox potentials of the redox chemicals screened (down) and cellular respiratory components (top). The effective redox chemicals for electron transfer are marked in bold.

In a different study by the Rosenbaum group (Schmitz et al. 2015), a mediator-based electron transfer route was established in *P. putida* KT24440 using genetic engineering. Production of phenazines, which were reported as electron mediators secreted by *Pseudomonas aeruginosa* (Rabaey et al. 2005; Pham et al. 2008), was introduced into KT2440. Phenazines are a group of redox compounds with formal redox potentials of about -110 to 0 mV (vs. SHE) at a neutral condition (Price-Whelan et al. 2007; Seviour et al. 2015). Under microaerobic conditions, *P. putida* KT2440 could catabolize glucose, secrete phenazines and subsequently use them as electron mediators to shuttle the electron transfer from the cells to the anode. A slight accumulation of gluconic acid, 2-ketogluconic acid and acetic acid were observed, however, the majority of carbon ended up in biomass and CO_2 that was similar to the aerobic metabolism of *P. putida*.

4.2 Anode-Driven Carbon Metabolism of *Pseudomonas putida*

In contrast to the microaerobic behavior using phenazines as mediators, the glucose metabolism was completely rearranged under the absence of oxygen in the BES. The anodic power redirected the carbon flux toward 2-ketogluconic acid as the end-product (with a molar yield of over 90% from glucose) (Lai et al. 2016). The 2-ketogluconic acid is an industrial precursor for the production of isoascorbic acid (Zorn et al. 2014), a preservative used in the food industry. The conversion rate was highly dependent on the redox potential of the mediator used. The higher the formal redox potential, the faster the glucose is oxidized. This demonstrates that the anoxic metabolism of *P. putida* is completely anode-dependent. The periplasmic oxidation pathway of glucose to 2-ketogluconic acid is normally not favored for *P. putida*, and about 90% of carbon flux would bypass this pathway into cytoplasmic glycolysis under aerobic condition (Nikel et al. 2015).

In addition to the periplasmic metabolism, some of the carbon also flowed into the intracellular metabolism. With glucose as substrate, the remaining carbon apart from 2-ketogluconic acid ended up in acetate. Acetate is commonly observed as a fermentation by-product, and microorganisms can gain excess electrons and ATP from acetate production. This can possibly be the reason for *P. putida* cells to produce acetate in BES reactor to improve the ATP yield for cell maintenance. A schematic of the process is given in Figure 4.

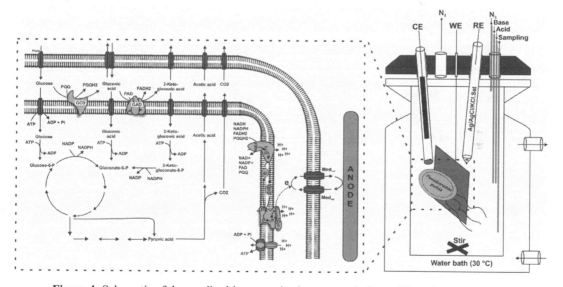

Figure 4: Schematic of the anodic-driven anoxic glucose metabolism of *Pseudomonas putida*.

In another study, it could also be demonstrated that *Pseudomonas putida* can use other carbon sources, like citric acid with an anode (Hintermayer et al. 2016). A para-hydroxybenzoic acid, overproducing mutant of *P. putida* KT2440, was using citric acid as the carbon source in a BES. In this process, the redox potential of the solution was maintained through the dosing of the redox mediator. In this process, an improvement of 69% of para-hydroxybenzoic acid yield could be observed compared to the aerobic batch process.

4.3 Metabolic Constraints Restricting the Anoxic Metabolism of *Pseudomonas putida*

Pseudomonas putida could use the anode as the electron sink to catabolize different carbon sources and gain energy from this process. In the case of using glucose as substrate and ferricyanide as a mediator, the intracellular ATP level was about four times higher than the anaerobic controls that increased from about 0.07 $\mu mol/g_{CDW}$ to about 0.28 $\mu mol/g_{CDW}$. Moreover, flux balance analysis showed that over 90% of the energy was generated through membrane-bound proton-dependent ATP synthase (i.e., the electron balance from anode) (Lai et al. 2016). However, cell metabolism was still largely constrained in the BES reactor.

The most obvious constraint was that no anaerobic growth was achieved with the anode in all studies. Although the cells cultivated with anode were showing improved survival compared to the normal anaerobic controls, the cell numbers were still decreasing significantly. This was the same as observed in the studies without electrodes for the wild type as well as the recombinant strains reported before (Nikel and de Lorenzo 2013; Steen et al. 2013). Most likely the cell growth is still limited by the restricted carbon metabolism. As discussed above, while feeding glucose as substrate, the majority of the carbon was only undergoing oxidation in the periplasm and only a small fraction was further catabolized in the intracellular central metabolism that provides the basic precursors for biomass formation (Noor et al. 2010). The reason for this restricted glucose metabolism remains unknown. One possibility could be the imbalance of intracellular redox pools. The periplasmic oxidation of glucose would feed the PQQ/ $PQQH_2$ and FAD/$FADH_2$ pools but might not balance the NAD/NADH and NADP/NADPH pools. When analyzing the redox ratio of the nicotinamide redox cofactor pools, it was observed that the NAD/NADH ratio was similar with or without an active anode, but the NADP/NADPH ratio was significantly increased while ferricyanide and/or anode were present. The shift in NADP/NADPH might be due to the demand of NADPH for further reducing 2-ketogluconic acid (Figure 1) that requires further investigation.

In addition to the redox unbalance, energy might also be a limiting factor, although as stated above, the cells are able to gain net ATP from the anode. According to the flux balance analysis results, the average ATP regeneration rate was calculated to be only 0.69 $\mu mol/g_{CDW}/h$ for using ferricyanide and glucose in the reactor. This number only accounts for 17-75% of the aerobic nongrowth associated maintenance (NGAM) requirement reported for *P. putida* KT2440 in the literature (Ebert et al. 2011; van Duuren et al. 2013). Since NGAM refers to the minimum ATP requirement for cell maintenance without growth, the observation of zero growth (even death) in the BES experiments seems plausible.

Furthermore, it is possible that regulatory factors responding to the absence of oxygen are present in *P. putida*, but this needs further investigation. The anoxic physiology of wild-type *P. putida* is largely unknown to date; but by using the anode as an electron sink to partially reactivate the cellular metabolism, quantitative approaches can now be applied to study and identify the metabolic constraints for *P. putida* under anaerobic conditions.

4.4 Mass Transfer Limitations to Electrode as the Rate Limiting Factor

Unlike many other microbial electrosynthesis systems, the anode-driven chemical production using *Pseudomonas putida* is not biofilm-based. The planktonic process of using *P. putida* in a BES can be simplified into two phases: the biologic phase (where *P. putida* cells uptake substrate and oxidized mediator) and the electrochemical phase (on the anode surface to regenerate the oxidized mediator). To determine the limiting factor for the whole process, the mass transfer flux of carbon substrate and mediators toward the planktonic cells and the electrode were accessed by mathematical calculations.

The calculations were done for the defined stir-tank bioreactor (Hintermayer et al. 2016) using ferricyanide as the mediator. The calculation of the mass transfer flux towards the planktonic cells followed the approach described in Chapter 10 from the book *Bioreaction Engineering Principles* (Nielsen et al. 2003), and the mass transfer towards the anode was calculated based on the empirical equation determined by Kato et al. (Kato et al. 2007), while the electrochemical reaction rate was calculated according to the Nicholson theory (Nicholson 1965).

Two basic assumptions for the calculations were taken i) the planktonic cells were evenly distributed in all the examined conditions and only diffusion had taken place between the liquid phase and cell pellets; ii) the boundary layer on the electrode surface where the ion migration took place was not considered because migration of ionic species could be negligible in a well-mixed system, and the convection force from bulk liquids phase to electrode surface played a major role (Agar 1947; Tobias et al. 1952; Goodridge and Scott 1995). Also, the kinetics of cross-cell membrane transportation was not taken into account due to the lack of data.

A list of the variables' values taken for calculation is presented as Table 1, and the calculated mass transfer fluxes can be seen in Table 2. According to the calculation, it was clear that the mass transfer of mediator toward the anode was the limiting factor which was far below others. This was largely due to the small electrode surface ($\sim$0.0062 m^2) compared to the specific surface area of planktonic cells ($\sim$5.46 m^2). In addition, the measured glucose uptake rate showed a linear relationship with the mass transfer of ferrocyanide to the anode (Figure 5), again demonstrating that the mass transfer to the electrode was the limiting factor.

Table 1: List of variables and their values to calculate k_1, k_2, k_3 and k_b. The values listed in this table are standard values or determined by experiments other than specified.

Variables	Description	Values	Units
A_e	Surface area of electrode (rod shape, graphite)	0.00628	m^2
b	Thickness of the stirrer	0.011	m
B_w	Diameter of electrode	0.01	m
d	Diameter of the stirrer	0.054	m
D	Diameter of the reactor	0.15	m
d_p	Length scale of single cell	1.5E-06 (Harwood et al. 1989)	m
d_e	Length of electrode	0.1	m
$D_{ferricyanide}$	Diffusion coefficient of ferricyanide in water	8.33E-10 (Bazán and Arvia 1965)	m^2/s
$D_{ferrocyanide}$	Diffusion coefficient of ferrocyanide in water	7.07E-10 (Bazán and Arvia 1965)	m^2/s
$D_{glucose}$	Diffusion coefficient of glucose in water	6.73E-10 (Koch 1996)	m^2/s
F	Faraday constant	96485	C/mol
g	Acceleration of gravity	9.81	m/s^2
n	Stirring speed	11.667 (700) 8.333 (500) 5 (300)	s^{-1} (rpm)
n_e	Performance factor of the stirrer	1	
OD_{600}	Optical density of cell resuspension	0.5 (equal to 0.243 g/L)	
R	Gas constant	8.314	J/K/mol
T	Temperature	303 (30)	K (°C)
v	Kinematic viscosity of the liquid	8.01E-07 (Kestin et al. 1978)	m^2/s
V	Working volume of reactor	2.5	L
v_c	Scanning rate in cycle voltammetry	50	mV/s
w	Fractional water content	0.6 (Bratbak and Dundas 1984)	
α	Electron transfer efficiency	0.5 (Hamelers et al. 2011; Rousseau et al. 2014)	
ΔE	Overpotential applied on working electrode	0.281	V
ΔE_p	Peak potential difference in cycle voltammetry	88.8	mV
η_l	Viscosity of liquid	0.000797 (Kestin et al. 1978)	kg/m/s, Pas
ρ_l	Liquid phase density	995.68 (Kestin et al. 1978)	kg/m^3
ρ_p	Cell pellet density	1100 (Bratbak and Dundas 1984)	kg/m^3

Table 2: Kinetic and molar flux calculated at different stirring speeds in the stir-tank reactor. Units of substance concentrations are all in mmol/L. k_1: mass transfer coefficient of ferrocyanide to anode; k_2: mass transfer coefficient of ferricyanide to planktonic cells; k_3: mass transfer coefficient of glucose to planktonic cells; k_b/k_f: backforward/forward rate constant for electrochemical reaction on anode.

Stirring speed	300 rpm	500 rpm	700 rpm
k_1, [cm/s]	$4.82 * 10^{-4}$	$6.45 * 10^{-4}$	$7.82 * 10^{-4}$
$J_{ferrocyanide}$ [mmol/h]	$0.109 * C_{ferrocyanide}$	$0.146 * C_{ferrocyanide}$	$0.177 * C_{ferrocyanide}$
k_2, [cm/s]	0.111	0.111	0.111
$J_{ferricyanide}$ [mmol/s]	$6.46 * C_{ferrocyanide}$		
k_3, [cm/s]	0.090	0.090	0.090
$J_{glucose}$, [mmol/s]	$5.22 * C_{glucose}$		
k_b, [cm/s]	1.26	1.26	1.26
k_f, [cm/s]	$2.67 * 10^{-5}$	$2.67 * 10^{-5}$	$2.67 * 10^{-5}$
J_{anode}, [mmol/s]	$0.08 * C_{ferrocyanide} - 1.68 * 10^{-6} C_{ferricyanide}$		

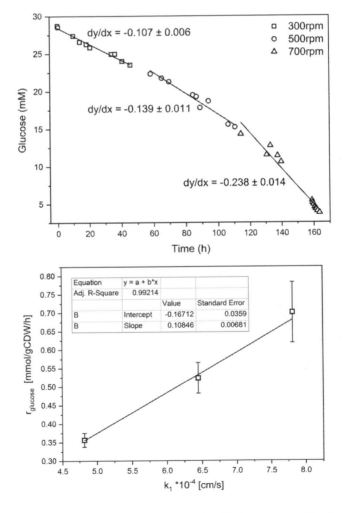

Figure 5: Glucose uptake rate (A) and the linear relationship between specific glucose uptake rate and mass transfer coefficient of ferrocyanide to anode (B) at different stirring speeds. Experiments and calculations were done using a defined stir-tank BES reactor (Hintermayer et al. 2016).

5. Conclusion and Future Perspectives

5.1 Conclusions

The bioelectrochemical technique provides a unique approach to balance the cellular redox status and thus steer the cell metabolism. This unique feature enables the microbial host to do 'redox unbalance' biosynthesis of high-value chemicals. *Pseudomonas putida*, a promising host for industrial biotechnology, is the first obligate aerobe that was confirmed to be capable of using an electrode as electron sink via a mediator-based route. This chapter summarizes the research efforts in the past few years regarding this topic and addresses the benefits and remaining bottlenecks. *Pseudomonas putida* can perform anode-dependent anoxic glucose catabolism and produce carbon product in high yield. The electron transfer rate and carbon turnover rate can also be engineered by targeted strain engineering. In addition, the anode-driven biosynthesis, using *Pseudomonas putida,* is nonbiofilm-based process that is of special benefits for process scaling up demonstrated by theoretical and experimental examinations. However, the anoxic metabolism of *Pseudomonas putida* triggered by anode is also restricted in terms of the low carbon turnover rate, imbalanced intracellular redox pool and major carbon flux limited in periplasmic space. These will need to be addressed and solved in the future by systematic characterization of cell phenotypes and strain development.

5.2 Future Perspectives

The anode-driven biochemical production using *P. putida* is promising but needs to be improved significantly before reaching the industrial application. The glucose consumption rates for *P. putida* in BES were below 0.5 mmol/g_{CDW}/h (Lai et al. 2016; Yu et al. 2018) that were less than 10% of those observed in aerobic cultures (Blank et al. 2008; Nikel et al. 2015). This can be improved by strain and process engineering that would ultimately require a more fundamental understanding of the physiology of *P. putida* under anaerobic conditions. In the following sections, the importance of rational strain design to enhance *P. putida* performance in BES would be addressed. Efficient approaches include physiology studies using omics techniques, computational simulation of metabolic networks and strain engineering with molecular tools. In addition, the feasibility of process scale-up will also be discussed from the aspect of nonbiofilm based microbial electrosynthesis.

5.2.1 Rational Strain Design

There is a large body of work regarding reactor design, electrode materials and interfacial interaction between microbes and solid electrodes (Zhou et al. 2011; Patil et al. 2012; Krieg et al. 2014). However, the cellular metabolic response to electrodes is largely unknown, and the product spectrum from bioelectrochemical reactors is still largely unpredictable and uncontrollable. This is currently a major limitation for industrial application. In mainstream synthetic biology, the concept of rational strain design has been used for decades for targeted production of a large variety of compounds (Kim et al. 2008; Chen and Nielsen 2013). It means to design a defined bioprocess in the microorganism from a given substrate to a target product with optimized performance (in terms of yield, titers, rates, etc.). While the microbial electrochemical technology enables a solution to decouple the carbon balance from the redox balance, which is a major limiting factor for rational strain design, this benefit has not been effectively used in the current microbial electrosynthesis research. One of the reasons is due to the widespread use of mixed culture communities that can hardly be quantitatively characterized and engineered.

However, with *P. putida* as the sole host for electro-biosynthesis, it would be feasible to apply this concept to rational design of the *P. putida* metabolism for targeted chemical production. A simplified workflow is presented in Figure 6, and it consists of an integrative loop of systems-level characterization and strain reconstruction (Kracke et al. 2018). In brief, multi-level omics techniques should be applied to quantitatively characterize the phenotypic change, and computational tools can be used to simulate the

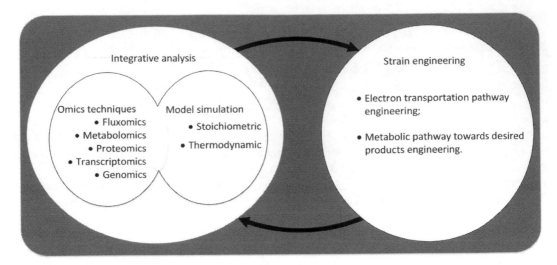

Figure 6: A systems biotechnology approach for strain development for the purpose of high efficient microbial electrosynthesis.

potential effects of electricity toward different target products, while metabolic engineering targets could subsequently be proposed for strain engineering based on these analyses.

This approach has been successfully applied to improve the *P. putida* performance in BES. Based on quantitative characterization, the membrane-bound glucose dehydrogenase was confirmed to be an effective metabolic engineering target. The *P. putida* mutant with glucose dehydrogenase presented dramatic improvements in the glucose uptake rate, the electron transfer rate and also the production rate that were over 470%, 470% and 640%, respectively, as compared to the wild type control (Yu et al. 2018). For future optimization, it will be important to identify the electron transfer route (transporter for mediators, redox proteins interacting with the mediator, etc.) and intracellular carbon flux constraints to enable rational design of the cellular metabolism for production at high yields, rates and titres.

5.2.2 Scaling Up

Apart from being accessible via rational strain design to improve the electro-biosynthesis performance, another advantage of using *P. putida* in BES is that it is relatively easy to be scaled up and compared to other microbial electrosynthesis systems.

For biofilm-based microbial electrosynthesis (also for other BES application fields), the physical contact between the microbe and the solid electrode is essential for the extracellular electron transfer. Furthermore, a thick biofilm is normally desired to improve the output of the electrons, but when the biofilm reaches a certain thickness, the mass transfer and electron transfer within the biofilm will become critical factors. Moreover, thoroughly mixing the bulk phase is also not feasible because i) the friction would disrupt the biofilm formation on the electrode surface and ii) the stress from mobile fluid would change the structure of biofilm, and the biofilm would become denser (Beyenal and Lewandowski 2002) that consequently would increase the resistance for mass and electron transfer within the biofilm. All these issues intrinsically determine the limits for scaling up the biofilm-based process. Therefore, a more feasible approach to reach the large scale is to stack up multiple small reactors rather than to scale-up a single reactor that, however, increases the capital cost and weakens the economic feasibility of the microbial electrosynthesis.

As mentioned above, *P. putida* does not normally form biofilms and through the use of mediators, the anode-driven biosynthesis process is also not biofilm-based that means the scaling up will not be limited by the physical contact between the *P. putida* cells and an electrode surface. In addition, the mass transfer and electron transfer limits cross biofilm will not exist for *Pseudomonas putida*-based process as well. The mass transfer of substrate as well as mediators to the cells could be solved by thoroughly mixing the bulk phase. In addition, the glucose uptake rate was almost identical in the 2.5 L scale stir-tank reactor used by Hintermayer et al. (2016) (0.357 mmol/g_{CDW}/h at 300 rpm stirring speed) compared to a 0.35L cylinder reactor (0.272 mmol/g_{CDW}/h at 400 rpm stirring speed) for a glucose dehydrogenase overexpression strain (Yu et al. 2018). The difference in the rates of these two cases can possibly be attributed to better mixing with stirring using mechanical coupling in the in stir-tank compared to magnetic coupling in the 0.35L reactor. Furthermore, as discussed in section 4.4, the glucose uptake rate increased with the increasing stirring speed in a stir-tank electro-bioreactor and mass transfer of mediator to the electrode is the limiting factor (Figure 4.2). Take together the microbial electrosynthesis using *P. putida* seems scalable and the future focus should be on increasing the specific surface area of electrode during the scale-up, while maintaining the mixing of the bulk phase.

Acknowledgment

BL would like to thank Prof. Dr.-Ing. Dirk Weuster-Botz at Technical University of Munich, Germany for hosting the research visit to his lab and supervising the calculations of mass transfer coefficients addressed in this chapter.

References

Agar JN (1947). Diffusion and convection at electrodes. Discuss Faraday Soc 1(0): 26-37.

Baez A, Shiloach J (2014). Effect of elevated oxygen concentration on bacteria, yeasts, and cells propagated for production of biological compounds. Microb Cell Fact 13: 181.

Bazán JC, Arvia AJ (1965). The diffusion of ferro- and ferricyanide ions in aqueous solutions of sodium hydroxide. Electrochim Acta 10(10): 1025-1032.

Belda E, van Heck RG, Jose Lopez-Sanchez M, Cruveiller S, Barbe V, Fraser C, Klenk HP, Petersen J, Morgat A, Nikel PI, Vallenet D, Rouy Z, Sekowska A, Martins Dos Santos VA, de Lorenzo V, Danchin A, Medigue C (2016). The revisited genome of *Pseudomonas putida* KT2440 enlightens its value as a robust metabolic chassis. Environ Microbiol 18(10): 3403-3424.

Beyenal H, Lewandowski Z (2002). Internal and external mass transfer in biofilms grown at various flow velocities. Biotechnol Prog 18(1): 55-61.

Blank LM, Ionidis G, Ebert BE, Buhler B, Schmid A (2008). Metabolic response of *Pseudomonas putida* during redox biocatalysis in the presence of a second octanol phase. FEBS J 275(20): 5173-5190.

Bollag JM, Henninger NM (1978). Effects of nitrite toxicity on soil bacteria under aerobic and anaerobic conditions. Soil Biol Biochem 10(5): 377-381.

Bratbak G, Dundas I (1984). Bacterial dry matter content and biomass estimations. Appl Environ Microbiol 48(4): 755-757.

Chavarría M, Nikel PI, Pérez-Pantoja D, de Lorenzo V (2013). The Entner–Doudoroff pathway empowers *Pseudomonas putida* KT2440 with a high tolerance to oxidative stress. Environ Microbiol 15(6): 1772-1785.

Chen X, Alonso AP, Allen DK, Reed JL, Shachar-Hill Y (2011). Synergy between ^{13}C-metabolic flux analysis and flux balance analysis for understanding metabolic adaptation to anaerobiosis in *E. coli*. Metab Eng 13(1): 38-48.

Chen Y, Nielsen J (2013). Advances in metabolic pathway and strain engineering paving the way for sustainable production of chemical building blocks. Curr Opin Biotechnol 24(6): 965-972.

Diaz-Ricci JC, Hitzmann B, Rinas U, Bailey JE (1990). Comparative studies of glucose catabolism by Escherichia coli grown in a complex medium under aerobic and anaerobic Conditions. Biotechnol Prog 6(5): 326-332.

Ebert BE, Kurth F, Grund M, Blank LM, Schmid A (2011). Response of *Pseudomonas putida* KT2440 to increased NADH and ATP demand. Appl Environ Microbiol 77(18): 6597-6605.

Erickson B, Nelson, Winters P (2012). Perspective on opportunities in industrial biotechnology in renewable chemicals. Biotechnol J 7(2): 176-185.

Escapa I, Garcia J, Buhler B, Blank L, Prieto M (2012). The polyhydroxyalkanoate metabolism controls carbon and energy spillage in *Pseudomonas putida*. Environ Microbiol 14: 1049 - 1063.

Fernandez M, Conde S, de la Torre J, Molina-Santiago C, Ramos JL, Duque E (2012). Mechanisms of resistance to chloramphenicol in *Pseudomonas putida* KT2440. Antimicrob Agents Chemother 56(2): 1001-1009.

Ferreira MT, Manso AS, Gaspar P, Pinho MG, Neves AR (2013). Effect of oxygen on glucose metabolism: utilization of lactate in *Staphylococcus Aureus* as revealed by in vivo NMR studies. PLoS One 8(3): e58277.

Goodridge F, Scott K (1995). Electrochemical process engineering : a guide to the design of electrolytic plant. Plenum Press, New York, USA.

Hamelers HV, Ter Heijne A, Stein N, Rozendal RA, Buisman CJ (2011). Butler-Volmer-Monod model for describing bio-anode polarization curves. Bioresour Technol 102(1): 381-387.

Hannon J, Bakker A, Lynd L, Wyman C (2007). Comparing the scale-up of anaerobic and aerobic processes. Annual Meeting of the American Institute of Chemical Engineers, Salt Lake City, Utah.

Harwood CS, Fosnaugh K, Dispensa M (1989). Flagellation of *Pseudomonas putida* and analysis of its motile behavior. J Bacteriol 171(7): 4063-4066.

Harwood CS, Rivelli M, Ornston LN (1984). Aromatic acids are chemoattractants for *Pseudomonas putida*. J Bacteriol 160(2): 622-628.

Hintermayer S, Yu S, Krömer JO, Weuster-Botz D (2016). Anodic respiration of *Pseudomonas putida* KT2440 in a stirred-tank bioreactor. Biochem Eng J 115: 1-13.

Hosseini R, Kuepper J, Koebbing S, Blank LM, Wierckx N, de Winde JH (2017). Regulation of solvent tolerance in *Pseudomonas putida* S12 mediated by mobile elements. Microbial Biotechnology 10(6): 1558-1568.

IEA. (2014). World energy outlook 2014. from http://www.worldenergyoutlook.org/publications/weo-2014/.

Ikeda T, Kurosaki T, Takayama K, Kano K, Miki K (1996). Measurements of oxidoreductase-like activity of intact bacterial cells by an amperometric method using a membrane-coated electrode. Anal Chem 68(1): 192-198.

Inoue A, Horikoshi K (1989). A Pseudomonas thrives in high concentrations of toluene. Nature 338(6212): 264-266.

Isken S, de Bont JA (1996). Active efflux of toluene in a solvent-resistant bacterium. J Bacteriol 178(20): 6056-6058.

Jimenez JI, Minambres B, Garcia JL, Diaz E (2002). Genomic analysis of the aromatic catabolic pathways from *Pseudomonas putida* KT2440. Environ Microbiol 4(12): 824-841.

Kato Y, Kamei N, Tada Y, Iwasaki Y, Nagatsu Y, Iwata S, Lee Y-S, Koh S-T (2007). Transport phenomena around cylindrical baffles in an agitated vessel measured by an electrochemical method. J Chem Eng Jpn 40(8): 611-616.

Kestin J, Sokolov M, Wakeham WA (1978). Viscosity of liquid water in the range $-8\,°C$ to $150\,°C$. J Phys Chem Ref Data 7(3): 941-948.

Kim T, Sohn S, Kim H, Lee S (2008). Strategies for systems-level metabolic engineering. Biotechnol J 3(5): 612-623.

Koch AL (1996). What size should a bacterium be? A question of scale. Annual Review Microbiology 50: 317-348.

Kolev NI (2012). Solubility of O_2, N_2, H_2 and CO_2 in water. In Kolev NI (eds) Multiphase flow dynamics 4: turbulence, gas adsorption and release, diesel fuel properties, Springer, Berlin, Germany, pp 209-239.

Kracke F, Lai B, Yu S, Krömer JO (2018). Balancing cellular redox metabolism in microbial electrosynthesis and electro fermentation – a chance for metabolic engineering. Metab Eng 45: 109-120.

Krieg T, Sydow A, Schröder U, Schrader J, Holtmann D (2014). Reactor concepts for bioelectrochemical syntheses and energy conversion. Trends Biotechnol 32(12): 645-655.

Lai B, Yu S, Bernhardt PV, Rabaey K, Virdis B, Krömer JO (2016). Anoxic metabolism and biochemical production in *Pseudomonas putida* F1 driven by a bioelectrochemical system. Biotechnol Biofuels 9: 39.

Lang K, Zierow J, Buehler K, Schmid A (2014). Metabolic engineering of *Pseudomonas* sp. strain VLB120 as platform biocatalyst for the production of isobutyric acid and other secondary metabolites. Microb Cell Fact 13(1): 2.

Lewis NS, Nocera DG (2006). Powering the planet: Chemical challenges in solar energy utilization. Proc Natl Acad Sci U S A 103(43): 15729-15735.

Marsili E, Baron DB, Shikhare ID, Coursolle D, Gralnick JA, Bond DR (2008). *Shewanella* secretes flavins that mediate extracellular electron transfer. Proc Natl Acad Sci U S A 105(10): 3968-3973.

Matsushita K, Ohno Y, Shinagawa E, Adachi O, Ameyama M (1980). Membrane-bound D-glucose dehydrogenase from *Pseudomonas* sp.: solubilization, purification and characterization. Agric Biol Chem 44(7): 1505-1512.

Matsushita K, Shinagawa E, Adachi O, Ameyama M (1979). Membrane-bound D-gluconate dehydrogenase from *Pseudomonas aeruginosa*: its kinetic properties and a reconstitution of gluconate oxidase. J Biochem 86(1): 249-256.

Meijnen JP, Verhoef S, Briedjlal AA, de Winde JH, Ruijssenaars HJ (2011). Improved p-hydroxybenzoate production by engineered *Pseudomonas putida* S12 by using a mixed-substrate feeding strategy. Appl Microbiol Biotechnol 90(3): 885-893.

Muir M, Williams L, Ferenci T (1985). Influence of transport energization on the growth yield of *Escherichia coli*. J Bacteriol 163(3): 1237-1242.

Nelson KE, Weinel C, Paulsen IT, Dodson RJ, Hilbert H, Martins dos Santos VA, Fouts DE, Gill SR, Pop M, Holmes M, Brinkac L, Beanan M, DeBoy RT, Daugherty S, Kolonay J, Madupu R, Nelson W, White O, Peterson J, Khouri H, Hance I, Chris Lee P, Holtzapple E, Scanlan D, Tran K, Moazzez A, Utterback T, Rizzo M, Lee K, Kosack D, Moestl D, Wedler H, Lauber J, Stjepandic D, Hoheisel J, Straetz M, Heim S, Kiewitz C, Eisen JA, Timmis KN, Dusterhoft A, Tummler B, Fraser CM (2002). Complete genome sequence and comparative analysis of the metabolically versatile *Pseudomonas putida* KT2440. Environ Microbiol 4(12): 799-808.

Ng CY, Farasat I, Maranas CD, Salis HM (2015). Rational design of a synthetic Entner-Doudoroff pathway for improved and controllable NADPH regeneration. Metab Eng 29: 86-96.

Nicholson RS (1965). Theory and application of cyclic voltammetry for measurement of electrode reaction kinetics. Anal Chem 37(11): 1351-1355.

Nielsen DR, Leonard E, Yoon S-H, Tseng H-C, Yuan C, Prather KLJ (2009). Engineering alternative butanol production platforms in heterologous bacteria. Metab Eng 11(4–5): 262-273.

Nielsen JH, Villadsen J, Lidén G (2003). Bioreaction engineering principles. Springer, New York, USA.

Nijkamp K, Luijk N, Bont J, Wery J (2005). The solvent-tolerant *Pseudomonas putida* S12 as host for the production of cinnamic acid from glucose. Appl Microbiol Biotechnol 69: 170 - 177.

Nijkamp K, Westerhof RG, Ballerstedt H, de Bont JA, Wery J (2007). Optimization of the solvent-tolerant *Pseudomonas putida* S12 as host for the production of p-coumarate from glucose. Appl Microbiol Biotechnol 74(3): 617-624.

Nikel PI, Chavarria M, Fuhrer T, Sauer U, de Lorenzo V (2015). *Pseudomonas putida* KT2440 metabolizes glucose through a cycle formed by enzymes of the Entner-Doudoroff, Embden-Meyerhof-Parnas, and Pentose Phosphate pathways. J Biol Chem 290(43): 25920-25932.

Nikel PI, de Lorenzo V (2013). Engineering an anaerobic metabolic regime in *Pseudomonas putida* KT2440 for the anoxic biodegradation of 1,3-dichloroprop-1-ene. Metab Eng 15: 98-112.

Nikel PI, Martinez-Garcia E, de Lorenzo V (2014). Biotechnological domestication of pseudomonads using synthetic biology. Nat Rev Micro 12(5): 368-379.

Noor E, Eden E, Milo R, Alon U (2010). Central carbon metabolism as a minimal biochemical walk between precursors for biomass and energy. Mol Cell 39(5): 809-820.

Park DH, Zeikus JG (2000). Electricity generation in microbial fuel cells using neutral red as an electronophore. Appl Environ Microbiol 66(4): 1292-1297.

Patil SA, Hägerhäll C, Gorton L (2012). Electron transfer mechanisms between microorganisms and electrodes in bioelectrochemical systems. Bioanal Rev 4(2): 159-192.

Pham TH, Boon N, Aelterman P, Clauwaert P, De Schamphelaire L, Vanhaecke L, De Maeyer K, Hofte M, Verstraete W, Rabaey K (2008). Metabolites produced by *Pseudomonas* sp. enable a gram-positive bacterium to achieve extracellular electron transfer. Appl Microbiol Biotechnol 77(5): 1119-1129.

Price-Whelan A, Dietrich LEP, Newman DK (2007). Pyocyanin alters redox homeostasis and carbon flux through central metabolic pathways in *Pseudomonas aeruginosa* PA14. J Bacteriol 189(17): 6372-6381.

Rabaey K, Boon N, Hofte M, Verstraete W (2005). Microbial phenazine production enhances electron transfer in biofuel cells. Environ Sci Technol 39(9): 3401-3408.

Ramos JL, Duque E, Gallegos MT, Godoy P, Ramos-Gonzalez MI, Rojas A, Teran W, Segura A (2002). Mechanisms of solvent tolerance in gram-negative bacteria. Annu Rev Microbiol 56: 743-768.

Ramos JL, Duque E, Rodríguez-Herva J-J, Godoy P, Haïdour A, Reyes F, Fernández-Barrero A (1997). Mechanisms for solvent tolerance in bacteria. J Biol Chem 272(7): 3887-3890.

Rich PR (2003). The molecular machinery of Keilin's respiratory chain. Biochem Soc Trans 31(6): 1095-1105.

Rousseau R, Delia M-L, Bergel A (2014). A theoretical model of transient cyclic voltammetry for electroactive biofilms. Energy & Environmental Science 7(3): 1079-1094.

Rühl J, Schmid A, Blank LM (2009). Selected *Pseudomonas putida* strains able to grow in the presence of high butanol concentrations. Appl Environ Microbiol 75(13): 4653-4656.

Saravolac EG, Taylor NF, Benz R, Hancock RE (1991). Purification of glucose-inducible outer membrane protein OprB of *Pseudomonas putida* and reconstitution of glucose-specific pores. J Bacteriol 173(16): 4970-4976.

Schmitz S, Nies S, Wierckx N, Blank LM, Rosenbaum MA (2015). Engineering mediator-based electroactivity in the obligate aerobic bacterium *Pseudomonas putida* KT2440. Front Microbiol 6: 284.

Seviour T, Doyle LE, Lauw SJ, Hinks J, Rice SA, Nesatyy VJ, Webster RD, Kjelleberg S, Marsili E (2015). Voltammetric profiling of redox-active metabolites expressed by *Pseudomonas aeruginosa* for diagnostic purposes. Chem Commun (Camb) 51(18): 3789-3792.

Silby MW, Winstanley C, Godfrey SA, Levy SB, Jackson RW (2011). Pseudomonas genomes: diverse and adaptable. FEMS Microbiol Rev 35(4): 652-680.

Singh R, Mailloux RJ, Puiseux-Dao S, Appanna VD (2007). Oxidative stress evokes a metabolic adaptation that favors increased NADPH synthesis and decreased NADH production in *Pseudomonas fluorescens*. J Bacteriol 189(18): 6665-6675.

Spaans SK, Weusthuis RA, van der Oost J, Kengen SW (2015). NADPH-generating systems in bacteria and archaea. Front Microbiol 6: 742.

Steen A, Ütkür FÖ, Borrero-de Acuña JM, Bunk B, Roselius L, Bühler B, Jahn D, Schobert M (2013). Construction and characterization of nitrate and nitrite respiring *Pseudomonas putida* KT2440 strains for anoxic biotechnical applications. J Biotechnol 163(2): 155-165.

Sudarsan S, Dethlefsen S, Blank LM, Siemann-Herzberg M, Schmid A (2014). The functional structure of central carbon metabolism in *Pseudomonas putida* KT2440. Appl Environ Microbiol 80(17): 5292-5303.

Tiquia-Arashiro SM, Mormile M (2013) Sustainable Technologies: Bioenergy and Biofuel from Biowaste and Biomass. Environ Technol 34 (13): 1637-1805

Tobias CW, Eisenberg M, Wilke CR (1952). Diffusion and convection in electrolysis - a theoritcal review. Electrochem Ionic Crystals 99(12): 359-365.

Tran QH, Unden G (1998). Changes in the proton potential and the cellular energetics of *Escherichia coli* during growth by aerobic and anaerobic respiration or by fermentation. Eur J Biochem 251(1-2): 538-543.

Tribe LA, Briens CL, Margaritis A (1995). Determination of the volumetric mass transfer coefficient (k_La) using the dynamic "gas out-gas in" method: Analysis of errors caused by dissolved oxygen probes. Biotechnol Bioeng 46(4): 388-392.

Udaondo Z, Duque E, Fernandez M, Molina L, de la Torre J, Bernal P, Niqui JL, Pini C, Roca A, Matilla MA, Molina-Henares MA, Silva-Jimenez H, Navarro-Aviles G, Busch A, Lacal J, Krell T, Segura A, Ramos JL (2012). Analysis of solvent tolerance in *Pseudomonas putida* DOT-T1E based on its genome sequence and a collection of mutants. FEBS Lett 586(18): 2932-2938.

United Nations. (2017). World population prospects: the 2017 revision. from https://esa.un.org/unpd/wpp/.

van Duuren J, Puchalka J, Mars A, Bucker R, Eggink G, Wittmann C, dos Santos VA (2013). Reconciling in vivo and in silico key biological parameters of *Pseudomonas putida* KT2440 during growth on glucose under carbon-limited condition. BMC Biotechnol 13(1): 93.

Varma A, Palsson BO (1994). Stoichiometric flux balance models quantitatively predict growth and metabolic by-product secretion in wild-type *Escherichia coli* W3110. Appl Environ Microbiol 60(10): 3724-3731.

Verhoef S, Ballerstedt H, Volkers RJM, de Winde JH, Ruijssenaars HJ (2010). Comparative transcriptomics and proteomics of p-hydroxybenzoate producing *Pseudomonas putida* S12: novel responses and implications for strain improvement. Appl Microbiol Biotechnol 87(2): 679-690.

Verhoef S, Wierckx N, Westerhof RG, de Winde JH, Ruijssenaars HJ (2009). Bioproduction of p-hydroxystyrene from glucose by the solvent-tolerant bacterium *Pseudomonas putida* S12 in a two-phase water-decanol fermentation. Appl Environ Microbiol 75(4): 931-936.

Vickers CE, Blank LM, Krömer JO (2010). Grand Challenge Commentary: Chassis cells for industrial biochemical production. Nat Chem Biol 6: 875.

Wierckx N, Ballerstedt H, de Bont J, Wery J (2005). Engineering of solvent-tolerant *Pseudomonas putida* S12 for bioproduction of phenol from glucose. Appl Environ Microbiol 71: 8221-8227.

Wu X, Monchy S, Taghavi S, Zhu W, Ramos J, van der Lelie D (2011). Comparative genomics and functional analysis of niche-specific adaptation in *Pseudomonas putida*. FEMS Microbiol Rev 35(2): 299-323.

Wylie JL, Worobec EA (1995). The OprB porin plays a central role in carbohydrate uptake in *Pseudomonas aeruginosa*. J Bacteriol 177(11): 3021-3026.

Yu S, Lai B, Plan MR, Hodson MP, Lestari EA, Song H, Kromer JO (2018). Improved performance of *Pseudomonas putida* in a bioelectrochemical system through overexpression of periplasmic glucose dehydrogenase. Biotechnol Bioeng 115(1): 145-155.

Zhou M, Chi M, Luo J, He H, Jin T (2011). An overview of electrode materials in microbial fuel cells. J Power Sources 196(10): 4427-4435.

Zorn H, Czermak P, Lipinski G-WvR (2014). Biotechnology of food and feed additives. Springer, Heidelberg, Germany.

Monitoring Electron Transfer Rates of Electrode-Respiring Cells

Ozlem Istanbullu, Jerome Babauta, Ryan Renslow and Haluk Beyenal*

The Gene and Linda Voiland School of Chemical Engineering and Bioengineering, Washington State University, Pullman, WA, The United States of America

1. Introduction

It is well known that certain bacteria can respire anaerobically on solid conducting materials using extracellular electron transfer (Borole et al. 2011; Gralnick and Newman 2007; Logan 2009). This ability is important for subsurface biogeochemical processes, microbiologically influenced corrosion and microbial fuel cell research (Renslow et al. 2015). For instance, cells respiring on electrodes are used in microbial fuel cells to generate electrical current (Logan and Regan 2006; Lovley 2006, 2008; Potter 1911), cells in sediments can dissolve metals and play a critical role in subsurface biogeochemistry (Coleman et al. 1993; Lovley et al. 2004; Nevin and Lovley 2002) and cells growing on metal surfaces can influence corrosion (Coetser and Cloete 2005; Herrera and Videla 2009). Cells capable of extracellular electron transfer start respiring on the relevant surfaces as soon as they attach to the surface. Most of the time attached cells mature and develop into biofilms and, therefore, respiration of biofilms on the electrode surfaces has generally been the chosen method to investigate extracellular electron transfer (Babauta et al. 2015a). However, cellular respiration and electrode cell interactions at the initial step of cell attachment have been primarily ignored due to its very small electron transfer rates and difficulties associated with *in situ* imaging of cells on solid electrodes (Renslow et al. 2013).

We have developed a method to quantify electron transfer rates to an electrode through the stages of biofilm formation from single cells up to multicellular communities (McLean et al. 2010). In this previous work we used *Shewanella oneidensis* strain MR-1 p519nGFP (constitutively expressed as green fluorescent protein), a model species capable of respiration on solid surfaces serving as an electron acceptor. The significant drawback of our system was that it operated as a microbial fuel cell where the anode and the cathode were connected through a resistor, allowing the studied electrode (anode) potential to change over time. A new method that uses the electrode as the sole electron acceptor (without any other soluble electron acceptor) and allows *in situ* imaging is required. A better system should be able to connect to a potentiostat to control the working electrode potential, quantify the electron transfer rates as current and allow for real-time imaging of the individual cells.

When an electrode is exposed to microorganisms, the electrode shows a different potential profile over time when compared to a sterile growth medium (Renslow et al. 2011; Xu et al. 2010). The cells

---

*Corresponding author: beyenal@wsu.edu

attached to the surface change the potential of the electrode. For *S. oneidensis* MR-1, the depression of the electrode potential is dependent upon the mediators expressed to transfer electrons. These mediators, known to be bind to the cell surfaces, have higher concentration as more cells colonize the electrode surface. In other words, cell numbers on the surface can be linked to the development of the electrode potential over time. In a simplistic approach, the electrode potential can be described as a pseudo-Nernst response where electrode potential is determined by the cell surface coverage. Equation 1 introduces the cell coverage into the electrode potential dependence by assuming that mediators are only present on the cell surfaces.

$$E_{OCP} = (1 - \phi)E_{iniatial} - (\phi)\frac{59}{n}\log\frac{[Med_{RED}]}{Med_{OX}} \tag{1}$$

This assumption can only be true for young biofilms with minimal EPS production since it is known that mediators are present in the biofilm matrix. This type of mechanistic approach has not been done before. However, steady state measurements of electrode potential after *S. oneidensis* biofilm growth revealed a limiting potential of -470 mV$_{SCE}$ (Babauta et al. 2015b). The typical electrode potential in sterile growth medium is approximately +200 mV$_{SCE}$. What is missing is the critical cell surface coverage that is required to reach the limiting electrode potential. This information is crucial in order to understand how electron flux to a solid conductor develops after initial attachment.

The goals of this study are 1) to quantify the *in situ* single cell electron transfer rates on an electrode surface when the electrode are the electron acceptor and 2) to quantify open circuit potential of the electrode and correlate it with the cell growth. We developed a well-controlled bioelectrochemical flow cell that allowed us to image the cells on the surface, quantify the open-circuit potential of an electrode and polarize another electrode to quantify electron transfer all in real-time. The bioelectrochemical flow cell we used was analogous to a microbial fuel cell, but the potential of the working electrode was controlled by a potentiostat. We used an inverted microscope to monitor *in situ* cell growth on the surface and counted the cells. We measured the potential of a nonpolarized electrode against an Ag/AgCl reference electrode located in the cathodic compartment of the bioelectrochemical cell.

2. Materials and Methods

2.1 Preparation of Growth Medium and Inocula

Shewanella oneidensis MR-1 stock samples, stored in 1 mL vials at -70°C, were thawed at room temperature (approximately 25°C) and used to inoculate sterilized 100 mL M1 minimal medium (described below). The culture was incubated in a shaker (Lab-line Instruments, Inc. Melrose Park, IL) at 150 rpm at room temperature. After 24 hours of incubation, 1 mL of the culture was transferred into a quartz cuvette in order to measure the optical density at 600 nm (OD$_{600}$). The culture was then used to create a set of diluted inocula in 50 mL pre-sterilized test tubes with sterilized M1 minimal medium. The set consisted of inocula with optical densities was fixed to 0.005, 0.013, 0.036, 0.301 and 0.402 OD$_{600}$.

M1 minimal medium was prepared with deionized water in an autoclavable carboy and included PIPES (piperazine-N,N'-bis [2-ethanesulfonic acid]) buffer 0.91 g/L, sodium hydroxide (NaOH) 0.3 g/L, ammonium chloride (NH$_4$Cl) 1.5 g/L, potassium chloride (KCl) 0.1 g/L, sodium phosphate monobasic (NaH$_2$PO$_4$H$_2$O) 0.6 g/L, sodium chloride (NaCl) 1.75 g/L and mineral solution 10 mL/L. The pH of the medium was set to 7.0 using HCl or NaOH as needed. A 0.2 µm polytetrafluoroethylene membrane filter was spliced to one of the ports of the carboy lid to allow for aseptic venting. The medium was autoclaved at 121°C for 20 minutes. After allowing the medium to cool, 10 mL/L vitamin solution, 10 mL/L amino acid solution and 1.1206 g/L sodium lactate [60% (w/w)] were added in distilled water.

Mineral Solution: Nitrilotriacetic acid 1.5 g/L, magnesium sulfate heptahydrate 3 g/L, magnesium sulfate monohydrate 0.5 g/L, ferrous sulfate heptahydrate 0.1 g/L, calcium chloride dihydrate 0.1 g/L, cobalt chloride hexahydrate 0.1 g/L, zinc chloride 0.13 g/L, cupric sulfate pentahydrate 0.01 g/L, aluminum potassium disulfate dodecahydrate 0.01 g/L, boric acid 0.01 g/L, sodium molybdate dihydrate

0.025 g/L, nickel chloride 0.024 g/L and sodium tungstate 0.025 g/L were added to deionized water. The pH was set to 7.0 by using HCl or NaOH as needed. The solution was filter-sterilized and kept refrigerated away from light.

Vitamin solution: Biotin 0.002 g/L, folic acid 0.002 g/L, pyridoxine HCl 0.01 g/L, riboflavin 0.005 g/L, thiamine HCl monohydrate 0.005 g/L, nicotinic acid 0.005 g/L, d-pantothenic acid hemicalcium salt 0.005 g/L, B12 0.0001 g/L, p-aminobenzoic acid 0.005 g/L and thioctic acid 0.005 g/L were added to deionized water. The pH was set to 7.0 by using HCl or NaOH as needed. The solution was filter sterilized and kept refrigerated and away from light.

Amino acid solution: L-glutonic acid 2 g/L, L-arginine 2 g/L and DL-serine 2 g/L were added to distilled water. The pH was set to 7.0 by using HCl or NaOH as needed. The solution was filter sterilized and kept refrigerated.

2.2 Preparation and Operation of Bioelectrochemical Flow Cell

For this study, a custom-built autoclavable bioelectrochemical flow cell (Figure 1) was designed to allow for simultaneous *in situ* monitoring of cell growth, quantification of metabolic rates per cell measured as current and measuring of electrode open circuit potential (OCP). It consisted of two adjacent parallel flow chambers: an anodic compartment for the working (polarized) and OCP-measuring (nonpolarized) electrodes and a cathodic compartment for the reference and counter electrodes. The flow cell was constructed out of polycarbonate, and the chambers were separated by a cation exchange membrane (CMI-7000, Membranes International, Ringwood, NJ, USA). Glassy carbon electrodes (1×1 cm), used for the working and OCP-measuring electrodes, were prepared by washing with deionized water and sonicating for 1 minute in 1 M HCl solution and then air dried. Platinum wires were connected to the electrodes by conductive epoxy that was subsequently covered with silicone rubber sealant. An Ag/AgCl wire was used as a pseudo-reference electrode and a graphite plate acted as the counter electrode. The flow cell assembly (electrodes, membrane and feed/waste lines) was sterilized by autoclave.

Once cooled, the flow cell assembly was placed on the microscope stage and the feed line was aseptically connected to the growth medium vessel. The medium was pumped to the anodic compartment of the flow cell at 0.4 mL/minute. In the cathodic compartment, sterile catholyte solution (3 M NaCl) was pumped at the same flow rate (0.4 mL/minutes). High purity nitrogen was fed into the M1 minimal medium prior to the inoculation to ensure anaerobic conditions. The flow cell was primed for at least 1 hour to purge trapped air bubbles and ensure anaerobic conditions.

After priming, the OCP of both anodic compartment electrodes were monitored using two standard three-digit digital voltmeters. After the OCP of both anodic compartment electrodes reached steady values and the system was completely anaerobic, the feed pump was stopped for inoculation. For each individual experiment, the flow cell was inoculated with 10 mL of one of the inocula (0.402, 0.301, 0.036, 0.013 or 0.005 OD_{600}) using a pre-sterilized syringe. After 2 hours of initial attachment, the medium was again pumped through the system. A Gamry Reference 600™ potentiostat was then used to polarize the working electrode to +400 $mV_{Ag/AgCl}$. The current values were continuously monitored and recorded by commercial software.

2.3 Cell Imaging and Analysis

We were able to image cells attached to the polarized and nonpolarized electrodes through a glass cover slip located at the base of the flow cell. Images were captured by Nikon DS-Qi1Mc camera mounted on a brightfield inverted microscope (Nikon Eclipse Ti-S) with a 40x extra long working distance objective. Images were taken every 24 hours, after inoculation, for four days. After four days, cell counts on the surface were unreliable due to biofilm thickness.

2.4 Cell Density

We manually counted the number of cells in each image and then calculated the number of cells per field of view. The size of the field of view is obtained from the NIS Element® software. Once a single cell counted, it was marked with the pointer.

2.5 Cell Respiration Rate

The cell respiration rates were determined from the experimentally measured current values while the working electrode was polarized to +400 $mV_{Ag/AgCl}$. The current values were continuously monitored and recorded. A Gamry Reference 600™ potentiostat was used to polarize the working electrode. The average current values were used at the time of cell counting. Then the cell reparation rates were calculated by dividing experimentally measured current to total cell numbers on the electrode and reported as current per cell with a unit of fA/cell.

2.6 Statistical Analysis

At least 20 images were taken during each imaging event. In our previous work (Ica et al. 2012), we found that taking 10 images generated statistically reproducible data. The average of these 10 images was considered as one biological replicate. The experiments were replicated biologically.

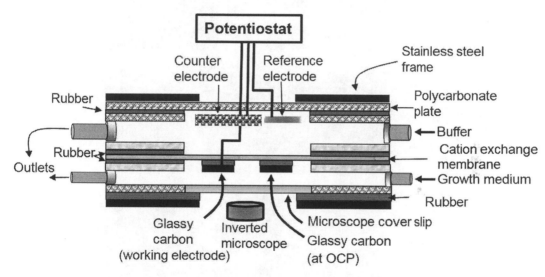

Figure 1: A schematic of the experimental configuration and bioelectrochemical flow cell. The figure is not drawn to scale.

3. Results and Discussion

3.1 Cell Attachment and Growth on the Electrodes

Using our bioelectrochemical flow cell imaging system, we successfully and repeatedly grew *S. oneidensis* biofilms and monitored their growth and metabolic activity. In this paper we define the measured current per cell as the cellular metabolic respiration rate since current shows the number of electrons transferred per unit cell that is equivalent to their respiration rates on the electrodes. Experiments were reproducible in every run; we obtained statistically similar numbers of cells attached to each electrode, and the statistical tests demonstrated that the results were also statistically similar. One example data set is shown in Figure 2 that demonstrates attached cells on the polarized and nonpolarized electrodes. This figure shows that the numbers of attached cells on the surface increases by time. The reported current values refer to the average respiration rates of single cells on the surface. During the growth of the cells in Figure 2, the current was ~0.3 $\mu A/cm^2$.

Figure 3 quantifies how the number of cells increased by time for the polarized and nonpolarized electrodes. The cell numbers were very small after day 1 ($6 \cdot 10^4$ cells/cm^2 for polarized and $5 \cdot 10^4$ for nonpolarized; there was no statistically significant difference P=0.2702). After day 1, the number of

Polarized   Non-polarized

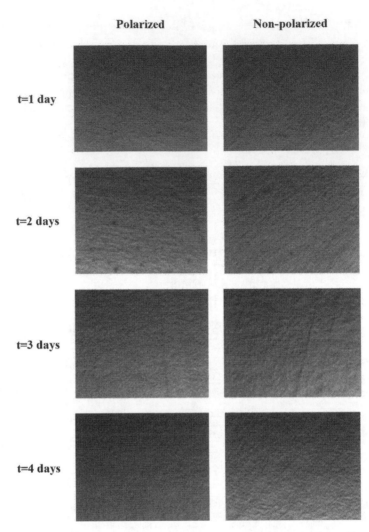

Figure 2: Representative images of the electrodes demonstrating cell
attachment and growth over time.

cells started to increase exponentially. During the duration of the experiment, the number of cells on
the polarized and nonpolarized electrodes were statistically similar (P=0.1168) and demonstrated that
the cells grew similarly, regardless of whether the electrode was polarized or not. This was unexpected
as *S. oneidensis* can utilize lactate only if there is an electron acceptor available (Babauta et al. 2012;
Babauta et al. 2011). For the polarized electrode, the electrode was the electron acceptor. However, for
the nonpolarized electrode, there was no electron acceptor directly provided. Possible explanations for
this include 1) electrons were somehow deposited on the non-polarized surface, 2) small amounts of
oxygen permeated into the system or 3) endogenously produced electron acceptors were used, such as
flavins or biomass. We know that nonpolarized electrodes can build up charge acting as a capacitor. If
the bacteria are able to take advantage of this, we expect that the OCP of the nonpolarized electrode will
decrease with cell growth. However, it is unlikely that even an initially forming biofilm would be able to
utilize this as a sole electron acceptor due to the amount of charge produced and the limited capacity of
the electrode as a capacitor (Lewandowski and Beyenal 2013). To tests these ideas, we measured open
circuit potential (OPC) of the nonpolarized electrode and determined cell respiration as function of the
inoculation density.

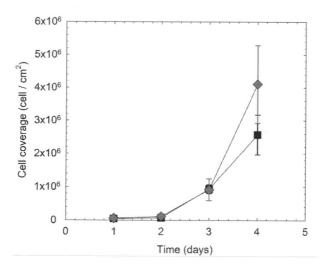

Figure 3: Number of cells on the polarized (♦) and nonpolarized (■) surfaces (inoculation OD$_{600}$=0.005).

Figure 4 shows that the open circuit potential (OPC) of the nonpolarized electrode decreases with the cell growth and that its value depends on the inoculation density. When inoculation density was low, the variation of OCP with time was slow. For example, at the lowest optical density of 0.005, the OCP on day 4 varied only slightly when compared to the initial value. Note on day 2, the OCP value was surprisingly high (asterisks in Figure 4). However, this value was not reproducible and was treated as an outlier. The OCP trend at an optical density of 0.013 was similar to that of optical density 0.005. However, when the flow cell was inoculated with a culture with an optical density of 0.036, the OCP decreased from ~200 mV$_{Ag/AgCl}$ to 50 mV$_{Ag/AgCl}$ within four days. The change in OCP was around 150 mV. At the higher cell densities (OD=0.301 and 0.402), the OCP decreased from ~300 mV$_{Ag/AgCl}$ to ~-180 mV$_{Ag/AgCl}$ within four days. This change of ~ 480 mV was almost three times higher than the change for low cell density inoculum. Interestingly, the OCP value reached ~-180 mV$_{Ag/AgCl}$ within two days for OD=0.301 and remained constant indicating that the OCP reached steady state. OCP is only sensitive to cell attachment at optical densities greater than 0.013.

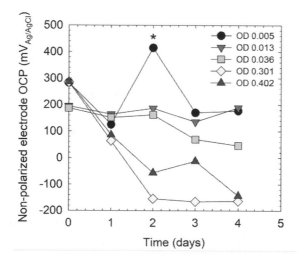

Figure 4: Open circuit potential (OCP) of the nonpolarized electrode decreases with the cell growth and depends on the inoculation density.

We also compared the OCP response of the nonpolarized electrode against the cell counts, except we normalized the cell counts to the fraction of electrode surface covered by cells. This was done by estimating the projected surface area of an individual cell lying flat to be approximately 10 μm². Figure 5 shows the aggregate OCP data of the nonpolarized data vs. the calculated percent surface coverage. OCP is insensitive to cell coverage until ~35%. Above this threshold, the OCP trends negative toward the OCP value typically measured for mature biofilm, ~-400 mV$_{Ag/AgCl}$ (Babauta et al. 2011). The dashed line in Figure 5 is centered on the average OCP value for all points below 35% excluding the outlier marked with an asterisk (same data point as shown in Figure 4). Below the threshold, the number of cells respiring on the electrode is not high enough to depress the OCP of the electrode. We hypothesize that this is due to the leakage current of the double layer capacitance being larger than the cell respiration rate. Thus, no charge is accumulated, and OCP is not depressed. With sufficient cell coverage, the total respiration rate can overcome the leakage rate. Charge then accumulates and the OCP shifts negatively (i.e., electron rich or highly reduced condition).

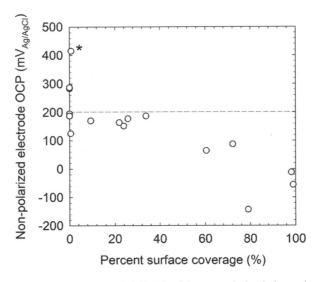

Figure 5: Open circuit potential (OCP) of the nonpolarized electrode across all optical densities and growth.

The fact that the OCP values measured here did not reach the limiting OCP value highlights an important difference between studying newly-attached cells vs. mature biofilm. The single cell monitoring methods, such as what is described herein, may not adequately capture the electron transfer rates of single cells respiring within a mature biofilm. This is because mature biofilms take advantage of the redox active extracellular matrix that takes time to develop. The redox active extracellular matrix provides further electron pathways for cell respiration other than direct electron transfer to the electrode surface. However, the single cell monitoring method allows researchers to study a one direction electron transfer between the attached cells and the electrode.

3.2 Practical Implications

Our results demonstrated that *S. oneidensis* MR-1 cells can grow on electrodes that are not polarized even under anaerobic conditions. After the cells are attached to the surface, the OCP of the electrode decreases as the cells grow. This decrease happens because the electrons generated as a result of cell respiration are deposited on the electrode surface, which acts as a capacitor, or more likely due to the change in concentration of redox couples near the electrode surface. The presented system can be used to monitor current. When we measured the current generated by the cells, we found that the calculated single cell respiration rate varied between 130 fA/cell and 8 fA/cell.

4. Conclusions

In the work we developed a bioelectrochemical system that monitored both the *in situ* cell growth and potential of the electrodes against a reference electrode and quantified the current of a polarized electrode. We concluded that the growth of the cells on the polarized and nonpolarized electrodes was statistically similar. The average cell respiration rate is critically dependent on the inoculum density. The denser the cell inoculation, the lesser is the average cell respiration. The potential of the nonpolarized electrode was dependent on the inoculation density. At the lowest inoculation density, the potential did not deviate from the initial value. Whereas the highest inoculation density, we observed a depression of the potential, indicating an accumulation of electrons at the electrode surface. This inoculum-dependent response of the potential highlights the complexities of single cell measurements. Single cells may respire on the electrode surface utilizing the double layer capacitance.

Acknowledgment

This work was supported by NSF awards # 0954186 and #706889.

References

Babauta JT, Renslow R, Lewandowski Z, Beyenal H (2012) Electrochemically active biofilms: facts and fiction. A review. Biofouling 28:789-812.

Babauta J.T, Beyenal H (2015a) Biofilm electrochemistry. In: Beyenal H, JT Babauta (eds) Electrochemically active biofilms in microbial fuel cells and bioelectrochemical systems: From laboratory practice to data interpretation, Wiley and Sons, New York, USA, pp 121-176.

Babauta J.T, Beyenal H (2015b) Introduction to electrochemically active biofilms. In: Beyenal H, JT Babauta (eds) Electrochemically active biofilms in microbial fuel cells and bioelectrochemical systems: From laboratory practice to data interpretation, Wiley and Sons, New York, USA, pp 1-136.

Babauta JT, Hung Duc N, Beyenal H (2011) Redox and pH microenvironments within *Shewanella oneidensis* MR-1 biofilms reveal an electron transfer mechanism. Environmental Science & Technology 45: 6654-6660.

Borole AP, Reguera G, Ringeisen B, Wang ZW, Feng YJ, Kim BH (2011) Electroactive biofilms: Current status and future research needs. Energy & Environmental Science 4: 4813-4834.

Coetser SE, Cloete TE, (2005) Biofouling and biocorrosion in industrial water systems. Critical Reviews in Microbiology 31: 213-232.

Coleman ML, Hedrick DB, Lovley DR, White DC, Pye K (1993) Reduction of Fe(III) in sediments by sulfate-reducing bacteria. Nature: 361, 436-438.

Gralnick JA, Newman DK (2007) Extracellular respiration. Molecular Microbiology 65:, 1-11.

Herrera LK, Videla HA (2009) Role of iron-reducing bacteria in corrosion and protection of carbon steel. International Biodeterioration & Biodegradation 63: 891-895.

Ica T, Caner V., Istanbullu O, Hung Duc N, Ahmed B, Call DR, Beyenal H (2012) Characterization of Mono- and Mixed-Culture *Campylobacter jejuni* Biofilms. Applied and Environmental Microbiology 78:, 1033-1038.

Lewandowski Z, Beyenal H (2013) Fundamentals of Biofilm Research, CRC Press, Boca Raton, FL, USA, pp 439-509.

Logan BE (2009) Exoelectrogenic bacteria that power microbial fuel cells. Nature Reviews Microbiology 7: 375-381.

Logan BE Regan JM (2006) Electricity-producing bacterial communities in microbial fuel cells. Trends in Microbiology 14: 512-518.

Lovley DR (2006) Bug juice: harvesting electricity with microorganisms. Nature Reviews Microbiology 4: 497-508.

Lovley DR (2008) The microbe electric: conversion of organic matter to electricity. Current Opinion in Biotechnology 19: 564-571.

Lovley DR, Holmes DE, Nevin KP (2004) Dissimilatory Fe(III) and Mn(IV) reduction. Advances in Microbial Physiology, 49: 219-286.

McLean, JS, Wanger G, Gorby YA, Wainstein M, McQuaid J, Ishii SI, Bretschger O, Beyenal H, Nealson KH (2010) Quantification of electron transfer rates to a solid phase electron acceptor through the stages of biofilm formation from single cells to multicellular communities. Environmental Science & Technology 44: 2721-2727.

Nevin KP, Lovley DR, (2002) Mechanisms for Fe(III) oxide reduction in sedimentary environments. Geomicrobiology Journal 19: 141-159.

Potter MC (1911) Electrical effects accompanying the decomposition of organic compounds. Proceedings of the Royal Society of London 84:, 260-276.

Renslow R, Babauta J, Ivory N, Beyenal H, Schenk J, Kuprat A, Fredrickson J, Beyenal H, Babauta JT (2015) Mathametical modeling of extracellular electron transfer in biofilms In: Beyenal H, JT Babauta (eds) Electrochemically active biofilms in microbial fuel cells and bioelectrochemical systems: From laboratory practice to data interpretation, Wiley and Sons, New York, USA, pp 281-344.

Renslow R, Babauta JT, Kuprat A, Schenk J, Ivory N, Fredrickson J, Beyenal H (2013) Modeling biofilms with dual extracellular electron transfer mechanisms. Physical Chemistry Chemical Physics 15: 19262-19283.

Renslow R, Donovan C, Shim M, Babauta JT, Nannapaneni S, Schenk J, Beyenal H (2011) Oxygen reduction kinetics on graphite cathodes in sediment microbial fuel cells. Physical Chemistry Chemical Physics 13: 21573-21584.

Xu F, Duan J, Hou B (2010) Electron transfer process from marine biofilms to graphite electrodes in seawater. Bioelectrochemistry 78: 92-95.

Electrode-Assisted Fermentations: Their Limitations and Future Research Directions

Veronica Palma-Delgado[1], Johannes Gescher[1,2] and Gunnar Sturm[1*]

[1] Institute for Applied Biosciences, Karlsruhe Institute for Technology (KIT), Karlsruhe, Germany
[2] Institute for Biological Interfaces, Karlsruhe Institute of Technology (KIT), Eggenstein-Leopoldshafen, Germany

1. Introduction

The metabolism of microorganisms is highly diverse. This gives rise to an astonishing variety of valuable compounds that can be produced in biotechnological processes. Many valuable compounds are synthesized using microbial fermentation reactions. In general, fermentation reactions refer to a metabolic process in which complex organic molecules are degraded into simpler substances under exclusion of an external terminal electron acceptor (TEA), e.g., oxygen, that distinguishes fermentation from respiration. Hence, the overall redox state of the substrate and the fermentation products will be equal. Complete oxidation of the substrate to CO_2 is hampered by the necessity to reoxidize NADH to NAD^+. In order to maintain the redox balance, electrons that are released during substrate oxidation have to be transferred to an internal electron acceptor. For instance, in lactic acid fermentation pyruvate represents the internal electron acceptor derived from glucose oxidation and serves as the target for NAD^+ recovery, resulting in the redox-balanced production of lactic acid and the recovery of the coenzyme NAD^+. Fermentations generally offer the benefit of low anabolism to catabolism ratio, meaning that the amount of substrate that is used to produce biomass, and not the desired end product, is rather low. Limitations arise if the desired end product is supposed to be more oxidized than the substrate. This necessitates the addition of an electron acceptor that is most often oxygen. Still, this leads to a process that is not energy limited anymore. This leads to a loss of carbon in the form of biomass and a loss of energy in the form of heat. Moreover, the input of oxygen into the system is accompanied by high energy demand and causes the problem of rather limited solubility. Here, the concept of electrode-assisted fermentation comes into play as it allows producing substances that are more oxidized than the substrate with an electrode as electron acceptor (Flynn et al. 2010). A scheme depicting the basic concept of electrode-assisted fermentations is given in Figure 1.

Since the process is anaerobic and the potential of the electrode can be steered according to the desired process kinetics, the electrode-assisted fermentation offers an undepletable electron acceptor in the form of an electrode, a low biomass production due to a limited energy gain for the organisms and a way to use the oxidation energy in the form of an electrical current. Still, the efficient connection

---

*Corresponding author: gunnar.sturm@kit.edu

of the cellular metabolism to the electrode as an electron acceptor is only possible with a limited number of microorganisms. Exoelectrogenic bacteria, like *Shewanella* or *Geobacter* species, developed mechanisms to transfer respiratory electrons to the outer membrane and thereby uncouple the process of NAD^+ recovery from the reduction of intracellular fermentation intermediates (Flynn et al. 2010; Bursac et al. 2017; Beblawy et al. 2018). As will be briefly described in the next section, their terminal reductases that are localized to the cell surface are rather nonspecific which allows for the catalysis of the electron transfer to the electrode surface. Besides these natural exoelectrogenic organisms, several researchers are trying to integrate electron transport pathways into *E. coli* strains in a synthetic biology approach.

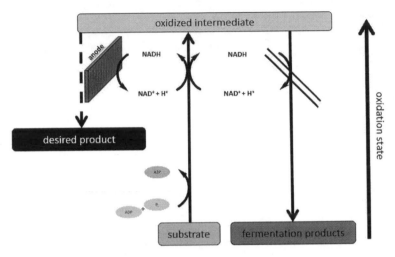

Figure 1: Scheme of electrode-assisted fermentation. NAD^+ is recovered via anodic re-oxidation of NADH instead of reducing oxidized intermediates of the metabolism. This allows fermentation end products that are higher oxidized than the substrate.

This review summarizes research efforts of the last decade leading to strains able to efficiently perform unbalanced as well as electrode-assisted fermentations. Furthermore, this review strikingly underpins future perspectives of this research area.

2. Extracellular Electron Transfer

A detailed description and comparison of the electron transfer mechanisms of *Shewanella* and *Geobacter* are given later in this book by Tender and Yates (2019) that is why we will only briefly describe our current understanding of the electron transport pathways in this chapter.

2.1 *Geobacter sulfurreducens*

Since *c*-type cytochromes were identified as key player proteins enabling extracellular electron transport (EET), it is not surprising that *G. sulfurreducens*—with over 100 *c*-type cytochromes encoded in its genome—is an effective EET organism. To perform EET, the electrons have to pass two membranes and the periplasm in between. Starting with electrons released during oxidative carbon metabolism, they are transported to the quinol pool in the inner membrane. The electrons are taken up by the inner membrane hepta-heme cytochrome ImcH for further supply to the periplasm. Zacharoff and colleagues recently suggested a second pathway depending on the redox potential of the respective electron acceptor. For low potential TEAs ($\leq$ - 0.1 mV), quinol oxidation is facilitated by Cbcl, a *b*-type and multiheme *c*-type cytochrome domains containing protein (Levar et al. 2014; Zacharoff et al. 2016). It is still under debate whether a third protein at the inner membrane, MacA, has a pivotal role in electron transfer reactions. Its importance seems to be rather based on its influence on OmcB expression or stability since an in *trans* expression of OmcB restored the ability to reduce ferric iron in a Δ*macA* strain (Kim and Lovley 2008; Levar et al. 2012).

PpcA is a tri-heme cytochrome that was found to be important for periplasmic electron transport in several *Geobacter* species and is well characterized in terms of structure and thermodynamics (Lloyd et al. 2003; Pokkuluri et al. 2004; Pessanha et al. 2006; Shelobolina et al. 2007). PpcA serves as a periplasmic link to the outer membrane cytochromes (OMCs) such as OmcB, OmcC, OmcE, OmcS, OmcT and OmcZ. These cytochromes play an important role in the reduction of ferric iron and electrode reduction, respectively (Butler et al. 2004; Lovley 2006; Qian et al. 2007, 2011). OmcB and OmcC form a transmembrane spanning porin cytochrome complex (Pcc) by associating with OmaB and OmaC as well as OmbB and OmbC, respectively. In these complexes, OmbB and OmbC represent the porin-like proteins, whereas OmaB and OmaC serve as the periplasmic electron entry site transferring the electrons to the respective OMC (Liu et al. 2014). The particular role of every single OMC is somewhat confusing since deletions in single *omc* genes result in very particular phenotypes with regard to respective electron acceptors. For instance, the deletion of the gene coding for the cytochrome OmcB provokes a strong decrease of the ferric citrate reduction rate and impairs growth on ferric oxide as electron acceptor, whereas it shows similar electron transfer rates compared to the wild type using an anode as TEA (Holmes et al. 2006; Reguera et al. 2006; Nevin et al. 2009). A deletion-mutant in *omcC* does not develop significant differences in growth compared to wild type levels (Leang et al. 2003). Inoue and co-workers demonstrated that the cytochrome OmcZ promotes the electron transfer in biofilms. Subsequent studies revealed that OmcZ accumulates at the interface of *Geobacter* biofilm and a graphite anode surface and are considered to catalyze the final electron transfer process from the cell to the anode (Inoue et al. 2010, 2011). However, OmcZ seems not to be involved in the reduction of insoluble Fe(III) oxides (Nevin et al. 2009). A deletion in the *omcE* gene leads to the inability to reduce Fe(III) oxide but soluble TEAs like ferric citrate do not seem to be affected (Mehta et al. 2005). The hexaheme *c*-type cytochrome OmcS was shown to co-localize with the pili of *Geobacter* and was suggested to play a role in final electron transfer to Fe(III) oxides particles (Leang et al. 2010). Purified OmcS was demonstrated to reduce a variety of substrates, including anthraquinone-2,6-disulfonate (AQDS), Fe(III) oxide, manganese and Cr(VI) (Qian et al. 2011). Since the distance between the single OmcS molecules at the pili surface seems to be rather wide, OmcS may not be responsible for facilitating electron transport over the pili itself (Leang et al. 2010). A long range electron transport along the pili was suggested to be based on overlapping π-orbitals of aromatic amino acids that would allow for a metallic-like conductivity of the pili, but the details of electron transfer and whether it is metallic-like or based on hopping between distinct electron transfer proteins is a matter of ongoing research (Malvankar et al. 2011, 2012, 2015; Strycharz-Glaven et al. 2011; Strycharz-Glaven and Tender 2012; Vargas et al. 2013).

In fact, the sheer number of cytochromes and their partially overlapping functions in electron transport processes gain redundancy of the overall process and make it difficult to assign certain functions. It appears as *Geobacter* acquired many different cytochromes in evolution and that these proteins share—at least to some extent— functional redundancy making it difficult for researchers to unravel their particular role in *Geobacter* EET. Nevertheless, the similarity of the Pcc formed by OmaB/C, OmbB/C and OmcB/C, respectively, to the outer membrane cytochrome complexes MtrCAB and MtrDEF from *Shewanella oneidensis* is of particular interest since it suggests that these complexes share the same role in conferring EET.

2.2 *Shewanella oneidensis*

In *Shewanella oneidensis* MR-1, the particular role of the single *c*-type cytochromes involved in EET is rather clear. The genome harbors the information for 41 cytochromes of which a significant number does not seem to be expressed (Beliaev et al. 2002; Meyer et al. 2004; Romine et al. 2008; Schuetz et al. 2009). During anaerobic respiration, electrons are transferred to the quinone pool. The tetraheme *c*-type cytochrome and quinol oxidase CymA reoxidizes the menaquinol by accepting the electrons and relays them to the periplasmic site of the inner membrane (Hartshorne et al. 2007; Firer-Sherwood et al. 2008).

The fundamental role of CymA was discovered early in *S. oneidensis* EET research due to the severe growth phenotypes developed by the deletion mutant. The deletion of the *cymA* gene disables the use of ferric iron, fumarate, nitrate, nitrite, DMSO and also an anode as TEA (Myers and Myers 1997; Schwalb et al. 2003; Lies et al. 2005; Gao et al. 2009). From there, CymA transports the electrons to a network of periplasmic redox proteins that is dominated by *c*-type cytochromes (Gralnick and Newman 2007). Especially two cytochromes, the periplasmic tetraheme *c*-type flavocytochrome and fumarate reductase FccA, and the small tetratheme *c*-type cytochrome STC seem to have an elementary function of an electron transfer hub. They are highly abundant in the periplasm, can be reduced by CymA and can also deliver the electrons to a number of different ET pathways (Gralnick et al. 2006; Schuetz et al. 2009; Coursolle and Gralnick 2010; Fonseca et al. 2013; Alves et al. 2015; Sturm et al. 2015). With regard to the distribution of electrons through and within the periplasmic space, both proteins share functional redundancy (Fonseca et al. 2013; Sturm et al. 2015). Electrode respiration depends on the delivery of electrons to the outer membrane heterotrimeric protein complex consisting of MtrA (periplasmic electron acceptor), MtrB (membrane-associated β-barrel protein) and the terminal reductase MtrC (Shi et al. 2012). The deletion of any of these proteins results in the inability or at least impairment of *S. oneidensis* cells to respire extracellular ferric iron (Beliaev et al. 2001; Bretschger et al. 2007; Gao et al. 2010; Schicklberger et al. 2011).

Previous studies discovered that the *c*-type cytochromes that are involved in periplasmic and extracellular electron transport processes share overlapping redox potential windows. These overlaps most likely facilitate electron transfer that is directed by the redox potential gradient between CymA and the TEA (Firer-Sherwood et al. 2008). For efficiently maintaining EET also with respect to longer distances and solid electron acceptors, *Shewanella* uses two different mechanisms. It was shown that cellular appendices of *Shewanella* are conductive and that this property is due to the functional expression of MtrC and OmcA (Gorby et al. 2006; El-Naggar et al. 2010; Subramanian et al. 2018). *Shewanella* nanowires resemble extensions of the outer membrane in the form of interconnected membrane vesicles. These vesicles contain in their lumen the typical periplasmic protein content and show similar surface decoration with MtrCAB complexes that can also be detected at the outer membrane (Pirbadian et al. 2014).

The second mechanism involves electron transfer via extracellular shuttling components. Flavins were identified to have a crucial role in *Shewanella* EET (Covington et al. 2010; Brutinel and Gralnick 2012; Kotloski and Gralnick 2013). Flavins seem not to be essential for the reduction of soluble electrons acceptors like Fe(III)-citrate because soluble TEAs can have direct contact with the cells. Nevertheless, the secretion of flavins can contribute to the acceleration of the reduction process (von Canstein et al. 2008). However, endogenously expressed and secreted flavins facilitate long-range electron transfer and account for ~70% of the electron transfer rate of *S. oneidensis* respiring electrodes (Marsili et al. 2008). In the presence of insoluble TEAs like graphite electrodes, endogenous flavins are actively exported as soluble redox molecules in order to improve the electron transfer. The essential genes for the synthesis of flavins have been identified to be *ribBA, ribD, ribH* and *ribE* (Brutinel et al. 2013). Flavin adenine dinucleotide (FAD) is exported into the periplasm through the bacterial FAD exporter Bfe (Kotloski and Gralnick 2013). Once in the periplasm, FAD can be incorporated into proteins as a cofactor or hydrolyzed to FMN and AMP by the periplasmic 5' nucleotidase UshA (Covington et al. 2010). Flavins are accumulated in supernatants of *S. oneidensis* cultures, and the growth conditions seem to have a great impact on their concentration. Studies conducted so far report flavin secretion resulting in concentrations ranging from 25 nM to 500 nM (Marsili et al. 2008; von Canstein et al. 2008; Velasquez-Orta et al. 2010; Zhai et al. 2016). The secretion of recyclable redox shuttles allows the reduction of solid electron acceptors at a distance and is potentially beneficial in intercellular electron transport in *Shewanella* biofilms. All the mechanisms facilitating EET described above make *Shewanella* an excellent organism for investigation of the particular pathways leading to efficient electron transport to extracellular electron acceptors. Still, other lines of evidence from differential pulse voltammetry suggest that flavins also have a role as cofactors of OMCs and facilitate extracellular electron transfer processes in their semiquinone status (Okamoto et al. 2013, 2014a, b; Xu et al. 2016).

2.3 *Escherichia coli*

The restriction limiting the use of *S. oneidensis* as target organism for unbalanced fermentation reactions is constituted by the rather small substrate spectrum that can be used as an electron source by the organism.

In contrast, the physiology of *E. coli* allows using a broad spectrum of fermentable substrates. Researchers are challenged to implement mechanisms that turn *E. coli* into an EET facilitating organism. The prime advantage of using *E. coli* in fermentative applications lies in the multitude of different fermentative processes that can be performed by the organism, its broad knowledge in the physiology of the organism as well as in its excellent genetic tractability.

The concept of electrode-assisted fermentation requires an alternative destination for electrons released by substrate oxidation to allow for the generation of products more oxidized than the substrate. Solid electrodes are undepletable electron acceptors that can readily take the role as the primary electron sink for electrode-assisted fermentations with *E. coli*. Unfortunately, until now, the specifically designed production strains use soluble mediators during the process due to the inability of *E. coli* to perform EET. Current research, therefore, aims at establishing biochemical pathways that facilitate EET in *E. coli* and is discussed later on. Also, examples of research attempts resulting in the transformation of *E. coli* into an organism capable of transferring electrons to the outside of the cell are also given later in this chapter.

3. Microbial Fuel Cells (MFCs) and Microbial Electrolysis Cells (MECs)

Electrode-assisted fermentations necessitate the electrode as an electron acceptor. Hence, two general modes of operating a bioelectrochemical system are possible. In MFC mode, the electrons are delivered into the electrical conduit and may power a device between the anode and cathode before the electrons are finally transferred to the respective TEA (e.g., oxygen) and form the end product (e.g., water). A detailed description of MFCs is examined by Kerzenmacher as well as Rahimnejad and his colleagues (Kerzenmacher 2012; Rahimnejad et al. 2015). Still, it is questionable whether it will be possible to design a robust electrode-assisted fermentation process in an MFC as it is typically characterized by rather low anode potentials that result from usually limiting performance of the cathode. Moreover, the low voltage of the system also limits the amount of electrical energy that can be produced. To gain more efficiency, researchers try to steer the anode potentials to optimal values and use the cathodic reduction reaction to produce other valuable compounds. In microbial electrolysis cells (MECs) the current produced by the anodic cells is used to reduce protons to elemental hydrogen (H_2). The reaction needs to be powered externally due to the low redox potential of the H^+/H_2 couple ($E_0^{'} = -414$ mV) and the necessary overpotential to establish the process. Therefore, additional energy is needed to lower the cathode potential even though MECs can lower the energy demand for hydrogen production by 62% to 81% (Zhang and Angelidaki 2014). Hence, the production of a valuable chemical can be combined with an efficient hydrogen formation. Hydrogen can then be either stored or used in further processes as, for instance, to generate methane. Kadier and his colleagues give more detailed examples later in this book (see Chapter 8).

4. Examples for Established Electrode-Assisted Fermentation

The concept of electrode-assisted fermentations was introduced by Flynn and his colleagues in 2010. They engineered a *Shewanella* strain by introducing glycerol consumption as well as an ethanol production module from *E. coli* and *Zymomonas mobilis*, respectively. This strain was able to stoichiometrically convert glycerol into ethanol by eliminating two surplus electrons by means of an electrode offered as terminal electron acceptor (Flynn et al. 2010). The additional deletion in the gene coding for phosphate acetyltransferase (*pta*) increased the carbon conversion rate from glycerol to ethanol from 75% to 85% (Flynn et al. 2010). Electrode-assisted fermentations were also established for the production of other products. One potential substance is 2,3-butanediol or its precursor acetoin. Acetoin (3-hydroxy-2-butanon) was rated as one of the top 30 most promising platform chemicals by the US Department of Energy in 2004 (Werpy and Petersen 2004). Consequently, a great deal of effort was put into the optimization of its production process (e.g., Sun et al. 2012; Zhang et al. 2013, 2016; Wang et al. 2013; Chen et al. 2013). Microbial fermentations, particularly bacterial fermentations, serve as the main source for the biotechnological production of acetoin (Xiao and Lu 2014). Sun et al. showed that a production yield of roughly 75 g l^{-1} at a rate of 1.88 g l^{-1} h^{-1} can be achieved by using *Serratia marcescens* as producing host and sucrose as substrate (Sun et al. 2012). An engineered strain of *Bacillus subtilis* was able to produce an average amount of ~62 g l^{-1} at a rate of 0.864 g l^{-1} h^{-1} but used a combination of monosaccharides derived from lignocellulosic hydrolysates (Zhang et al. 2016). Until now, acetoin was produced almost exclusively under oxic conditions as it is more oxidized than glucose (Wang et al. 2013). Recently, Bursac and colleagues established a strain variant of *S. oneidensis* capable of acetoin production. They introduced a plasmid-based genetic module consisting of codon-optimized versions of *B. subtilis* derived acetolactate synthase (*alsS*) and acetolactate decarboxylase (*alsD*). This strain was capable of converting 40% of the catabolically consumed lactate into acetoin while roughly 60% was further converted into the natural end product of the strain (Bursac et al. 2017). Further strain development was carried out to increase the amount of acetoin produced by considerably reducing the amount of acetate released from the reaction. Knockouts in phosphate acetyltransferase and acetate kinase (e.g., *pta, ackA*) further increased the ratio of acetoin conversion from lactate to ~86% of the theoretical maximum. Of note these experiments were conducted in cell suspensions using fumarate as the terminal electron acceptor (Bursac et al. 2017).

S. oneidensis harbors three prophages in its genome (Gödeke et al. 2011). Since the authors could show that a deletion strain deficient in the λ-prophage produces 1.34x more current in an MFC setup, it was chosen as host carrying the mutations and modifications described above. This strain was also subjected to anode-respiring conditions. In this setup, the strain produced 78% of the theoretical maximum in 72 hours. Furthermore, the system produced a current density of roughly 20 μA cm^{-2}. So far, this work is the first description of the anoxic conversion leading to the production of acetoin close to the benchmark levels with regard to carbon efficiency (Bursac et al. 2017).

Since a major drawback of using *Shewanella* strains as producing hosts is their limited substrate spectrum, a great deal of effort was invested either to broaden the spectrum of carbon sources of the organism or to convert organisms with a diverse substrate spectrum into exoelectrogens. The Ajo-Franklin group was recently successful in establishing extracellular electron transfer in *E. coli* by implementing the key proteins of the pathway from *S. oneidensis* facilitating EET. They were able to show that expression of the MtrCAB complex was sufficient to enable extracellular electron transport, and the engineered cells reduced Fe$_2$O$_3$ particles 4x faster than the wild type strain (Jensen et al. 2010). Additional integration of the cytoplasmic membrane-bound quinol oxidase CymA from *S. oneidensis* enabled extracellular electron transfer and led to an increase of biomass (Teravest et al. 2014; Jensen et al. 2016). As proof of principle, this work shows the possibility of facilitating EET in organisms that are by nature not capable of it. Nevertheless, the results also show that the productivity of electron transfer rates cannot compete with EET-inherent organisms like *S. oneidensis*. In another approach Sturm-Richter et al. showed that EET in *E. coli* is possible without emulating the whole protein cascade from *S. oneidensis* by substituting the outer membrane complex MtrCAB through the supply of the membrane permeable electron shuttle methylene blue (Sturm-Richter et al. 2015). The organism harbors the genetic information for CymA, MtrA and STC but not MtrB and MtrC. It means that the electron transport to the periplasm is guaranteed but transport over the outer membrane has to be facilitated artificially – in this case, via the addition of methylene blue.

The strain from that study served as a chassis for further strain developments by Förster and colleagues who deleted the genes *frdA-D, adhE, ldhA, pta* and *ack* in order to increase the pyruvate production and avoid undesired by-products, and afterward inserted the gene module coding *alsS* and *alsD* for acetoin production. Thus, the metabolism of the strain was forced to accumulate a high amount of the precursors of acetoin, and the authors were able to show that the overall acetoin yield from glucose could be raised up to 90% under nitrate-reducing conditions and up to 79% under electrode respiring conditions with no detectable by-product in the fermentation broth (Förster et al. 2017).

5. Necessities to Realize Electrode-Assisted Fermentations on an Industrial Level

The question could be whether this technology of electrode-assisted fermentation can be—judged from the current stage of research and development—competitive at some point. As discussed above, the so far established strains allow gaining carbon recoveries that are either better or at least on the same level compared to the traditional biotechnological production systems. Still, besides the carbon recovery, the space-time yield is another important parameter and so far, the bottleneck for electrode-assisted fermentation systems. Let us conduct the following theoretical experiment in order to recognize the key aspects of an industrial realization.

Our research revealed that a model bioelectrochemical community (composed of exoelectrogenic wild type strains *S. oneidensis*, *G. sulfurreducens* and *G. metallireducens*) can produce current densities of 600 µA cm^{-2} in a very simple reactor setup (this current density is 30x higher compared to the engineered *Shewanella* strain mentioned above) (Prokhorova et al. 2017). To gain this current we used a graphite electrode material that was not improved in its characteristics by any further steps. Let us assume that we work with a reactor that contains 0.5 m^2 of electrode surface area per liter of reactor volume that is theoretically possible but is an engineering challenge at the same time. This means that we will produce a current of 3 A that can be correlated to a biotechnological activity. Using the elementary charge and the Avogadro number, we can translate this to 112 mmol of electrons per hour and liter. Let us assume that we try to produce acetoin from glucose that cannot be metabolized by the model organism community so far. Per mol of acetoin produced i.e. 4 mol of electrons will have to be transferred to the electrode surface. Consequently, it could be possible to produce 2.5 g of acetoin per hour and a liter of reactor volume that is more than the typical minimum threshold for biotechnological production routines of 1 g l^{-1} h^{-1}. Moreover, this process would be combined with the concomitant production of 1.34 liters of hydrogen per liter reactor volume and hour. In summary, it could be possible to use bioelectrochemical technologies for biotechnological productions.

6. Conclusions and Future Perspectives

As described in the previous sections, the field of electrode-assisted fermentations is of great interest regarding the production of platform chemicals like acetoin or other highly valuable chemicals. From today's view, it is conceivable that electrode-assisted fermentations will contribute to a noteworthy amount of industrially-produced platform chemicals in the future. For this it will be necessary to establish production strains that meet the necessary requirements for current densities and reactor systems that will allow exploiting exoelectrogenic biofilms in continuous production systems. These two focus areas determine the need for future research directions that should establish ways:

- To raise current densities in a combined approach of synthetic engineering of the organisms; for instance, by increasing the biofilm densities and conductivities and the materials that allow optimal connection of the biofilm to the electrode material.
- To design reactor systems to exploit the enormous potential of productive exoelectrogenic biofilms as they are natural retentostat systems that offer a strategy to design continuous processes without the need to separate the biocatalyst from the desired product.

References

Alves MN, Neto SE, Alves AS, Fonseca BM, Carrêlo A, Pacheco I, Paquete CM, Soares CM, Louro RO (2015) Characterization of the periplasmic redox network that sustains the versatile anaerobic metabolism of *Shewanella oneidensis* MR-1. Front Microbiol 6. doi: 10.3389/fmicb.2015.00665

Beblawy S, Bursac T, Paquete C, Louro R, Clarke TA, Gescher J (2018) Extracellular reduction of solid electron acceptors by *Shewanella oneidensis*. Mol Microbiol 109: 571–583 doi: 10.1111/mmi.14067

Beliaev AS, Saffarini DA, McLaughlin JL, Hunnicutt D (2001) MtrC, an outer membrane decahaem *c* cytochrome required for metal reduction in *Shewanella putrefaciens* MR-1. Mol Microbiol 39: 722–730. doi: 10.1046/j.1365-2958.2001.02257.x

Beliaev AS, Thompson DK, Khare T, Lim H, Brandt CC, Li G, Murray AE, Heidelberg JF, Giometti CS, Yates J, Nealson KH, Tiedje JM, Zhou J (2002) Gene and Protein Expression Profiles of *Shewanella oneidensis* during Anaerobic Growth with Different Electron Acceptors. Omics A J Integr Biol 6: 39–60. doi: 10.1089/15362310252780834

Bretschger O, Obraztsova A, Sturm CA, In SC, Gorby YA, Reed SB, Culley DE, Reardon CL, Barua S, Romine MF, Zhou J, Beliaev AS, Bouhenni R, Saffarini D, Mansfeld F, Kim BH, Fredrickson JK, Nealson KH (2007) Current production and metal oxide reduction by *Shewanella oneidensis* MR-1 wild type and mutants. Appl Environ Microbiol 73: 7003–7012. doi: 10.1128/AEM.01087-07

Brutinel ED, Dean AM, Gralnick JA (2013) Description of a riboflavin biosynthetic gene variant prevalent in the phylum proteobacteria. J Bacteriol 195: 5479–5486. doi: 10.1128/JB.00651-13

Brutinel ED, Gralnick JA (2012) Shuttling happens: Soluble flavin mediators of extracellular electron transfer in *Shewanella*. Appl Microbiol Biotechnol 93: 41–48. doi: 10.1007/s00253-011-3653-0

Bursac T, Gralnick JA, Gescher J (2017) Acetoin production via unbalanced fermentation in *Shewanella oneidensis*. Biotechnol Bioeng 114: 1283–1289. doi: 10.1002/bit.26243

Butler JE, Kaufmann F, Coppi M V, Nunez C, Lovley DR (2004) MacA, a Diheme *c*-Type Cytochrome Involved in Fe(III) Reduction by *Geobacter sulfurreducens*. J Bacteriol 186: 4042–4045. doi: 10.1128/JB.186.12.4042

Chen T, Liu WX, Fu J, Zhang B, Tang YJ (2013) Engineering *Bacillus subtilis* for acetoin production from glucose and xylose mixtures. J Biotechnol 168: 499–505. doi: 10.1016/j.jbiotec.2013.09.020

Coursolle D, Gralnick JA (2010) Modularity of the Mtr respiratory pathway of *Shewanella oneidensis* strain MR-1. Mol Microbiol 77: 995–1008. doi: 10.1109/ICEPT.2013.6756496

Covington ED, Gelbmann CB, Kotloski NJ, Gralnick JA (2010) An essential role for UshA in processing of extracellular flavin electron shuttles by *Shewanella oneidensis*. Mol Microbiol 78: 519–532. doi: 10.1111/j.1365-2958.2010.07353.x

El-Naggar MY, Wanger G, Leung KM, Yuzvinsky TD, Southam G, Yang J, Lau WM, Nealson KH, Gorby YA (2010) Electrical transport along bacterial nanowires from *Shewanella oneidensis* MR-1. Proc Natl Acad Sci 107: 18127–18131. doi: 10.1073/pnas.1004880107

Firer-Sherwood M, Pulcu GS, Elliott SJ (2008) Electrochemical interrogations of the Mtr cytochromes from *Shewanella*: opening a potential window. J Biol Inorg Chem 13: 849–854. doi: 10.1007/s00775-008-0398-z

Flynn JM, Ross DE, Hunt KA, Bond DR, Gralnick JA (2010) Enabling unbalanced fermentations by using engineered electrode-interfaced bacteria. mBio 1: e00190-10. doi: 10.1128/mBio.00190-10

Fonseca BM, Paquete CM, Neto SE, Pacheco I, Soares CM, Louro RO (2013) Mind the gap: cytochrome interactions reveal electron pathways across the periplasm of *Shewanella oneidensis* MR-1. Biochem J 449: 101–108. doi: 10.1042/BJ20121467

Förster AH, Beblawy S, Golitsch F, Gescher J (2017) Electrode-assisted acetoin production in a metabolically engineered *Escherichia coli* strain. Biotechnol Biofuels 10: 65. doi: 10.1186/s13068-017-0745-9

Gao H, Barua S, Liang Y, Wu L, Dong Y, Reed S, Chen J, Culley D, Kennedy D, Yang Y, He Z, Nealson KH, Fredrickson JK, Tiedje JM, Romine M, Zhou J (2010) Impacts of *Shewanella oneidensis* *c*-type cytochromes on aerobic and anaerobic respiration. Microb Biotechnol 3: 455–466. doi: 10.1111/j.1751-7915.2010.00181.x

Gao H, Yang ZK, Barua S, Reed SB, Romine MF, Nealson KH, Fredrickson JK, Tiedje JM, Zhou J (2009) Reduction of nitrate in *Shewanella oneidensis* depends on atypical NAP and NRF systems with NapB as a preferred electron transport protein from CymA to NapA. ISME J 3: 966–976. doi: 10.1038/ismej.2009.40

Gödeke J, Paul K, Lassak J, Thormann KM (2011) Phage-induced lysis enhances biofilm formation in *Shewanella oneidensis* MR-1. ISME J 5: 613–626. doi: 10.1038/ismej.2010.153

Gorby YA, Yanina S, McLean JS, Rosso KM, Moyles D, Dohnalkova A, Beveridge TJ, Chang IS, Kim BH, Kim KS, Culley DE, Reed SB, Romine MF, Saffarini DA, Hill EA, Shi L, Elias DA, Kennedy DW, et al. (2006) Electrically conductive bacterial nanowires produced by *Shewanella oneidensis* strain MR-1 and other microorganisms. Proc Natl Acad Sci 103: 11358–11363. doi: 10.1073/pnas.0604517103

Gralnick JA, Newman DK (2007) Extracellular respiration. Mol Microbiol 65: 1–11. doi: 10.1111/j.1365-2958.2007.05778.x

Gralnick JA, Vali H, Lies DP, Newman DK (2006) Extracellular respiration of dimethyl sulfoxide by *Shewanella oneidensis* strain MR-1. Proc Natl Acad Sci 103: 4669–4674. doi: 10.1073/pnas.0505959103

Hartshorne RS, Jepson BN, Clarke TA, Field SJ, Fredrickson J, Zachara J, Shi L, Butt JN, Richardson DJ (2007) Characterization of *Shewanella oneidensis* MtrC: A cell-surface decaheme cytochrome involved in respiratory electron transport to extracellular electron acceptors. J Biol Inorg Chem 12: 1083–1094. doi: 10.1007/s00775-007-0278-y

Holmes DE, Chaudhuri SK, Nevin KP, Mehta T, Methé BA, Liu A, Ward JE, Woodard TL, Webster J, Lovley DR (2006) Microarray and genetic analysis of electron transfer to electrodes in *Geobacter sulfurreducens*. Environ Microbiol 8: 1805–1815. doi: 10.1111/j.1462-2920.2006.01065.x

Inoue K, Leang C, Franks AE, Woodard TL, Nevin KP, Lovley DR (2011) Specific localization of the *c*-type cytochrome OmcZ at the anode surface in current-producing biofilms of *Geobacter sulfurreducens*. Environ Microbiol Rep 3: 211–217. doi: 10.1111/j.1758-2229.2010.00210.x

Inoue K, Qian X, Morgado L, Kim BC, Mester T, Izallalen M, Salgueiro CA, Lovley DR (2010) Purification and characterization of OmcZ, an outer-surface, octaheme *c*-type cytochrome essential for optimal current production by *Geobacter sulfurreducens*. Appl Environ Microbiol 76: 3999–4007. doi: 10.1128/AEM.00027-10

Jensen HM, Albers AE, Malley KR, Londer YY, Cohen BE, Helms BA, Weigele P, Groves JT, Ajo-Franklin CM (2010) Engineering of a synthetic electron conduit in living cells. Proc Natl Acad Sci 107: 19213–19218. doi: 10.1073/pnas.1009645107

Jensen HM, TerAvest MA, Kokish MG, Ajo-Franklin CM (2016) CymA and Exogenous Flavins Improve Extracellular Electron Transfer and Couple It to Cell Growth in Mtr-Expressing *Escherichia coli*. ACS Synth Biol 5: 679–688. doi: 10.1021/acssynbio.5b00279

Kerzenmacher S (2012) Dissimilatory metal reducers producing electricity: Microbial fuel cells. In: Microbial Metal Respiration: From Geochemistry to Potential Applications. Springer Berlin Heidelberg, Berlin, Heidelberg, pp 203–230. doi: 10.1007/978-3-642-32867-1_8

Kim BC, Lovley DR (2008) Investigation of direct vs. indirect involvement of the *c*-type cytochrome MacA in Fe(III) reduction by *Geobacter sulfurreducens*. FEMS Microbiol Lett 286: 39–44. doi: 10.1111/j.1574-6968.2008.01252.x

Kotloski NJ, Gralnick JA (2013) Flavin electron shuttles dominate extracellular electron transfer by *Shewanella oneidensis*. mBio 4. doi: 10.1128/mBio.00553-12

Leang C, Coppi MV, Lovley DR (2003) OmcB, a *c*-type polyheme cytochrome, involved in Fe(III) reduction in *Geobacter sulfurreducens*. J Bacteriol 185: 2096–2103. doi: 10.1128/JB.185.7.2096-2103.2003

Leang C, Qian X, Mester T, Lovley DR (2010) Alignment of the *c*-type cytochrome OmcS along pili of *Geobacter sulfurreducens*. Appl Environ Microbiol 76: 4080–4084. doi: 10.1128/AEM.00023-10

Levar CE, Chan CH, Mehta-Kolte MG, Bond DR (2014) An inner membrane cytochrome required only for reduction of high redox potential extracellular electron acceptors. mBio 5: e02034-14. doi: 10.1128/mBio.02034-14

Levar CE, Rollefson JB, Bond DR (2012) Energetic and molecular constraints on the mechanism of environmental fe(III) reduction by *Geobacter*. In: Microbial Metal Respiration: From Geochemistry to Potential Applications. pp 29–48. doi: 10.1007/978-3-642-32867-1_2

Lies DP, Hernandez ME, Kappler A, Mielke RE, Gralnick JA, Newman DK (2005) *Shewanella oneidensis* MR-1 Uses Overlapping Pathways for Iron Reduction at a Distance and by Direct Contact under Conditions Relevant for Biofilms. Appl Environ Microbiol 71: 4414–4426. doi: 10.1128/AEM.71.8.4414-4426.2005

Liu Y, Wang Z, Liu J, Levar C, Edwards MJ, Babauta JT, Kennedy DW, Shi Z, Beyenal H, Bond DR, Clarke TA, Butt JN, Richardson DJ, Rosso KM, Zachara JM, Fredrickson JK, Shi L (2014) A trans-outer membrane porin-cytochrome protein complex for extracellular electron transfer by *Geobacter sulfurreducens* PCA. Environ Microbiol Rep 6: 776–785. doi: 10.1111/1758-2229.12204

Lloyd JR, Leang C, Myerson ALH, Coppi MV, Cuifo S, Methe B, Sandler SJ, Lovley DR (2003) Biochemical and genetic characterization of PpcA, a periplasmic *c*-type cytochrome in *Geobacter sulfurreducens*. Biochem J 369: 153–161. doi: 10.1042/bj20020597

Lovley DR (2006) Bug juice: Harvesting electricity with microorganisms. Nat Rev Microbiol 4: 497–508. doi: 10.1038/nrmicro1442

Malvankar NS, Tuominen MT, Lovley DR (2012) Comment on "on electrical conductivity of microbial nanowires and biofilms" by S. M. Strycharz-Glaven, R. M. Snider, A. Guiseppi-Elie and L. M. Tender, Energy Environ. Sci., 2011, 4, 4366. Energy Environ. Sci. 5: 6247–6249. doi: 10.1039/C2EE02613A

Malvankar NS, Vargas M, Nevin K, Tremblay PL, Evans-Lutterodt K, Nykypanchuk D, Martz E, Tuomine MT, Lovley DR (2015) Structural basis for metallic-like conductivity in microbial nanowires. mBio 6. doi: 10.1128/mBio.00084-15

Malvankar NS, Vargas M, Nevin KP, Franks AE, Leang C, Kim BC, Inoue K, Mester T, Covalla SF, Johnson JP, Rotello VM, Tuominen MT, Lovley DR (2011) Tunable metallic-like conductivity in microbial nanowire networks. Nat Nanotechnol 6: 573–579. doi: 10.1038/nnano.2011.119

Marsili E, Baron DB, Shikhare ID, Coursolle D, Gralnick JA, Bond DR (2008) *Shewanella* secretes flavins that mediate extracellular electron transfer. Proc Natl Acad Sci 105: 3968–3973. doi: 10.1073/pnas.0710525105

Mehta T, Coppi MV, Childers SE, Lovley DR (2005) Outer membrane *c*-type cytochromes required for Fe(III) and Mn(IV) oxide reduction in *Geobacter sulfurreducens*. Appl Environ Microbiol 71: 8634–8641. doi: 10.1128/AEM.71.12.8634-8641.2005

Meyer TE, Tsapin AI, Vandenberghe I, De Smet L, Frishman D, Nealson KH, Cusanovich MA, Van Beeumen JJ (2004) Identification of 42 Possible Cytochrome *c* Genes in the *Shewanella oneidensis* Genome and Characterization of Six Soluble Cytochromes. Omi A J Integr Biol 8: 57–77. doi: 10.1089/153623104773547499

Myers CR, Myers JM (1997) Cloning and sequence of *cymA*, a gene encoding a tetraheme cytochrome *c* required for reduction of iron(III), fumarate, and nitrate by *Shewanella putrefaciens* MR-1. J Bacteriol 179: 1143–1152. doi: 10.1128/jb.179.4.1143-1152.1997

Nevin KP, Kim BC, Glaven RH, Johnson JP, Woodward TL, Methé BA, Didonato RJ, Covalla SF, Franks AE, Liu A, Lovley DR (2009) Anode biofilm transcriptomics reveals outer surface components essential for high density current production in *Geobacter sulfurreducens* fuel cells. PLoS One 4. doi: 10.1371/journal.pone.0005628

Okamoto A, Hashimoto K, Nealson KH (2014a) Flavin redox bifurcation as a mechanism for controlling the direction of electron flow during extracellular electron transfer. Angew Chemie - Int Ed 53: 10988–10991. doi: 10.1002/anie.201407004

Okamoto A, Hashimoto K, Nealson KH, Nakamura R (2013) Rate enhancement of bacterial extracellular electron transport involves bound flavin semiquinones. Proc Natl Acad Sci 110: 7856–7861. doi: 10.1073/pnas.1220823110

Okamoto A, Kalathil S, Deng X, Hashimoto K, Nakamura R, Nealson KH (2014b) Cell-secreted flavins bound to membrane cytochromes dictate electron transfer reactions to surfaces with diverse charge and pH. Sci Rep 4. doi: 10.1038/srep05628

Pessanha M, Morgado L, Louro RO, Londer YY, Pokkuluri PR, Schiffer M, Salgueiro CA (2006) Thermodynamic characterization of triheme cytochrome PpcA from *Geobacter sulfurreducens*: Evidence for a role played in e⁻/H⁺ energy transduction. Biochemistry 45: 13910–13917. doi: 10.1021/bi061394v

Pirbadian S, Barchinger SE, Leung KM, Byun HS, Jangir Y, Bouhenni RA, Reed SB, Romine MF, Saffarini DA, Shi L, Gorby YA, Golbeck JH, El-Naggar MY (2014) *Shewanella oneidensis* MR-1 nanowires are outer membrane and periplasmic extensions of the extracellular electron transport components. Proc Natl Acad Sci 111: 12883–12888. doi: 10.1073/pnas.1410551111

Pokkuluri PR, Londer YY, Duke NEC, Long WC, Schiffer M (2004) Family of Cytochrome *c* 7 -Type Proteins from *Geobacter sulfurreducens*: Structure of One Cytochrome *c* 7 at 1.45 Å Resolution. Biochemistry 43: 849–859. doi: 10.1021/bi0301439

Prokhorova A, Sturm-Richter K, Doetsch A, Gescher J (2017) Resilience, dynamics, and interactions within a model multispecies exoelectrogenic-biofilm community. Appl Environ Microbiol 83: e03033-16. doi: 10.1128/AEM.03033-16

Qian X, Mester T, Morgado L, Arakawa T, Sharma ML, Inoue K, Joseph C, Salgueiro CA, Maroney MJ, Lovley DR (2011) Biochemical characterization of purified OmcS, a *c*-type cytochrome required for insoluble Fe(III) reduction in *Geobacter sulfurreducens*. Biochim Biophys Acta - Bioenerg 1807: 404–412. doi: 10.1016/j.bbabio.2011.01.003

Qian X, Reguera G, Mester T, Lovley DR (2007) Evidence that OmcB and OmpB of *Geobacter sulfurreducens* are outer membrane surface proteins. FEMS Microbiol Lett 277: 21–27. doi: 10.1111/j.1574-6968.2007.00915.x

Rahimnejad M, Adhami A, Darvari S, Zirepour A, Oh S-E (2015) Microbial fuel cell as new technology for bioelectricity generation: A review. Alexandria Eng J 54: 745–756. doi: 10.1016/J.AEJ.2015.03.031

Reguera G, Nevin KP, Nicoll JS, Covalla SF, Woodard TL, Lovley DR (2006) Biofilm and nanowire production leads to increased current in *Geobacter sulfurreducens* fuel cells. Appl Environ Microbiol 72: 7345–7348. doi: 10.1128/AEM.01444-06

Romine MF, Carlson TS, Norbeck AD, McCue LA, Lipton MS (2008) Identification of mobile elements and pseudogenes in the *Shewanella oneidensis* MR-1 genome. Appl Environ Microbiol 74: 3257–65. doi: 10.1128/AEM.02720-07

Schicklberger M, Bücking C, Schuetz B, Heide H, Gescher J (2011) Involvement of the *Shewanella oneidensis* decaheme cytochrome MtrA in the periplasmic stability of the beta-barrel protein MtrB. Appl Environ Microbiol 77: 1520–3. doi: 10.1128/AEM.01201-10

Schuetz B, Schicklberger M, Kuermann JA, Spormann M, Gescher J (2009) Periplasmic electron transfer via the *c*-type cytochromes MtrA and FccA of *Shewanella oneidensis* MR-1. Appl Environ Microbiol 75: 7789–7796. doi: 10.1128/AEM.01834-09

Schwalb C, Chapman SK, Reid GA (2003) The tetraheme cytochrome CymA is required for anaerobic respiration with dimethyl sulfoxide and nitrite in *Shewanella oneidensis*. Biochemistry 42: 9491–9497. doi: 10.1021/bi034456f

Shelobolina ES, Coppi MV, Korenevsky AA, DiDonato LN, Sullivan SA, Konishi H, Xu H, Leang C, Butler JE, Kim BC, Lovley DR (2007) Importance of *c*-type cytochromes for U(VI) reduction by *Geobacter sulfurreducens*. BMC Microbiol 7: 16. doi: 10.1186/1471-2180-7-16

Shi L, Rosso KM, Clarke TA, Richardson DJ, Zachara JM, Fredrickson JK (2012) Molecular underpinnings of Fe(III) oxide reduction by *Shewanella oneidensis* MR-1. Front Microbiol 3: 50. doi: 10.3389/fmicb.2012.00050

Strycharz-Glaven SM, Snider RM, Guiseppi-Elie A, Tender LM (2011) On the electrical conductivity of microbial nanowires and biofilms. Energy Environ Sci 4: 4366–4379 doi: 10.1039/C1EE01753E

Strycharz-Glaven SM, Tender LM (2012) Reply to the "Comment on 'on electrical conductivity of microbial nanowires and biofilms'" by N. S. Malvankar, M. T. Tuominen and D. R. Lovley, Energy Environ Sci, 2012, 5, DOI: 10.1039/c2ee02613a. Energy Environ Sci 5: 6250–6255.

Sturm-Richter K, Golitsch F, Sturm G, Kipf E, Dittrich A, Beblawy S, Kerzenmacher S, Gescher J (2015) Unbalanced fermentation of glycerol in *Escherichia coli* via heterologous production of an electron transport chain and electrode interaction in microbial electrochemical cells. Bioresour Technol 186: 89–96. doi: 10.1016/j.biortech.2015.02.116

Sturm G, Richter K, Doetsch A, Heide H, Louro RO, Gescher J (2015) A dynamic periplasmic electron transfer network enables respiratory flexibility beyond a thermodynamic regulatory regime. ISME J 9: 1802–1811. doi: 10.1038/ismej.2014.264

Subramanian P, Pirbadian S, El-Naggar MY, Jensen GJ (2018) Ultrastructure of *Shewanella oneidensis* MR-1 nanowires revealed by electron cryotomography. Proc Natl Acad Sci 115: E3246–E3255. doi: 10.1073/pnas.1718810115

Sun JA, Zhang LY, Rao B, Shen YL, Wei DZ (2012) Enhanced acetoin production by *Serratia marcescens* H32 with expression of a water-forming NADH oxidase. Bioresour Technol 119: 94–98. doi: 10.1016/j.biortech.2012.05.108

Tender L and Yates M (2019) Electron transfer mechanisms in *Geobacter* and *Shewanella*: A comparative analysis. In: Microbial Electrochemical Technologies. Tiquia-Arashiro SM and Pant D (eds). Springer Publishers. Accepted for publication.

Teravest MA, Zajdel TJ, Ajo-Franklin CM (2014) The Mtr Pathway of *Shewanella oneidensis* MR-1 Couples Substrate Utilization to Current Production in *Escherichia coli*. ChemElectroChem 1: 1874–1879. doi: 10.1002/celc.201402194

Vargas M, Malvankar NS, Tremblay PL, Leang C, Smith JA, Patel P, Synoeyenbos-West O, Nevin KP, Lovley DR (2013) Aromatic amino acids required for pili conductivity and long-range extracellular electron transport in *Geobacter sulfurreducens*. mBio 4. doi: 10.1128/mBio.00105-13

Velasquez-Orta SB, Head IM, Curtis TP, Scott K, Lloyd JR, von Canstein H (2010) The effect of flavin electron shuttles in microbial fuel cells current production. Appl Microbiol Biotechnol 85: 1373–1381. doi: 10.1007/s00253-009-2172-8

von Canstein H, Ogawa J, Shimizu S, Lloyd JR (2008) Secretion of flavins by *Shewanella* species and their role in extracellular electron transfer. Appl Environ Microbiol 74: 615–623. doi: 10.1128/AEM.01387-07

Wang X, Lv M, Zhang L, Li K, Gao C, Ma C, Xu P (2013) Efficient bioconversion of 2,3-butanediol into acetoin using *Gluconobacter oxydans* DSM 2003. Biotechnol Biofuels 6: 155. doi: 10.1186/1754-6834-6-155

Werpy T, Petersen G (2004) Top Value Added Chemicals From Biomass: Volume I – Results of Screening for Potential Candidates from Sugars and Synthesis Gas. US DOE, doi: 10.2172/15008859

Xiao Z, Lu JR (2014) Strategies for enhancing fermentative production of acetoin: A review. Biotechnol Adv 32: 492–503. doi: 10.1016/J.BIOTECHADV.2014.01.002

Xu S, Jangir Y, El-Naggar MY (2016) Disentangling the roles of free and cytochrome-bound flavins in extracellular electron transport from *Shewanella oneidensis* MR-1. Electrochim Acta 198: 49–55. doi: 10.1016/j.electacta.2016.03.074

Zacharoff L, Chan CH, Bond DR (2016) Reduction of low potential electron acceptors requires the CbcL inner membrane cytochrome of *Geobacter sulfurreducens*. Bioelectrochemistry 107: 7–13. doi: 10.1016/j.bioelechem.2015.08.003

Zhai DD, Li B, Sun JZ, Sun DZ, Si RW, Yong YC (2016) Enhanced power production from microbial fuel cells with high cell density culture. Water Sci Technol 73: 2176–2181. doi: 10.2166/wst.2016.059

Zhang B, Li XL, Fu J, Li N, Wang Z, Tang YJ, Chen T (2016) Production of acetoin through simultaneous utilization of glucose, xylose, and arabinose by engineered *Bacillus subtilis*. PLoS One 11: 1–14. doi: 10.1371/journal.pone.0159298

Zhang X, Zhang R, Yang T, Zhang J, Xu M, Li H, Xu Z, Rao Z (2013) Mutation breeding of acetoin high producing *Bacillus subtilis* blocked in 2,3-butanediol dehydrogenase. World J Microbiol Biotechnol 29: 1783–1789. doi: 10.1007/s11274-013-1339-8

Zhang Y, Angelidaki I (2014) Microbial electrolysis cells turning to be versatile technology: Recent advances and future challenges. Water Res. 56: 11–25. doi: 10.1016/j.watres.2014.02.031

Challenges and Opportunities of Microbial Symbiosis for Electrosynthesis Applications

Oskar Modin[1] and Nikolaos Xafenias[2]

[1] Division of Water Environment Technology, Department of Architecture and Civil Engineering, Chalmers University of Technology, Gothenburg, 41296, Sweden

[2] Division of Industrial Biotechnology, Department of Biology and Biological Engineering, Chalmers University of Technology, Gothenburg 41296, Sweden

1. Introduction

Microbial electrosynthesis (MES) refers to the generation of desirable products using electrochemical reactions catalyzed by microorganisms (Schröder et al. 2015). MES systems could, for instance, be used to synthesize a range of chemicals from electricity and carbon dioxide. In such systems, the biocathode plays an important role. At a biocathode, living microorganisms harvest electrons and carry out reduction reactions. Examples include reduction of hydrogen ions into hydrogen gas (Rozendal et al. 2008) and carbon dioxide into methane (Cheng et al. 2009), acetate (Nevin et al. 2010), butyrate (Choi et al. 2012), alcohols or other organic chemicals (Arends et al. 2017; Vassilev et al. 2018). Both pure and mixed culture biocathodes are described in the literature. Undefined mixed cultures are typically less sensitive to disturbances, can use a broad range of substrates and produce a large variety of products. However, pure cultures and defined co-cultures can generate products with higher specificity (ter Heijne et al. 2017). In mixed cultures, the metabolic capabilities of different microorganisms are combined and can be used to naturally develop biocathodes for producing target molecules, e.g., to enhance the methane yield in anaerobic digesters (Sasaki et al. 2010). However, the performance of a mixed culture biocathode is very much dependent on the dynamics of the microorganisms that make up the community and can, therefore, be troublesome to handle. For practical applications, biocathodes should be efficient and contain electroactive microorganisms that can deliver high current densities at low overpotentials. They should also be specific and contain a specialized consortium that generates target products while avoiding undesired side products.

MES has been reviewed in several recent articles such as the possibilities to integrate MES in biorefineries (Sadhukhan et al. 2016), the reduction of multi-carbon substrates into alcohols and solvents (Mostafazadeh et al. 2017), the reduction of CO_2 with mixed cultures (ter Heijne et al. 2017) and the use of pure and defined cultures (Rosenbaum et al. 2017) were discussed. Previous reviews have covered process engineering aspects, the microorganisms involved, the substrates used and the products generated. However, to our knowledge, there is no review focusing specifically on methods to control microbial community dynamics in mixed culture systems. As the favorable selection and control of target

*Corresponding author: oskar.modin@chalmers.se, xafenias@chalmers.se

microorganisms are vital for the successful production of target chemicals and fuels, this knowledge should be seriously considered when designing process systems.

The goal of this chapter is to review strategies researchers have used to control the selection of efficient and specific biocathode communities and to give an updated overview of this rapidly evolving research area. We start with an overview of the thermodynamics that set the boundaries for biocathodes before proceeding to the microorganisms used and the products generated from electricity and CO_2. Then we go through the selection strategies that have been applied to control the community composition and the process, and finally, we highlight some areas that need further research to improve our understanding of how biocathode communities actually develop and function.

2. Thermodynamic Boundaries

All reactions taking place in a microbial electrochemical system are constrained by thermodynamics expressed mainly by the Nernst equation (Equation 1). The cathode potential, the concentrations of electron acceptors (reactants) and their products in the electrode vicinity as well as the temperature are the selective forces that set the boundaries for processes that microorganisms can use to grow. Since water must be present in the cathode for the bioelectrochemical processes to take place, reduction of H^+ to H_2 is a very fundamental and possible reaction given a suitable temperature, pH and sufficiently low cathode potentials as determined by the Nernst equation (lower than -0.41 V vs. SHE at 25°C and pH 7). When CO_2 and its soluble form HCO_3^- (dominant at neutral pH) are provided as the only external electron acceptors, reduction to methane or acetate are the most likely reactions. Indeed, studies on mixed culture MES have often observed product mixes of H_2, CH_4 and acetate (Marshall et al. 2012). For microorganisms to be able to derive energy from harvesting electrons from a cathode and catalyzing any of these reduction reactions, the cathode potential must be lower than the reduction potential of the redox couple. For near-neutral pH, which most microorganisms favor, reduction of CO_2 to CH_4 gives the best energy return because of the potential difference between the electrode and the redox couple being the highest. Consequently, methanogens are easily selected in mixed culture biocathode communities, and CH_4 is usually a major product (Saheb Alam et al. 2018; van Eerten-Jansen et al. 2013). Although considered as an opportunity for CH_4-targeted applications, this can also be an obstacle when more valuable products are desired. On the other hand, although the reduction of H^+ to H_2 is thermodynamically the least favorable, it can occur abiotically on the cathode surface. This means that H_2 may serve as an electron shuttle between the cathode and the microorganisms if the cathode is operated at a sufficiently low potential.

$$E = E^0 - (RT/nF)*\ln([products]/[reactants]) \tag{1}$$

Where:

E: half-reaction redox potential

E^0: standard redox potential

R: global gas constant (8.314 J K^{-1} mol^{-1})

T: temperature (°C)

n: number of electrons required

F: Faraday's constant; 96,485.3 C (mol e$^-$)$^{-1}$

3. Microorganisms and Products in Microbial Electrosynthesis Applications

To date, most MES and electrofermentation studies have focused on the use of either undefined mixed populations (Table 1) or pure cultures (Table 2) with studies on undefined mixed cultures pointing toward the different genera that dominate MES applications and are capable of autotrophic H_2, CH_4 and acetate production (e.g., *Desulfovibrio* sp., *Methanobacterium* sp. and *Acetobacterium* sp. respectively; Table 1). Research on undefined mixed populations has offered us valuable advancements and helped understand and expand our possibilities for MES applications by setting the starting point for selecting microbes that thrive in particular environments and perform desired functions (Rosenbaum et al. 2017). Undefined mixed populations have shown some important advantages compared to pure cultures (Table 3), including higher current production and production rates, increased robustness and also a plethora of metabolic pathways and microbial interactions that allow the use of a broad range of substrates and products. They also offer simplicity in operation, lower operational costs and less need for special handling when it comes to decontamination of undefined wastewater sources, and the use of undefined mixed populations can be the most efficient and cost-effective option. However, when the process requires production of more complex and valuable chemicals than hydrogen, methane or acetate in combination with demands for high specificity, rates and yields, MES biotechnology has a lot to gain from the use and development of specialized pure and defined co-cultures that minimize the production of unwanted secondary products and reduce the downstream processing costs.

Table 1: MES using mixed cultures

Microbial Source	Key-Microbes Detected	Major Products	Source	Notes
Enriched microbial anodes	*Desulfovibrio vulgaris, Geobacter sulfurreducens*	H_2	(Croese et al. 2011)	
Two different marine sediments	*Eubacterium limosum, Desulfovibrio* sp., *Gemmata obscuriglobus, Mesorhizobium* sp., *Rhodococcus* sp. and *Azospirillum* sp.	H_2, CH_4	(Pisciotta et al. 2012)	
Municipal wastewater and anaerobic digester sludge (9:1)	*Methanobacterium* sp., *Acetobacterium* sp. (after inhibition of the methanogens)	CH_4, acetate	(Saheb Alam et al. 2018)	Acetate observed after inhibition of the methanogens.
River sediment and anaerobic digestate	*Acetobacter* sp., *Methanosaeta* sp., *Methanobacterium* sp., *Methanomassiliicoccus* sp.	H_2, CH_4	(Mateos et al. 2018)	Cathodic biofilm was very diverse
TCE and cis-DCE dechlorinating cultures established by enrichment from contaminated lake sediments	*Desulfitobacterium* spp.	H_2	(Villano et al. 2011)	No methane was detected.
A mixed methanogenic culture was compared to an enriched hydrogenotrophic methanogenic culture	*Methanobrevibacter arboriphilus*	CH_4	(Dykstra and Pavlostathis 2017)	A 3.8-fold more methane was produced in the enriched hydrogenotrophic culture.
Activated sludge	*Acetobacterium* sp.	H_2, acetate	(Su and Jiang 2013)	Methanogens inhibition with 2-BES.
Granular sludge from a beer brewery	*Acetobacterium* sp. and *Desulfovibrio* sp.	H_2, acetate	(Xiang et al. 2017)	Methanogens inhibition with 2-BES.

Contd.

Laboratory-scale reactor inoculated with sulfate-reducing bacteria	*Curvibacter* sp., *Desulfovibrio* sp., *Desulfobacter* sp., *Syntrophobacter* sp.	Methanol, ethanol, propanol, butanol, acetone	(Sharma et al. 2013)	Acetic and butyric acid were used as substrates.
Freshwater bog sediment	*Trichococcus palustris, Oscillibacter* sp., *Clostridium* sp.	H_2, acetate, propionate, butyrate, ethanol, butanol	(Zaybak et al. 2013)	
Anodes of laboratory-scale microbial electrolysis reactors	*Desulfovibrio* sp., *Promicromonospora* sp.	H_2	(Croese et al. 2014; Jeremiasse et al. 2012)	
Anaerobic sludge from a UASB reactor treating distillery wastewater	*Methanobacterium* sp., *Desulfovibrio putealis, Hydrogenophaga caeni*, and *Methylocystis* sp.	H_2, CH_4	(van Eerten-Jansen et al. 2013)	
An acetogenic/methanogenic microbiome	*Acetobacterium* sp.	H_2, formate, acetate	(LaBelle et al. 2014)	Methanogens inhibition with 2-BES.
A retention basin for brewery wastewater treatment	*Methanobacterium* sp. and *Acetobacterium* sp.	H_2, CH_4, acetate	(Marshall et al. 2012)	
Mesophilic anaerobic sludge	*Acetobacterium* sp. and *Acetoanaerobium* sp.	H_2, CH_4, acetate	(Xafenias and Mapelli 2014)	

Table 2: MES using pure and defined co-cultures

Microorganism(s)	Product	Source	Notes
Sporomusa ovata	Acetate	(Nevin et al. 2010; Nie et al. 2013)	
Sporomusa sphaeroides	Acetate	(Nevin et al. 2011)	
Sporomusa silvacetica	Acetate	(Nevin et al. 2011)	
Clostridium ljungdahlii	Acetate, 2-oxobutyrate, formate	(Nevin et al. 2011)	
Clostridium aceticum	Acetate, 2-oxobutyrate	(Nevin et al. 2011)	
Moorella thermoacetica	Acetate	(Nevin et al. 2011)	
Cupriavidus necator (Former *Ralstonia eutropha*) Re2133-pEG12	isopropanol	(Torella et al. 2015)	
Cupriavidus necator (Former *Ralstonia eutropha*) LH74D	3-methyl-1-butanol, Isobutanol	(Li et al. 2012)	Formate was the electron mediator, produced abiotically from CO_2 by a cathode.
Propionibacterium freudenreichii	Propionate	(Croese et al. 2011)	From lactate.

Contd.

(*Contd.*)

Microorganism(s)	Product	Source	Notes
Methanobacterium palustre	CH_4	(Cheng et al. 2009)	
Methanobacterium thermoautotrophicus	CH_4	(Sato et al. 2013)	
IS4 and *Methanococcus maripaludis*	CH_4	(Deutzmann and Spormann 2016)	Hydrogenotrophic methanogenesis with *IS4* as the H_2 producer.
Methanococcus maripaludis	CH_4, H_2 or formate after inhibition of methanogenesis	(Lohner et al. 2014)	Hydrogenase-independent electron uptake.
IS4 and *Acetobacterium woodii*	Acetate	(Deutzmann and Spormann 2016)	Hydrogenotrophic acetogenesis with *IS4* as the H_2 producer.
Pyrococcus furiosus	3-hydroxypropionic acid	(Keller et al. 2013)	Hydrogen as electron donor, no electricity involved; thermophilic conditions- 73° C; heterologous expression of five genes of the carbon cycle of *Metallosphaera sedula*.

Table 3: Comparative potential advantages and disadvantages for application of different microbial cultivation strategies in MES applications

vs.	Undefined, mixed cultures		Pure cultures		Defined co-cultures	
	Pure Cultures	Defined Co-cultures	Defined Co-cultures	Undefined, Mixed cultures	Undefined, Mixed cultures	Pure Cultures
Current produced under given potential and production rates	↑	↑	↓	↓	↓	↑
Robustness and O_2 scavenging	↑	↑	↓	↓	↓	↑
Plethora of metabolic pathways, microbial interactions and range of potential substrates	↑	↑	↓	↓	↓	↑
Unwanted by-products and intermediates	↑	↑	-	↓	↓	-
Yields of high value products	↓	↓	-	↑	↑	-
Operation costs	↓	↓	↑	↑	↑	↓
Downstream processing costs	↑	↑	-	↓	↓	-

Research on MES using pure cultures has been performed using mainly a number of well-known acetogens like *Sporomusa ovata* and *Clostridium ljungdahlii* for acetate production while methanogens capable of autotrophic CO_2 conversion like *Methanobacterium* sp. have also been studied to a good degree (Table 2). The use of defined co-cultures opens up possibilities for not only combining metabolic features from different species and expanding our portfolio of substrates and products but also of different electron transfer mechanisms. One recent example is the co-application of *IS4* together with *Methanococcus maripaludis* for CH_4 electrosynthesis to that of *IS4* together with *Acetobacterium woodii* for acetate electrosynthesis (Deutzmann and Spormann 2016). Although it is very promising, successful application of defined co-culture cultivations is still a very challenging chapter of microbial biotechnology. Like operation with pure cultures, the defined co-cultures require a strictly sterile environment with conditions optimized for the particular application while satisfying the metabolic needs of the individual members. Small changes in these conditions can have significant effects on the individual growth profiles and, therefore, the production rates and may even lead to a chaotic biological dynamics (Becks et al. 2005). As defined co-cultures are very dynamic systems, microbial evolution and sensitivity analysis will also be needed for optimization (Höffner and Barton 2014). Control of the biotic and abiotic factors could make the interaction between the individual members more predictable and modifiable, however, this task can be very difficult to accomplish depending on the complexity of the biological system. The physiology of the individual members would also need to be studied in detail, including the flux analyses and the building of metabolic models that can later on be used to engineer new microbial strains and defined co-culture communities (Rosenbaum et al. 2017). Application of these strictly controlled conditions will increase capital and operational costs compared to operation with undefined mixed cultures; however, it is the only way to open up opportunities for producing high-value products at competitive rates.

Metabolic pathway engineering can also offer powerful tools for tailor-made MES by allowing the transfer of pathways from electroactive microbes to production hosts and vice versa (Rosenbaum and Henrich 2014). Autotrophic production of more valuable compounds has been made possible via metabolic engineering and the production of 3-hydroxypropionic acid using *Pyrococcus furiosus* (Keller et al. 2013), but also isopropanol, isobutanol and 3-methyl-1-butanol using *Cupriavidus necator* (Li et al. 2012; Torella et al. 2015) have been demonstrated. Although in the first stage, these developments are paving the way for expanding MES toward specialty and fine chemicals from CO_2.

4. Strategies to Control Mixed Microbial Community Composition and Function of Biocathodes

Although mixed, undefined microbial cultures have been proven robust and sufficient for bioelectrochemical wastewater treatment where a complex mixture of organic components is converted into CO_2 and biomass, control of the microbial community composition is essential for biotechnological applications. The microbial communities on biocathodes as well as in other engineered and natural environments are shaped by a range of processes that ecologists have categorized as selection, dispersal, drift and diversification (Nemergut et al. 2013; Vellend 2010). Selection refers to changes in the community caused by fitness differences between species; dispersal refers to the movement of microorganisms; drift refers to changes due to random death and reproduction of individual cells; diversification refers to factors such a genetic mutations and horizontal gene transfer that give rise to new species or species with new functions. All four processes likely affect how biocathode communities develop and function over time. Selection and dispersal are processes that to some extent can be engineered and controlled. As MES is developing to compete with the state-of-the-art, researchers have tried to control the microbial community composition of biocathodes by designing reactors and operational strategies that cause selection of microorganisms performing desired functions (Jourdin et al. 2014; LaBelle et al. 2014). The electrode potential is, of course, a very important selection force that sets the thermodynamic boundaries for the microbial communities; so important that it has even been demonstrated that different electrode potentials (anodic or cathodic) lead to the enrichment of different microbial communities even when both electrodes are located in the same reactor chamber (Xafenias and Mapelli 2014).

Microbial communities that are capable of catalyzing cathodic reactions have generally been more challenging to enrich than their anodic counterparts. A possible reason for this is that cathodic communities in MES reactors are typically autotrophic whereas bioanodes are typically fed with organic substrates leading to the enrichment of heterotrophs. With more energy needed to be invested in the synthesis of organic compounds from CO_2, autotrophs have lower growth rates than heterotrophs and are thus challenging to cultivate. Another challenge is to ensure that mixed culture MES systems produce a defined set of products. In Figure 1 we can see that the reduction potentials for H^+/H_2, CO_2/CH_4 and CO_2/acetate are very close to each other, and therefore, for a given cathode potential there are similar thermodynamic driving forces for all reactions. Thus, kinetic limitations and the relative abundance of different functional groups in the community will determine not only the composition of the product spectrum but also the development of the microbial communities. Since CH_4 is the energetically most favorable end-product, methanogens tend to dominate biocathode biofilms unless specific inhibition strategies are employed (Karthikeyan et al. 2017). Several different strategies to influence the development and function of microbial communities in MES systems have been investigated and some of the studies are reviewed below.

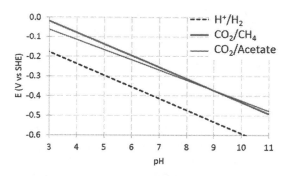

Figure 1: Reduction potentials for different redox couples as functions of pH (except for the H^+ activity, standard conditions were assumed in Equation 1, e.g., 25°C, 1 atm and 1 M).

4.1 Biological Anodes as a Pre-Enrichment Step

A method for the rapid start-up of biocathodes, by first pre-enriching them as bioanodes, was first proposed by Rozendal et al. (2008). In that study, the goal was to enrich a biocathode catalyzing H_2 generation, and their strategy was to exploit the reversible nature of hydrogenases. By first enriching a hydrogenotrophic bioanode and then lowering the electrode potential, so that the electrode would work as a cathode, the same biofilm that catalyzed H_2 oxidation could be used to catalyze H_2 generation with a cathodic current density of 1.2 A/m² at -0.7 V vs. SHE (Rozendal et al. 2008). In another study with mixed culture biocathodes, pre-enriched as bioanodes, a cathodic current of 1.2 A/m² at -0.7 V vs. SHE was produced (Croese et al. 2011). Microbial community analysis showed that *Desulfovibrio* sp. was the most abundant member of the biocathode while *Geobacter sulfurreducens* was also enriched. Both these species seemed to have an active role in the biocathodes, and in studies with pure cultures, both *Desulfovibrio* sp. and *Geobacter sulfurreducens* have been shown to catalyze H_2 generation (Croese et al. 2011; Geelhoed and Stams 2011). Anodic pre-enrichment was also tested by Pisciotta et al. (2012) who enriched bioanodes in MFCs using two different sediment inocula and a nutrient medium with acetate as the carbon source. After switching to biocathode mode, the 0.22 m² brush-type electrode could produce current at a potential as high as -0.439 V vs. SHE while the maximum current normalized to the surface area was about 0.002 A/m² produced under the same potential. Further subculturing of the biocathode cultures on different electrode materials resulted in current densities ranging from 0.001 to 0.052 A/m². A steep cathodic current polarization curve was observed when applied potentials were lower than -0.35 V vs. SHE along with a maximum current density of 0.02 A/m² under -0.7 V vs. SHE. The dominant members of the biocathode community enriched from one type of sediment were classified as *Eubacterium limosum*, *Desulfovibrio* sp. and *Gemmata obscuriglobus*. The methanogen *Methanocorpusculum labreanum* was also detected and both H_2 and CH_4 were produced. The biocathode enriched from another sediment type contained *Gemmata obscuriglobus*, *Mesorhizobium* sp., *Rhodococcus* sp. and *Azospirillum* sp. (Pisciotta et al. 2012).

The transition in microbial community composition, when bioanodes are converted to biocathodes, was studied using high-throughput sequencing by Saheb Alam et al. (2018). The authors fed bioanodes with acetate and observed communities dominated by *Geobacter* sp. When the electrodes were converted to biocathodes, a cathodic current density of 0.016 ± 0.007 A/m² was generated at -0.65 V vs. SHE that is comparable to the current densities observed by Pisciotta et al. (2012) under similar pH and electrode potentials of -0.439 V vs. SHE. After prolonged enrichment for more than 170 days, the current density increased to 0.6-3.6 A/m². During that time, the community composition had completely changed and was dominated by *Methanobacterium* sp. with CH_4 as the only observed product. After inhibition of the methanogens using 2-bromoethane sulfonate (2-BES), acetate was the main final product and *Acetobacterium* sp. was dominating the community (Figure 2). Start-up of biocathodes directly from anaerobic sludge and sewage in a parallel reactor resulted in a biocathode with very similar performance and microbial community composition. These results suggested that the reactor design and operation were more important selection forces for the biocathode than the founding community. Although the bioanodes did produce some cathodic current when they converted to biocathodes, the cathodic current was >200 times lower than the current generated after a new microbial community developed on the cathode surfaces (Saheb Alam et al. 2018). In another study that followed a similar methodological concept, a biocathode pre-enriched as bioanode was compared to direct biocathode start-up using two different inocula: river mud and anaerobic digestate. The bioanodes were enriched in a mix of acetate, propionate and glucose and showed an improved start-up in terms of consistency in operation, current and products produced and specialization of the biocathodes. Mixtures of acetate, H_2 and CH_4 at current densities, ranging from 0 to 1 A/m² at -0.6 V vs. SHE, were produced in that study. The authors also observed a shift in community structure when the electrodes changed from anodes to cathodes and the community composition varied depending on inoculum and start-up procedure (Mateos et al. 2018).

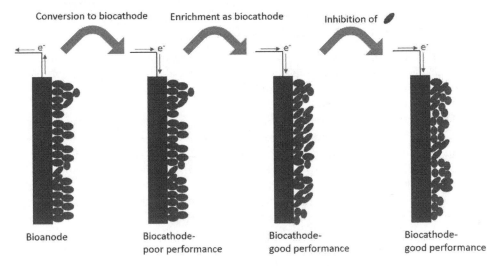

Figure 2: Transition of microbial community from bioanode to biocathode.

4.2 Direct and Indirect (Electrochemical) H₂ Supply

As H_2 is a major electron and energy carrier in biocathodes, supplementation and control of H_2 levels have been another pre-enrichment approach to support the growth of autotrophic microbes. In one study, an H_2-utilizing mixed methanogenic culture produced methane at current densities of about 0.75 A/m² and cathodic potentials lower than about -0.75 V vs. SHE (Villano et al. 2010). In another study, a trichloroethylene-dechlorinating culture was enriched in H_2 and was able to produce cathodic current densities of 4.4 A/m² under a potential of -0.9 V vs. SHE that was higher than the current produced in the abiotic control. The mixed culture was dominated by *Desulfitobacterium* spp. and was able to catalyze the H_2 generation (Villano et al. 2011). Dykstra and Pavlostathis (2017) compared a mixed methanogenic culture and methanogens pre-enriched in H_2 as inocula. The hydrogenotrophic enriched culture generated higher current density ($\sim$1 A/m²) than the mixed methanogens ($\sim$0.2 A/m²) at a potential of -0.8 V vs. SHE. *Methanobrevibacter arboriphilus* was enriched in both reactors, but the bacterial communities differed depending on the inoculum (Dykstra and Pavlostathis 2017). Su and Jiang (2013) supplied flushes of an H_2/CO_2 gas mixture during the first five batch cycles of a biocathode enrichment. Methanogens were inhibited with 2-BES that resulted in an acetate-producing biocathode dominated by *Acetobacter* sp. (Su and Jiang 2013). In another study, an acetogenic culture was also pre-enriched on H_2/CO_2 with 2-BES in the culture medium (Xiang et al. 2017). Sulfate-reducing bacteria played a key role for MES in that study, and reactors operated with 6 mM sulfate had more biomass and higher acetate production rate than reactors operated without sulfate. Sulfate-reducing bacteria have also been used to reduce acetate and butyrate to alcohols and acetone via MES (Sharma et al. 2013). As those two substrates are common fermentation residuals, the latter study paved the way for the valorization of industrial side streams.

4.3 Supply of Organic Substrate in Biocathodes

Using CO_2 as a carbon source is very energy-demanding for the microorganisms. To enhance microbial yield and speed up the start-up time, a few studies have investigated the addition of an organic substrate before or during the enrichment of the biocathode. Srikanth et al. (2018) pre-enriched a culture collected from corroding metals on glucose and CO_2, and the culture was used as inoculum in a MES reactor after glucose concentration had gradually decreased to zero. The reactor was operated for 90 days, and in the end, ethanol and butanol made up 61% of the products generated from CO_2. A gas diffusion cathode was used and assisted to achieve high CO_2 mass transfer and high current densities (9-11 A/m²) (Srikanth et al. 2018). In another study, freshwater sediment samples were pre-enriched with glucose as carbon source before starting a biocathode at -0.4 V vs. SHE along with glucose in the growth medium for the first two weeks of enrichment. The biocathode generated 0.034 A/m² and produced a mixture of butanol, ethanol, acetate, propionate, butyrate and hydrogen gas although fermentation of the added glucose may have partially contributed to the product mix. *Trichococcus* sp, *Oscillibacter* sp. and *Clostridium* sp. were dominant in that biocathode (Zaybak et al. 2013). Addition of 1 mM acetate also had an impact and resulted in reduced start-up times of H_2-producing biocathodes in two other studies (Croese et al. 2014; Jeremiasse et al. 2012).

4.4 Effects of Pre-Enrichment on Microbial Selection

Pre-enrichment of H_2-utilizing microorganisms appears to be a successful strategy to select for a community with the capacity to function in a biocathode. Interestingly, pre-enrichment of bioanodes on acetate was a less successful strategy for start-up biocathodes. Two studies observed relatively low cathodic current densities (<0.02 A/m²) with acetate-enriched electrodes (Pisciotta et al. 2012; Saheb Alam et al. 2018), whereas 1.2 A/m² was generated in another study when H_2 was used as feed during the pre-enrichment (Rozendal et al. 2008). Similar current densities were obtained in two other studies that did not use an electrode as an electron acceptor during the pre-enrichment with H_2 (Dykstra and Pavlostathis 2017; Villano et al. 2010). These results suggest that for a biocathode to generate high current densities, enrichment with H_2-utilizing microorganisms can be more important than having an electrochemically active biofilm on the electrode.

Several studies have demonstrated that hydrogenotrophic methanogens and homoacetogens will dominate in biocathodes and provide important functions (Su and Jiang 2013; Villano et al. 2010). Among known hydrogenotrophic methanogens, *Methanobacterium* sp. have been commonly observed (Marshall et al. 2012; Mateos et al. 2018; Saheb Alam et al. 2018; van Eerten-Jansen et al. 2013), while *Acetobacter* sp. are among the homoacetogens that are frequently selected for biocathodes (LaBelle et al. 2014; Saheb Alam et al. 2018; Xafenias and Mapelli 2014) (Table 1). Another group of bacteria often found in biocathodes is that of sulfate-reducing bacteria. This group of bacteria seems to play a key role in catalyzing H_2 generation that can subsequently be used as an electron donor by homoacetogens and methanogens. *Desulfovibrio* sp. has been found in several biocathode studies (LaBelle et al. 2014; Mateos et al. 2018; van Eerten-Jansen et al. 2013; Xiang et al. 2017) and has been shown to catalyze H_2 generation in biocathodes (Aulenta et al. 2012; Croese et al. 2011).

4.5 Methods to Prevent Methanogenesis

Methane is often an unwanted product in MES processes catalyzed by undefined microbial consortia. To avoid methane formation, a number of measures can be taken (Figure 3). A common and effective method that has been used in several studies is the addition of 2-bromoethanosulfonate (2-BES) a chemical that interferes with methyl co-enzyme M in methanogens (Marshall et al. 2012; Saheb Alam et al. 2018). Another method is the pH control that may be applied in combination with 2-BES addition. In the study of LaBelle et al. (2014), a low pH of about 6.4 was applied in three replicate biocathode reactors for a period of six days, and 2-BES was also added in the medium. Afterward, 2-BES was removed and the reactors were left to operate for another 24 days. While one of the reactors started producing methane after 11 days of operation, no methane production was observed in the other two reactors. Acetate and hydrogen were produced by the biocathodes that were dominated by *Acetobacter* sp.; however, acetate production ceased upon a pH drop to values below 5 (LaBelle et al. 2014). In another study, methanogenesis was inhibited using 2-BES at a pH of 6.8. Acetate was the main product and the production was also observed at a pH as low as 5.8 (Batlle-Vilanova et al. 2016).

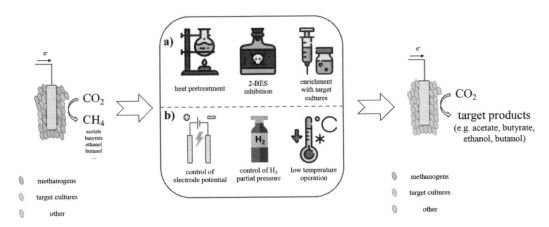

Figure 3: Summary of methods to prevent methanogenesis mainly taken before (a) and during (b) biocathodes operation for MES applications.

Another strategy for methanogenesis inhibition is pre-enrichment measures to obtain a mixed culture inoculum free of methanogens (Modestra et al. 2015). Bajracharya et al. (2016) pre-heated anaerobic sludge at 90°C for 1 hour to kill methanogens. Afterward, the sample was pre-enriched in the presence of 2-BES, first with fructose and then with H_2/CO_2. Finally, the culture was mixed with a pure culture of *Clostridium ljungdahlii* and used as inoculum for a MES reactor. 2-BES was no longer added in the reactor that could operate for nearly a year without methanogenesis observed. Acetate was the major product while butyrate and ethanol were co-produced (Bajracharya et al. 2016). A similar pre-enrichment strategy was also used in another study (Modestra and Mohan 2017).

Applied cathode potentials also affect product composition. In one study, the cathode potentials below ca. -0.8 V vs. SHE led to methane formation as the dominant process. However, at -0.85 V vs. SHE, acetate production rate was higher (Jiang et al. 2013). This suggests that acetogens may outcompete methanogens under certain conditions that also has been previously observed for certain strains at high H_2 concentration and low temperature (Kotsyurbenko et al. 2001). Xafenias and Mapelli (2014), and thus confirming these conclusions and also showing that higher anodic potentials do not always lead to higher removal of chemical oxygen demand (COD) in reactors where the anode and the cathode are hydraulically connected. This is due to acetogens growth and acetate production on the cathode when the potential drops below a certain value (i.e., below ca. -0.92 V vs. SHE in that study). In addition, higher net power production was produced (mainly as CH_4) in a membraneless system compared to a nonbioelectrochemical anaerobic digestion system, both fed with acetate when the cathode potential was higher (but below -0.86 V vs. SHE in that study). The study concluded that both anode and cathode potentials seem to be crucial when the purpose is to remove COD and produce net power in membraneless systems treating water contaminated with organics (Xafenias and Mapelli 2014).

5. Conclusions and Future Perspectives

Research on undefined mixed cultures in MES biocathodes has advanced rapidly during the last couple of years, contributing significantly to our understanding of the microbial communities in such systems. Based on these studies a number of strategies have been developed for selecting microbes that can be applied for biotechnology applications. Methods to encourage growth and dominance of acetogens have been a focus for MES applications, including measures to alter the biocathodes' physicochemical (e.g., pH, cathode potential) and biological environment (e.g., by pre-enrichment and inhibition of methanogens). Although the proof-of-concept has been demonstrated and we can now tell to some degree how specific parameters affect bioproduction, more detailed studies are needed to understand the microbial ecology in depth in order to design effective MES systems. The systematic characterization and categorization of the imposed causes with the observed effects is needed, and these linkages need to be reproduced in different laboratories. The study of time series on the development of the biofilm and the suspended microbial consortia would offer valuable information toward that direction, although it is by nature problematic as it involves invasive methods. The effect of other ecological processes such as drift and dispersal also need to be studied. Biocathodes operated under nonsterile conditions are continuously exposed to external microorganisms that may affect how the community composition and function change over time. Progress in our understanding of how mixed culture biocathodes communities develop and function also goes hand in hand with advancements in genetic and systems biology tools that are necessary to enhance bioproduction in a systematic way. These would lead to the development of more advanced microorganisms than the ones currently known, allowing the production and use of monocultures and defined co-cultures that can compete with the current state-of-the-art.

References

Arends JBA, Patil SA, Roume H, Rabaey K (2017). Continuous long-term electricity-driven bioproduction of carboxylates and isopropanol from CO_2 with a mixed microbial community. J CO$_2$ Util 20: 141–149.

Aulenta F, Catapano L, Snip L, Villano M, Majone M (2012). Linking bacterial metabolism to graphite cathodes: electrochemical insights into the H(2)-producing capability of *Desulfovibrio* sp. ChemSusChem 5: 1080–5.

Bajracharya S, Yuliasni R, Vanbroekhoven K, Buisman CJN, Strik DPBTB, Pant D (2016). Long-term operation of microbial electrosynthesis cell reducing CO_2 to multi-carbon chemicals with a mixed culture avoiding methanogenesis. Bioelectrochemistry 113: 26–34.

Batlle-Vilanova P, Puig S, Gonzalez-Olmos R, Balaguer MD, Colprim J (2016). Continuous acetate production through microbial electrosynthesis from CO_2 with microbial mixed culture. J. Chem Technol Biotechnol 91: 921–927.

Becks L, Hilker FM, Malchow H, Jürgens K, Arndt H (2005). Experimental demonstration of chaos in a microbial food web. Nature 435: 1226–1229.

Cheng S, Xing D, Call DF, Logan BE (2009). Direct Biological Conversion of Electrical Current into Methane by Electromethanogenesis. Environ Sci Technol 43: 3953–3958.

Choi O, Um Y, Sang B-I (2012). Butyrate production enhancement by *Clostridium tyrobutyricum* using electron mediators and a cathodic electron donor Biotechnol Bioeng 109: 2494–2502.

Croese E, Jeremiasse AW, Marshall IP, Spormann AM, Euverink GJ, Geelhoed JS, Stams AJ, Plugge CM (2014). Influence of setup and carbon source on the bacterial community of biocathodes in microbial electrolysis cells. Enzym Microb Technol 61-62: 67–75.

Croese E, Pereira MA, Euverink G-JW, Stams AJM, Geelhoed JS (2011). Analysis of the microbial community of the biocathode of a hydrogen-producing microbial electrolysis cell. Appl Microbiol Biotechnol 92: 1083–1093.

Deutzmann JS, Spormann AM (2016). Enhanced microbial electrosynthesis by using defined co-cultures. ISME J 11: 704–714.

Dykstra CM, Pavlostathis SG (2017). Methanogenic Biocathode Microbial Community Development and the Role of Bacteria. Environ Sci Technol 51: 5306–5316.

Geelhoed JS, Stams AJ (2011). Electricity-assisted biological hydrogen production from acetate by *Geobacter sulfurreducens*. Environ Sci Technol 45: 815–820.

Höffner K, Barton PI (2014). Design of Microbial Consortia for Industrial Biotechnology. Proc 8th Int Conf Found Comput Process Des 34: 65–74.

Jeremiasse AW, Hamelers HVM, Croese E, Buisman CJN (2012). Acetate enhances startup of a H-producing microbial biocathode. Biotechnol Bioeng 109: 657–664.

Jiang Y, Su M, Zhang Y, Zhan G, Tao Y, Li D (2013). Bioelectrochemical systems for simultaneously production of methane and acetate from carbon dioxide at relatively high rate. Int J Hydrog Energy 38: 3497–3502.

Jourdin L, Freguia S, Donose BC, Chen J, Wallace GG, Keller J, Flexer V (2014). A novel carbon nanotube modified scaffold as an efficient biocathode material for improved microbial electrosynthesis. J Mater Chem 2: 13093–13102.

Karthikeyan R, Cheng KY, Selvam A, Bose A, Wong JWC (2017). Bioelectrohydrogenesis and inhibition of methanogenic activity in microbial electrolysis cells - A review. Biotechnol Adv 35: 758–771.

Keller MW, Schut GJ, Lipscomb GL, Menon AL, Iwuchukwu IJ, Leuko TT, Thorgersen MP, Nixon WJ, Hawkins AS, Kelly RM, Adams MWW (2013). Exploiting microbial hyperthermophilicity to produce an industrial chemical, using hydrogen and carbon dioxide. Proc Natl Acad Sci United States Am 110: 5840–5845.

Kotsyurbenko OR, Glagolev MV, Nozhevnikova AN, Conrad R (2001). Competition between homoacetogenic bacteria and methanogenic archaea for hydrogen at low temperature. FEMS Microbiol Ecol 38: 153–159.

LaBelle EV, Marshall CW, Gilbert JA, May HD (2014). Influence of acidic pH on hydrogen and acetate production by an electrosynthetic microbiome. PLoS One 9: e109935.

Li H, Opgenorth PH, Wernick DG, Rogers S, Wu T-Y, Higashide W, Malati P, Huo Y-X, Cho KM, Liao JC (2012). Integrated electromicrobial conversion of CO_2 to higher alcohols. Science 335: 1596.

Lohner ST, Deutzmann JS, Logan BE, Leigh J, Spormann AM (2014). Hydrogenase-independent uptake and metabolism of electrons by the archaeon *Methanococcus maripaludis*. ISME J 8: 1673–1681.

Marshall CW, Ross DE, Fichot EB, Norman RS, May HD (2012). Electrosynthesis of commodity chemicals by an autotrophic microbial community. Appl Environ Microbiol 78: 8412–8420.

Mateos R, Sotres A, Alonso RM, Escapa A, Moran A (2018). Impact of the start-up process on the microbial communities in biocathodes for electrosynthesis. Bioelectrochemistry 121: 27–37.

Modestra JA, Mohan SV (2017). Microbial electrosynthesis of carboxylic acids through CO_2 reduction with selectively enriched biocatalyst: Microbial dynamics. J CO_2 Util 20: 190–199.

Modestra JA, Navaneeth B, Mohan SV (2015). Bio-electrocatalytic reduction of CO_2: Enrichment of homoacetogens and pH optimization towards enhancement of carboxylic acids biosynthesis. J CO_2 Util 10: 78–87.

Mostafazadeh et al. (2017). Journal of Environmental Chemical Engineering 5(1): 940–954.

Nemergut DR, Schmidt SK, Fukami T, O'Neill SP, Bilinski TM, Stanish LF, Knelman JE, Darcy JL, Lynch RC, Wickey P, Ferrenberg S (2013). Patterns and processes of microbial community assembly. Microbiol Mol Biol Rev 77: 342–356.

Nevin KP, Hensley SA, Franks AE, Summers ZM, Ou J, Woodard TL, Snoeyenbos-West OL, Lovley DR (2011). Electrosynthesis of organic compounds from carbon dioxide is catalyzed by a diversity of acetogenic microorganisms. Appl Environ Microbiol 77: 2882–2886.

Nevin KP, Woodard TL, Franks AE, Summers ZM, Lovley DR (2010). Microbial electrosynthesis: feeding microbes electricity to convert carbon dioxide and water to multicarbon extracellular organic compounds. mBio 1: e00103-10.

Nie H, Zhang T, Cui M, Lu H, Lovley DR, Russell TP (2013). Improved cathode for high efficient microbial-catalyzed reduction in microbial electrosynthesis cells. Phys Chem Chem Phys 15: 14290.

Pisciotta JM, Zaybak Z, Call DF, Nam JY, Logan BE (2012). Enrichment of microbial electrolysis cell biocathodes from sediment microbial fuel cell bioanodes. Appl Environ Microbiol 78: 5212–5219.

Rosenbaum MA, Berger C, Schmitz S, Uhlig R (2017). Microbial Electrosynthesis I: Pure and Defined Mixed Culture Engineering. In: Advances in Biochemical Engineering/Biotechnology. Springer, Berlin, Germany, pp 1-22.

Rosenbaum MA, Henrich AW (2014). Engineering microbial electrocatalysis for chemical and fuel production. Curr Opin Biotechnol 29: 93–98.

Rozendal R, Jeremiasse AW, Hamelers HVM, Buisman CJN (2008). Hydrogen production with a microbial biocathode. Environ Sci Technol 42: 629–634.

Sadhukhan J, Lloyd JR, Scott K, Premier GC, Yu EH, Curtis T, Head IM (2016). A critical review of integration analysis of microbial electrosynthesis (MES) systems with waste biorefineries for the production of biofuel and chemical from reuse of CO_2. Renew Sustain Energy Rev 56: 116–132.

Saheb Alam S, Singh A, Hermansson M, Persson F, Schnürer A, Wilén B-M, Modin O (2018). Effect of start-up strategies and electrode materials on carbon dioxide reduction on biocathodes. Appl Environ Microbiol 84: e02242–17.

Sasaki K, Sasaki D, Morita M, Hirano S, Matsumoto N, Ohmura N, Igarashi Y (2010). Bioelectrochemical system stabilizes methane fermentation from garbage slurry. Bioresour Technol 101: 3415–3422.

Sato K, Kawaguchi H, Kobayashi H (2013). Bio-electrochemical conversion of carbon dioxide to methane in geological storage reservoirs. Energy Convers Manag 66: 343–350.

Schröder U, Harnisch F, Angenent LT (2015). Microbial electrochemistry and technology: terminology and classification. Energy Environ Sci 8: 513–519.

Sharma M, Aryal N, Sarma PM, Vanbroekhoven K, Lal B, Benetton XD, Pant D (2013). Bioelectrocatalyzed reduction of acetic and butyric acids via direct electron transfer using a mixed culture of sulfate-reducers drives electrosynthesis of alcohols and acetone. Chem Commun 49: 6495–6497.

Srikanth S, Singh D, Vanbroekhoven K, Pant D, Kumar M, Puri SK, Ramakumar SSV (2018). Electro-biocatalytic conversion of carbon dioxide to alcohols using gas diffusion electrode. Bioresour Technol 265: 45-51.

Su M, Jiang Y (2013). Production of Acetate from Carbon Dioxide in Bioelectrochemical Systems Based on Autotrophic Mixed Culture. J Microbiol Biotechnol 23: 1140–1146.

Ter Heijne A, Geppert F, Sleutels THJA, Batlle-Vilanova P, Liu D, Puig S (2017). Mixed Culture Biocathodes for Production of Hydrogen, Methane, and Carboxylates. In: Advances in Biochemical Engineering/Biotechnology. Springer, Berlin, Germany, pp 1-27.

Torella JP, Gagliardi CJ, Chen JS, Bediako DK, Colón B, Way JC, Silver PA, Nocera DG (2015). Efficient solar-to-fuels production from a hybrid microbial–water-splitting catalyst system. Proc Natl Acad Sci 112: 2337–2342.

Van Eerten-Jansen MC, Veldhoen AB, Plugge CM, Stams AJ, Buisman CJ, Ter Heijne A (2013). Microbial community analysis of a methane-producing biocathode in a bioelectrochemical system. Archaea 2013.

Vassilev I, Hernandez PA, Batlle-Vilanova P, Freguia S, Krömer JO, Keller J, Ledezma P, Virdis B (2018). Microbial Electrosynthesis of Isobutyric, Butyric, Caproic Acids, and Corresponding Alcohols from Carbon Dioxide. ACS Sustain Chem Eng 6: 8485–8493.

Vellend M (2010). Conceptual synthesis in community ecology. Q Rev Biol 85: 183–206.

Villano M, Aulenta F, Ciucci C, Ferri T, Giuliano A, Majone M (2010). Bioelectrochemical reduction of CO_2 to CH_4 via direct and indirect extracellular electron transfer by a hydrogenophilic methanogenic culture. Bioresour Technol 101: 3085–3090.

Villano M, De Bonis L, Rossetti S, Aulenta F, Majone M (2011). Bioelectrochemical hydrogen production with hydrogenophilic dechlorinating bacteria as electrocatalytic agents. Bioresour Technol 102: 3193–3199.

Xafenias N, Mapelli V (2014). Performance and bacterial enrichment of bioelectrochemical systems during methane and acetate production. Int J Hydrog Energy 39: 21864-21875.

Xiang Y, Liu G, Zhang R, Lu Y, Luo H (2017). Acetate production and electron utilization facilitated by sulfate-reducing bacteria in a microbial electrosynthesis system. Bioresour Technol 241: 821–829.

Zaybak Z, Pisciotta JM, Tokash JC, Logan, BE (2013). Enhanced start-up of anaerobic facultatively autotrophic biocathodes in bioelectrochemical systems. J Biotechnol 168: 478–485.

PART-III

Applications of Microbial Electrochemical Systems in Wastewater Treatment, Bioenergy, Biosensors and Electrosynthesis

Microbial Electrolysis Cell (MEC): A Versatile Technology for Hydrogen, Value-added Chemicals Production and Wastewater Treatment

Abudukeremu Kadier[1,2*]**, Piyush Parkhey**[3]**, Ademola Adekunle**[4]**, Pankaj Kumar Rai**[5]**, Mohd Sahaid Kalil**[1,2]**, S. Venkata Mohan**[3] **and Azah Mohamed**[6]

[1] Department of Chemical and Process Engineering, Faculty of Engineering and Built Environment, National University of Malaysia (UKM), 43600 UKM Bangi, Selangor, Malaysia

[2] Research Centre for Sustainable Process Technology (CESPRO), Faculty of Engineering and Built Environment, National University of Malaysia (UKM), 43600 UKM Bangi, Selangor, Malaysia

[3] Bioengineering and Environmental Sciences Laboratory, Environmental Engineering and Fossil Fuel (EEFF) Department, CSIR-Indian Institute of Chemical Technology (CSIR-IICT), Hyderabad 500 007, India

[4] Bioresource Engineering Department, McGill University, 21111 Lakeshore Rd., Ste-Anne-de-Bellevue, QC H9X 3V9, Canada

[5] Department of Biotechnology and Bioinformatics Center, Barkatullah University (BU), Bhopal 462 026, India

[6] Research Centre for Integrated Systems Engineering and Advanced Technologies (INTEGRA), Faculty of Engineering and Built Environment, National University of Malaysia (UKM), Bangi 43600, Selangor, Malaysia

1. Introduction

The present scenario of the energy crisis is arising due to the depletion of fossil fuels (FFs) and increasing global energy demand. Scientists around the world are engaged to work on various alternative energy sources. Also, the rapid combustion of FFs generates greenhouse gasses, such as CO_2, that contributes to global climate change (Kadier et al. 2016a). The ever-increasing population produce an immense volume of wastewater creating a nuisance to the environment. Collectively, these two major issues, namely energy crisis and wastewater disposal problem requires immediate attention for world's sustainable development. Wastewater usually contains a complex mixture of organic substrates that has to be removed before being discharged into the surroundings (Lu and Ren 2016). In general, the removal of these contaminants comprises a group of biological, chemical and physical treatment process. Wastewater treatment plants usually use large pumps and blowers that are energy intensive and raise treatment costs. Thus, the operating costs of treating wastewater are sure to rise despite the very fact that a major quantity of the energy inputs can be recovered as biogas via anaerobic digestion (AD). On the other hand, aerobic

*Corresponding author: abudoukeremu@163.com

treatments make matters worse and produce a large amount of sludge that is needed to be disposed of that only adds more to the operating cost (Yu et al. 2018). Nevertheless, the operating costs of treatment plants could be greatly decreased if wastewater treatment is accompanied by green energy and value-added biochemical production (Angenent et al. 2004). It is worth mentioning that wastewater exploitation requires versatile and robust technologies due to its complex composition.

Among the biochemicals or fuels that may be extracted from wastewaters, H_2 occupies a leading position as a result of its attention-grabbing characteristics as a fuel; it is clean and provides CO_2 and other pollutant emissions-free energy carrier (Rai 2016). On burning, H_2 does not produce greenhouse gas (GHG), ozone depletion chemicals or acid rain. It is a high calorific value fuel, and has the highest energy content per unit weight in comparison to other gaseous fuels (Kadier et al. 2016a; Kadier et al. 2018a). H_2 can be generated biologically by various methods, including dark fermentation (DF), photo fermentation and biophotolysis (Nikolaidis and Poullikkas, 2017; Kadier et al. 2017b). However, in the said processes the complete utilization of substrate is not possible as the metabolic pathway of the microorganisms directs the synthesis of alternative by-products (Rai and Singh 2016). Besides, it had been reported that lower H_2 yield achieved in conventional biological hydrogen production processes is due to the thermodynamic barrier. This thermodynamic barrier may be overcome by application of a small input of electrical energy (Rozendal et al. 2006) in a bioelectrochemical system (BES) called microbial electrolysis cell (MEC). MEC is comparatively a new and sustainable approach for H_2 production from organic waste matter, including wastewater (Kumar et al. 2017). H_2 production from wastewater using MECs is a promising approach toward wastewater treatment and green energy generation as it has the potential to overcome the bottlenecks of conventional hydrogen production technologies (Escapa et al. 2016; Zhen et al. 2017). Recent developments demonstrate that MECs represent a promising technology for coupling wastewater treatment and energy recovery by using the wastewater as a source of free electron (Logan et al. 2008; Rozendal et al. 2008, Kadier et al. 2017a). MECs are BESs in which electrochemical reactions are microbially catalyzed. They utilize domestic and industrial wastewater as a feedstock to generate H_2 through the catalytic action of microorganisms in the presence of electric current and absence of oxygen (O_2) (Kadier et al. 2014; Khan et al. 2017). In MECs, external power is supplied to drive thermodynamically favorable reactions at the cathode (Kadier et al. 2015a; Kadier et al. 2015b). The application of MECs is not limited to only H_2 production utilizing wastewater. Breakthrough researches are also going on the application of MECs in the field of value-added product synthesis, biosensor, resource recovery and waste treatment.

The present chapter provides a brief introduction to MEC, including the working principles, thermodynamics, possible electron transfer mechanisms at the anode and current application of MEC. In addition, a comprehensive overview of different types of conventional technologies for waste-to-energy (WTE) generation and WTE-MEC integrated approaches is presented. Finally, the advantages of MECs over other conventional ones are also discussed in detail. Thus, this chapter is a first comprehensive review of current knowledge in comparing and integrating MEC and other conventional WTE technologies.

2. Microbial Electrolysis Cell (MEC) Technology

2.1 Working Principles

The working principle of the MEC is similar to that of the microbial fuel cell (MFC) as they both rely on oxidation of organic material at the anode and complimentary reduction at the cathode. This oxygen dependence in MFCs implies the non-spontaneity of the process and could be a serious limitation. To overcome this cathodic limitation, Liu et al. (2005) and Rozendal et al. (2006) developed a novel system called MEC where the entire process is assisted with a small voltage. In an MEC, electrochemically active bacteria (EAB) or electrogenic bacteria are the dominant populations at the anode and oxidize organic matters to protons, electrons and CO_2. Finally, H_2 is evolved at the cathode by reducing the produced protons and electrons. This is subsequently reflected in the presence of obligate anaerobes (Logan et al. 2008; Kadier et al. 2018b).

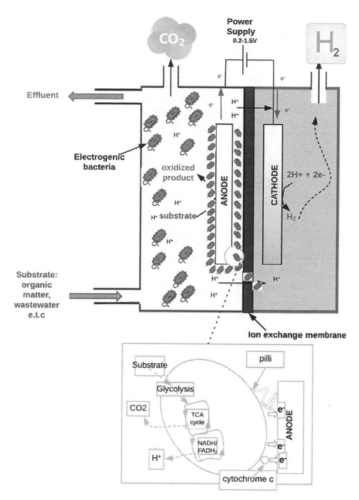

Figure 1: A schematic of a typical two-chamber MEC especially as it relates to oxidation at the anode, reduction at the cathode and electron transfer.

This electrically assisted operation provides robustness to potential applications of this technology. In general, the essential components for the construction and operation of an MEC include a power supply with constant DC source typically between 0.2-0.8V, a microbial enriched anode, a catalyzed cathode (e.g., Pt catalyzed cathode) and a well-designed gas collection system. Different architecture exists (Kadier et al. 2016b; Kadier et al. 2017c) and they are basically optimized for the process/application that the MECs is built on but important parameters considered include anode-cathode distance, flow rates and configuration and presence/absence of a membrane. Recent studies (Call and Logan 2008; Wang and Ren 2013; Kadier et al. 2017c) have confirmed that a single-chamber MEC is cost-effective and lead to an increase in the coulombic efficiency (CE) of the entire process.

2.2 Thermodynamics of MEC

The reactions that lead to H_2 evolution using acetate as an example (Equation 1) in an MEC typically have a positive Gibbs free energy (ΔG), hence will not proceed unless external energy is added (usually a small voltage). This added energy must be greater than the equilibrium voltage that can be calculated as the negative ratio of the Gibbs free energy of reaction and the product of the number of electrons involved and Faraday's constant as shown in Equation 2 (Logan et al. 2008; Harnisch and Schroder 2010; Kadier et al. 2016a). An alternative method is to use the Nernst equation to determine the difference between the theoretical cathode and anode potential. However, overpotentials at both electrodes could mean that the energy requirement might be greater than required. Thus for H_2 evolution using acetate under standard biological conditions (T = 25 °C, P = 1 bar, pH = 7), the following equation can be derived:

$$CH_3COO^- + 4H_2O \rightarrow 4HCO_3^- + H^+ + 4H_2 \left(\Delta G_r^\circ = +104.6 \frac{kJ}{mol} \right) \qquad (1)$$

$$E_{eq} = \frac{\Delta G_r^\circ}{nF} = -\frac{104.6 \times 10^3}{8 \times 96485} = -0.14 \text{ V} \qquad (2)$$

This thermodynamic calculation is important as it eventually affects both the success of the forward reaction as well as the cost. The requirement for H_2 evolution using acetate with the (Liu et al. 2005) minimum required energy supplied or minimum overpotential was estimated at 0.29 kWh/m³ (Logan et al. 2008). There are theoretical thermodynamic losses in the MECs relating to anodic over-potential, cathodic overpotential and ohmic losses, and although some of the energy is converted into chemical energy in the H_2, operating close to the (Equation 2) has been shown to minimize losses.

2.3 Electron Transfer Mechanism in an MEC

The electron transfer in MECs is accomplished mainly by direct electron transfer (DET) or mediated electron transfer (MET) (Kitching et al. 2017). Direct electron transfer (DET), which employs redox proteins on the membranes as the linking species, has been documented for electroactive bacteria such as *Geobacter sulfurreducens* and *Shewanella putrefaciens* (Lovely 2006, 2011; Logan 2009) and occurs via direct contact with the criteria for outer *c*-type cytochromes and other heme protein (Chaudhuri and Lovely 2003; Holmes et al. 2004). It has also been documented that protrusions from cells (pili) also aid in this electron transfer with a propensity to increase under anaerobic conditions. DET has also been documented to be carried out with nanowires that conductive appendages present when cytochromes MtrC and OmcA are explored. Micrographs of some of the microorganism identified in working MFCs have shown that the transfer of the electrons from the microbe to the anode can occur through cell surface protrusions. Mediated electron transfer (MET), which employs dissolved redox species as the linking species (Schroder, 2007), involves no contact between the microbes and the electrode as implied. Rather, electron transfer is achieved with mediators that can be endogenous or exogenous and is capable of reversible oxidation at the cathode.

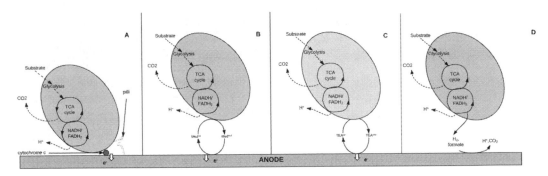

Figure 2: A schematic picture of the various electron transfer pathways that may occur in an MEC: (A) DET through membrane-bound *c*-type cytochromes or nanowires (B) MET through artificial mediators (C) MET via primary respiratory metabolites and (D) MET via secondary respiratory metabolites.

2.4 Current Applications of MEC

The applications of an MEC are becoming increasingly diverse with the advancements in technology being achieved in reactor design and operation techniques. In general, at least four distinct classifications can be identified on the possible applications of the use of microbial electrochemical systems i.e., synthesis, waste treatment, resource recovery and biosensing. This is in addition to its very versatile use for electrochemical research and integration with other electrochemical reactors. As reviewed by Kadier et al. (2014), Zhang and Angelidaki (2014), Kadier et al. (2016a), Lu and Ren (2016), Zhen et al. (2017) and many other works, these current application categories are summarized diagrammatically in Figure 3.

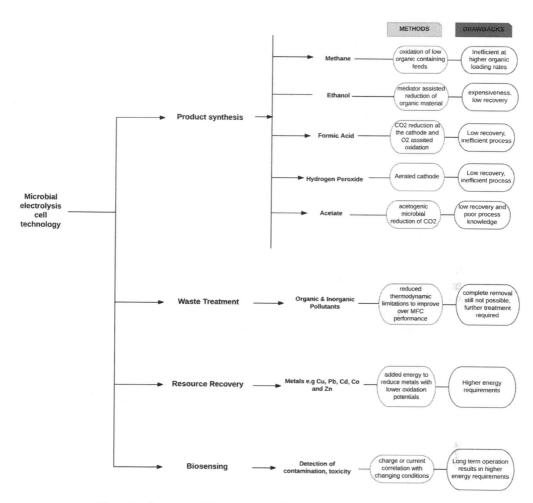

Figure 3: Summary of the current applications of MEC technology

3. Conventional Waste-To-Energy (WTE) Technologies and MEC-WTE Integrated Approaches

In the past couple of decades, the world has witnessed a tremendous rate of industrial development, growth and urbanization. This rapid rate of development along with the ever-increasing population has led to indiscriminate usage of the energy sources of the world. The demand for energy is constantly on the rise, and the unrestrained utilization of fossil fuel reserves has led to their dwindling stocks. This has put a serious question looming over the research fraternity over the identification of non-conventional sources of energy that can act as substitutes for the conventional fossil fuels. Additionally, over the course of modernization and industrialization not only has the societal and economic status of mankind has elevated, but there has also been a manifold increase in the waste generation all over the world. According to a world economic forum report, the per capita municipal solid waste (MSW) generation was expected to reach 1.42 kg/day by 2015 and was 0.64 kg/day in 2015 (Kalyani and Pandey 2014). It is thus quite evident that if the increase in waste generation at this rate goes unchecked, then its accumulation and disposal would be a major concern in future times. Therefore, to prevent any damage to the environment or cause any negative effect on the pace of the economy, it is utmost necessary that the waste generated because of domestic or industrial activities is judiciously managed. The concept of waste to energy holds tremendous potential in reducing the waste and production of renewable sources of energy. It provides a sustainable means of decomposition of waste along with energy generation. Herein, we describe the five most frequently and thoroughly studied conventional methodologies for the conversion of WTE and MEC-WTE technology integrated systems.

3.1 Incineration and MEC-Based Treatment of Incineration Leachate

Incineration is one of the most extensive methods utilized for waste-to-energy conversion and is essentially a technique of combustion of solid wastes at high temperatures of about 750-1,000 °C (Kalyani and Pandey 2014) in excess of O_2, thereby bringing complete oxidation of waste. The waste materials after incineration are mostly converted to ash that constitutes the inorganic component of the waste, flue gases that are composed of oxides of sulfur, carbon, nitrogen and heat. The energy from the heat produced from incineration is recovered in the form of either steam or electricity when it is coupled to an efficient energy recovery system (Singh et al. 2011). Several factors contribute toward the popularity of incineration as the most common waste-to-energy technology. One of the foremost reasons is the reduction in waste mass by 70% and volume by 90% (Kalyani and Pandey 2014). This reduction in volume and mass of waste helps in reducing the amount of waste that goes for landfilling. Therefore, incineration can be a useful method for energy recovery from waste in countries that have limited land availability. Additionally, incineration is also an ideal method for converting high calorific value wastes and deriving energy from them.

The major drawback with incineration is the production of noxious pollutants or humic substances such as furans and dioxins apart from ash and flue gasses. Moreover, the flue gasses containing the oxides of sulfur, carbon and nitrogen contribute to the greenhouse emissions and further aggravating the situation. In order to overcome some of these hindrances and to enhance biomethanogenic treatment of fresh incineration leachate, MECs were incorporated into AD (Gao et al. 2017). The results demonstrated that AD-MECs were more effective and inexpensive at the treatment of incineration leachate.

3.2 Gasification for H_2 Production

Gasification is a thermochemical process of waste-to-energy conversion wherein carbonaceous materials, most commonly biomass is subjected to partial combustion at high temperatures of 700 °C to produce CO, CO_2 and H_2. The chemical reactions occurring during the gasification can be summarized as illustrated in Figure 4.

The most significant advantage of gasification is that the energy recovery using syngas is higher than the direct combustion of fuel or biomass. Syngas can either be directly burned or converted to methanol and hydrogen or it can also be converted into synthetic fuel that is a mixture of CO and H_2. The second advantage of gasification is its simpler control of emissions as compared to combustion. Since the process of gasification is carried out at a comparatively higher temperature and pressure conditions, the removal of oxides of sulfur and nitrogen along with other pollutants such as mercury, selenium, arsenic and cadmium becomes easier. The water requirements in gasification are also much less than other technologies. Nevertheless, the process of gasification also suffers from certain bottlenecks. Primarily, the cost of construction of a gasifier plant exceeds that of a natural gas plant (National Energy Technology Laboratory, US Department of Energy, n.d.; Vreugdenhil et al. 2014). Additionally, complex and sensitive operational procedures, difficulty in maintenance and continuous operation of the plant are the other disadvantages associated with the technology.

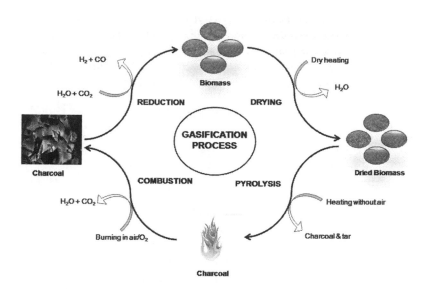

Figure 4: The four steps of the gasification process in the production of CO_2 and H_2.

3.3 Pyrolysis and Pyrolysis-MEC Coupled Novel Process

Pyrolysis is another thermochemical waste decomposition method that is carried out at high temperatures in the absence of O_2. Few scientists opine that pyrolysis is the precursor step in both gasification and solid fuel combustion (Singh and Gu 2010). It can also be called as an indirect gasification method with inert gasses being the external energy source (Belgiorno et al. 2003). The process of pyrolysis has manifold advantages as the waste disposal method. Principally, it can convert waste and other feedstocks to various solid, liquid or gaseous fuels. Secondly, the technology is easy and economical to operate and is also self-sustaining. Thirdly, there is a significant reduction in the volume of waste post-pyrolysis with a minimum undesired by-product formation (Serio et al. 2000; Singh et al. 2011). Depending upon the operative criteria, three types of pyrolysis have been reported. These are conventional, fast and flash pyrolysis. The conditions for each type are tabulated in Table 1.

Table 1: Types of pyrolysis depending upon the operational parameter conditions

Parameter	Temperature (°C)	Rate of Heating (K/s)	Particle Size (mm)	Solid Residence Time (s)	References
Conventional pyrolysis	277-627	0.1-1	5-50	300-3600	Katyal, 2007; Mohan et al. 2006
Fast pyrolysis	577-977	10-200	<1	0.5-10	Demirbas 2009,
Flash pyrolysis	777-1027	>1000	<0.2	<0.5	2007

As can be seen from Table 1, conventional pyrolysis is a slower method with longer residence time. Fast pyrolysis and flash pyrolysis that occur at higher temperatures have lower residence time and therefore are the preferred technology at present (Singh et al. 2011). Even though pyrolysis is a better technique than other thermal methods of waste-to-energy conversions, there still are certain issues that need diligent attention. Firstly, the initial investments for this technique are still high that limit its economic viability. Moreover, the ash produced upon the process completion is contaminated with toxic compounds or heavy metals. A new integrated pyrolysis-MEC process to handle the toxic compounds and the generating H_2 utilizing a switchgrass pyrolysis-derived aqueous phase is reported (Lewis et al. 2015). In this integrated system, the aqueous stream produced during pyrolysis of switchgrass was employed as a fuel source for H_2 production in the MECs; a maximum hydrogen production rate (HPR) of 4.3 L H_2/L anode-day with a loading of 10 g COD/L anode-day was achieved. H_2 yields ranged from $50 \pm 3.2\%$ to $76 \pm 0.5\%$ while CE ranged from $54 \pm 6.5\%$ to $96 \pm 0.21\%$, respectively (Lewis et al. 2015).

3.4 Landfilling and MEC-Based Integrated Process for Resource Recovery From Landfill Leachate

Landfilling is one of the most common waste disposal strategies followed in most of the developing nations and includes dumping of waste in pits followed by covering them with soil. In landfills, the objective is to keep the waste separated from groundwater and air. The microorganisms present in the landfill site would digest the waste and produce a mixture of gasses most commonly CH_4 and CO_2. This gas mixture is termed as landfill gas. This landfill gas can be used directly on-site as a fuel in boilers for providing heat. Or else, they can be used for electricity production through turbines or fuel cells (Sullivan, 2010). Among waste management approaches, landfilling is a simple and an inexpensive process for waste disposal and does not require trained manpower for complex plant operations unlike in incineration or pyrolysis (Renou et al. 2008). However, the safety and environmental issues associated with using landfill gas far exceed its advantages as a waste-to-energy alternative. Landfill leachate is one of the most complex waste streams to treat due to the presence of a wide range of organic and inorganic compounds, such as toxic compounds, chlorinated organic, nutrients, inorganic salts and heavy metals in leachate (Foo and Hameed 2009). Therefore, there is a demand for environmentally friendly approaches for landfill leachate treatment. Leachate treatment technologies such as advanced oxidation, membrane separation, chemical treatment, and biological processes have been practiced so far (Greenman et al. 2009). To investigate the potential of resource recovery from landfill leachate, an MEC and Forward Osmosis (FO) coupled system for treating leachate samples were applied (Qin et al. 2016). This MEC-FO process was first developed by Qin and He (2014) to utilize MECs to recover NH_4^+ from wastewater and use the recovered NH_4^+ as a draw solute in FO for water recovery.

3.5 Anaerobic Digestion (AD) and AD-MEC Hybrid Process for CH_4 and H_2 Production

Anaerobic digestion is a biochemical process wherein complex organic wastes are broken down by the action of microorganisms in absence of O_2 forming gasses mostly composed of methane (CH_4) and CO_2 through the four major steps, including hydrolysis, acidogenesis, acetogenesis and methanogenesis (Yu et al. 2018). The process of AD can be divided into four major steps as shown in Figure 5. The biogas produced because of the complete process can be utilized directly as a source of heat energy by burning, converting to methanol or using it for electricity production.

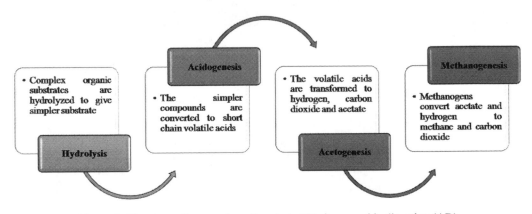

Figure 5: The steps of conversion of waste to CH_4 in anaerobic digestion (AD).

The single most important advantage of AD is the conversion of waste-to-energy in an environment-friendly manner. The process of AD is carried out in controlled conditions in a biogas plant. There are no extreme conditions required for its operation, and the emission of greenhouse gasses during and after the process is also minimal. Additionally, a wide range of solid wastes and wastewater such as effluents from industrial, household or municipal sources and sewage treatment can be effectively converted to biogas. AD also reduces the amount of waste that otherwise would be carried to landfills or incinerators for decomposition. However, the process is ineffective in converting wastes with low organic load as the microbial activity is mostly dependent upon the organic content. Further, the process has other bottlenecks such as the requirement of stringent control conditions for bacterial growth and gas production efficiencies.

Recently numerous studies have shown that MECs and AD can be integrated for enhancing the efficiency of AD (Sadhukhan et al. 2016; Lee et al. 2017; Yu et al. 2018). Bo et al. (2014) inserted a pair of MEC electrodes in an AD reactor. The MEC generated H_2 at the cathode that could react with the CO_2 to produce CH_4 *in situ* by hydrogenotrophic methanogens. Compared with the conventional AD reactor, the coupled MEC-AD reactor achieved very high content of CH_4 up to 98%. It was also reported that CH_4 yield, COD removal and carbon recovery was increased by 24-230%, 130-300% and 55-56%, respectively (Bo et al. 2014). Furthermore, a novel combination of AD and MEC was designed by Cai et al. (2016) in which two ADs were separated by an anion exchange membrane each working as anode and cathode, respectively. With sludge fermentation liquid, 0.247 mL CH_4/mL reactor/day methane was generated at cathodic AD and raised by 51.53% compared to AD control reactor. Moreover, it was demonstrated by Yin et al. (2016) that by co-cultivating *Geobacter* and *Methanosarcina* in an AD-MEC integrated system, the CH_4 yield was further enhanced by 24.1%, obtaining 360.2 mL/g-COD; these results were comparable to the theoretical CH_4 yield of an AD.

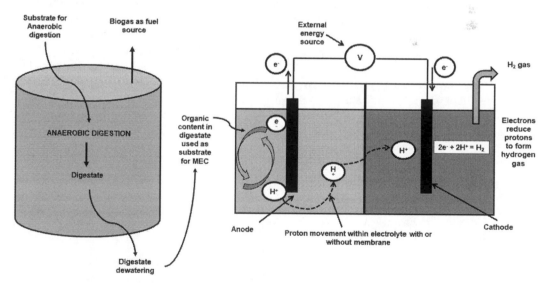

Figure 6: Schematic diagram of the integration scheme of anaerobic digestion (AD) and microbial electrolysis cell (MEC)

3.6 Dark-Fermentations (DF), Photo-Fermentation and Integrated Approaches for Bio-H_2 Production

Dark fermentation is a biochemical process in which anaerobic bacteria in a series of steps like AD produce H_2 from organic substrates. The process of DF of conversion of organic substrates into bio-H_2 mainly takes place in four steps (Mohan, 2010). In the first step, the complex organic substrates are hydrolyzed into simpler monomers by the action of fermentative hydrolytic bacteria. In the second step, the monomers such as simple sugars, peptides, amino acids are used as a carbon source by the same fermentative microorganisms to produce a mixture of organic acids and alcohols. The fermentation end products are then converted into H_2 and acid intermediates by acidogenesis bacteria.

The DF mode of H_2 production is a more efficient process as compared to the photosynthetic route that is carried out by algae and photosynthetic bacteria (Mohanakrishna and Mohan 2013). The net hydrogen yield in DF is higher than the photosynthetic mode of H_2 production. Furthermore, the DF approach is simpler to operate since the bacteria do not have a specific prerequisite of light for growth and H_2 production. They have a higher growth rate and in addition to that the process of DF does not offer any concerns with the maintenance of purity of culture as it can be operated with mixed microflora. However, the most significant advantage with the fermentative process is its flexibility with substrates, specifically the wastewaters that reduce the input costs thereby making the process cost-effective. The process of DF hydrogen production also suffers certain drawbacks, the primary of them are low conversion efficiency of feedstock and residue organics in fermentation effluent (Marone et al. 2017).

One probable solution is to integrate the process with either photo-fermentation (Venkata Mohan et al. 2008) or MEC to realize higher organic removal efficiencies (Wang et al. 2011; Rivera et al. 2015; Sivagurunathan et al. 2017). DF of glucose is limited to production of 4 moles H_2 and 2 mole acetate per mole glucose. Aided by MEC, acetate can also be converted into H_2. The combination of DF-MEC technologies results in a theoretical production of 12 moles hydrogen per mole glucose (Clauwaert et al. 2008). In another study, DF of corn stalk was integrated with an MEC (at 0.8 V), thereby tripling the HPR (387.1 mL H_2/g-corn stalk) (Li et al. 2014). This concept of utilizing unused energy of DF through MECs was also demonstrated by Rivera et al. (2015) by using DF effluent as the substrate for MEC operation. The effluent containing various volatile fatty acids produced 81 mLH$_2$/L/d with 85% organic removal rate (Rivera et al. 2015). Furthermore, Wang et al. (2015) used a novel strategy to extract maximum H_2 from cellulose. They connected two MFCs in series to an MEC to produce a maximum of 0.43 V using fermentation effluent as a substrate, obtained a HPR from the MEC of 0.48 m^3 H_2/m^3/d, and a H_2 yield of 33.2 mmol H_2/g COD removed in the MEC. The overall H_2 production for the integrated system (DF, MFC and MEC) was increased by 41% compared with fermentation alone to 14.3 mmol H_2/g cellulose, with a total HPR of 0.24 m^{3}H$_2$/m^3/d and an overall energy recovery efficiency of 23% (Wang et al. 2015). The main advantage of this combined fermentation and MFC-MEC system is that H_2 is generated at higher yields than DF alone, without the need for exogenous electrical power input.

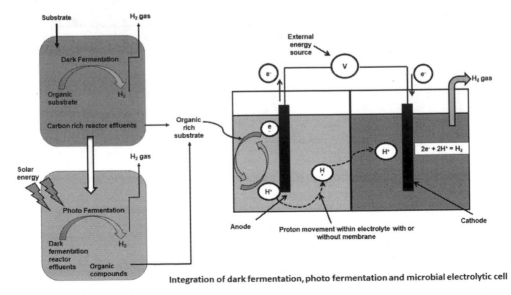

Integration of dark fermentation, photo fermentation and microbial electrolytic cell

Figure 7: Schematic diagram of the integration scheme of dark fermentation (DF), photo fermentation and microbial electrolysis cell (MEC)

4. The Advantages of MEC over Other Potential Waste-To-Energy (WTE) Technologies

MECs are cutting edge technique for treatment of wastewater because (1) it produces the energy output in the form of H_2, (2) it can decrease solids production and thus, reduce sludge handling or treatment costs, (3) it can be useful to recover value-added products from waste, and (4) possibly decrease the release of odors. MEC system has numerous advantages over other biohydrogen processes as microbial population present in an inoculum can oxidize a wide variety of substrates ranging from simple to complex wastewater as well as industrial and lignocellulosic waste. Comparing to the DF process, the MEC has a higher H_2 recovery and wider substrate diversity (Escapa et al. 2016; Sivagurunathan et al. 2017). The HPR in the MEC was about fivefold higher than that of DF (Call and Logan, 2008).

Considering the amount of energy required for input voltage supply, there is a need to make more cost-effective and economical process to make MEC comparable with existing conventional wastewater treatment technologies (Logan et al. 2008). The energy required to maintain the external applied potential in MEC is approximately equivalent to energy consumption for operating the aerator in activated sludge process (Tchobanoglous et al. 2003). To make the MEC system economical and cost-effective, the efficiency should be improved in terms of hydrogen production and other value-added products. In this regard, process optimization and use of more efficient processes would help to improve the environmental impact as well as the economic balance. Additionally, MEC offers advantage of nutrient removal and recovery over anaerobic digestion system (Haddadi et al. 2014). Also, the gas produced in MEC is more valuable over AD.

Moreover, on the basis of chemical oxygen demand (COD) loading rates, MECs could be comparable to other wastewater treatment technologies such as activated sludge process and AD (Rozendal et al. 2007). Also, compared to conventional water electrolysis, the applied voltage energy investment is significantly low and thus reduced overall cost. Recent developed microbial electrochemical technology i.e., MEC offers effective wastewater treatment along with simultaneous hydrogen and other value-added product recovery. This article mainly focused on detailed comprehensive review of the various substrates used for oxidation in MEC varying from simple sugars to complex industrial wastewater along with different aspects of MEC to make it implementable for field scale applications.

5. Conclusions and Future Perspectives

Since the inception of the MECs, substantial progress has been made to improve the technology by applying various designs, membranes, electrodes materials, microbes and substrates. MECs are not only best suited for H_2 production from wastewater, but it could also be utilized to recover value-added products from wastewater. Compared to conventional biohydrogen production technologies i.e., photo-fermentation and DF, MEC proves to be a potential technology as it is able to overcome the main bottleneck which is the thermodynamic barrier present in the conventional approach. Operation/construction cost of MECs is the major hurdle for the application of MECs to various objectives. To reduce the operation cost of MECs various strategies had been employed, including the use of less expensive electrodes, membraneless design having added advantage of high H_2 recovery and production rates. Although MECs seem to be more promising in comparison to other H_2 production and wastewater treatment technologies, but further improvement in the MECs is desired to cope with the operation hurdles. Utilizing wastewater for H_2 production in MECs seem good, but since the composition of wastewater is not always well balanced in nutrients, supplementation is required prior to use in MECs. Under these conditions, it is very difficult for MECs to compete with other technologies. In recent years, MECs capable of removing nutrients from the wastewater have been developed but their operating cost is very high. The production of valuable biochemicals including acetate, ethanol, formic acid or hydrogen peroxide in MECs utilizing wastewater made this technology more attractive. However, the production of fine chemicals supposed to be used in the food and pharma industry might give rise to health issues. Scaling up of MECs means a flow of large electric currents. Voltage fluctuation/losses in MECs are directly depending on current, so this parameter should be taken care of for the development of efficient MECs. Reports on MECs indicated that efficiency of MECs is also affected by various operating conditions like inoculums type, applied voltage, ionic strength, temperature, pH, hydraulic retention time (HRT) and organic loading. Thus, the integration of H_2 and other value-added biochemicals production from wastewater treatment using MECs seem to be a feasible approach as the high operating cost of MECs could be compensated from other products. Thus, we can hope that in the future more efficient and cost-effective MECs will be developed, and these improved MECs will have the capacity to operate at a commercial level.

Acknowledgments

This work was supported by the National University of Malaysia (UKM), Project No: DIP-2017-019 and the Malaysian Ministry of Higher Education (MOHE). Special thanks to the Head of Project DIP-2017-019, Prof. Dr. Mohd Sahaid Kalil & Members.

References

Angenent LT, Karim K, Al-Dahhan MH, Wrenn BA, Domíguez-Espinosa R (2004) Production of bioenergy and biochemicals from industrial and agricultural wastewater. Trends Biotechnol 22: 477-485.

Belgiorno V, De Feo G, Della Rocca C, Napoli, RMA (2003) Energy from gasification of solid wastes. Waste Manag 23(1): 1–15.

Bo T, Zhu X, Zhang L, Tao Y, He X, Li D, Yan Z (2014) A new upgraded biogas production process: Coupling microbial electrolysis cell and anaerobic digestion in single-chamber, barrel-shape stainless steel reactor. Electrochem Commun 45: 67–70.

Cai W, Han T, Guo Z, Varrone C, Wang A, Liu W (2016) Methane production enhancement by an independent cathode in integrated anaerobic reactor with microbial electrolysis. Bioresour Technol 208: 13–18.

Call D, Logan BE (2008) Hydrogen production in a single chamber microbial electrolysis cell lacking a membrane. Environ Sci Technol 42: 3401-3406.

Chaudhuri SK, Lovley DR (2003) Electricity generation by direct oxidation of glucose in mediatorless microbial fuel cells. Nat Biotechnol 21: 1229.

Clauwaert P, Toledo R, Van Der Ha D, Crab R, Verstraete W, Hu H, Udert KM, Rabaey K (2008) Combining biocatalyzed electrolysis with anaerobic digestion. Water Sci Technol 57: 575–579.

Demirbas A (2007) Producing bio-oil from olive cake by fast pyrolysis. Energy source A, Recovery Util Environ Effects 30: 38-44.

Demirbas A (2009) Biorefineries: current activities and future developments. Energy Convers Manag 50: 2782-2801.

Escapa A, Mateos R, Martínez EJ, Blanes J (2016) Microbial electrolysis cells: An emerging technology for wastewater treatment and energy recovery. From laboratory to pilot plant and beyond. Renew Sustain Energy Rev 55: 942-956.

Foo K, Hameed B, (2009) An overview of landfill leachate treatment via activated carbon adsorption process. J Hazard Mater 171: 54-60.

Gao Y, Sun D, Dang Y, Lei Y, Ji J, Lv T, Bian R, Xiao Z, Yan L, Holmes DE (2017) Enhancing biomethanogenic treatment of fresh incineration leachate using single chambered microbial electrolysis cells. Bioresour Technol 231: 129-137.

Greenman J, Gálvez A, Giusti L, Ieropoulos I (2009) Electricity from landfill leachate using microbial fuel cells: comparison with a biological aerated filter. Enzyme Microb Technol 44: 112-119.

Haddadi S, Nabi-Bidhendi G, Mehrdadi N (2014) Nitrogen removal from wastewater through microbial electrolysis cells and cation exchange membrane. J Environ Health Sci Eng 12: 48.

Harnisch F, Schröder U (2010) From MFC to MXC: chemical and biological cathodes and their potential for microbial bioelectrochemical systems. Chem Soc Rev 39: 4433-4448.

Holmes DE, Nicoll J S, Bond DR, Lovley DR (2004) Potential role of a novel psychrotolerant member of the family *Geobacteraceae, Geopsychrobacter electrodiphilus* gen. nov., sp. nov., in electricity production by a marine sediment fuel cell. Appl and Environ Microbiol 70: 6023-6030.

Hu H, Fan Y, Liu H (2008) Hydrogen production using single-chamber membrane-free microbial electrolysis cells. Water Res 42: 4172-4178.

Kadier A, Jiang Y, Lai B, Rai PK, Chandrasekhar K, Mohamed A, Kalil MS (2018a) Biohydrogen production in microbial electrolysis cells from renewable resources. In: O Konur (eds), Bioenergy Biofuels, Taylor & Francis Group, Boca Raton: CRC Press. London, UK; pp. 331-356.

Kadier A, Kalil MS, Chandrasekhar K, Mohanakrishna G, Saratale GD, Saratale RG, Kumar G, Arivalagan P, Sivagurunathan P (2018b) Surpassing the current limitations of high purity H_2 production in microbial electrolysis cell (MECs): Strategies for inhibiting growth of methanogens. Bioelectrochem 19: 211–219.

Kadier A, Logroño W, Rai PK, Kalil MS, Mohamed A, Hasan HA, Hamid AA (2017a) None-platinum electrode catalysts and membranes for highly efficient and inexpensive H_2 production in microbial electrolysis cells (MECs): A review. Iran J Catal 7: 89-102.

Kadier A, Kalil MS, Mohamed A, Hasan AA, Abdeshahian P, Fooladi T, Hamid, AA (2017b) Microbial electrolysis cells (MECs) as innovative technology for sustainable hydrogen production: Fundamentals and perspective applications. In: M Sankir, ND Sankir (eds), Hydrogen Production Technologies, Wiley-Scrivener Publishing LLC, USA; pp 407–458.

Kadier A, Kalil MS, Mohamed A, Hamid AA (2017c) A new design enhances hydrogen production by *G. sulfurreducens PCA* strain in a single-chamber microbial electrolysis cell (MEC). J Technol 79: 71–79.

Kadier A, Kalil MS, Abdeshahian P, Chandrasekhar K, Mohamed A, Azman NF, Logroño W, Simayi Y, Hamid AA (2016a) Recent advances and emerging challenges in microbial electrolysis cells (MECs) for microbial production of hydrogen and value-added chemicals. Renew Sustain Energy Rev 61: 501–525.

Kadier A, Simayi Y, Abdeshahian P, Azman NF, Chandrasekhar K, Kalil MS (2016b) A comprehensive review of microbial electrolysis cells (MEC) reactor designs and configurations for sustainable hydrogen gas production. Alexandria Eng J 55: 427-443.

Kadier A, Simayi Y, Chandrasekhar K, Ismail M, Kalil MS (2015a) Hydrogen gas production with an electroformed Ni mesh cathode catalysts in a single-chamber microbial electrolysis cell (MEC). Int J Hydrogen Energy 40: 14095-14103.

Kadier A, Abdeshahian P, Simayi Y, Ismail M, Hamid AA, Kalil MS (2015b) Grey relational analysis for comparative assessment of different cathode materials in microbial electrolysis cells. Energy 90: 1556-1562.

Kadier A, Simayi Y, Kalil MS, Abdeshahian P, Hamid AA (2014) A review of the substrates used in microbial electrolysis cells (MECs) for producing sustainable and clean hydrogen gas. Renew Energy 71: 466-472.

Kalyani KA, Pandey KK (2014) Waste to energy status in India: a short review. Renew Sustain Energy Rev 31: 113-120.

Katyal S (2007) Effect of carbonization temperature on combustion reactivity of bagasse char. Energy Sourc A, Recovery Util Environ Effects 29: 1477-1485.

Khan MZ, Nizami AS, Rehan M, Ouda OKM, Sultana S, Ismail IM, Shahzad K (2017) Microbial electrolysis cells for hydrogen production and urban wastewater treatment: A case study of Saudi Arabia. Appl Energ 185: 410-420.

Kitching M, Butler R, Marsili E (2017) Microbial bioelectrosynthesis of hydrogen: current challenges and scale-up. Enzyme Microb Technol 96: 1-13.

Kumar G, Saratale RG, Kadier A, Sivagurunathan P, Zhen G, Kim SH, Saratale GD (2017) A review on bio-electrochemical systems (BESs) for the syngas and value-added biochemicals production. Chemosphere 177: 84-92.

Lee B, Park JG, Shin WB, Tian DJ, Jun HB (2017) Microbial communities change in an anaerobic digestion after application of microbial electrolysis cells. Bioresour Technol 234: 273–280.

Li X, Liang D, Bai Y, Fan Y (2014) Enhanced H_2 production from corn stalk by integrating dark fermentation and single chamber microbial electrolysis cells with double anode arrangement. Int J Hydrogen Energy 39: 2–7.

Lin CY, Leu HJ, Lee KH (2016) Hydrogen production from beverage wastewater via dark fermentation and room-temperature methane reforming. Int J Hydrogen Energy 41: 21736-21746.

Liu H, Grot S, Logan BE (2005) Electrochemically assisted microbial production of hydrogen from acetate. Environ Sci Technol 39: 4317-4320.

Lewis AJ, Ren S, Ye X, Kim P, Labbe N, Borole AP (2015) Hydrogen production from switchgrass via an integrated pyrolysis-microbial electrolysis process. Bioresour Technol 195: 231–41.

Logan BE (2009) Exoelectrogenic bacteria that power microbial fuel cells. Nat Rev Microbiol 7: 375.

Logan BE, Call D, Cheng S, Hamelers HV, Sleutels TH, Jeremiasse AW, Rozendal, RA (2008) Microbial electrolysis cells for high yield hydrogen gas production from organic matter. Environ Sci Technol 42: 8630-8640.

Lovley DR (2006) Bug juice: harvesting electricity with microorganisms. Nat Rev Microbiol 4: 497.

Lovley DR (2011) Powering microbes with electricity: direct electron transfer from electrodes to microbes. Environ Microbiol Rep 3: 27-35.

Lu L, Ren ZJ (2016) Microbial electrolysis cells for waste biorefinery: A state of the art review. Bioresour Technol 215: 254-264.

Marone A, Ayala-Campos OR, Trably E, Carmona-Martínez AA, Moscoviz R, Latrille E, Steyer JP, Alcaraz-Gonzalez V, Bernet N (2017) Coupling dark fermentation and microbial electrolysis to enhance bio-hydrogen production from agro-industrial wastewaters and by-products in a bio-refinery framework. Int J Hydrogen Energy 42: 1609-1621.

Mohan D, Pittman CU, Steele PH (2006) Pyrolysis of wood/biomass for bio-oil: a critical review. Energy & Fuels 848-889.

Mohan SV (2010) Waste to renewable energy: a sustainable and green approach towards production of biohydrogen by acidogenic fermentation. In Sustain Biotechnol 129-164.

Mohan SV, Srikanth S, Dinakar P, Sarma PN (2008) Photo-biological hydrogen production by the adopted mixed culture: Data enveloping analysis. Int J Hydrogen Energy 33: 559-569.

Mohanakrishna G, Mohan SV (2013) Multiple process integrations for broad perspective analysis of fermentative H_2 production from wastewater treatment: Technical and environmental considerations. Appl Energy 107: 244-254.

National Energy Technology Laboratory. Advantage of Gasification. URL - https://netl.doe.gov/research/coal/energy-systems/gasification/gasifipedia/advantage-of-gasification. US Department of Energy.

Nikolaidis P, Poullikkas A (2017) A comparative overview of hydrogen production processes. Renew Sustain Energy Rev 67: 597–611.

Qin M, He Z (2014) Self-supplied ammonium bicarbonate draw solute for achieving wastewater treatment and recovery in a microbial electrolysis cell-forward osmosis coupled system. Environ Sci Technol Lett 1: 437-441.

Qin M, Molitor H, Brazil B, Novak JT, He Z (2016) Recovery of nitrogen and water from landfill leachate by a microbial electrolysis cell-forward osmosis system. Bioresour Technol 200: 485-92.

Rai PK, Singh SP (2016) Integrated dark-and photo-fermentation: Recent advances and provisions for improvement. Int J Hydrog Energy 41: 19957-19971.

Rai PK (2016) Recent Advances in substrate utilization for fermentative hydrogen Production. J Appl Biol and Biotechnol 4: 059-067.

Renou S, Givaudan J, Poulain S, Dirassouyan F, Moulin P (2008) Landfill leachate treatment: Review and opportunity. J Hazard Mater 150: 468-493.

Rivera I, Buitrón G, Bakonyi P, Nemestóthy N, Bélafi-Bakó K (2015) Hydrogen production in a microbial electrolysis cell fed with a dark fermentation effluent. J Appl Electrochem 45: 1223-1229.

Rozendal RA, Hamelers HV, Euverink GJ, Metz SJ, Buisman CJ (2006) Principle and perspectives of hydrogen production through biocatalyzed electrolysis. Int J Hydrogen Energy 31: 1632-1640.

Rozendal RA, Hamelers HV, Molenkamp RJ, Buisman, CJ (2007) Performance of single chamber biocatalyzed electrolysis with different types of ion exchange membranes. Water Res 41: 1984-1994.

Rozendal RA, Jeremiasse AW, Hamelers HV, Buisman CJ (2008) Hydrogen production with a microbial biocathode. Environ Sci and Technol 42: 629-634.

Sadhukhan J, Lloyd JR, Scott K, Premier GC, Yu EH, Curtis T, Head IM (2016) A critical review of integration analysis of microbial electrosynthesis (MES) systems with waste biorefineries for the production of biofuel and chemical from reuse of CO_2. Renew Sustain Energy Rev 56: 116–132.

Schröder U (2007) Anodic electron transfer mechanisms in microbial fuel cells and their energy efficiency. Physl Chem Chem Phys 9: 2619-2629.

Serio MA, Chen Y, Wójtowicz MA, Suuberg EM (2000) Pyrolysis processing of mixed solid waste streams. ACS Div of Fuel Chem Prepr 45: 466-474.

Singh J, Gu S (2010) Biomass conversion to energy in India critique. Renew and Sustain Energy Rev 14: 1367-1378.

Singh RP, Tyagi VV, Allen T, Ibrahim MH, Kothari R (2011) An overview for exploring the possibilities of energy generation from municipal solid waste (MSW) in Indian scenario. Renew and Sustain Energy Rev 15: 4797-4808.

Sivagurunathan P, Kuppam C Mudhoo A, Saratale GD, Kadier A, Zhen G, Chatellard L, Trably E, Kumar G (2017) A comprehensive review on two-stage integrative schemes for the valorization of dark fermentative effluents. Crit Rev Biotechnol 21: 1-15.

Sullivan P, SENIOR VICE PRESIDENT, S E (2010) The Importance of Landfill Gas Capture and Utilization in the US. Biocycle Magazine, Earth Engineering Center, Columbia University.

Tchobanoglous G, Burton FL, Stensel HD (2003) Adsorption. In: Metcalf & Eddy Inc. (Ed), Wastewater Engineering: Treatment and Reuse, 4th Edition. McGraw-Hill, New York; pp 1138-1162.

Vreugdenhil BJ, Van Der Drift A, Van Der Meijden CM, Grootjes, AJ (2014) Indirect gasification: A new technology for a better use of Victorian Brown Coal. Melbourne.

Wang A, Sun D, Cao G, Wang H, Ren N, Wu WM , Logan BE (2011) Integrated hydrogen production process from cellulose by combining dark fermentation, microbial fuel cells, and a microbial electrolysis cell. Bioresour Technol 102: 4137-4143.

Wang H, Ren ZJ (2013) A comprehensive review of microbial electrochemical systems as a platform technology. Biotechnol Adv 31: 1796-1807.

Wang A, Sun D, Cao G, Wang H, Ren N, Wu W, Logan BE (2015) Integrated hydrogen production process from cellulose by combining dark fermentation microbial fuel cells, and a microbial. Bioresour Technol 102: 4137–4143.

Yin Q, Zhu X, Zhan G, Bo T, Yang Y, Tao Y, He X, Li D, Yan Z (2016) Enhanced methane production in an anaerobic digestion and microbial electrolysis cell coupled system with co-cultivation of *Geobacter* and *Methanosarcina*. J Environ Sci 42: 210–214.

Yu Z, Leng X, Zhao S, Ji J, Zhou T, Khan A, Kakde A, Liu P, Li X (2018) A review on the applications of microbial electrolysis cells in anaerobic digestion. Bioresour Technol 255: 340-348.

Zhang Y, Angelidaki I (2014) Microbial electrolysis cells turning to be versatile technology: recent advances and future challenges. Water Res 56: 11-25.

Zhen G, Lu X, Kumar G, Bakonyi P, Xu K, Zhao Y (2017) Microbial electrolysis cell platform for simultaneous waste biorefinery and clean electrofuels generation: Current situation, challenges and future perspectives. Prog Energy Combust Sci 63: 119-145.

Mechanisms of Heavy Metal Separation in Bioelectrochemical Systems and Relative Significance of Precipitation

Hui Guo and Younggy Kim[*]

McMaster University, Department of Civil Engineering, 1280 Main St. W., JHE 301, Hamilton, Ontario, L8S 4L8, Canada

1. Introduction

Heavy metals are the elements having high atomic weights or high densities (Fergusson and Erric 1990). Some heavy metals such as copper, iron and zinc are important trace elements in natural waters (Manahan 2017). These heavy metals are essential nutrients for plant and animals at low levels but toxic at high concentration. The toxicity of heavy metals to microorganisms and plants were reported by Giller et al. (1998) and Nagajyoti et al. (2010). Some heavy metals are of particular concern because of their toxicities to humans. For instance, cadmium can affect several enzymes and cause kidney damage and bone disease (Manahan 2017). Exposure to uranium can increase the risk of cancer (Achparaki et al. 2012). Lead adversely affects the central and peripheral nervous systems (Hu 2002; Tchounwou et al. 2012). The industrial wastewater from metal plating facilities, mining operations and pesticide-producing establishment are the main sources of heavy metal pollution (Srivastava and Majumder 2008; Fu and Wang 2011). Because heavy metals are nonbiodegradable (Kurniawan et al. 2006) and can be accumulated in soils and living organisms, the effective treatments of the waste sources should be conducted. Many methods, which include chemical precipitation, coagulation-flocculation, flotation, membrane filtration, ion exchange, electrodialysis and adsorption, have been developed to separate heavy metal ions from aqueous solution (Fu and Wang 2011; Kurniawan et al. 2006; He and Chen 2014). However, the conventional methods require high energy consumption and chemical cost. Therefore, an innovative method that can separate heavy metals effectively and sustainably is in need.

Bioelectrochemical systems (BES) have emerged as a novel technology for wastewater treatment and energy production (Rabaey et al. 2010; Wang et al. 2015; ElMekawy et al. 2015). In literature, BES is used to present both in microbial fuel cells (MFCs) and microbial electrolysis cells (MECs) (Nancharaiah et al. 2015). MFCs can oxidize the substrates with current generation (Liu et al. 2005; He et al. 2005; Logan et al. 2006), while MECs can produce hydrogen gas with an addition of a small voltage (Call and Logan 2008; Logan et al. 2008). During the process, the electrons produced from the substrates are transferred to the anode and flow to the cathode. The oxidized heavy metal ions can gain the electrons from the cathode and be reduced to metallic metals. This reduction process makes the separation of heavy metals achievable in BES. There are many studies that reported the utility of BES for separating heavy metals. Therefore, in this review, we give a summary of previous studies that focuses on the heavy metal separation in BES.

*Corresponding author: younggy@mcmaster.ca

Nancharaiah et al. (2015), Wang and Ren (2014) and Dominguez-Benetton et al. (2018) reviewed the removal and recovery of metals in BES. In the study of Wang and Ren (2014), the BES was classified into four categories based on the cathode type (abiotic cathode or biocathode) and reactor type (MEC or MFC). The category of the reactor with bipolar membrane was added in the study conducted by Nancharaiah et al. (2015). Dominguez-Benetton et al. (2018) reported a division of BES into four categories based on the mechanisms for metal transformation and recovery. In these three reviews, the performance of previous reactors was summarized and the reduction of heavy metal ions was discussed. However, the precipitation of heavy metal ions, which is possible in BES, was not included in their study. Some precipitates such as cadmium hydroxide, cadmium carbonate and cobalt hydroxide were detected in previous studies (Colantonio and Kim 2016; Huang et al. 2015). Thus, in this study, we review the mechanism of heavy metal separation with an emphasis on precipitation of metals in BES. In summary, the objectives of this review are to: summarize the previous studies related to the heavy metal separation in BES; report the fundamentals for separating heavy metal in BES; discuss the mechanisms that include reduction, precipitation and adsorption of heavy metal separation in BES.

2. Fundamentals of Heavy Metal Separation Using Microbial Electrochemistry

2.1 Reduction Potential

In most of the microbial electrochemistry systems, organic substrates are oxidized at the anode to provide the electrons. The electrons flowed from the anode to the cathode where most of the reduction of heavy metals occurs. The reduction potential represents the possibility of a chemical species gaining the electrons. The typical cathode potential in MFCs is from 0.1 to 0.3 V vs. SHE (Logan et al. 2006). For the metal ions whose reduction potential is higher than the typical cathode potential, they can be reduced in MFCs and the reduction is spontaneous. For instance, the reduction of Co^{3+} to Co^{2+} (1.82 V vs. SHE), Au^{3+} to Au^{0} (1.00 V vs. SHE) and Cr^{6+} to Cr^{3+} (1.33 V vs. SHE) have been reported in MFCs (Huang et al. 2013; Choi and Hu 2013; Gangadharan and Nambi 2015). The cathode potential in MECs depends on the applied voltage and varies from -0.25 to -1.5 V (Logan et al. 2016; Huang et al. 2014; Colantonio and Kim 2016). The broad range of the cathode potentials makes the heavy metals, such as Cd^{2+}, Co^{2+} and Ni^{2+} whose reduction potential is lower than -0.4 V (Table 1), reduced in MECs (Wang et al. 2016; Jiang et al. 2014; Qin et al. 2012).

Table 1: Standard reduction potential of redox couples and theoretical reduction potential of redox couples when $[M_{ox}] = 1$ mM or $10 \mu M^{a}$

Redox couples	Standard (V vs. SHE)	1 mM (V vs. SHE)	10 μM (V vs. SHE)
Co^{3+}/Co^{2+}	1.82	1.88[b]	1.88[b]
Cr^{6+}/Cr^{3+}	1.33	0.39[b]	0.41[b]
Au^{3+}/Au^{0}	1.00	0.94	0.90
V^{6+}/V^{4+}	0.991	0.55[b]	0.55[b]
Hg^{2+}/Hg^{0}	0.855	0.77	0.71
Ag^{+}/Ag^{0}	0.799	0.62	0.51
Fe^{3+}/Fe^{2+}	0.77	0.74[b]	0.74[b]
Se^{4+}/Se^{0}	0.74	0.29	0.26
Cu^{2+}/Cu^{0}	0.34	0.25	0.19
U^{6+}/U^{4+}	0.327	-0.53[b]	-0.53[b]
Ni^{2+}/Ni^{0}	-0.257	-0.34	-0.40
Co^{2+}/Co^{0}	-0.28	-0.37	-0.43
Cd^{2+}/Cd^{0}	-0.4	-0.49	-0.55
Zn^{2+}/Zn^{0}	-0.76	-0.85	-0.91

a: $[M_{ox}]$: the concentration of oxidized heavy metals; pH=7; at 20°C
b: $M_{ox} = 0.1 M_{red}$ (the concentration of reduced heavy metals)

The redox potential, which is affected by different conditions such as pH and temperature, can be calculated using the Nernst equation (Equation 1, Sawyer et al. 1994).

$$E = E^0 - \frac{RT}{nF} \ln Q \tag{1}$$

In Equation 1, E is the half-cell redox potential at the operating conditions; E^0 is the standard half-cell redox potential; R is the gas constant (8.314 J/K/mol); T is the temperature (K); n is the number of electrons transferred in the half-cell reaction; F is the Faraday constant (9.6485×10^4 C/mol); Q is the reaction quotient. Table 1 shows the theoretical reduction potential of heavy metals with different concentrations at pH 7.

2.2 Precipitation

The oxygen reduction in MFCs (Equation 2) and the hydrogen production (Equation 3) in MECs can result in the high pH near the cathode (Van Phuong et al. 2011, 2012; Cusick and Logan 2012). Some heavy metal ions can easily form the precipitation with hydroxide (OH^-) at high pH. The equilibrium concentration of the heavy metal ions ($[M_{eq}]$) can be calculated based on the solubility product (K_{sp}). Table 2 shows a summary of the pK_{sp} value of hydroxide precipitation and $[M_{eq}]$ at pH 7 and 12. From Table 2, it can be seen that the formation of some heavy metal hydroxide precipitation requires a high pH condition. For example, $10^{-0.3}$ M (or 56 mg/L) Cd^{2+} can dissolve in the solution at pH 7, while only $10^{-11.7}$ M (or 1.12×10^{-10} mg/L) can dissolve at pH 12. For some other heavy metals such as Cu^{2+}, Hg^{2+} and Cr^{3+}, the dissolved concentration is low even at natural pH.

$$O_2 + 4H^+ + 4e^- \rightarrow 2H_2O \tag{2}$$

$$2H^+ + 2e^- \rightarrow H_2 \tag{3}$$

Table 2: Equilibrium concentration (M) of heavy metal ions $[M_{eq}]$ at pH 7 and pH 12 with hydroxide

Reactions	pK_{sp}[a]	$[M_{eq}]$ at pH = 7	$[M_{eq}]$ at pH = 12
$Co^{2+} + 2OH^- \rightarrow Co(OH)_2$ (s)	15.7	$10^{-1.7}$	$10^{-11.7}$
$Ni^{2+} + 2OH^- \rightarrow Ni(OH)_2$ (s)	17.2	$10^{-3.2}$	$10^{-13.2}$
$Cu^{2+} + 2OH^- \rightarrow Cu(OH)_2$ (s)	20.4	$10^{-6.4}$	$10^{-16.4}$
$Zn^{2+} + 2OH^- \rightarrow Zn(OH)_2$ (s)	16.8	$10^{-2.8}$	$10^{-12.8}$
$Hg^{2+} + 2OH^- \rightarrow Hg(OH)_2$ (s)	25.4	$10^{-11.4}$	$10^{-21.4}$
$Cd^{2+} + 2OH^- \rightarrow Cd(OH)_2$ (s)	14.3	$10^{-0.3}$	$10^{-10.3}$
$Ag^+ + OH^- \rightarrow Ag(OH)$ (s)	7.7	$10^{-0.7}$	$10^{-5.7}$
$Cr^{3+} + 3OH^- \rightarrow Cr(OH)_3$ (s)	30.0	10^{-9}	$10^{-24.0}$

a: Werener and Morgan (2012)

The carbonate precipitation is also possible in BES. The oxidation of organics at the anode can produce bicarbonate (HCO_3^-) (Equation 4). The HCO_3^- can convert to CO_3^{2-} with OH^- in the cathode chamber (Equation 5). The concentration of HCO_3^- and CO_3^{2-} depend on the pH. When the concentration of total carbonate ($C_{T, CO_3^{2-}}$) equals to 10 mM, the concentration of CO_3^{2-} is $10^{-5.4}$ M (pH 7) and 10^{-2} M (pH 12). With these CO_3^{2-}, some heavy metals are easy to form carbonate precipitation. Table 3 shows the summary of pK_{sp} of carbonate precipitation and $[M_{eq}]$ at different pH. From Table 3, it can be seen that some heavy metal ions such as Ni^{2+} and Ag^+ require a higher pH or higher $C_{T, CO_3^{2-}}$ to form the carbonate precipitation. Some heavy metal ions such as Cd^{2+} and Hg^{2+} can easily form the precipitation even with the low concentration at neutral pH.

$$Organics \rightarrow HCO_3^- + e^- \qquad (4)$$

$$HCO_3^- + OH^- \rightarrow CO_3^{2-} + H_2O \qquad (5)$$

Table 3: Equilibrium concentration (M) of heavy metal ions at pH 7 and pH 12 with $C_{T, CO_3^{2-}} = 10$ mM

Reactions	pK_{sp}[a]	pH = 7 ($[CO_3^{2-}] = 10^{-5.4}$ M)	pH = 12 ($[CO_3^{2-}] = 10^{-2}$ M)
$Co^{2+} + CO_3^{2-} \rightarrow CoCO_3$ (s)	10.0	$10^{-4.6}$	10^{-8}
$Ni^{2+} + CO_3^{2-} \rightarrow NiCO_3$ (s)	6.9	$10^{-1.5}$	$10^{-4.9}$
$Cu^{2+} + CO_3^{2-} \rightarrow CuCO_3$ (s)	9.6	$10^{-4.2}$	$10^{-7.6}$
$Zn^{2+} + CO_3^{2-} \rightarrow ZnCO_3$ (s)	10	$10^{-4.6}$	$10^{-8.0}$
$Hg^{2+} + CO_3^{2-} \rightarrow HgCO_3$ (s)	16.1	$10^{-10.7}$	$10^{-14.1}$
$Cd^{2+} + CO_3^{2-} \rightarrow CdCO_3$ (s)	13.7	$10^{-8.3}$	$10^{-11.7}$
$2Ag^+ + CO_3^{2-} \rightarrow Ag_2CO_3$ (s)	11.1	$10^{-2.85}$	$10^{-4.55}$

a: Werener and Morgan (2012)

3. Mechanisms of Heavy Metal Separation in BES

3.1 Overview

Reduction, precipitation and adsorption have been reported as the three mechanisms for heavy metal separation in BES (Table 4). The heavy metal ions can be reduced at the cathode (Figure 1A). They can also be reduced at the anode by the microorganisms (Figure 1B). Figure 1C shows the formation of hydroxide precipitation of heavy metals. In addition, the heavy metal ions can also combine with carbonate to form carbonate precipitation (Figure 1D). The adsorption by the electrode or biosorption by the microorganisms contributes to the heavy metal separation in BES as well.

Table 4: The summary of heavy metal separation in BES

Heavy metals	Concentration (mM)	Reactor mode	Removal efficiency	Mechanisms	Reference
Cadmium Cd^{2+}	0.1	MEC	71-91% in 48 hours	Cathodic reduction, Precipitation,	Colantonio and Kim (2016)
	0.45-1.8	MFC	89-93% in 60 hours	Cathodic reduction	Choi et al. (2014)
	0.45	MEC	39-47% in 4 hours	Cathodic reduction, Adsorption	Wang et al. (2016)
	10^{-4}-10^{-1} μM 0.02 mM	MEC	69% in 168 hours	Cathodic reduction, biosorption	Colantonio et al. (2016)
Cobalt (Co^{2+} or Co^{3+})	Co^{2+} 0.847	MEC	92% in 6 hours	Cathodic reduction, Adsorption	Jiang et al. (2014)
	$LiCoO_2$ (s) and Co^{2+} 0.34	MFC-MEC	88% in 6 hours	Cathodic reduction, Adsorption	Huang et al. (2014)
	$LiCoO_2$ (s)	MFC	62-70% in 48 hours	Cathodic reduction	Huang et al. (2013)
	Co^{2+} 0.36	MFC	93% in 6 hours	Cathodic reduction, Precipitation	Huang et al. (2015)

(Contd.)

Table 4: (*Contd.*)

Heavy metals	Concentration (mM)	Reactor mode	Removal efficiency	Mechanisms	Reference
Copper (Cu^{2+})	15.6	MFC	~100% in 168 hours	Cathodic reduction, Precipitation	Heijne et al. (2010)
	0.79-100	MFC	> 99% in 144 hours	Cathodic reduction	Tao et al. (2011a)
	31.25	MFC	90% in 24 hours	Cathodic reduction	Rodenas et al. (2015)
	0-100	MFC	70% in 144 hours	Cathodic reduction, Precipitation	Tao et al. (2011b)
	9.4, 31.3	MFC	92% in 480 hours	Cathodic reduction	Tao et al. (2011c)
	Cu(NH$_3$)$_4$$^{2+}$	MFC	84% in 8 hours	Cathodic reduction	Zhang et al. (2012)
	5-15	MFC	98% in 24 hours	Cathodic reduction	Wu et al. (2018)
Nickle (Ni^{2+})	0.85-17	MEC	33-99% in 19.8 hours	Cathodic reduction, Adsorption	Qin et al. (2012)
Mercury (Hg^{2+})	0.125-0.5	MFC	>94% in 5 hours	Cathodic reduction, Precipitation	Wang et al. (2011)
Gold (Au^{4+})	0-10.1	MFC	~100% in 12 hours	Cathodic reduction, Precipitation	Choi and Hu (2013)
Silver (Ag^{+})	0.46-1.84	MFC	>98% in 8 hours	Cathodic reduction	Choi and Cui (2012)
	AgNO$_3$ or AgS$_2$O$_3$	MFC	95% in 35 hours	Cathodic reduction	Tao et al. (2012)
	9.26	MFC	~100% in 21 hours	Cathodic reduction	Wang et.al (2013)
Vanadium (V^{5+})	9.8	MFC	25.3% in 72 hours	Cathodic reduction	Zhang et al. (2009)
	4-8	MFC	~100% in 168 hours	Cathodic reduction	Qiu et al. (2017)
	4.9, 9.8	MFC	67.9% in 240 hours	Cathodic reduction	Zhang et al. (2010)
	4.9-19.6	MFC	26.1% in 72 hours	Cathodic reduction	Zhang et al. (2010)
Zinc (Zn^{2+})	0.23-0.63	MEC	17-99% in 48 hours	Precipitation	Teng et al. (2016)
	1.54 -12.3	MEC	60%-99% in 23 hours	Cathodic reduction, Precipitation	Modin et al. (2017)
	0.2-0.5	MFC	94-99% in 48 hours	Precipitation. Biosorption	Abourached et al. (2014)
Selenium (Se^{4+})	0-5.1	MFC	~99%	Bio-reduction	Catal et al. (2009)
Chromium (Cr^{6+})	0.96-9.6	MFC	99.5% in 25 hours	Cathodic reduction	Li et al. (2008)
	0.48-3.84	MFC	~100% in 150 hours	Cathodic reduction	Wang et al. (2008)
	1.44-5.77	MFC	~100% in 48 hours	Cathodic reduction	Gangadharan and Nambi (2015)
	0.5	MFC	97% in 26 hours	Cathodic reduction	Li et al. (2009)
	1.8-19.2	MFC	19.2~100% in 14 days	Cathodic reduction, Precipitation	Kim et al. (2017)
	0.76	MFC	>80% in 4 hours	Cathodic reduction, precipitation	Huang et al. (2011)
	0.38-0.77	MFC	~100% in 7 hours	Cathodic reduction	Huang et al. (2010)
	0.42-1.2	MFC	~100% in 120 hours	Cathodic reduction, precipitation	Tandukar et al. (2009)
	0.19-0.57	MFC	97.5% in 4.5 hours	Cathodic reduction	Shi et al. (2017)
	0.19	MFC	~100% or 42.5% in 3.5 hours	Chemical reduction, Cathodic reduction	Liu et al. (2011)
	0-2.88	MFC	65.6% in 3 hours	Chemical reduction, Cathodic reduction	Wang et al. (2017)

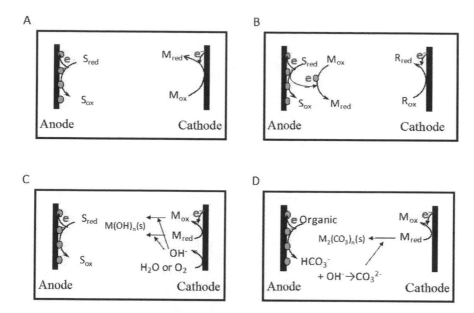

Figure 1: Mechanisms of heavy metal separation in BES. A. reduction at the cathode; B. reduction at the anode; C. precipitation with OH⁻; D. precipitation with CO_3^{2-}

3.2 Cobalt

Cobalt can be found in some minerals and they are mostly produced as a by-product of nickel refining. Cobalt has many industrial and medical applications. It is widely used in lithium-ion batteries where the cathode is made by lithium cobalt oxide ($LiCoO_2$). Various technologies such as ion exchange, chemical precipitation and solvent extraction have been used to recover cobalt (Marafi and Stanislaus 2008). BES was also proposed to separate the cobalt from aqueous solution.

Cobalt can be reduced at the cathode in BES. The reduction of Co^{3+} to Co^{2+} at the cathode is spontaneous since the redox potential of Co^{3+}/Co^{2+} couple is high (1.81 V vs. SHE). In the work conducted by Huang et al. (2013), the Co^{3+} from insoluble $LiCoO_2$ (s) was reduced to the soluble Co^{2+} in two chambers MFCs (Equation 6). The recovery efficiencies from 62.5% to 70.5% were achieved under different conditions such as initial pH and external resistor with 48 hours of operation (Table 4). To further recover the cobalt, the soluble Co^{2+} needs to be reduced to insoluble Co^0 (Equation 7). This reduction demands an external power supply due to the low redox potential of Co^{2+}/Co^0 (-0.232 V SHE). The reduction of Co^{2+} to Co^0 was enhanced by using biocathode in the studies of Huang et al. (2014) and Jiang et al. (2014). In addition, the applied voltage can affect the reduction of Co^{2+} and a higher applied voltage condition results in a higher reduction rate. However, when the applied voltage was larger than 0.5 V, more electrons were provided for hydrogen evolution instead of cobalt reduction. Therefore, the optimal applied voltage of Co^{2+} reduction was 0.3 to 0.5 V (Jiang et al. 2014). A self-drive system that the $LiCoO_2/Co^{2+}$ MFC was used to provide energy for the Co^{2+}/Co^0 MEC was proposed by Huang et al. (2014). In their study, 46 ± 2 mg/L·h of cobalt leaching rate was achieved in the MFC when 7 ± 0 mg/L·h cobalt reduction rate was shown in the MEC. The overall cobalt yield was 0.15 ± 0.01 g Co/ g Co.

$$LiCoO_2 + 4H^+ + e^- \rightarrow Li^+ + Co^{2+} + 2H_2O \qquad (6)$$

$$Co^{2+} + 2e^- \rightarrow Co \qquad (7)$$

Co^{2+} can be separated by precipitation in BES as well. Co^{2+} ions can easily combine with OH⁻ and CO_3^{2-} to form the precipitation of $Co(OH)_2$ and $Co(CO)_3$ because of the low K_{sp} (Table 2 and

Table 3). Huang et al. (2015) reported that 93.3% of cobalt was removed by $Co(OH)_2$ precipitation in MFCs within 6 hours operation. The cobalt precipitation was formed on bacterial surfaces that demonstrated the contribution of the oxygen-reducing biocathode. There was no $Co(CO)_3$ formed in these MECs because the cation exchange membrane limited the transfer of HCO_3^- between the anode chamber and the cathode chamber. However, the precipitation of $Co(CO)_3$ in BES without the membrane is possible. It can be seen from Table 3 that only $10^{-4.6}$ M Co^{2+} (1.5 mg-Co/L) can dissolve in the water with $C_{T, CO_3^{2-}}$ of 10 mM at pH 7. The concentration decrease to 10^{-8} M (0.59 μg-Co/L) at pH 12.

The mechanism of adsorption has been reported by Huang et al. (2014) and Jiang et al. (2014). 46.1% and 27% of cobalt were removed under open circuit conditions in their studies.

3.3 Cadmium

Cadmium is toxic and has been listed as a group-B1 carcinogen element by the US Environmental Protection Agency (EPA; Purkayastha et al. 2014). The sources of cadmium in the environment include coal combustion, iron and steel production, usage of NiCd batteries and electroplating. A number of technologies such as precipitation, coagulation and membrane filtration have been applied to remove cadmium (Purkayastha et al. 2014). BES was also used to separate cadmium from aqueous solution.

Cadmium removal in BES can be achieved by the cathodic reduction (Equation 8). The reduction of Cd^{2+} to Cd^0 requires an external power supply because of the low redox potential of Cd^{2+}/Cd^0 (-0.403 V vs. SHE). Choi et al. (2014) used Cr^{6+} as a cathodic reactant to remove Cd^{2+} in an MFC. 94.5% of Cd^{2+} was removed within 60 hours of operation in the study. An enhanced reduction rate was reported in the study of Wang et al. (2016) by using the deposited Cu cathodes. Cd^{2+} removal rates in the MECs with deposited Cu (4.96-5.86 mg/L·h with different cathode materials) were 1.8-4.2 times higher than that in the MECs without deposited Cu (1.18-3.26 mg/L·h).

$$Cd^{2+} + 2e^- \rightarrow Cd \tag{8}$$

Precipitation of cadmium hydroxide ($Cd(OH)_2$) and cadmium carbonate ($CdCO_3$) was reported by Colantonio and Kim (2016). They found that the precipitation was responsible for more than 60% of cadmium removal under the applied voltage of 0.4 V in a single chamber MEC. To avoid the dissociation of $Cd(OH)_2$ and $CdCO_3$, sufficient amounts of substrates were suggested to feed to the reactor in their study. From Table 2 and Table 3, $10^{-8.3}$ M Cd^{2+} (0.56 μg-Cd/L) can result in the formation of $CdCO_3$, while $Cd(OH)_2$ can be formed with $10^{-0.3}$ M Cd^{2+} (56 mg-Cd/L) at pH 7. At pH 12, the equilibrium concentration of Cd^{2+} decreases to $10^{-10.3}$ M (5.6×10^{-3} μg/L).

In addition, the separation of cadmium by biosorption at the anode was demonstrated in an MFC by Abourached et al. (2014). The precipitation process of Cd^{2+} on the cathode is difficult when oxygen exists in the MFC because the standard redox potential of Cd^{2+} is much lower than O_2. However, more than 89% of cadmium was removed in the study of Abourached et al. (2014) by using the air-cathode MFCs. This result can be explained by the biosorption of Cd^{2+} at the anode. The performance of the MFC was affected by the initial cadmium concentration. The maximum tolerable concentration of cadmium was 200 μM in MFCs. Cadmium removal by anode biosorption was also reported in MECs with low cadmium concentration in the study conducted by Colantonio et al. (2016). In their study, 59.3%, 6.3% and 4.4% of cadmium were removed by anode biosorption with different initial Cd^{2+} concentration.

3.4 Copper

Copper is an essential element to living organisms because it is a key constituent of the respiratory enzyme complex. However, a high dose of copper is toxic to all life forms. Copper commonly exists in the effluent from electronics plating, wire drawing, copper polishing and paint (Zamani et al. 2007). Numerous treatment technologies, which include adsorption, biosorption and co-precipitation with calcium carbonate, are available for copper removal and recovery (Aston et al. 2010; Khosravi and Alamdari 2009).

Copper can be separated by reduction at the cathode in BES. The reduction of Cu^{2+} to Cu^0 at the cathode occurs spontaneously because of the positive standard redox potential of Cu^{2+}/Cu^0 couple (0.337 V vs. SHE). There are two major ways for the reduction of Cu^{2+}: Cu^{2+} directly reduces to Cu^0 at the cathode (Equation 9); Cu^{2+} first reduces to Cu_2O and then Cu_2O reduces to Cu^0 (Equation 10 and 11). Although the standard reduction potential of Cu^{2+}/Cu^0 (0.337 V vs. SHE) is higher than that of Cu^{2+}/Cu_2O (0.207 V vs. SHE), the reduction of Cu^{2+} to Cu_2O is more favorable when pH > 4.7 based on the Nernst equations (Tao et al. 2011a). Both metallic copper Cu^0 and cuprous oxide (Cu_2O) appeared on the cathode in studies conducted by Tao et al. (2011a) and Wu et al. (2018). The formation of Cu_2O can be governed by pH. It was shown that only metallic Cu was formed on the cathode by controlling the pH < 3 in studies of Heijne et al. (2010) and Rodenas et al. (2015). In addition to the pH, the formation of Cu_2O can be controlled by providing enough electrons (substrates) since the formation of Cu requires one more electron than the formation of Cu_2O (Tao et al. 2011c).

$$Cu^{2+} + 2e^- = Cu(s) \tag{9}$$

$$2Cu^{2+} + 2H_2O + 2e^- = Cu_2O + 2H^+ \tag{10}$$

$$Cu_2O + 2H^+ + 2e^- = 2Cu + H_2O \tag{11}$$

Apart from Cu^{2+}, some copper complexes can also be reduced at the cathode. For instance, ammonia-copper complexes ($Cu(NH_3)_4^{2+}$) was fed to a dual chamber MFC in the study of Zhang et al. (2012). The reduction of $Cu(NH_3)_4^{2+}$ can be achieved in two ways: Cu^{2+} that released from $Cu(NH_3)_4^{2+}$ is reduced to Cu or Cu_2O directly; $Cu(NH_3)_4^{2+}$ accepted an electron to form $Cu(NH_3)_4^+$ and then $Cu(NH_3)_4^+$ is deposited as Cu or Cu_2O on the cathode. These two ways were affected by the pH. $Cu(NH_3)_4^{2+}$ was dissociated to Cu^{2+} at pH < 5.34, while it was reduced to $Cu(NH_3)_4^+$ at pH > 8.83. In the study conducted by Zhang et al. (2012), 84% of copper was removed with the initial concentration of 350 mg-Cu/L at pH 3.0 after 8 hours.

Copper can also be separated by precipitation in BES because of the low K_{sp} of $Cu(OH)_2$ and $CuCO_3$ (Table 2 and Table 3). For instance, $10^{-6.4}$ M Cu^{2+} (25.5 µg-Cu/L) can result in the formation of $Cu(OH)_2$ at pH 7. In addition, the formation of insoluble copper complex brochantite ($CuSO_4 \cdot 3Cu(OH)_2$; Equation 12) also contributed to the separation of Cu^{2+}. This complex was detected on the cathode in the studies of Tao et al. (2011a, 2011b). Brochantite can be formed at the high initial Cu^{2+} concentration (~6400 mg/L). With the high concentration of Cu^{2+}, low removal efficiencies of 18.6% and 28.1% were shown in a dual chamber MFC and a membrane-free MFC, respectively (Tao et al. 2011a, Tao et al. 2011b). The precipitation of copper sulfide was also observed in the study of Miran et al. (2017).

$$4Cu^{2+} + 6H_2O + SO_4^{2-} = CuSO_4 \cdot 3Cu(OH)_2 + 6H^+ \tag{12}$$

The mechanism of anode biosorption was reported by Tao et al. (2011a). In this study, the Cu^{2+} concentration in the anode chamber decreased from 5.3 mg/L to 0.17 mg/L.

3.5 Mercury

Mercury is harmful and toxic to human beings. The major sources of mercury contaminations include dental practice wastes, fertilizers, pulp paper wastes and coal combustors used in electricity generation (Baeyens et al. 2016; Bender 2008; Morimoto et al. 2005). Many technologies such as activated carbon adsorption, ion exchange and precipitation have been used to remove the mercury (Monteagudo and Ortiz 2000; Hutchison et al. 2008).

Similar to cobalt and cadmium, the separation of Hg^{2+} was also achieved in BES. The reduction of Hg^{2+} to Hg^0 in BES is spontaneous since the Hg^{2+}/Hg^0 couple has a high redox potential (0.851 V vs. SHE). The reduction at the cathode has two ways: Hg^{2+} is reduced to Hg^0 directly (Equation 13); Hg^{2+} is reduced to Hg_2^{2+} first and then Hg_2^{2+} is reduced to Hg^0 (Equation 14 and Equation 15). In the study of Wang et al. (2011), the removal efficiency larger than 94% was achieved under different conditions (e.g., initial pH, initial Hg concentration).

$$Hg^{2+} + 2e^- \rightarrow Hg\,(l)\,(0.851\,V\,vs.\,SHE) \tag{13}$$

$$2Hg^{2+} + 2e^- \rightarrow Hg_2^{2+}\,(l)\,(0.911\,V\,vs.\,SHE) \tag{14}$$

$$Hg_2^{2+} + 2e^- \rightarrow 2Hg\,(l)\,(0.796\,V\,vs.\,SHE) \tag{15}$$

Precipitation also contributed to the separation of Hg^{2+} since Hg_2Cl_2 was detected on the cathode in the study of Wang et al. (2011). There were no $Hg(OH)_2$ and $HgCO_3$ formed in the system because of the low initial pH. However, the precipitation of these two precipitates is possible in BES because of the low K_{sp} of $Hg(OH)_2$ and $HgCO_3$ (Table 2 and Table 3). For instance, less than $10^{-10.7}$ M Cu^{2+} (4×10^{-3} μg-Hg/L) can dissolve in the solution at pH 7. The mechanism of precipitation of $Hg(OH)_2$ and $HgCO_3$ can be investigated in the future study.

3.6 Gold

Gold is widely used in the electronics industry because of its great electrical conductivity and outstanding corrosion resistance. It is commonly present in leach solutions and electroplating wastes (Flores and Okeefe 1995). Many methods such as ion exchange and biosorption can be used to separate gold from wastewater (Gomes et al. 2001; Das 2010).

Gold ions (Au^{3+}) can deposit on the cathode by reducing Au^{3+} in BES (Equation 16). The transfer of electrons from the cathode to Au^{3+} is spontaneous because of the high standard redox potential of Au^{3+}/Au^0 (1.002 V vs. SHE). Choi and Hu (2013) reported that 99.89% of gold was recovered in a cubical dual chamber MFC within 12 hours. Also, pH and initial Au^{3+} concentration can affect the removal efficiency.

$$AuCl^{4-} + 3e = Au(s) + 4Cl^- \tag{16}$$

Au^{3+} is likely to form precipitates of $Au(OH)_{3(s)}$ at high pH conditions, allowing separation of gold. In the study of Choi and Hu (2013), the power density of MFC decreased from 1.37 to 0.78 W/m² with the pH increasing from 2 to 5. The low current density was due to the low conductivity of the catholyte with the $Au(OH)_3$ formation.

3.7 Silver

Silver is widely used in jewelry, electronics and photographic industries. It is a precious metal and only exists in nature with limited amounts. The high concentration of some silver compounds is toxic to aquatic life (Naddy et al. 2007). Several methods such as ion exchange, chemical reduction and electrolysis are available to recover or remove silver from aqueous solution (Blondeau and Veron 2010).

Separation of Ag^+ can be achieved by reducing Ag^+ to Ag^0 at the cathode in BES. The reduction of Ag^+ to Ag is spontaneous because of the high redox potential of Ag^+/Ag^0 (0.799 V vs. SHE; Equation 17). In the study of Choi and Cui (2012), more than 98% of silver was recovered in the MFC with different initial concentrations of $AgNO_3$ (50 to 200 mg/L) in 8 hours. High removal efficiencies (> 89%) were shown using MFCs that started with 1 mM $AgNO_3$ (170 mg/L) with pH ranging from 2 to 6.6 (Tao et al. 2012). In addition to $AgNO_3$, some silver complex compounds such as silver thiosulfate($[AgS_2O_3]^-$) and diamine silver ($[Ag(NH_3)_2^+]$) were used as cathodic solutions in the studies of Tao et al. (2012) and Wang et al. (2013). The slower removal rates (95% in 35 hours and 99.9% in 21 hours; Table 4) were shown for complexes reduction since the redox potentials of $[AgS_2O_3]^-/Ag\,(0)$ (0.25 V vs. SHE) and $[Ag(NH_3)_2^+]/Ag^0$ (0.373 V vs. SHE) were lower (Equation18, Equation 19 and Equation 20) than that of Ag^+/Ag^0. Through the SEM-EDS analysis, Ag^0 crystals were shown on the cathode. A small fraction of Ag_2S was also detected at pH 4.0 and 6.5 with $[AgS_2O_3]^-$ reduction (Equation 19).

$$Ag^+ + e^- = Ag(s) \tag{17}$$

$$[AgS_2O_3]^- + e^- = Ag(s) + S_2O_3^{3-} \tag{18}$$

$$[AgS_2O_3]^- + 8e^- + 6H^+ = 2Ag2S(s) + 3H_2O + 3S_2O^{3-} \tag{19}$$

$$[Ag(NH_3)_2]^+ + e^- = Ag(s) + 2NH_3 \tag{20}$$

The precipitation of Ag_2CO_3 and AgOH requires high pH conditions. At pH 7, the solubility of Ag^+ is $10^{-2.85}$ M (152 mg-Ag/L) and $10^{-0.7}$ M (21.5 g-Ag/L) with CO_3^{2-} and OH^-, respectively. Therefore, the formation of Ag_2CO_3 and AgOH is difficult at pH 7. The solubility of Ag^+ is smaller than $10^{-5.7}$ M (0.2 mg-Ag/L) at pH 12. The low solubility makes the formation of Ag_2CO_3 and AgOH possible. There were no previous BES related to the results of AgOH or Ag_2CO_3 precipitation. The precipitation can be investigated in the future as a potential method to separate silver in BES.

3.8 Vanadium

Vanadium is the main pollutant found in wastewater from vanadium mining and vanadium pentoxide (V_2O_5) production (Bauer et al. 2000). Most of the vanadium is used as a steel additive and vanadium-steel alloys for tools, piston rods and armor plates. Vanadium is a trace element in living organisms. However, some vanadium compounds are harmful to people and sometimes can be even fatal. V^{5+} is more toxic than V^{4+} that is insoluble at high pH (neutral or alkaline pH). Therefore, the reduction of V^{5+} to V^{4+} is an applicable method to remove or recover vanadium from aqueous solution.

Reduction of V^{5+} to V^{4+} is spontaneous because the V^{5+}/V^{4+} couple has a high positive redox potential (0.991 V vs. SHE; Equation 21). The removal efficiency ranged from 25 to 100% was achieved in previous studies (Table 4; Zhang et al. 2009; Zhang et al. 2012; Zhang et al. 2010; Qiu et al. 2017). The color of cathodic electrolyte changed from yellow-brown to sky-blue during the reduction process of V^{5+} to V^{4+}. In order to separate V^{4+} from cathodic solution, the pH of the cathodic electrolyte was adjusted to 6 by using $NH_3 \cdot H_2O$. During the pH increasing process, the color changed from sky-blue to dark grey due to the reoxidization of some V^{4+}. Both Cr^{6+} and V^{5+} were reduced in the study of Zhang et al. (2012) that contributed to a better MFC performance. The recovery of Cr^{6+} and V^{5+} can be separated since chromium was mainly deposited on the cathode surface, while vanadium stayed in the catholyte. The vanadium can be removed later by increasing the pH.

$$VO_2^{2+} + e^- + 2H^+ = VO^{2+} + H_2O \tag{21}$$

3.9 Chromium

The main sources of hexavalent chromium in the environment are from electroplating, leather tanning and wood product processes (Jadhav et al. 2012). Methods such as chemical precipitation, ion exchange, membrane filtration and biosorption have been developed to treat chromium in aqueous solution (Kurniawan et al. 2006; Quintelas et al. 2006). Cr(III) is the form of chromium with less toxicity and less solubility compared to Cr^{6+}. Thus, the reduction of Cr(VI) to Cr(III) is regarded as a safe and efficient process for Cr(VI) removal.

The reduction of Cr(VI) is achievable at the cathode of MFCs since the redox potential of Cr(VI)/Cr(III) couple is high (1.33 V vs. SHE; Equation 22). There are three different ways for Cr(VI) reduction at the cathode: Cr(VI) is reduced at the abiotic cathode directly; Cr(VI) is reduced by microorganisms at the biocathode; the cathode products such as H_2O_2 can also reduce Cr(VI). 99.5% of Cr(VI) that was removed in a dual chamber MFC with abiotic cathode after 25 hours treatment in the study of Li et al. (2008). The removal efficiency was affected by the pH and initial concentration of Cr(VI). The similar removal efficiency was shown in previous studies by Wang et al. (2008) and Gangadharan and Nambi (2015). The cathode materials also affect the reduction rates at the cathode. Li et al. (2009) improved the reduction rates and power generation by using the rutile-coated cathode. In addition various removal efficiencies (19.2-100%) were shown in MFCs fed with chromium wastewater (Kim et al. 2017).

$$Cr^2O_7^{2-} + 4e^- + 8H^+ = Cr^2O_3 + 4H_2O \tag{22}$$

The reduction of Cr(VI) was also investigated in several MFCs with biocathode. In these MFCs, the cathode chambers were inoculated with various mediums that included healthy MFC effluent (Huang et al. 2011a), primary clarifier effluent (Huang et al. 2011b), Cr(VI) reducing bacteria cultures from Cr^{6+} contaminated soil (Huang et al. 2010) and a mixture of denitrifying and anaerobic mixed cultures (Tandukar et al. 2009). The removal efficiencies of Cr(VI) ranged from 60 to 100% (i.e., 0.46-20.4 mg/VSS·h) in these MFCs. In addition to the inoculated mediums, many other factors can also affect the removal efficiency of Cr(VI) in MFCs. For instance, Huang et al. (2010) evaluated the effects of different biocathode materials, i.e., graphite fibers, graphite felt and graphite granules. The highest removal efficiency of 90.2% was achieved in the MFC with graphite fibers. The removal efficiency of 97.5% within 4.5 hours was achieved in the study of Shi et al. (2017) by using the natural pyrrhotite-coated cathode. The effects of pH and initial concentration of Cr(VI) were studied by Huang et al. (2011b) and Huang et al. (2010). In addition, Liu et al. (2011) evaluated the effects of different inoculums for the anode on Cr(VI) removal efficiency. Among the MFCs that inoculated with *Shewanella decolorationis* S12, *Klebsiella pneumonia* L17 and anaerobic activated sludge, the MFC inoculated with anaerobic activated sludge had the best performance with 97% Cr(VI) removal efficiency after 3 hours operation.

Cr(VI) can also be reduced by hydrogen peroxide that was produced at the cathode with oxygen reduction (Equation 23). The mechanism was demonstrated in the study of Liu et al. (2011) with an air-bubbling-cathode MFC and a nitrogen-bubbling-cathode MFC. A complete reduction of Cr(VI) was achieved in the air-bubbling-cathode MFC after 4 hours of operation, while only 42.5% of Cr(VI) was reduced in the MFC without air. The faster removal rate of Cr(VI) with the air-bubbling cathode MFC can be explained by the production of H_2O_2 that was electrochemically generated via the reaction of O_2 reduction (Equation 24). After adding H_2O_2, the increased Cr(VI) removal rates of the nitrogen-bubbling cathode MFC indicated the contribution of H_2O_2. Similar to H_2O_2, Fe^{3+} was used as an electron shuttle mediator to enhance the reduction of Cr(VI) in the study of Wang et al. (2017).

$$O_2 + 2H^+ + 2e^- \rightarrow H_2O_2 \tag{23}$$

$$2HCrO^{4-} + 3H_2O_2 + 8H^+ = 2Cr^{3+} + 8H_2O + 3O_2 \tag{24}$$

After reduction, Cr^{3+} ions can easily combine with OH⁻ to form the precipitates in BES (Table 2). $Cr(OH)_3$ was detected in many previous studies (Tandukar et al. 2009; Kim et al. 2017; Huang et al. 2011a). The formation of $Cr(OH)_3$ was highly pH dependent, and it was usually generated at pH from 6.5 to 10.

3.10 Nickel

Nickel is the fifth most common element on the earth. The dominated use of nickel is the production of ferronickel for stainless steel (Reck et al. 2008). It is also widely used for producing batteries, alloy steels and non-ferrous alloys. A high dose of nickel can cause various pathological effects such as kidney diseases, lung fibrosis and even cancer in humans (Denkhaus and Salnikow 2002). Various technologies, including chemical precipitation, ion exchange, membrane filtration and adsorption, were investigated for nickel recovery and removal. (Papadopoulos et al. 2004; Chen et al. 2009; Landaburu-Aguirre et al. 2012).

BES can also be used to separate nickel from aqueous solution by reducing Ni^{2+} to Ni^0. Direct reduction of soluble Ni^{2+} to metal Ni^0 at the cathode requires an external power supply because of the negative standard redox potential (-0.25 V vs. SHE; Equation 25). Ni^{2+} recovery from a nickel sulfate solution was studied in an MEC with different applied voltages from 0.5 V to 1.1 V (Qin et al. 2012). The maximum removal efficiency of $67 \pm 5.3\%$ was achieved under the applied voltage of 1.1 V. The pH and initial Ni^{2+} concentration can affect the performance of MEC. 87% of Ni^{2+} was removed in the single test with the initial concentration of 530 mg-Ni^{2+}/L when the MEC fed with artificial acid mine drainage in the study of Luo et al. (2014). For the mixed metal (nickel, copper and iron) test, copper deposited on the cathode first and was followed by nickel and ferric.

$$Ni^{2+} + 2e^- \rightarrow Ni \tag{25}$$

Precipitation is the potential mechanism of nickel separation in BES although it has not been reported in previous studies. The precipitation of $Ni(OH)_2$ and $NiCO_3$ requires high pH conditions. At pH 7, $10^{-3.2}$ M Ni^{2+} (37 mg/L) can dissolve in the water based on the K_{sp} of $Ni(OH)_2$ (Table 2). The high solubility of $10^{-1.5}$ M Ni^{2+} (1.86 g/L) was shown for $NiCO_3$ because of the high Ksp. At pH 12, the negligible amount of Ni^{2+} ($10^{-13.2}$ M Ni^{2+} or $10^{-5.4}$ µg/L) can dissolve in the water because of the formation of $Ni(OH)_2$.

Adsorption at the cathode also contributes to the nickel removal that was proved by Qin et al. (2012). The nickel removal efficiency of $9 \pm 0.1\%$ was achieved in an MEC under open circuit conditions.

3.11 Zinc

Zinc is an essential trace element for living things, but it can be carcinogenic in excess. Various technologies that include ion exchange, precipitation and adsorption are used to remove or recover zinc from aqueous solution (Alyüz et al. 2009; Chen et al. 2011).

Direct reduction of zinc at the cathode in a BES system requires an external power supply because of the low standard redox potential (-0.764 vs. SHE; Equation 26). Reduction of zinc at the cathode was investigated in an MEC using sodium acetate as the substrate (Modin et al. 2017). 60-99% of zinc was removed with different operating conditions, such as initial zinc concentration and catholyte type. The removal efficiency can be affected by the hydrogen generation since Zn^{2+} was reduced at the cathode potential of -1.0 V that was close to the potential when hydrogen generation occurred.

$$Zn^{2+} + 2e^- \rightarrow Zn \tag{26}$$

The separation of zinc by precipitation of $Zn(OH)_2$ and ZnS has been reported in previous studies (Modin et al.2017; Teng et al. 2016; Abourached et al. 2014). The formation of $Zn(OH)_2$ contributed to the bulk removal of zinc in the study of Teng et al. (2016) since the reduction of Zn^{2+} was difficult at the cathode when sulfide existed in the system. Up to 99% of Zn^{2+} was removed with the initial concentration ranged from 10 to 40 mg-Zn/L. This result is consistent with the study of Abourached et al. (2014) that up to 99% of Zn^{2+} was removed by the precipitation of ZnS with the initial concentration ranged from 13 to 32.5 mg-Zn/L. In addition, the formation of $ZnCO_3$ is possible in BES because of the low K_{sp}. At pH 7, $10^{-4.6}$ M Zn^{2+} (1.4 mg/L) can result in the precipitation of $ZnCO_3$, while 10^{-8} M Zn^{2+} (0.56 µg/L) can result in the precipitation at pH 12.

Zinc also can be separated from aqueous solution by biosorption at the anode. In the study of Abourached et al. (2014), the removal of zinc was studied in an air-cathode MFC. When the oxygen existed in the system, Zn^{2+} was difficult to reduce because of the low standard redox potential (-0.764 V vs. SHE). However, more than 94% of zinc was removed in this MFC, and this high removal efficiency was due to the biosorption at the anode. In addition, the lower removal efficiency was shown in the autoclave MFC with nonliving microbial cells that indicated that the microorganisms played an important role in zinc removal in MFCs.

3.12 Selenium

Selenium is an essential trace element for some species, but large amounts of selenium are toxic. Selenium is used in various industrial products and processes, such as pigments, electronics, photoelectric cells and glass manufacturing. The current methods for removing selenium include precipitation, adsorption, ion exchange and reduction (Twidwell et al. 1999).

The separation of selenium was studied in the single chamber MFCs by reducing Se^{4+} to Se^0. Se^{4+} can be reduced spontaneously since the redox potential of Se^{4+}/Se^0 is +0.41 V vs. SHE. However, there was no electric current generated when the MFC was converted from aerobic condition to anaerobic condition in the study of Catal et al. (2009). This result indicated that oxygen was reduced on the cathode, while the reduction of Se^{4+} to Se^0 was caused by the respiring microorganisms at the anode. 99% of 50 mg/L selenite was removed in 48 hours with acetate as the substrate. Also, 99% of 200 mg/L selenite was removed in 72 hours with glucose as the substrate. In addition, Lee et al. (2007) and Banuelos et al.

(2013) illustrated that *Shewanella* species, well-known for their capability to generate current in MFCs, can use selenite as the electron acceptor.

4. Conclusions

BES shows great efficiency for separating heavy metals from aqueous solution. Many factors govern the removal efficiency, such as the operation time, pH, initial concentration and cathode potential (MFC) or applied voltage (MEC). The mechanism of cathodic reduction has been demonstrated for metal separations in many previous studies. In addition, the mechanism of precipitation can also contribute to the metal separation in BES since some heavy metals are easy to form precipitates with OH^- or CO_3^{2-}. For instance, Hg^{2+}, Cu^{2+}, Au^{3+}, Cd^{2+} and Cr^{3+} can form the precipitants even at neutral pH with low metal ion concentration. Some heavy metals such as Co^{2+}, Ni^{2+}, Zn^{2+} and Ag^+ require a high pH condition. Thus, it is still possible for the metals to form precipitants near the BES cathode since the oxygen reduction in MFC and hydrogen production in MEC can result in high pH conditions near the BES cathode. The mechanism of precipitation has been investigated in some previous studies, but more future studies are needed to understand this mechanism.

5. Future Perspectives

There were many previous studies focused heavily on the reduction mechanism of the heavy metal separation in BES. However, only few studies discussed the mechanism of precipitation. The electrode reactions such as hydroxide production at the cathode and carbonate production at the anode enhance the separation of heavy metal by precipitation. Therefore, we suggest that future studies should investigate the mechanism of precipitation for heavy metal separation in BES.

Acknowledgement

This study was supported by Early Researcher Awards (ER16-12-126, Ontario Ministry of Research and Innovation), Canada Research Chairs Program (950-2320518, Government of Canada), Discovery Grants (435547-2013, Natural Sciences and Engineering Research Council of Canada), Leaders Opportunity Fund (31604, Canada Foundation for Innovation), and Ontario Research Fund-Research Infrastructure (31604, Ontario Ministry of Economic Development and Innovation).

References

Abourached C, Catal T, Liu H (2014) Efficacy of single chamber microbial fuel cells for removal of cadmium and zinc with simultaneous electricity production. Water Research 51:228-33.

Achparaki M, Thessalonikeos E, Tsoukali H, Mastrogianni O, Zaggelidou E, Chatzinikolaou F, Vasilliades N, Raikos N (2012) Heavy metals toxicity. Aristotle University Medical Journal 39(1):29-34.

Alyüz B, Veli S (2009) Kinetics and equilibrium studies for the removal of nickel and zinc from aqueous solutions by ion exchange resins. Journal of Hazardous Materials 167(1-3):482-8.

Aston JE, Apel WA, Lee BD, Peyton BM (2010) Effects of cell condition, pH, and temperature on lead, zinc, and copper sorption to Acidithiobacillus caldus strain BC13. Journal of Hazardous Materials 184(1-3):34-41.

Baeyens W, Ebinghaus R, Vasiliev O, editors (2016) Global and regional mercury cycles: sources, fluxes and mass balances. Springer Science & Business Media.

Bauer G, Güther V, Hess H, Otto A, Roidl O, Roller H, Sattelberger S, Köther-Becker S, Beyer T (2000) Vanadium and vanadium compounds. Ullmann's Encyclopedia of Industrial Chemistry.

Banuelos GS, Lin ZQ, Yin X (2013) Selenium in the environment and human health. CRC Press.

Bender M (2008) Facing Up to the Hazards of Mercury Tooth Fillings. Mercury.

Blondeau JP, Veron O (2010) Precipitation of silver nanoparticles in glass by multiple wavelength nanosecond laser irradiation. Journal of Optoelectronics and Advanced Materials 445-50.

Call D, Logan BE (2008) Hydrogen production in a single chamber microbial electrolysis cell lacking a membrane. Environmental Science & Technology 42(9):3401-6.

Catal T, Bermek H, Liu H (2009) Removal of selenite from wastewater using microbial fuel cells. Biotechnology Letters 31(8):1211-6.

Chen C, Hu J, Shao D, Li J, Wang X (2009) Adsorption behavior of multiwall carbon nanotube/iron oxide magnetic composites for Ni (II) and Sr (II). Journal of Hazardous Materials 164(2-3):923-8.

Chen X, Chen G, Chen L, Chen Y, Lehmann J, McBride MB, Hay AG (2011) Adsorption of copper and zinc by biochars produced from pyrolysis of hardwood and corn straw in aqueous solution. Bioresource Technology 102(19):8877-84.

Choi C, Cui Y (2012) Recovery of silver from wastewater coupled with power generation using a microbial fuel cell. Bioresource Technology 107:522-5.

Choi C, Hu N (2013) The modeling of gold recovery from tetrachloroaurate wastewater using a microbial fuel cell. Bioresource Technology 133:589-98.

Choi C, Hu N, Lim B (2014) Cadmium recovery by coupling double microbial fuel cells. Bioresource Technology 170:361-9.

Colantonio N, Guo H, Kim Y (2016) Effect of Low Cadmium Concentration on the Removal Efficiency and Mechanisms in Microbial Electrolysis Cells. ChemistrySelect 1(21):6920-4.

Colantonio N, Kim Y (2016) Cadmium (II) removal mechanisms in microbial electrolysis cells. Journal of Hazardous Materials 311:134-41.

Cusick RD, Logan BE (2012) Phosphate recovery as struvite within a single chamber microbial electrolysis cell. Bioresource Technology 107:110-5.

Das N (2010) Recovery of precious metals through biosorption—a review. Hydrometallurgy 103(1-4):180-9.

Denkhaus E, Salnikow K. Nickel essentiality, toxicity, and carcinogenicity (2002) Critical reviews in oncology/hematology 42(1):35-56.

Dominguez-Benetton X, Varia JC, Pozo G, Modin O, ter-Heijne A, Fransaer J, Rabaey K (2018) Metal recovery by microbial electro-metallurgy. Progress in Materials Science.

ElMekawy A, Srikanth S, Bajracharya S, Hegab HM, Nigam PS, Singh A, Mohan SV, Pant D (2015). Food and agricultural wastes as substrates for bioelectrochemical system (BES): the synchronized recovery of sustainable energy and waste treatment. Food Research International 73:213-25.

Fergusson JE (1990). Heavy elements: chemistry, environmental impact and health effects. Pergamon.

Flores C, OKeefe TJ (1995) Gold recovery from organic solvents using galvanic stripping. Minerals, Metals and Materials Society, Warrendale, PA (United States).

Fu F, Wang Q (2011) Removal of heavy metal ions from wastewaters: a review. Journal of Environmental Management 92(3):407-18.

Gangadharan P, Nambi IM (2015) Hexavalent chromium reduction and energy recovery by using dual chambered microbial fuel cell. Water Science and Technology 71(3):353-8.

Giannopoulou I, Panias D (2008) Differential precipitation of copper and nickel from acidic polymetallic aqueous solutions. Hydrometallurgy 90(2-4):137-46.

Giller KE, Witter E, Mcgrath SP (1998) Toxicity of heavy metals to microorganisms and microbial processes in agricultural soils: a review. Soil Biology and Biochemistry 30(10-11):1389-414.

Gomes CP, Almeida MF, Loureiro JM (2001) Gold recovery with ion exchange used resins. Separation and Purification Technology 24(1-2):35-57.

Heijne AT, Liu F, Weijden RV, Weijma J, Buisman CJ, Hamelers HV (2010) Copper recovery combined with electricity production in a microbial fuel cell. Environmental Science & Technology 44(11):4376-81.

He J, Chen JP (2014) A comprehensive review on biosorption of heavy metals by algal biomass: materials, performances, chemistry, and modeling simulation tools. Bioresource Technology 160:67-78.

He Z, Minteer SD, Angenent LT (2005) Electricity generation from artificial wastewater using an upflow microbial fuel cell. Environmental Science & Technology 39(14):5262-7.

Huang L, Chai X, Chen G, Logan BE (2011a) Effect of set potential on hexavalent chromium reduction and electricity generation from biocathode microbial fuel cells. Environmental Science & Technology 45(11):5025-31.

Huang L, Chai X, Cheng S, Chen G (2011b) Evaluation of carbon-based materials in tubular biocathode microbial fuel cells in terms of hexavalent chromium reduction and electricity generation. Chemical Engineering Journal 166(2):652-61.

Huang L, Chen J, Quan X, Yang F (2010) Enhancement of hexavalent chromium reduction and electricity production from a biocathode microbial fuel cell. Bioprocess and Biosystems Engineering 33(8):937-45.

Huang L, Jiang L, Wang Q, Quan X, Yang J, Chen L (2014) Cobalt recovery with simultaneous methane and acetate production in biocathode microbial electrolysis cells. Chemical Engineering Journal 253:281-90.

Huang L, Li T, Liu C, Quan X, Chen L, Wang A, Chen G (2013) Synergetic interactions improve cobalt leaching from lithium cobalt oxide in microbial fuel cells. Bioresource Technology 128:539-46.

Huang L, Liu Y, Yu L, Quan X, Chen G (2015) A new clean approach for production of cobalt dihydroxide from aqueous Co (II) using oxygen-reducing biocathode microbial fuel cells. Journal of Cleaner Production 86:441-6.

Huang L, Yao B, Wu D, Quan X (2014) Complete cobalt recovery from lithium cobalt oxide in self-driven microbial fuel cell–microbial electrolysis cell systems. Journal of Power Sources 259:54-64.

Hu H (2002) Human health and heavy metals. Life Support: The Environment and Human Health; MIT Press: Cambridge, MA, USA.

Hutchison A, Atwood D, Santilliann-Jiminez QE (2008) The removal of mercury from water by open chain ligands containing multiple sulfurs. Journal of Hazardous Materials 156(1-3):458-65.

Jadhav UU, Hocheng H (2012) A review of recovery of metals from industrial waste. Journal of Achievements in Materials and Manufacturing Engineering 54(2):159-67.

Jiang L, Huang L, Sun Y (2014) Recovery of flakey cobalt from aqueous Co (II) with simultaneous hydrogen production in microbial electrolysis cells. International Journal of Hydrogen Energy 39(2):654-63.

Khosravi J, Alamdari A (2009) Copper removal from oil-field brine by coprecipitation. Journal of Hazardous Materials 166(2-3):695-700.

Kim C, Lee CR, Song YE, Heo J, Choi SM, Lim DH, Cho J, Park C, Jang M, Kim JR (2017) Hexavalent chromium as a cathodic electron acceptor in a bipolar membrane microbial fuel cell with the simultaneous treatment of electroplating wastewater. Chemical Engineering Journal 15(328):703-7.

Kurniawan TA, Chan GY, Lo WH, Babel S (2006) Physico–chemical treatment techniques for wastewater laden with heavy metals. Chemical Engineering Journal 118(1-2):83-98.

Landaburu-Aguirre J, Pongrácz E, Sarpola A, Keiski RL (2012) Simultaneous removal of heavy metals from phosphorus rich real wastewaters by micellar-enhanced ultrafiltration. Separation and Purification Technology 88:130-7.

Lee JH, Han J, Choi H, Hur HG (2007) Effects of temperature and dissolved oxygen on Se (IV) removal and Se (0) precipitation by Shewanella sp. HN-41. Chemosphere 68(10):1898-905.

Liu H, Cheng S, Logan BE (2005) Production of electricity from acetate or butyrate using a single chamber microbial fuel cell. Environmental Science & Technology 39(2):658-62.

Liu L, Yuan Y, Li FB, Feng CH (2011) *In situ* Cr (VI) reduction with electrogenerated hydrogen peroxide driven by iron-reducing bacteria. Bioresource Technology 102(3):2468-73.

Li Y, Lu A, Ding H, Jin S, Yan Y, Wang C, Zen C, Wang X (2009) Cr(VI) reduction at rutile-catalyzed cathode in microbial fuel cells. Electrochemistry Communications 11(7):1496-9.

Li Z, Zhang X, Lei L (2008) Electricity production during the treatment of real electroplating wastewater containing Cr6+ using microbial fuel cell. Process Biochemistry 43(12):1352-8.

Logan BE, Call D, Cheng S, Hamelers HV, Sleutels TH, Jeremiasse AW, Rozendal RA (2008) Microbial electrolysis cells for high yield hydrogen gas production from organic matter. Environmental Science & Technology 42(23):8630-40.

Logan BE, Hamelers B, Rozendal R, Schröder U, Keller J, Freguia S, Aelterman P, Verstraete W, Rabaey K (2006) Microbial fuel cells: methodology and technology. Environmental Science & Technology 40(17):5181-92.

Luo H, Liu G, Zhang R, Bai Y, Fu S, Hou Y (2014) Heavy metal recovery combined with H2 production from artificial acid mine drainage using the microbial electrolysis cell. Journal of Hazardous Materials 270:153-9.

Manahan S (2017) Environmental chemistry. CRC press.

Marafi M, Stanislaus A (2008) Spent hydroprocessing catalyst management: A review: Part II. Advances in metal recovery and safe disposal methods. Resources, Conservation and Recycling 53(1-2):1-26.

Miran W, Jang J, Nawaz M, Shahzad A, Jeong SE, Jeon CO, Lee DS (2017). Mixed sulfate-reducing bacteria-enriched microbial fuel cells for the treatment of wastewater containing copper. Chemosphere. 189:134-42.

Modin O, Fuad N, Rauch S (2017) Microbial electrochemical recovery of zinc. Electrochimica Acta 248:58-63.

Monteagudo JM, Ortiz MJ (2000) Removal of inorganic mercury from mine waste water by ion exchange. Journal of Chemical Technology and Biotechnology 75(9):767-72.

Morimoto T, Wu S, Uddin MA, Sasaoka E (2005) Characteristics of the mercury vapor removal from coal combustion flue gas by activated carbon using H2S. Fuel 84(14-15):1968-74.

Naddy RB, Gorsuch JW, Rehner AB, McNerney GR, Bell RA, Kramer JR (2007) Chronic toxicity of silver nitrate to Ceriodaphnia dubia and Daphnia magna, and potential mitigating factors. Aquatic Toxicology 84(1):1-0.

Nagajyoti PC, Lee KD, Sreekanth TV (2010) Heavy metals, occurrence and toxicity for plants: a review. Environmental Chemistry Letters 8(3):199-216.

Nancharaiah YV, Mohan SV, Lens PN (2015) Metals removal and recovery in bioelectrochemical systems: a review. Bioresource Technology 195:102-14.

Papadopoulos A, Fatta D, Parperis K, Mentzis A, Haralambous KJ, Loizidou M (2004) Nickel uptake from a wastewater stream produced in a metal finishing industry by combination of ion-exchange and precipitation methods. Separation and Purification Technology 39(3):181-8.

Purkayastha D, Mishra U, Biswas S (2014) A comprehensive review on Cd (II) removal from aqueous solution. Journal of Water Process Engineering 2:105-28.

Qin B, Luo H, Liu G, Zhang R, Chen S, Hou Y, Luo Y (2012) Nickel ion removal from wastewater using the microbial electrolysis cell. Bioresource Technology 121:458-61.

Qiu R, Zhang B, Li J, Lv Q, Wang S, Gu Q (2017) Enhanced vanadium (V) reduction and bioelectricity generation in microbial fuel cells with biocathode. Journal of Power Sources. 359: 379-83.

Quintelas C, Sousa E, Silva F, Neto S, Tavares T (2006) Competitive biosorption of ortho-cresol, phenol, chlorophenol and chromium (VI) from aqueous solution by a bacterial biofilm supported on granular activated carbon. Process Biochemistry 41(9):2087-91.

Rabaey K, Bützer S, Brown S, Keller J, Rozendal RA (2010) High current generation coupled to caustic production using a lamellar bioelectrochemical system. Environmental Science & Technology 44(11):4315-21.

Rabaey K (2006) Microbial fuel cells: methodology and technology. Environmental Science & Technology 40(17):5181-92.

Reck BK, Müller DB, Rostkowski K, Graedel TE (2008) Anthropogenic nickel cycle: Insights into use, trade, and recycling. Environmental Science & Technology 42(9):3394-400.

Rodenas Motos P, Ter Heijne A, van der Weijden R, Saakes M, Buisman CJ, Sleutels TH (2015) High rate copper and energy recovery in microbial fuel cells. Frontiers in microbiology 6:527.

Sawyer CN, McCarty PL, Parkin GF (1994) Chemistry for environmental engineers. New York. Mc Graw-Hill Book Company.

Shi J, Zhao W, Liu C, Jiang T, Ding H (2017) Enhanced Performance for Treatment of Cr (VI)-Containing Wastewater by Microbial Fuel Cells with Natural Pyrrhotite-Coated Cathode. Water. 9(12):979.

Srivastava NK, Majumder CB (2008) Novel biofiltration methods for the treatment of heavy metals from industrial wastewater. Journal of Hazardous Materials 151(1):1-8.

Stumm W, Morgan JJ (2012) Aquatic chemistry: chemical equilibria and rates in natural waters. John Wiley & Sons.

Tandukar M, Huber SJ, Onodera T, Pavlostathis SG (2009) Biological chromium (VI) reduction in the cathode of a microbial fuel cell. Environmental Science & Technology 43(21):8159-65.

Tao HC, Gao ZY, Ding H, Xu N, Wu WM (2012) Recovery of silver from silver (I)-containing solutions in bioelectrochemical reactors. Bioresource Technology 111:92-7.

Tao HC, Liang M, Li W, Zhang LJ, Ni JR, Wu WM (2011a) Removal of copper from aqueous solution by electrodeposition in cathode chamber of microbial fuel cell. Journal of Hazardous Materials 189(1-2):186-92.

Tao HC, Li W, Liang M, Xu N, Ni JR, Wu WM (2011b) A membrane-free baffled microbial fuel cell for cathodic reduction of Cu (II) with electricity generation. Bioresource Technology 102(7):4774-8.

Tao HC, Zhang LJ, Gao ZY, Wu WM (2011c) Copper reduction in a pilot-scale membrane-free bioelectrochemical reactor. Bioresource Technology 102(22):10334-9.

Tchounwou PB, Yedjou CG, Patlolla AK, Sutton DJ (2012) Heavy metal toxicity and the environment. In: Molecular, clinical and environmental toxicology. Springer, Basel.

Teng W, Liu G, Luo H, Zhang R, Xiang Y (2016) Simultaneous sulfate and zinc removal from acid wastewater using an acidophilic and autotrophic biocathode. Journal of Hazardous Materials 304:159-165.

Twidwell LG, McCloskey J, Miranda P, Gale M (1999) Technologies and potential technologies for removing selenium from process and mine wastewater. In: Proceedings of the TMS Fall Extraction and Processing Conference Vol. 2, pp. 1645-1656.

Van Phuong N, Kwon SC, Lee JY, Lee JH, Lee KH (2012). The effects of pH and polyethylene glycol on the Cr (III) solution chemistry and electrodeposition of chromium. Surface and Coatings Technology 206(21):4349-55.

Van Phuong N, Kwon SC, Lee JY, Shin J, Lee YI (2011) Mechanistic study on the effect of PEG molecules in a trivalent chromium electrodeposition process. Microchemical Journal 99(1):7-14.

Wang G, Huang L, Zhang Y (2008) Cathodic reduction of hexavalent chromium [Cr (VI)] coupled with electricity generation in microbial fuel cells. Biotechnology Letters 30(11):1959.

Wang H, Luo H, Fallgren PH, Jin S, Ren ZJ (2015) Bioelectrochemical system platform for sustainable environmental remediation and energy generation. Biotechnology Advances 33(3):317-34.

Wang H, Ren ZJ (2014) Bioelectrochemical metal recovery from wastewater: a review. Water Research 66:219-32.

Wang Q, Huang L, Pan Y, Quan X, Puma GL (2017) Impact of Fe (III) as an effective electron-shuttle mediator for enhanced Cr (VI) reduction in microbial fuel cells: Reduction of diffusional resistances and cathode overpotentials. Journal of Hazardous Materials 321:896-906.

Wang Q, Huang L, Pan Y, Zhou P, Quan X, Logan BE, Chen H (2016) Cooperative cathode electrode and *in situ* deposited copper for subsequent enhanced Cd (II) removal and hydrogen evolution in bioelectrochemical systems. Bioresource Technology 200:565-71.

Wang YH, Wang BS, Pan B, Chen QY, Yan W (2013) Electricity production from a bio-electrochemical cell for silver recovery in alkaline media. Applied Energy 112:1337-41.

Wang Z, Lim B, Choi C (2011) Removal of Hg2+ as an electron acceptor coupled with power generation using a microbial fuel cell. Bioresource Technology 102(10):6304-7.

Wu Y, Zhao X, Jin M, Li Y, Li S, Kong F, Nan J, Wang A (2018) Copper removal and microbial community analysis in single chamber microbial fuel cell. Bioresource Technology.

Zamani HA, Rajabzadeh G, Firouz A, Ganjali MR (2007) Determination of copper (II) in wastewater by electroplating samples using a PVC-membrane copper (II)-selective electrode. Journal of Analytical Chemistry 62(11):1080-7.

Zhang B, Feng C, Ni J, Zhang J, Huang W (2012) Simultaneous reduction of vanadium (V) and chromium (VI) with enhanced energy recovery based on microbial fuel cell technology. Journal of Power Sources 204:34-9.

Zhang BG, Zhou SG, Zhao HZ, Shi CH, Kong LC, Sun JJ, Yang Y, Ni JR (2010) Factors affecting the performance of microbial fuel cells for sulfide and vanadium (V) treatment. Bioprocess and Biosystems Engineering 33(2):187-94.

Zhang B, Zhao H, Shi C, Zhou S, Ni J (2009) Simultaneous removal of sulfide and organics with vanadium (V) reduction in microbial fuel cells. Journal of Chemical Technology and Biotechnology 84(12):1780-6.

Zhang LJ, Tao HC, Wei XY, Lei T, Li JB, Wang AJ, Wu WM (2012) Bioelectrochemical recovery of ammonia–copper (II) complexes from wastewater using a dual chamber microbial fuel cell. Chemosphere 89(10):1177-82.

Photosynthetic Algal Microbial Fuel Cell for Simultaneous NH_3-N Removal and Bioelectricity Generation

Chaolin Tan[1,2], Ming Li[1,2], Minghua Zhou[1,2]*, Xiaoyu Tian[1,2], Huanhuan He[1,2] and Tingyue Gu[1,3]*

[1] Key Laboratory of Pollution Process and Environmental Criteria, Ministry of Education, College of Environmental Science and Engineering, Nankai University, Tianjin 300350, P. R. China
[2] Tianjin Key Laboratory of Urban Ecology Environmental Remediation and Pollution Control, College of Environmental Science and Engineering, Nankai University, Tianjin 300350, P. R. China
[3] Department of Chemical and Biomolecular Engineering, Ohio University, Athens, OH 45701, USA

1. Introduction

The effluent of a sewage treatment plant typically contains a lot of nutrients, such as nitrogen and phosphorus (Chiu et al. 2015; Yang et al. 2018). Nitrogen, if beyond the water's ability to self-regulate, will lead to eutrophication of water. Therefore, proper treatment is needed before discharge. There are two main methods for N removal: one, is the conversion to N_2 and the other is stripping of NH_3 (Feng et al. 2008; Kuntke et al. 2012). The traditional wastewater treatment method is often expensive and results in a lot of sludge. Microalgae technology is attractive. Markl (1997) proposed that the use of microalgae could effectively remove nitrogen, phosphorus and other nutrients. However, it is difficult to harvest microalgae biomass, and this limits wastewater treatment by microalgae (Wang et al. 2016). Immobilized *Chlorella vulgaris* has been applied to solve this problem (Gao et al. 2011). In a previous study, it was confirmed that immobilized *C. vulgaris* can be used in microbial fuel cells (MFCs) effectively. The autotrophic growth of *C. vulgaris* uses ammonium salts in the culture medium as the nitrogen source and converts it into energy through anabolism (He et al. 2014; Zhou et al. 2012). This means that, theoretically, immobilized *C. vulgaris* can remove ammonia nitrogen in the treatment of wastewater.

Ammonia nitrogen wastewater treatment is mainly achieved through the following three ways. One way is the assimilation of immobilized *C. vulgaris* since immobilized *Chlorella* uses ammonia nitrogen as the nitrogen source for its growth and its conversion to organic nitrogen. Given different forms of nitrogen sources, algal growth preferentially utilizes ammonia nitrogen (Razzak et al. 2017). The second way is the adsorption by immobilized algae matrix (Gao et al. 2011). Calcium alginate is often used for algae immobilization and it has certain adsorption ability. During the initial interaction, the outer surface of immobilized *C. vulgaris* preferentially adsorbs some ammonia nitrogen. The third way is ammonia volatilization that can be caused by changes in culture medium pH. The photosynthesis by algal cells will utilize CO_2, and thus change the pH value of ammonia wastewater causing volatilization of ammonia.

---

*Corresponding author: zhoumh@nankai.edu.cn (M. Zhou), gu@ohio.edu (T. Gu)

Ammonia in wastewater mostly exists in the form of free ammonia and ammonium ion with a dynamic balance is as shown below:

$$NH_4^+ + OH^- \rightleftharpoons NH_3 + H_2O \tag{1}$$

As pH increases, this equilibrium shifts to the right and the free ammonia amount increases. Ammonia is mostly in the form of ammonium ion at a pH around 7. When pH is 9, 40% of ammonia nitrogen in ammonia nitrogen wastewater is converted into ammonia due to volatilization (Aosiman et al. 2014).

In situ separation of suspended algal biomass from a culture medium is more difficult than that from immobilized algae biomass. The latter also offers a stronger resistance to toxic matters with stable MFC operations (Jin et al. 2011).

A photosynthetic algal microbial fuel cell (PAMFC) works in the following way: microorganisms in the anode chamber oxidizes an organic substrate to release electrons, protons and CO_2 as shown in Figure 1. Under light illumination, microalgae in the cathode chamber uses CO_2 fed from the anode chamber as the carbon source for growth and oxygen production via photosynthesis. The biogenic oxygen serves as the electron acceptor for bioelectricity generation. Therefore, in theory, a PAMFC can simultaneously realize wastewater treatment and bioenergy production without CO_2 discharge due to its sequestration by microalgae.

In this study, the feasibility of using immobilized *C. vulgaris* as the cathode oxygen supplier for simultaneous NH_3-N removal from the catholyte and bioelectricity generation is demonstrated using a laboratory-scale PAMFC system. Comparisons of power output and NH_3-N removal efficiency are carried out between a PAMFC using immobilized *C. vulgaris* and a PAMFC using suspended *C. vulgaris*.

2. Materials and Methods

2.1 Microbes and chemicals

C. vulgaris (strain FACHB-24) was purchased from the Freshwater Algae Culture Collection at the Institute of Hydrobiology (Chinese Academy of Sciences, Wuhan, China). The anodic biofilm was acclimated from an anaerobic sludge (Teda Sewage Treatment Plant, Tianjin, China). The anolyte was 50 mM phosphate buffered nutrient solution (NH_4Cl 0.31g/L, KCl 0.13g/L, $NaH_2PO_4 \cdot 2H_2O$ 3.32 g/L, $Na_2HPO_4 \cdot 12H_2O$ 10.36 g/L, trace minerals 12.5 mL/L and vitamins 5 mL/L) supplemented with 1.0 g/L glucose medium as substrate. The BG11 culture medium for pre-culturing *C. vulgaris* before immobilization had the following composition (g/L): 1.5 $NaNO_3$, 0.04 K_2HPO_4, 0.075 $MgSO_4 \cdot 7H_2O$, 0.036 $CaCl_2 \cdot 2H_2O$, 0.006 citric acid, 0.006 ferric ammonium citrate, 0.001 $EDTANa_2$, 0.02 Na_2CO_3 and 1 mL trace metal solution as previously described (Zhou et al. 2012). The catholyte was artificial wastewater constituted by diluting an ammonia solution. Its pH was adjusted using NaOH solution. All chemicals were analytical reagent-grade chemicals.

2.2 PAMFC Reactor Configuration and Operation

A PAMFC reactor was constructed with an anodic chamber and a cathodic chamber separated by a cation exchange membrane (CEM) (80 cm^2, Ultrex CMI7000, AnKeTech, Membrane Separation Engineering & Technology Co., Ltd., Beijing, China) as shown in Figure 1 (Zhou et al. 2012). The PAMFC anolyte and catholyte volumes in this work were both 200 mL. The anode was made of carbon felt, while the cathode was carbon fiber cloth containing 0.1 mg/cm^2 Pt catalyst (6 cm × 6 cm; Jilin Carbon Plant, Jilin, China). The cathode was inoculated with *C. vulgaris*. Immobilized *C. vulgaris* was prepared using the following steps. The algal cells in exponential growth phase were harvested by centrifugation (model TGL-16C, Jiangsu Jinyi Instrument Technology Co., Ltd., Jintan, Jiangsu, China) at 3,500 rpm for 10 minutes. The cell pellet was rinsed with sterilized water. They were resuspended in BG11 culture medium to grow a concentrated cell suspension. Using the previous procedure (Zhou et al. 2012) algae alginate beads was prepared.

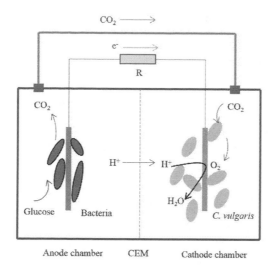

Figure 1: Schematic illustrations of a PAMFC bioreactor with CO_2 recycle.

The PAMFC was operated in batch mode in an illuminated incubator at 25 °C. The headspace of the anode chamber was vented to the headspace of the cathode chamber to allow CO_2 released from organic carbon degradation in the anode chamber for utilization by immobilized *C. vulgaris* in the cathode chamber.

2.3 Data Analysis

The cell mass of suspended algae was determined using optical density at 683 nm wavelength on UV759 UV-Vis spectrophotometer (Shanghai Precision Scientific Instruments Corporation, Shanghai, China). Voltage output was recorded twice every hour using a data acquisition system (PISO-813, ICPDAS Co., Ltd., Shanghai, China). Polarization curves were obtained by adjusting the external resistance from 1,000 to 50 Ω. The experimental pH and initial concentration of NH$_3$-N were optimized, and the effects of pH and initial concentration of ammonia in the catholyte on the removal of ammonia and the power generation of PAMFC were investigated. The same PAMFC bioreactor was also tested with suspended *C. vulgaris* cells for comparison to prove the advantages of algae immobilization.

3. Results and Discussion

3.1 Effect of pH on Ammonia Nitrogen Removal and Bioelectricity Generation

3.1.1 Effect of pH on Ammonia Nitrogen Removal

Figure 2a shows ammonia nitrogen removal efficiency from the catholyte at different pH values. The PAMFC achieved nearly 100% removal at initial pH values of 5, 7 and 9 on day 3. Moreover, the pH values in all three cases with different initial pH values varied from 5.5 to 8 during the six days operation (Figure 2b). After three days, the ammonia nitrogen concentration increased and then dropped slightly, and the ammonia nitrogen removal efficiency for each pH reached around 90% when the experiment was terminated (Figure 2a). For the phenomenon of ammonia nitrogen concentration take-off after day 3, the reason could be that ammonia might be converted into other forms upon absorption and utilization (Razzak et al. 2017). Nitrogen that existed in the form of ammonium ion was not detected on the third day, but then ammonia was regenerated from nitrification-denitrification or the adsorption by the immobilized algae. The concentration of ammonia nitrogen on the first day dropped to 2.5 mg/L that was below the concentration (5 mg/L) of ammonia nitrogen specified in the national first class 'A' standard in China. Thus, the wastewater could be discharged at the end of day 1 or before the end of day 3.

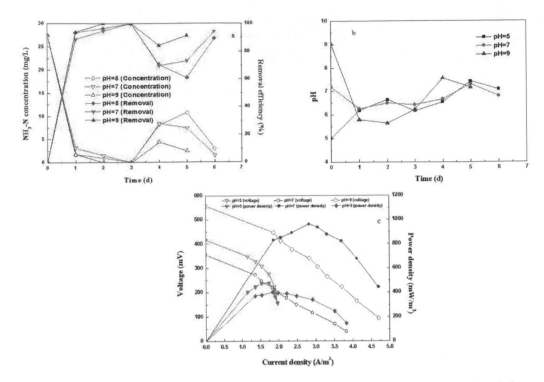

Figure 2: Effect of catholyte pH on ammonia nitrogen removal (a) change of pH in catholyte during PAMFC operation (b) polarization curves and power density curves at different initial pH values (c).

3.1.2 Effect of Initial pH on Power Generation in PAMFC

When the initial pH was neutral (pH 7), the PAMFC voltage output was relatively stable and the cycle time could last 153 hours with an average maximum voltage output of 361 mV. When the pH was acidic (pH 5), the voltage graph was mainly divided into two phases. During the first three days of operation, the PAMFC operated smoothly at a lower voltage output (around 215 mV) and then quickly rose to 270 mV on day 3. The continuing operation for another 1.5 days, the voltage dropped to below 100 mV.

With an alkaline pH (pH 9), the maximum voltage output was higher than that at pH 5, but the overall power output was less. Polarization curves and power density curves were measured at these three different initial pH values when the voltage was relatively stable. As shown in Figure 2c, the power density under pH 7 was significantly higher than at initial pH 5 and pH 9. The maximum power density at pH 7 was 963 mW/m³ that was 101% and 140% higher than that at pH 5 (478 mW/m³) and at pH 9 (402 mW/m³), respectively.

When the initial pH value was 5 or 7, it changed to 6.2 after 1 day and then increased slightly afterward. However, when the initial pH was 9, it dropped rapidly to about 5.5 after 1 day. At this time, the voltage output was at a lower level. Then the pH started to rise gradually and the voltage output began to increase. Therefore, 6-8 were considered favorable pH values for bioenergy generation.

3.2 Effects of Ammonia Nitrogen Concentration on the Removal Efficiency and Bioelectricity Generation

3.2.1 Effect of Ammonia Nitrogen Concentration on the Removal Efficiency

During the initial stage of the PAMFC operation, the ammonia nitrogen concentration was higher and its removal was faster. Regardless of the initial concentration (30-120 mg/L), the ammonia nitrogen concentration rapidly decreased on the first day, but it did not change significantly afterward. With an increased initial concentration of ammonia nitrogen, the removal efficiency of ammonia nitrogen first increased and then decreased. The removal efficiency values at initial ammonia nitrogen concentrations of 30, 60, 90 and 120 mg/L were 29.4%, 55.9%, 53.2% and 46%, respectively after one day (Figure 3a). After five days, the removal efficiency values all reached above 75%. Bian (2010) showed that when suspended *C. vulgaris* was used, the higher the concentration of ammonia nitrogen was the more obvious was the inhibition of the algal growth. Here, the removal efficiency was 93.8% at the initial ammonia nitrogen concentration of 30 mg/L after three days. Therefore, it can be concluded that immobilized *C. vulgaris* can tolerate a higher concentration of ammonia nitrogen, and the removal of ammonia nitrogen was better than the suspended state.

Apart from utilization by algae, there were other factors that enhanced the removal of ammonia nitrogen in this system. The pH value of wastewater changed during the experiment (Figure 3b). When the wastewater had an alkaline pH, some ammonia nitrogen was converted into ammonia gas. Some ammonia could be adsorbed on the surface of the algae beads (Tam and Wong 2000). Thus, the initial ammonia nitrogen concentration will affect the removal of ammonia nitrogen if it is too high or too low. At the end of the four days test, the maximum combined ammonia nitrogen removal efficiency values of were 76.0%, 95.3%, 98.8% and 78.8% at initial ammonia nitrogen concentrations of 30, 60, 90 and 120 mg/L, respectively (Figure 3a).

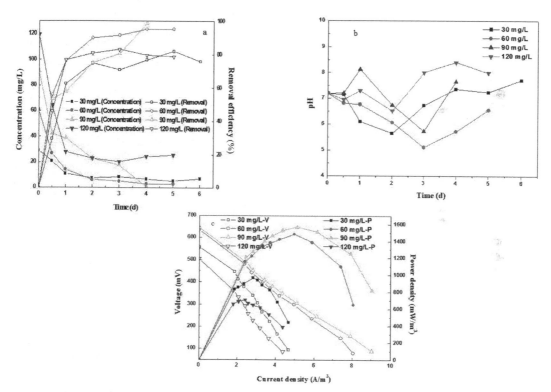

Figure 3: Residual ammonia nitrogen concentration in catholyte and removal efficiency (a) catholyte pH (b) polarization curves and power density curves (c) under different initial ammonia nitrogen concentrations.

3.2.2 *Effect of Initial Ammonia Nitrogen Concentration on Power Output*

The power output using ammonia nitrogen wastewater as the culture medium was lower than that using the BG11 medium. Its maximum voltage output was below 400 mV (Figure 3c). When the initial ammonia nitrogen concentration was between 30-60 mg/L, the voltage output was relatively stable. When the initial concentration was greater than 60 mg/L, the voltage output dropped rapidly in the early stage because this high initial concentration of ammonia nitrogen inhibited the growth of the immobilized *C. vulgaris*. After a period of adaptation by the cells, the voltage output began to rise and became stabilized for about one day. Then it dropped quickly below 100 mV. When the initial ammonia nitrogen concentration was 120 mg/L, the plateau voltage had the shortest duration (Figure 3c).

The polarization curves and power density curves in Figure 3c clearly indicate that the initial ammonia nitrogen concentration in the catholyte had an important effect on bioelectricity production. A high concentration (120 mg/L) and a low concentration (30 mg/L) both resulted in a power density of less than 1 W/m³. The inhibition effect of the high concentration (120 mg/L) was especially pronounced. The initial ammonia nitrogen concentrations of 60 mg/L and 90 mg/L were preferred, yielding maximum power densities of 1.48 W/m³ and 1.57 W/m³, respectively.

3.3 Nitrogen Removal and Bioelectricity Generation in PAMFC with Immobilized and Suspended *C. vulgaris*

3.3.1 Nitrogen Removal and Bioelectricity Generation with Cell-Free Sodium Alginate Beads

Cell-free sodium alginate beads were used to absorb ammonia nitrogen in a control test using the conditions the same as in the PAMFC test with immobilized algae beads. In Figure 4a, the cell-free beads had a maximum ammonia nitrogen removal efficiency of 16.7% at 30 mg/L initial ammonia nitrogen. The removal of ammonia nitrogen by cell-free beads was mainly owing to adsorption. When the adsorption saturation was reached, the ammonia nitrogen concentration in the culture medium remained basically unchanged. The removal efficiency of 16.7% was relatively small compared to the aforementioned overall maximum removal efficiency values of 76-98.8% using immobilized algae beads in the PAMFC.

Figure 4b shows that the MFC using cell-free alginate beads produced a negligible voltage output. Without additional oxygen supplied from an external feed or from *C. vulgaris* metabolism, the dissolved oxygen in the catholyte was gradually depleted. This control test also confirmed that algae produced oxygen indeed served as the electron acceptor to sustain PAMFC bioelectricity generation.

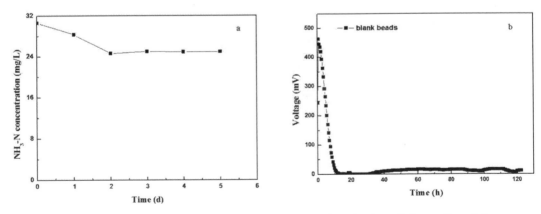

Figure 4: Efficiency of ammonia nitrogen removal due to adsorption by cell-free sodium alginate beads (a) voltage output using cell-free sodium alginate beads in PAMFC (b).

3.3.2 Comparison of Ammonia Nitrogen Removal Efficiency Using Immobilized *C. vulgaris* vs. Suspended *C. vulgaris* in a Flask and in PAMFC Operations

Suspended and immobilized *C. vulgaris* in Erlenmeyer flasks filled with 100 mL ammonia nitrogen wastewater were found to yield very different ammonia nitrogen removal efficiency values at an initial ammonia nitrogen concentration of 30 mg/L. The ammonia nitrogen concentration decreased to 21 mg/L with suspended *C. vulgaris* after 0.5 days (Figure 5a). Then it began to rise and remained at around 26 mg/L, yielding a removal efficiency of only 13%. The pH of the wastewater changed from the initial value of 7.2 to above 8 after 0.5 days (Figure 5b). Bian (2010) simulated MFC removal of ammonia nitrogen in urine. In the initial stage, the removal efficiency by microalgae was only 4.1-7.8% due to the inhibition of a high initial ammonia nitrogen concentration of 50 mg/L, but the removal efficiency reached 95% on the eighth day.

Compared with the suspended *C. vulgaris* PAMFC, the immobilized *C. vulgaris* PAMFC showed a much better removal efficiency of ammonia nitrogen. The ammonia nitrogen concentration continued to decrease and reached the lowest point of 9.74 mg/L on the seventh day and then remained basically unchanged, yielding a maximum ammonia nitrogen removal efficiency of 67.5% (Figure 5a).

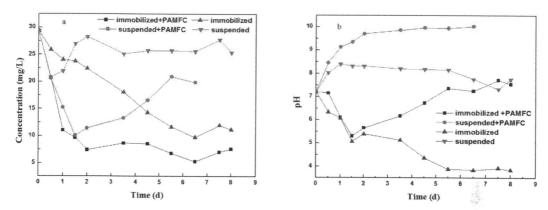

Figure 5: Catholyte residual ammonia nitrogen values (a) and pH values (b) with immobilized *C. vulgaris* and suspended *C. vulgaris* in PAMFC and flask operations.

The removal efficiency of ammonia nitrogen in PAMFC was better than that in an Erlenmeyer flask for both immobilized and suspended *C. vulgaris* cases. The concentration of ammonia nitrogen decreased to 10.1 mg/L after 1.5 days with suspended *C. vulgaris* in PAMFC, yield a removal efficiency of 66.4 %. Then the ammonia nitrogen content began to increase. The pH of the wastewater varied between 9.6 and 10 from day 2 to day 6.5. It is known that when wastewater is alkalinic, ammonium ion is easily converted to ammonia gas for direct volatilization and removal. The ammonia nitrogen content in the liquid phase should further decrease. However, the data in this work showed the opposite. This phenomenon could be explained using the working principle of the PAMFC in this work. Its anolyte contained a large amount of ammonia. When the pH of the wastewater in the cathode chamber was too high, the anion/cation ratio was unbalanced between the anode chamber and the cathode chamber. Then, some of the ammonium ions in the anode chamber diffused to the cathode chamber through the cation exchange membrane that increased the ammonia nitrogen content in the catholyte (Hu et al. 2015). In addition, suspended *C. vulgaris* tended to attach to the cathode chamber wall or deposited at the bottom of the chamber resulting in an increase of mass transfer resistance from the bulk liquid to the cells. Furthermore, some cells grew on the cathode surface that increased the proton and oxygen transfer resistances. These unfavorable factors led to a lower power performance of the PAMFC with suspended *C. vulgaris* compared to that of immobilized *C. vulgaris* (Zhou et al. 2012).

The concentration of ammonia nitrogen in the PAMFC with immobilized *C. vulgaris* reached its lowest point after two days. It did not increase afterward, unlike in the PAMFC with suspended *C. vulgaris*. The pH of the wastewater varied between 5.5 and 8, and the maximum ammonia nitrogen removal efficiency was 75.3% that was achieved on the second day. Therefore, after two days, the wastewater could be discharged with a good ammonia nitrogen removal efficiency.

Table 1 summarized ammonia nitrogen removal efficiency values under different conditions in this work and compared them with other MFC systems in the literature. It shows that using immobilized *C. vulgaris* in PAMFC achieved the 98.8% efficiency that was among the highest removal efficiency values in the MFC literature.

Table 1: Ammonia nitrogen removal efficiency comparison

Species/position	Wastewater type	Cultivation system	Removal efficiency	Reference
Chlorella sp.	Municipal wastewater	Reactor	93%	Li et al. 2011
Chlorella sp.	Animal wastewater	Flasks	98%	Chen et al. 2012
C. vulgaris	Dairy manure	Flasks	99.7%	Wang et al. 2010
C. vulgaris	Piggery wastewater	Unknown	99%	Molinuevo-Salces et al. 2016
Scenedesmus obliquus	Urban wastewater	Photo-bioreactor	98%	Gouveia et al. 2016
Consortia algae/ diatoms	Dairy wastewater	Tanks	96%	Woertz et al. 2009
Consortia	Domestic wastewater	Unknown	75%	Sutherland et al. 2014
Consortia bacteria/ microalgae	Municipal wastewater	Bottles	83%	Delgadillo-Mirquez et al. 2016
C. vulgaris	Piggery wastewater	Flasks	60%	
Anode	Synthetic wastewater	MFC	71.4%	Abou-Shanab et al. 2013
Anode	Swine-farming wastewater	MFC	83%	Liu et al. 2017
Anode	Synthetic wastewater	MDC	88.3%	Colombo et al. 2017
Cathode (*C. vulgaris*)	Synthetic wastewater	MFC	98.8%	Zhang and Angelidaki 2015 This work

3.3.3 Bioelectricity Generation in PAMFC with Immobilized or Suspended C. vulgaris

The polarization curves and power density curves of PAMFC operations with immobilized and suspended *C. vulgaris* for the treatment of ammonia nitrogen wastewater were compared in Figure 6. The maximum voltage output for immobilized *C. vulgaris* was at 378 mV that was 15% higher than that for suspended *C. vulgaris* state (328 mV). A power density curve can provide the internal resistance at the peak of the

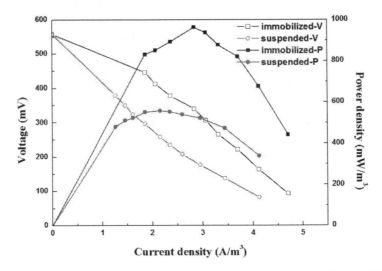

Figure 6: Polarization and power density curves of PAMFC operations with immobilized and suspended *C. vulgaris* using an external resistance of 1000 Ω.

power output when the internal resistance is the same as the external resistance according to Jacobi's law. The suspended and immobilized *C. vulgaris* PAMFC operations did not show much difference in internal resistance or an open circuit potential despite the fact that the two operations exhibited considerable differences in the output voltage.

The maximum power density values were 963 mW/m^3 and 557 mW/m^3 for the immobilized and suspended *C. vulgaris* PAMFC operations, respectively, reflecting an increase of 73% for the immobilized *C. vulgaris* PAMFC. Coulombic efficiency depends not only on the output voltage but also the COD (chemical oxygen demand) removal efficiency. In this work, the COD removal efficiency values in immobilized and suspended *C. vulgaris* PAMFC operations were 80.9% and 73.1%, respectively, corresponding to Coulombic efficiency values of 8.3% and 6.7%, respectively. Therefore, in terms of bioelectricity production efficiency, immobilized *C. vulgaris* PAMFC was a better choice. It is possible to further improve the efficiency of power generation and wastewater treatment through optimization.

4. Conclusion

The removal of ammonia nitrogen from wastewater in the cathode chamber by immobilized *C. vulgaris* PAMFC was primarily achieved by biological assimilation, while the adsorption of immobilized algae beads and ammonia volatilization were small. When the catholyte pH was neutral, the power generation of PAMFC and the removal efficiency of ammonia nitrogen were the highest. Thus, the pH of wastewater should be maintained between 6 and 8 in practice. The removal efficiency values of ammonia nitrogen under different conditions were ranked as follows: immobilized *C. vulgaris* PAMFC > suspended *C. vulgaris* PAMFC > immobilized *C. vulgaris* in flask > suspended *C. vulgaris* in a flask. The maximum power density obtained with immobilized *C. vulgaris* in PAMFC was 963 mW/m^3 that was 73% higher than that of suspended *C. vulgaris* in PAMFC. The Coulombic efficiency and the COD removal efficiency were also higher with immobilized *C. vulgaris* in PAMFC.

Acknowledgment

This work was financially supported by the National Natural Science Foundation of China (Nos. 91545126, 21773129 and 21273120) and the Fundamental Research Funds for the Central Universities.

References

Abou-Shanab RAI, Ji MK, Kim HC, Paeng KJ, Jeon BH (2013) Microalgal species growing on piggery wastewater as a valuable candidate for nutrient removal and biodiesel production. J Environ Manage 115: 257-264.

Aosiman T, Yang L, An D, Wang T (2014) Progresses in Air Stripping for Treatment of Ammonia Wastewater. Petrochem Technol 43(11): 1348-1353.

Bian L (2010) The removal and recycle of nitrogen and phosphorus nutrients and wastewater purification by microalgae.

Chen R, Li R, Deitz L, Liu Y, Stevenson RJ, Liao W (2012) Freshwater algal cultivation with animal waste for nutrient removal and biomass production. Biomass Bioenerg 39: 128-138.

Chiu SY, Kao CY, Chen TY, Chang YB, Kuo CM, Lin CS (2015) Cultivation of microalgal *Chlorella* for biomass and lipid production using wastewater as nutrient resource. Bioresour Technol 184: 179-89.

Colombo A, Marzorati S, Lucchini G, Cristiani P, Pant D, Schievano A (2017) Assisting cultivation of photosynthetic microorganisms by microbial fuel cells to enhance nutrients recovery from wastewater. Bioresour Technol 237: 240-248.

Delgadillo-Mirquez L, Lopes F, Taidi B, Pareau D (2016) Nitrogen and phosphate removal from wastewater with a mixed microalgae and bacteria culture. Biotechnol Rep (Amst) 11: 18-26.

Feng D, Wu Z, Xu S (2008) Nitrification of human urine for its stabilization and nutrient recycling. Bioresour Technol 99(14): 6299-304.

Gao QT, Wong YS, Tam NF (2011) Removal and biodegradation of nonylphenol by immobilized *Chlorella vulgaris*. Bioresour Technol 102(22): 10230-10238.

Gouveia L, Graca S, Sousa C, Ambrosano L, Ribeiro B, Botrel EP, Neto PC, Ferreira AF, Silva CM (2016) Microalgae biomass production using wastewater: Treatment and costs Scale-up considerations. Algal Res 16: 167-176.

He H, Zhou M, Yang J, Hu Y, Zhao Y (2014) Simultaneous wastewater treatment, electricity generation and biomass production by an immobilized photosynthetic algal microbial fuel cell. Bioprocess Biosyst Eng 37(5): 873-80.

Hu X, Liu B, Zhou J, Jin R, Qiao S, Liu G (2015) CO_2 fixation, lipid production, and power generation by a novel air-lift-type microbial carbon capture cell system. Environ Sci Technol 49(17): 10710-10717.

Jin J, Yang LH, Chan SMN, Luan TG, Li Y, Tam NFY (2011) Effect of nutrients on the biodegradation of tributyltin (TBT) by alginate immobilized microalga, *Chlorella vulgaris*, in natural river water. J Hazard Mater 185(2-3): 1582-1586.

Kuntke P, Smiech KM, Bruning H, Zeeman G, Saakes M, Sleutels TH, Hamelers HV, Buisman CJ (2012) Ammonium recovery and energy production from urine by a microbial fuel cell. Water Res 46(8): 2627-2636.

Li YC, Chen YF, Chen P, Min M, Zhou WG, Martinez B, Zhu J, Ruan R (2011) Characterization of a microalga *Chlorella* sp. well adapted to highly concentrated municipal wastewater for nutrient removal and biodiesel production. Bioresour Technol 102(8): 5138-5144.

Liu S, Li L, Li H, Wang H, Yang P (2017) Study on ammonium and organics removal combined with electricity generation in a continuous flow microbial fuel cell. Bioresour Technol 243: 1087-1096.

Markl H (1997) CO_2 Transport and photosynthetic productivity of a continuous culture of algae. Biotechnol Bioeng 19: 1851-1862.

Molinuevo-Salces B, Mandy A, Ballesteros M, Gonzalez-Fernandez C (2016) From piggery wastewater nutrients to biogas: Microalgae biomass revalorization through anaerobic digestion. Renew Energ 96: 1103-1110.

Razzak SA, Ali SAM, Hossain MM, deLasa H (2017) Biological CO_2 fixation with production of microalgae in wastewater – A review. Renew Sust Energ Rev 76: 379-390.

Sutherland DL, Turnbull MH, Broady PA, Craggs RJ (2014) Effects of two different nutrient loads on microalgal production, nutrient removal and photosynthetic efficiency in pilot-scale wastewater high rate algal ponds. Water Res 66: 53-62.

Tam NFY, Wong YS (2000) Effect of immobilized microalgal bead concentrations on wastewater nutrient removal. Environ Pollut 107: 145–151.

Wang LA, Wang YK, Chen P, Ruan R (2010) Semi-continuous cultivation of *Chlorella vulgaris* for treating undigested and digested dairy manures. Appl Biochem Biotech 162(8): 2324-2332.

Wang Y, Ho SH, Cheng CL, Guo WQ, Nagarajan D, Ren NQ, Lee DJ, Chang JS (2016) Perspectives on the feasibility of using microalgae for industrial wastewater treatment. Bioresour Technol 222: 485-497.

Woertz I, Feffer A, Lundquist T, Nelson Y (2009) Algae grown on dairy and municipal wastewater for simultaneous nutrient removal and lipid production for biofuel feedstock. J Environ Eng-Asce 135(11): 1115-1122.

Yang Z, Pei H, Hou Q, Jiang L, Zhang L, Nie C (2018) Algal biofilm-assisted microbial fuel cell to enhance domestic wastewater treatment: Nutrient, organics removal and bioenergy production. Chem Eng J 332: 277-285.

Zhang Y, Angelidaki I (2015) Submersible microbial desalination cell for simultaneous ammonia recovery and electricity production from anaerobic reactors containing high levels of ammonia. Bioresour Technol 177: 233-239.

Zhou M, He H, Jin T, Wang H (2012) Power generation enhancement in novel microbial carbon capture cells with immobilized *Chlorella vulgaris*. J Power Sources 214: 216-219.

Nutrients Removal and Recovery in Bioelectrochemical Systems

Nan Zhao[1], Qun Yan[1] and Zhen He[2*]

[1] School of Environmental and Civil Engineering, Jiangnan University, Wuxi 214122, China
[2] Department of Civil and Environmental Engineering, Virginia Polytechnic Institute and
  State University, Blacksburg, VA 24061, USA

1. Introduction

Nitrogen and phosphorus are two essential elements for life on the earth. Nitrogen is one of the most important elements for nitrogen-containing biomolecules such as amino acids that can be used to form different kinds of proteins. Phosphorus—broadly existing in cells, proteins and bones in the form of phosphate—takes part in many crucial metabolic processes of various living organisms. For thousands of years, the balance of nitrogen and phosphorus has been maintained in nature via nitrogen and phosphorus cycles, respectively. The balances were recently affected by anthropogenic activities for demand of chemical fertilizers in modern agriculture to replenish soil fertility and to maintain soil nutrients reserve (Nancharaiah et al. 2016a).

Although nitrogen gas (N_2) is abundant in the atmosphere (78.0%), only a little of such kind of unreactive form can be used. Most organisms can make use of nitrogen only in the reactive forms (e.g., NO_3^- and NH_4^+) because they are unable to metabolize the inert N_2 (Hoffman et al. 2014). In nature, the fixation of inert nitrogen gas to ammonia nitrogen and ammonium nitrogen is conducted through lightning and a select group of microorganisms containing nitrogenase (e.g. prokaryotes) (Nancharaiah et al. 2016a). Industrial transformation of N_2 to NH_3 has been dominant for the production of chemical fertilizers by the Haber–Bosch process (Fowler et al. 2013). It is reported that only about 42–47% of the nitrogen added to croplands globally is harvested as crop products (Zhang 2017), and the residue contributes to increasing eutrophication of water bodies that deteriorates the water quality and causes acidification of atmosphere that is also hazardous to human beings. Moreover, the discharge of effluent, with insufficient removal of ammonia nitrogen from wastewater treatment plants, aggravates such situations (Feng and Sun 2015). Hence, the significance of nitrogen removal from wastewater prior to discharge has been universally recognized.

Phosphorus, which receives public attention as both a crucial nutrient for living organisms and an important aquatic pollutant, is a nonrenewable resource that exists in the form of phosphate rocks in nature. Mining from phosphate deposits is the most viable way to extract phosphorus for fertilizer production. It is predicted that commercial phosphate rock reserves will be depleted in 70-140 years if there is a lack of proper management (Li et al. 2017). On the other hand, discharge of phosphorus contributes to the

*Corresponding author: zhenhe@vt.edu

process of eutrophication in lakes, reservoirs, estuaries and oceans, leading to degradation of freshwater ecosystems and reducing the possibility of water reclamation (Sukačová et al. 2015). Phosphorus released to the environment from point sources (e.g. due to inefficient phosphorus removal in wastewater treatment plants) and nonpoint sources (e.g., soil erosion and agricultural runoff with unused chemical fertilizers) account for a great deal of the mined phosphorus (Rittmann et al. 2011). Therefore, phosphorus recovered from wastewater streams should be considered as a potentially important phosphorus resource for industry and agriculture.

Compared to conventional nutrient removal technologies, such as nitrification and denitrification for nitrogen removal, and biological processes and chemical precipitation for phosphorus removal, BES has a potential to be an energy-efficient approach for nutrients removal and recovery though it is mostly still in the stage of bench-scale studies. BES can achieve wastewater treatment with simultaneous electricity and/or chemicals (e.g. methane) production and have been extensively studied and developed in the past decade via integrating microbiology, electrochemistry, materials science and engineering, etc. (Wang and Ren 2013). BES has also been proposed as a novel technology to separate and recover nitrogen and phosphorus from wastewaters (Kelly and He 2014). However, the focus of most existing treatment approaches is the removal of nutrients. When it comes to recovery, we need to take a further step. The objective of this chapter is to provide a comprehensive introduction of the nitrogen and phosphorus removal and recovery in BES and to discuss the perspectives and challenges that warrant further investigation.

2. Nitrogen Removal and Recovery

2.1 Background

Nitrogen removal by conventional technologies (aerobic and anaerobic processes) in wastewater treatment is usually energy intensive (3.4-6 kWh kg N^{-1}) compared to the energy consumption of BES-based nitrogen removal (estimated at 2.22-3.17 kWh kg N^{-1}) (McCarty et al. 2011; Eusebi et al. 2009; Wang et al. 2016; Molognoni et al. 2017). This is mostly because of aeration that is required to supply oxygen for nitrification (ammonium is oxidized subsequently to nitrate by ammonium and nitrite-oxidizing bacteria). In addition, COD (chemical oxygen demand) is needed to provide electrons for denitrification (nitrate is reduced subsequently to nitrogen gas by denitrifying bacteria). The implementation of converting nitrogen in the form of reactive ammonium to inert harmless nitrogen gas accounts for a major cost of wastewater treatment (Batstone et al. 2015).

To reduce energy consumption, several more energy-efficient processes have been developed as alternatives to conventional technologies for nitrogen removal. For instance, anaerobic ammonium oxidation (ANAMMOX) can achieve nitrogen removal by converting ammonium and nitrite to nitrogen gas by ANAMMOX bacteria (Lackner et al. 2014). In the Single reactor system for High-rate Ammonium Removal Over Nitrite (SHARON), ammonium is partially oxidized to nitrite by the nitrite-oxidizing bacteria instead of complete oxidation to nitrate, and then ammonium and nitrite are oxidized to N_2 (Arredondo et al. 2015). The completely autotrophic nitrogen removal over nitrite (CANON) process was reported to simultaneously accomplish the aerobic and anaerobic oxidation of ammonium with both aerobic and anaerobic ammonium-oxidizing microorganisms coexisting in the same reactor (Third et al. 2001). Another technology is so-called coupled aerobic–anoxic nitrous decomposition operation (CANDO) in which three steps is completed: 1. partial nitrification of NH_4^+ to NO_2^-, 2. partial anoxic reduction of NO_2^- to N_2O and 3. decomposition of N_2O to N_2, O_2 and energy (Scherson et al. 2013). However, these technologies still focus on removal of nitrogen instead of recovery while treating wastewater at the same time.

2.2 Nitrogen Removal and Recovery in BES

2.2.1 Bioelectrochemical Nitrogen Transformation and Ammonia Transport

BES is capable of removing and recovering nitrogen via bioelectrochemical, chemical and biological processes (Figure 1). The ammonia nitrogen contained in wastewater can migrate from the anode to the cathode where it can be either recovered or transformed to nitrate by microbiological oxidation. Nitrate can be reduced to nitrogen gas as an electron acceptor through either bioelectrochemical denitrification or heterotrophic denitrification (Kelly and He 2014). Bioelectrochemical denitrification occurs in the cathode whereby accepting electrons from the oxidation of acetate or glucose at the anode (Ghosh Ray and Ghangrekar 2014) nitrate is converted to nitrogen gas as a terminal electron acceptor. Bioelectrochemical denitrification can reduce nitrite to nitrogen by autotrophic denitrifying bacteria in either pure culture (Gregory et al. 2004) or mixed culture (Park et al. 2005) via accepting electrons from a cathode electrode. The microbial community in such biocathode has been studied by many researchers, revealing that various species were active and dominant in the denitrifying biofilm affected by operational factors or enrichment approaches (Kelly and He 2014). It was found that denitrifying bacteria containing a special *nirS* gene were dominant in the community that played a crucial role in the denitrification process (Vilar-Sanz et al. 2013).

In most wastewaters, the predominant form of nitrogen is ammonium (Peccia et al. 2013). Ammonium can be removed and recovered in BES via several mechanisms (Figure 1). The first mechanism is based on nitrification and denitrification of ammonium. It can be transformed to nitrate by microbiological oxidation that is subsequently reduced to inert nitrogen gas by denitrifiers through accepting electrons from the cathode electrode. Direct conversion of ammonium to nitrogen gas may also be possible in a BES via an ANAMMOX-like process that uses ammonium as an electron donor in the anode, and there was nearly a linear relationship between ammonium removal rate and current generation (Figure 2) (Vilajeliu-Pons et al. 2017). In a tubular MFC, complete nitrification in the outer cathode and denitrification in the

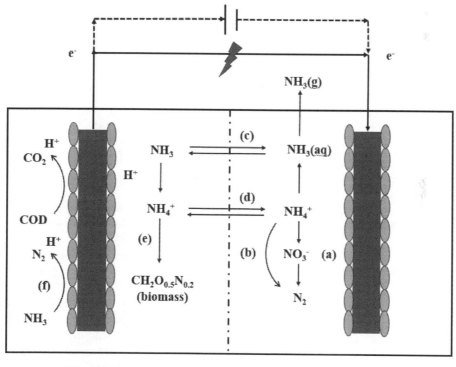

Figure 1: Mechanisms of nitrogen transformation processes in BES: (a) nitrification and denitrification of ammonium nitrogen in a catholyte; (b) anaerobic ammonium oxidation; (c) ammonia diffusion driven by the concentration gradient between the anolyte and catholyte; (d) ammonium migration induced by the electric field; (e) ammonium utilized for biomass during bacterial growth; and (f) direct oxidation of ammonium to nitrogen gas at the anode.

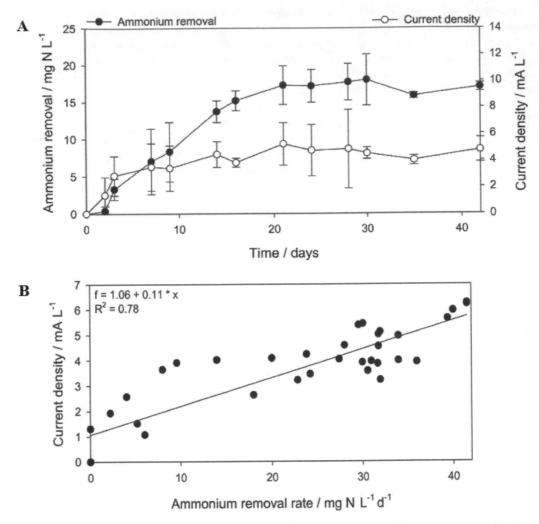

Figure 2: Performance of BES reactors over 42 days. (A) Ammonium removal and current density in the anode compartment of BES that were continuously operated at +0.8 V vs. SHE with a constant influent flow of 1 L d^{-1}. (B) Correlation between the ammonium removal rate and the current density (Vilajeliu-Pons et al. 2017). Copyright © Water Research (Elsevier). Reproduced with permission.

inner cathode was accomplished by removing more than 96.0% of ammonium, and the total nitrogen removal efficiency was between 66.7 and 89.6% (Zhang and He 2012a). Simultaneous nitrification and denitrification (SND) were reported in an MFC with a nitrogen removal efficiency of about 94.0% (Virdis et al. 2010). A BES containing membrane-aerated nitritation-anammox in the cathode was developed to achieve total nitrogen removal efficiency of ~ 95.0% using a loop operation and under an externally applied voltage (Yang et al. 2017).

The second mechanism focuses on ammonia recovery via the transport (diffusion and migration) of ammonia and ammonium ions. Ammonium ions, induced by the electric field, are capable of migrating from the anode to the cathode across a CEM (cation exchange membrane) to maintain the charge balance. Driven by the concentration gradient between the anolyte and catholyte, ammonium ions can diffuse to the cathode as well. Due to the alkaline pH, ammonium is converted to ammonia gas that can then be stripped out of the catholyte. Ammonium transport also acts as a proton/cation carrier; it was reported in an MFC that the NH_4^+/NH_3 migration accounted for approximately 90.0% of the ionic flux in the

BES (Cord-Ruwisch et al. 2011). In a two-chamber MFC with a high concentration of ammonium, the charge transport was found to be proportional to the concentration of ions and would be influenced by a concentration gradient and charge exchange processes (Kuntke et al. 2011). In addition to MFCs, ammonium transport was also reported in MECs and MDCs. Instead of stripping ammonia from the catholyte by air or nitrogen gas, Cheng et al. applied hydrogen evolved at the cathode to volatilize NH_3 and recycled it to the anode chamber in an MEC, successfully refraining the anolyte from acidification. This has provided the feasibility of pH control via ammonium recycle and has given a novel pathway for nitrogen recovery from wastewater (Cheng et al. 2013). Inspired by the MDC concept, Chen et al. constructed a microbial nutrient recovery cell (MNRC) to achieve simultaneous nitrogen and phosphorus recovery by exchanging the relative positions of AEM (anion exchange membrane) and CEM (Chen et al. 2015).

In addition to those main mechanisms above, there is also utilization of ammonium for biomass during bacterial growth and the direct oxidation of ammonium to nitrogen gas at the anode that lacks solid proof so far (Arredondo et al. 2015). Some researchers believed that ammonium was involved in electricity generation either directly as the anodic fuel or indirectly as substrates for bacteria (Zhen et al. 2009), while others did not consider the direct oxidation of ammonium was a source of electricity generation (Kim et al. 2008; Zang et al. 2012b). Herein, we do not elaborate in detail.

2.2.2 Nitrogen Removal and Recovery from Different Water Streams Using BES

Nitrogen compounds in different forms exist broadly in various types of water streams. BES has been applied for nitrogen removal and recovery from groundwater as well as wastewaters such as landfill leachate and urine that rejects water and so on (Table 1). BES has gained its popularity recently with nitrogen removal and recovery for its numerous advantages compared to traditional technologies. Firstly, there is no need for additional chemicals for increasing the pH of the catholyte. Both the oxygen reduction and hydrogen evolution are capable of elevating the catholyte pH that favors the ammonium conversion to ammonia. Secondly, recycling ammonia from a cathode to an anode can control the pH of the anolyte from acidifying and thus sustain current (e.g. pH < 5.5 severely inhibits current generation) without the dosage of alkali or buffers (Cheng et al. 2013). Thirdly, compared to traditional energy-intensive ammonia recovery technologies (such as stripping or electrodialysis), the BES enables ammonia recovery in energy-efficient way as well as current generation and wastewater treatment. In addition, the hydrogen produced in the cathode chamber of MEC can be utilized to strip ammonia out that then can be absorbed in acid bottles.

Table 1: The nitrogen removal and/or recovery performance in different BES reactors. N/A: Not applicable.

BES type	Waste source	Removal efficiency or rate	Recovery efficiency or rate	References
MEC	Groundwater	77.3%	N/A	(Tong and He 2013)
MEC	Landfill leachate	65.7% (with aeration) 54.1% (without aeration)	N/A	(Qin et al. 2016)
MEC-FO	Landfill leachate	63.7%	53.8%	
MFC	Urine	N/A	3.3 g_N d^{-1} m^{-2}	(Ieropoulos et al. 2012)
MEC	Urine	27.8-34.3%	N/A	(Kuntke et al. 2014)
MEC	Reject water	N/A	79.0% (real reject water) 94.0% (synthetic reject water)	(Wu and Modin 2013)

Nitrate-containing groundwater has become a serious threat to human health as groundwater is a major source of supply for drinking water, especially in rural areas where surface water supply is limited. It is reported that groundwater is the primary source of drinking water for 44.0% of the population in the United States. Hence, nitrate removal from groundwater has been broadly studied by using BES. Nitrate can be removed from groundwater through bioelectrochemical denitrification in the cathode. Energy consumption of such an approach is highly variable, depending on the factors such as where BES is installed (e.g., *in situ* vs. *ex situ*) and whether an external power supply will be provided. A preliminary analysis of energy consumption reports specific energy consumption based on the mass of NO_3^--N of 0.341 and 1.602 kWh kg NO_3^--N^{-1} obtained from *in situ* and *ex situ* treatments with MFCs (Cecconet et al. 2018). A new approach was developed for *in situ* nitrate removal in a BES by reducing nitrate to nitrogen gas in the anode (Tong and He 2013). The highest nitrate removal rate of about 208.0 g NO_3^--N m^{-3} d^{-1} was achieved and another undesired ion exchange to groundwater was inhibited when applying an electrical potential of 0.8 V. Further study of nitrate reduction and anion transport competition were conducted in a BES consisting of two tubes, similar to a tubular MDC, that were submerged in groundwater containing nitrate acting as the cathode chamber (Figure 3A). It was concluded that nitrate ions migration driven by electricity was dominant in nitrogen removal rather than biological reduction, and OH$^-$ ions were main competitors with nitrate ions among the anions migration (Tong and He 2014).

Landfill leachate with high concentrations of various contaminants is formed in landfill during the degradation of solid wastes and permeation of wastewater and can result in groundwater and soil pollution. Leachate has been studied as an anode substrate of BES for current generation and resources recovery (Iskander et al. 2016). Because of its high ammonium nitrogen concentration, leachate has also created a great opportunity for nitrogen recovery. A microbial electrolysis cell (MEC)–forward osmosis (FO) system was designed (Figure 3B) to recover ammonium and water from synthetic solutions (Qin et al. 2016) and could accomplish ammonium and water recovery from actual landfill leachate. The recovered ammonium was converted to ammonium bicarbonate that was then used as the draw solution in the FO for water recovery from the treated leachate used as the feed solution. It was also demonstrated that aeration, though increasing energy consumption of the BES, was critical for recovering ammonium.

Urea ($(NH_2)_2CO$) contained in urine is the main form of nitrogen. Urine can be applied as a feed for BES by offering possibilities for nitrogen reduction and ammonium recovery due to its high conductivity (20.0 mS/cm) and urea concentration (20.0 g/L) (Nancharaiah et al. 2016b). Both MFCs and MECs have been proposed to treat urine and recover ammonium as well. Ieropoulos et al. (Ieropoulos et al. 2012) investigated the feasibility of producing electricity and recovering ammonia nitrogen from urine using an MFC for the first time. In an MFC equipped with a gas diffusion cathode, urine was used as both ammonium and energy source. Ammonium transporting into the cathode chamber was converted to volatile ammonia due to high pH produced in catholyte and then was recovered via absorption in an acid bottle. This MFC obtained an ammonium recovery rate of around 3.3 g$_N$ d^{-1} m^{-2} and a surplus of energy 3.5 kJ g$_N^{-1}$. Thus, simultaneous ammonium and energy recovery can be achieved (Kuntke et al. 2012). The current density produced in MFCs is limited by the internal resistance, while MECs can achieve higher current with an applied voltage that also drives the production of hydrogen gas that may compensate the additional energy input (Arredondo et al. 2015). This was demonstrated in an MEC treating diluted urine for ammonium recovery (Figure 3C) with simultaneous COD removal and hydrogen production (Kuntke et al. 2014). It should be noted that MECs may be capable of obtaining a higher ammonia recovery rate than MFCs. The researchers have demonstrated that with efficient recovery the source-separated urine could supply 20.0% of current macronutrient usage and remove 50.0–80.0% of nutrients present in wastewater (Ledezma et al. 2015). It was reported that the energy input of ammonia recovery using microbial electrochemical technologies were lower than that required for processes such as ammonia stripping and electrodialysis and nearly the same as struvite crystallization (Maurer et al. 2003). Detailed economic analysis for ammonium recovery from urine using MFCs and MECs has been carried out by Rodriguez Arredondo and his colleagues (Arredondo et al. 2015).

Nitrogen removal and recovery have been investigated using various wastewaters. Reject water produced from sludge treatment process was treated in an MEC to realize simultaneous hydrogen

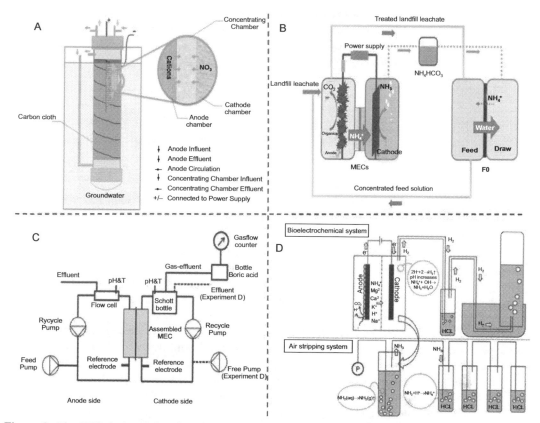

Figure 3: The BES designed for nitrogen removal and recovery from different wastewater streams. (A) A tubular reactor submerged in groundwater containing nitrate acting as the cathode chamber. (B) A microbial electrolysis cell (MEC)-forward osmosis (FO) system for ammonium and water recovery from the treated landfill leachate. (C) An MEC for ammonium recovery, COD removal and hydrogen production from diluted urine. (D) An MEC for simultaneous hydrogen production and ammonia recovery from reject water (Kuntke et al. 2014; Tong and He 2014; Wu and Modin 2013; Qin et al. 2016). (A) Copyright © RSC Advances (Royal Society of Chemistry). Reproduced with permission. (B) Copyright © Bioresource Technology (Elsevier). Reproduced with permission. (C) Copyright © International Journal of Hydrogen Energy (Elsevier). Reproduced with permission. (D) Copyright © Bioresource Technology (Elsevier). Reproduced with permission.

production and ammonia recovery. In this system, ammonia was stripped away by the hydrogen produced in the cathode chamber, and ammonium recovery efficiencies of 79.0% from real reject water and 94.0% from synthetic reject water was achieved (Figure 3D) (Wu and Modin 2013). Desloover et al. used an MEC to understand the NH_4^+ flux from an anode to a cathode and to recover ammonia from anaerobic digestate with produced hydrogen that could volatilize ammonia out of the catholyte; their results showed that a NH_4^+ charge transfer efficiency of 96.0% and NH_4^+ flux of 120.0 g N m^{-2} d^{-1} could be obtained (Desloover et al. 2012). In an MFC coupled with a stripping/absorption unit, pig slurry was applied as the feed solution (Sotres et al. 2015). It was found that the highest nitrogen flux was 7.0 g N d^{-1} m^{-2} when using buffer as a catholyte that then increased to 10.0 g N d^{-1} m^{-2} when shifting the MFC mode to the MEC mode; further improvement to 25.0 g N d^{-1} m^{-2} was realized by using NaCl solution as a catholyte.

2.2.3. *Factors Influencing Nitrogen Removal and Recovery in BES*

Several parameters in BES, such as external resistance, electrolyte pH, membrane type, C/N ratio and other operating conditions, are critical to system performance and nitrogen removal. The interactions between these parameters and their influences should be explored to improve nitrogen removal efficiency.

A high current generation under a low external resistance means more electrons produced from the anode substrate that will benefit the bioelectrochemical denitrification. In a dual-cathode MFC, nitrate and total nitrogen removal efficiency were improved from 52.1% and 51.9% to 66.4% and 68.0%, respectively, by reducing the external resistance from 712 to 10 Ω to increase current generation (Zhang and He 2012b). Haddadi et al. demonstrated that current density was of significant importance to the ammonia recovery in an MEC (microbial electrochemical cell) because of electricity-driven ammonium migration that accounted for 61.0% of ammonium transport (Haddadi et al. 2013).

Both anolyte and catholyte pH are of great importance to nitrogen removal. With anolyte pH varied from 6.0 to 9.0 in a novel three-dimensional BES (3D-BES) (Chen et al. 2016), it was found that the nitrate removal efficiencies of about 98.0% and 96.0% were obtained at pH 7.0 and 8.0, and the dominant bacterial phylum *Firmicutes* and class *Clostridia* decreased under pH 6.0 and 9.0 conditions that led to low nitrate removal efficiencies. Clauwaert et al. revealed that nitrogen removal rate was increased more than twice when maintaining the pH in the biocathode at 7.2 compared to that without pH adjustment, suggesting the proton supply had limited the nitrate reduction in the cathode (Clauwaert et al. 2009).

The membrane type could influence the performance of a BES to some degree. The researchers have investigated the nitrogen removal in MFCs with anion exchange membrane and cation exchange membrane, respectively. The results showed that there was a significant improvement for the total nitrogen removal efficiency from 8.0% in the MFC based on a CEM to 57.0% in the MFC containing an AEM. This phenomenon was contributed to the nitrate migration across AEM and heterotrophic denitrification in the anode (Li and He 2015).

A proper C/N ratio in the anode is considered as a critical parameter for nitrogen removal because the organic substrates are the source of electrons that will be accepted in the cathode for bioelectrochemical denitrification. However, heterotrophic denitrification may be stimulated and bioelectrochemical denitrification may be postponed when substrates are oversupplied (Zhang and He 2013). The effects of different C/N ratios on nitrogen removal in a BES have been investigated (Figure 4), and it was concluded that an increased C/N ratio benefited nitrate removal and depressed nitrite accumulation but did not increase autotrophic denitrification. High C/N ratios inhibited soluble microbial products excretion and increased electrogenesis without anode transformation efficiency improvement (Huang et al. 2013).

Ammonia recovery is greatly dependent upon the factors, such as catholyte pH, current density, type of membrane, the concentration of ammonium in wastewater and aeration. Optimizing these parameters is expected to enhance ammonia recovery efficiency. As previously described, ammonium transported to

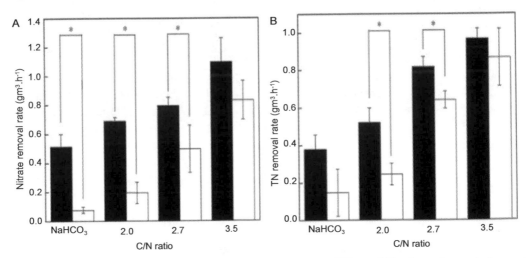

Figure 4: (A) Nitrate. (B) Total nitrogen (TN) removal performance at different C/N ratios and a constant current of 5 mA. Filled bars represent the BESs, unfilled bars represent control reactors. The asterisk (*) indicates that the difference was significant (p < 0.05) (Huang et al. 2013). Copyright © Bioresource Technology (Elsevier). Reproduced with permission.

the cathode can be converted to ammonia gas due to the alkaline pH in the catholyte and be stripped out by air or N_2. In this case, the catholyte pH plays a crucial role in the ammonia recovery. In an MEC using NaCl solution as catholyte, ammonium recovery in a subsequent stripping and absorption process could be favored because of the high pH as a result of the cathode reaction (Cerrillo et al. 2016). Some factors are inter-related and influence the ammonia recovery in different ways. For instance, a high current density is believed to be in favor of the transport of ammonium ions, and thus ammonia recovery will be decreased via decreasing external resistance or increasing applied voltage. It was noted that under low current densities (e.g. I= 1.0 A m^{-2}), diffusion was domination of the ammonium transport, while migration induced by an electric field dominated the transport at high current densities (e.g. I> 5.0 A m^{-2}) (Liu et al. 2016). In addition, the current density generated in a BES can be affected by the type of membrane to some extent. Compared to 7.2 A m^{-2} produced in an MEC with a CEM, a higher current density of 10.2 A m^{-2} (applied voltage 1.0 V) was achieved with an AEM (Kuntke et al. 2011). It could be explained that the internal resistance of the AEM-MEC was lower than that of the CEM-MEC (Sleutels et al. 2013). Concentration gradient across the CEM partly induces ammonia diffusion and migration. In a study of different ammonia concentration of an MFC, although the current density was not significantly affected, the charge transport was proportional to the concentration gradients and therefore influenced ion recovery (Kuntke et al. 2011). Aeration is another important parameter for ammonia recovery. Qin et al. have examined the influence of aeration in the MEC under two modes: aeration in the cathode (oxygen reduction) and no aeration (hydrogen evolution) and found that higher ammonium removal efficiency and ammonia recovery from the leachate could be achieved in the presence of cathode aeration (Figure 5) (Qin et al. 2016).

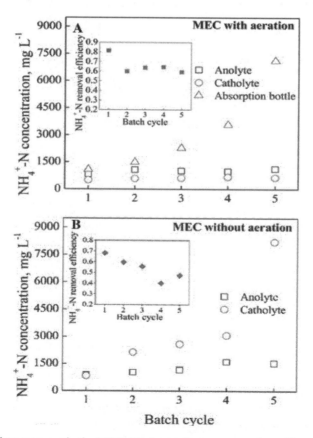

Figure 5: Ammonium recovery in the MEC. (A) Ammonium concentration with aeration. (B) Ammonium concentration without aeration. The insets in (A) and (B) show the NH_4^+-N removal efficiency ($\times$100%) (Qin et al. 2016). Copyright © Bioresource Technology (Elsevier). Reproduced with permission.

3. Phosphorus Removal and Recovery

3.1 Background

Addressing the phosphorus pollution is of great importance. Traditional phosphorus removal technologies include biological processes, chemical precipitation and membrane filtration, etc. In the enhanced biological phosphorus removal (EBPR), polyphosphate accumulating organisms (PAOs) are enriched through aerobic and anaerobic processes during which such special bacteria are capable of storing phosphate within their cells at a level higher than normal and then are removed as waste sludge. Chemical precipitation by adding precipitates, such as lime ($Ca(OH)_2$), alum ($Al_2(SO_4)_3 \cdot 18H_2O$) and ferric chloride ($FeCl_3$), is considered as an easy and low cost way to remove phosphorus from wastewater (De-Bashan and Bashan 2004). Membrane filtration systems are applied to achieve phosphorus removal in the form of both dissolved and solid phosphorus, for instance, reverse osmosis (RO) systems and membrane bioreactors (MBRs) (Sengupta et al. 2015).

3.2 Phosphorus Removal and Recovery using BES

Phosphorus removal/recovery has not been investigated as broadly as nitrogen removal/recovery in BES. Nevertheless, as both valuable resource and waste, phosphorus has stimulated a strong interest in its removal and recovery in MFCs and MECs. Some of the studies are summarized in Table 2. Struvite mineral ($MgNH_4PO_4 \cdot 6H_2O$) is the most common form of phosphorus recovered in BES and can form in the presence of magnesium, ammonium nitrogen and phosphorus under an alkaline environment (>9.2). It is believed that BES can facilitate struvite formation due to the oxygen reduction in the cathode compartment that would elevate the catholyte pH without additional adjustment.

Table 2: The phosphorus removal and/or performance in various BES reactors. MESC: Microbial electrolysis struvite-precipitation cell. IPB: Integrated photobioelectrochemical. N/A: Not applicable.

BES type	Form	Removal efficiency (%)	Recovery efficiency (%)	References
MFC	Struvite	N/A	82	(Fischer et al. 2011)
MFC	Struvite	70-82	27	(Ichihashi and Hirooka 2012)
MESC	Struvite	20-40	N/A	(Cusick and Logan 2012)
MFC MEC	Struvite	N/A	67	(Happe et al. 2016)
IPB	Algal biomass	82	N/A	(Xiao et al. 2012)
MFC	Algal biomass	58	N/A	(Jiang et al. 2012)
MFC + external photobioreactor		92		

Recovering phosphorus as struvite was first studied in an MFC with two chambers. The electrons and protons produced from metabolic activity were used to mobilize phosphorus from an insoluble form to a soluble form and then were precipitated with $MgCl_2$ and NH_4OH to form struvite during which 82.0% of orthophosphate was recovered from digester sludge (Fischer et al. 2011). In another study, the possibility of recovering electrical power and phosphorus simultaneously from swine wastewater was investigated in a single chamber MFC with an air cathode that achieved about 70.0-82.0% of phosphorus removal; it was demonstrated that phosphorus in the form of suspended solid was first dissolved and then precipitated on the surface of cathode (Ichihashi and Hirooka 2012). A single chamber microbial electrolysis struvite-precipitation cell (MESC) was developed to produce hydrogen and struvite concurrently. The results showed the phosphorus removal rate, the struvite crystallization rate as well as hydrogen production rate were highly dependent on the applied voltage and cathode material (stainless steel mesh or flat plates)

(Cusick and Logan 2012). Recently, a 3-L BES with three chambers that was operated in an MFC mode and then in an MEC mode (Figure 6) achieved a phosphorus recovery efficiency of 67.0% from sewage sludge; in this system, diffusion from the sewage sludge layer was considered as the rate-limiting step and phosphorus release was found faster in the MEC mode (Happe et al. 2016).

The photosynthetic processes in combination with BES can also accomplish phosphorus removal and recovery. Many photoautotrophs (such as algae and cyanobacteria) are able to efficiently take nutrients for biomass synthesis from wastewater (Li et al. 2014). Algal growth can be realized either within a BES or in an algal bioreactor coupled with a BES externally. The researchers designed an integrated photobioelectrochemical (IPB) system by installing an MFC inside an algal bioreactor (Figure 7A) and achieved removal efficiencies of 92.0% for COD (in the MFC) and 82.0% for phosphorus (in the algal

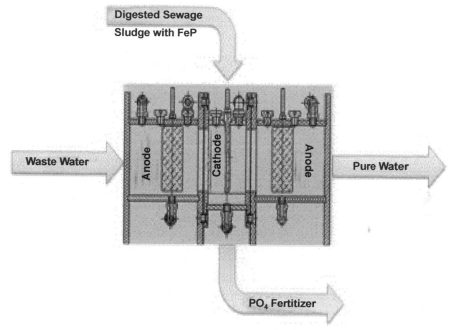

Figure 6: A 3-L BES with three chambers operated in an MFC and MEC mode to achieve phosphorus recovery (Happe et al. 2016). Copyright © Bioresource Technology (Elsevier). Reproduced with permission.

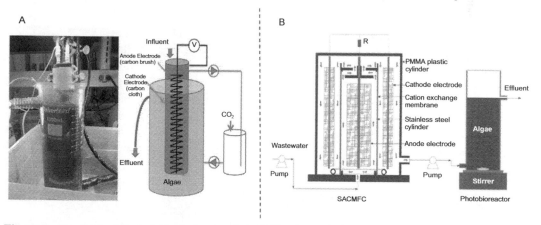

Figure 7: The integrated photobioelectrochemical (IPB) system designed for phosphorus recovery. (A) An MFC installed inside an algal bioreactor. (B) An MFC with an external photobioreactor (Xiao et al. 2012; Jiang et al. 2012). (A) Copyright © Environmental Science & Technology (American Chemical Society). Reproduced with permission; (B) Copyright © Biotechnology Letters (Springer). Reproduced with permission.

bioreactor) (Xiao et al. 2012). Jiang et al. fed wastewater to an MFC for COD removal and then fed the effluent from the cathode chamber to an external photobioreactor (Figure 7B) for nutrients removal (Jiang et al. 2012). Total phosphorus removal efficiency was increased to 92.0% compared with that in a standalone MFC. Phosphorus recovered simultaneously along with nitrogen is of great interest since many types of wastewaters contain both of them that we will discuss in the next section.

4. Simultaneous Removal and Recovery of Nitrogen and Phosphorus in BES

Phosphorus and nitrogen can be recovered simultaneously from high strength wastewaters (such as swine wastewater and urine) by forming magnesium ammonium phosphate (MAP) or struvite at a high pH range (8.5-10) (Kumar and Pal 2015; Iskander 2016; Kumar and Pal 2015). An MFC with MAP precipitation as a pre-treatment process was proposed for reclaiming both electricity and a slow-release fertilizer from urine in which high concentrations of nitrogen and phosphorus from urine were recovered via MAP precipitation process, and the urine was subsequently fed to the MFC to generate electricity. About 95.0% phosphorus and 29.0% nitrogen were recovered and 65.0% COD was removed (Zang et al. 2012a).

A microbial nutrient recovery cell (MNRC) has been developed to recover nutrient ions and purify wastewater simultaneously and also taking advantage of recovering the energy contained in wastewater. In this MNRC, more than 82% of COD was removed and the concentrations of NH_4^+ and PO_4^{3-} in the recovery chamber increased to more than 1.5 and 2.2 times, respectively, compared to the initial concentrations when wastewater was circulated between the anode and cathode chambers (Chen et al. 2015). In another study, Zhang et al. developed a bioelectrochemical system named 'R²-BES' for removing and possibly recovering nutrients from wastewater in which about 83.4% of ammonium nitrogen and 52.4% of phosphate were removed under an applied voltage of 0.8 V by integrating both CEM (for ammonium Migration) and AEM (for phosphate transport) into the BES (Zhang et al. 2014). The concept of source separation, which means separating different kinds of waste streams at their first point of collection, has been proposed over the past few years. Nitrogen and phosphorus can be recovered from source-separated urine. In a review (Ledezma et al. 2015), the researchers represented a business case 'value from urine' that was favored by the nutrients recovery and calculated the capital and energy costs in detail. However, reconcentrating nutrients is energy-intensive. An MEC sandwiched by three AME/CEM pairs, so-called nutrient separation microbial electrolysis cell (NSMEC), was constructed in which nitrogen and phosphorus were removed and reconcentrated from source-separated human urine. During a batch cycle, 54.0% COD was removed and ammonium and phosphate ions were concentrated by 4.5 and 3.0 times, respectively (Tice and Kim 2014).

As previously described, photosynthetic systems are also capable of removing and recovering nitrogen and phosphorus at the same time via the photosynthetic processes with algal growth (Kelly and He 2014). Zhang et al. developed a sediment-type photomicrobial fuel cell (PFC) for removing organic contaminants and nutrients from wastewater through a synergistic interaction between microalgae and electrochemically active bacteria and obtained 99.0% organic carbon removal along with 88.0% nitrogen removal and 70.0% phosphorus removal (Zhang et al. 2011). The results showed that algae biomass uptake accounted for most of the nutrients removal. Luo et al. have presented a review on utilizing the integrated photobioelectrochemical systems for energy recovery and nutrient remediation to which one can refer for more information in detail (Luo et al. 2017).

5. Conclusions and Future Perspectives

Bioelectrochemical systems are promising for nutrients removal and recovery with multiple advantages over conventional technologies. Nevertheless, there exist some challenges to be addressed for further development of BES for nutrients removal and recovery. As previously described, ammonium transport is affected by different parameters, such as current generation, electrolyte pH, membrane type, C/N ratio, the concentration of ammonium in wastewater and aeration. Some of these factors are inter-related to some extent; hence, there is a need to optimize them for improving nitrogen removal and/or recovery efficiency. Some concerns emerging in other types of BES are also of particular importance for the BES with nitrogen removal and/or recovery. For instance, realization of long-term operation and system stability, better understanding of the critical factors influencing microbiological processes (microorganisms are sensitive to temperature, pH, substrates, etc.), difficulties of scaling up of the reactors, transformation from bench research to practical utilization (in most of the previous studies, synthetic wastewater was used as substrates or raw wastewater was used with organic compounds and/ or buffer applied into the system) and reduction of the capital and operational costs. The characteristics of wastewater can obviously affect nutrient removal and recovery in BES. Removal has been applied for most wastewater treatment processes with low-strength nitrogen concentration, while recovery as a sustainable process is more suitable for those containing high-strength nitrogen concentrations (Kelly and He 2014).

Although the studies providing phosphorus removal and recovery are fewer than those of nitrogen removal and recovery, phosphorus as a limited resource is critically important. Because most of the phosphorus removal and recovery technologies are realized through precipitation, the electrolyte pH is crucial. In addition, due to the precipitation formation on the cathode, how to effectively collect those precipitates and how to reuse the cathode are of great challenges that need to be investigated in the future. The economic analysis should be performed and the long-term performance and stability of BES should be better understood.

Acknowledgement

This study was financially supported by Postgraduate Research & Practice Innovation Program of Jiangsu Provice (KYCX17_1484). The authors would like to thank Mr. Shiqiang Zou and Mr. Han Wang for helpful suggestions.

References

Arredondo MR, Kuntke P, Jeremiasse A, Sleutels T, Buisman C, ter Heijne A (2015) Bioelectrochemical systems for nitrogen removal and recovery from wastewater. Environ Sci: Water Res & Technol 1: 22-33.

Batstone D, Hülsen T, Mehta C, Keller J (2015) Platforms for energy and nutrient recovery from domestic wastewater: A review. Chemosphere 140: 2-11.

Cecconet D, Zou, S, Capodaglio, AG, He Z (2018) Evaluation of energy consumption of treating nitrate-contaminated groundwater by bioelectrochemical systems. Sci Total Enviro 636: 881-890.

Cerrillo M, Oliveras J, Viñas M, Bonmatí A (2016) Comparative assessment of raw and digested pig slurry treatment in bioelectrochemical systems. Bioelectrochem 110: 69-78.

Chen D, Wei L, Zou Z, Yang K, Wang H (2016) Bacterial communities in a novel three-dimensional bioelectrochemical denitrification system: the effects of pH. Appl Microbiol Biotechnol 100: 6805-6813.

Chen X, Sun D, Zhang X, Liang P, Huang X (2015) Novel self-driven microbial nutrient recovery cell with simultaneous wastewater purification. Sci Rep 5: 15744.

Cheng KY, Kaksonen AH, Cord-Ruwisch R (2013) Ammonia recycling enables sustainable operation of bioelectrochemical systems. Bioresour Technol 143: 25-31.

Clauwaert P, Desloover J, Shea C, Nerenberg R, Boon N, Verstraete W (2009) Enhanced nitrogen removal in bio-electrochemical systems by pH control. Biotechnol Lett 31: 1537-1543.

Cord-Ruwisch R, Law Y Cheng KY (2011) Ammonium as a sustainable proton shuttle in bioelectrochemical systems. Bioresour Technol 102: 9691-9696.

Cusick RD, Logan BE (2012) Phosphate recovery as struvite within a single chamber microbial electrolysis cell. Bioresour Technol 107: 110-115.

De-Bashan LE, Bashan Y (2004) Recent advances in removing phosphorus from wastewater and its future use as fertilizer (1997–2003). Water Res 38: 4222-4246.

Desloover J, Abate Woldeyohannis A, Verstraete W, Boon N, Rabaey K (2012) Electrochemical resource recovery from digestate to prevent ammonia toxicity during anaerobic digestion. Environ Sci Technol 46: 12209-12216.

Eusebi A, Nardelli P, Gatti G, Battistoni P, Cecchi F (2009) From conventional activated sludge to alternate oxic/anoxic process: the optimisation of winery wastewater treatment. Water Sci Technol 60: 1041-1048.

Feng Z, Sun T (2015) A novel selective hybrid cation exchanger for low-concentration ammonia nitrogen removal from natural water and secondary wastewater. Chem Eng J 281: 295-302.

Fischer F, Bastian C, Happe M, Mabillard E, Schmidt N (2011) Microbial fuel cell enables phosphate recovery from digested sewage sludge as struvite. Bioresour Technol 102: 5824-5830.

Fowler D, Coyle M, Skiba U, Sutton MA, Cape JN, Reis S, Sheppard LJ, Jenkins A, Grizzetti B, Galloway JN (2013) The global nitrogen cycle in the twenty-first century. Phil Trans R Soc B 368: 20130164.

Ghosh Ray S, Ghangrekar M (2014) Evaluation of Electrical Properties under Different Operating Conditions of Bio-electrochemical System Treating Thin Stillage. Recent Advances in Bioenergy Research: 381.

Gregory KB, Bond DR, Lovley DR (2004) Graphite electrodes as electron donors for anaerobic respiration. Environ Microbiol 6: 596-604.

Haddadi S, Elbeshbishy E, Lee H-S (2013) Implication of diffusion and significance of anodic pH in nitrogen-recovering microbial electrochemical cells. Bioresour Technol 142: 562-569.

Happe M, Sugnaux M, Cachelin CP, Stauffer M, Zufferey G, Kahoun T, Salamin P-A, Egli T, Comninellis C, Grogg A-F (2016) Scale-up of phosphate remobilization from sewage sludge in a microbial fuel cell. Bioresour Technol 200: 435-443.

Hoffman BM, Lukoyanov D, Yang Z-Y, Dean DR, Seefeldt LC (2014) Mechanism of nitrogen fixation by nitrogenase: the next stage. Chem Rev 114: 4041-4062.

Huang B, Feng H, Wang M, Li N, Cong Y, Shen D (2013) The effect of C/N ratio on nitrogen removal in a bioelectrochemical system. Bioresour Technol 132C: 91.

Ichihashi O, Hirooka K (2012) Removal and recovery of phosphorus as struvite from swine wastewater using microbial fuel cell. Bioresour Technol 114: 303-307.

Ieropoulos I, Greenman J, Melhuish C (2012) Urine utilisation by microbial fuel cells; energy fuel for the future. PCCP 14: 94-98.

Iskander SM, Brazil B, Novak JT, He Z (2016) Resource recovery from landfill leachate using bioelectrochemical systems: opportunities, challenges, and perspectives. Bioresour Technol 201: 347-354.

Jiang H, Luo S, Shi X, Dai M, Guo R-b (2012) A novel microbial fuel cell and photobioreactor system for continuous domestic wastewater treatment and bioelectricity generation. Biotechnol Lett 34: 1269-1274.

Kelly PT, He Z (2014) Nutrients removal and recovery in bioelectrochemical systems: a review. Bioresour Technol 153: 351-60.

Kim JR, Zuo Y, Regan JM, Logan BE (2008) Analysis of ammonia loss mechanisms in microbial fuel cells treating animal wastewater. Biotechnol Bioeng 99: 1120–1127.

Kumar R, Pal P (2015) Assessing the feasibility of N and P recovery by struvite precipitation from nutrient-rich wastewater: a review. Environ Sci and Pollut Res 22: 17453-17464.

Kuntke P, Geleji M, Bruning H, Zeeman G, Hamelers H, Buisman C (2011) Effects of ammonium concentration and charge exchange on ammonium recovery from high strength wastewater using a microbial fuel cell. Bioresour Technol 102: 4376-4382.

Kuntke P, Sleutels T, Saakes M, Buisman C (2014) Hydrogen production and ammonium recovery from urine by a Microbial Electrolysis Cell. Int J Hydrogen Energy 39: 4771-4778.

Kuntke P, Śmiech K, Bruning H, Zeeman G, Saakes M, Sleutels T, Hamelers H, Buisman C (2012) Ammonium recovery and energy production from urine by a microbial fuel cell. Water Res 46: 2627-2636.

Lackner S, Gilbert EM, Vlaeminck SE, Joss A, Horn H, van Loosdrecht MC (2014) Full-scale partial nitritation/anammox experiences–an application survey. Water Res 55: 292-303.

Ledezma P, Kuntke P, Buisman CJ, Keller J, Freguia S (2015) Source-separated urine opens golden opportunities for microbial electrochemical technologies. Trends Biotechnol 33: 214-220.

Li B, Boiarkina I, Young B, Yu W, Singhal N (2017) Prediction of Future Phosphate Rock: A Demand Based Model. J Environ Info.

Li J, He Z (2015) Optimizing the performance of a membrane bio-electrochemical reactor using an anion exchange membrane for wastewater treatment. Environ Sci: Water Res & Technol 1: 355-362.

Li W-W, Yu H-Q, He Z (2014) Towards sustainable wastewater treatment by using microbial fuel cells-centered technologies. Energy Environ Sci 7: 911-924.

Liu Y, Qin M, Luo S, He Z, Qiao R (2016) Understanding Ammonium Transport in Bioelectrochemical Systems towards its Recovery. Sci Rep 6: 22547.

Luo S, Berges JA, He Z, Young EB (2017) Algal-microbial community collaboration for energy recovery and nutrient remediation from wastewater in integrated photobioelectrochemical systems. Algal Res 24: 527-539.

Maurer M, Schwegler P, Larsen T (2003) Nutrients in urine: energetic aspects of removal and recovery. Water Sci Technol 48: 37-46.

McCarty PL, Bae J, Kim J (2011) Domestic wastewater treatment as a net energy producer–can this be achieved? ACS Publications.

Molognoni D, Devecseri M, Cecconet D, Capodaglio AG (2017) Cathodic groundwater denitrification with a bioelectrochemical system. J Water Process Eng 19: 67-73.

Nancharaiah Y, Mohan SV, Lens P (2016a) Recent advances in nutrient removal and recovery in biological and bioelectrochemical systems. Bioresour Technol 215: 173-185.

Nancharaiah YV, Venkata Mohan S, Lens PNL (2016b) Recent advances in nutrient removal and recovery in biological and bioelectrochemical systems. Bioresour Technol 215: 173-185.

Park HI, kun Kim D, Choi Y-J, Pak D (2005) Nitrate reduction using an electrode as direct electron donor in a biofilm-electrode reactor. Process Biochem 40: 3383-3388.

Peccia J, Haznedaroglu B, Gutierrez J, Zimmerman JB (2013) Nitrogen supply is an important driver of sustainable microalgae biofuel production. Trends Biotechnol 31: 134-138.

Qin M, Molitor H, Brazil B, Novak JT, He Z (2016) Recovery of nitrogen and water from landfill leachate by a microbial electrolysis cell-forward osmosis system. Bioresour Technol 200: 485-92.

Rittmann BE, Mayer B, Westerhoff P, Edwards M (2011) Capturing the lost phosphorus. Chemosphere 84: 846-853.

Scherson YD, Wells GF, Woo S-G, Lee J, Park J, Cantwell BJ, Criddle CS (2013) Nitrogen removal with energy recovery through N 2 O decomposition. Energy Environ Sci 6: 241-248.

Sengupta S, Nawaz T, Beaudry J (2015) Nitrogen and Phosphorus Recovery from Wastewater. Curr Pollut Rep 1: 155-166.

Sleutels TH, Ter Heijne A, Buisman CJ, Hamelers HV (2013) Steady-state performance and chemical efficiency of Microbial Electrolysis Cells. Int J Hydrogen Energy 38: 7201-7208.

Sotres A, Cerrillo M, Viñas M, Bonmatí A (2015) Nitrogen recovery from pig slurry in a two-chambered bioelectrochemical system. Bioresour Technol 194: 373-382.

Sukačová K, Trtílek M, Rataj T (2015) Phosphorus removal using a microalgal biofilm in a new biofilm photobioreactor for tertiary wastewater treatment. Water Res 71: 55-63.

Third K, Sliekers AO, Kuenen J, Jetten M (2001) The CANON system (completely autotrophic nitrogen-removal over nitrite) under ammonium limitation: interaction and competition between three groups of bacteria. Syst Appl Microbiol 24: 588-596.

Tice RC, Kim Y (2014) Energy efficient reconcentration of diluted human urine using ion exchange membranes in bioelectrochemical systems. Water Res 64: 61-72.

Tong Y, He Z (2013) Nitrate removal from groundwater driven by electricity generation and heterotrophic denitrification in a bioelectrochemical system. J Hazard Mater 262: 614-619.

Tong Y, He Z (2014) Current-driven nitrate migration out of groundwater by using a bioelectrochemical system. RSC Adv 4: 10290.

Vilajeliu-Pons A, Koch C, Balaguer MD, Colprim J, Harnisch F, Puig S (2017) Microbial electricity driven anoxic ammonium removal. Water Res 130: 168-175.

Vilar-Sanz A, Puig S, García-Lledó A, Trias R, Balaguer MD, Colprim J, Bañeras L (2013) Denitrifying bacterial communities affect current production and nitrous oxide accumulation in a microbial fuel cell. PLoS One 8: e63460.

Virdis B, Rabaey K, Rozendal RA, Yuan Z, Keller J (2010) Simultaneous nitrification, denitrification and carbon removal in microbial fuel cells. Water Res 44: 2970-2980.

Wang H, Hang Q, Crittenden J, Zhou Y, Yuan Q, Liu H (2016) Combined autotrophic nitritation and bioelectrochemical-sulfur denitrification for treatment of ammonium rich wastewater with low C/N ratio. Environ Sci and Pollut Res 23: 2329-2340.

Wang H, Ren ZJ (2013) A comprehensive review of microbial electrochemical systems as a platform technology. Biotechnol Adv 31: 1796-1807.

Wu X, Modin O (2013) Ammonium recovery from reject water combined with hydrogen production in a bioelectrochemical reactor. Bioresour Technol 146: 530-536.

Xiao L, Young EB, Berges JA, He Z (2012) Integrated photo-bioelectrochemical system for contaminants removal and bioenergy production. Environ Sci Technol 46: 11459-11466.

Yang Y, Li X, Yang X, He Z (2017) Enhanced nitrogen removal by membrane-aerated nitritation-anammox in a bioelectrochemical system. Bioresour Technol 238: 22-29.

Zang G-L, Sheng G-P, Li W-W, Tong Z-H, Zeng RJ, Shi C, Yu H-Q (2012a) Nutrient removal and energy production in a urine treatment process using magnesium ammonium phosphate precipitation and a microbial fuel cell technique. PCCP 14: 1978-1984.

Zang GL, Sheng GP, Li WW, Tong ZH, Zeng RJ, Shi C, Yu HQ (2012b) Nutrient removal and energy production in a urine treatment process using magnesium ammonium phosphate precipitation and a microbial fuel cell technique. PCCP 14: 1978-84.

Zhang F, He Z (2012a) Integrated organic and nitrogen removal with electricity generation in a tubular dual-cathode microbial fuel cell. Process Biochem 47: 2146-2151.

Zhang F, He Z (2012b) Simultaneous nitrification and denitrification with electricity generation in dual-cathode microbial fuel cells. J Chem Technol Biotechnol 87: 153-159.

Zhang F, He Z (2013) A cooperative microbial fuel cell system for waste treatment and energy recovery. Environ Technol 34: 1905.

Zhang F, Li J, He Z (2014) A new method for nutrients removal and recovery from wastewater using a bioelectrochemical system. Bioresour Technol 166: 630-4.

Zhang X (2017) Biogeochemistry: A plan for efficient use of nitrogen fertilizers. Nature 543: 322.

Zhang Y, Noori JS, Angelidaki I (2011) Simultaneous organic carbon, nutrients removal and energy production in a photomicrobial fuel cell (PFC). Energy Environ Sci 4: 4340-4346.

Zhen H, Kan JJ, Wang YB, Huang YL, Mansfeld F, Nealson KH (2009) Electricity production coupled to ammonium in a microbial fuel cell. Environ Sci Technol 43: 3391-7.

Microbial Fuel Cells: Treatment Efficiency and Comparative Bioelectricity Production from Various Wastewaters

Andrea G. Capodaglio* and Silvia Bolognesi

Department of Civil Engineering and Architecture (D.I.C.Ar), University of Pavia, Pavia 27100, Italy

1. Introduction

Global depletion of fossil reserves makes it necessary to develop alternative energy sources, including renewable bioenergy from wastes, with a neutral or negative carbon footprint. Current wastewater treatment processes (WWTPs) are energy-intensive (0.5 - 2 kWh m^{-3} treated) depending on the process applied and waste composition. This way about 3-4% of all US energy demand (similarly to other developed countries) is attributable to water sanitation ($\approx$110 TWh/year) (Gude 2016). Greenhouse gases (GHGs) such as carbon dioxide (CO_2) and nitrous oxide (N_2O) are released, in the process, to the atmosphere; for every ton of wastewater treated, 1.5 tons of GHGs are released (Wang et al. 2010). Furthermore, treatment processes generally produce residuals needing further disposal at high additional energy cost (Li et al. 2014).

In this context, microbial electrochemical technologies (METs) represent a wastewater energy and resources recovery technology class, including microbial fuel cells (MFCs) which is a promising technology for bioelectricity production (Capodaglio et al. 2013, 2016) capable of converting chemical energy embedded in organic substrates directly into electrical energy and exploiting the biocatalytic effect of specific electroactive bacteria (EAB) acting on reactions of substrate oxidation and/or reduction. Practical MFCs application in WWTPs, however, has been long delayed by the instability of the engineered systems and by low power and voltage outputs achievable so far. Anodic side-reactions (e.g. methanogenesis) and effects of microorganism competition represent process drawbacks, partly mitigated by appropriate strategies (Molognoni et al. 2014, 2016). For these reasons, it is essential to keep a close eye on the ongoing development of this technology.

Various substrates are suitable for degradation in MFCs with electricity production from simple compounds to complex organic matter mixtures; however, substrate typology critically affects the production of electricity. The bioelectricity generation in MFCs depends on many operational and construction parameters, including architecture configuration more than biodegradation efficiency (Capodaglio et al. 2015). The scope of this review is to establish the current state-of-the-art in MFC

---

*Corresponding author: capo@unipv.it

applications with the most varied types of substrates and to discuss critically the possible future of the technology.

2. Energy Consumption and Recovery in Wastewater Treatment

So far, the aims of WWTPs were to meet effluent standards for discharge (removing at least carbonaceous and nutrient compounds) and stabilize biosolids for land application. Current advances in WWTPs scope have expanded to a larger array of pollutants removal (e.g. emerging contaminants) to energy consumption minimization and energy and resource recovery optimization. In fact, WWTPs are now often addressed as WRRFs (Water Resources and Recovery Facilities).

Typical configurations of civil WWTPs present aerobic biotreatment followed by anaerobic biosolids digestion. Unit power requirements depend upon process and topographic factors: aeration and hydraulic profile are determinants. Energy needs for typical WWTPs are about 0.6 kWhm^{-3} of wastewater treated, about half thereof for supplying oxygen and 20% for auxiliary processes (e.g., pumping, headworks, lighting, post-aeration, heating and sludge handling) (McCarty et al. 2011). Carbon removal energy demand is lower than nitrogen removal. Specific energy requirements usually decrease with plant size (Gude 2016). Generally, energy represents 15 to 40% of the total operational cost of WWTPs and only lagging behind workforce (Gikas 2017).

Up to 50% of WWTP's energy requirements may be satisfied by recovering biogas (CH_4) produced during biosolids digestion, and targeted facilities modifications may further reduce them considerably (EPA 2006). Most of the energy content of raw wastewater is attributable to settleable volatile solids (Gude 2016). Residual biosolids are rich in organic carbon and nutrients with energy content 4 to 5 times higher than that of raw wastewater; Shizas and Bagley (2004) quantified energy content of primary sludge as 15.9 MJ/kg and of secondary sludge as 12.4 MJ/kg (dry weight basis), respectively.

The key issues for an energy balance in WWTPs, however, are that wastewater is generated in excess, regardless of resource scarcity (water for dilution and conveyance of waste), and its intrinsic energy content generally has a lower exploitation level than its full potential due to thermodynamic and technological constraints (Gikas 2017). Today, anaerobic digestion (AD) is the most common energy recovery strategy from WWTPs; organic matter in primary and secondary treatment solids is converted into biogas that could provide between 39-76% of all energy consumed in a WWTP (Soares et al. 2017) with electric energy yield depending on employed technology. The EPA estimated that for each MG (3,785 m^3) of wastewater treated approximately 491 kWh may be produced with a microturbine following AD or 525 kWh could be produced with internal combustion engines. Stillwell et al. (2010) calculated that WWTPs with capacity < 5 MGD (18,900 m^3 d^{-1}) do not produce enough biogas to make electricity generation feasible or cost-effective.

Other options for energy recovery from wastewater come from chemical and thermochemical processes, including gasification, liquefaction and pyrolysis. Biosolids may constitute an excellent feedstock for biodiesel production due to their high lipid concentration (Tiquia-Arashiro and Mormile 2013). Algae-based processes could be energy positive as they produce biomass that might be used as feedstock for high-value biofuels. Several lines of research are attempting to effectively convert WWTPs into WRRFs. Among these, optimization and scale-up of MFC technology is certainly important.

The purpose of this review is to analyze and compare bioelectricity generation and treatment efficiency of several wastewaters with MFC, based on the experiences of many authors so far, in order to bring researches one step closer to successful scaling up of the technology.

3. Microbial Fuel Cells

MFCs are bioelectrochemical systems (BESs) that rely on the catalytic action of EABs to oxidize organic substrate in anodic conditions and release electrons and protons. Electrons travel through an external circuit from the anode to cathode, while protons (or other charge-balancing ions) reach the cathode through ionic selective membranes. There electrons and protons combine with a terminal electron acceptor (TEA), usually oxygen (Logan and Rabaey 2012). Extracellular electron transfer (EET) by EABs may be achieved by different paths as illustrated in Figure 1a. MFCs may also be equipped with biocathodes, where electrotrophic biomass acts as a catalyzer of reduction reactions, improving cell sustainability compared to metal catalysts (He and Angenent 2006; Nikhil et al. 2017). The main advantage of MFCs, besides removal of organic matter, is electric energy generation that could be harvested by low-power management systems (Dallago et al. 2016). Dual chamber is a conventional architecture for this technology, but also air-cathode (cathode exposed to air) (see Figure 1) and single chamber systems are common.

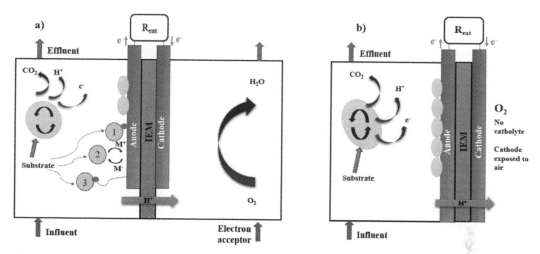

Figure 1: a) Scheme of a two-chamber MFC; the possible electron transfer mechanisms: (1) direct electron transfer; (2) electron transfer through mediators and (3) electron transfer through nanowires. b) Scheme of single chamber MFC with open air cathode.

Electrochemical reactions in MFCs are exergonic as they possess negative Gibbs' free energy, indicating spontaneously proceeding reactions with electrical energy release. Typical redox reactions involved in the energetic balance are reported in Equations 1-3, considering acetate as anodic substrate, electron donor (ED), and oxygen as TEA at the cathode (Rozendal et al. 2008). Different EDs and TEAs (i.e. NO_3^-) lead to different bioelectrochemical reactions with different yields in bioelectricity production.

$$\text{Anode: } CH_3COO^- + 4H_2O \rightarrow 2HCO_3^- + 9H^+ + 8e^- \quad (E^0 = -0.289 \text{ V vs. SHE}) \tag{1}$$

$$\text{Cathode: } 2O_2 + 8H^+ + 8e^- \rightarrow 4H_2O \quad (E^0 = 0.805 \text{ V vs. SHE}) \tag{2}$$

$$\text{Total: } CH_3COO^- + 2O_2 \rightarrow 2HCO_3^- + H^+ \quad (\Delta E^0 = 1.094 \text{ V}) \tag{3}$$

where E^0 is anode/cathode potential, and ΔE^0 the electromotive force.

4. Wastewater as MFCs Substrate

MFCs are a promising technology to achieve both wastewater treatment and energy production (ElMekawy et al. 2015), and therefore, researchers are investigating combinations of wastewaters and MFC designs. Results are commonly reported according to parameters such as coulombic efficiency (CE, %), COD removal efficiency (ηCOD) and electrical energy production in terms of current (Cd) or power density (Pd). The substrate is one of the most important factors affecting bioelectricity generation (Pant et al. 2010). In general, MFCs produce very low amounts of residuals, much lower than biosolids with high levels (about 66%), while organics produced by conventional processes require further treatment. This often overlooked detail constitutes an additional economic and energetic advantage of MFCs over conventional, non-EAB based biological processes. Table 1 summarizes the

Table 1: Physicochemical characterization of several wastewaters.

Substrate	Organic matter		Alkalinity	Kjeldahl nitrogen	Ammonium	Nitrite and nitrate	Total susp sol	Conductivity	pH
	BOD_5 mgO_2L^{-1}	COD mgO_2L^{-1}	Alk mgL^{-1}	TKN $mgN\,L^{-1}$	NH_4^+ $mgN\,L^{-1}$	NO_x^- $mgN\,L^{-1}$	TSS $mg\,L^{-1}$	Cond $mS\,cm^{-1}$	pH -
Domestic WW	508±187	376±133	653±154	87.2±21.0	68.8±11.7	0.91±0.82	176	0.68	6-8
Swine manure	2200±665	1302±174	4745±575	650±40	540±50	n.a.	425±140	8.6±0.8	n.a.
Agricultural	n.a.	9700	n.a.	n.a.	677	n.a.	600	5.6	6.5
Dairy WW	300-1400	650-3000	257-657	20-160	10-20	10-20	250-2700	2.000 ± 1.500	4-12
Paper WW	n.a.	600	316		<2	n.a.	20	1.390	6.3
Distillery WW	35000-50000	90000±10000	n.a.	n.a.	n.a.	n.a.	18460-20200	33.3±3	4-4.5
Brewery WW	n.a.	3000-5000	0	n.a.	16±5	8	500±50	702	3.5-7
Food WW	1300	1900	380	n.a.	120	n.a.	800	n.a.	7

main physicochemical characteristics of wastewaters tested (at laboratory scale) for MFC treatment with bioelectricity production. The following paragraphs summarize the literature results of MFC applications according to five main wastewater origin classifications.

4.1 Domestic (Municipal) Wastewater

Municipal wastewaters may include a multitude of pollutants with reduced toxicity, so they generally are a suitable substrate for MFCs and shown to respond positively to EAB's metabolism. Ammonium recovery and electricity generation from source-separated urine were also shown possible. Santoro et al. (2013) demonstrated single compartment MFC treating raw human urine as a cost-effective process for power production, contaminant removal and nutrient recovery. Suitability of separated human feces (blackwater) for electricity production with MFCs was demonstrated by Fangzhou et al. (2011). Table 2 summarizes results of municipal wastewater-fed MFC's operation efficiency and energy recovery. It should be observed that power densities are commonly reported in terms of electrodes unit volume or unit surface, depending on cell architecture and constituent materials, and that often a direct comparison of these values is not possible.

Table 2: Domestic wastewater used as substrate in MFCs and their respective performances.

Substrate	MFC type	Substrate concentration	COD removal (ηCOD)	Power density (max)	CE	Reference
Domestic wastewater	Air-cathode	300 mgCODL^{-1}	71%	103 mWm^{-2}	18.4%	(You et al. 2006)
Domestic wastewater	Upflow membraneless + photobioreactor	238.7 mgCODL^{-1}	77.9	481 mWm^{-3}	14%	(Jiang et al. 2013)
Domestic wastewater	Air-cathode	345 mgCODL^{-1}	83%	22.5 Whm^{-3}	18%	(Cusick et al. 2010)
Human urine	Single compartment	17 gCODL^{-1}	75%	*	*	(Santoro et al. 2013)
Human feces	Two chambered	650 mgCODL^{-1}	71%	70.8 mWm^{-2}	*	(Fangzhou et al. 2011)

*indicates data not available from the cited reference

4.2 Food Industry Wastewater

Food industry wastewater is considered an ideal substrate for electricity generation in MFCs due to the high organic content, high biodegradability and large availability (ElMekawy et al. 2015). Production of food wastes (FW) involves the entire food supply chain, starting from industrial processing up to wholesale trade. FW may be coincinerated with other urban waste or codigested with biosolids for energy production. Throughout the world, huge quantities of food wastes rich in carbohydrate content are produced and about 27% of MSW (Municipal Solid Waste) is composed of food waste (Parfitt et al. 2010; Behera et al. 2012). Globally about 33% of food is wasted, prompting investigations on food waste as a potential substrate in MFCs (Pandey et al. 2016).

Food wastes have two main origins: vegetal (i.e., cereals, potatoes, oil and citrus crops) or animal (i.e., meat by-products and cheese whey) (Galanakis 2012). One of the leading food industries worldwide is the dairy industry that is known for processing raw milk into consumer products. Dairy wastewater is complex in nature and contains high concentrations of fermentable substrates that can be regarded as an efficient anolyte in MFCs (He et al. 2017). In addition, the dairy industry is one of the most water-intensive processes; hence, considerably high interest has developed on MFC technology for dairy waste processing (Mahdi Mardanpour et al. 2012; Nimje et al. 2012; Cecconet et al. 2018). Treatment of cheese whey (a by-product of cheesemaking characterized by 4-5% carbohydrates, <1% proteins, 0.4-0.5% fat, <1% lactic acid and 1-3% salts content) in MFCs has also been successfully attempted (Pandey et al. 2016).

Palm oil is one of the most relevant and high yield crops in tropical countries and its processing yields high amounts of palm oil mill effluent (POME) that is characterized by complex substrates inclusive of amino acids, inorganic nutrients and a mixture of carbohydrates ranging from hemicelluloses to simple sugars. More than 2.5 tons of POME can be produced by processing 1-ton palm oil (Ahmad et al. 2003), with COD and BOD up to 50,000 mg L^{-1} and 25,000 mg L^{-1}, respectively, and high acidic content cause environmental concern. Conventional POME processing requires hydrolysis and is highly energy consuming, while MFCs processing, of which several positive experiences are reported, could balance energy requirement with power generation (Cheng et al. 2010; Leaño et al. 2012; Baranitharan et al. 2015).

Apart from the abovementioned specific streams, various food industry wastewaters have been investigated as MFC substrate. The protein industry wastewater is characterized as nontoxic with few hazardous compounds, high BOD and high organics content of simple sugars and starch (Mansoorian et al. 2016). Acidogenic food waste leachate, characterized by a complex structure and high organic load, was also evaluated (Rikame et al. 2012; Min et al. 2013). Vegetable market waste is also a potential candidate for bioelectricity generation due to high biodegradability of the organic fraction within (Mohanakrishna et al. 2010a; Venkata Mohan et al. 2010). Canteen food waste was also fed to MFCs by Cercado-Quezada et al. (2010). Mustard tuber processing wastewater (MTWW) is characterized by high-strength and salinity and also generated in large volumes. Guo et al. (2013) reported treating MTWW with dual chamber MFCs, obtaining Pd_{max} = 246 mW m^{-2}, 67% CE and 85% COD removal. Fermented apple juice, a brewery product, was tested as MFC substrate by Cercado-Quezada et al. (2010). Table 3 summarizes literature-reported experiences with MFC application to food industry wastewaters.

Table 3: List of food processing wastewater used as substrate in MFCs and their respective performances.

Substrate	MFC type	Substrate concentration	COD removal (ηCOD)	Power density (max)	CE	Reference
Acidogenic food waste leachate	Dual chambered	5000 mgCODL^{-1}	90%	15.14 Wm^{-3}	*	(Rikame et al. 2012)
Acidogenic food waste leachate	Dual chambered	1000 mgCODL^{-1}	87%	432 mWm^{-3}	20%	(Min et al. 2013)
Canteen based food waste	Single chambered	3200 mgCODL^{-1}	86.4%	556 mWm^{-2}	23.5%	(Cercado-Quezada et al. 2010)
Canteen based food waste	Single chambered	1.13 kgCODm^{-3}d^{-1}	64.8%	5.13 mWm^{-2}	*	(Goud et al. 2011)
Cassava mill wastewater	Two chambered	16000 mgCODL^{-1}	72%	1771 mWm^{-2}	20%	(Kaewkannetra et al. 2011)
Cereal processing wastewater	Dual chambered	595 mgCODL^{-1}	95%	81 mWm^{-2}	40.5%	(Oh and Logan 2005)
Chocolate industry wastewater	Dual chambered	1459 mgCODL^{-1}	74.8%	1500 mWm^{-2}	*	(Patil et al. 2009)

(Contd.)

Table 3: (*Contd.*)

Substrate	MFC type	Substrate concentration	COD removal (ηCOD)	Power density (max)	CE	Reference
Composite vegetable waste	Single chambered mediator-less	52 gCODL^{-1}	62.86%	57.38 mWm^{-2}	*	(Venkata Mohan et al. 2010)
Fermented canteen waste	Single chambered	823 mgCODL^{-1}	44.3%	530 mAm^{-2}	*	(Goud et al. 2011)
Dairy industry wastewater	Annular single chamber spiral	1000 mgCODL^{-1}	91%	20.2 Wm^{-3}	26.9%	(Mahdi Mardanpour et al. 2012)
Dairy industry wastewater	catalyst-less and mediator-less membrane	53.22 kgCODm^{-3}	90.46%,	621.13 mWm^{-2}	37.16%	(Mansoorian 2016)
Dairy industry wastewater	Two chamber	1500 mgCODL^{-1}	85%	27 Wm^{-3}	36%	(Cecconet et al. 2017)
Cheese whey	Dual chambered tubular	1134 mgCODL^{-1}	74.8%	1.3±0.5 Wm^{-3}	*	(Kelly and He 2014)
Cheese whey	Dual chambered	6.7 gCODL^{-1}	94%	46 mWm^{-2}	11.3%	(Tremouli et al. 2013)
Yogurt waste	Dual chambered	8159 mgCODL^{-1}	*	53.8 mWm^{-2}	*	(Cercado-Quezada et al. 2010)
Fermented apple juice	Dual chambered	3501 mgCOD L^{-1}	*	43.7 mWm^{-2}	*	(Cercado-Quezada et al. 2010)
Fermented vegetable waste	Single chambered	0.93 kgCODm^{-3}d^{-1}	80%	111.76 mWm^{-2}	*	(Mohanakrishna et al. 2010b)
Food processing wastewater	Single chamber mediatorless	1900 mgCODL^{-1}	86%	230 mWm^{-2}	21%	(Mansoorian et al. 2013)
Mustard tuber wastewater	Dual chambered	550 mgCODL^{-1}	57.1%	246 mWm^{-2}	67.7%	(Guo et al. 2013)
Palm oil mill effluent	Dual chambered	2680 mgCODL^{-1}	32%	107.35 mWm^{-2}	74%	(Baranitharan et al. 2015)
Palm oil mill effluent	Cylindrical	8000-10500 mgCODL^{-1}	90%	44.5 mWm^{-2}	*	(Cheng et al. 2010)
Palm oil mill effluent	Dual chambered	38400 mgCODL^{-1}	54%	18.92 mWm^{-2}	*	(Leaño et al. 2012)
Protein food industry wastewater	Dual chambered	1900 mgCODL^{-1}	86%	45 mWm^{-2}	15%	(Mansoorian et al. 2013)
Vegetable waste	Single chambered	0.70 kgCODm^{-3}d^{-1}	62.9%	57.38 mWm^{-2}	*	(Venkata Mohan et al. 2010)

*indicates the data not available from the cited reference

4.3 Agricultural By-products

Abundance and renewability of cellulosic and lignocellulosic agricultural residues make them cheap renewable energy and carbon resource. Lignocellulosic biomass is often used for biofuel or biochar production (Plecha et al. 2013) but cannot be directly used by MFCs for electricity generation. However, fermentable sugars derived from these materials by acid or enzymatic treatments are suitable for MFC processing (ElMekawy et al. 2015).

Starch products industry use large amounts of process water, generating large amounts of process wastewater (SPW) characterized by high COD (16,870-22,800 mgL^{-1}). SPW also contains relatively high amounts of carbohydrates (2,300-3,500 mgL^{-1}), sugars (0.65-1.18%), proteins (0.12-0.15%) and starch (1,500-2,600 mgL^{-1}), representing an energy-packed resource that could impose heavy loads on the environment but may also be converted into a wide variety of final products (Jin et al. 2002). Cassava mill wastewater is carbohydrate-rich with high COD, BOD, TSS and low ammonium concentrations (Kaewkannetra et al. 2011).

Wheat straw is a quite common agricultural residue. Organic carbon in wheat straw consists of about 34-40% cellulose, 21 to 26% hemicellulose and 11 to 23% lignin; this may be hydrolyzed to obtain a carbohydrate-rich liquid substrate 'wheat straw hydrolysate (Khan and Mubeen 2012) that has been used in MFC studies for electricity generation (Zhang et al. 2009). Although recovered power density was low, it proved suitable as MFC substrate.

Rice straw, on the other hand, is one of the cheapest and most abundant agricultural wastes, mainly composed of cellulose, hemicellulose and lignin, and better suited as MFC substrate. Using rice milling industry wastewater as substrate, Pd_{max} = 2.3 Wm^{-3} and ηCOD = 96.5%, were reported in earthen-pot MFCs with additional 84% lignin and 81% phenol removal (Behera et al. 2010). Gurung and Oh (2015) used powdered rice straw without pretreatment and inoculated a mixed culture of cellulose-degrading bacteria in MFCs; at an initial concentration of 1 gL^{-1}, Pd_{max} = 190 mWm^{-2} was reported. So far, MFCs have proven effective in the treatment of rice waste materials and bioelectricity harvesting.

Corn stover is an agricultural by-product mainly composed of cellulose and hemicellulose (70%) and is converted into sugars by cellulosic enzyme treatment or steam explosion (ElMekawy et al. 2015). Raw corn stover was fed as the substrate for electricity generation in single chambered MFCs. Although the treatment was effective (X. Wang et al. 2009), power output was much lower than other substrates.

Molasses from sugarcane mills are widely used in the fermentation industry, representing one of the most important raw materials for ethanol production due to low cost and wide availability. Its use as raw material for fermentation products (i.e., alcohol and amino acids) produces large amounts of high strength wastewater that still needs treatment. Several methods, chemical and biological, are applied. MFCs processing has also been successful; however, due to the complex nature of this substrate, it is difficult to accomplish both wastewater treatment and energy recovery at desired levels (B. Zhang et al. 2009; Mohanakrishna et al. 2010a; Sevda et al. 2013).

4.3.1 Cattle Manure

Livestock industry includes feedlots (manure production) and slaughterhouses. These produce high-strength wastewaters, mainly constituted of biodegradable organic carbon, fats and proteins (Zheng and Nirmalakhandan 2010). Use of manure and slaughterhouse wastewater as MFCs substrate has been reported.

Animal carcass wastewater is conventionally disposed of through alkaline hydrolysis; the resulting sterile solution is a coffee-colored alkaline solution characterized by high amounts of BOD (70 gL^{-1}), COD (105 gL^{-1}), ammonia (1 gL^{-1}), organic nitrogen (8 gL^{-1}) and total phosphorus (0.4 gL^{-1}) (Das 2008). Manure wastewater is the main waste of farming activity. Causes of concern during treatment are the generation of methane gas, ammonia and odors (due to volatile organic acids) (Pandey et al. 2016). Animal manure is normally high strength, rich in nitrates and phosphates (Tam et al. 1996; Garrison et al. 2001; Tiquia 2003) and must be treated to specific regulations. Alkaline-thermally pretreated swine wastewater seems to be suitable in enhancing MFC operation compared to raw swine wastewater enhancing resources and energy recovery (Guo and Ma 2015).

Table 4 lists agricultural residuals and livestock industry process wastewater used as anodic MFC substrate and shows their respective performances.

Table 4: List of agricultural by-products and livestock industry process wastewater used as anodic substrate in MFCs, and their respective performances.

Substrate	MFC type	Substrate concentration	COD removal (ηCOD)	Power density (max)	CE	Reference
Beet-sugar wastewater	Upflow anaerobic sludge blanket reactor	*	53.2%	1410 mWm^{-2}	*	(Cheng et al. 2016)
Beet-sugar wastewater	Anaerobic baffled stacking (ABSMFC)	*	70%	115.5 mWm^{-2}	*	(Zhao et al. 2013)
Distillery wastewater (molasses based)	Single chamber open air-cathode	15.2 kgCOD m^{-3} d^{-1}	72.8%	124.35 mWm^{-2}	*	(Mohanakrishna et al. 2010a)
Molasses wastewater	Single chambered cuboid	127500 mgCODL^{-1}	53.2%	1410 mWm^{-2}	1%	(B. Zhang et al. 2009)
Molasses wastewater mixed with sewage	Single chambered	9958 mgCODL^{-1}	59%	382 mWm^{-2}	*	(Sevda et al. 2013)
Powdered rice straw	Two chambered	1000 mgCODL^{-1}	*	190 mWm^{-2}	37%	(Gurung and Oh 2015)
Powdered rice straw	H-type	1000 mgCODL^{-1}	*	145 mWm^{-2}	54.3 - 45.3%	(Hassan et al. 2014)
Raw corn stover	Bottle-type air-cathode	*	42±8% (cellulose) 17±7% (hemicellulose)	331 mWm^{-2}	*	(Wang et al. 2009)
Rice Milling	Earthen pot	2250 mgCODL^{-1}	96.5%	2.3 Wm^{-3}	21%	(Behera et al. 2010)
Rice straw hydrolysate	Air-cathode single chamber	400 mgCODL^{-1}	49-72%	137.6 ± 15.5 mWm^{-2}	8.5-17%	(Wang et al. 2014)
Starch processing wastewater	Air-cathode single chamber	4852 mgCODL^{-1}	98.0%	239.4 mWm^{-2}	8%	(Lu et al. 2009)
Steam exploded corn stover residue	Bottle-type air-cathode	*	60±4% (cellulose) 15±4% (hemicellulose)	406 mWm^{-2}	*	(Wang et al. 2009)
Wheat straw hydrolysate	H-type dual chamber	250-2000 mgCODL^{-1}	*	123 mWm^{-2}	15.5-37.1%	(Y. Zhang et al. 2009)
Animal carcass wastewater	Up-flow tubular air-cathode	11180 mgCODL^{-1}	50.7%	2.19 Wm^{-3}	0.25%	(Li et al. 2013)

(Contd.)

Cattle manure slurry	Air cathode cassette electrode	*	41.9-56.7%	765 mWm^{-2}	28.8%	(Inoue et al. 2013)
Manure wash water	Air-cathode single chamber	*	*	215 mWm^{-2}	*	(Zheng and Nirmalakhandan 2010)
Slaughterhouse wastewater	Dual chambered	4850 mgCODL^{-1}	93±1%	578 mWm^{-2}	64±2%	(Katuri et al. 2012)
Swine wastewater	Single chambered	8320 mgCODL^{-1}	*	182 mWm^{-2}	*	(Min et al. 2005)
Swine wastewater	Large-scale single chambered	*	85.6%	382 mWm^{-2}	*	(Cheng et al. 2014)

*indicates the data not available from the cited reference

4.4 Beverage Industry Wastewater

Brewery wastewater has been a favorite substrate in earlier MFCs studies primarily because of high organic content and a substantial lack of inhibitory substances. Large amounts of brewery wastewater are produced from different industrial operations (saccharification, fermentation, cooling, washing, etc.), typically with COD of 3,000--5,000 mgL^{-1} (about a tenfold of domestic wastewater), characterized by the presence of sugar, starch and protein components (ElMekawy et al., 2015). Aerobic sequencing batch reactors, cross-flow ultrafiltration (UF) membrane anaerobic reactors and up-flow UASBs are common biological methods used for its treatment; aerobic treatment is effective but requires energy for aeration, and anaerobic treatment needs to operate at high temperature (35-45°C) for maximum efficiency. MFCs have, therefore, been tested; brewery wastewater treatment using air-cathode MFC was investigated by Feng et al. (2008) that achieved Pd$_{max}$ = 528 mW/m^2. Other air--cathode MFCs achieved 669 mWm^{-2} (24.1 Wm^{-3}) running continuously on a similar substrate (Wen et al. 2010). Dong et al. (2015) tested a full-scale (90 L), 5 modules, stacked system fed with brewery wastewater that achieved ηCOD = 88% and Pd$_{ave}$ = 171 ± 8.4 mWm^{-2}.

Winery wastewater was also evaluated as MFCs substrate due to its high strength. Unbalanced nutrients/COD ratios are the major challenge in winery wastewater treatment. With this substrate, maximum energy recovery of 31.7 Whm^{-3}, 65% COD removal and 18% CE were reported (Cusick et al. 2010) down to 3.82 Wm^{-3} Pd, 41% ηCOD and 45% CE in two chamber MFCs with reject wine as substrate and *Acetobacter aceti* and *Gluconobacter roseus* as biocatalysts (Rengasamy and Berchmans 2012). Table 5 summarizes published MFC applications to beverage industry wastewaters.

4.5 Other Industrial Wastewaters

Acid-mine drainage (AMD) is characterized by low pH and solubilized metals such as lead, copper, cadmium and arsenic that are dangerous for aquatic environments. AMD is generated by biological oxidation of contained metal sulfides to sulfates. Processed with MFC technology, AMD generated Pd$_{max}$ = 290 mWm^{-2} with 97% CE. In addition, removal and recovery of metals and other by-products from AMD are possible using MFCs (S. Cheng et al. 2007). Coal tar and coke wastewaters are also successfully treated in a similar way.

Pharmaceutical industry wastewater containing penicillin was fed to an air--cathode, single chamber MFC by Wen et al. (2011) that included the mixture of 1 gL^{-1} glucose, and 50 mgL^{-1} penicillin yielded Pd$_{max}$ = 101.2Wm^{-3}. Recalcitrant pharmaceutical effluents, characterized by complex composition and high toxicity (Velvizhi and Venkata Mohan 2012), were also tested: paracetamol-containing effluent was successfully treated (L. Zhang et al. 2015); steroidal drug wastewater turned highly toxic after acid hydrolysis and wastewater from hydrocortisone production, intermediate of steroidal drugs, were fed to MFCs by R. Liu et al. (2012).

Table 5: List of beverage industry wastewater used as anodic substrate in MFCs and their respective performances.

Substrate	MFC type	Substrate concentration	COD removal (ηCOD)	Power density (max)	CE	Reference
Beer brewery processing wastewater	Air-cathode	2240 mgCODL^{-1}	87%	483 mWm^{-2}	38%	(Feng et al. 2008)
Beer brewery wastewater	One-chamber air-cathode	625 mgCODL^{-1}	43%	264 mWm^{-2}	19.75%	(Wen et al. 2009)
Beer brewery wastewater	One-chamber air-cathode	1501 mgCODL^{-1}	47.6	669 mWm^{-2}	2.5%	(Wen et al. 2010)
Brewery wastewater	Tubular air-cathode	2125 mgCODL^{-1}	93%	96 mWm^{-2}	28%	(Zhuang et al. 2010)
Brewery wastewater	Serpentine type stack	2120 mgCODL^{-1}	86.4%	97.2 mWm^{-2}	7.6%	(Zhuang et al. 2012)
Brewery wastewater	Stackable baffled MFC	800 mgCODL^{-1}	87.6%	181 ± 21 mWm^{-2}	19.1%	(Dong et al. 2015)
Reject wine	Two-chambered	7.8 gCODL^{-1}	41%	3.82 Wm^{-3}	45%	(Rengasamy and Berchmans 2012)
Winery wastewater	Two-chambered	2200 mgCODL^{-1}	65%	31.7 Whm^{-3}	18%	(Cusick et al. 2010)

*indicates the data not available from the cited reference

Paper recycling wastewater contains soluble organics and particulate matter (cellulose), ineffectively treated with traditional technologies, while sustainable agriculture and bio-based industries have indicated other efficient methods for cellulose-containing wastewater treatment and recovery. Treatment efficiency of these wastewaters in MFCs is limited by low conductivity and yielding power densities, Pd_{max} = 1,070 mWm^{-2} and 880 mWm^{-2} in single and two chamber air-cathode MFCs, respectively (S. Cheng et al. 2011).

Textile industries are among the most complex wastewaters, such as dyes containing recalcitrant organics, toxic, mutagenic and carcinogenic chemicals. Azo dyes constitute the largest class of synthetic dyes and are extensively present in effluents from dye-manufacturing and textile industries (Pant et al. 2010). Physical, chemical and electrochemical methods are conventionally used for their treatment, but the development of toxic intermediates, low removal efficiency and high specificity are some of the limitations of these methods. Decolorization capacity of MFCs for certain dyes has been investigated. Kalathil et al. (2011) built a granular activated carbon-based cell (GACB-MFC) that achieved Pd = 1.7 Wm^{-3}, ηCOD = 71% anodic and 76% cathodic with virtually nontoxic cathode effluent and a threefold less toxic anodic effluent within 48 hrs. In a subsequent study, a higher power density (8 Wm^{-3}) was achieved (Kalathil et al. 2012). Fang et al. (2015) demonstrated electric production from azo wastewater using an MFC-constructed wetland (CW-MFC) coupled system with highest Pd = 0.852 Wm^{-3}. Sun et al. (2009) tested simultaneous treatment of azo dye and readily biodegradable organic wastewater and observed accelerated decolorization of active brilliant red X-3B (ABRX3) in MFCs with glucose and confectionary wastewater as co-substrates. Electric production was affected by high concentrations of ABRX3 (>300 mgL^{-1}) due to competition between dye and anode for electrons from carbon sources. The study, however, exemplified how mixtures of different substrates may amplify treatment results.

Petroleum hydrocarbons contaminated sites, and refinery effluents treatment by MFCs is appealing. Studies have been carried out for the possible *in situ* applications. Petroleum sludge contamination treatment has been studied on site by Chandrasekhar and Venkata Mohan (2012) leading to a power generation of 53.11 mWm^{-2}. Morris and Jin (2012) reported 24% TPH (total petroleum hydrocarbon) removal and 2162 mWm^{-3} Pd in sediment-MFCs containing TPH. Foad Marashi et al. (2013) studied raw PTA (purified terephthalic acid) which is a raw material for petrochemical manufacturing with high organic strength wastewater treatment from a petrochemical plant in a membraneless single chamber MFC. Table 6 summarizes published results of MFCs treatment of different industrial wastewaters.

Table 6: List of industrial wastewaters used as anodic substrate in MFCs and their respective performances.

Substrate	MFC type	Substrate concentration	COD removal (ηCOD)	Power density (max)	CE	Reference
Acid-mine drainage	AMD fuel cell			290 mWm^{-2}	97%	(Cheng et al. 2007)
Hydrocortisone production wastewater	Air-cathode single chambered	1340 mgCODL^{-1}	82%	22.3 Wm^{-3}	30%	(Liu et al. 2012)
Recalcitrant pharmaceutical industrial effluent	Single chambered air-cathode	7.98 kgCODL^{-1}	85%	205.61 mWm^{-2}	*	(Velvizhi and Venkata Mohan 2012)
Synthetic penicillin wastewater	Air--cathode single chamber	*	87.1%	101.2 Wm^{-3}	*	(Wen et al. 2011)
Cellulose	Two chamber air-cathode	*	70%	880 mWm^{-2}	50%	(S. Cheng et al. 2011)
Paper recycling wastewater	Single chamber	480 mgCODL^{-1}	60%	506 mWm^{-2}	24%	(Huang and Logan 2008)
Azo dye wastewater	granular activated carbon-based MFC (GACB-MFC)	2080 mgCODL^{-1}	71% (anode) 76% (cathode)	1.7 Wm^{-3}	*	(Kalathil et al. 2011)
Real dye wastewater	granular activated carbon-based MFC (GACB-MFC)	2000-2200 mgCODL^{-1}	71%	8.0 Wm^{-3}	*	(Kalathil et al. 2012)
Azo dye wastewater	Constructed wetland (CW-MFC)	180-500 mgCODL^{-1}	*	0.852 Wm^{-3}	0.037-1.890%	(Fang et al. 2015)
Petroleum refinery wastewater	Dual chambered GC-packing-type	250 mgCODL^{-1}	64%	330.4 mWcm^{-3}	*	(Guo et al. 2016)
Purified terephtalic acid wastewater	Single chambered	8000 mgCODL^{-1}	74%	31.8 mWm^{-2}	2.05%	(Foad Marashi et al. 2013)
Real-field petroleum sludge	Single chambered open air-cathode	3gL^{-1}	*	53.11 mWm^{-2}	*	(Chandrasekhar and Venkata Mohan 2012)
Total petroleum hydrocarbons	Sediment MFC	18590 mgTPHkg^{-1}	24% (TPH)	2162 mWm^{-3}	*	(Morris and Jin 2012)

*indicates the data not available from the cited reference

It is difficult to draw specific conclusions related to the performance of MFCs treating different substrates based on results in Tables 3 to 5 due to many variables, besides wastewater type, affecting the performance of the systems analyzed, such as inoculum type and procedures, materials and architecture, reactor volume, electrodes type, hydraulic retention time (HRT), loading rates, etc.

Nature and composition of wastewater have a significant influence on MFC performance and energy generation efficiencies. Many highly biodegradable substrates could be effectively processed by MFCs with discrete energy recovery. On the other hand, the more complex substrates are, the more difficult is their treatment that results in lower organic matter removal and power production.

5. Challenges for MFC Performance

Evolution of MFC technology on many fronts (reactor designs, materials, biocatalysts, etc.) is slowly bringing it much closer to full industrial potential and actual application for bioenergy production and simultaneous wastewater treatment. The use of wastewater as an electron donor is desirable due to the growing demand for sustainable treatment with minimum carbon impact and energy recovery. The MFC's extractable power densities are gradually rising by orders of magnitude, and some companies started launching industrialized MFC-based treatment systems, however, many challenges still remain. High cost and low power output still impair the commercial success of MFC technologies, both as standalone or combined technology.

Besides the current intrinsic technological limitations, the selection of appropriate substrates in terms of molecular complexity is paramount; composition and strength are among the main factors affecting MFC's performance. Low conductivity or out of range pH could impair microbial activity, therefore, their initial values and online control must be considered.

This chapter considered many MFC feedstocks in terms of bioelectricity production and organic removal. Among these, distillery along with agro-processing industry effluents was treated at a higher efficiency due to the presence of methanogenic inhibitors and electron transferring mediators (i.e. lignin) within. Also, food and dairy wastewaters showed good performance that was limited, however, by the presence of other electron acceptors and of non-exoelectrogenic microorganisms, such as fermenters and methanogens. Animal processing wastewaters are especially suitable for the presence of blood (proteins) and organic compounds.

Low CE has been a general issue in MFCs fed with real wastewaters due to competitive non-exoelectrogenic biomass growth at electrodes, substrate consumption via competing metabolic pathways (fermentation and methanogenesis), presence of toxins/inhibitors for EABs, large number of electrons locked in substrates by other electron acceptors and low electron transfer efficiency.

MFC scaling up (from laboratory to full-scale) issues are a major economic obstacle to systems that could be easily maintained and produce satisfactory power levels. Scaling up MFC technology from milliliters to liters generally led to higher electrode potential losses that considerably reduced current densities obtained (ElMekawy et al. 2015). Scaling up may be achieved in two ways: the interconnection of multiple small cells or geometric enlargement of a cell up to the desired volume. The cathode surface area has a critical role in scaling up; the performance is negatively influenced, with power density drop, by decrease of the area/volume ratio, so that enlargement of cells has a very low chance to succeed. Interconnected MFCs stacks may be an effective alternative to overcome problems of electrode spacing, orientation and surface area to improve power output. The application of stacked MFCs in parallel or series would be essential to significantly increase bioelectricity generation. Specifically stacked MFCs in parallel seem to have greater potential to increase MFC's performance parameters compared to serial connection.

Nowadays, MFCs capital costs are about 30 times higher than traditional technology for domestic wastewater considering the configuration and treatment capacity (He et al. 2017). These are due to the use of an expensive electrode, catalyst and membrane materials and, certainly, to general uncertainty about design criteria. New materials are being constantly explored and developed to improve MFC's economic feasibility and performance.

6. Conclusions

Microbial fuel cells (MFCs) are recognized as an innovative technology with certain potential advantages in the field of wastewater treatment. This chapter summarized MFC's application results on substrates from different sources and compositions. Almost all these applications were limited to the laboratory or pilot-scale as there are no reports yet about the application of this technology to real-scale WRRFs. This is due to a still limited technical understanding of the technology, especially concerning the design parameters and material's effects on full-scale system development. Initial costs and the limited levels of energetic recovery achieved so far compared to the initial and theoretical expectations seem to have deflected industrial interest away from this technology, even though scientific interest in it is still high as demonstrated by intensive publication activities on the subject.

It should be, however, considered that MFC technology already offers, even at this development stage, several important advantages over conventional aerobic and also anaerobic technologies. They require low input energy due to lack of aeration and produce extremely low amounts of residual biosolids. Even compared to anaerobic digestion, MFCs have the advantage of direct power generation without intermediates, such as biogas at low temperatures and concentrations. So far, expensive solutions have been imagined to maximize energy recovery from these processes. However, their final balance could just as well turn out positive by forfeiting significant energy harvesting in favor of more economic constructive solutions with sufficient pollutants removal capacity and low operating cost and the lack of high cost for the disposal of excess biosolids.

7. Future Perspectives

MFC technology is still in need of an evolution to compete directly with current wastewater treatment technologies. This competition may come along with two parallel pathways: recognition and exploitation of current limitations and drastic technological improvement. The former would lead to simpler full-scale systems where electric energy generation is forfeited or purposefully limited to achieve desired pollutant removal levels. The aforementioned residual advantages of MFCs would likely make these systems sustainable. The latter would require heavy investments in new materials for the electrochemical components and perhaps the development of MFCs "superbugs" through mutagenesis and rDNA technology. It is, in fact, conceivable that these techniques could be used for this purpose in the future as it is currently happening in the field of biofuels (biobutanol) fermentation with the development of specially engineered algae. For instance, development of new anodophilic microbes with improved characteristics could vastly enhance internal electron transport rates and thus the power density output from MFCs.

Even if the generation of high power from MFCs may be a long way off, understanding the inner mechanisms connecting the organic matter degradation to electrons transfer is likely to bring in significant insights into the microbial respiratory capabilities and might lead to unforeseen applications in this, and also in unrelated fields, such as nanoelectronics.

References

Ahmad AL, Ismail S, Bhatia S (2003) Water recycling from palm oil mill effluent (POME) using membrane technology. Desalination 157 (May): 87-95.

Baranitharan E, Khan MR, Yousuf A, Fei W, Teo A, Yuan G, Tan A (2015) Enhanced power generation using controlled inoculum from palm oil mill effluent fed microbial fuel cell. Fuel 143: 72-79. https://doi.org/10.1016/j.fuel.2014.11.030

Behera M, Jana PS, More TT, Ghangrekar MM (2010) Rice mill wastewater treatment in microbial fuel cells fabricated using proton exchange membrane and earthen pot at different pH. Bioelectrochemistry 79(2): 228-233. https://doi.org/10.1016/j.bioelechem.2010.06.002

Behera SK, Park JM, Kim KH, Park H (2012) Methane production from food waste leachate in laboratory-scale simulated landfill. Waste Manag 30(8-9): 1502-1508. https://doi.org/10.1016/j.wasman.2010.02.028

Capodaglio AG, Molognoni D, Dallago E, Liberale A, Cella R, Longoni P, Pantaleoni L (2013) Microbial fuel cells for direct electrical energy recovery from urban wastewaters. Sci World J, Hindawi, 2013: 634738. https://doi.org/10.1155/2013/634738

Capodaglio AG, Molognoni D, Puig S, Balaguer MD, Colprim J (2015) Role of operating conditions on energetic pathways in a Microbial Fuel Cell. Energy Procedia 74: 728-735. https://doi.org/10.1016/j.egypro.2015.07.808

Capodaglio AG, Molognoni D, Pons AV (2016) A multi-perspective review of microbial fuel-cells for wastewater treatment: bio-electro-chemical, microbiologic and modeling aspects. AIP Conf. Proc. 1758: 030032. https://doi.org/10.1063/1.4959428

Cecconet D, Molognoni D, Callegari A, Capodaglio AG (2018) Agro-food industry wastewater treatment with microbial fuel cells: Energetic recovery issues. Int J Hydrogen Energy 43: 500-511. https://doi.org/10.1016/j.ijhydene.2017.07.231

Cercado-Quezada B, Delia M, Bergel A (2010) Testing various food-industry wastes for electricity production in microbial fuel cell. Bioresour Technol 101(8): 2748-2754. https://doi.org/10.1016/j.biortech.2009.11.076

Chandrasekhar K, Venkata Mohan S (2012) Bio-electrochemical remediation of real field petroleum sludge as an electron donor with simultaneous power generation facilitates biotransformation of PAH: Effect of substrate concentration. Bioresour Technol 110: 517-525. https://doi.org/10.1016/j.biortech.2012.01.128

Cheng CY, Li CC, Chung YC (2014) Continuous electricity generation and pollutant removal from swine wastewater using single chambered air-cathode microbial fuel cell. Adv Mat Res 953-954: 158-162. https://doi.org/10.4028/www.scientific.net/AMR.953-954.158

Cheng J, Zhu X, Ni J, Borthwick A (2010) Palm oil mill effluent treatment using a two-stage microbial fuel cells system integrated with immobilized biological aerated filters. Bioresour Technol 101(8): 2729-2734. https://doi.org/10.1016/j.biortech.2009.12.017

Cheng S, Dempsey BA, Logan BE (2007) Electricity generation from synthetic acid-mine drainage (AMD) water using fuel cell technologies. Environ Sci Technol 41(23): 8149-8153.

Cheng S, Kiely P, Logan BE (2011) Pre-acclimation of a wastewater inoculum to cellulose in an aqueous - cathode MEC improves power generation in air - cathode MFCs. Bioresour Technol, Elsevier Ltd 102(1): 367-371. doi: 10.1016/j.biortech.2010.05.083.

Cheng X, He L, Lu H, Chen Y, Ren L (2016) Optimal water resources management and system benefit for the Marcellus shale-gas reservoir in Pennsylvania and West Virginia. J Hydrol 540: 412-422. https://doi.org/10.1016/j.jhydrol.2016.06.041

Cusick RD, Kiely PD, Logan BE (2010) A monetary comparison of energy recovered from microbial fuel cells and microbial electrolysis cells fed winery or domestic wastewaters. Int J Hydrogen Energy, Elsevier Ltd 35(17): 8855-8861. doi: 10.1016/j.ijhydene.2010.06.077.

Dallago E, Barnabei AL, Liberale A, Torelli G, Venchi G (2016) A 300-mV low-power management system for energy harvesting applications. IEEE Trans Power Electron. 31(3): 2273-2281. https://doi.org/10.1109/TPEL.2015.2431439

Das KC (2008) Co-composting of alkaline tissue digester effluent with yard trimmings. Waste Manag 28: 1785-1790. doi: 10.1016/j.wasman.2007.08.027.

Dong Y, Qu Y, He W, Du Y, Liu J, Han X, Feng Y (2015) A 90-liter stackable baffled microbial fuel cell for brewery wastewater treatment based on energy self-sufficient mode. Bioresour Technol 195: 66-72. https://doi.org/10.1016/j.biortech.2015.06.026

ElMekawy A, Srikanth S, Bajracharya S, Hegab HM, Nigam PS, Singh A, Pant D (2015) Food and agricultural wastes as substrates for bioelectrochemical system (BES): The synchronized recovery of sustainable energy and waste treatment. Food Res Int 73: 213-225. https://doi.org/10.1016/j.foodres.2014.11.045

EPA (2006) Wastewater Management Fact Sheet: Energy Conservation. Available at: nepis.epa.gov/Exe/ZyPURL.cgi?Dockey=P100IL6T.TXT.

Fang Z, Song H, Cang N, Li X (2015) Electricity production from Azo dye wastewater using a microbial fuel cell coupled constructed wetland operating under different operating conditions. Biosens Bioelectron 68: 135-141. https://doi.org/10.1016/j.bios.2014.12.047

Fangzhou D, Zhenglong L, Shaoqiang Y, Beizhen X, Hong L (2011) Electricity generation directly using human feces wastewater for life support system. Acta Astronaut 68(9-10): 1537-1547. https://doi.org/10.1016/j.actaastro.2009.12.013

Feng Y, Wang X, Logan BE, Lee H (2008) Brewery wastewater treatment using air-cathode microbial fuel cells. Appl Microbiol and Biotechnol 78(5): 873-880. https://doi.org/10.1007/s00253-008-1360-2

Foad Marashi SK, Kariminia HR, Pour Savizi IS (2013) Bimodal electricity generation and aromatic compounds removal from purified terephthalic acid plant wastewater in a microbial fuel cell. Biotechnol Lett 35: 197-203. doi: 10.1007/s10529-012-1063-8.

Galanakis CM (2012) Recovery of high components from food wastes: Conventional, emerging technologies and commercialized applications. Trends Food Sci Technol, Elsevier Ltd 26(2): 68-87. doi: 10.1016/j.tifs.2012.03.003.

Garrison M, Richard TL, Tiquia SM, Honeyman MS (2001) Nutrient losses from unlined bedded swine hoop structures and an associated windrow composting site. ASAE Paper No. 012238. American Society of Agricultural Engineers (ASAE) International Meeting. Sacramento, California, U.S.A. July 29-August 1, 2001.

Gikas P (2017) Towards energy positive wastewater treatment plants. J. Environ. Manage., Elsevier Ltd 203: 621-629. doi: 10.1016/j.jenvman.2016.05.061.

Goud RK, Babu PS, Mohan SV (2011) Canteen based composite food waste as potential anodic fuel for bioelectricity generation in single chambered microbial fuel cell (MFC): Bio-electrochemical evaluation under increasing substrate loading condition. Int J Hydrogen Energy, Elsevier Ltd 36(10): 6210-6218. doi: 10.1016/j.ijhydene.2011.02.056.

Gude VG (2016) Wastewater treatment in microbial fuel cells - An overview. J Clean Prod, Elsevier Ltd, 122 (December 2016): 287-307. doi: 10.1016/j.jclepro.2016.02.022.

Guo F, Fu G, Zhang Z, Zhang C (2013) Mustard tuber wastewater treatment and simultaneous electricity generation using microbial fuel cells. Bioresour Technol 136: 425-430. https://doi.org/10.1016/j.biortech.2013.02.116

Guo J, Ma J (2015) Bioflocculant from pre-treated sludge and its applications in sludge dewatering and swine wastewater pretreatment. Bioresour Technol 196: 736-740. https://doi.org/10.1016/j.biortech.2015.07.113

Guo X, Zhan Y, Chen C, Cai B, Wang Y, Guo S (2016) Influence of packing material characteristics on the performance of microbial fuel cells using petroleum refinery wastewater as fuel. Renew Energy 87: 437-444. https://doi.org/10.1016/j.renene.2015.10.041

Gurung A, Oh SE (2015) Rice straw as a potential biomass for generation of bioelectrical energy using microbial fuel cells (MFCs). Energy Sources, Part A: Recovery, Utilization, and Environmental Effects. Taylor & Francis 37(24): 2625-2631. doi: 10.1080/15567036.2012.728678.

Hassan SHA, Gad El-Rab SMF, Rahimnejad M, Ghasemi M, Joo J, Sik-ok Y, Kim IS (2014) Electricity generation from rice straw using a microbial fuel cell. Int J Hydrogen Energy 39(17): 9490-9496. https://doi.org/10.1016/j.ijhydene.2014.03.259

He L, Du P, Chen Y, Lu H, Cheng X, Chang B, Wang Z (2017) Advances in microbial fuel cells for wastewater treatment. Renew Sustain Energy Rev 71 (December 2016): 388-403. https://doi.org/10.1016/j.rser.2016.12.069

He Z, Angenent LT (2006) Application of bacterial biocathodes in microbial fuel cells. Electroanalysis 18(19-20): 2009-2015. doi: 10.1002/elan.200603628.

Huang L, Logan BE (2008) Electricity generation and treatment of paper recycling wastewater using a microbial fuel cell. Appl Microbiol Biotechnol 80(2): 349-355. doi: 10.1007/s00253-008-1546-7.

Inoue K, Ito T, Kawano Y, Iguchi A, Miyahara M, Suzuki Y, Watanabe K (2013) Electricity generation from cattle manure slurry by cassette-electrode microbial fuel cells. J Biosc Bioeng 116(5): 610-615. https://doi.org/10.1016/j.jbiosc.2013.05.011

Jiang H, Luo S, Shi X, Dai M, Guo R (2013) A system combining microbial fuel cell with photobioreactor for continuous domestic wastewater treatment and bioelectricity generation. J Cent South Univ 20(2): 488-494. https://doi.org/10.1007/s11771-013-1510-2

Jin B, Yan XQ, Yu Q, Van Leeuwen JH (2002) A comprehensive pilot plant system for fungal biomass protein production and wastewater reclamation. Adv. Environ. Res. 6: 179-189.

Kaewkannetra P, Chiwes W, Chiu TY (2011) Treatment of cassava mill wastewater and production of electricity through microbial fuel cell technology. Fuel, Elsevier Ltd 90(8): 2746-2750. doi: 10.1016/j.fuel.2011.03.031.

Kalathil S, Lee J, Cho MH (2011) Granular activated carbon based microbial fuel cell for simultaneous decolorization of real dye wastewater and electricity generation. N Biotechnol Elsevier B.V. 29(1): 32-37. doi: 10.1016/j.nbt.2011.04.014.

Kalathil S, Lee J, Cho MH (2012) Efficient decolorization of real dye wastewater and bioelectricity generation using a novel single chamber biocathode-microbial fuel cell. Bioresour Technol, Elsevier Ltd 119: 22-27. doi: 10.1016/j.biortech.2012.05.059.

Katuri KP, Enright A, Flaherty VO, Leech D (2012) Microbial analysis of anodic biofilm in a microbial fuel cell using slaughterhouse wastewater. Bioelectrochemistry 87: 164-171. https://doi.org/10.1016/j.bioelechem.2011.12.002

Kelly PT, He Z (2014) Understanding the application niche of microbial fuel cells in a cheese wastewater treatment process. Bioresour Technol, Elsevier Ltd 157: 154-160. doi: 10.1016/j.biortech.2014.01.085.

Khan TS, Mubeen U (2012) Wheat straw: A pragmatic overview. Curr Res J Biol Sci. 4(6): 673-675.

Leaño EP, Anceno AJ, Babel S (2012) Ultrasonic pretreatment of palm oil mill effluent: Impact on biohydrogen production, bioelectricity generation, and underlying microbial communities. Int J Hydrogen Energy 37: 12241-12249 doi: 10.1016/j.ijhydene.2.012.06.007.

Li WW, Yu HQ, He Z (2014) Towards sustainable wastewater treatment by using microbial fuel cells-centered technologies. Energy Environ Sci 7(3): 911-924. doi: 10.1039/C3EE43106A.

Li X, Zhu N, Wang Y, Li P, Wu P, Wu J (2013) Animal carcass wastewater treatment and bioelectricity generation in up-flow tubular microbial fuel cells: Effects of HRT and non-precious metallic catalyst. Bioresour Technol 128: 454-460. https://doi.org/10.1016/j.biortech.2012.10.053

Liu R, Gao C, Zhao Y, Wang A, Lu S, Wang M (2012) Biological treatment of steroidal drug industrial effluent and electricity generation in the microbial fuel cells. Bioresour Technol 123: 86-91. https://doi.org/10.1016/j.biortech.2012.07.094

Logan BE, Rabaey K (2012) Conversion of wastes into bioelectricity and chemicals by using microbial electrochemical technologies. Science. American Association for the Advancement of Science 337(6095): 686–690. doi: 10.1126/science.1217412.

Lu N, Zhou S, Zhuang L, Zhang J, Ni J. (2009) Electricity generation from starch processing wastewater using microbial fuel cell technology. Biochem Eng J 43(3): 246-251. https://doi.org/10.1016/j.bej.2008.10.005

Mahdi Mardanpour M, Esfahani MN, Behzad T, Sedaqatvand R (2012) Single chamber microbial fuel cell with spiral anode for dairy wastewater treatment. Biosens Bioelectron 38(1): 264-269. https://doi.org/10.1016/j.bios.2012.05.046

Mansoorian HJ (2016) Evaluation of dairy industry wastewater treatment and simultaneous bioelectricity generation in a catalyst-less and mediator-less membrane microbial fuel cell. J Saudi Chem Soc. King Saud University 20(1): 88-100. doi: 10.1016/j.jscs.2014.08.002.

McCarty PL, Bae J, Kim J (2011) Domestic wastewater treatment as a net energy producer-can this be achieved? Environ Sci Technol 45(17): 7100-7106. doi: 10.1021/es2014264.

Min B, Kim JR, Oh SE, Regan JM, Logan BE (2005) Electricity generation from swine wastewater using microbial fuel cells. Water Res 39(20): 4961-4968. https://doi.org/10.1016/j.watres.2005.09.039

Min X, Yu K, Selvam A, Wong JWC (2013) Bioelectricity production from acidic food waste leachate using microbial fuel cells: Effect of microbial inocula. Process Biochem 48(2): 283-288. https://doi.org/10.1016/j.procbio.2012.10.001

Mohanakrishna G, Venkata Mohan S, Sarma PN (2010a) Bio-electrochemical treatment of distillery wastewater in microbial fuel cell facilitating decolorization and desalination along with power generation. J Hazard Mater, Elsevier B.V. 177(1-3): 487-494. doi: 10.1016/j.jhazmat.2009.12.059.

Mohanakrishna G, Venkata Mohan S, Sarma PN (2010b) Utilizing acid-rich effluents of fermentative hydrogen production process as substrate for harnessing bioelectricity: An integrative approach. Int J Hydrogen Energy, Elsevier Ltd 35(8): 3440-3449. doi: 10.1016/j.ijhydene.2010.01.084.

Molognoni D, Puig S, Balaguer MD, Liberale A, Capodaglio AG, Callegari A, Colprim J (2014) Reducing start-up time and minimizing energy losses of Microbial Fuel Cells using Maximum Power Point Tracking strategy. J Power Sources 269: 403-411. https://doi.org/10.1016/j.jpowsour.2014.07.033

Molognoni D, Puig S, Balaguer MD, Capodaglio AG, Callegari A, Colprim J (2016) Multiparametric control for enhanced biofilm selection in microbial fuel cells. J Chem Technol Biotechnol 91(6): 1720-1727. https://doi.org/10.1002/jctb.4760

Morris JM, Jin S (2012) Enhanced biodegradation of hydrocarbon-contaminated sediments using microbial fuel cells. J Hazard Mater, Elsevier B.V. 213-214: 474-477. doi: 10.1016/j.jhazmat.2012.02.029.

Nikhil GN, Suman P, Venkata Mohan S, Swamy YV (2017) Energy-positive nitrogen removal of pharmaceutical wastewater by coupling heterotrophic nitrification and electrotrophic denitrification. Chem Eng J 326: 715-720. https://doi.org/10.1016/j.cej.2017.05.165

Nimje VR, Chen CY, Chen HR, Chen CC, Huang YM, Tseng MJ, Chang YF (2012) Comparative bioelectricity production from various wastewaters in microbial fuel cells using mixed cultures and a pure strain of *Shewanella oneidensis*. Bioresour Technol 104: 315-323. https://doi.org/10.1016/j.biortech.2011.09.129

Oh S, Logan BE (2005) Hydrogen and electricity production from a food processing wastewater using fermentation and microbial fuel cell technologies. Water Res 39: 4673-4682. doi: 10.1016/j.watres.2005.09.019.

Pandey P, Shinde VN, Deopurkar RL, Kale SP, Patil SA, Pant D (2016) Recent advances in the use of different substrates in microbial fuel cells toward wastewater treatment and simultaneous energy recovery. Appl Energy 168: 706-723. https://doi.org/10.1016/j.apenergy.2016.01.056

Pant D, Van Bogaert G, Diels L, Vanbroekhoven K (2010) A review of the substrates used in microbial fuel cells (MFCs) for sustainable energy production. Bioresour Technol 101(6): 1533-1543. https://doi.org/10.1016/j.biortech.2009.10.017

Parfitt J, Barthel M, Macnaughton S (2010) Food waste within food supply chains: Quantification and potential for change to 2050. Phil Trans R Soc B 365: 3065-3081. doi: 10.1098/rstb.2010.0126.

Patil SA, Surakasi VP, Koul S, Ijmulwar S, Vivek A, Shouche YS, Kapadnis BP (2009) Electricity generation using chocolate industry wastewater and its treatment in activated sludge based microbial fuel cell and analysis of developed microbial community in the anode chamber. Bioresour Technol 100(21): 5132-5139. https://doi.org/10.1016/j.biortech.2009.05.041

Plecha S, Hall D, Tiquia-Arashiro SM (2013) Screening and characterization of soil microbes capable of degrading cellulose from switchgrass (*Panicum virgatum* L.). Environ Technol 34: 1895-1904.

Rengasamy K, Berchmans S (2012) Simultaneous degradation of bad wine and electricity generation with the aid of the coexisting biocatalysts *Acetobacter aceti* and *Gluconobacter roseus*. Bioresour Technol, Elsevier Ltd 104: 388-393. doi: 10.1016/j.biortech.2011.10.092.

Rikame SS, Mungray AA, Mungray AK (2012) Electricity generation from acidogenic food waste leachate using dual chamber mediator less microbial fuel cell. Int Biodeterior Biodegradation, Elsevier Ltd 75: 131-137. doi: 10.1016/j.ibiod.2012.09.006.

Rozendal RA, Hamelers HVM, Rabaey K, Keller J, Buisman CJN (2008) Towards practical implementation of bioelectrochemical wastewater treatment. Trends Biotechnol 26(8): 450-459. https://doi.org/10.1016/j.tibtech.2008.04.008

Santoro C, Ieropoulos I, Greenman J, Cristiani P, Vadas T (2013) Power generation and contaminant removal in single chamber microbial fuel cells (SCMFCs) treating human urine. IntJ Hydrogen Energy 38: 11543-11551. https://doi.org/10.1016/j.ijhydene.2013.02.070

Sevda S, Dominguez-Benetton X, Vanbroekhoven K, Wever HD (2013) High strength wastewater treatment accompanied by power generation using air cathode microbial fuel cell. Appl Energy 105: 194-206. https://doi.org/10.1016/j.apenergy.2012.12.037

Shizas I, Bagley DM (2004) Experimental determination of energy content of unknown organics in municipal wastewater streams. J Energy Eng 130(2): 45-53. doi: 10.1061/(ASCE)0733-9402(2004)130:2(45).

Soares RB, Santos Memelli M, Pereira Roque R, Gonçalves RF (2017) Comparative analysis of the energy consumption of different wastewater treatment plants. Int J Archit Arts Appl 33(611): 79-86. https://doi.org/10.11648/j.ijaaa.20170306.11

Stillwell AS, Hoppock DC, Webber ME (2010) Energy recovery from wastewater treatment plants in the United States: A case study of the energy-water nexus. Sustainability 2(4): 945-962. https://doi.org/10.3390/su2040945

Tam NFY, Tiquia SM, Vrijmoed LLP (1996) Nutrient transformation of pig manure under pig-on-litter system. *In*: M De Bertoldi, P Sequi, B Lemmes, T Papi (eds). Chapman and Hall, London, U.K. pp. 96-105.

Tiquia SM (2003) Evaluation of organic matter and nutrient composition of partially decomposed and composted spent pig-litter. Environ Technol 24: 97-108.

Tiquia-Arashiro SM, Mormile M (2013) Sustainable technologies: Bioenergy and biofuel from biowaste and biomass. Environ Technol 34(13): 1637-1805.

Tremouli A, Antonopoulou G, Bebelis S, Lyberatos G (2013) Operation and characterization of a microbial fuel cell fed with pretreated cheese whey at different organic loads. Bioresour Technol 131: 380-389. https://doi.org/10.1016/j.biortech.2012.12.173

Velvizhi G, Venkata Mohan S (2012) Electrogenic activity and electron losses under increasing organic load of recalcitrant pharmaceutical wastewater. Int J Hydrogen Energy, Elsevier Ltd 37(7): 5969-5978. doi: 10.1016/j.ijhydene.2011.12.112.

Venkata Mohan S, Mohanakrishna G, Sarma PN (2010) Composite vegetable waste as renewable resource for bioelectricity generation through non-catalyzed open-air cathode microbial fuel cell. Bioresour Technol, Elsevier Ltd 101(3): 970-976. doi: 10.1016/j.biortech.2009.09.005.

Wang X, Feng J, Wang H, Qu Y, Yu Y, Ren N, Lee HE (2009) Bioaugmentation for electricity generation from corn stover biomass using microbial fuel cells. Environ Sci Technol 43(15): 6088-6093.

Wang X, Feng Y, Liu J, Lee H, Li C, Li N, Ren N (2010) Sequestration of CO_2 discharged from anode by algal cathode in microbial carbon capture cells (MCCs). Biosens Bioelectron 25(12): 2639-2643. https://doi.org/10.1016/j.bios.2010.04.036

Wang Z, Lee T, Lim B, Choi C, Park J (2014) Microbial community structures differentiated in a single chamber air-cathode microbial fuel cell fueled with rice straw hydrolysate. Biotechnol Biofuels 7(9): 1-10.

Wen Q, Wu Y, Cao D, Zhao L, Sun Q (2009) Electricity generation and modeling of microbial fuel cell from continuous beer brewery wastewater. Bioresour Technol 100(18): 4171-4175. https://doi.org/10.1016/j.biortech.2009.02.058

Wen Q, Wu Y, Zhao L, Sun Q (2010) Production of electricity from the treatment of continuous brewery wastewater using a microbial fuel cell. Fuel 89(7): 1381-1385. https://doi.org/10.1016/j.fuel.2009.11.004

Wen Q, Kong F, Zheng H, Cao D, Ren Y, Yin J (2011) Electricity generation from synthetic penicillin wastewater in an air-cathode single chamber microbial fuel cell. Chem Eng J 168(2): 572-576. https://doi.org/10.1016/j.cej.2011.01.025

You SJ, Zhao QL, Jiang JQ, Zhang JN (2006) Treatment of domestic wastewater with simultaneous electricity generation in microbial fuel cell under continuous operation. Chem Biochem Eng Q 20(4): 407-412.

Zhang B, Zhao H, Zhou S, Shi C, Wang C, Ni J (2009) A novel UASB - MFC - BAF integrated system for high strength molasses wastewater treatment and bioelectricity generation. Bioresour Technol 100(23): 5687-5693. https://doi.org/10.1016/j.biortech.2009.06.045

Zhang L, Yin X, Fong S, Li Y (2015) Bio-electrochemical degradation of paracetamol in a microbial fuel cell-Fenton system. Chem Eng J 276: 185-192. https://doi.org/10.1016/j.cej.2015.04.065

Zhang Y, Min B, Huang L, Angelidaki I (2009) Generation of electricity and analysis of microbial communities in wheat straw biomass-powered microbial fuel cells. Appl Environ Microbiol 75(11): 3389-3395. https://doi.org/10.1128/AEM.02240-08

Zhao Y, Collum S, Phelan M, Goodbody T, Doherty L, Hu Y (2013) Preliminary investigation of constructed wetland incorporating microbial fuel cell: Batch and continuous flow trials. Chem Eng J 229: 364-370. https://doi.org/10.1016/j.cej.2013.06.023

Zheng X, Nirmalakhandan N (2010) Cattle wastes as substrates for bioelectricity production via microbial fuel cells. Biotechnol Lett 32(12): 1809-1814. doi: 10.1007/s10529-010-0360-3.

Zhuang L, Feng C, Zhou S, Li Y, Wang Y (2010) Comparison of membrane- and cloth-cathode assembly for scalable microbial fuel cells: Construction, performance and cost. Process Biochem 45(6): 929-934. https://doi.org/10.1016/j.procbio.2010.02.014

Zhuang L, Zheng Y, Zhou S, Yuan Y, Yuan H, Chen Y (2012) Scalable microbial fuel cell (MFC) stack for continuous real wastewater treatment. Bioresour Technol 106: 82-88. https://doi.org/10.1016/j.biortech.2011.11.019

Bioelectrochemical Systems for Sustainable Groundwater Denitrification and Removal of Arsenic and Chromium

Andrea G. Capodaglio* and Daniele Cecconet

Department of Civil Engineering and Architecture (D.I.C.Ar.), University of Pavia, Pavia 27100, Italy

1. Introduction

Groundwater resources constitute, on the average, about 80% of drinking water sources in Europe and 50% in the U.S. On a worldwide scale, groundwater is still the primary source of drinking water, and among the many ubiquitous pollutants that can limit its use, aside from nitrates (Cecconet et al. 2018), and perchlorates (Urbansky and Schock 1999; Molognoni et al. 2017), arsenic (As) is one of the most dangerous contaminants. Chromium from both natural and anthropogenic sources may also be present in groundwater (CSWRCB 2009).

Various researchers have identified bioelectrochemical systems (BESs) as an environmental technology that could provide a sustainable and effective solution for the removal of these (and other) pollutants from groundwater (Modin and Aulenta 2017) due to interaction of different redox conditions during the process such as As(III) oxidation at the anode, reduction of Cr(VI) and nitrates at the cathode (Figure 1). The scope of this chapter is to illustrate known applications of bioelectrochemical systems for removal of these pollutants and discuss issues related to their wider implementation and future extension of BES technologies to solve other groundwater-related problems.

2. Groundwater Contamination Issues

The presence of nitrate, detected worldwide, may constitute a threat to human health, especially to infants and seniors. The main sources of nitrate in groundwater are anthropogenic due to intensive use of fertilizers (Bouchard et al. 1992), but other important sources are via leakages from sewage systems, on-site wastewater disposal systems and cattle feedlots (Wakida and Lerner 2005; Tiquia et al. 2007; Tiquia et al. 2008; Tiquia 2010). Nitrate can also naturally occur in groundwater with background natural concentrations reported to be higher than 3 mgN-NO_3^- L^{-1} (Cho et al. 2012; Menció et al. 2016). Nitrate has been identified as toxic (Pawełczyk 2012), and its intake can lead to *methaemoglobinemia* in infants, hence, guidelines have been issued worldwide; the USA and Canada have limits of 10 mgN-NO_3^- L^{-1} and 1 mgN-NO_2^- L^{-1} (Health Canada 2017; USEPA 2010), respectively, the former with addition of a combined limit of 10 mgN L^{-1} as sum of both forms. WHO guidelines recommend values of 11 mg

---

*Corresponding author: capo@unipv.it

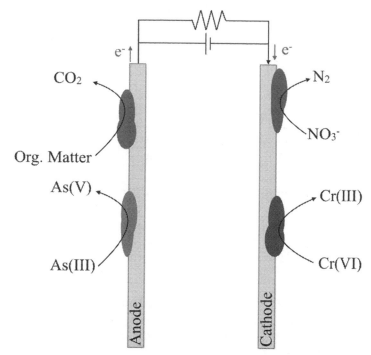

Figure 1: Scheme of electrochemical reactions for the contaminants considered
in the chapter.

N-NO$_3^-$ L^{-1} and 0.9 mg N-NO$_2^-$ L^{-1}, respectively, and a combined limit (sum of the ratios of nitrate and nitrite concentrations to their guideline value < 1) (WHO 2017); the EU adopted a limit of 11.3 mg N-NO$_3^-$ L^{-1} and 0.5 mg N-NO$_2^-$ L^{-1} for nitrate and nitrite, respectively, in addition to the WHO combined limit (CEC 1998).

The presence of arsenic in groundwater is threatening to human health as well due to its toxicity and carcinogenicity (Jomova et al. 2011). A long-term contaminated drinking water exposure can cause skin, lung, bladder and kidney cancer as well as other disorders (Agusa et al. 2014), affecting particularly women and children (Ahmed et al. 2006). It is naturally present in the environment in different oxidation states such as As(V), As(III), As(0) and As(-III). In water, its most common valences are As(V) (arsenate), stable in aerobic surface waters, and As(III) (arsenite) stable in anaerobic groundwater (Jain and Ali 2000). Biologically, As(III) is about 60 times more toxic than As(V) (USEPA 2011). Its presence in natural water could be related to leaching from rocks and sediments (Hering and Elimelech 1995) or to anthropic activities such as agricultural or industrial practices (Jadhav et al. 2015, Liao et al. 2016). In the first case, rocks naturally containing As can cause its dissolution in groundwater (Barringer et al. 2010); in the others, cause of the pollution may lie in the use (or overuse) of As-containing materials such as dyes, wood preservatives, pesticides, pharmaceutical substances, additives, etc., (Mandal and Suzuki 2002) that may also accumulate in the groundwater (Khaska et al. 2015), sediments and surface waters (Barringer et al. 2010, Hu et al. 2013). Arsenic poisoning is widespread in Asia, Africa and South America, however, its occurrence has also been reported worldwide and is not unknown in developed countries, including Canada (McGuigan et al. 2010) and Italy (Dalla Libera et al. 2017).

Recently, the World Health Organization (WHO) had established a guideline value of 10 μg L^{-1} for Arsenic in drinking water (WHO 2017), the same value is applied by EPA (USEPA 2010). Hexavalent chromium [Cr(VI)] is a toxic heavy metal and is found in effluents from industries, such as electroplating, steel and metal alloys production, leather tanning, cement, dye/pigment manufacturing, mining, photographic materials, paints and fungicide production (Saha et al. 2011). Its accumulation in the environment is a serious threat since Cr(VI) is a known mutagen, teratogen and carcinogen besides being also highly corrosive (Zhitkovich 2011). Its high solubility makes it a highly mobile element in the environment. Perhaps the most famous Cr(VI) groundwater contamination was the 'Hinkley case' caused by the Pacific Gas and Electric Company (PG&E) around the town of Hinkley in the Mojave Desert (California). From 1952 to 1966, PG&E used Cr(VI) as rust suppressors in its natural gas transmission pipelines, dumping roughly 1.5 million m^3 of chromium-tainted wastewater into unlined ponds (Egilman 2006). The U.S.EPA identified Cr(VI) as one of the 17 chemicals posing greatest threats to human health (U.S.EPA 1990), and thus establishing a limit Cr(VI) concentration $< 50\ \mu g\ L^{-1}$.

3 NO_3^--Contaminated Groundwater Treatment

3.1 Common Nitrate Removal Processes

A wide range of technologies have been applied for groundwater denitrification that includes chemical techniques, such as electrodialysis (El Midaoui et al. 2002), reverse osmosis (RO) in combination with nano-filtration (Epsztein et al. 2015), electrodeionization (Zhang and Chen 2016), electrocatalytic reduction (Duca and Koper 2012), photocatalytic degradation (Anderson 2011) and chemical reduction (Fu et al. 2014), and physical techniques such as adsorption (Bhatnagar and Sillanpää 2011, Capodaglio et al. 2015) that have been tested with varying results. A valuable alternative to chemical and physical techniques may consist of the application of biological processes. Heterotrophic (Capodaglio et al. 2016) and autotrophic (Molognoni et al. 2017) denitrification can be suitable for the removal of nitrate from a liquid matrix; the former is principally applied in the case of wastewaters with an abundance of organic matter, while the latter may be applicable for groundwater denitrification in which there is generally a lack thereof (Wu et al. 2005; Gentile et al. 2006).

Autotrophic nitrate removal has, in fact, the advantage of not requiring an organic carbon source; however, the slow growth rate of autotrophic bacteria and low nitrate removal rate have so far contributed to a relatively scarce number of full-scale plants in operation at the present time. The addition of a carbon source (methanol, ethanol or acetic acid) to groundwater under heterotrophic denitrification is, in fact, expensive and may result in effluent turbidity increase due to bacterial growth and excessive organic carbon. Nevertheless, this process is currently applied extensively because of high efficiency (high degradation rates) and the existing know-how on this type of reactors (Mohseni-Bandpi et al. 2013).

3.2 BESs for Nitrate Removal

BESs have been proposed in the last decade for autotrophic denitrification. The most commonly adopted strategy is to perform denitrification in a biocathode, where nitrate will act as an electron acceptor, and consequently reduce through intermediate steps to $N_2\uparrow$ as shown in Equations 1-4:

$$NO_3^- + 2H^+ + 2e^- \rightarrow NO_2^- + H_2O \qquad (1)$$

$$NO_2^- + 2H^+ + e^- \rightarrow NO + H_2O \qquad (2)$$

$$+ H_2O \qquad (3)$$

$$N_2O + 2H^+ + 2e^- \rightarrow N_2 + H_2O \qquad (4)$$

An appropriate electron supply for the reaction can be guaranteed by oxidation of organic matter at a bioanode by external application of a power supply or a combination of both. In case of incomplete denitrification, reduced N-forms such as nitrite can appear in the effluent as reported by Srinivasan et al. (2016) and Cecconet et al. (2018). N_2O may also accumulate (Desloover et al. 2011, Molognoni et al. 2017), while fast kinetics of NO reduction does not allow its accumulation (Virdis et al. 2009).

The feasibility of autotrophic denitrification was first demonstrated by Gregory et al. (2004), reporting a simultaneous reduction of nitrate and nitrite. The first MFC application of biocathode denitrification was described by Clauwaert et al. (2007), achieving a nitrate removal rate of 16 gN m^{-3} NCC* d^{-1}. Several similar examples have been reported since using different anolytes such as glucose (Jia et al.

---

* NCC: Net Cathodic Chamber

2008), acetate (Puig et al. 2012; Oon et al. 2016) and domestic wastewater (Fang et al. 2011). The application of Pt-catalyst at the cathode showed to enhance MFC performance both in terms of higher energy production and better denitrification (Fang et al. 2011).

Energy production in MFCs is dependent on the availability of electron acceptors, in this case nitrate, at the biocathode (Jia et al. 2008; An et al. 2016). Since oxidation of organic matter at the anode is the source of electrons, increasing organic load and COD removal at the bioanode leads to denitrification increase at the cathode due to the larger amount of moving electrons. This is also coupled with simultaneous higher power output (Oon et al. 2016). The use of an external power source (potentiostat) can help overcome the main limiting factor of groundwater denitrification in MFCs and oxidation at the anode. Abiotic anodes without oxidation could be used, coupled with a biotic cathode, similarly to what was described by Cecconet et al. (2018) and Nguyen et al. (2015).

The operational parameters can strongly influence the performance of denitrifying BESs. The pH of catholyte can enhance or limit denitrification and cause the presence of reduced intermediate N-forms; Clauwaert et al. (2009) reported incomplete denitrification at cathode effluent pH = 8.3, while complete denitrification was observed at pH = 7.2. The values of pH > 8 showed to stop denitrification, resulting in N_2O accumulation (Molognoni et al. 2017). Cecconet et al. (2018) reported influent pH=8 as a cause for nitrite accumulation and subsequent pH reduction to 7 leading to complete denitrification. The optimal pH for denitrification was reported as 7.5 (Kurt et al. 1987).

Hydraulic retention time (HRT) can also influence denitrification; Pous et al. (2017) suggested diverse bacterial responses at different HRTs. Interactions with the availability of influent nitrate were noticed as its removal rates increased with its concentration increase at low HRTs (1.2-1.6 h). The accumulation of intermediate N-forms has been reported in BESs with different configurations at the decrease of HRT (Cecconet et al. 2018).

Most available studies have dealt with laboratories or *ex situ* treatments, nevertheless, some attempts of *in situ* treatments were also made; Nguyen et al. (2016a) described the denitrifying performance of a biocathode buried in the sand. The decrease of nitrate reduction rates occurred at increasing sand depths, obtaining 38.7% denitrification performance in a fully buried biocathode compared to liquid phase only.

A submerged microbial desalination-denitrification cell (SMDDC) for *in situ* nitrate removal was developed by Zhang and Angelidaki (2013); the BES-generated current was able to attract nitrate inside the anode and transferred it subsequently to the cathode where it was autotrophically reduced.

The influence of other electron acceptors (e.g. perchlorate) in solution was examined by Butler et al. (2010). The authors reported that successful combined treatment could be possible and that presence of nitrate would enhance power production compared to the sole presence of perchlorate as the electron acceptor. A high value of pH at the cathode (pH = 8.5) seemed to enhance the removal of perchlorate in this case. Jiang et al. (2017) obtained high simultaneous removals of perchlorate and nitrate (40.97% for perchlorate and 86.03% for nitrate) jointly with electric production (3.10 A m^{-3}) in MFCs; the best results in terms of energy production and contaminants removal were obtained at the NO_3^-/ClO_4^- ratio of 1:1 with denitrifying bacteria observed as predominant species.

Due to the occasional accumulation of nitrite in BES processes, research on NO_2^- reduction was carried out. The accumulation of nitrite was initially highlighted by Srinivasan et al. (2016). The power production in the nitrate/nitrite reductive step was higher than in nitrite reduction, and no process modifications were able to completely remove nitrite. Puig et al. (2011) demonstrated that in autotrophic denitrification processes, nitrate and nitrite could be considered 'interchangeable' as electron acceptors by exoelectrogenic bacteria while still obtaining a measurable power production.

4. As-Contaminated Groundwater Treatment

4.1 Common Arsenic Removal Processes

Several technologies were developed for arsenic removal from water solutions, such as oxidation, aluminum coagulation, lime softening, membrane nano-filtration and adsorption (Joshi and Chaudhuri 1996; Zouboulis and Katsoyiannis 2005; Çiftçi et al. 2011).

Arsenic can be present in water in different forms, mainly inorganic, andarsenate [As(V)] and arsenite [As(III)] are the two most common forms (Masscheleyn et al. 1991). Due to a predominantly reducing environment, arsenic is present generally in the As(III) form in groundwater and is less soluble and more toxic than As(V) (Jain and Ali 2000). Due to the different properties, As(V) has been reported easier to remove in physical processes (Lin and Wu 2001; Wickramasinghe et al. 2004; Jiang 2001).

A two-step approach could, therefore, be hypothesized, involving preliminary oxidation of As(III) to As(V) that is followed by physical removal. Traditional physicochemical techniques to oxidize arsenite to arsenate, however, suffer typically from high chemical costs and generation of toxic by-products. Oxidation can, however, be achieved using different techniques (Bissen and Frimmel 2003). Microbially-catalyzed oxidation, based on self-regenerating catalysts, received considerable attention as possible environmentally friendly pre-treatment for arsenic removal; in general, biological processes are preferred to physical and chemical methods due to inferior cost and environmental impact (Bahar et al. 2013; Nguyen et al. 2017). Bioremediation, unlike organic compounds removal, is accomplished by a complex chain of biologically-mediated transformation, accumulation, sorption and volatilization processes; specifically, arsenic bioremediation relies on microbes to detoxify, mobilize/immobilize this compound through redox, bio-methylation, sorption and complexation reactions (Wang and Zhao 2009; Pal and Paknikar 2012).

Heterotrophic and autotrophic As(III)-oxidizing microorganisms have been described; since the first report (Green 1918), strains from approximately 21 genera of As(III)-oxidizing prokaryotes have been reported, phylogenetically grouped under α-, β- and γ-Proteobacteria, *Deinocci* (i.e. *Thermus*) and *Crenarchaeota* (Stolz et al. 2002). These are physiologically diverse and include heterotrophs and the recently described *chemolithoautotrops* (bacteria obtaining energy from autotrophic oxidation of inorganic compounds). In contrast to heterotrophic As(III)-oxidizing bacteria, which is widely present in soils, autotrophic oxidizers utilize arsenite as electron donor, oxygen as the electron acceptor and CO_2 as carbon source (Zhang et al. 2015). Methylation of arsenic (oxidative addition of methyl groups producing compounds as monomethylarsonic and other acids to trimethylarsine) is a detoxification strategy occurring in living organisms from bacteria to humans. Arsenic methylation by fungi and other eukaryotes is well recognized, while less is known concerning this reaction in bacterial systems. *Methanobacterium formicicum* was found to be very efficient at producing methylated arsines and arsine, while anaerobes *Clostridium collagenovorans* (fermentative) and sulfate-reducing *Desulfovibrio gigas* and *Desulfovibrio vulgaris* can produce small amounts of trimethylarsine.

Once As(III) is oxidized to As(V), cost-effective and efficient adsorbents are required for complete removal. Adsorption on biological materials seems to be the most promising as they are naturally available, cheap and renewable. Some limitations with this technology include the possible necessary addition of some external carbon source for bacterial growth and the injection of oxygen into the aquifer (Agarwal et al. 2005). The injection of organic matter, necessary in heterotrophic oxidation, could originate groundwater contamination. A feasible solution could be the use of a virtually-unlimited electron acceptor (electrode) and consequently a BES for groundwater arsenic remediation (Pous et al. 2018). The use of BESs to remove arsenic from groundwater (Pous et al. 2015) could be achieved alone or in conjunction to the removal of other observed pollutants, like nitrate (Molognoni et al. 2017). These are based on functional modifications of MFCs (Capodaglio et al. 2013; Molognoni et al. 2014). Although seemingly offering a quite promising alternative to other methods, with already established proof of concept in laboratory studies, they have not still originated full-scale applications so far.

4.2 Hybrid MFC-ZVI System

Removal of arsenic using MFCs was first proved by Xue et al. (2013) by coupling an MFC and a zero-valent iron (ZVI) process. ZVI has many applications in groundwater remediation; one being the "permeable iron wall" technique in which ZVI is implemented as a permeable reactive barrier (PRB) that filters out contaminants in the aquifer during its natural flow (Figure 2).

ZVI technology, releasing Fe^{2+} oxidated to Fe^{3+} during a sacrificial corrosion process, produces strong oxidants (H_2O_2 and radical •OH) that are capable of oxidizing As(III) to As(V). Generated Fe(II), however, can react with H_2O_2 and consume it; hence, the addition of 2,2'-bipyridyl (BPY) is necessary to prevent such reaction. To control the corrosion process and thus the release of Fe^{2+}, it is necessary to have current flowing in the electrolyte of the system as illustrated in Figure 3.

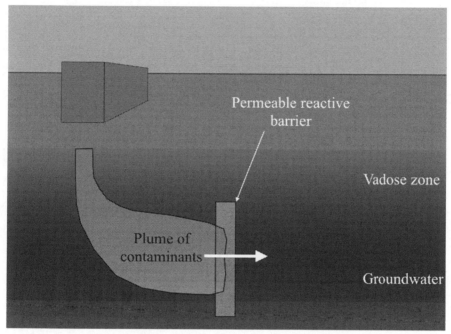

Figure 2: Permeable reactive barrier concept.

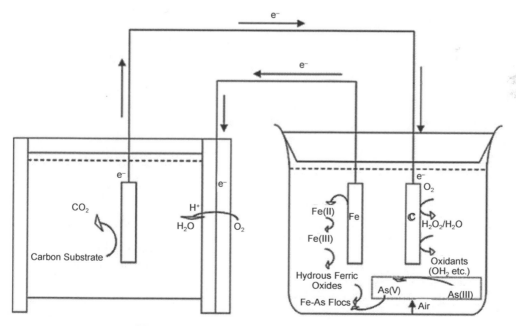

Figure 3: The MFC/ZVI process proposed by Xue et al. (2013). Reproduced from Xue et al. (2013) with permission from Elsevier.

The electricity produced by an MFC can be used for this purpose whereby arsenic is not removed by the MFC directly but by the associated ZVI system. Other authors (Tandukar et al. 2009; Ter Heijne et al. 2010) have postulated that MFCs can also be used to treat water containing CrO_4^{2-} (see Section 4) in similar ways. In the arsenic case, the hybrid process proved to obtain better results compared to the conventional ZVI process, generating more H_2O_2-derived oxidants under BPY addition, facilitating arsenic oxidation and achieving concentrations below EPA and WHO standards.

Arsenic in the MFC-ZVI process is divided into three fractions: the first remains in solution, the second retains on the electrodes and the third gets removed by flocs formed during precipitation. Pre-oxidation with external oxidants is a requisite to guarantee effective arsenic removal; however, *in situ* oxidant generation could simplify process operation and reduce its cost.

Compared to other techniques, the MFC-ZVI hybrid process does not show the highest arsenic removals, however, it is relatively efficient and capable of successfully removing arsenic to levels compliant with current standards.

4.3 Anaerobic Arsenite Oxidation

BESs were successfully applied for bioremediation of subsurface contaminants, including perchlorate (Butler et al. 2010), trichloroethene (Aulenta et al. 2008, 2009, 2010), hexavalent chromium (Xafenias et al. 2013), benzene (Rakoczy et al. 2013), nitrate (Pous et al. 2013) and sulfate (Coma et al. 2013) in laboratory settings. So far, the potential for remediation of As-contaminated groundwater was only casually explored. Brastad and He (2013), who applied microbial desalination cells to separate hardness from water, found that the process separated 89% of the arsenic contained in the solution.

The first study proposing direct arsenic removal with BES technology was published by Pous et al. (2015) that was based on work by Wang and Zhao (2009) and Pal and Paknikar (2012). The proposed process would provide As(III)-oxidizing microorganisms with virtually unlimited, low-cost and low-maintenance electron acceptors and was tested in an anaerobic BES, using a polarized (+0.497 V vs. SHE) graphite electrode as electron acceptor, based on the formal redox potential of arsenite oxidase of NT-26 autotrophic bacteria reported by Bernhardt and Santini (2006). The cyclic voltammetry (CV) showed that As(III) oxidation potential occurred at +0.500 V vs. SHE, close to the applied one, and no peaks relative to the presence of redox mediators were observed thus suggesting direct extracellular electron transfer. Microbiological analysis showed a predominance of *gamma-Proteobacteria* and *delta-Proteobacteria* on the electrode where the former dominated the microbial population at the anode.

Figure 4 shows trends of arsenic forms during the test under polarization potential of +0.497 V. When electrode polarization is applied, an almost linear decrease in As(III) concentration mirrored by stoichiometric increase of As(V) is observed with practically identical rates of removal and formation of 420 ± 38 and 428 ± 59 µg L^{-1} d^{-1}. The total arsenic (As$_{TOT}$) remained stable, indicating the absence of precipitation and/or adsorption onto graphite surfaces.

This approach would allow sustainable groundwater treatment since it requires only a polarized graphite electrode to induce arsenite oxidation with no need for chemicals. The process could be applied either *in situ* or on-site and several strategies (e.g., increasing area of the electrode, increasing biomass density, optimizing potential of the electrode) could be applied to increase its rates. This could lead to more effective bio(electro)remediation of As-contaminated groundwater; the introduction of electrodes in an aquifer (similarly to PBRs) could allow the development of localized treatment zones. With appropriate adsorbing material next to the anode, a complete treatment sequence would thus be achieved. Adsorbers, replaced upon saturation, could typically consist of conductive metals or metal oxides, serving as a cathode in the process. In this case, a cathodic polarization would reduce the corrosion of materials while increasing its lifetime and adsorption capacity.

An important step forward was achieved by Nguyen et al. (2016b) in a BES coupling anodic As(III) oxidation and cathodic denitrification that showed electrons produced by anaerobic oxidation of arsenite could be recovered for denitrification. Reactors using carbon paper as electrode material and bipolar membranes as separators between chambers were operated under DC supply; a complete oxidation of

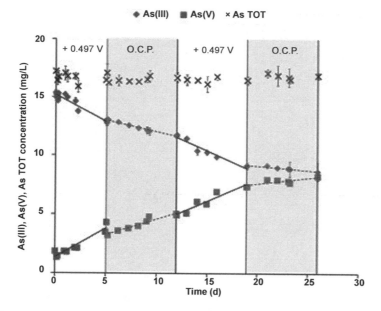

Figure 4: Evolution of As(III), As(V) and As$_{TOT}$ concentrations (Pous et al. 2015). Reproduced from Pous et al. (2015) with permission from Elsevier.

As(III) was observed in reactors under power supply (1 V) and potentiostat (anode at +0.5 V vs. SHE), while only incomplete reaction was reported in MFC mode or open circuit that confirmed bioelectrochemical oxidation is controlled by the polarized electrode serving as electron acceptor. A higher removal rates compared to Pous et al. (2015) (29.6 and 0.42 mg L^{-1}d^{-1}, respectively) due to increased electrode effective area and a higher density of microbial species present were reported. However, adsorption on surface of electrodes or precipitation was not reported. No differences in oxidation rates were reported between the use of power supply and potentiostat, but complete denitrification occurred only in the potentiostat-operated system. Nitrite accumulation occurred under DC supply likely due to less stable cathode potential. Microbiological analysis of anode biofilm showed the predominance of *alpha-Proteobacteria* and *delta-Proteobacteria*, two species different from those observed by Pous et al. (2015).

Under this approach, a virtually unlimited, low-cost and low-maintenance electron acceptor as well as a physical support for attachment could be provided. Cheap carbon-based materials such as graphite or even renewable substitutes currently studied, like biochar (Callegari and Capodaglio 2018), could be used as anodes and cathodes, required minimal maintenance and were not being prone to corrosion. Furthermore, solar accumulation systems could be used to drive the process. Furthermore, a bioelectrochemical approach would eliminate needs for chemicals and oxygen to stimulate microbial activity, minimizing the generation of by-products.

4.4 Spontaneous As(III) Oxidation with Bioelectricity Generation in Single-Chamber MFCs

Li et al. (2016) investigated spontaneous As(III) oxidation at the anode of single-cell MFCs with carbon-fiber felt anode and plain carbon paper cathode. Additions of 200 μg As(III) L^{-1} in the form of NaAsO$_2$ to the anolytes of MFCs (MFC-As) occurred, while an equal number of cells were operated without addition as control (MFC-C). Corresponding abiotic systems without inoculation (FC-As and FC-C) were run to exclude purely electrochemical effects. Significant As(III) removal was observed during the operation of MFC-As, with corresponding As(V) generation, as shown in Figure 5. As(III) was removed nearly to completion during a 7-day operating cycle, while it was hardly removed during the same time in abiotic FC-As (not shown).

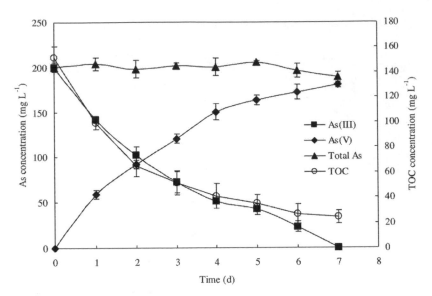

Figure 5: As concentrations (As(III), As(V), total As) and TOC in MFC-As during a 7 days operating cycle (Li et al. 2016). Reproduced from Li et al. (2016) with permission from Elsevier.

Anaerobic As(III) oxidation occurred with glucose as organic carbon source according to the reactions:

$$AsO_2^- + 2H_2O \rightarrow AsO_4^{3-} + 4H^+ + 2e^- \qquad E'_0 = -0.56 \text{ V} \qquad (5)$$

$$C_6H_{12}O_6 + 6H_2O \rightarrow 6CO_2 + 24H^+ + 24e^- \qquad E'_0 = 0.014 \text{ V} \qquad (6)$$

In these conditions, As(III) was oxidized faster compared to pure anaerobic cultures as mixed cultures could better handle complex conditions. Moreover, the removal efficiency of As(III) was comparable with those from fixed anode potential MFCs, observed by Pous et al. (2015), while the present technology could be considered more practical as spontaneous As(III) oxidation occurs without additional energy. Tests indicated that most bacteria could still survive with As(III)-spiked anolyte and thus properly work in the proposed system. Since a sufficient amount of organics may not be present in groundwater, the addition of exact amounts of carbon could be necessary to support effective As(III) removal and preventing induced organic pollution of treated water.

Cell inoculum contained 27 genotypes of bacterial phylum in MFC-As where remarkable changes occurred, with many original genotypes disappearing, while the types *Chlorobi*, as well as *Firmicutes*, increased significantly. *Actinobacteria* also appeared, suggesting that the structure of the bacterial community had evolved, adapting to the new operating conditions. The results indicated that As(III) oxidation in MFCs could be an effective strategy for remediation of arsenic-polluted groundwater environment. Polarization curves and power outputs also confirmed these findings with a maximum power density of 752.6 ± 17 mW m^{-2} in MFC-As and minimal voltage/power density in the abiotic cells.

5. Biological Chromium VI Reduction with MFCs

5.1 Common Cr(VI) Removal Processes

Removal of Cr(VI) is a major challenge in water and wastewater treatment. Hexavalent chromium is usually treated by physicochemical processes (e.g., adsorption, electrochemical separation, ion exchange, filtration, membrane processes, precipitation and solvent extraction), typically involving pH adjustment and addition of reducing agents to the solution, making them expensive with the production of toxic sludge. Microbiologically catalyzed reduction of Cr(VI) to Cr(III), first described by Romanenko and Koren'kov

(1977), is viewed as a potentially attractive option. A wide range of aerobic and anaerobic microorganisms can reduce Cr(VI) to Cr(III): phylum *proteobacteria,* such as *Pseudomonas dechromaticans, Escherichia coli, Desulfovibrio vulgaris, Shewanella oneidensis, Aeronomas dechromatica* and *Enterobacter cloacae* besides other species from classes *Bacilli* and *Clostridia* (Chen and Hao 1998; Bae et al. 2000). Biological chromium reduction processes always need the addition of external carbon sources, increasing cost and reducing process efficiency since the latter is also used by other microbial communities and does not contribute to chromium reduction.

5.2 Cathodic Chromium Reduction with Microbial Fuel Cells

The reducing environment of a cathode is an advantage for environmental biotechnology since it can be used for the treatment of oxidized pollutants. In an MFC biocathode there is no need for metal catalysts or mediators, unlike in conventional MFCs using abiotic and aerated cathodes. Based on these principles, Tandukar et al. (2009) first attempted to biologically reduce Cr(VI) by means of an MFC biocathode. The system used is schematized in Figure 6.

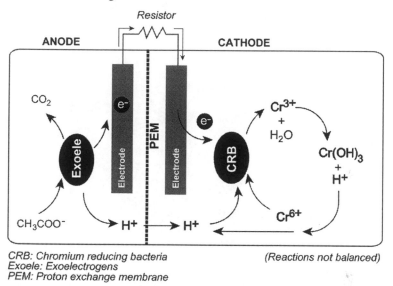

CRB: *Chromium reducing bacteria*
Exoele: *Exoelectrogens*
PEM: *Proton exchange membrane*

(Reactions not balanced)

Figure 6: Schematics of the biocathode MFC used by Tandukar et al. (2009). Reprinted with permission from Tandukar et al. (2009). Copyright 2009 American Chemical Society.

H-type borosilicate glass MFCs were operated with a mixture of denitrifying anaerobic mixed cultures with the same nutrient solution used for anode and cathode, except for the presence of 0.1 g L^{-1} of $MgCl_2$ in the catholyte, to avoid sulfate reduction and abiotic Cr(VI) reduction. The cathode compartment was also supplied with 0.2 g$NaHCO_3$ L^{-1} as an inorganic carbon source for Cr(VI)-reducing microorganisms.

Four consecutive Cr(VI) spikes were introduced into the system to investigate the effect on microorganisms' reductive activity and power generation (Figure 7). Cr(VI) reduction followed close to zero-order kinetics rather than first-order as initially imagined.

Direct relationships were observed between Cr(VI) concentration and specific reduction rate while also affecting power generation; maximum specific reduction rate was 0.46 mgCr(VI) gVSS^{-1} h^{-1} at initial Cr(VI) concentration of 63 mg L^{-1}, comparable to that of conventional biological reduction processes, varying from 0.20 to 0.62 mgCr(VI) gVSS^{-1} h^{-1} for different substrates (Xafenias et al. 2013).

When spiked with a higher Cr(VI) concentration (about 80 mg L^{-1}), observed bio-reduction rates were much lower. This decrease could be attributed to microbiological inhibition by Cr(VI) and/or accumulated Cr(III). The total chromium in filtered and unfiltered catholyte was absent after each test indicating the absence of any form of soluble chromium, complete reduction of Cr(VI) and that reduced

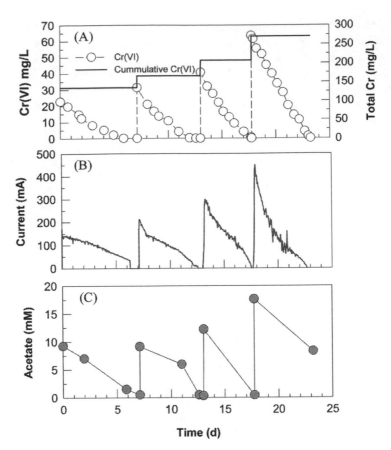

Figure 7: (A) cycles of Cr(VI) addition and reduction; (B) current production; (C) acetate concentration in the anode compartment. Reprinted with permission from Tandukar et al. (2009). Copyright 2009 American Chemical Society.

Cr(III) was precipitated (as $Cr(OH)_3$) and/or adsorbed on biomass or electrode surfaces. In the anodic compartment, CH_4 produced an amount equivalent to 18% of total initial COD that increased to over 50% in the final phases. A suppression of anode methanogenesis is one of the typical challenges of MFCs operation in view of its negative effects on their coulombic efficiency.

In standard conditions, reduction of Cr(VI) to Cr(III) occurs at potential of 1.33 V higher than reduction of oxygen to water (1.23 V). Therefore, cathodic Cr(VI) reduction has potential to generate higher electricity than an aerated cathode with similar anode. Half-reduction reactions at anode and cathode are shown, respectively:

Anode:

$$CO_2 + HCO_3^- + 8H^+ + 8e^- \rightarrow CH_3COO^- + 3H_2O \qquad E'_0 = 0.284 \text{ V} \qquad (7)$$

Cathode:

$$Cr_2O_7^{2-} + 14H^+ + 6e^- \rightarrow 2Cr(III) + 7H_2O \qquad E'_0 = 0.365 \text{ V} \qquad (8)$$

Precipitation:

$$2Cr(III) + 7H_2O \rightarrow 2Cr(OH)_{3(s)} + 6H^+ + H_2O \qquad (6.5 < pH < 10) \qquad (9)$$

The stoichiometry of reactions (Equations 7-9) shows that protons released by the oxidation of acetate are not sufficient for Cr(VI) reduction, which could limit the reaction, raising the pH of the cathodic compartment. Nevertheless, pH in this compartment is near neutral, facilitating the precipitation of Cr(III) as $Cr(OH)_3$ (Equation 7). Three moles of protons are generated per mole of Cr(III) that will be utilized for Cr(VI) reduction. The generation of protons via $Cr(OH)_3$ formation compensates 75% of the proton deficiency.

The calculation of the MFC's theoretical EMF_{max} (maximum electromotive force) shows that molar concentration of oxidized acetate should be 0.38 times the molar concentration of reduced Cr(VI), neglecting biomass growth and other competing processes. Reduction of 1 mol of Cr(VI) to Cr(III) has a reduction potential of 0.365 V, and complete oxidation of 0.38 mols of acetate has a reduction potential of -0.287 V (vs. SHE). Consequently, a combination of the two half-reactions could generate a total EMF of 0.652 V, however, its actual value is affected by numerous factors, such as biomass activity, an internal resistance of the system and others.

The maximum current and power densities obtained by Tandukar et al. (2009) were 123.4 mA m^{-2} and 55.5 mW m^{-2}, respectively, at an initial Cr(VI) concentration of 63 mg L^{-1}. These values are slightly lower than those reported for conventional MFCs with an aerated cathode. The coulombic balance and total coulombs transferred from the anode were compared with coulombs required for Cr(VI) reduction (Table 1). The study confirmed that Cr(VI) can be efficiently reduced in an MFC biocathode under mostly autotrophic conditions. Any suitable organic wastewater could be used in the anode compartment, facilitating simultaneous treatment of different substrates in a single system; however, pH control in both compartments may be necessary. A complete removal of chromium can be obtained by settling and removing biomass from the cathode, however, separation of adsorbed and non-adsorbed Cr(III) from biomass could be a major issue for the practical application of this process. Retention of biomass in the biocathode will also be another major aspect for the design of a continuous-flow MCF based on this concept.

Table 1: Summary of experimental results (from Tandukar et al. 2009). Reprinted with permission from Tandukar et al. (2009). Copyright 2009 American Chemical Society.

Parameter		Initial Cr(VI) Concentration (mg L^{-1})			
		22	**31**	**40**	**63**
Specific reduction rate	mgCr(VI) gVSS^{-1} h^{-1}	0.18 ± 0.11^a	0.22 ± 0.13	0.30 ± 0.12	0.46 ± 0.17
I_{max}	mA	0.17	0.20	0.30	0.45
Resistance	kΩ	1	1	1	1
Power density	mW m^{-2}	7.9	11	24.7	55.5
Power volume^{-1}	mA m^{-3}	115.6	160	360	810
Current density	mA m^{-2}	46.6	54.8	83.2	123.4
Coulombs required (C_R)	C	29.3	40.2	52.3	81.4
Coulombs transferred (C_t)	C	25.9	55.9	61.5	68.9

[a]: Mean $\pm$ standard deviation ($n \geq 10$).

6. Conclusions

BESs have been proposed in recent years as an environmental technology that could provide a sustainable and effective solution for the removal of several pollutants from water and groundwater. In particular, this chapter has addressed the state-of-the-art in the removal of nitrate, arsenic and chromium by their means. So far, the results seem positive although some achieved transformation rates are still lower than those of traditional processes. This issue is often not due to microbiological constraints but to practical issues relating to construction and materials in these systems. Also, the majority of published studies dealt with these systems in laboratory settings and pilot *ex situ* treatments, with very few real-scale applications, notwithstanding the many and repeated proofs-of-concept achieved by these technologies.

In the authors' opinion, BESs technologies are an interesting alternative approach to groundwater remediation problems and are due for full-scale applications as soon as some eminently practical issues will be tested and resolved.

7. Future Perspectives

The analysis of the described technologies has found that the majority of BES applications for groundwater bioremediation are relative to pilot or *ex situ* treatments. Although the most real application will probably be implemented on site (water treated as needed after pumping), higher attention should also be focused on *in situ* applications where the aquifer itself is decontaminated passing through a PBR-like system. In particular, specific BES designs should be investigated due to the expected differences in operating conditions in the aquifer rather than on the ground.

In some described applications a mixture of two contaminants were treated, but the complexity of groundwater contamination situations suggests that more complex conditions could occur and should be examined for possible combined treatment. In these cases, the resilience of BES applications toward potentially harmful contaminants for biomass should be assessed.

References

Agarwal N, Semmens MJ, Novak PJ, Hozalski RM (2005) Zone of influence of a gas permeable membrane system for delivery of gases to groundwater. Water Resources Research 41(5): 1–10.

Agusa T, Trang PTK, Lan VM, Anh DH, Tanabe S, Viet PH, Berg M (2014) Human exposure to arsenic from drinking water in Vietnam. Science of the Total Environment 488–489(1): 562–569.

Ahmed MF, Ahuja S, Alauddin M, Hug SJ, Lloyd JR, Pfaff A, Pichler T, Saltikov C, Stute M, van Geen A (2006) Epidemiology: Ensuring Safe Drinking Water in Bangladesh. Science 314(5806): 1687–1688.

An BM, Song YH, Shin JW, Park JY (2016) Two-chamber microbial fuel cell to simultaneously remove ethanolamine and nitrate. Desalination and Water Treatment 57(17): 7866–7873.

Anderson JA (2011) Photocatalytic nitrate reduction over Au/TiO2. Catalysis Today 175(1): 316–321.

Aulenta F, Canosa A, Majone M, Panero S, Reale P, Rossetti S (2008) Trichloroethene dechlorination and H2 evolution are alternative biological pathways of electric charge utilization by a dechlorinating culture in a bioelectrochemical system. Environmental Science and Technology 42(16): 6185–6190.

Aulenta F, Canosa A, Reale P, Rossetti S, Panero S, Majone M (2009) Microbial reductive dechlorination of trichloroethene to ethene with electrodes serving as electron donors without the external addition of redox mediators. Biotechnology and Bioengineering 103 (1): 85–91.

Aulenta F, Reale P, Canosa A, Rossetti S, Panero S, Majone M (2010) Characterization of an electro-active biocathode capable of dechlorinating trichloroethene and cis-dichloroethene to ethene. Biosensors and Bioelectronics 25(7): 1796–1802.

Bae WC, Kang TG, Kang IK, Won YJ, Jeong BC (2000) Reduction of Hexavalent Chromium by Escherichia coli ATCC 33456 in Batch and Continuous Cultures. The Journal of Microbiology 38(1): 36–39.

Bahar MM, Megharaj M, Naidu R (2013) Bioremediation of arsenic-contaminated water: Recent advances and future prospects. Water, Air, and Soil Pollution 224(12): 1–20.

Barringer JL, Mumford A, Young LY, Reilly PA, Bonin JL, Rosman R (2010) Pathways for arsenic from sediments to groundwater to streams: Biogeochemical processes in the Inner Coastal Plain, New Jersey, USA. Water Research 44(19): 5532–5544.

Bernhardt PV, Santini JM (2006) Protein film voltammetry of arsenite oxidase from the chemolithioautotrophic arsenite-oxidizing bacterium NT-26. Biochemistry 45(9): 2804–2809.

Bhatnagar A, Sillanpää M (2011) A review of emerging adsorbents for nitrate removal from water. Chemical Engineering Journal 168(2): 493–504.

Bissen M, Frimmel FH (2003) Arsenic— a Review. Part II: Oxidation of Arsenic and its Removal in Water Treatment. Acta hydrochimica et hydrobiologica 31(2): 97–107.

Bouchard DC, Williams MK, Surampalli RY (1992) Nitrate Contamination of Groundwater: Sources and Potential Health Effects. Journal (American Water Works Association) 84(9): 85–90.

Brastad KS, He Z (2013) Water softening using microbial desalination cell technology. Desalination 309: 32–37.

Butler CS, Clauwaert P, Green SJ, Verstraete W, Nerenberg R (2010) Bioelectrochemical perchlorate reduction in a microbial fuel cell. Environmental Science & Technology 44(12): 4685–4691.

Callegari A, Capodaglio A (2018) Properties and beneficial uses of (bio)chars, with special attention to products from sewage sludge pyrolysis. Resources 7(1): 20.

Capodaglio AG, Hlavínek P, Raboni M (2015) Physico-chemical technologies for nitrogen removal from wastewaters: a review. Ambiente e Agua - An Interdisciplinary Journal of Applied Science 10(3): 445–458.

Capodaglio AG, Hlavínek P, Raboni M (2016) Advances in wastewater nitrogen removal by biological processes: state of the art review. Ambiente e Agua - An Interdisciplinary Journal of Applied Science 11(2): 250.

Capodaglio AG, Molognoni D, Dallago E, Liberale A, Cella R, Longoni P, Pantaleoni L (2013) Microbial fuel cells for direct electrical energy recovery from urban wastewaters (Article number 634738). The Scientific World Journal 2013.

CEC (1998) Council Directive 98/83/EC of 3 November 1998 on the quality of water intended for human consumption. Official Journal L 330, 05/12/1998: 0032–0054.

Cecconet D, Devecseri M, Callegari A, Capodaglio AG (2018) Effects of process operating conditions on the autotrophic denitrification of nitrate-contaminated groundwater using bioelectrochemical systems. Science of The Total Environment 613–614: 663–671.

Chen JM, Hao OJ (1998) Microbial Chromium (VI) Reduction. Critical Reviews in Environmental Science and Technology 28(3): 219–251.

Cho K, Zholi A, Frabutt D, Flood M, Floyd D, Tiquia SM (2012) Linking bacterial diversity and geochemistry of uranium-contaminated groundwater. Environ Technol 33: 1629-1640.

Çiftçi TD, Yayayürük O, Henden E (2011) Study of arsenic(III) and arsenic(V) removal from waters using ferric hydroxide supported on silica gel prepared at low pH. Environmental Technology 32(3): 341–351.

Clauwaert P, Desloover J, Shea C, Nerenberg R, Boon N, Verstraete W (2009) Enhanced nitrogen removal in bio-electrochemical systems by pH control. Biotechnology Letters 31(10): 1537–1543.

Clauwaert P, Rabaey K, Aelterman P, De Schamphelaire L, Pham TH, Boeckx P, Boon N, Verstraete W (2007) Biological denitrification in microbial fuel cells. Environmental Science & Technology 41(9): 3354–3360.

Coma M, Puig S, Pous N, Balaguer MD, Colprim J (2013) Biocatalysed sulphate removal in a BES cathode. Bioresource Technology 130: 218–223.

CSWRCB (2009) Groundwater Information Sheet. Chromium VI. Rev. 2009/09.

Dalla Libera N, Fabbri P, Mason L, Piccinini L, Pola M (2017) Exceedance probability map: a tool helping the definition of arsenic Natural Background Level (NBL) within the Drainage Basin to the Venice Lagoon (NE Italy). *In*: 19th EGU General Assembly, EGU2017, proceedings from the conference held 23-28 April, 2017 in Vienna, Austria., p.6388. 6388.

Desloover J, Puig S, Virdis B, Clauwaert P, Boeckx P, Verstraete W, Boon N (2011) Biocathodic nitrous oxide removal in bioelectrochemical systems. Environmental Science & Technology 45(24): 10557–10566.

Duca M, Koper MTM (2012) Powering denitrification: the perspectives of electrocatalytic nitrate reduction. Energy & Environmental Science 5(12): 9726.

Egilman D (2006) Corporate Corruption of Science—The Case of Chromium(VI). *International* Journal of Occupational and Environmental Health 12(2): 169–176.

El Midaoui A, Elhannouni F, Taky M, Chay L, Menkouchi Sahli MA, Echihabi L, Hafsi M (2002) Optimization of nitrate removal operation from ground water by electrodialysis. Separation and Purification Technology 29(3): 235–244.

Epsztein R, Nir O, Lahav O, Green M (2015) Selective nitrate removal from groundwater using a hybrid nanofiltration–reverse osmosis filtration scheme. Chemical Engineering Journal 279: 372–378.

Fang C, Min B, Angelidaki I (2011) Nitrate as an Oxidant in the Cathode Chamber of a Microbial Fuel Cell for Both Power Generation and Nutrient Removal Purposes. Applied Biochemistry and Biotechnology 164(4): 464–474.

Fu F, Dionysiou DD, Liu H (2014) The use of zero-valent iron for groundwater remediation and wastewater treatment: A review. Journal of Hazardous Materials 267: 194–205.

Gentile M, Yan T, Tiquia SM, Fields MW, Nyman J, Zhou J and Criddle CS (2006) Stability and resilience in a denitrifying fluidized bed reactor. Microbial Ecol 52(2): 311-321.

Green HH (1918) Description of a bacterium which oxidizes arsenite to arsenate, and of one which reduces arsenate to arsenite, isolated from a cattle-dipping Etank. South Afr J Sci 14: 465–467.

Gregory KB, Bond DR, Lovley DR (2004) Graphite electrodes as electron donors for anaerobic respiration. Environmental Microbiology 6(6): 596–604.

Health Canada (2017) Guidelines for Canadian Drinking Water Quality Summary Table Federal-Provincial-Territorial Committee on Drinking Water of the Federal-Provincial-Territorial Committee on Health and the Environment February 2017, (February 2017).

Hering JG, Elimelech M (1995) International perspective on arsenic in groundwater: problems and treatment strategies. *In*: Proceedings of the American Water Works Association, Annual Conference.

Hu B, Li J, Zhao J, Yang J, Bai F, Dou Y (2013) Heavy metal in surface sediments of the Liaodong Bay, Bohai Sea: Distribution, contamination, and sources. Environmental Monitoring and Assessment 185(6): 5071–5083.

Jadhav SV, Bringas E, Yadav GD, Rathod VK, Ortiz I, Marathe KV (2015) Arsenic and fluoride contaminated groundwaters: A review of current technologies for contaminants removal. Journal of Environmental Management 162: 306–325.

Jain CK, Ali I (2000) Arsenic: Occurrence, toxicity and speciation techniques. Water Research 34(17): 4304–4312.

Jia YH, Tran HT, Kim DH, Oh S-J, Park D-H, Zhang R-H, Ahn D-H (2008) Simultaneous organics removal and bio-electrochemical denitrification in microbial fuel cells. Bioprocess and Biosystems Engineering 31(4): 315–321.

Jiang C, Yang Q, Wang D, Zhong Y, Chen F, Li X, Zeng G, Li X, Shang M (2017) Simultaneous perchlorate and nitrate removal coupled with electricity generation in autotrophic denitrifying biocathode microbial fuel cell. Chemical Engineering Journal 308: 783–790.

Jiang JQ (2001) Removing arsenic from groundwater for the developing world—a review. Water science and technology : a journal of the International Association on Water Pollution Research 44(6): 89–98.

Jomova K, Jenisova Z, Feszterova M, Baros S, Liska J, Hudecova D, Rhodes CJ, Valko M (2011) Arsenic: Toxicity, oxidative stress and human disease. Journal of Applied Toxicology 31(2): 95–107.

Joshi A, Chaudhuri M (1996) Removal of arsenic from ground water by iron oxide-coated sand. Journal of Environmental Engineering 122(8): 769–771.

Khaska M, Le Gal La Salle C, Verdoux P, Boutin R (2015) Tracking natural and anthropogenic origins of dissolved arsenic during surface and groundwater interaction in a post-closure mining context: Isotopic constraints. Journal of Contaminant Hydrology 177–178: 122–135.

Kurt M, Dunn IJ, Bourne JR (1987) Biological denitrification of drinking water using autotrophic organisms with H2 in a fluidized-bed biofilm reactor. Biotechnology and Bioengineering 29(4): 493–501.

Li YY, Zhang B, Cheng M, Li YY, Hao L, Guo H (2016) Spontaneous arsenic (III) oxidation with bioelectricity generation in single-chamber microbial fuel cells. Journal of Hazardous Materials 306: 8–12.

Liao L, Jean JS, Chakraborty S, Lee MK, Kar S, Yang HJ, Li Z (2016) Hydrogeochemistry of groundwater and arsenic adsorption characteristics of subsurface sediments in an alluvial plain, SW Taiwan. (Article number 1305). Sustainability (Switzerland) 8(12).

Lin TF, Wu JK (2001) Adsorption of arsenite and arsenate within activated alumina grains: Equilibrium and kinetics. Water Research 35(8): 2049–2057.

Mandal BK, Suzuki KT (2002) Arsenic round the world: A review. Talanta 58(1): 201–235.

Masscheleyn PH, Delaune RD, Patrick WH 1991. Effect of redox potential and pH on arsenic speciation and solubility in a contaminated soil. Environmental Science and Technology 25(8): 1414–1419.

McGuigan CF, Hamula CLA, Huang S, Gabos S, Le XC 2010. A review on arsenic concentrations in Canadian drinking water. Environmental Reviews 18: 291–307.

Menció A, Mas-Pla J, Otero N, Regàs O, Boy-Roura M, Puig R, Bach J, Domènech C, Zamorano M, Brusi D, Folch A (2016) Nitrate pollution of groundwater; all right…, but nothing else? Science of The Total Environment 539: 241–251.

Modin O, Aulenta F (2017) Three promising applications of microbial electrochemistry for the water sector. Environ Sci: Water Res Technol 3(3): 391–402.

Mohseni-Bandpi A, Elliott DJ, Zazouli MA 2013. Biological nitrate removal processes from drinking water supply-a review. Journal of Environmental Health Science & Engineering 11 (1): 35.

Molognoni D, Devecseri M, Cecconet D, Capodaglio AG 2017. Cathodic groundwater denitrification with a bioelectrochemical system. Journal of Water Process Engineering 19 (April), 67–73.

Molognoni D, Puig S, Balaguer MD, Liberale A, Capodaglio AG, Callegari A, Colprim J (2014) Reducing start-up time and minimizing energy losses of microbial fuel cells using maximum power point tracking strategy. Journal of Power Sources 269: 403–411.

Nguyen VK, Hong S, Park Y, Jo K, Lee T (2015) Autotrophic denitrification performance and bacterial community at biocathodes of bioelectrochemical systems with either abiotic or biotic anodes. Journal of Bioscience and Bioengineering 119(2): 180–187.

Nguyen VK, Park Y, Yu J, Lee T (2016a) Bioelectrochemical denitrification on biocathode buried in simulated aquifer saturated with nitrate-contaminated groundwater. Environmental Science and Pollution Research 23(15): 15443–15451.

Nguyen VK, Park Y, Yu J, Lee T (2016b) Simultaneous arsenite oxidation and nitrate reduction at the electrodes of bioelectrochemical systems. Environmental Science and Pollution Research 23(19): 19978–19988.

Nguyen VK, Tran HT, Park Y, Yu J, Lee T (2017) Microbial arsenite oxidation with oxygen, nitrate, or an electrode as the sole electron acceptor. Journal of Industrial Microbiology and Biotechnology 44(6): 857–868.

Oon Y-S, Ong S-A, Ho L-N, Wong Y-S, Oon Y-L, Lehl HK, Thung W-E (2016) Long-term operation of double chambered microbial fuel cell for bio-electro denitrification. Bioprocess and Biosystems Engineering 39(6): 893–900.

Pal A, Paknikar KM (2012) Bioremediation of arsenic from contaminated water. *In*: T. Satyanarayana, Bhavdish Narain Johri, Anil Prakash (eds.). Microorganisms in Environmental Management. Dordrecht: Springer Netherlands, pp. 477–523.

Pawełczyk A (2012) Assessment of health hazard associated with nitrogen compounds in water. Water Science and Technology 66(3): 666–672.

Pous N, Balaguer MD, Colprim J, Puig S (2018) Opportunities for groundwater microbial electro-remediation. Microbial Biotechnology 11(1): 119–135.

Pous N, Casentini B, Rossetti S, Fazi S, Puig S, Aulenta F (2015) Anaerobic arsenite oxidation with an electrode serving as the sole electron acceptor: A novel approach to the bioremediation of arsenic-polluted groundwater. Journal of Hazardous Materials 283: 617–622.

Pous N, Puig S, Balaguer MD, Colprim J (2017) Effect of hydraulic retention time and substrate availability in denitrifying bioelectrochemical systems. Environmental Science: Water Research & Technology 3(5): 922–929.

Pous N, Puig S, Coma M, Balaguer MD, Colprim J (2013) Bioremediation of nitrate-polluted groundwater in a microbial fuel cell. Journal of Chemical Technology & Biotechnology 88(9): 1690–1696.

Puig S, Coma M, Desloover J, Boon N, Colprim J, Balaguer MD 2012. Autotrophic denitrification in microbial fuel cells treating low ionic strength waters. Environmental Science and Technology 46(4): 2309–2315.

Puig S, Serra M, Vilar-Sanz A, Cabré M, Bañeras L, Colprim J, Balaguer MD (2011) Autotrophic nitrite removal in the cathode of microbial fuel cells. Bioresource Technology 102(6): 4462–4467.

Rakoczy J, Feisthauer S, Wasmund K, Bombach P, Neu TR, Vogt C, Richnow HH (2013) Benzene and sulfide removal from groundwater treated in a microbial fuel cell. Biotechnology and Bioengineering 110(12): 3104–3113.

Romanenko VI, Koren'kov VN (1977) Pure culture of bacteria using chromates and bichromates as hydrogen acceptors during development under anaerobic conditions. Mikrobiologiia 46(3): 414–7.

Saha R, Nandi R, Saha B (2011) Sources and toxicity of hexavalent chromium. Journal of Coordination Chemistry 64(10): 1782–1806.

Srinivasan V, Weinrich J, Butler C (2016) Nitrite accumulation in a denitrifying biocathode microbial fuel cell. Environmental Science: Water Research & Technology 2(2): 344–352.

Stolz J, Basu P, Oremland R (2002) Microbial transformation of elements: the case of arsenic and selenium. International Microbiology 5(4): 201–207.

Tandukar M, Huber SJ, Onodera T, Pavlostathis SG 2009. Biological chromium(VI) reduction in the cathode of a microbial fuel cell. Environmental Science and Technology 43(21): 8159–8165.

ter Heijne A, Strik DPBTB, Hamelers HVM, Buisman CJN (2010) Cathode potential and mass transfer determine performance of oxygen reducing biocathodes in microbial fuel cells. Environmental Science and Technology 44(18): 7151–7156.

The Council of the European Union, 1998. Council Directive 98/83/EC of 3 November 1998 on the quality of water intended for human consumption. Official Journal of the European Communities.

Tiquia SM (2010) Metabolic diversity of the heterotrophic microorganisms and potential link to pollution of the Rouge River. Environ Pollut 158: 1435-1443.

Tiquia SM (2011) Extracellular hydrolytic enzyme activities of the heterotrophic microbial communities of the Rouge River: An approach to evaluate ecosystem response to urbanization. Microbial Ecol 62: 679-689.

Tiquia SM, Davis D, Hadid H, Kasparian S, Ismail M, Sahly R, Shim J, Singh S, Murray KS (2007) Halophilic and halotolerant bacteria from river waters and shallow groundwater along the Rouge River of southeastern Michigan. Environ Technol 28: 297-307.

Tiquia SM, Schleibak M, Schlaff J, Floyd C, Hadid H, Murray KS (2008) Microbial community profiling and characterization of some heterotrophic bacterial isolates from river waters and shallow groundwater wells along the Rouge River, Southeast Michigan. Environ Technol 29: 651-663.

Urbansky ET, Schock MR (1999) Issues in managing the risks associated with perchlorate in drinking water. Journal of Environmental Management 56(2): 79–95.

USEPA, 1990. The drinking water criteria document on chromium; EPA 440/5-84-030. Washington D.C.

USEPA, 2010. Part 141 - National Primary Drinking Water regulations.

USEPA, 2011. Costs of Arsenic Removal Technologies for Small Water Systems. Washington D.C.

Virdis B, Rabaey K, Yuan Z, Rozendal RA, Keller J (2009) Electron fluxes in a microbial fuel cell performing carbon and nitrogen removal. Environmental Science and Technology 43(13): 5144–5149.

Wakida FT, Lerner DN (2005) Non-agricultural sources of groundwater nitrate: A review and case study. Water Research 39(1): 3–16.

Wang S, Zhao X (2009) On the potential of biological treatment for arsenic contaminated soils and groundwater. Journal of Environmental Management 90(8): 2367–2376.

WHO, 2017. Guidelines for Drinking-water quality: fourth edition incorporating the first addendum.

Wickramasinghe SR, Han B, Zimbron J, Shen Z, Karim MN (2004) Arsenic removal by coagulation and filtration: comparison of groundwaters from the United States and Bangladesh. Desalination 169(3): 231–244.

Wu W, Gu B, Fields MW, Gentile M, Ku Y-K, Yan H, Tiquia SM, Yan T, Nyman J, Zhou J, Jardine PM, Criddle CS (2005) Uranium (VI) reduction by denitrifying biomass. Bioremediation J 9: 1-13.

Xafenias N, Zhang Y, Banks CJ (2013) Enhanced performance of hexavalent chromium reducing cathodes in the presence of *Shewanella oneidensis* MR-1 and lactate. Environmental Science and Technology 47(9): 4512–4520.

Xue A, Shen ZZ, Zhao B, Zhao HZ 2013. Arsenite removal from aqueous solution by a microbial fuel cell-zerovalent iron hybrid process. Journal of Hazardous Materials 261: 621–627.

Zhang J, Zhou W, Liu B, He J, Shen Q, Zhao FJ 2015. Anaerobic arsenite oxidation by an autotrophic arsenite-oxidizing bacterium from an arsenic-contaminated paddy soil. Environmental Science and Technology 49(10): 5956–5964.

Zhang Y, Angelidaki I (2013) A new method for in situ nitrate removal from groundwater using submerged microbial desalination–denitrification cell (SMDDC). Water Research 47(5): 1827–1836.

Zhang Z, Chen A (2016) Simultaneous removal of nitrate and hardness ions from groundwater using electrodeionization. Separation and Purification Technology 164: 107–113.

Zhitkovich A (2011) Chromium in drinking water: Sources, metabolism, and cancer risks. Chemical Research in Toxicology 24(10): 1617–1629.

Zouboulis AI, Katsoyiannis IA (2005) Recent advances in the bioremediation of arsenic-contaminated groundwaters. Environment International 31(2): 213–219.

An Overview of Carbon and Nanoparticles Application in Bioelectrochemical System for Energy Production and Resource Recovery

Shiv Singh[1,2*], Kshitij Tewari[2] and Deepak Pant[3*]

[1] Lightweight Metallic Materials, Council of Scientific and Industrial Research-Advanced Materials and Processes Research Institute, Hoshangabad Road, Bhopal, Madhya Pradesh 462026, India

[2] Nanomaterial Toxicology Group, CSIR-Indian Institute of Toxicology Research (CSIR-IITR), Vishvigyan Bhawan, 31 Mahatma Gandhi Marg, Lucknow-226001, Uttar Pradesh, India

[3] Separation & Conversion Technology, Flemish Institute for Technological Research (VITO), Boeretang 200, 2400 Mol, Belgium

1. Introduction

Bioelectrochemical systems (BESs) have gained much attention in the field of energy sources due to their unique potential to recover different nutrients, inorganic and organic materials (Colombo et al. 2017; Modi et al. 2016; Singh et al. 2018; Sun et al. 2018). BESs are a unique, sustainable, eco-friendly microbial system for converting electrical energy into chemical energy and vice versa. From the past few decades, BESs are well-known for contribution in the field of developing sustainable technology for wastewater treatment, producing electricity and useful chemicals. Wastewater containing biodegradable organic compounds and other contaminants are being used as fuels in BESs that could otherwise move to water streams and become responsible for water pollution (ElMekawy et al. 2015; Venkata Mohan et al. 2010). In BESs, electrochemically active microorganisms act as biocatalysts deposited on solid electrodes and forming bioanodes/or biocathodes. These electroactive bacteria (EABs) known as electrogens, grown on electrodes, derive their food from organic waste and act as bioelectrocatalyst releasing electrons and protons during their metabolism. Predictably, they were being used for the treatment of wastewater and bioenergy production through bioelectrochemical reactions (Modi et al. 2017; Singh and Verma 2015a, b). In recent years, another set of EABs known as electrotrophs (electron eater) are being used to obtain useful compound from waste materials in a process described as microbial electrosynthesis (MES) (Gupta et al. 2017; Kadier et al. 2016; Lu et al. 2015; Marshall et al. 2013; Modi et al. 2016; Mook et al. 2013; Nevin et al. 2010; Singh et al. 2018; Singh and Verma 2015a, b; Venkata Mohan et al. 2014b). The basic advantage of BESs is generating energy and producing high efficient and valuable chemicals/products at low costs (Clauwaert et al. 2008; Khare et al. 2016; Venkata Mohan et al. 2008).

Apart from BESs, there are limited approaches that can reduce CO_2 without emission of carbon to date although researchers are working on it since a long time (Bajracharya et al. 2017a). Currently, researchers

*Corresponding authors: sshiv.singh@ampri.res.in, sshiviitk@gmail.com, deepak.pant@vito.be, pantonline@gmail.com

are approaching BESs in different ways to make them efficient, including different configurations, cation/anion exchange membranes (CEM/AEM), compositions of anolyte and catholyte, types of biocatalyst and electrodes materials (Modi et al. 2016, 2017; Noori et al. 2017; Singh and Verma 2015a, b). Among different BES components, the electrodes/electrode materials are considered as the most critical one that could directly affect the efficiency of prepared BESs. The presence of metal/metal ions has a significant impact on BESs for enhancing/reducing the performances of BESs because most of the components of BESs are directly or indirectly affected by metal/metal ions (Lu et al. 2015). It is believed that the existence of metal ions mostly enhances the performance of BESs. The reason behind is in many cases the metal ions act as electron transfer mediators and facilitate the activity of enzymes, biocatalysts and oxygen reduction reaction (ORR) (Ehrlich 1997; Kilpin and Dyson 2013; Lovley 2006; Lu et al. 2015; Singh et al. 2016; Singh and Verma 2015a, b; van der Maas et al. 2005).

With this background, this contemporary review emphasizes on the role of carbon and metal NPs in BESs particularly for (i) energy generation in a microbial fuel cell (MFC) and (ii) chemical production in MES. Also, in brief, other types of BESs are discussed below. In this chapter, we provide the most up-to-date information of nanomaterials used in MFC for electricity generation and MES for producing useful chemical/products along with CO_2 conversion. Firstly, we have described the nanomaterials used in MFC and MES and how they have enhanced the performances of electrodes and its system for energy recovery and chemical production. It is also discussed in what way useful interactions between electrode materials and bacteria are favorable for the production rate of value-added products.

2. Categories of BESs

In a typical BES, there exists an anode compartment and a cathode compartment that is filled with anolyte and catholyte, respectively. There is an optional arrangement for the cation exchange membrane (CEM)/anion exchange membrane (AEM) or separator (Modi et al. 2016; Pant et al. 2012). The microorganisms oxidize organic materials of wastewater under anaerobic conditions and generate electrons and protons at the anode chamber. The electrons move to the counter electrode via an external circuit, while the protons diffuse to the cathode through proton exchange membrane (PEM) resulting in the bioenergy generation (MFC mode). On the other hand, the electron accepting bacteria at the cathode is used for CO_2 reduction into value-added products, like formic acid, acetic acid, butyric acid, methanol, ethanol, glycerol, hydrogen peroxide, methane, etc. (MES mode). The prime objectives of BESs are to treat wastes/wastewaters and produce electricity useful products/chemicals (Modi et al. 2017; Pant et al. 2012; Singh and Verma 2015a, b).

There are two key approaches for electron transfer from bacteria in a typical BESs: (i) direct electron transfer where physical interaction between the electrode and bacteria for electron transfer and (ii) indirect electron transfer where extracellular electron transfer via electrochemically excretion of electron shuttling redox molecules (Modi et al. 2017; Pant et al. 2012; Singh and Verma 2015a, b; Velasquez-Orta et al. 2010).

Depending on applications, BESs are classified into four domains: (i) MFC, (ii) MEC/MES, (iii) microbial desalination cells (MDC) and (iv) microbial solar cells (MSC) (Pant et al. 2012). Among these, MFC and MEC/MES are primarily known for the production of bioelectricity and value-added chemicals/products, respectively, from the wastes. Whereas MDC and MSC are used for energy recovery with desalination and without feeding food and CO_2, respectively (Pant et al. 2012). A comprehensive schematic is shown in Figure 1. Nonetheless, the focus of the current review is only MFC and MES.

Characteristically, an MFC oxidizes organic matter for producing bioenergy. Basic fundamentals, theories, working principles and types of MFCs (single and double) are very well-known (Bond et al. 2002; Gupta et al. 2017; Kim et al. 2002; Min and Logan 2004; Modi et al. 2017), therefore, these points are not discussed in current review. However, a schematic (Figure 2) is being produced here for better understanding. There are several factors that are directly/indirectly responsible for affecting the performance of MFCs. Among them, it is observed that mixed cultures are favorable for high output of MFC instead of pure cultures and it has been extensively used. Also for maximum power generation,

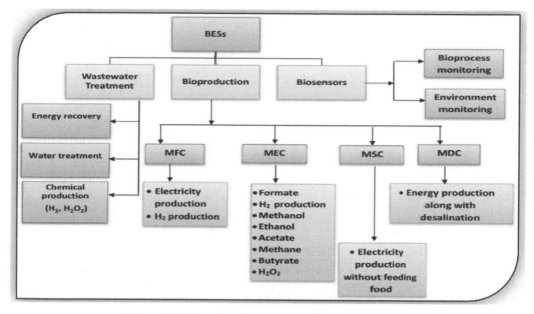

Figure 1: Schematic of division and sub division of BESs

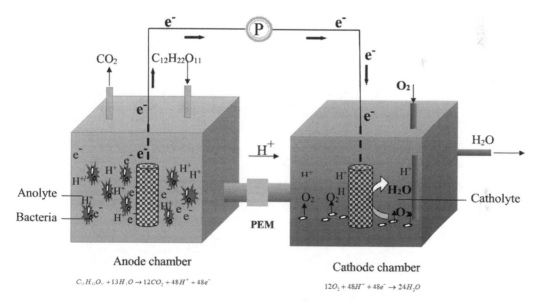

Figure 2: Schematic description of MFC.

the total internal resistance particularly charge transfer resistance should be very low (Singh and Verma 2015b). In few studies, it is explained that if potassium ferricyanide is applied from different existed electron acceptor at the cathode chamber, the power density enhances 1.5-1.8 folds (Logan et al. 2006; Oh and Logan 2006). However, the use of such chemical mediators is not sustainable and has been discouraged in recent years. Different cost estimation studies have also been done on MFCs (Singh et al. 2018; Singh et al. 2016). Considering, for instance, a food-based industry can produce organic waste 7,500 kg/day, it may produce 950 kW or 350 kW (if efficiency is 30%) of power. It is also defined for 1 kW/m³ energy generation, 350 m³ volume of reactor is needed to conquer energy, and its current cost is approximately 3.9 M Euros (Logan 2004; Logan et al. 2006; Rabaey and Verstraete 2005).

MEC is another version of nonspontaneous MFC that needs external energy to run. If there is the unavailability of oxygen in catholyte/cathode chamber, then MFC starts to work as MEC shown in Figure 3. In MEC, there is the existence of exoelectrogen (electron donor) and exoelectrotroph (electron acceptor) at anode and cathode chamber, respectively. Generally, during oxidation of the organic matter, electrons are propelled to the anode surface by exoelectrogen and after that transported to the cathode where electrons linked with exoelectrotroph to generate/facilitate value-added products depending on available electrons i.e., H_2, methane and ethanol (Min et al. 2013). It has been designed for spending electricity to accomplish electrochemical reactions at the cathode for generating useful chemicals in anaerobic conditions. Villano et al. have proposed that methane is dominating product of MEC instead of hydrogen and also confirmed that it deals with the gas and liquid wastes (Villano et al. 2011).

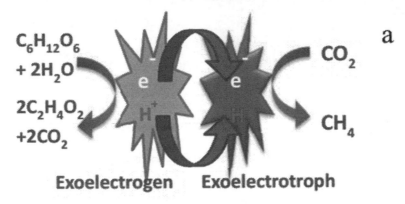

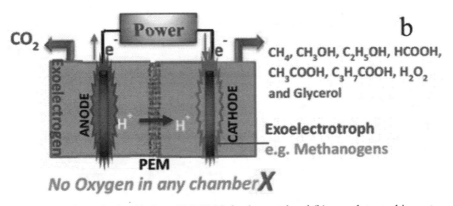

Figure 3: Schematic description of MEC (a) fundamental and (b) complete working set-up.

3. Electrocatalyst in Different BESs

As discussed above, NPs are playing a significant role for electricity generation and facilitate the conversion of CO_2 into useful products/value-added chemicals (Centi and Perathoner 2012; Centi et al. 2003; Perathoner et al. 2007). The catalyst that participates in the electrochemical reaction is known as an electrocatalyst. In the past few years, several studies have been reported on CO_2 reduction via electrochemical route using a different catalyst (DuBois 2006; Oloman and Li 2008). However, the conversion efficacy of electrons is larger in comparison to high overpotentials greater than 1.5 V. To reduce this overpotential there is another way to reduce CO_2 i.e. bioelectrochemical approach. Furthermore, in this context, it is important to mention that there are several amendments done in BES to modify the physicochemical properties of bioelectrocatalyst to make the process efficient (Marshall et al. 2012). In previous years, there have been a lot of research efforts made toward the making and modifying of the electrode materials (electrocatalyst) for both the anodic and cathodic chamber in different types of BES. These modified electrocatalysts are different types of nanomaterials, including carbon NPs, metal NPs and carbon-based materials. These different types of electrocatalyst can enhance the performance in terms of electron transfer, bacteria adhesion, active surface area, biocompatibility, stability, inertness and electrochemical efficiency. The applications of BESs mainly lie in two different routes: (i) spontaneous (electricity) (Singh and Verma 2015a, b) and (ii) nonspontaneous (value-added products) (Centi et al. 2013; Sasaki et al. 2012).

4. Bioelectro Synthesis of $C_xH_yO_z$ from CO_2 in MES

In MESs, biocatalysts play a vital role for reducing and/or oxidizing organic foods into useful products/ chemicals, like formic acid, acetic acid, butyric acid, methanol, ethanol, glycerol, hydrogen peroxide, methane, bioelectricity generation, etc. (Aulenta et al. 2008; Lovley and Nevin 2011; Modi et al. 2017; Ren 2013; Singh et al. 2016; Singh and Verma 2015a; Venkata Mohan et al. 2014a). In a typical MES, the anaerobic and untreated culture was preferred at anode chamber while pretreated and enriched homocultures were preferred at the cathode for getting pure and useful value-added products. The efficacy of settling the electrons in the microbial catalyzed electrochemical systems have been studied for various mixed type and pure biocatalysts. Some of the researchers have reported that high power densities at the anode chamber could be achieved on the enhancement of electrochemically active bacteria (EAB) and mixed culture (Aelterman et al. 2008; Srikanth et al. 2010; Wang and Han 2009). Several EAB and metal-reducing microorganisms, including *Shewanella haliotis, S. oneidensis* (Carmona-Martinez et al. 2011; Carmona-Martínez et al. 2013), *Geopsychronacter electrodiphilus* (Holmes et al. 2004), *Desulfobulbus propionicus* (Holmes et al. 2004), *E. coli* (Modi et al. 2017; Raghavulu et al. 2011; Singh and Verma 2015a, b), *Rhodoferax ferrireducens* (Chaudhuri and Lovley 2003), *Geobacter sulfurreducens* (Bond and Lovley 2003; Richter et al. 2009), *P. aeruginosa, Pseudomonas otitidis* (Rabaey et al. 2008; Raghavulu et al. 2011), *Rhodopseudomonas palustris DX-1* (Xing et al. 2008), *Aeromonas hydrophila* (Pham et al. 2003), etc., have been reported in different BESs. Many authors have studied *Acetobacterium spp.* and *Methanobacterium* biocatalysts that are extensively used for the production of acetate and methane (Marshall et al. 2013). Min et al. (2013) observed that the addition of methanogenic inhibitor in the presence of *Acetobacterium woodi* had endorsed the microbial population. Nevin and her co-workers have studied *Acetobacterium woodi* and concluded the interesting fact that these bacteria are unable to generate current/electricity (Nevin et al. 2011; Nevin et al. 2010). Cheng and his co-worker have suggested that the electromethanogens may convert renewable energy sources, including solar energy, biomass energy and wind energy into biofuels (Cheng et al. 2009). The authors reported that the overall yields of methanol/ methane (used as primary products) were less because the surface of electrodes showed poisoning in absenteeism of high overpotentials. These above studies have been done for planar-type electrodes (Centi et al. 2007). Some of the researchers have been reported a high production efficacy on the formation of acetone and isopropanol at higher temperature ~ 60°C using Fe/Pt CNT-based electrocatalyst (Gangeri et al. 2009). There are three different types of electrocatalysts based on iron-CNT or platinum-CNT and with 10% Fe loaded N-CNT. In this, the value-added product isopropanol is the major product in all reactions and found different production efficacy. They found better results in N-CNT in comparison to bare CNTs. Also, they reported isopropanol is the main product and other value-added chemicals such as ethanol and acetaldehyde as side products with low production rate. Some authors reported that Pt electrocatalyst is more effective in comparison with Fe electrocatalyst for the production of methanol. The overall efficiency of electrocatalytic conversion of electrocatalyst is found better at 60°C and 1 bar pressure, and electrocatalytic conversion of carbon monoxide to higher alcohols and hydrocarbons are possible at higher temperatures (above 200°C and pressures ~30 bars or more). Carbon black has also used as nanomaterials for production of acetone as a leading product and isopropanol as a side product. The use of CNTs instead of carbon black is desirable in terms of value-added products, such as isopropanol and acetone. The productivity is better in CNT-based materials in comparison to carbon black (Centi and Perathoner 2010). Marshall and his co-workers have first used a carbon source as a

CO_2 that is reduced into to a mixture of hydrogen, methane and acetate using brewery wastewater and its potential is -590 mV (vs. SHE) (Diekert and Wohlfarth 1994). Several biocatalysts and respective electrodes based on metal and/or carbon cited in the current review and used in BESs for the conversion of value-added products are tabulated in Table 1.

5. Reported Statistics on the Bioelectro Synthesis of $C_xH_yO_z$ from CO_2

5.1 Formate/Formic acid

Formic acid is an intermediary chemical compound that is used in various chemical industries. Some of the authors have studied that the use of BESs as a substitute for CO_2 fixation has an advanced strategy for using formic acid as an energy carrier. Li and his co-workers have studied the transformation of CO_2 into formate/ formic acid using *Ralstonia eutropha* as a biocatalyst. Reda and his co-workers reported the production of formate/formic acid using pyrolytic graphite edge as an electrode and *Syntrophobacter funaroxidans* as a substrate for the conversion of CO_2 at cathodic potential -0.4 V/-0.8V (vs. NHE). The enzyme named formate dehydrogenases has used for interconversion of CO_2 and formate electrochemically. This enzyme increases the rate of CO_2 conversion that increases by two times. According to thermodynamics, formate/ formic acid and hydrogen have different oxidation potentials (Reda et al. 2008). Zhao and his co-workers have used an MFC stack connected in series for the conversion of CO_2 into formic acid by degrading the carbonaceous substances. They reported the production rate of formate from CO_2 is nearly 0.09 mM/L/h and the efficacy of electron recovery is 64.8% using carbon paper coated with Pt as an electrode in single chambered MFCs (Zhao et al. 2012). The substrate formate dehydrogenase used as a biocatalyst follows Wood-Ljungdahl (WL) pathway for reversible reduction of CO_2 to formate. In brief, the reduction of two moles of CO_2 into acetate and other by-products are generally known as WL pathway (Ragsdale and Pierce 2008). Zhou and his co-workers have reported the reduction of CO_2 into formic acid at the cathode. They used five MFCs in series that generates a voltage of 2.73 V. The production rate of formic acid was found to be 4.27 mg /L/h in double chambered MFCs using non-wet proofed carbon fiber as an electrode and corresponding 64.8% faraday efficacy was achieved. However, the concentration of formic acid and the rate of production are less; the authors suggested that this methodology will be helpful for recycling the carbon from various types of wastewaters and effluents without using external energy input via carbon fiber based electrodes (Zhao et al. 2012).

5.2 Acetate/Acetic acid

In BESs, production of acetate/acetic acid from CO_2 regarded as a primary intermediate product/chemical. Thereafter, reduction of acetate/acetic acid into value-added chemical such as butyric acid/butanol, ethanol, etc. is considered. Recently, some authors have reported the enriched microbial communities and most of the acetogenic microorganisms are used for the production of acetate as the main product at the cathode (Bajracharya et al. 2016; May et al. 2016; Sharma et al. 2013; Steinbusch et al. 2010). The main use of *Sporomusa spp.* and *Clostridium spp.* as a biocatalyst for CO_2 reduction into acetate/acetic acid using graphite electrodes has been demonstrated (Schuchmann and Muller 2014). Nevin et al. found that *Sporomusa ovata* has the capability for electrosynthesis in the cathodic chamber at potential -0.4 V using graphite stick as an electrode in the anode and cathode chamber. In cathode chamber, electrons are capable to reduce CO_2 into acetate, 2-oxobutyrate and formate/formic acid. A similar observation has been reported when potential is found less than or equal to -0.59 V (vs. NHE) in the cathodic chamber. In this work, they also reported the miscellany of acetogenic bacteria present in culture for verifying acetogens other than *S. ovata* are effective or not for electrosynthesis (Nevin et al. 2011). Marshall et al. have studied that the mechanism of biocatalysis occur at the interface of an electrode which is supported by a catalytic wave (0.460 V vs. SHE) in cyclic voltammetry images of the biocathode and produced value-added chemicals. They reported production rate of acetate was increased from 0.18 mM/day (-0.4 V vs. SHE) to 4.0 mM/day (0.-59 V vs. SHE) using granular graphite bed as an electrode connected with

Table 1: Synopsis of conversion of CO_2 into value-added chemicals through different biocatalyst and electrodes in MES.

Substrate/Biocatalyst	Chamber	Reaction	Electrode materials	Operation	Product	Reference
Sporomusa ovata	Dual chamber	-	Graphite electrodes	Batch mode	Acetate, oxybutyrate	(Nevin et al. 2010)
Mixed acetogen from brewery waste water sludge	Dual chamber	-	Graphite granules	Batch mode	Acetate, hydrogen	(Marshall et al. 2013)
Acetobacterium woodii	Dual chamber	$H^+ + e^- \rightarrow H_2$ $H_2 + CO_2 \rightarrow CH_3COOH$	Titanium (anode) and carbon felt (cathode)	Batch mode	Hydrogen, acetate	(Min et al. 2013)
Cheese whey	Single chamber	$2H^+ + 2e^- \rightarrow H_2$		Continuous mode	Hydrogen, butyric acid, acetic acid	(Davila-Vazquez et al. 2009)
Brewery waste water: Acetobacterium spp.	Dual chamber	$H^+ + e^- \rightarrow H_2$ $CO_2 + 8H^+ + 8e^- \rightarrow CH_4 + 2H_2O$	Graphite granules	Semi-batch mode	Hydrogen, methane, acetate	(Marshall et al. 2012)
Mixed enriched culture biocatalyst	Dual chamber	-	Carbon paper contains Pt (cathode) and carbon fiber (anode)	Batch mode	Formic acid	(Zhao et al. 2012)
C. ljungdahlii and Moorella thermoacetica	Dual chamber	-	graphite electrodes	Batch mode	Acetate, oxybutyrate, formate	(Nevin et al. 2011)
Methanobacterium palustre	Dual chamber	$CO_2 + 8H^+ + 8e^- \rightarrow CH_4 + 2H_2O$	Carbon cloth (cathode) and graphite fiber (anode)	Batch mode	Methane	(Cheng et al. 2009)
Clostridium ljungdahlii	Dual chamber	-	Titanium with an Iridium coated electrode (anode) and graphite (cathode)	Batch mode	Acetate, ethanol, butyrate	(Bajracharya et al. 2017b)
Cyanobacteria	Dual chamber	$8H^+ + 8e^- + N_2 \rightarrow 2NH_3 + H_2$ $2H^+ + 2e^- \rightarrow H_2$		Batch mode	Hydrogen	(Bothe et al. 2010)

(Contd.)

Table 1: (Contd.)

Substrate/Biocatalyst	Chamber	Reaction	Electrode materials	Operation	Product	Reference
Enterobacter aerogenes	Dual chamber	-	Carbon cloth; electrodes	Batch mode	Glycerol, hydrogen, ethanol	(Sakai and Yagishita 2007)
Mixed enriched culture biocatalyst	Dual chamber	-	Gas diffusion; electrode	Batch mode	Ethanol, butyrate, acetate	(Bajracharya et al. 2016)
mixed culture waste water	Dual chamber	-	Gas diffusion; electrode	Batch mode	Acetate, hydrogen, methane	(Bajracharya et al. 2016)
Clostridium tyrobutyricum	Dual chamber	$\frac{1}{4}CO_2 + H^+ + e^- \rightarrow \frac{1}{48}C_{12}H_{22}O_{11} + \frac{13}{48}H_2O$ $C_{12}H_{22}O_{11} + 0.22\,NH_4^+ + 1.92C_4H_7O_2^- + 3.06CO_2 \rightarrow 0.4C_2H$ $0.22C_5H_7O_2N + 1.05H^2 +$ $1.04H_2O + 2.5H^+$	Graphite felt (cathode), Pt anode electrodes	Batch mode	Butyrate	(Choi et al. 2012)
Mixed enriched culture biocatalyst	Dual chamber	-	Commercial carbon cloth (cathode) and Ti rod (anode)	Batch mode	Butyrate	(Ganigue et al. 2015)
Mixed culture waste water treatment plant sludge	Dual chamber	-	Carbon felt	Batch mode	Methane, acetate	(Jiang et al. 2013)
Hydrogenophilic methanogenic culture	Dual chamber	$CO_2 + 8H^+ + 8e^- \rightarrow CH_4 + 2H_2O$	Carbon paper (cathode) and glassy carbon rod (anode)	Batch mode	Methane	(Villano et al. 2010)

(Contd.)

Biocatalyst	Chamber	Equation	Electrode	Mode	Product	Reference
Pond sediments and water treatment plant sludge sludge	Dual chamber	-	Nano web reticulated vitrous carbon (RVC), unmodified RVC, Graphite Plate	Batch mode	Acetate	(Jourdin et al. 2014)
Mixed enriched culture biocatalyst	Single chamber	-	Graphite brushes (anode)	Continuous mode	Methane, Acetate	(Rader and Logan 2010)
Mixed enriched culture biocatalyst	Dual chamber	-	Graphite rod, graphite felt, graphite foam	Continuous mode	Methane	(Chaudhuri and Lovley 2003)
Hydrogenotrophic methanogens	Dual chamber	$CO_2 + 2H_2O \rightarrow CH_4 + 2O_2$ (E_{cell} = −1.05 to −1.22 V vs NHE)	Pt coated Ti mesh (anode) and graphite felt (cathode)	Continuous mode	Methane	(Van Eerten-Jansen et al. 2012)
Enriched culture from bog sediment	Dual chamber	-	Carbon rods attached with Ti wire	Batch mode	Butyrate/butanol, propionate, acetate, hydrogen, ethanol	(Zaybak et al. 2013)
Mixed enriched culture Biocatalyst	Single chamber	$CO_2 + 2H_2O \rightarrow CH_4 + 2O_2$	Plain carbon felt attached with Ti wire	Batch mode	Methane	(Kobayashi et al. 2013)
Methanobacterium Thermoautotrophicus	Single chamber	$CO_2 + 2H_2O \rightarrow CH_4 + 2O_2$ $CH_3COOH \rightarrow CH_4 + CO_2$	Carbon paper	Batch mode	Methane	(Sato et al. 2013)
Enriched culture from bog sediment	Dual chamber	-	Granular Graphite	Continuous mode	Acetate	(Battle-Vilanova et al. 2016)
Sporomusa ovata	Dual chamber	-	Graphite coated with Ni nanowires	Batch mode	Acetate	(Nie et al. 2013)
Clostridiales	Dual chamber	-	Carbon felt (cathode) and SS (anode)	Batch mode	Acetate	(Gildemyn et al. 2017)
Methanobacterium sp.	Dual chamber	-	Pt coated Ti mesh (anode) and graphite felt (cathode)	Batch mode	Methane	(Van Eerten-Jansen et al. 2013)

titanium wire (Marshall et al. 2012). Gong et al. have used *Desulfobulbus propionicus* and *Sporomusa ovata* as biocatalysts for oxidizing sulfur into sulfate effectively that generates extra six electrons during the process at the anodic chamber in the microbial electrosynthesis for efficient conversion of CO_2 into acetate/acetic acid at the cathode chamber using graphite electrodes. The production rate of acetate was 49.9 mmol/day.m² and the coulombic efficacy was found to be greater than 90% (Gong et al. 2013). Some of the researchers also observed that when BES operated for a long time interval using graphite rod connected with Ti wire, the autotrophs increased the production rate of acetate/acetic acid (Marshall et al. 2013). Min and coworkers have observed that while producing acetate from CO_2, the cathode potential plays a very critical role for enhancing the production rate of acetate in the presence of *Acetobacterium woodi*. They used Ti plate in anode and carbon felt in cathode as an electrode (assuming Ag/AgCl as reference electrodes). Thereafter, they increased the production rate of acetate from 0.38 mM/day to 2.35 mM/day, i.e. the efficacy of CO_2 reduction has been improved from 53.6% to 89.5% (Min et al. 2013). Jiang and his coworkers reported the larger production rates of acetate and methane from CO_2 using mixed culture as a biocatalyst and carbon felt electrodes. In this, they used an electroactive biofilm that is attached with an electrode along with acetogenic and methanogenic bacteria at the cathode. Therefore, this method enhances the surface area for a higher rate of production of acetate and methane. They found the high production rate of acetate is 94.73 mg/d and efficacy for capturing the current reached up to 97%. Subsequently, acetogenic bacteria was used as an electron donor, i.e. reduction of CO_2 into useful products (Jiang et al. 2013). There are numerous types of acetogenic bacteria named *Clostridium aceticum*, *Sporomusa sphaeroides*, *Moorella thermoacetica* and *Clostridium ljungdahlii*, etc., that are capable of consuming electric current using graphite stick electrodes for generating some valuable chemicals/products, such as 2-oxobutyrate acetate and formate (Nevin et al. 2011). Bajracharya et al. proposed that mixed culture inoculum used in BES for restricting methane generation to produce acetate/acetic acid. They used rectangular titanium with iridium coated electrode at the anodic chamber, and graphite stick is sandwiched between the pieces of graphite felts at the cathode. They reported the maximum production of acetate is 400 mg/l/d at the potential of -1 V (Bajracharya et al. 2017b). Batlle-Vilanova et al. have modified the BES in which they increase the partial pressure of H_2 that shifts the spectrum of the final product toward useful products. They reported the production of acetate by mixed culture in continuous mode from CO_2 and the production rate was found to be 0.98 mmol/L/day at the cathode potential -0.6 V in the presence of graphite granules in the anode and cathode chambers (Batlle-Vilanova et al. 2016). They found that conversion of CO_2 into acetate occurs where the potential requirement is lower due to the resistance of mass and charge transfers and the thermodynamical essentials for cathodic potential is -0.28V vs. SHE. It is found that there are some factors which affects the performance of the process such as mediators that are helpful in transferring of electrons, type of electrode material, reactor design, as well as the availability and supply of electricity to the MES, etc. (Desloover et al. 2012).

5.3. Butyrate/Butyric acid

Butyric acid is a chemical compound, and it is extensively used in the production of biofuels by biotransformation processes. It is also used as an industrial solvent and is mainly used in pharmaceutical industries (Dwidar et al. 2012). During the production of biohydrogen from cheese whey in CSTR, butyric acid and acetic acid also generated along with hydrogen generation as main constituents. The production ratio of acetic acid to butyric acid is found to be 2.4 (Davila-Vazquez et al. 2009). Some of the authors have suggested that the absence of the mediator methyl viologen at cathode compartment using graphite felt electrode with reference Ag/AgCl electrode that produce an adequate amount of by-products such as butyric acid/butyrate instead of ethanol (Steinbusch et al. 2010). Choi and his co-workers have reported the significant production of butyric acid/butyrate in BES using *Clostridium tyrobutyricum* at cathode chamber with cathode potential have been found -0.4V in the presence of Pt (anode) and graphite felt (cathode) electrodes. They used neutral red as a mediator for enhancing the production of butyrate and its rate of production is 8.8 g/L while the yield of production was approximately 0.44 g/g (Choi et al. 2012). Ganigue and his co-workers have reported the highest production rate and the highest concentration of butyrate/butyric acid was 1.82 mMC/d and 20 mMC respectively, using carbon cloth as a commercial electrode and Ti rods are used as a counter electrode. They investigated two different methodologies: (i) the production of butyrate through the WL pathway that directly converts CO_2 and (ii) the production of butyrate through the chain-elongation (reverse β–oxidation) using acetate and ethanol (Ganigue et al. 2015). Vilanova et al. have validated the enhancement of *Megasphaera sueciensis* at the biocathode that enhances the production of butyrate using commercial carbon cloth electrode at the cathode and Ti used as an anode. They reported the production rate of butyrate was found 7.2 mMC/day with simultaneous production of other products, such as acetate, ethanol and butanol at small rates. In this, they achieved the butyrate to acetate ratio is 16:4 and achieved a higher concentration of butyrate is 252.4 mMC (Batlle-Vilanova et al. 2017).

5.4. Methanol

Methanol is widely used as an organic solvent in various industries. Azuma and his co-workers reported the production of methanol using copper as a metal catalyst by reduction of CO_2 and also found methane and ethylene as side products. They observed the production of methane and ethylene in the presence of all metal electrodes and found efficacy is too low except Cu. This process has found plenty of energy but is not sustainable (Azuma et al. 1990). Torella and his coworkers used engineered *R. eutropha* for the conversion of CO_2 into methanol by electrolysis process using either NiMoZn or stainless steel as the cathode electrode. This process has an ability to produce isopropanol and the concentration found was approximately 216 mg/L (Torella et al. 2015).

5.5. Ethanol

The availability of lignocellulosic biomass is in abundance for the production of ethanol that is so far the most preferred route to make it (Singh et al. 2010). However, in recent years, gaseous substrate like CO_2 has also been used to make ethanol. The *Clostridium* species has an ability to utilize the mixture of CO, CO_2 and H_2 for producing ethanol. Bajracharya et al. (2017b) used rectangle titanium with iridium coated anode electrode and graphite stick sandwiched between the pieces of graphite felts as the cathode electrode in an MES. In the cathode compartment, the mediator methyl viologen promotes acetic acid/acetate into ethanol and its concentration was 13.5 mM using graphite felt electrode with reference of Ag/AgCl electrode. The work of methyl viologen was to avoid side reactions like methanogenesis reaction and found the ethanol production in plenty. If the mediator such as methyl viologen was not present, then several bi-products such as butyrate/ butyric acid were found (Steinbusch et al. 2010). Blanchet et al. have reported the concentration of ethanol is 35 mM using *Sporomusa ovata* as biocatalyst. The acetogenic bacteria *S. ovata* was primarily known for the production of acetate in large amount by CO_2 reduction using carbon cloth electrode connected with Pt wire. Subsequently, some additional products are also produced, mainly ethanol in a little amount approximately below 1 mM (Blanchet et al. 2014). Younesi and his co-workers have reported the maximum concentration of ethanol to be 13 mM that is equivalent to 600 mg/L using *Clostridium ljungdahlii* as (Younesi et al. 2005). Some of the authors have reported that the various acetogens such as *C. ljungdahlii*, *Clostridium autoethanogenum* or *Clostridium ragsdalei* have the ability to produce a significant amount of ethanol via CO_2 reduction (Schiel-Bengelsdorf and Dürre 2012).

5.6. Glycerol

There is limited study found on the reduction of CO_2 to glycerol using carbon or metal nanoparticles. Sakai and Yagishita (2007) reported the production of glycerol in large quantity by biodiesel wastes using *Enterobacter aerogenes* as a substrate in BESs. The concentration of glycerol was found 154 mM at the potential of 0.2 V using carbon electrodes. Hydrogen and ethanol were also produced along with glycerol from biodiesel wastes in small extent (Sakai and Yagishita 2007).

5.7. Methane

Methane is known as an energy carrier, and it is a very noble source of renewable energy. It is easy to transport as well as easy to store. Most of the authors reported that the recovery of methane has been done by anaerobic digestion (Khanal 2009; Pant et al. 2012). Cheng et al. have proposed methane is simply produced by biological conversion using *Methanobacterium palustre* with electromethanogens in BES. They used a dual chambered reactor; the anodic potential was required to -0.7V (vs. Ag/AgCl) for producing methane via CO_2 reduction using carbon cloth as a cathode electrode and graphite fiber as an anode electrode. In this, they used an exoelectrogenic biofilm for generating current at the anodic chamber for the production of methane. Subsequently, reaching at -1.0V (vs. Ag/AgCl), the efficacy of the storing current is 96% (Cheng et al. 2009). Marshall et al. (2012) have reported the production of value-added chemicals/products like hydrogen, acetate and methane in brewery wastewater sludge at -0.59 V vs. SHE in presence of *Acetobacterium* biocatalyst. In this, the occurrence of electrosynthetic biocatalysis occurs at the interface of the electrode that is supported by a catalytic wave (460 mV vs. SHE) in cyclic voltammetry images of the biocathode and produced value-added chemicals. In this way, the production of methane at the cathodic chamber was increased from 1 mM/day to 7 mM/day using graphite granules as an electrode. The efficacy of electron recovery in the chamber was found nearly 54.8% (Marshall et al. 2012). Clauwaert and Verstraete reported the production of methane in a single chamber at a rate of 0.75/L/d using graphite as electrode under the potential of -0.8 V vs. SHE. In anaerobic reactors, the production rate of methane is greater than 1 g COD /L fed with concentrated organic substrates that is the order of several liters methane per liter reactor per day. On the other hand, in our lower-loaded MECs, the production rate of methane is less than 1 g COD /L that is below 0.7 L methane per liter reactor per day (Clauwaert and Verstraete 2009). Moreover, Feng and his co-workers have observed at anode that the modified graphite fiber has been used for increasing the production of biomethane in BES. In MECs, the utilization of wastes for anaerobic digestion processes of methane operates at low temperatures and does not require any aeration costs, and the production rate of methane is larger that makes this method robust and cost-effective (van Eerten-Jansen et al. 2015; Wagner et al. 2009). Some of the authors have used the garbage slurry at neutral pH for production of methane using carbon sheets as a working electrode at the potential in the range between -0.39 to -0.59 V vs. SHE (Sasaki et al. 2012). They found that electromethanogenesis is an important and effective process with good conversion efficiency using biocathode (Rader and Logan 2010). Thereafter, Liu and his co-workers have reported bioelectrochemical anaerobic digestion system operates at lower temperatures to be beneficial for enhancing the production of methane using organic wastes in presence of granular activated carbon electrodes (Liu et al. 2016). This system operates at the cathodic potential of -0.90 V vs. Ag/AgCl, and the increased generation of methane was 31 mg methane-COD/g. Furthermore, they reported the current density at cathode is about 2.9 A/m^2 and rate of production of methane was about 5.2 L methane/m^2/day at a cathodic potential of -0.7 V vs. NHE using platinum-coated titanium mesh as an anode and graphite felt as a cathode (van Eerten-Jansen et al. 2015; van Eerten-Jansen Mieke et al. 2011). Most of the researchers (Ditzig et al. 2007; Wagner et al. 2009) have studied that while using animal wastewater as a substrate in BESs, they found the maximum yield for methane in comparison to hydrogen. Some of the researchers have observed the production of methane via EAB in MES is efficient and economical. Thereafter, some significant factors have also been considered for enhancing the methane production, like properties of used material, electrode design, reactor design and other operational conditions (Liu et al. 2016; van Eerten-Jansen et al. 2015).

6. Conclusions

This chapter summarizes the development of nanomaterials used in different types of BES for electricity generation and producing useful/valuable products. Here, we presented a wide range of substrate, such as various industrial wastewater sludge, animal waste, microorganisms and carbon-based nanomaterials as a biocatalyst for bioenergy generation/electricity production. Also, conversion of CO_2 into useful products/

or chemicals, such as formate, methanol, ethanol, acetate, methane and hydrogen in BES. The doped NPs have the capability to enhance the efficacy of BESs for the alternative generation of energy and value-added CxHyOz products. From the substantial number of published papers reviewed here, it is observed that different types of carbon-based electrodes have high electrical conductivity, high specific surface area, biocompatibility, hierarchical porous and are economical as well. This review will be very useful for quick reference to the reader and will help them to select appropriate products obtained for fixing substrate for their anodic or cathodic electroactive bacteria for further studies and modifications.

7. Future Perspectives

From the above discussion, it is evident that an extensive range of value-added products and energy carriers such as electricity and/or hydrogen are being produced from different BESs using various nanoparticles dispersed carbon electrodes and also involves CO_2 reduction in several instances. BESs propose a versatile podium and show vast potential for simultaneous generation of energy and value-added products. However, so far it is not very clear which electrode is best for value-added CxHyOz products from CO_2. The CO_2 reduction still requires substantial expertise in multiple disciplines, viz. microbiology, biochemistry, design engineering, kinetics, bioengineering, material science, environmental science, bioelectrochemistry, separation and purification technology, downstream processing, enzymology, etc., to get higher yields of value-added CxHyOz products. One can also implement combinations of efficient anodes and cathodes for facilitating the production of value-added products. There is still a need to realize synergistic interaction between biocatalyst and electrodes surface. Current density, output power and nature of synthesized value-added products also depend on the mechanism of transferring electrons between the biocatalyst and the electrode. So, it is necessary to focus on electron transfer mechanisms so that losses can be reduced either by improved materials of electrodes or catalyst for reactions. Overall, it can be considered that BESs have demonstrable advantages over some existing technologies and could be incorporated into industrial wastewater treatment and resource recovery plans.

Acknowledgment

The authors are grateful for the support of the Department of Science and Technology (DST), New Delhi, India in the form of research grant (DST/INSPIRE/04/2015/001869). The authors are also obliged to CSIR-Indian Institute of Toxicology Research for providing necessary infrastructural facilities.

References

Aelterman P, Versichele M, Marzorati M, Boon N, Verstraete W (2008) Loading rate and external resistance control the electricity generation of microbial fuel cells with different three-dimensional anodes. Bioresource technology 99(18): 8895-8902.

Aulenta F, Canosa A, Majone M, Panero S, Reale P, Rossetti S (2008) Trichloroethene Dechlorination and H2 Evolution Are Alternative Biological Pathways of Electric Charge Utilization by a Dechlorinating Culture in a Bioelectrochemical System. Environmental Science & Technology 42(16): 6185-6190.

Azuma M, Hashimoto K, Hiramoto M, Watanabe M, Sakata T (1990) Electrochemical Reduction of Carbon Dioxide on Various Metal Electrodes in Low-Temperature Aqueous $KHCO_3$ Media. Journal of The Electrochemical Society 137(6): 1772-1778.

Bajracharya S, Srikanth S, Mohanakrishna G, Zacharia R, Strik DP, Pant D (2017a) Biotransformation of carbon dioxide in bioelectrochemical systems: State of the art and future prospects. Journal of Power Sources 356: 256-273.

Bajracharya S, Vanbroekhoven K, Buisman CJN, Pant D, Strik DPBTB (2016) Application of gas diffusion biocathode in microbial electrosynthesis from carbon dioxide. Environmental Science and Pollution Research 23(22): 22292-22308.

Bajracharya S, Yuliasni R, Vanbroekhoven K, Buisman CJN, Strik DPBTB, Pant D (2017b) Long-term operation of microbial electrosynthesis cell reducing CO2 to multi-carbon chemicals with a mixed culture avoiding methanogenesis. Bioelectrochemistry 113: 26-34.

Batlle-Vilanova P, Ganigué R, Ramió-Pujol S, Bañeras L, Jiménez G, Hidalgo M, Balaguer MD, Colprim J, Puig S (2017) Microbial electrosynthesis of butyrate from carbon dioxide: Production and extraction. Bioelectrochemistry 117: 57-64.

Batlle-Vilanova P, Puig S, Gonzalez-Olmos R, Balaguer MD, Colprim J (2016) Continuous acetate production through microbial electrosynthesis from CO2 with microbial mixed culture. Journal of Chemical Technology & Biotechnology 91(4): 921-927.

Blanchet E, Pécastaings S, Erable B, Roques C, Bergel A (2014) Protons accumulation during anodic phase turned to advantage for oxygen reduction during cathodic phase in reversible bioelectrodes. Bioresource Technology 173: 224-230.

Bond DR, Holmes DE, Tender LM, Lovley DR (2002) Electrode-Reducing Microorganisms That Harvest Energy from Marine Sediments. Science 295(5554): 483-485.

Bond DR, Lovley DR (2003) Electricity Production by Geobacter sulfurreducens Attached to Electrodes. Applied and Environmental Microbiology 69(3): 1548-1555.

Bothe H, Schmitz O, Yates MG, Newton WE (2010) Nitrogen Fixation and Hydrogen Metabolism in Cyanobacteria. Microbiology and Molecular Biology Reviews 74(4): 529-551.

Carmona-Martinez AA, Harnisch F, Fitzgerald LA, Biffinger JC, Ringeisen BR, Schröder U (2011) Cyclic voltammetric analysis of the electron transfer of Shewanella oneidensis MR-1 and nanofilament and cytochrome knock-out mutants. Bioelectrochemistry 81(2): 74-80.

Carmona-Martínez AA, Harnisch F, Kuhlicke U, Neu TR, Schröder U (2013) Electron transfer and biofilm formation of Shewanella putrefaciens as function of anode potential. Bioelectrochemistry 93: 23-29.

Centi G, Perathoner S (2010) Problems and perspectives in nanostructured carbon-based electrodes for clean and sustainable energy. Catalysis Today 150(1): 151-162.

Centi G, Perathoner S, 2012. Reduction of greenhouse gas emissions by catalytic processes. Handbook of Climate Change Mitigation, pp. 1849-1890. Springer.

Centi G, Perathoner S, Rak ZS (2003) 58 Gas-phase electrocatalytic conversion of CO 2 to fuels over gas diffusion membranes containing Pt or Pd nanoclusters. Studies in Surface Science and Catalysis 145: 283-286.

Centi G, Perathoner S, Winè G, Gangeri M (2007) Electrocatalytic conversion of CO 2 to long carbon-chain hydrocarbons. Green Chemistry 9(6): 671-678.

Centi G, Quadrelli EA, Perathoner S (2013) Catalysis for CO2 conversion: a key technology for rapid introduction of renewable energy in the value chain of chemical industries. Energy & Environmental Science 6(6): 1711-1731.

Chaudhuri SK, Lovley DR (2003) Electricity generation by direct oxidation of glucose in mediatorless microbial fuel cells. Nat Biotech 21(10): 1229-1232.

Cheng S, Xing D, Call DF, Logan BE (2009) Direct Biological Conversion of Electrical Current into Methane by Electromethanogenesis. Environmental Science & Technology 43(10): 3953-3958.

Choi O, Um Y, Sang B-I (2012) Butyrate production enhancement by Clostridium tyrobutyricum using electron mediators and a cathodic electron donor. Biotechnology and Bioengineering 109(10): 2494-2502.

Clauwaert P, Aelterman P, Pham TH, De Schamphelaire L, Carballa M, Rabaey K, Verstraete W (2008) Minimizing losses in bio-electrochemical systems: the road to applications. Applied Microbiology and Biotechnology 79(6): 901-913.

Clauwaert P, Verstraete W (2009) Methanogenesis in membraneless microbial electrolysis cells. Applied Microbiology and Biotechnology 82(5): 829-836.

Colombo A, Marzorati S, Lucchini G, Cristiani P, Pant D, Schievano A (2017) Assisting cultivation of photosynthetic microorganisms by microbial fuel cells to enhance nutrients recovery from wastewater. Bioresource Technology 237: 240-248.

Davila-Vazquez G, Cota-Navarro CB, Rosales-Colunga LM, de León-Rodríguez A, Razo-Flores E (2009) Continuous biohydrogen production using cheese whey: Improving the hydrogen production rate. International Journal of Hydrogen Energy 34(10): 4296-4304.

Desloover J, Arends Jan BA, Hennebel T, Rabaey K (2012) Operational and technical considerations for microbial electrosynthesis. Biochemical Society Transactions 40(6): 1233-1238.

Diekert G, Wohlfarth G (1994) Metabolism of homoacetogens. Antonie van Leeuwenhoek 66(1): 209-221.

Ditzig J, Liu H, Logan BE (2007) Production of hydrogen from domestic wastewater using a bioelectrochemically assisted microbial reactor (BEAMR). International Journal of Hydrogen Energy 32(13): 2296-2304.

DuBois DL (2006) Electrochemical reactions of carbon dioxide. Encyclopedia of electrochemistry.

Dwidar M, Park J-Y, Mitchell RJ, Sang B-I (2012) The Future of Butyric Acid in Industry. The Scientific World Journal 2012: 10.

Ehrlich HL (1997) Microbes and metals. Applied Microbiology and Biotechnology 48(6): 687-692.

ElMekawy A, Srikanth S, Bajracharya S, Hegab HM, Nigam PS, Singh A, Mohan SV, Pant D (2015) Food and agricultural wastes as substrates for bioelectrochemical system (BES): The synchronized recovery of sustainable energy and waste treatment. Food Research International 73: 213-225.

Gangeri M, Perathoner S, Caudo S, Centi G, Amadou J, Begin D, Pham-Huu C, Ledoux MJ, Tessonnier J-P, Su DS (2009) Fe and Pt carbon nanotubes for the electrocatalytic conversion of carbon dioxide to oxygenates. Catalysis Today 143(1): 57-63.

Ganigue R, Puig S, Batlle-Vilanova P, Balaguer MD, Colprim J (2015) Microbial electrosynthesis of butyrate from carbon dioxide. Chemical Communications 51(15): 3235-3238.

Gildemyn S, Verbeeck K, Jansen R, Rabaey K (2017) The type of ion selective membrane determines stability and production levels of microbial electrosynthesis. Bioresource Technology 224: 358-364.

Gong Y, Ebrahim A, Feist AM, Embree M, Zhang T, Lovley D, Zengler K (2013) Sulfide-Driven Microbial Electrosynthesis. Environmental Science & Technology 47(1): 568-573.

Gupta S, Yadav A, Singh S, Verma N (2017) Synthesis of Silicon Carbide-Derived Carbon as an Electrode of a Microbial Fuel Cell and an Adsorbent of Aqueous Cr(VI). Industrial & Engineering Chemistry Research 56(5): 1233-1244.

Holmes DE, Bond DR, Lovley DR (2004) Electron Transfer by Desulfobulbus propionicus to Fe(III) and Graphite Electrodes. Applied and Environmental Microbiology 70(2): 1234-1237.

Jiang Y, Su M, Zhang Y, Zhan G, Tao Y, Li D (2013) Bioelectrochemical systems for simultaneously production of methane and acetate from carbon dioxide at relatively high rate. International Journal of Hydrogen Energy 38(8): 3497-3502.

Jourdin L, Freguia S, Donose BC, Chen J, Wallace GG, Keller J, Flexer V (2014) A novel carbon nanotube modified scaffold as an efficient biocathode material for improved microbial electrosynthesis. Journal of Materials Chemistry A 2(32): 13093-13102.

Kadier A, Kalil MS, Abdeshahian P, Chandrasekhar K, Mohamed A, Azman NF, Logroño W, Simayi Y, Hamid AA (2016) Recent advances and emerging challenges in microbial electrolysis cells (MECs) for microbial production of hydrogen and value-added chemicals. Renewable and Sustainable Energy Reviews 61: 501-525.

Khanal SK, 2009. Bioenergy Generation from Residues of Biofuel Industries. Anaerobic Biotechnology for Bioenergy Production, pp. 161-188. Wiley-Blackwell.

Khare P, Ramkumar J, Verma N (2016) Carbon Nanofiber-skinned Three Dimensional Ni/Carbon Micropillars: High Performance Electrodes of a Microbial Fuel Cell. Electrochimica Acta 219: 88-98.

Kilpin KJ, Dyson PJ (2013) Enzyme inhibition by metal complexes: concepts, strategies and applications. Chemical Science 4(4): 1410-1419.

Kim HJ, Park HS, Hyun MS, Chang IS, Kim M, Kim BH (2002) A mediator-less microbial fuel cell using a metal reducing bacterium, Shewanella putrefaciens. Enzyme and Microbial Technology 30(2): 145-152.

Kobayashi H, Saito N, Fu Q, Kawaguchi H, Vilcaez J, Wakayama T, Maeda H, Sato K (2013) Bio-electrochemical property and phylogenetic diversity of microbial communities associated with bioelectrodes of an electromethanogenic reactor. Journal of Bioscience and Bioengineering 116(1): 114-117.

Liu D, Zhang L, Chen S, Buisman C, ter Heijne A (2016) Bioelectrochemical enhancement of methane production in low temperature anaerobic digestion at 10 °C. Water Research 99: 281-287.

Logan BE (2004) Peer Reviewed: Extracting Hydrogen and Electricity from Renewable Resources. Environmental Science & Technology 38(9): 160A-167A.

Logan BE, Hamelers B, Rozendal R, Schröder U, Keller J, Freguia S, Aelterman P, Verstraete W, Rabaey K (2006) Microbial Fuel Cells: Methodology and Technology. Environmental Science & Technology 40(17): 5181-5192.

Lovley DR (2006) Bug juice: harvesting electricity with microorganisms. Nat Rev Micro 4(7): 497-508.

Lovley DR, Nevin KP (2011) A shift in the current: New applications and concepts for microbe-electrode electron exchange. Current Opinion in Biotechnology 22(3): 441-448.

Lu Z, Chang D, Ma J, Huang G, Cai L, Zhang L (2015) Behavior of metal ions in bioelectrochemical systems: A review. Journal of Power Sources 275: 243-260.

Marshall CW, Ross DE, Fichot EB, Norman RS, May HD (2012) Electrosynthesis of Commodity Chemicals by an Autotrophic Microbial Community. Applied and Environmental Microbiology 78(23): 8412-8420.

Marshall CW, Ross DE, Fichot EB, Norman RS, May HD (2013) Long-term Operation of Microbial Electrosynthesis Systems Improves Acetate Production by Autotrophic Microbiomes. Environmental Science & Technology 47(11): 6023-6029.

May HD, Evans PJ, LaBelle EV (2016) The bioelectrosynthesis of acetate. Current Opinion in Biotechnology 42: 225-233.

Min B, Logan BE (2004) Continuous Electricity Generation from Domestic Wastewater and Organic Substrates in a Flat Plate Microbial Fuel Cell. Environmental Science & Technology 38(21): 5809-5814.

Min S, Yong J, Da Ping L (2013) Production of Acetate from Carbon Dioxide in Bioelectrochemical Systems Based on Autotrophic Mixed Culture. Journal of Microbiology and Biotechnology 23(8): 1140-1146.

Modi A, Singh S, Verma N (2016) In situ nitrogen-doping of nickel nanoparticle-dispersed carbon nanofiber-based electrodes: Its positive effects on the performance of a microbial fuel cell. Electrochimica Acta 190: 620-627.

Modi A, Singh S, Verma N (2017) Improved performance of a single chamber microbial fuel cell using nitrogen-doped polymer-metal-carbon nanocomposite-based air-cathode. International Journal of Hydrogen Energy 42(5): 3271-3280.

Mook WT, Aroua MKT, Chakrabarti MH, Noor IM, Irfan MF, Low CTJ (2013) A review on the effect of bio-electrodes on denitrification and organic matter removal processes in bio-electrochemical systems. Journal of Industrial and Engineering Chemistry 19(1): 1-13.

Nevin KP, Hensley SA, Franks AE, Summers ZM, Ou J, Woodard TL, Snoeyenbos-West OL, Lovley DR (2011) Electrosynthesis of Organic Compounds from Carbon Dioxide Is Catalyzed by a Diversity of Acetogenic Microorganisms. Applied and Environmental Microbiology 77(9): 2882-2886.

Nevin KP, Woodard TL, Franks AE, Summers ZM, Lovley DR (2010) Microbial electrosynthesis: feeding microbes electricity to convert carbon dioxide and water to multicarbon extracellular organic compounds. MBio 1(2): e00103-00110.

Nie H, Zhang T, Cui M, Lu H, Lovley DR, Russell TP (2013) Improved cathode for high efficient microbial-catalyzed reduction in microbial electrosynthesis cells. Physical Chemistry Chemical Physics 15(34): 14290-14294.

Noori MT, Mukherjee CK, Ghangrekar MM (2017) Enhancing performance of microbial fuel cell by using graphene supported V2O5-nanorod catalytic cathode. Electrochimica Acta 228: 513-521.

Oh S-E, Logan BE (2006) Proton exchange membrane and electrode surface areas as factors that affect power generation in microbial fuel cells. Applied Microbiology and Biotechnology 70(2): 162-169.

Oloman C, Li H (2008) Electrochemical processing of carbon dioxide. ChemSusChem 1(5): 385-391.

Pant D, Singh A, Van Bogaert G, Irving Olsen S, Singh Nigam P, Diels L, Vanbroekhoven K (2012) Bioelectrochemical systems (BES) for sustainable energy production and product recovery from organic wastes and industrial wastewaters. RSC Advances 2(4): 1248-1263.

Perathoner S, Passalacqua R, Centi G, Su DS, Weinberg G (2007) Photoactive titania nanostructured thin films: synthesis and characteristics of ordered helical nanocoil array. Catalysis today 122(1): 3-13.

Pham CA, Jung SJ, Phung NT, Lee J, Chang IS, Kim BH, Yi H, Chun J (2003) A novel electrochemically active and Fe(III)-reducing bacterium phylogenetically related to Aeromonas hydrophila, isolated from a microbial fuel cell. FEMS Microbiology Letters 223(1): 129-134.

Rabaey K, Read ST, Clauwaert P, Freguia S, Bond PL, Blackall LL, Keller J (2008) Cathodic oxygen reduction catalyzed by bacteria in microbial fuel cells. ISME J 2(5): 519-527.

Rabaey K, Verstraete W (2005) Microbial fuel cells: novel biotechnology for energy generation. Trends in Biotechnology 23(6): 291-298.

Rader GK, Logan BE (2010) Multi-electrode continuous flow microbial electrolysis cell for biogas production from acetate. International Journal of Hydrogen Energy 35(17): 8848-8854.

Raghavulu SV, Goud RK, Sarma PN, Mohan SV (2011) Saccharomyces cerevisiae as anodic biocatalyst for power generation in biofuel cell: Influence of redox condition and substrate load. Bioresource Technology 102(3): 2751-2757.

Ragsdale SW, Pierce E (2008) Acetogenesis and the Wood–Ljungdahl pathway of CO2 fixation. Biochimica et Biophysica Acta (BBA) - Proteins and Proteomics 1784(12): 1873-1898.

Reda T, Plugge CM, Abram NJ, Hirst J (2008) Reversible interconversion of carbon dioxide and formate by an electroactive enzyme. Proceedings of the National Academy of Sciences 105(31): 10654-10658.

Ren Z, 2013. The Principle and Applications of Bioelectrochemical Systems. In: Gupta, V.K., Tuohy, M.G. (Eds.), Biofuel Technologies: Recent Developments, pp. 501-527. Springer Berlin Heidelberg, Berlin, Heidelberg.

Richter H, Nevin KP, Jia H, Lowy DA, Lovley DR, Tender LM (2009) Cyclic voltammetry of biofilms of wild type and mutant Geobacter sulfurreducens on fuel cell anodes indicates possible roles of OmcB, OmcZ, type IV pili, and protons in extracellular electron transfer. Energy & Environmental Science 2(5): 506-516.

Sakai S, Yagishita T (2007) Microbial production of hydrogen and ethanol from glycerol-containing wastes discharged from a biodiesel fuel production plant in a bioelectrochemical reactor with thionine. Biotechnology and Bioengineering 98(2): 340-348.

Sasaki K, Morita M, Matsumoto N, Sasaki D, Hirano S-i, Ohmura N, Igarashi Y (2012) Construction of hydrogen fermentation from garbage slurry using the membrane free bioelectrochemical system. Journal of Bioscience and Bioengineering 114(1): 64-69.

Sato K, Kawaguchi H, Kobayashi H (2013) Bio-electrochemical conversion of carbon dioxide to methane in geological storage reservoirs. Energy Conversion and Management 66: 343-350.

Schiel-Bengelsdorf B, Dürre P (2012) Pathway engineering and synthetic biology using acetogens. FEBS Letters 586(15): 2191-2198.

Schuchmann K, Muller V (2014) Autotrophy at the thermodynamic limit of life: a model for energy conservation in acetogenic bacteria. Nat Rev Micro 12(12): 809-821.

Sharma M, Aryal N, Sarma PM, Vanbroekhoven K, Lal B, Benetton XD, Pant D (2013) Bioelectrocatalyzed reduction of acetic and butyric acids via direct electron transfer using a mixed culture of sulfate-reducers drives electrosynthesis of alcohols and acetone. Chemical Communications 49(58): 6495-6497.

Singh A, Pant D, Korres NE, Nizami A-S, Prasad S, Murphy JD (2010) Key issues in life cycle assessment of ethanol production from lignocellulosic biomass: Challenges and perspectives. Bioresource Technology 101(13): 5003-5012.

Singh S, Bairagi PK, Verma N (2018) Candle soot-derived carbon nanoparticles: An inexpensive and efficient electrode for microbial fuel cells. Electrochimica Acta 264: 119-127.

Singh S, Modi A, Verma N (2016) Enhanced power generation using a novel polymer-coated nanoparticles dispersed-carbon micro-nanofibers-based air-cathode in a membrane-less single chamber microbial fuel cell. International Journal of Hydrogen Energy 41(2): 1237-1247.

Singh S, Verma N (2015a) Fabrication of Ni nanoparticles-dispersed carbon micro-nanofibers as the electrodes of a microbial fuel cell for bio-energy production. International Journal of Hydrogen Energy 40(2): 1145-1153.

Singh S, Verma N (2015b) Graphitic carbon micronanofibers asymmetrically dispersed with alumina-nickel nanoparticles: A novel electrode for mediatorless microbial fuel cells. International Journal of Hydrogen Energy 40(17): 5928-5938.

Srikanth S, Venkata Mohan S, Sarma PN (2010) Positive anodic poised potential regulates microbial fuel cell performance with the function of open and closed circuitry. Bioresource Technology 101(14): 5337-5344.

Steinbusch KJJ, Hamelers HVM, Schaap JD, Kampman C, Buisman CJN (2010) Bioelectrochemical Ethanol Production through Mediated Acetate Reduction by Mixed Cultures. Environmental Science & Technology 44(1): 513-517.

Sun D, Gao Y, Hou D, Zuo K, Chen X, Liang P, Zhang X, Ren ZJ, Huang X (2018) Energy-neutral sustainable nutrient recovery incorporated with the wastewater purification process in an enlarged microbial nutrient recovery cell. Journal of Power Sources 384: 160-164.

Torella JP, Gagliardi CJ, Chen JS, Bediako DK, Colón B, Way JC, Silver PA, Nocera DG (2015) Efficient solar-to-fuels production from a hybrid microbial–water-splitting catalyst system. Proceedings of the National Academy of Sciences 112(8): 2337-2342.

van der Maas P, Peng S, Klapwijk B, Lens P (2005) Enzymatic versus Nonenzymatic Conversions during the Reduction of EDTA-Chelated Fe(III) in BioDeNOx Reactors. Environmental Science & Technology 39(8): 2616-2623.

Van Eerten-Jansen MCAA, Heijne AT, Buisman CJN, Hamelers HVM (2012) Microbial electrolysis cells for production of methane from CO2: long-term performance and perspectives. International Journal of Energy Research 36(6): 809-819.

van Eerten-Jansen MCAA, Jansen NC, Plugge CM, de Wilde V, Buisman CJN, ter Heijne A (2015) Analysis of the mechanisms of bioelectrochemical methane production by mixed cultures. Journal of Chemical Technology & Biotechnology 90(5): 963-970.

Van Eerten-Jansen MCAA, Ter Heijne A, Grootscholten TIM, Steinbusch KJJ, Sleutels THJA, Hamelers HVM, Buisman CJN (2013) Bioelectrochemical Production of Caproate and Caprylate from Acetate by Mixed Cultures. ACS Sustainable Chemistry & Engineering 1(5): 513-518.

Van Eerten-Jansen Mieke CAA, Heijne Annemiek T, Buisman Cees JN, Hamelers Hubertus VM (2011) Microbial electrolysis cells for production of methane from CO2: long-term performance and perspectives. International Journal of Energy Research 36(6): 809-819.

Velasquez-Orta SB, Head IM, Curtis TP, Scott K, Lloyd JR, von Canstein H (2010) The effect of flavin electron shuttles in microbial fuel cells current production. Applied Microbiology and Biotechnology 85(5): 1373-1381.

Venkata Mohan S, Lalit Babu V, Sarma PN (2008) Effect of various pretreatment methods on anaerobic mixed microflora to enhance biohydrogen production utilizing dairy wastewater as substrate. Bioresource Technology 99(1): 59-67.

Venkata Mohan S, Mohanakrishna G, Velvizhi G, Babu VL, Sarma PN (2010) Bio-catalyzed electrochemical treatment of real field dairy wastewater with simultaneous power generation. Biochemical Engineering Journal 51(1): 32-39.

Venkata Mohan S, Velvizhi G, Annie Modestra J, Srikanth S (2014a) Microbial fuel cell: Critical factors regulating bio-catalyzed electrochemical process and recent advancements. Renewable and Sustainable Energy Reviews 40: 779-797.

Venkata Mohan S, Velvizhi G, Vamshi Krishna K, Lenin Babu M (2014b) Microbial catalyzed electrochemical systems: A bio-factory with multi-facet applications. Bioresource Technology 165: 355-364.

Villano M, Aulenta F, Ciucci C, Ferri T, Giuliano A, Majone M (2010) Bioelectrochemical reduction of CO2 to CH4 via direct and indirect extracellular electron transfer by a hydrogenophilic methanogenic culture. Bioresource Technology 101(9): 3085-3090.

Villano M, Monaco G, Aulenta F, Majone M (2011) Electrochemically assisted methane production in a biofilm reactor. Journal of Power Sources 196(22): 9467-9472.

Wagner RC, Regan JM, Oh S-E, Zuo Y, Logan BE (2009) Hydrogen and methane production from swine wastewater using microbial electrolysis cells. Water Research 43(5): 1480-1488.

Wang B, Han J-I (2009) A single chamber stackable microbial fuel cell with air cathode. Biotechnology Letters 31(3): 387-393.

Xing D, Zuo Y, Cheng S, Regan JM, Logan BE (2008) Electricity Generation by *Rhodopseudomonas palustris* DX-1. Environmental Science & Technology 42(11): 4146-4151.

Younesi H, Najafpour G, Mohamed AR (2005) Ethanol and acetate production from synthesis gas via fermentation processes using anaerobic bacterium, Clostridium ljungdahlii. Biochemical Engineering Journal 27(2): 110-119.

Zaybak Z, Pisciotta JM, Tokash JC, Logan BE (2013) Enhanced start-up of anaerobic facultatively autotrophic biocathodes in bioelectrochemical systems. Journal of Biotechnology 168(4): 478-485.

Zhao H-Z, Zhang Y, Chang Y-Y, Li Z-S (2012) Conversion of a substrate carbon source to formic acid for carbon dioxide emission reduction utilizing series-stacked microbial fuel cells. Journal of Power Sources 217: 59-64.

Opportunities for Hydrogen Production from Urban/Industrial Wastewater in Bioelectrochemical Systems

Albert Guisasola[1*], Juan Antonio Baeza[1], Antonella Marone[1], Éric Trably[2] and Nicolas Bernet[2]

[1] Departament d'Enginyeria Química, Escola d'Enginyeria, Universitat Autònoma de Barcelona, 08193, Bellaterra (Barcelona), Spain

[2] LBE, Univ Montpellier, INRA, 102 Avenue des Etangs, 11100, Narbonne, France

1. Introduction

Wastewater treatment is necessary in modern times for the development of our society. Despite the recent outstanding progress, wastewater treatment is still an energy consuming process. However, the new paradigm in environmental engineering is to consider wastewater as not a mere waste but a resource in a circular economy scenario. Thus, researchers' approaches to wastewater treatment technologies are aiming to not only treat wastewater but to also recover as many resources as possible (e.g., phosphorus, nitrogen, cellulose, PHA, energy, etc.). Focusing on the energy content, the chemical energy in the wastewater is stored as organic matter. If this energy could be technically exploited, wastewater treatment would not be a sink of energy but a source instead. Calculating the energy content of the wastewater is not an easy task and different authors have recently taken the challenge of estimating this energy content to commonly measured values, such as chemical oxygen demand (COD), total organic carbon (TOC) or biological oxygen demand (BOD). Despite this, values may be variable among different wastewaters, and a correlation of 13-15 kJ·g^{-1} COD seems a conservative approach (Shizas et al. 2005; Logan 2008; Heidrich et al. 2011; Korth et al. 2017). This means that for a 'model' wastewater of a medium-sized wastewater treatment plant (WWTP), Q = 10^5 m^3·d^{-1}, with an average COD inlet of 0.4 g COD·L^{-1}, the energy contained is 1.56·10^5 kWh·d^{-1}. This would mean that for average consumption of a 9 kWh·home^{-1}·d^{-1}, more than 17,400 homes could be served with this power. This number is certainly overestimated since these calculations assume 100% of energy recovery that is certainly far from reality. A sensitive goal is to recover only 25-50% of this energy (Logan 2008).

How can we recover this energy? Energy carriers are moving toward more hydrogen-rich fuels (C, coal → -CH$_2$-, oil → CH$_4$, natural gas → H$_2$, hydrogen), which together with the necessity to avoid carbon dioxide emissions, converts hydrogen to the energy carrier of the future. Hydrogen is a clean and renewable energy carrier that does not influence greenhouse gas emissions in its energy generation process and has a high combustion heat (122 kJ g^{-1}) when compared to other possible fuels (methane 50.1 kJ g^{-1} or ethanol 26.5 kJ g^{-1}). Finally, it is expected that the price of hydrogen will not be very

---

*Corresponding author: albert.guisasola@uab.cat

fluctuating if it is produced from natural sources. Hydrogen, nowadays, is mainly produced by natural gas steam reforming, and therefore it cannot be considered either renewable or carbon-neutral fuel. Water electrolysis is an alternative, but it is energy consuming and remains a promising technology for a future with more available renewable energy.

Biological hydrogen production offers the possibility to upgrade wastes via the generation of hydrogen, a renewable energy vector with no net contribution to the greenhouse effect. Hydrogen can be produced biologically by photosynthesis, dark fermentation or bioelectrochemistry (Drapcho et al. 2008; Lee et al. 2010b; Züttel et al. 2011). The potential maximum hydrogen yield that can be obtained from wastewater can be estimated from its COD content. One mole of COD requires one mole of oxygen for its oxidation and produces four electrons that potentially would produce two moles of hydrogen. Thus, one gram of COD would produce stoichiometrically 0.125 g H_2 that is a maximum value without considering bacterial growth, and the adequate biomass growth yield should be used for a more precise evaluation. Hence, our 'model' wastewater would allow maximum production of 5 Tn H_2/d that is $6.2 \cdot 10^4$ m³/d (at 298K and 1 atm). This would mean an energy production of $1.71 \cdot 10^5$ kWh·d⁻¹ that is very similar to the value obtained based on the experimental calculations showed above. Moreover, if we compare this value with the potential energy obtainment via methane (assuming 0.35 $L_{CH4} \cdot g^{-1}COD$), we would obtain $1.28 \cdot 10^5$ kWh·d⁻¹. This comparison is only in terms of energy recovery potential since both technologies have very different reactor configurations and a deeper comparison, such as a life-cycle analysis would be required (Rozendal et al. 2008; Foley et al. 2010; Escapa et al. 2016). The previous discussion has been only focused on urban wastewater. However, an important niche exists for bioelectrochemical hydrogen production from industrial wastewaters where potential higher COD concentrations can be found.

In practice, the values of maximum molar hydrogen production described above are almost impossible to obtain. The differences between reality and theory are quantified through several widely used indexes, such as coulombic efficiency (CE), cathodic gas recovery (r_{CAT}) or energy recovery (r_{E+S}) (Logan 2008; Selembo et al. 2009; Ruiz et al. 2013). CE defines the ratio of coulombs contained in the substrate consumed that are recovered as current intensity. r_{CAT} is calculated as the ratio of moles of hydrogen measured and moles of hydrogen that can be produced based on current intensity measured. Finally, r_{E+S} compares the energy contained in the hydrogen with the sum of the energy contained in the initial substrate and energy needed. Bioelectrochemical systems (BES) comprise emerging technologies that combine electrochemistry with the metabolism of microorganisms. In the case of bioelectrochemical hydrogen production, the microorganisms are located at the anode whereas hydrogen production occurs, usually abiotically, at the cathode (Figure 1) (Liu et al. 2005a; Rozendal et al. 2007; Logan et al. 2008; Cheng and Logan 2011). Microorganisms whose metabolism is related to anodic processes are called anode-respiring bacteria (ARB), and they are capable to oxidize the organic matter available under anaerobic conditions by using an insoluble anode as the terminal electron acceptor. ARB is also known as exoelectrogenic bacteria since they can transfer the electron extracellularly to the anode. Thus, organic matter is oxidized on the anode, while protons are reduced on the cathode. These devices are known as microbial electrolysis cells (MECs). Since the flow of electrons is favored toward more positive potentials, external energy input is required to drive the process.

This chapter reviews the current state-of-the-art of the MEC technology for the treatment of real wastewaters at both lab and pilot-scale. MEC technology has, nowadays, proven to be successful at lab-scale; however, its scale-up has several limitations and a successful MEC pilot plant has not been reported yet. This chapter critically reviews for the first time the ten attempts to scale-up these systems, and based on the results found at lab-scale, the chapter aims at providing guidelines and perspectives for a more successful design of future MEC plants.

2. Microbiology of MEC for Hydrogen Production

The microbial communities enriched at the anode, including biofilm and planktonic microorganisms, are diverse and depend on the composition of the inoculum and substrates—the type of MEC being used (i.e., one or two chamber)—electrode material and the operational conditions. In the case of complex

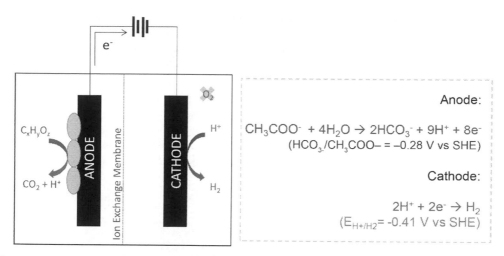

Figure 1: Schematic design of an MEC for bioelectrochemical hydrogen formation from organic sources.

substrates such as wastewaters, a very diverse microbial community is present. The anodic microbial community includes exoelectrogens (ARB) and hydrolytic and fermentative bacteria that are able to convert complex organic molecule to simpler molecules such as volatile fatty acids (VFAs) which can be easily used as electron donors by ARB. Thus, these populations have positive interactions favoring an efficient substrate conversion to electron and therefore hydrogen at the cathode. Depending on the composition of the wastewater used to feed the MEC, alternative electron acceptors to the anode can be present, such as nitrate or sulfate, that are responsible for a decrease in the CE. Potential oxygen leakages could also favor heterotrophic bacteria and decrease the CE.

ARB is naturally enriched from any inoculum in the presence of an anode. In many studies using acetate as an electron donor, the biofilm community is dominated by bacteria from the *Geobacter* genus (*δ-Proteobacteria*), generally *Geobacter sulfurreducens*. When more complex substrates are used, such as wastewates, the diversity increases but the strong selection pressure generally favors *Proteobacteria* and even *δ-Proteobacteria* (Flayac et al. 2018). Hydrolytic and fermentative bacteria are present when complex substrates basically waste or wastewater containing polymeric carbohydrates, fats or proteins is used to feed the MEC (Kokko et al. 2018). These complex substrates cannot be directly used by ARB, and this hydrolysis/fermentation step is, therefore, necessary to provide a substrate for the current generation. Montpart et al. (2015) fed single chambers MEC with glycerol, milk and starch and observed *Bacteroidetes* and *Actinobacteria* as dominant in the biofilm.

In a one-chamber membraneless system, the fact that both electrolytes are not physically separated, contributes to the growth of microorganisms that do not only compete with ARB for the substrate but also for hydrogen. If the hydrogen that evolves in the cathode of an MEC is not rapidly evacuated into the gas collection, there is a risk of hydrogen scavenging (Lee and Rittmann 2010; Lu et al. 2011; Parameswaran et al. 2011a; Lu et al. 2012; Ruiz et al. 2013; Rago et al. 2015a). Hydrogen scavenging reactions include homoacetogenesis and hydrogenotrophic methanogenesis. Moreover, when working with fermentable substrates, the acetate or the hydrogen formed in fermentation can also be used for methanogenesis as electron donor that can account for important electron losses at the anodic compartment (Lee et al. 2009; Parameswaran et al. 2009). Hydrogenotrophic methanogens are more often detected in anodic microbial communities than acetoclastic methanogens. One hypothesis is that acetoclastic methanogens could be outcompeted by ARB due to a higher affinity of the latter for acetate. Homoacetogenesis is a competitive reaction to hydrogenotrophic methanogenesis. However, due to thermodynamic and kinetic advantages, hydrogenotrophic methanogens have been shown to outcompete homoacetogens so that homoacetogenesis only demonstrates when methanogenesis is inhibited (Parameswaran et al. 2010; Ruiz et al. 2013; Tiquia-Arashiro 2014).

Figure 2 presents the main routes of syntrophic interactions in MECs as proposed by Kadier et al. (2018). This figure only considers the H_2 produced by fermentative bacteria and not the potential H_2 produced at the cathode. It is important to highlight that in the case of complex organic substrates, the growth of H_2-scavengers is indispensable to lower H_2 partial pressure and makes the acetate-forming reactions thermodynamically favorable (Gao et al. 2014). If the H_2-scavengers are methanogens, the CE is negatively impacted, whereas the development of homo-acetogens seems to build a positive syntrophy with ARB (Parameswaran et al. 2011b). Thus, Gao et al. (2014) explain the good results obtained in a continuous MEC fed with digestate by efficient syntrophic interactions among fermenters, homoacetogens and ARB. Their MEC showed a high current density of 14.6 A/m² and high CE of 98% ± 35% and 87% ± 13% at HRT 4 days and 8 days, respectively. Methane accounted for only 3.4% of the COD removed.

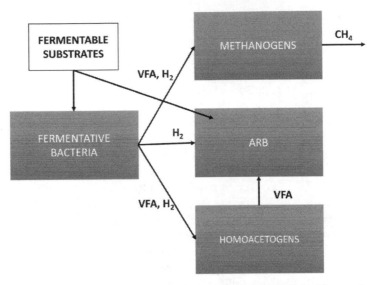

Figure 2: The three main routes of syntrophic interactions in MECs fed with organic wastes

3. State-Of-The-Art of Hydrogen Production from Wastewater by MEC at Lab-scale

So far, lab studies with MEC have been successful and have demonstrated them to be a biological technology with a higher yield of hydrogen produced per unit of organic matter. However, the majority of the studies were conducted in small reactors (i.e. reactor volumes below 100 mL) with the use of simple substrates and optimized and buffered culture media (Lu and Ren 2016). Acetate, among all the substrates tested, is recognized to be the best candidate in terms of H_2 production performances, leading to H_2 production rate up to 6.3–50 m³/m³/d with high CE (~85%) and overall energy efficiency in most cases even exceeding 100% (Cheng and Logan 2011; Jeremiasse et al. 2010; Liang et al. 2011; Tartakovsky et al. 2009). In general, smaller reactors showed higher system performance, indeed declines in current density, hydrogen production rate and efficiency were observed with the increase in the reactor's dimension (Lu and Ren 2016). Up to now, the maximum hydrogen rate reported has been of 50 m³/m³/d, it was obtained with a 40mL two-chamber reactor, with a Ni-foam cathode and applied voltage of 1V using acetate as substrate (Jeremiasse et al. 2010). However, this production rate was calculated based on the measured current density by assuming a cathodic H_2 of 90% but not experimentally measured. The calculated values often represent an overestimation compared to the actual H_2 production rate that could be experimentally measured.

Besides these promising results, certain issues need to be addressed and resolved to operate MECs in real conditions. As can be seen from the data in Table 1, the rate of hydrogen production using wastewater

Table 1: Electrode half-reactions in MEC and standard reduction potentials for acetate as carbon source.

Electrode	Reaction	E^0 (V) *vs* SHE
Anode	1) $CH_3COO^- + 4H_2O \rightarrow 2HCO_3^- + 9H^+ + 8e^-$	0.187
Cathode	2a) $2H^+ + 2e^- \rightarrow H_2$	0
	2b) $2H_2O + 2e^- \rightarrow H_2 + 2OH^-$	-0.828

substrates ranges from 0.01 m³/m³/d to 2.27 m³/m³/d and the CE in wastewater treatment is lower (10–87%) than that of pure substrates (Parkhey and Gupta 2017). Since not all substrates can be directly used by ARB, a consortium is needed where fermenters and ARB coexist (see Section 2).

The first attempt to produce hydrogen from real domestic wastewater was reported by Ditzig et al. (2007). This study reported a maximum hydrogen recovery of 42%, obtained in packed-bed mode, at an applied voltage of 0.5V with a CE of 23%. Only 154 ml H_2/g COD consumed was obtained although COD degradation was 90% that theoretically could give 1,414 ml H_2/g COD consumed. The authors hypothesized that the main reason for the low hydrogen recovery was appreciable gas loss via diffusion through the tubing. Lately, Escapa et al. (2012) studied the impact of different organic loading rates (OLR) and applied voltages on MEC performances during continuous unbuffered (pH of approximately 6.7 and conductivity of 0.9 mS/cm) domestic wastewater treatment. Hydrogen was produced only at OLRs above 448 mg COD/L/d (HRTs below 24 hours), reaching a maximum of 0.30 m³/m³/d at the OLR of 1,994 mg COD/L/d. Further increases (up to 3,120 mg COD/L/d) in the OLR did not increase hydrogen production significantly, indicating a maximum of hydrogen production and current production when the OLR is set above 2,000 mg COD/L/d. Moreover, the energy consumption was also found to be highly dependent on the OLR. When the OLR was set at 3,120 mg COD/L/d, the energy consumption per unit of COD removed was 0.77 Wh/g COD, and then it increased almost linearly until 2.20 Wh/g COD when the lowest OLR was imposed. These latest findings were partially explained by the fact the internal resistance decreased with increasing OLR from approximately 100 Ω, when low OLRs were imposed, to 48 Ω at an OLR of 3,120 mg COD/L/d. Despite not being optimized, these first attempts showed MEC platform holds great potentials for future wastewater treatment. Indeed, Cusick et al. (2010) obtained a similar hydrogen production rate (i.e. 0.28 m³/m³/d) estimated a cost of $3.01/kg H_2 for the hydrogen produced from domestic wastewater that was less than the estimated merchant value of hydrogen ($6/kg-$H_2$).

Apart from domestic wastewater, different types of wastewater have been tested for H_2 production in MECs. Methanol-rich industrial wastewater (4070 mg COD/L) from a chemical manufacturing facility was used in a one-chamber MEC to evaluate molybdenum disulfide (MoS_2) and stainless steel (SS) as alternative cathode catalysts to platinum (Pt) (Tenca et al. 2012). As expected MEC performance highly varied depending on the cathode catalyst, showing H_2 production rates from 0.26 to 0.58 m³/m³/d and COD removal rates from 1.8 to 2.8 kg COD/m³/day. The use of the MoS_2 catalyst generally resulted in better performance than the SS cathodes although the use of the Pt catalyst provided the best performance. However, the produced gas was mainly composed of methane for all the conditions. Interestingly, swine wastewater was shown to be suitable to be treated by MEC technology. Using a single-chamber MEC with a graphite-fiber brush anode, hydrogen gas was generated at a maximum rate of 1.0 m³/m³/d using a full-strength (12,825 mg COD/L) with COD removals of 72% in long tests (184 hours). However, appreciable amounts of methane gas were also produced in this process, resulting in biogas with several components (Wagner et al. 2009).

Illustratively, MEC technology was proposed to treat spent yeast issued from alcoholic fermentation (Sosa-hernández et al. 2016). The maximum hydrogen production rate was increased from 0.64 ± 0.06 to 1.24 ± 0.46 m³/m³/d by increasing the organic load from 0.75 to 1.5 g COD/L, and the highest hydrogen production rate of 2.18 ± 0.66 m³/m³/d was achieved when ethanol was added to keep high the organic load (2.7 g COD/L). The use of dairy wastewater for hydrogen production in MECs was suggested by Montpart et al. (2015) since it seemed to have the capacity to inhibit methanogenesis without the use of a chemical inhibitor (Montpart et al. 2015; Moreno et al. 2015). H_2 production in an MEC feed with cheese whey and diluted to 2g COD/L, ranged from 0.6 $m^3/m^3/d$ to 0.8 $m^3/m^3/d$. Even though no methane was detected inhibitor, about 71% of the energy invested in the cheese whey wastewater treatment was recovered as hydrogen gas (Rago et al. 2016). These results represent a significant improvement of the performances obtained with milk powder feed at half the organic load (1 g COD/L) by the same team (Montpart et al. 2015) where only 0.086 $m^3/m^3/d$ was observed. Besides the higher organic load such improvement was likely resulting from the enrichment of an efficient syntrophic consortium between fermentative and electroactive bacteria.

4. Enhancing Hydrogen Recovery from Dark Fermentation with MECs

Among all the biotechnologies that can produce hydrogen, dark fermentation (DF) processes have gained increased attention over the past decades mostly because of the wide capacity of fermentative microbial communities to convert organic matter to valuable metabolic products. By using complex and unsterilized substrates, such as organic-rich waste and wastewaters, the costs for producing hydrogen are lower than other biological or more conventional technologies (electrosynthesis), i.e. down to 2-3 €/kgH$_2$, and with the lowest environmental impacts in terms of global warming and acidification potentials (Dincer and Acar 2014). The other main advantages of the DF processes are: (1) the possibility to be integrated into conventional waste/wastewater treatment streams to furnish an additional source of energy, (2) the use of inexpensive and well-known technology of fermentation and (3) the wide potential of adaptation of microbial mixed cultures to various organic materials (Guo et al. 2010). In particular, bioH$_2$ production by DF of 'low valuable; organic waste streams (OWS) is of particular interest since it combines the objectives of sustainable waste management by treating pollution, and the generation of a valuable clean energy product (Hallenbeck 2009).

Although such systems present high productivities for a biological process, in a maximal order of magnitude of few L of hydrogen/L reactor/h (Hawkes et al. 2007), the conversion yields in terms of a mole of hydrogen produced per mole of the substrate are low with regard to other technologies. Indeed, hydrogen production by fermentative bacteria is limited by the intrinsic capacity of the microorganisms to oxidize organic substrates mainly because of metabolic restrictions. As an illustration, a maximum of 77 g H_2 per kg of glucose or 640 g per ton of olive effluent can be expected (Rozendal et al. 2006). Indeed, carbon and hydrogen molecules issued from the substrate are mainly converted to other microbial metabolic by-products in forms of short-chain acids (e.g., acetate, butyrate) or alcohols (e.g. ethanol). In fact, fermentative microorganisms mainly produce hydrogen from simple carbohydrates (e.g., sucrose, glucose) or from more complex polysaccharides, such as molasses up to hemicellulosic compounds mainly following the acetate and butyrate pathways (Guo et al. 2010). The fermentation of glucose produces a maximum of 4 moles of hydrogen per mole of glucose through the acetate pathway, while only 2 moles of hydrogen per mole of glucose is produced through the butyrate pathway. The amounts of hydrogen per mass unit from the acetate and butyrate pathways are equivalent to 44 g H_2 and 22 g H_2 per kg of glucose, respectively, over 133 g/kg as the theoretical maximal yield of 12 moles H_2 per mole glucose. In terms of COD, the conversion yield presents a maximum of 33% of conversion with the acetate pathway. In practice, fermentation H_2 processes yield only 2.4 to 3.2 moles of hydrogen per mole of glucose (Kovács et al. 2006; Ghimire et al. 2015). However, environmental sustainability and economic feasibility can only be reached if the overall H_2 yields on dissolved organic material can approach 60-80% (Benemann 1996). For that reason, bioH$_2$ production by DF is likely to be industrially viable only if it is integrated within a process that can utilize the fermentation end-products. Therefore, it can be proposed to convert organic acids from fermentation into hydrogen using a minimum of electrical energy through MECs (Liu et al. 2005b).

Up to now, only a low number of publications have dealt with direct coupling of DF processes with MEC to produce hydrogen. Moreover, the range of substrates, microbial inocula and operating conditions are extended, making comparisons difficult between the studies (Bakonyi et al. 2018). As operating conditions, most of the studies are dealing with batch tests for the DF process. MEC is mostly inoculated with anaerobic sludge or domestic wastewater. Anode material is often a carbon, graphite felt or brush to favor the electron transfer with the electroactive bacteria. Cathode material corresponds to stainless steel, graphite or Pt mesh. The operating temperature ranges between 25 and 37°C with an average value of 30°C. Finally, the applied voltage ranges from 0.44 to 1 V (DC supply or potentiostat). If considering the type of substrate in DF, single carbon sources have first been used as model substrates, such as glucose, cellulose, sucrose or molasses (Lu et al. 2009; Lalaurette et al. 2009; Bakonyi et al. 2018). As an illustration, on glucose-based synthetic effluents Babu et al. (2013) observed a hydrogen yield in DF of 11.2 moleH$_2$/kg COD equivalent to approximately 2 moles hydrogen per mole of glucose (17.8% of the maximum theoretical value or 86% of the H$_2$ yield in mixed cultures). The following effluent, mainly composed of acetate, butyrate and propionate, was further used in MEC to reach an H$_2$ yield of 2 moles hydrogen per kg VFA. Overall, the coupled system performances could be estimated at around 2.52 moles hydrogen per mole of initial glucose (21% of the maximum theoretical yield) (Babu et al. 2013).

With more complex substrate, Lalaurette et al. (2009) reported first a two-stage system with acid-pretreated corn stover as a substrate but finally observed methane generation from the MEC. Using a pure culture of *Clostridium thermocellum*, they observed a hydrogen yield of 1.6 mole H$_2$ per mole glucose$_{equivalent}$ (13% of the max conversion) with productivity ranging from 0.25 to 1.65 m^3/m^3/d according to the substrate, corn stover or cellobiose. Interestingly, the overall yield was estimated at around 9.95 mole H$_2$ per mole of glucose equivalent (83% of the max conversion). More recently, there is a clear tendency to use raw sources of substrates, such as real wastewaters, by-products and lingo-cellulose materials coming from different industrial processes. Marone et al. (2017) reviewed a compilation of published works that report the use of acid-pretreated corn stalk, crude glycerol, cheese whey or waste activated sludge for the production of hydrogen by coupling dark fermentation and microbial electrolysis (Marone et al. 2017). Moreover, Marone et al. (2017) evaluated six different wastewaters and industrial by-products coming from cheese, fruit juice, paper, sugar, fruit processing and spirits factories for the feasibility of hydrogen production in a two-step process. These authors reported high performances from complex industrial wastewaters particularly when using carbohydrate-rich effluents. Using fruit juice factory wastewaters, they showed a maximal hydrogen yield around 10 moles hydrogen per mole glucose$_{equivalent}$ degraded (more than 80% of the theoretical value) in batch tests with a COD removal of 72% and only 8% of the H$_2$ coming from the DF process. Mainly the hydrogen conversion efficiency ranged between 2.3 and 10 moles H$_2$ per mole glucose equivalent depending on the carbohydrate content of the effluents (Marone et al. 2017). Other parameters than yield and productivities must also be considered such as the overall energy efficiency of the system (Bakonyi et al. 2018). In Lalaurette et al. (2009), the overall energy efficiency was estimated to reach around 270%, on the basis of the electricity needed to recover H$_2$, making the process highly beneficial. CE and r$_{CAT}$ are highly variable, depending on the operating conditions, ranging 8-92% and 22-101%, respectively (Bakonyi et al. 2018).

5. Scaling-Up MEC Systems: Pilot Plant Configurations and Performance

The results of hydrogen production with MEC at different lab-scale have led to a greater focus on the application of this technology on a larger scale. Despite the large number of manuscripts published about hydrogen production by MEC, research on the scale-up of these systems is limited. Currently, most MEC studies are conducted at lab-scale in 10-100 mL reactors. Therefore, the scale-up of MEC seems to be an unsolved problem that hinders its real applicability as a clean energy production system. However, the number of real pilot-scale MEC applications is scarce as reflected in only 10 experiences of these systems with four or more liters of volume reported in the literature. Table 2 summarizes the main characteristics reported on these pilot systems, such as substrate and load used, volume, area vs. volume for the anode, materials for anode and cathode, membrane or separator used, performance as COD removal and H$_2$ productivity or maximum current density.

Table 2. Studies on hydrogen production from wastewater (WW) in bench-scale MECs.

Wastewater (g COD/L)	Reactor	V (mL)	Anode	Cathode	Membrane/separator	E_{app} (mV)	H_2 rate (m^3/m^3 day)	R_{cat} (%)	H_2 Yield (%)	H_2 (%)	CA_{Max} (A/m^2)	ER (kWh/KgCOD)	COD_{rem} (%)	CE (%)	Ref.
Urban WW amended with PBS (0.204-0.481)	Two-chamber packed-bed	135	Graphite-granule	Carbon paper with a Pt loading	CEM; NafionT	500	0.01	42.7	9.8	n.a.	0.471	n.a.	87–100%	23.2	(Ditzig et al., 2007) and produced a maximum Coulombic efficiency of 26% (applied voltage of 0.41 V
Urban WW (0.345)	Single chamber cube MEC	28	Graphite fiber brushes pretreated using an ammonia gas process	Carbon cloth with platinum catalyst layer	no	-900	0.28 ± 0.04	104 ±10	n.a.	70 ±1	n.a.	0.14 ± 0.04	58 ± 3	64 ± 9	Cusick et al., 2010
Urban WW (ORL 1.944 g /L$_a$·d)*	Continuous-flow single-chamber MEC	90	Graphite felt	Ni-based gas diffusion electrode (GDE)	0.7mm porous, cellulosic, non-woven fabric (J-cloth®)	1000	0.30 L/ (Lr d)	43	11	n.a.	n.a.	-0.89 (consumed)	51	54	Escapa et al., 2012
Urban WW (0.45)	Single chamber cube MEC	32	Carbon fiber brush	Stainless steel mesh with platinum catalyst layer on carbon black, using Nafion as a binder	no	900	n.a.	<10	>10	n.a.	<1 (I_{avg-90})	-4	80	>100	Ullery and Logan, 2014

(Contd.)

Chemical manufacturing facility WW (4.07)	Single chamber cube MEC	28	Carbon fiber brush	Three different types of cathode catalysts were examined: SS, MoS2, and Pt	no	700	0.26-0.58	n.a.	n.a.	32 ± 4	1.2-2.1	3.1-3.8	1.8-2.8 kg COD/ m3/d	10-12	Tenca et al., 2012
De-oiled refinery WW (0.33-0.62)	Single chambered mini-MECs were	5	Graphite plates	Stainless steel mesh	no	700	n.a.	n.a.	79	73	2.1 ± 0.2	n.a.	79	n.a.	Ren et al., 2013
Polymer and chemical production WW (1.4)	Single chamber cube MEC	32	Carbon fiber brush	Sainless steel mesh with platinum catalyst layer on carbon black, using Nafion as a binder	no	900	n.a.	>40	>40	n.a.	>1 (I_{avg-90})	-3.5	75	>100	Ullery and Logan, 2014
Swine WW (12.825)	Single chamber cube MEC	28	Graphite-fiber brush anode	Carbon cloth with platinum catalyst layer	no	500	1 ± 0.1	29 ± 2	20 ± 1	64 ± 1	112 ± 25 A/ m^3	0.8	72 ± 4	70 ± 2	Wagner et al., 2009
Winery WW (2.2)	Single chamber cube MEC	28	Graphite fiber brushes pretreated using an ammonia gas process	Carbon cloth with platinum catalyst layer	no	-900	0.17 ± 0.01	78 ± 20	n.a.	70 ± 8	n.a.	-0.32 ± 0.3	47 ± 3	50 ± 8	Cusick et al., 2010

(Contd.)

Table 2. (*Contd.*)

Wastewater (g COD/L)	Reactor	V (mL)	Anode	Cathode	Membrane/ separator	E_{app} (mV)	H_2 rate (m^3/m^3 day)	R_{cat} (%)	H_2 Yield (%)	H_2 (%)	CA_{Max} (A/m^2)	ER (kWh/ KgCOD)	COD_{rem} (%)	CE (%)	Ref.
Potato WW (2.1)	Single chamber cube MEC	28	Graphite fiber brushes pretreated using an ammonia gas process	Carbon cloth with platinum catalyst layer	no	900	0.74	n.a.	83 ± 4	73	n.a.	n.a.	79	80	Kiely et al., 2011
Food processing WW (1.83)	Single chamber cube MEC	28	Carbon fiber brush	Three different types of cathode catalysts were examined: SS, MoS2, and Pt	no	700	0.10-0.35	n.a.	n.a.	86	2.4	1.2 ± 0.2	0.9 COD/ m3/d	26-35	Tenca et al., 2012
Molasses WW (diluted to 2)	Single chamber cube MEC	25	Carbon fiber brush	Biocathode on carbon cloth	no	800	1.3	59.7	50	45	70 A/ m^3	n.a.	n.a.	92.6	Wang et al., 2014
Cheese whey (diluted to 1)	Single chamber cube MEC	32	Carbon fiber brush	Graphite fibre cloth	no	800	0.8	49 ± 2	n.a.	n.a.	0.01	n.a.	n.a.	> 100	Rago et al., 2017

V: working volume; E_{app}: Voltage applied; H_2 rate: H_2 production rate; R_{cat}: Cathodic H_2 recovery; H_2 Yield: Overal H_2 Yield; H_2: H_2 purity; CA_{Max}: Max current density; ER: Energy recovery; COD_{rem}: COD removal.
n.a.: not available
*ORL with best performances

Table 3. Studies on hydrogen production from wastewater (WW) in pilot-scale MECs under continuous operation, ordered by publication date

Wastewater (g COD/L)	HRT (h)	organic loading rate (g/L/d)	Reactor type	V (L)	Area:volume anode ratio (m²/m³)	Anode	Cathode	Anolyte	Catholyte	T (°C)	Membrane/separator	E_{app} (V)	H_2 rate (m³/m³ d)	R_{cat} (%)	H_2 (%)	CA_{Max} (A/m²)	:COD removal (%)	CE (%)	
Winery WW diluted with boiler water (0.7-2)	24	n.a.	Single chamber with 144 electrodes pair	1000	7.2	Graphite fibre brush	Stainless steel mesh	WW	WW	31	Separator: strips of glass fibre matting	0.9	0.19	No H_2 recovered	CH_4 86%, no H_2	1	62	n.a.	Cusick et al., 2011
Urban WW (0.054-0.112)	4-25	0.62-1.32	Single chamber with separator	4	24	Carbon felt	Carbon paper	WW	WW	20	Porous cellulosic non-woven fabric	1	0.045	10-94	CH_4 1-11%	0.22	86	9-24	Gil-Carrera et al., 2013b
Urban WW (0.93-0.135)	4-10	0.46-1.59	Single chamber with separator	4	24	Carbon felt	Carbon paper with Ni	WW	WW	20	Porous cellulosic non-woven fabric	1	0.022	5-24	82	0.21	80	9-31	Gil-Carrera et al., 2013a
Urban WW (0.45)	24	0.14	Double chamber with 6 cassette cells	120	16.4	Sheet of carbon felt	Stainless steel wire wool	WW	50 mM PBS	16	Rhinohide non-selective battery separator	1.1	0.015	70	100	0.3	34	55	Heidrich et al, 2013
Urban WW after primary settling (0.64)	29	0.54	Double chamber	46	6.3	Graphite plates	Graphite plates	WW	HCO_3^- buffer 0.5M	25-36	CEM	Anode potential set to 0.2 V vs. Ag/AgCl	n.a.	n.a.	n.a.	0.72	67	11	Heidrich et al., 2014
Urban WW (0.15-2)	24		Double chamber with 6 cassette cells	100	16.4	Carbon felt	Stainless steel wool	WW	50 mM PBS	1-22	Rhinohide non-selective battery separator	1.1	0.007	49	98-99	0.25	33	41.2	Heidrich et al., 2014
Urban WW (0.070-0 40)	7-42	0.04-0.56	Single chamber membrane-less	6	38	Thick graphite felt	Ni-based gas diffusion electrodes	WW	WW	19	Porous geotextile	0.7	0.005	20	87	0.36	10-70	97-460	Escapa et al., 2015

(Contd.)

Table 3. (*Contd.*)

Wastewater (g COD /L)	HRT (h)	organic loading rate (g/L/d)	Reactor type	V (L)	Area:volume anode ratio (m²/m³)	Anode	Cathode	Anolyte	Catholyte	T (°C)	Membrane/ separator	E_{app} (V)	H₂ rate (m³/m³ d)	R_{cat} (%)	H₂ (%)	CA_{Max} (A/ m²)	:COD removal. (%)	CE (%)	
Synthetic + acetate + 35 g/L NaCl (6.42)	12	12.0	Double chamber	4	19	Graphite felt	Empty stainless steel cylindrical tube	Saline water	Saline water	37	AEM	Anode potential set to 0.44 V vs. SHE	0.9	20	>90	10.6	50-70	20	Carmona-Martinez et al., 2015
Urban WW (0.3-0.5)	24-48	0.25 - 0.5	Double chamber with 10 cassette cells	130	12.6	Stainless steel mesh and graphite fibres	Stainless steel wool	WW	NaCl 4 g/L	22	AEM	1	0.031	82	94	0.3	25	28	Baeza et al., 2017
Urban WW (0.3-0.5)	5	1.6	Double chamber with 3 cassette cells	175	34	Graphite felt	Stainless steel wire wool	WW	NaCl 0.1M	11.5	Rhinohide non-selective battery separator	0.9	0.005	10	93	0.29	63	21	Cotterill et al., 2017

V: Working volume; E_{app}: Voltage applied; H₂ rate: H₂ production rate; R_{cat}: Cathodic H₂ recovery; H₂ Yield: Overal H₂ Yield; H₂: H₂ purity; CA_{Max}: Max current density; COD_{rem}: COD removal. n.a.: Not available

Cusick et al. (2011) was the first published attempt of an MEC for producing hydrogen at a pilot scale. A pilot plant with 1,000 L of volume and single chamber configuration was designed for treating winery wastewater. The anodes were carbon fiber brushes and the cathodes were made of stainless steel mesh, and they were separated by strips of glass fiber (Table 3). Some hydrogen production was observed on days 22-43 when operating at 15-22 °C: 0.09 ± 0.04 $m^3/m^3/d$ with $33\pm22\%$ H_2, $21\pm12\%$ CH_4 and $51\pm11\%$ CO_2. When a heating system was installed and the temperature increased to 31 ± 1 °C, biogas production increased up to 0.28 $m^3/m^3/d$ but finally decreased to 0.16 $m^3/m^3/d$ and, most importantly, methane concentration increased up to $86\pm6\%$ and H_2 was only detected at trace concentrations. The period with higher hydrogen productivity was coincident with the fed of sugar-rich wastewater at sub-mesophilic temperatures. Hence, part of the hydrogen recovered could be produced by sugar fermentation as well as electrohydrogenesis (Cusick et al. 2011). The amount of hydrogen in the biogas decreased progressively, and the hydrogen produced at the cathode or by fermentation was converted to methane primarily by hydrogenotrophic methanogens, as demonstrated by the percentage of methane in the biogas (86%), higher than that produced under acetoclastic methanogenesis (<70%). Overall, there was a COD removal of $62\pm20\%$ from the wastewater once the reactor was acclimated. One of the clear lessons from this study was the need for anode and cathode separation (i.e. double chamber configuration) to prevent hydrogenotrophic methanogenesis. As a positive remark for this pilot, although the biogas produced was mainly methane, its energy content was larger than the energy applied to the electrodes, making this process potentially useful for net energy recovery.

A semi-pilot tubular MEC configuration was proposed by Gil-Carrera et al. (2013a, b) although in this case a much lower volume of 4 L was used. This was also a single chamber configuration with a carbon felt anode separated from the cathode (carbon paper with Ni cathode) by means of a porous cellulosic non-woven fabric. Urban wastewater was used, obtaining COD removal higher than 80%. Hydrogen production in the range 0.022-0.045 $m^3/m^3/d$ was obtained, in this case, with H_2 content up to 82% but also with methane presence 1-11%, probably due to the single chamber configuration. In any case, the use of MEC for urban wastewater treatment was demonstrated feasible in terms of energy consumption and COD removal efficiency

Heidrich et al. (2013, 2014) showed a 120 L pilot plant based on a novel scalable cassette configuration that was built using low-cost alternatives to the standard lab materials used for the cathode and membrane. The anode was made of a sheet of carbon felt, the cathode with stainless steel wire wool and a non-selective battery separator was used as a membrane. The plant had a double chamber configuration with six cassette cells with wastewater as anolyte and 50 mM PBS as catholyte. COD removal was 34%, lower than in previous systems, but the operation was at 16°C and treating real urban wastewater. It produced 0.015 $m^3/m^3/d$ but with much higher purity (around 100%) due to the double chamber configuration.

Another double-chamber configuration was proposed by Brown et al. (2014) while focusing on the comparison of lab-scale MEC with the prototype and a performance assessment model for comparing different systems. The prototype was a 46 L double chamber configuration with graphite plates for both anode and cathode and separated by a CEM. The anolyte was the wastewater and the catholyte was carbonate buffer, and each one was recirculated with a pump. The prototype performed better when operated with wastewater as opposed to synthetic wastewater. Removal efficiencies around 67% for COD and 45% for nitrogen were observed with wastewater although no information about hydrogen production was reported. Finally, it was found that in addition to optimizing the performance of electrodes and the area/volume ratio in the anode, establishing an optimal flow regime within the anode chamber was also necessary for maintaining high performance at a high loading rate.

The MEC reactor built by Escapa et al. (2015) was designed as a modular pilot-scale system with independent elements. Two twin membraneless MEC units of 6 L each were studied in this work. Each flat module had an integrated anode and cathode in a single chamber configuration. The anode was based on thick graphite felt, while the cathode was a Ni-based gas diffusion electrode and both were separated by a porous geotextile element. Urban wastewater was treated, obtaining COD removal between 10 and 70% depending on the load applied. H_2 productivity around 0.005 $m^3/m^3/d$ was observed, with a purity of 87%. However, H_2 recycling phenomenon was detected, decreasing the energy efficiency of the system. The authors suggested that for a proper scale-up of MEC systems, special attention should be paid to hydrogen management, fluid dynamics inside the anodic chamber, and the adaptation of microbial communities during the start-up.

Carmona-Martínez et al. (2015) designed and operated a novel scalable MEC configuration. The 4L reactor had a double chamber cylindrical configuration with graphite felt as an anode, an empty cylindrical stainless steel tube as a cathode and an AEM as the separator. The catholyte was saline water (35g NaCl/L) and the anolyte had the same composition plus acetate. COD removal in the range 50-70% was obtained, with 0.9 $m^3/m^3/d$ and H_2 content > 90%. Although this is the highest hydrogen production observed at this scale, the conditions are not comparable to the rest of works presented, because of the high salinity used and the fact that no real wastewater was tested in this system.

Another cassette-based configuration was designed and tested by Baeza et al. (2017). The 130 L pilot was made with 10 cassette cells, each one with two flat anodes made of stainless steel mesh wrapped with graphite fibers and a cathode of stainless steel wool separated by two AEM. The catholyte was water with NaCl 4g/L and the anolyte was urban wastewater. COD removal was around 25%, but it could be improved at lower organic loading rates. The hydrogen production was 0.031 $m^3/m^3/d$ with a hydrogen content higher than 94%. The amount of hydrogen recovered improved the previous results obtained by Heidrich et al. (2013) in a cassette configuration, probably due to the improved cell design, with the tightening of 34 wing nuts that held the PVC frames to the rest of cell materials.

Cotterill et al. (2017) presented the last MEC pilot plant reported nowadays in the literature. This 175 L pilot plant up-scales the cassette design previously evaluated by the same group (Heidrich et al. 2013; Heidrich et al. 2014). Each cassette, with a size of 0.5 × 1.2 m, was composed of two graphite felt anodes, a stainless steel wire wool cathode, and a non-selective battery separator as a membrane. A high COD removal of 63% was obtained, although the H_2 recovery was not very high (0.005 $m^3/m^3/d$) but with 93% of purity. The important result of this work is the scale-up of the previous configuration, with a 16-fold increase in electrode size and the reduction in HRT to 5 h.

Comparing the COD removal of these works, values between 10 and 86% have been reported, although the different wastewater composition, biodegradability and load applied do not allow a fair comparison. Typical organic loading rates in systems treating urban wastewater are in the range 0.04-1.6 g COD/L/d, while reported hydraulic retention time is between 4 and 48 h. Achieving high COD removal efficiency is interesting in view of treating wastewater in which nutrient removal is not required or as a way to implement a high rate system for only COD removal (Baeza et al. 2017). However, low oxidation rates in the anode are typically found due to low mixing. This is translated in the need for systems with higher volume and higher hydraulic residence time. To avoid mass transfer limitation when no mixing is used, conventional stirring or recycling pumps are being implemented, although this translates into an increased need for energy and can make economically unfeasible the treatment.

6. Conclusions and Future Perspectives

The results with MEC at lab scale show that to optimize an MEC in terms of hydrogen yield and CE, it is important to control the metabolic pathways to favor ARB in the anodic microbial community. One of the main challenges will be to prevent electron losses mainly by methanogenesis (Lalaurette et al. 2009; Karthikeyan et al. 2017; Kadier et al. 2018). When treating wastewater by MEC, several strategies have been proposed to prevent methanogenesis besides the use of chemical inhibitors (Montpart et al. 2015; Rago et al. 2016; Marone et al. 2017). In fact, Rago et al. (2015b) showed that methanogenesis could not be completely suppressed in a membraneless MEC, despite the use of concentrations of BES as high as 200 mM, probably because of the resistance of some methanogens or the protective effect of the biofilm. The proposed strategies at industrial scale include: (i) the use of low operational temperature (i.e. 9°C) (Wang et al. 2014), (ii) a low acetate concentration and hydraulic retention time (Sosa-hernández et al.

2016), (iii) air exposure of the cathode and reduction of the time for a fed-batch cycle (Lu et al. 2009), (iv) the use of a particular substrate (Montpart et al. 2014; Rago et al. 2016) and (v) to rapidly extract H_2 using a gas-permeable hydrophobic membrane and vacuum that prevented methanogenesis during the 51 days of operation. As suggested by Gao et al. (2014), the way to succeed is to develop syntrophic interactions from the start-up of the MEC among fermenters, homoacetogens and ARB to build a stable and efficient network, thus outcompeting methanogens from the systems. It seems more effective to prevent methanogens from growing rather than trying to eliminate them when they are already installed.

When scaling up the systems, the plant performances have not been high yet since the bioelectrochemical parameters are still far from lab-scale systems although high COD removal values have been reported in some of the pilot plant attempts. CEs are typically below 30%, indicating many processes that are not driven by the electrical current are also taking place in these pilot plants. There are also cases with electron recycling that decrease energy efficiency (Escapa et al. 2015). The maximum current density is typically around $0.3 A/m^2$. The highest value when treating wastewater was $1 A/m^2$ (Cusick et al. 2011) although in this case it was linked to methane production. These values are far from the $10.6 A/m^2$ obtained by Carmona-Martínez et al. (2015), but in this case, the conditions are not comparable as saline water with acetate was used as anolyte instead of urban wastewater. On the other hand, there seems to be a consensus about using a power source with an applied potential around 1 V, except two cases were a potentiostat was applied.

Regarding the architecture of the pilot plant, there is a clear trend to increase the anodic ratio of the projected area vs. volume within a range of reported values from 6.3 to $38 m^2/m^3$. In addition, it was found that establishing an optimal flow regime within the anode chamber is also needed for achieving high performance at a high loading rate (Brown et al. 2014). The use of double chamber configurations as the cassette modules seem to provide better scalability and improved recovery of gas with higher H_2 content. The appearance of H_2 scavengers as hydrogenotrophic methanogens or homoacetogens seems to discard the utilization of single chamber MECs. In any case, H_2 leakage is still one of the most important problems in MECs at pilot scale, and H_2 recovery is in general low. The cost of the materials is also a current focus of interest (Cotterill et al. 2017) in order to obtain systems that can compete cost-effectively with more conventional treatments. The utilization of materials as graphite felt or stainless steel wool seems then a clear trend nowadays although this could also lead to higher overpotential losses that make require higher applied voltage and hence higher power consumption.

Finally, the combination of MEC and dark fermentation has opened a new possibility for both systems. Much of the research effort in the past decades regarding dark fermentation has been focused on enhancing the hydrogen production rate by controlling the operation parameters, modifying the reactor design or the selection of specific electroactive microorganisms. However, efficient and controlled conversion of complex OWS into hydrogen requires a complex microbial community (mixed cultures) where ecological, as well as metabolic interactions between microorganisms, can occur. This makes a fermenter or an MEC similar to a microecosystem subject to the rules of natural systems, and the functions of hydrogen production and depollution may be regarded as 'ecosystem services' from which the operators benefit. Species' composition and characteristics and interactions among species are often more important than the species' richness itself in maintaining ecosystem processes and related services. The gain or loss of one or several very particular species, sometimes, present amplifying effects on both community and ecosystem processes. Those species with this effect, but unpredictable traits, are termed 'keystone species' or 'core microbiome'.

References

Babu ML, Subhash GV, Sarma PN, Mohan SV (2013) Bio-electrolytic conversion of acidogenic effluents to biohydrogen: an integration strategy for higher substrate conversion and product recovery. Bioresour Technol 133:322–31. doi: 10.1016/j.biortech.2013.01.029

Baeza JA, Martínez-Miró À, Guerrero J, Ruiz Y, Guisasola A (2017) Bioelectrochemical hydrogen production from urban wastewater on a pilot scale. J Power Sources 356:500–509. doi: 10.1016/j.jpowsour.2017.02.087

Bakonyi P, Kumar G, Koók L, Tóth G, Rózsenberszki T, Bélafi-Bakó K, Nemestóthy N (2018) Microbial electrohydrogenesis linked to dark fermentation as integrated application for enhanced biohydrogen production: A review on process characteristics, experiences and lessons. Bioresour. Technol. 251:381–389

Benemann J (1996) Hydrogen biotechnology: Progress and prospects. Nat Biotechnol 14:1101–1103. doi: 10.1038/nbt0996-1101

Brown RK, Harnisch F, Wirth S, Wahlandt H, Dockhorn T, Dichtl N, Schröder U (2014) Evaluating the effects of scaling up on the performance of bioelectrochemical systems using a technical scale microbial electrolysis cell. Bioresour Technol 163:206–213. doi: 10.1016/j.biortech.2014.04.044

Carmona-Martínez AA, Trably E, Milferstedt K, Lacroix R, Etcheverry L, Bernet N (2015) Long-term continuous production of H2 in a microbial electrolysis cell (MEC) treating saline wastewater. Water Res 81:149–156. doi: 10.1016/j.watres.2015.05.041

Cheng S, Logan BE (2011) High hydrogen production rate of microbial electrolysis cell (MEC) with reduced electrode spacing. Bioresour Technol 102:3571–4. doi: 10.1016/j.biortech.2010.10.025

Cotterill SE, Dolfing J, Jones C, Curtis TP, Heidrich ES (2017) Low Temperature Domestic Wastewater Treatment in a Microbial Electrolysis Cell with 1 m2 Anodes: Towards System Scale-Up. Fuel Cells 17:584–592. doi: 10.1002/fuce.201700034

Cusick RD, Bryan B, Parker DS, Merrill MD, Mehanna M, Kiely PD, Liu G, Logan BE (2011) Performance of a pilot-scale continuous flow microbial electrolysis cell fed winery wastewater. Appl Microbiol Biotechnol 89:2053–2063. doi: 10.1007/s00253-011-3130-9

Cusick RD, Kiely PD, Logan BE (2010) A monetary comparison of energy recovered from microbial fuel cells and microbial electrolysis cells fed winery or domestic wastewaters. Int J Hydrogen Energy 35:8855–8861. doi: 10.1016/j.ijhydene.2010.06.077

Dincer I, Acar C (2014) Review and evaluation of hydrogen production methods for better sustainability. Int J Hydrogen Energy 40:11094–11111. doi: 10.1016/j.ijhydene.2014.12.035

Ditzig J, Liu H, Logan BE (2007) Production of hydrogen from domestic wastewater using a bioelectrochemically assisted microbial reactor (BEAMR). Int J Hydrogen Energy 32:2296–2304. doi: 10.1016/j.ijhydene.2007.02.035

Drapcho CM, Nhuan NP, Walker TH (2008) Biofuels engineering process technology. McGraw-Hill New York

Escapa A, Gil-Carrera L, García V, Morán A (2012) Performance of a continuous flow microbial electrolysis cell (MEC) fed with domestic wastewater. Bioresour Technol 117:55–62. doi: 10.1016/j.biortech.2012.04.060

Escapa A, Mateos R, Martínez EJ, Blanes J (2016) Microbial electrolysis cells: An emerging technology for wastewater treatment and energy recovery. from laboratory to pilot plant and beyond. Renew Sustain Energy Rev 55:942–956. doi: 10.1016/j.rser.2015.11.029

Escapa A, San-Martín MII, Mateos R, Morán A (2015) Scaling-up of membraneless microbial electrolysis cells (MECs) for domestic wastewater treatment: Bottlenecks and limitations. Bioresour Technol 180:72–78. doi: 10.1016/j.biortech.2014.12.096

Flayac J-C, Trably E, Bernet N (2018) Microbial Ecology of Anodic Biofilms: From Species Selection to Microbial Interactions. In: Das D (ed) Microbial Ecology of Anodic Biofilms: From Species Selection to Microbial Interactions. Springer International Publishing, Cham, pp 63–85

Foley JM, Rozendal R a, Hertle CK, Lant P a, Rabaey K (2010) Life cycle assessment of high-rate anaerobic treatment, microbial fuel cells, and microbial electrolysis cells. Environ Sci Technol 44:3629–37. doi: 10.1021/es100125h

Gao Y, Ryu H, Santo Domingo JW, Lee H-S (2014) Syntrophic interactions between H2-scavenging and anode-respiring bacteria can improve current density in microbial electrochemical cells. Bioresour Technol 153:245–253. doi: http://dx.doi.org/10.1016/j.biortech.2013.11.077

Ghimire A, Frunzo L, Pirozzi F, Trably E, Escudie R, Lens PNL, Esposito G (2015) A review on dark fermentative biohydrogen production from organic biomass: Process parameters and use of by-products. Appl. Energy 144:73–95

Gil-Carrera L, Escapa a, Carracedo B, Morán a, Gómez X (2013a) Performance of a semi-pilot tubular microbial electrolysis cell (MEC) under several hydraulic retention times and applied voltages. Bioresour Technol 146:63–9. doi: 10.1016/j.biortech.2013.07.020

Gil-Carrera L, Escapa A, Moreno R, Morán A (2013b) Reduced energy consumption during low strength domestic wastewater treatment in a semi-pilot tubular microbial electrolysis cell. J Environ Manage 122:1–7. doi: 10.1016/j.jenvman.2013.03.001

Guo XM, Trably E, Latrille E, Carrère H, Steyer JP (2010) Hydrogen production from agricultural waste by dark fermentation: A review. Int. J. Hydrogen Energy 35:10660–10673

Hallenbeck PC (2009) Fermentative hydrogen production: Principles, progress, and prognosis. Int J Hydrogen Energy 34:7379–7389. doi: 10.1016/j.ijhydene.2008.12.080

Hawkes FR, Hussy I, Kyazze G, Dinsdale R, Hawkes DL (2007) Continuous dark fermentative hydrogen production by mesophilic microflora: Principles and progress. Int J Hydrogen Energy 32:172–184. doi: 10.1016/j.ijhydene.2006.08.014

Heidrich ES, Curtis TP, Dolfing J (2011) Determination of the Internal Chemical Energy of Wastewater. 45:827–832. doi: 10.1021/es103058w

Heidrich ES, Dolfing J, Scott K, Edwards SR, Jones C, Curtis TP (2013) Production of hydrogen from domestic wastewater in a pilot-scale microbial electrolysis cell. Appl Microbiol Biotechnol 97:6979–6989. doi: 10.1007/s00253-012-4456-7

Heidrich ES, Edwards SR, Dolfing J, Cotterill SE, Curtis TP (2014) Performance of a pilot scale microbial electrolysis cell fed on domestic wastewater at ambient temperatures for a 12month period. Bioresour Technol 173:87–95. doi: 10.1016/j.biortech.2014.09.083

Jeremiasse AW, Hamelers HVM, Saakes M, Buisman CJN (2010) Ni foam cathode enables high volumetric H2 production in a microbial electrolysis cell. Int J Hydrogen Energy 35:12716–12723. doi: 10.1016/j.ijhydene.2010.08.131

Kadier A, Kalil MS, Chandrasekhar K, Mohanakrishna G, Saratale GD, Saratale RG, Kumar G, Pugazhendhi A, Sivagurunathan P (2018) Surpassing the current limitations of high purity H2production in microbial electrolysis cell (MECs): Strategies for inhibiting growth of methanogens. Bioelectrochemistry 119:211–219. doi: 10.1016/j.bioelechem.2017.09.014

Karthikeyan R, Yu K, Selvam A, Bose A, Wong JWC (2017) Bioelectrohydrogenesis and inhibition of methanogenic activity in microbial electrolysis cells - A review. Biotechnol Adv 35:758–771. doi: 10.1016/j.biotechadv.2017.07.004

Kokko M, Epple S, Gescher J, Kerzenmacher S (2018) Effects of wastewater constituents and operational conditions on the composition and dynamics of anodic microbial communities in bioelectrochemical systems. Bioresour Technol 258:376–389. doi: 10.1016/j.biortech.2018.01.090

Korth B, Maskow T, Günther S, Harnisch F (2017) Estimating the Energy Content of Wastewater Using Combustion Calorimetry and Different Drying Processes. Front. Energy Res. 5:23

Kovács KL, Maróti G, Rákhely G (2006) A novel approach for biohydrogen production. Int J Hydrogen Energy 31:1460–1468. doi: 10.1016/j.ijhydene.2006.06.011

Lalaurette E, Thammannagowda S, Mohagheghi A, Maness P-C, Logan BE (2009) Hydrogen production from cellulose in a two-stage process combining fermentation and electrohydrogenesis. Int J Hydrogen Energy 34:6201–6210. doi: 10.1016/j.ijhydene.2009.05.112

Lee H-S, Torres CI, Parameswaran P, Rittmann BE (2009) Fate of H2 in an upflow single-chamber microbial electrolysis cell using a metal-catalyst-free cathode. Environ Sci Technol 43:7971–6. doi: 10.1021/es900204j

Lee H-S, Rittmann BE (2010) Significance of biological hydrogen oxidation in a continuous single-chamber microbial electrolysis cell. Environ Sci Technol 44:948–54. doi: 10.1021/es9025358

Lee H-S, Vermaas WFJ, Rittmann BE (2010) Biological hydrogen production: prospects and challenges. Trends Biotechnol 28:262–71. doi: 10.1016/j.tibtech.2010.01.007

Liang DW, Peng SK, Lu SF, Liu YY, Lan F, Xiang Y (2011) Enhancement of hydrogen production in a single chamber microbial electrolysis cell through anode arrangement optimization. Bioresour Technol 102:10881–10885. doi: 10.1016/j.biortech.2011.09.028

Liu H, Cheng S, Logan BE (2005a) Production of electricity from acetate or butyrate using a single-chamber microbial fuel cell. Environ Sci Technol 39:658–62

Liu H, Grot S, Logan BE (2005b) Electrochemically assisted microbial production of hydrogen from acetate. Environ Sci Technol 39:4317–4320

Logan BE (2008) Microbial Fuel Cells

Logan BE, Call D, Cheng S, Hamelers HVM, Sleutels THJA, Jeremiasse AW, Rozendal RA (2008) Microbial Electrolysis Cells for High Yield Hydrogen Gas Production from Organic Matter. Environ Sci Technol 42:8630–8640. doi: 10.1021/es801553z

Lu L, Ren N, Xing D, Logan BE (2009) Hydrogen production with effluent from an ethanol-H2-coproducing fermentation reactor using a single-chamber microbial electrolysis cell. Biosens Bioelectron 24:3055–60. doi: 10.1016/j.bios.2009.03.024

Lu L, Ren N, Zhao X, Wang H, Wu D, Xing D (2011) Hydrogen production, methanogen inhibition and microbial community structures in psychrophilic single-chamber microbial electrolysis cells. Energy Environ Sci 4:1329. doi: 10.1039/c0ee00588f

Lu L, Ren ZJ (2016) Microbial electrolysis cells for waste biorefinery: A state of the art review. Bioresour Technol 215:254–264. doi: 10.1016/j.biortech.2016.03.034

Lu L, Xing D, Ren N (2012) Pyrosequencing reveals highly diverse microbial communities in microbial electrolysis cells involved in enhanced H2 production from waste activated sludge. Water Res 46:2425–2434. doi: https://doi.org/10.1016/j.watres.2012.02.005

Marone A, Ayala-Campos OR, Trably E, Carmona-Martinez AA, Moscoviz R, Latrille E, Steyer J, Alcaraz-Gonzalez V, Bernet N (2017) Coupling dark fermentation and microbial electrolysis to enhance bio-hydrogen production from agro-industrial wastewaters and by-products in a bio-refinery framework. Int J Hydrogen Energy 42:1909–1621. doi: http://dx.doi.org/10.1016/j.ijhydene.2016.09.166

Montpart N, Rago L, Baeza JA, Guisasola A (2015) Hydrogen production in single chamber microbial electrolysis cells with different complex substrates. Water Res 68:601–615. doi: 10.1016/j.watres.2014.10.026

Moreno R, Escapa A, Cara J, Carracedo B, Gómez X (2015) A two-stage process for hydrogen production from cheese whey: Integration of dark fermentation and biocatalyzed electrolysis. Int J Hydrogen Energy 40:168–175. doi: 10.1016/j.ijhydene.2014.10.120

Parameswaran P, Torres CI, Lee H-S, Krajmalnik-Brown R, Rittmann BE (2009) Syntrophic interactions among anode respiring bacteria (ARB) and Non-ARB in a biofilm anode: electron balances. Biotechnol Bioeng 103:513–23. doi: 10.1002/bit.22267

Parameswaran P, Torres CI, Lee H-S, Rittmann BE, Krajmalnik-Brown R (2011a) Hydrogen consumption in microbial electrochemical systems (MXCs): the role of homo-acetogenic bacteria. Bioresour Technol 102:263–71. doi: 10.1016/j.biortech.2010.03.133

Parameswaran P, Torres CI, Lee HS, Rittmann BE, Krajmalnik-Brown R (2011b) Hydrogen consumption in microbial electrochemical systems (MXCs): The role of homo-acetogenic bacteria. Bioresour Technol 102:263–271. doi: 10.1016/j.biortech.2010.03.133

Parameswaran P, Zhang HS, Torres CI, Rittmann BE, Krajmalnik-Brown R (2010) Microbial Community Structure in a Biofilm Anode Fed With a Fermentable Substrate: The Significance of Hydrogen Scavengers. Biotechnol Bioeng 105:69–78. doi: 10.1002/bit.22508

Parkhey P, Gupta P (2017) Improvisations in structural features of microbial electrolytic cell and process parameters of electrohydrogenesis for efficient biohydrogen production: a review. Renew Sustain Energy Rev 69:1085–1099. doi: 10.1016/J.RSER.2016.09.101

Rago L, Baeza JA, Guisasola A (2016) Bioelectrochemical hydrogen production with cheese whey as sole substrate. J Chem Technol Biotechnol. doi: 10.1002/jctb.4987

Rago L, Guerrero J, Baeza JA, Guisasola A (2015a) 2-Bromoethanesulfonate degradation in bioelectrochemical systems. Bioelectrochemistry 105:44–49. doi: 10.1016/j.bioelechem.2015.05.001

Rago L, Ruiz Y, Baeza JA, Guisasola A, Cortés P (2015b) Microbial community analysis in a long-term membrane-less microbial electrolysis cell with hydrogen and methane production. Bioelectrochemistry 106:359–368. doi: 10.1016/j.bioelechem.2015.06.003

Rozendal R a, Hamelers HVM, Molenkamp RJ, Buisman CJN (2007) Performance of single chamber biocatalyzed electrolysis with different types of ion exchange membranes. Water Res 41:1984–94. doi: 10.1016/j.watres.2007.01.019

Rozendal R a, Hamelers HVM, Rabaey K, Keller J, Buisman CJN (2008) Towards practical implementation of bioelectrochemical wastewater treatment. Trends Biotechnol 26:450–9. doi: 10.1016/j.tibtech.2008.04.008

Rozendal RA, Hamelers HVM, Euverink GJW, Metz SJ, Buisman CJN (2006) Principle and perspectives of hydrogen production through biocatalyzed electrolysis. Int J Hydrogen Energy 31:1632–1640

Ruiz Y, Baeza JA, Guisasola A (2013) Revealing the proliferation of hydrogen scavengers in a single-chamber microbial electrolysis cell using electron balances. Int J Hydrogen Energy 38:15917–15927. doi: 10.1016/j.ijhydene.2013.10.034

Selembo P a., Merrill MD, Logan BE (2009) The use of stainless steel and nickel alloys as low-cost cathodes in microbial electrolysis cells. J Power Sources 190:271–278. doi: 10.1016/j.jpowsour.2008.12.144

Shizas I, Bagley DM, Asce M (2005) Experimental Determination of Energy Content of Unknown Organics in Municipal Wastewater Streams. 130:

Sosa-hernández O, Popat SC, Parameswaran P, Alemán-nava GS, Torres CI, Buitrón G, Parra-saldívar R (2016) Application of microbial electrolysis cells to treat spent yeast from an alcoholic fermentation. Bioresour Technol 200:342–349. doi: 10.1016/j.biortech.2015.10.053

Tartakovsky B, Manuel M, Wang H, Guiot S (2009) High rate membrane-less microbial electrolysis cell for continuous hydrogen production. Int J Hydrogen Energy 34:672–677. doi: 10.1016/j.ijhydene.2008.11.003

Tenca A, Cusick RD, Schievano A, Oberti R, Logan BE (2012) Evaluation of low cost cathode materials for treatment of industrial and food processing wastewater using microbial electrolysis cells. Int J Hydrogen Energy 38:1859–1865. doi: 10.1016/j.ijhydene.2012.11.103

Tiquia-Arashiro SM (2014) Thermophilic Carboxydotrophs and their Biotechnological Applications. Springerbriefs in Microbiology: Extremophilic Microorganisms. Springer International Publishing. 131 p.

Wagner RC, Regan JM, Oh SE, Zuo Y, Logan BE (2009) Hydrogen and methane production from swine wastewater using microbial electrolysis cells. Water Res 43:1480–1488. doi: 10.1016/j.watres.2008.12.037

Wang Y, Guo W, Xing D, Chang J (2014) Hydrogen production using biocathode single- chamber microbial electrolysis cells fed by molasses wastewater at low temperature. Int J Hydrogen Energy 39:19369–19375. doi: 10.1016/j.ijhydene.2014.07.071

Züttel A, Borgschulte A, Schlapbach L (2011) Hydrogen as a future energy carrier. John Wiley & Sons

Microbial Fuel Cell Sensors for Water and Wastewater Monitoring

Sharon Velasquez-Orta[1]*, Ekaete Utuk[1] and Martin Spurr[2]

[1] School of Engineering, Newcastle University, Newcastle upon Tyne, NE1 7RU, United Kingdom
[2] School of Natural and Environmental Sciences, Newcastle University, Newcastle upon Tyne, NE1 7RU, United Kingdom

1. Introduction

Within bioelectrochemical systems (BES), microbial fuel cells (MFCs) generate electricity through electrochemical reactions that convert the chemical energy in an organic substrate to electrical energy. The use of bacteria to generate electricity was first discovered in 1911 (Potter 1911). MFCs have progressed, over the last century, from application for electricity production from organic waste to use for environmental monitoring in the last fifteen years (Karube et al. 1977; Kim et al. 2003; Kim and Kwon 1999; Matsunaga et al. 1980). MFC biosensors can be made using simple architecture and low-cost materials. They could be self-sustaining and may not require an external power source. In water and wastewater monitoring, MFCs can offer the capability of simultaneous treatment of contaminated medium and generation of a measured concentration response within a short response time, in contrast with traditional methods (Schneider et al. 2016; Sun et al. 2015; Jiang et al. 2018). Most reviews focus on operating principles and designs/configurations of MFC biosensors. This review provides an overview of the application of MFC biosensors for water and wastewater monitoring using biological/chemical oxygen demand (BOD/COD) or toxicity. It will cover the operating principles of MFC biosensors for water quality monitoring with emphasis on microbial consortium in anode and cathode, electron transfer and current generation mechanism. Factors affecting performance, challenges of the technology as well as future perspectives will also be discussed.

In MFC, an organic substrate is oxidized by electroactive bacteria. This results in release of the electrons that are transferred from the anode, via an external circuit, to a terminal electron acceptor (TEA), such as oxygen, at the cathode, thereby generating electric current. Protons from the anodic chamber diffuse across a proton exchange membrane (PEM) into the cathodic chamber and react with the electrons and oxygen. A typical microbial fuel cell (Figure 1) consists of a vessel comprising two electrodes (anode and cathode) separated by an ion-exchange membrane and with a resistor connected across the electrodes. The electric current generated is measured by a voltammeter as the potential difference between the two electrodes. The data is recorded on a computer using a data logging software.

---

*Corresponding author: sharon.velasquez-orta@ncl.ac.uk

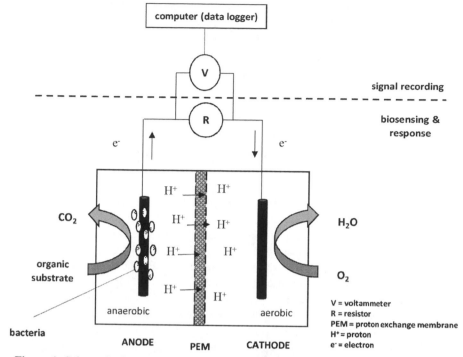

Figure 1: Schematic diagram of a microbial fuel cell showing the redox reactions occurring at the anode and cathode chambers.

2. Operating Principles

The biosensing element of MFC biosensors will be an electroactive microorganism at the anode or cathode electrode surface. In other words, redox reactions of the metabolic pathways of the microorganisms are responsible for the electric current or voltage generated. Any analyte that modifies these reactions will likely affect the current output. The nature of the analyte can either enhance or diminish current output, depending on the type of interaction that takes place between the analyte and bacteria or its metabolic pathway (Chang et al. 2005). Since the bacteria act as a biocatalyst, then any analyte that alters its ability to carry out its catalytic function will cause a change in the generated current or voltage. The biosensoric capabilities of the MFC are based on the ability to detect this change and involve correlating the analyte concentration to the current or voltage output.

2.1 Microbial Consortium

In MFC, biofilms are formed on an anodic surface after inoculation and a sustained substrate supply. Most electrodes employed, when using microbial biocatalysts, are made of carbon materials. Pure strains inoculated in the anodic side of MFC biosensors, include *Shewanella putrefaciens* IR-1, *Shewanella oneidensis, Geobacter sulfurreducens* (DSM 12127), *Serratia marsecens, Bacillus subtilis, Bacillus licheniformis, Trichosporon cutaneum, Klebsiella oxytoca, Hansennula anomala, Pseudomonas putida, Torulopsis candida, Proteus vulgaris, Clostridium butyricum, Aeromonas formicans, Psuedomonas syringae, Escherichia coli, Moraxella* and *Saccharomyces cerevisiae* (Liu et al. 2000; Riedel et al. 1988; Tan et al. 1993; Chee et al. 1999; Kim, Chang et al. 2003; Su et al. 2011; Kim and Kwon 1999; Dávila et al. 2011; Kim et al. 1999). Most of the strains previously listed require the use of mediators in order to facilitate electron exchange between microbes and electrodes. One of the first reported mediatorless MFC sensors was a lactate sensor using a pure culture at the anode (Kim et al. 1999). Later, it was found that when using pure cultures of bacteria for a specific substrate, bacteria could be responsive to different types of substrates that can give false positives. As a result, MFC sensors should be better suited to give information on broader parameters, such as BOD. This should also be adequate when using biofilms of mixed microbial culture. Mixed cultures yield higher power densities and are more stable than pure culture biofilms because of increased capacity to utilize a broader range of substrates as fuel (Du et al. 2007; Logan et al. 2006; Lei et al. 2006). Additionally, mixed culture inocula may be sourced naturally from the environment, including marine sediments, freshwater sediments, garden soil, activated sludge or anaerobic sludge. Biocathodes may also be employed in certain configurations, where the cathode serves as an electron donor for a microbial biofilm on its surface, such as *Thiobacillus ferrooxidans* (Du et al. 2007). Other phyla found in biocathode communities include *Proteobacteria, Bacteroidetes, Chlorobi* and *Actinobacteria* (Chen et al. 2008). Mixed microbial cultures, sourced from the environment, can also be used in MFC cathodic electrodes as the response element (Velasquez-Orta et al. 2017).

2.2 Analyte

A wide range of bacterial substrates has been employed in MFCs' biosensors as the analyte. These include different forms of carbohydrates (acetate, arabitol, mannitol, glucose, galatitol, sucrose, xylose and cellulose), chemicals (nitrilotriacetic acid, phenol, fumarate, lactate, pyruvate and butyrate), waste (farm manure and real or synthetic wastewaters) and algae biomass (Pant et al. 2010; Velasquez-Orta et al. 2009). In MFC biosensors reviewed for this communication, acetate, glucose and glutamate were the most common substrates utilized by electroactive bacteria. This is not surprising as these compounds are not only simple to oxidize but also have a high energy store. Furthermore, glucose and glutamic acid are the carbon sources used to calibrate the standardized BOD_5 test (APHA 2005). The substrate can be supplied in batch mode, where the reactor is fed at periodic intervals, or in continuous mode, where the substrate is fed into the system at a specified flow rate. Although batch mode operations have been most commonly studied, a decline in current can occur due to starvation caused by insufficient fuel supply. When the reactor is maintained in continuous mode a constant current can be maintained, thereby providing more stability to the system. The operation mode of MFC impacts on its performance as it can influence the electron transfer mechanism. For instance, electron shuttles can be lost in continuous mode that would be thermodynamically unfavorable for the bacteria producing shuttles, therefore bacteria capable of direct electron transfer to the anode surface would likely be the dominant species (Lovley 2006). This observation suggests some level of adaptation in MFC microbial consortiums to favor the growth and proliferation of bacteria that can thrive best in the environmental conditions, thereby highlighting the dynamic nature of the consortium to changes in the environment.

2.3 Electron Transfer Mechanism

Although there is still considerable debate on the specifics of electron transfer mechanisms in microbial fuel cells, most literature identifies two types of electron transfer (ET) mechanisms, namely mediated and non-mediated. Non-mediated ET involves direct contact between the bacteria and the electrode surface. Electrons are transferred directly from the bacteria to the electrode using nanowires (conductive pili) or membrane-associated cytochromes e.g., *Geobacter sulfurreducens*, *Shewanella oneidensis* and *Thermincola potens* (Logan 2008) (Philips et al. 2016; Logan 2008). In mediated ET, electron shuttles are used to transport electrons between the bacteria and the electrode surface. This contact is made by soluble electron shuttles such as natural mediators secreted by the electroactive bacteria or chemical reagents (artificial redox mediators) (Abrevaya et al. 2015; Logan et al. 2006). Flavins or redox endogenous mediators such as 2-amino-3-dicarboxy-1,4-naphthoquinone and pyocyanin can facilitate electron shuttling between the bacteria and the anode (Santoro et al. 2017; Velasquez-Orta et al. 2010). Artificial redox mediators such as flavins, thionine, neutral red, methylene blue and anthraquinone-2,6-disulfonate (AQDS) facilitate electron transfer, especially for bacteria that are unable to transfer electrons on their own; they are often used in small quantities (Abrevaya et al. 2015; Logan et al. 2006).

In MFC biosensors, redox mediators provide the link between the bacteria and the electrodes by harvesting electrons from the bacterial cells, during which they are reduced, to the anode surface where the electrons are released, and the mediators are reoxidized. They are then made available for the transport of more electrons. This mechanism facilitates electron transport in the organism that is unable to produce its own natural mediators or do not have nanowires. Bacteria such as *Shewanella putrefaciens* did not require added mediators for electron transfer, while thionine and hydrogen were used to facilitate electron transfer when *Proteus vulgaris*, *Clostridium butyricum* and *Aeromonas formicans* were utilized in MFC biosensors (Matsunaga et al. 1980; Thurston et al. 1985; Kim et al. 2002; Karube et al. 1977). The use of external mediators may not promote the stability of the biosensors (Chang et al. 2004). In several cases where anaerobic or activated sludge was used as the enrichment medium for the biosensors, mediators were not used (Chang et al. 2004; Kim, Hyun et al. 2007; Kim, Youn, et al. 2003). This may be due to the ability of the diverse microbial community present in the inocula to utilize various electron shuttle mechanisms. Due to the external mediator's potential toxicity and cost, there is currently little interest in adding redox compounds to MFC sensors.

2.4 Current or Voltage Generation

Current is the most commonly calibrated signal generated by MFC biosensors as it can be monitored and recorded in real-time, and it is an indirect measure of the concentration of a target analyte. A plot of the current generated vs. the concentration of the analyte is known as a calibration curve. It is a means of quantifying the concentration of the analyte and defines the nature of the relationship between the analyte concentration and the current output. The relationship between biological oxygen demand (BOD), a standard water industry method for quantifying unspecified or unknown concentrations of biodegradable organic substrates, and MFC current output has been shown to be linear up to an identified saturation concentration (upper limit). Beyond this limit, higher concentrations of the organic substrate cannot be measured as the bacteria are unable to oxidize further the substrate until concentration levels subside (Jiang et al. 2018). This sensing parameter is particularly relevant for water quality monitoring (in correlation with BOD or COD). It was recently demonstrated that the dynamic range could be significantly extended (2-3 fold) using a three-stage MFC reactor configuration assembled hydraulically in series (Spurr et al. 2018). A wide linear range is crucial for application to monitor wastewaters from urban or industrial sources or for volatile fatty acid (VFA) monitoring during anaerobic digestion to prevent VFA build up in the reactor that could lead to a system failure. Membrane fouling or unfavorable pH shifts can also cause the systematic failure of the MFCs.

Changes in the open circuit potential, or voltage, can also be used as the response element in MFC biosensors (Wang et al. 2018; Liu et al. 2014). In an ideal system, when there is no current flow through the system (open circuit potential), the potential difference between the anode and the cathode is referred to as the overall electromotive force or cell potential (E_{emf}) and is related to Gibbs free energy (ΔG). The relationship between the cell potential, temperature and concentration of the reactants can be expressed by the Nernst equation. In reality, with MFCs, including those used as biosensors, the actual potential (E_{MFC}) is less than E_{emf} because of energy losses through the energy used to start the reaction (activation losses, η_{act}), overcome internal resistance (ohmic losses, η_{ohm}) and energy losses due to mass transport within the system (concentration losses, η_{concn}) (Logan et al. 2006; Esfandyari et al. 2017). The anode potential must, therefore, be kept at a low potential, enough to drive the reactions but minimize activation losses. Activation losses can also be minimized by increasing the anode surface area and operating temperature (Logan et al. 2006). Use of membrane or electrode materials with low resistance, shortening the distance between the electrodes and increasing electrolyte conductivity can reduce ohmic overpotential and increase current output in the biosensor (Logan et al. 2006; Bard et al. 2008). Concentration losses, mainly due to mass transfer limitations, can also be a result of a limited removal of oxidized species from the anode or supply of reduced species. Concentration losses can be minimized by using buffer solutions to maintain the pH of the electrolyte within an acceptable range (Logan 2008) or by improving mixing near electrodes.

3. Performance Indicators

An efficient MFC biosensor must be capable of rapid and precise *in situ* monitoring of compounds, have long operational stability, low maintenance requirements and short recovery time (Kim, Hyun et al.

2007). The efficiency of an MFC biosensor is measured by parameters, such as response time, detection limit, sensitivity, recovery time and stability. These indicators can be expressed in terms of the current generated, coulombic efficiency or yield and power or current density with reference to the anode or cathode surface area. Electroanalytical techniques such as cyclic voltammetry, chronoamperometry, chronopotentiometry, electrical impedance spectroscopy as well as polarization curves, peak current and power density can be used to study and characterize MFC to optimize their performance (Logan 2008) and may be applicable to improve MFC biosensing performance.

The time duration for current to achieve 95% of the steady state current after a change in current output due to the presence of a target compound is one method used to calculate the response time of the biosensor (Di Lorenzo et al. 2009a). Short response time is vital for rapid monitoring. The minimum quantity of a compound that can be measured by the biosensor is its detection limit. The lower the quantity that can be detected, the higher the performance of the biosensor; this is especially important for monitoring drinking water resources where contaminants can be present in trace amounts. The sensitivity of an MFC biosensor is the change in current per unit change in the concentration of the compound and is determined by the anodic biofilm. It is measured in relation to the anode surface area as shown below:

$$\text{Sensitivity} = \frac{\Delta I}{\Delta C} \frac{1}{A} \tag{1}$$

where ΔI (μA) is a unit change in current, ΔC (mM) is a unit change in concentration and A (m^2) is the anode surface area. Biosensors with larger unit changes in current appear to be more sensitive (Stein et al. 2012). Sensitivity can be increased by improving electron recovery either by enhancing electroactive bacteria in the anode or by inhibiting the activity of other bacteria that could compete with the bacteria for substrate (Jiang et al. 2018). It has been reported that low external resistance can improve the response time and sensitivity (Pasternak et al. 2017).

A good biosensor should be able to recover rapidly after periods of starvation, non-usage, sudden disturbance or toxicity presence. Recovery time is determined by the nature of the compounds or periods without fuel, anodic biofilm and operational conditions in the sensor. Bacteria in the biofilm that are able to recover quickly from any stress or damage caused by such periods demonstrate resilience, and this may lengthen the operational stability of the biosensor in which they are utilized. A biosensor with long operational stability and low maintenance provides increased reliability for water or wastewater monitoring. Recovery time has been shown to increase with prolonged periods of starvation, but this can be improved by electrode modification (Chang et al. 2004; Kaur et al. 2014). Whereas low resistance and increased concentrations of the analyte can lengthen the time it takes for a biosensor to recover (Pasternak et al. 2017).

4. Environmental Monitoring

Organic compounds are the primary pollutants in wastewaters and are difficult to characterize, hence analytical techniques such as biological oxygen demand (BOD) and chemical oxygen demand (COD) are used to monitor the amount of organic matter in wastewater. BOD is a measure of the oxygen consumed from biological degradation of organic pollution in water and is widely regulated for assessment of water quality, while COD is a measure of the amount of oxygen required for complete chemical oxidation of organic matter to carbon dioxide. As previously mentioned, the current produced in the MFC biosensor is proportional to the concentration of the organic substrate. By measuring and calibrating this current, the amount of organic content in water or wastewater can be estimated.

Existing literature shows that MFCs have been widely validated against chemical oxygen demand (COD) or biological oxygen demand (BOD) sensors for use in monitoring of water quality (Ayyaru and Dharmalingam 2014; Chang et al. 2004; Di Lorenzo et al. 2009a; Di Lorenzo et al. 2009b; Di Lorenzo et al. 2014; Kang et al. 2003; Karube et al. 1977; Kim, Chang et al. 2003; Kim, Youn et al. 2003; Kumlanghan et al. 2007; Liu et al. 2014; Zhang and Angelidaki 2011; Pasternak et al. 2017). A recent review of MFC biosensors for environmental monitoring described the use of self-powered dualchambered MFC designs for monitoring biological oxygen demand (BOD) and toxicity with response times of five minutes to ten hours and high stability of up to five years in one instance (Sun et al. 2015). The electricity generation was directly related to the concentration of organic matter or toxins in the wastewater, whereas the presence of organic matter favored increased current generation, the introduction of toxins caused a decrease in the current generation. MFCs have been used as volatile fatty acid (VFA) sensors to monitor biological processes (e.g. liquid waste) in anaerobic digestion during wastewater treatment, as VFA accumulation inhibits microbial activity, that decreases the efficiency of COD removal and can cause system failure (Kaur et al. 2013; Kretzschmar et al. 2017). The most common VFAs analyzed are acetate, propionate and butyrate. MFC biosensors used to monitor toxicity in water rely on the metabolic activity of the biofilm to sense the presence of toxicants. The introduction of a toxic substance into the MFC reactor causes an inhibition in the metabolic activity of the electroactive bacteria, resulting in a decline in signal or current output. The performance indicators for toxicity biosensors are current density and power output (Sun et al. 2015; Stein et al. 2012c).

BOD and toxicity sensors are the most widely studied MFC biosensors for water quality monitoring (Table 1), utilizing mostly synthetic media containing acetate, glucose and/or glutamic acid as the fuel source. Continuous operation, which resembles the closest approach to the real-world application for real-time monitoring, has been the predominant feeding mode. Although response time (and method for calculating such) varied between the biosensors, the majority recorded response times within 1 hour (confirming the suitability of MFC biosensors for rapid water and wastewater monitoring) compared to the standard offline tests including BOD that requires five days sample incubation and COD that requires two hours heating and 30 minutes cooling of samples. Response time could be improved by reducing the anodic volume or employing high substrate concentrations at low feeding rates (Moon et al. 2004). A decrease in current density following the introduction of known concentrations of a toxicant (Nickel, Ni) was also observed in a microbial electrolysis cell (MEC) biosensor using different ion exchange membranes that suggested microbial activity was inhibited by the presence of chemical toxicants/ pollutants (Stein et al. 2012a). Addition of toxic substances such as organophosphorus compounds and mercury and cadmium decreased current generation by inhibiting electron transfer mechanism of electroactive microorganisms (Kim, Hyun et al. 2007). Biosensor response signals also decreased after the introduction of the chromium or iron (Liu et al. 2014) but increased when using nitrate or acetate. The magnitude of the response was linearly correlated to the concentration of the tested analytes, with a distinguishing voltage signal that changes between toxic and non-toxic analytes.

The double chamber is a common configuration used for MFC biosensors, and some studies utilize a single chamber MFC that enables low maintenance requirements. A singlechamber MFC was used to monitor COD removal and VFA concentrations in four types of industrial wastewater (Velasquez-Orta et al. 2011). Submersible MFCs were also tested to monitor changes within an activated sludge tank or in groundwater, giving an indication of microbial activity or organic matter loads (Zhang and Angelidaki 2011; Xu et al. 2014). More recently, various modifications of sediment MFC biosensors (sediment/bulk liquid, sediment/sediment, bulk liquid/air and bulk liquid/bulk liquid) have been investigated for *in situ* monitoring of crude fecal contamination in groundwater (Velasquez-Orta et al. 2017). Here, a cathodic electrode used as the sensing element was exposed to the analyte (water) producing a decrease in current output after fecal contamination.

5. Challenges

Limitations of the performance of MFC biosensors include substrate concentration, high internal resistance of the system, diffusion of oxygen into the anode chamber, the presence of alternate terminal electron acceptors at the anode, proton permeability across the PEM, oxygen supply and consumption in the cathode chamber as well as the interference of environmental factors, such as temperature, pH and ionic conductivity of the electrolyte used (Kim, Hyun et al. 2007; Larrosa-Guerrero et al. 2010; Li et al.

Table 1: Microbial fuel cell-based biosensors for water and wastewater monitoring

S/N	MFC configuration	Electrode material Anode	Cathode	Substrate	Analyte	Mode of operation	Current output (mA)	Response time (minutes)	Detection limit/range (mg L^{-1})	Stability	Reference
1	Double chamber with AEM	Lead	Carbon with platinum catalyst	Glucose and glutamic acid	BOD	Batch	ND	30-40	14-190	40 days	(Karube et al. 1977)
2	Double chamber using CEM	Graphite felt	Graphite felt	Wastewater from starch processing plant	COD	Batch	1.7	ND	50	3 yrs	(Gil et al. 2003)
3	Double chamber using CEM	Graphite felt	Graphite felt	Wastewater from starch processing plant	BOD	Batch	1.1*	30-60	2.5-206	5 yrs	(Kim, Chang, et al. 2003)
4	Double chamber using CEM	Graphite felt	Graphite felt	Glucose and glutamic acid	BOD	Continuous	ND	30	91-142	> 1 yr	(Kim, Youn, et al. 2003)
5	Double chamber using CEM	Graphite felt	Platinum-coated graphite felt	Glucose and glutamate in surface water and artificial wastewater	BOD	Continuous	0.01-0.02	ND	2-6	> 8 weeks	(Kang et al. 2003)
6	Double chamber using CEM	Graphite felt	Graphite felt	Glucose and glutamic acid	BOD	Continuous	3.7-5.2	60	20-200	> 1 yr	(Chang et al. 2004)
7	Double chamber using CEM	Graphite felt	Graphite felt	Glucose and glutamic acid	BOD	Continuous	1.9*	5-36	50-100	> 2 yrs	(Moon et al. 2004)
8	Double chamber using CEM	Graphite felt	Graphite felt	Glucose and glutamic acid	Mercury, cadmium, lead, PCBs (polychlorinated biphenyl), organophosphorus compound (diazinon solution in acetonitrite).	Continuous	0.026-0.040	20-120	1-10 (OrgP) 0.01-1 (Cd) 0.1-1 (Pb)	1 year	(Kim, Hyun et al. 2007)

(Contd.)

9	Double chamber (using PEM) coupled an anaerobic reactor	Graphite rod	Graphite roll	Glucose	BOD	Batch	0.0013	3-5	25	ND	(Kumlanghan et al. 2007)
10	Double chamber using CEM, AEM, monovalent cation selective membrane, bipolar membrane	Graphite	Graphite	Synthetic wastewater, Acetate	Nickel	Continuous	ND	120	13.2 – 187.6	ND	(Stein et al. 2012a; ter Heijne et al. 2008)
11	Single chamber	Graphite rod	Carbon fibre paper with carbon nanoparticles	Acetate	BOD	Batch	0.02-1	300 - 1200	32-1280	ND	(Modin and Wilén 2012)
12	Single chamber	Graphite granules	Carbon supported with platinum catalyst	Glucose, Artificial water	COD	Continuous	0.01-0.09	40	50-500	> 7 months	(Di Lorenzo et al. 2009b) (Di Lorenzo et al. 2009a)
13	Double chamber (submerged) using CEM	Toray carbon paper	Toray carbon paper/ Platinum (Pt)	Acetate, glucose, wastewater	Microbial activity, BOD, COD	Batch	0.1	40	10-250	> 5 months	(Zhang and Angelidaki 2011)
14	Single chamber (submerged) using CEM	Toray carbon paper	Wet proof carbon paper/ Platinum (Pt)	Domestic wastewater	BOD	Batch	0.2	30-60	17-183	ND	(Peixoto et al. 2011)
15	Single chamber	Carbon felt	Carbon cloth	Glucose and glutamic acid	BOD	Continuous	0.00047[#]	132	5-120	ND	(Yang et al. 2013)
16	Single chamber with air cathode using PEM	Carbon cloth	Carbon cloth	Potassium acetate	COD, cadmium (Cd)	Continuous	ND	2.8	3-164 (COD) 0.0001 – 0.1 (Cd)	ND	(Di Lorenzo et al. 2014)

(Contd.)

Table 1. (*Contd.*)

S/N	MFC configuration	Electrode material		Substrate	Analyte	Mode of operation	Current output (mA)	Response time (minutes)	Detection limit/range (mg L^{-1})	Stability	Reference
		Anode	Cathode								
17	Single chamber with air cathode	Carbon cloth	Platinum coated carbon cloth	ND	Chromium, iron, nitrate, sodium acetate	Batch	0.00023#	35 – 40	1-8 (Cr) 1-18 (Fe, NO$_3$)	ND	(Liu et al. 2014)
18	Single chamber with PEM	Carbon cloth	Carbon cloth coated with Pt	Artificial wastewater, Glucose	BOD	Batch	0.88	80	50-1000	ND	(Ayyaru and Dharmalingam 2014)
19	Double chamber with CEM	Granular carbon	K3[Fe(CN)6] / graphite felt	Acetate	COD	Continuous	21.6	120	200	ND	(Wu et al. 2015)
20	Double chamber	ND	ND	Glucose Methionine Acetate Glycerol	BOD	Continuous	0.0196	10-15	235	ND	(Hsieh et al. 2016)
21	Double chamber with CEM	Polished graphite rods	Polished graphite rods	Glucose and glutamic acid	BOD	Batch	0.031	60-1260	50-250	60 days	(Anam et al. 2017)
22	4 single chambers connected in parallel	Carbon fibre veil	Carbon fibre veil	Human urine	COD	Batch	18.1	3	15 - 150	5 months	(Pasternak et al. 2017)
23	3-stage Single chamber with CEM	Carbon cloth	Carbon with platinum catalyst	Glucose and glutamic acid	BOD	Continuous	0.58	150	25-750	2 years	(Spurr et al. 2018)

CEM. cation exchange membrane; PEM. proton exchange membrane; AEM. anion exchange membrane; ND. No data available;* as reported in (Peixoto et al. 2011);
Calculated from available data.

2017; Schneider et al. 2016; Gil et al. 2003). Other challenges of MFC biosensors include specificity, sensitivity, standardization and microorganisms used for the anodic biofilm and scalability for mass production.

5.1 Reactor Configuration

The internal resistance of an MFC biosensor depends on the design and configuration of the system and in turn, determines its performance. MFC performance can be enhanced by reducing the distance between electrodes and by using miniature reactor sizes (Ringeisen et al. 2006). A VFA MFC sensor using polypyrrole-modified carbon electrode as the working electrode favored bacterial attachment to the electrode surface and improved electron transfer rate, and the recovery time of the sensor was enhanced (<2-10 mins) when compared with the unmodified electrodes (10-30 minutes) (Kretzschmar et al. 2017). The use of natural polymers, such as agarose and polyacrylamide, in the presence of mediators also increased the start-up time and stability of the sensors (Kaur et al. 2014).

For electroactive bacteria to proliferate, the anode must be maintained under anaerobic conditions. Oxygen diffusion into the anode chamber leads to loss of the organic substrate through aerobic respiration and results in low MFC performance. Biosensor designs that minimize oxygen diffusion into the anodic chamber and lower the density of anodic biofilms have been shown to improve sensitivity. Lowering the flow or the shear rate also increases the sensitivity of the biosensor (Chang et al. 2004; Shen et al. 2013). The use of a sulfonated poly ether membrane, which prevented oxygen diffusion, was reported as being more effective than Nafion™, allowing detection of up to 650 ppm glucose concentration (Ayyaru and Dharmalingam 2014). However, its applicability in MFC biosensors for BOD measurements still needs to be determined. Zhang and Angelidaki (2011) demonstrated the use of a submersible MFC biosensor for monitoring BOD in groundwater where the anaerobic conditions were maintained in the anode by immersing it in a subsurface environment. The compact reactor design also minimized ohmic losses within the system. Although using a membraneless configuration promoted proton diffusion into the cathode, this did not improve performance as the process was inhibited by high concentrations of cations (Kim, Chang et al. 2007). It has also been reported that membraneless systems are advantageous because internal resistance is reduced, and no pH gradient is formed between the anode and the cathode. Nevertheless, such configurations are susceptible to biofouling of the cathode that reduces system performance (Logan 2010). Electrode and membrane fouling has reportedly led to diminished MFC performance after more than six months of operation (Kim, Youn et al. 2003).

From the above, it can be seen that the internal resistance is a major limitation in the operation and performance of an MFC biosensor. The lower the internal resistance of the system, the higher is its performance. The internal resistance of the reactor can be reduced by optimizing the configuration either by reducing the distance between electrodes or by using membranes or separators that permit the diffusion of protons to the cathode chamber while preventing the influx of oxygen into the anode chamber. It is also important to select electrodes with surface characteristics that enhance electron transfer to provide adequate support for the bacterial biofilm, and the cathode and membrane materials must be such that enhances oxygen reduction and proton diffusion while preventing its diffusion into the anode chamber, respectively. Although single chamber configurations are simpler and cheaper, oxygen diffusion into the anode and membrane fouling are ongoing challenges with this design. Not using a reactor, as proposed in sediment MFCs, results in a system difficult to control. In such systems, the interpretation of the current output would need the measurement of other variables.

5.2 Operational Conditions

Operating conditions such as pH, temperature and conductivity as well as the redox potentials of the electrodes and the feeding mode impact the performance of MFC biosensors. A change in temperature can influence both the reaction kinetics and the thermodynamics of an MFC biosensor. Temperature significantly affects the metabolism of the bacteria, which is one of the most important factors that affect performance, in addition to electrode potentials, activation energy, mass transfer process and conductivity of the electrolyte. Among all these, changes are seen in the current generation and BOD or COD removal of the biosensor. For instance, increase in temperature also increases COD and VFA removal as well as conductivity in a single chamber MFC biosensor that is used to monitor COD (Larrosa-Guerrero et al. 2010; Oliveira et al. 2013). Current output increases with increasing temperature and conductivity (Zhang and Angelidaki 2011). At pH values above 8, current output reduced that suggested microbial metabolism is influenced by the pH (Yang et al. 2013). The current output was, therefore, diminished under this condition. The most suitable pH range for MFCs has been reported as 6–8. The pH changes can be controlled by using phosphate, bicarbonate, borax or synthetic zwitterionic buffers (Oliveira et al. 2013; Gil et al. 2003). These examples indicate that acidic or very alkaline conditions in the anodic chamber have a negative impact on biofilm stability and electron transfer. The effect of environmental parameters on the performance of modified MFCs, used to monitor fecal contamination in groundwater, showed a decline in dissolved oxygen concentration and an increase in current output after a contamination event. Current output declined with an increase in temperature (Velasquez-Orta et al. 2017). Current output was, however, low when these MFCs were tested in real groundwater well. The difference in response was attributed to the different soil, water and microbial characteristics of the actual groundwater well when compared with the laboratory tests. This example illustrates how unpredictable environmental conditions may affect biosensor performance *in situ*. Further research is required to determine MFC performance with real wastewaters and design MFC biosensors that can measure response signal without the interference of environmental factors.

In a continuous mode single chambered MFC, used to monitor copper toxicity, Shen et al. (2013) showed that decreasing the flow rate of the reactor and maintaining an anoxic environment in the anodic compartment improved the biofilm density and enhanced its sensitivity. Nevertheless, Moon et al. (2004) reported that lowering the feeding rate improve response time but did not improve sensitivity; to improve sensitivity, the feeding was increased from 0.053 ml min^{-1} to 0.65 ml min^{-1}. This shows that improving one factor or indicator does not necessarily result in an overall improvement in performance. The key to an efficient biosensor is establishing a combination of conditions that allow the biosensor to perform at its best without compromising its sensitivity or stability. This is an example of some of the complexities required to optimize the performance of an MFC biosensor.

5.3 Microbial Consortium

When using mixed cultures, a diversity of microorganisms are enriched in microbial fuel cells most of which, if not all, are electroactive bacteria (Du et al. 2007; Kim et al. 2006; Logan 2008; Santoro et al. 2017). As earlier discussed, the electron transfer mechanism and proton diffusion across the anodic chamber can hinder efficient current generation. MFC performance is dependent on the electron transfer mechanism of the microbial biofilm in the anodic chamber; however, this mechanism has not been thoroughly understood. (Li et al. 2017; Schneider et al. 2016; Sun et al. 2015; Velasquez-Orta et al. 2017). For instance, although mixed cultures produced higher current output, when isolates from this consortium were used as pure culture they generated lower current than expected (Logan and Regan 2006). Most microorganisms in MFC reactors are unable to thrive in severe weather (e.g. low temperatures) or ionic conditions (e.g. starvation). There is a need to develop biosensors using bacteria that can flourish in extreme environments (Lei et al. 2006).

5.4 Standardization

Even though MFC biosensors can detect the presence of toxic substances, the quantification of these substances is still a challenge as the sensor only measures the signal response to the change caused by the toxic substance. In recent times, calibration curves have been produced to establish a relationship between the current output and the concentration of the toxic substance (Jiang et al. 2018). While this may be acceptable for laboratory investigations or use as early-warning detection systems, but *in situ* or online monitoring would require devices that can provide information on the precise amount of the target analyte present in the water being analyzed. Various configurations, conditions and methods of measurements have been employed to characterize MFC biosensors, hence it can sometimes be difficult to make a comprehensive comparison of the performance of these sensors. Although most literature uses current, current density or power density to describe MFC performance, this may not be suitable for MFC biosensors where the key focus is detection or monitoring of analyses rather than power generation. A common platform or standard would provide a means of establishing the specific minimum requirements of an efficient biosensor. In addition, the use of different synthetic calibrants and variation between validation methods including BOD$_5$, COD, DOC and substrate concentration make the comparison of sensing ranges difficult.

5.5 Specificity

This is also another area that is not clear. So far, the investigation of MFC biosensors for water and wastewater monitoring has predominantly been conducted in the laboratory with few field trials. Laboratory experiments are conducted under highly controlled conditions and measure very specific (mostly single) parameters to provide clarity on the reactions occurring and the influencing factors. In reality, the interaction of biotic and abiotic environmental factors with biosensor performance involves more complex reactions. MFC modeling can be used to design biosensors that are adaptable to these interactions. With the use of fixed anode potential for measurement of the response signal, the use of different anode potentials for sensing different compounds has been proposed. This concept requires further investigation.

6. Conclusions

MFC biosensors can be used for rapid monitoring of BOD and COD during anaerobic digestion in wastewater treatment plants. They can also be employed as toxicity sensors for monitoring chemical compounds such as nickel, cadmium, chromium and organophosphate compounds in water with the capacity to measure concentrations as low as 1mg L^{-1}. Limitations of the performance of MFC biosensors include electron transfer from the biofilm to the anode, substrate concentration, internal resistance of the system, proton permeability across the PEM, oxygen supply and consumption in the cathode chamber as well as the interference of environmental factors such as temperature, pH and ionic conductivity of the electrolyte. Practical applications have demonstrated that of modification of reactor design, configuration and materials are usually required to manage these limitations. There is indeed a plethora of practical evidence of the rapid, sensitive and *in situ* capabilities of MFC biosensors. However, a wide range of configurations and methods have been employed for these investigations, making it difficult to establish basic standards against which an efficient biosensor can be measured. Nevertheless, its usefulness for real-time water and wastewater monitoring is undeniable. Further investigations need to be conducted to develop biosensors using bacteria that can flourish in extreme environments and utilize a wider range of substrates as a fuel source as well as compact miniaturized MFC biosensors for field applications. Reactor designs that minimize internal resistance while improving response time, specificity and sensitivity are the focus of ongoing research efforts. A common platform or standard would provide a means of establishing the specific minimum requirements of an efficient biosensor.

7. Future Perspectives

Most biosensors, especially BOD and COD, are only able to detect overall response such as total organic carbon without distinguishing between the various forms of carbon present. Although recent studies have explored the development of sensors for target contaminants, the ability of a biosensor to detect a single analyte in a complex matrix is still a progressive area of research for MFC biosensors. A foremost concern for MFC biosensors is scalability, and membrane electrode assembly consisting of arrays of MFCs are more preferable for boosting performance. Simpler and cost-effective material remains the sensible choice for scaling up reactors. Genetic engineering of target genes of electroactive bacteria may be further explored as a means of enhancing the electrogenic activities of the microbial consortium. Although this is not the main focus of these biosensors, power management systems are also being developed for storing the energy generated by MFC in order to put it to other relevant uses, such as powering remote sensors or small lighting devices.

The promising potentials of MFC biosensors for water and wastewater monitoring as a low cost and low energy solution to water management problems will continue to drive further refinement of this technology. MFC biosensors could also be integrated with wastewater management treatment plants to monitor the effectiveness of the treatment process.

Acknowledgements

Dr S B Velasquez Orta appreciates the funds provided by the Natural Environment Research Council, the Economic and Social Research Council, and UKaid [grant number NE/L002108/1] to conduct research in MFC biosensors.

References

Abrevaya, Ximena C, Natalia J Sacco, Maria C Bonetto, Astrid Hilding-Ohlsson, Eduardo Cortón (2015) Analytical applications of microbial fuel cells. Part I: Biochemical oxygen demand. Biosens Bioelectron 63: 580-590.

Anam, Maira, Sameen Yousaf, Iqra Sharafat, Zargona Zafar, Kamran Ayaz, Naeem Ali (2017) Comparing natural and artificially designed bacterial consortia as biosensing elements for rapid non-specific detection of organic pollutant through microbial fuel cell. Int J Electrochem Sci 12: 2836-2851.

APHA. 2005. Biochemical Oxygen Demand (BOD) *In*: Standard Methods for the Examination of Water and Wastewater, edited by L.S.C.A.D. Eaton, E. W. Rice and A. E. Greenberg. American Public Health Association. Washington. 5(2-5): 13.

Ayyaru, Sivasankaran, Sangeetha Dharmalingam (2014) Enhanced response of microbial fuel cell using sulfonated poly ether ether ketone membrane as a biochemical oxygen demand sensor. Anal Chim Acta 818: 15-22.

Bard, Allen J., György Inzelt, Fritz Scholz (2008) Electrochemical Dictionary. Springer Science & Business Media. 991 pages. https://doi.org/10.1007/978-3-642-29551-5_7,

Chang, In Seop, Jae Kyung Jang, Geun Cheol Gil, Mia Kim, Hyung Joo Kim, Byung Won Cho, Byung Hong Kim (2004) Continuous determination of biochemical oxygen demand using microbial fuel cell type biosensor. Biosens Bioelectron 19(6): 607-613.

Chang, In Seop, Hyunsoo Moon, Jae Kyung Jang, Byung Hong Kim (2005) Improvement of a microbial fuel cell performance as a BOD sensor using respiratory inhibitors. Biosens Bioelectron 20(9): 1856-1859.

Chee, Gab-Joo, Yoko Nomura, Isao Karube (1999) Biosensor for the estimation of low biochemical oxygen demand. Anal Chim Acta 379(1): 185-191.

Chen, Guo-Wei, Soo-Jung Choi, Tae-Ho Lee, Gil-Young Lee, Jae-Hwan Cha, Chang-Won Kim (2008) Application of biocathode in microbial fuel cells: Cell performance and microbial community. Appl Microbiol Biotechnol 79(3): 379-388.

Dávila, D, JP Esquivel, N Sabaté, J Mas (2011) Silicon-based microfabricated microbial fuel cell toxicity sensor. Biosens Bioelectron 26(5): 2426-2430.

Di Lorenzo, Mirella, Tom P Curtis, Ian M Head, Sharon B Velasquez-Orta, Keith Scott (2009a) A single chamber packed bed microbial fuel cell biosensor for measuring organic content of wastewater. Water Sci Technol 60(11): 2879-2887.

Di Lorenzo, Mirella, Tom P Curtis, Ian M Head, Keith Scott (2009b) A single-chamber microbial fuel cell as a biosensor for wastewaters. Water Res 43(13): 3145-3154.

Di Lorenzo, Mirella, Alexander R Thomson, Kenneth Schneider, Petra J Cameron, Ioannis Ieropoulos (2014) A small-scale air-cathode microbial fuel cell for on-line monitoring of water quality. Biosens Bioelectron 62: 182-188.

Du, Zhuwei, Haoran Li, Tingyue Gu (2007) A state of the art review on microbial fuel cells: A promising technology for wastewater treatment and bioenergy. Biotechnol Adv 25(5): 464-482.

Esfandyari, Morteza, Mohmmad Ali Fanaei, Reza Gheshlaghi, Mahmood Akhavan Mahdavi (2017) Mathematical modeling of two-chamber batch microbial fuel cell with pure culture of *Shewanella*. Chem Eng Res Des 117: 34-42.

Gil, Geun-Cheol, In-Seop Chang, Byung Hong Kim, Mia Kim, Jae-Kyung Jang, Hyung Soo Park, Hyung Joo Kim (2003) Operational parameters affecting the performance of a mediator-less microbial fuel cell. Biosens Bioelectron 18(4): 327-334.

Hsieh, Min-Chi, Chiu-Yu Cheng, Man-Hai Liu, Ying-Chien Chung (2016) Effects of operating parameters on measurements of biochemical oxygen demand using a mediatorless microbial fuel cell biosensor. Sensors 16(1): 35.

Jiang, Yong, Xufei Yang, Peng Liang, Panpan Liu, Xia Huang (2018) Microbial fuel cell sensors for water quality early warning systems: Fundamentals, signal resolution, optimization and future challenges. Renew Sust Energ Rev 81: 292-305.

Kang, Kui Hyun, Jae Kyung Jang, The Hai Pham, Hyunsoo Moon, In Seop Chang, Byung Hong Kim (2003) A microbial fuel cell with improved cathode reaction as a low biochemical oxygen demand sensor. Biotechnol Lett 25(16): 1357-1361.

Karube, Isao, Tadashi Matsunaga, Satoshi Mitsuda, Shuichi Suzuki (1977) Microbial electrode BOD sensors. Biotechnol Bioeng 19(10): 1535-1547.

Kaur, Amandeep, Jung Rae Kim, Iain Michie, Richard M Dinsdale, Alan J Guwy, Giuliano C Premier (2013) Microbial fuel cell type biosensor for specific volatile fatty acids using acclimated bacterial communities. Biosens Bioelectron 47: 50-55.

Kaur, Amandeep, Saad Ibrahim, Christopher J Pickett, Iain S Michie, Richard M Dinsdale, Alan J Guwy, Guiliano C Premier (2014) Anode modification to improve the performance of a microbial fuel cell volatile fatty acid biosensor. Sens Actuators, B 201: 266-273.

Kim, Byung Hong, In Seop Chang, Geun Cheol Gil, Hyung Soo Park, and Hyung Joo Kim (2003) Novel BOD (biological oxygen demand) sensor using mediator-less microbial fuel cell. Biotechnol Lett 25(7): 541-545.

Kim, Byung Hong, In Seop Chang, Hyunsoo Moon (2006) Microbial fuel cell-type biochemical oxygen demand sensor. Encycl Sens 10: 1-12.

Kim, Byung Hong, In Seop Chang, Geoffrey M Gadd (2007a) Challenges in microbial fuel cell development and operation. Appl Microbiol Biotechnol 76(3): 485.

Kim, HJ, MS Hyun, IS Chang, BH Kim (1999a) A microbial fuel cell type lactate biosensor using a metal-reducing bacterium, *Shewanella putrefaciens*. J Microbiol Biotechnol 9(3): 365-367.

Kim, Hyung Joo, Hyung Soo Park, Moon Sik Hyun, In Seop Chang, Mia Kim, Byung Hong Kim (2002) A mediator-less microbial fuel cell using a metal reducing bacterium, *Shewanella putrefaciens*. Enzyme Microb Technol 30(2): 145-152.

Kim, Mal-Nam, Hee-Sun Kwon (1999b) Biochemical oxygen demand sensor using *Serratia marcescens* LSY 4. Biosens Bioelectron 14(1): 1-7.

Kim, Mia, Su Mi Youn, Sung Hye Shin, Ji Gu Jang, Seol Hee Han, Moon Sik Hyun, Geoffrey M Gadd, Hyung Joo Kim (2003b) Practical field application of a novel BOD monitoring system. J Environ Monit 5(4): 640-643.

Kim, Mia, Moon Sik Hyun, Geoffrey M. Gadd, Hyung Joo Kim (2007b) A novel biomonitoring system using microbial fuel cells. J Environ Monit 9(12): 1323-1328.

Kretzschmar, Jörg, Christin Koch, Jan Liebetrau, Michael Mertig, Falk Harnisch (2017) Electroactive biofilms as sensor for volatile fatty acids: Cross sensitivity, response dynamics, latency and stability. Sens Actuators, B 241: 466-472.

Kumlanghan, Ampai, Jing Liu, Panote Thavarungkul, Proespichaya Kanatharana, Bo Mattiasson (2007) Microbial fuel cell-based biosensor for fast analysis of biodegradable organic matter. Biosens Bioelectron 22(12): 2939-2944.

Larrosa-Guerrero, A, K Scott, IM Head, F Mateo, A Ginesta, C Godinez (2010) Effect of temperature on the performance of microbial fuel cells. Fuel 89(12): 3985-3994.

Lei, Yu, Wilfred Chen, Ashok Mulchandani (2006) Microbial biosensors. Anal Chim Acta 568(1–2): 200-210.

Li, Xiaojing, Xin Wang, Liping Weng, Qixing Zhou, Yongtao Li (2017) Microbial fuel cells for organic-contaminated soil remedial applications: A Review. Energ Tech 5(8): 1156-1164.

Liu, Bingchuan, Yu Lei, Baikun Li (2014) A batch-mode cube microbial fuel cell based "shock" biosensor for wastewater quality monitoring. Biosens Bioelectron 62: 308-314.

Liu, Jing, Lovisa Björnsson, Bo Mattiasson (2000) Immobilised activated sludge based biosensor for biochemical oxygen demand measurement. Biosens Bioelectron 14(12): 883-893.

Logan, Bruce E, John M Regan (2006) Microbial Fuel Cells—Challenges and Applications. Environ Sci Technol 40(17): 5172-5180.

Logan, Bruce E, Bert Hamelers, René Rozendal, Uwe Schröder, Jürg Keller, Stefano Freguia, Peter Alterman, Willy Verstraete, Korneel Rabaey (2006a) Microbial fuel cells: Methodology and technology. Environ Sci Technol 40(17): 5181-5192.

Logan, Bruce E (2008) Microbial Fuel Cells. John Wiley & Sons. New Jersey, USA.

Logan, Bruce E (2010) Scaling up microbial fuel cells and other bioelectrochemical systems. Appl Microbiol Biotechnol 85(6): 1665-1671.

Lovley, Derek R (2006) Microbial fuel cells: novel microbial physiologies and engineering approaches. Curr Opin Biotechnol 17(3): 327-332.

Matsunaga, Tadashi, Isao Karube, Shuichi Suzuki (1980) A specific microbial sensor for formic acid. Eur J Appl Microbiol Biotechnol 10(3): 235-243.

Modin, Oskar, Britt-Marie Wilén (2012) A novel bioelectrochemical BOD sensor operating with voltage input. Water Res 46(18): 6113-6120.

Moon, Hyunsoo, In Seop Chang, Kui Hyun Kang, Jae Kyung Jang, Byung Hong Kim (2004) Improving the dynamic response of a mediator-less microbial fuel cell as a biochemical oxygen demand (BOD) sensor. Biotechnol Lett 26(22): 1717-1721.

Oliveira, VB, M Simões, LF Melo, AMFR Pinto (2013) Overview on the developments of microbial fuel cells. Biochem Eng J 73 (Supplement C): 53-64.

Pant, Deepak, Gilbert Van Bogaert, Ludo Diels, Karolien Vanbroekhoven (2010) A review of the substrates used in microbial fuel cells (MFCs) for sustainable energy production. Bioresour Technol 101(6): 1533-1543.

Pasternak, Grzegorz, John Greenman, Ioannis Ieropoulos (2017) Self-powered, autonomous Biological Oxygen Demand biosensor for online water quality monitoring. Sens Actuators, B: Chemical 244 (Supplement C): 815-822.

Peixoto, Luciana, Booki Min, Gilberto Martins, Antonio G Brito, Pablo Kroff, Pier Parpot, Irini Angelidaki, Regina Nogueira (2011) *In situ* microbial fuel cell-based biosensor for organic carbon. Bioelectrochem 81(2): 99-103.

Philips, Jo, Kristof Verbeeck, Korneel Rabaey, JBA Arends (2016) Electron transfer mechanisms in biofilms. pp. 67-113. *In*: K Scott and E Yu (Eds). Microbial Electrochemical and Fuel Cells. Woodhead Publishing. Cambridge, UK.

Potter, Michael C (1911) Electrical effects accompanying the decomposition of organic compounds. Proc R Soc Lond B 84(571): 260-276.

Riedel, K, R Renneberg, M Kühn, F Scheller (1988) A fast estimation of biochemical oxygen demand using microbial sensors. Appl Microbiol Biotechnol 28(3): 316-318.

Ringeisen, Bradley R, Emily Henderson, Peter K. Wu, Jeremy Pietron, Ricky Ray, Brenda Little, Justin C Biffinger, Joanne M Jones-Meehan (2006) High power density from a miniature microbial fuel cell using *Shewanella oneidensis* DSP10. Environ Sci Technol 40(8): 2629-2634.

Santoro, C, C Arbizzani, B Erable, I Ieropoulos (2017) Microbial fuel cells: From fundamentals to applications. A review. J Power Sources 356: 225-244.

Schneider, György, Tamás Kovács, Gábor Rákhely, Miklós Czeller (2016) Biosensoric potential of microbial fuel cells. Appl Microbiol Biotechnol 100(16): 7001-7009.

Shen, Yujia, Meng Wang, In Seop Chang, How Yong Ng (2013) Effect of shear rate on the response of microbial fuel cell toxicity sensor to Cu (II). Bioresour Technol 136: 707-710.

Spurr, Martin WA, Eileen H Yu, Keith Scott, Ian M Head (2018) Extending the dynamic range of biochemical oxygen demand sensing with multi-stage microbial fuel cells. Environ Sci: Water Res Technol 4(12): 2029-2040.

Stein, Nienke Elisabeth, Hubertus MV Hamelers, Gerrit van Straten, Karel J Keesman (2012a) On-line detection of toxic components using a microbial fuel cell-based biosensor. J Process Control 22(9): 1755-1761.

Stein, Nienke E, Hubertus VM Hamelers, Cees NJ Buisman (2012b) Influence of membrane type, current and potential on the response to chemical toxicants of a microbial fuel cell based biosensor. Sens Actuators B: Chemical 163(1): 1-7.

Stein, Nienke E, Hubertus VM Hamelers, Cees NJ Buisman (2012c) The effect of different control mechanisms on the sensitivity and recovery time of a microbial fuel cell based biosensor. Sens Actuators B: Chemical 171: 816-821.

Su, Liang, Wenzhao Jia, Changjun Hou, Yu Lei (2011) Microbial biosensors: A review. Biosens Bioelectron 26(5): 1788-1799.

Sun, Jian-Zhong, Gakai Peter Kingori, Rong-Wei Si, Dan-Dan Zhai, Zhi-Hong Liao, De-Zhen Sun, Tao Zheng, Yang-Chun Yong (2015) Microbial fuel cell-based biosensors for environmental monitoring: A review. Water Sci Technol 71(6): 801.

Tan, TC, F Li, KG Neoh (1993) Measurement of BOD by initial rate of response of a microbial sensor. Sens Actuators, B: Chemical 10(2): 137-142.

ter Heijne, Annemiek, Hubertus VM Hamelers, Michel Saakes, Cees JN Buisman (2008) Performance of non-porous graphite and titanium-based anodes in microbial fuel cells. Electrochim Acta 53(18): 5697-5703.

Thurston, CF, HP Bennetto, GM Delaney (1985) Glucose metabolism in a microbial fuel cell. Stoichiometry of product formation in a thionine-mediated Proteus vulgaris fuel cell and its relation to coulombic yields. J Gen Microbiol 131(6): 1393-1401.

Velasquez-Orta, SB, IM. Head, TP Curtis, K Scott (2011) Factors affecting current production in microbial fuel cells using different industrial wastewaters. Bioresour Technol 102(8): 5105-5112.

Velasquez-Orta, SB, D Werner, JC Varia, S Mgana (2017) Microbial fuel cells for inexpensive continuous in-situ monitoring of groundwater quality. Water Res 117: 9-17.

Velasquez-Orta, Sharon B., Tom P. Curtis, and Bruce E. Logan (2009) Energy from algae using microbial fuel cells. Biotechnol Bioeng 103(6): 1068-1076.

Velasquez-Orta, Sharon B, Ian M Head, Thomas P Curtis, Keith Scott, Jonathan R Lloyd, Harald von Canstein (2010) The effect of flavin electron shuttles in microbial fuel cells current production. Appl Microbiol Biotechnol 85(5): 1373-1381.

Wang, Donglin, Peng Liang, Yong Jiang, Panpan Liu, Bo Miao, Wen Hao, Xia Huang (2018) Open external circuit for microbial fuel cell sensor to monitor the nitrate in aquatic environment. Biosens Bioelectron 111: 97-101.

Wu, Shijia, Peng Liang, Changyong Zhang, Hui Li, Kuichang Zuo, Xia Huang (2015) Enhanced performance of microbial fuel cell at low substrate concentrations by adsorptive anode. Electrochim Acta 161: 245-251.

Xu, Gui-Hua, Yun-Kun Wang, Guo-Ping Sheng, Yang Mu, Han-Qing Yu (2014) An MFC-Based Online Monitoring and Alert System for Activated Sludge Process. Sci Rep 4: 6779.

Yang, Gai-Xiu, Yong-Ming Sun, Xiao-Ying Kong, Feng Zhen, Ying Li, Lain-Hua Li, Ting-Zhou Lei, Zhen-Hong Yuan, Guan-Yi Chen (2013) Factors affecting the performance of a single-chamber microbial fuel cell-type biological oxygen demand sensor. Water Sci Technol 68(9): 1914-1919.

Zhang, Yifeng, Irini Angelidaki (2011) Submersible microbial fuel cell sensor for monitoring microbial activity and BOD in groundwater: Focusing on impact of anodic biofilm on sensor applicability. Biotechnol Bioeng 108(10): 2339-2347.

Bioelectrochemical Enhancement and Intensification of Methane Production from Anaerobic Digestion

Christy M. Dykstra[1,2] and Spyros G. Pavlostathis[2*]

[1] San Diego State University, San Diego, California, USA
[2] Georgia Institute of Technology, Atlanta, Georgia, USA

1. Introduction

As the world's energy demand continues to soar, renewable sources of carbon-neutral energy are needed to sustainably support population growth and a global increase in the standard of living. In order to achieve net zero energy goals, water resource recovery facilities (WRRFs) must be designed to take advantage of carbon recycling for energy recovery during wastewater treatment. Anaerobic digestion is used for waste volume reduction, and energy recovery from biogas rich in methane (CH_4), which is a carbon neutral fuel because its carbon does not originate from previously sequestered sources (i.e., fossil fuels). Degradable complex organics go through four stages during anaerobic digestion: hydrolysis, acidogenesis, acetogenesis and methanogenesis (Table 1). The resulting biogas, produced from anaerobic digestion, consists of approximately 30-40% carbon dioxide (CO_2) and 60-70% CH_4 (Petersson and Wellinger 2009), along with other trace gases (e.g., H_2S, N_2, H_2, etc.). Because of the relatively high CO_2 content of anaerobic digester biogas, energy recovery is often limited to specialized combined heat and power (CHP) equipment.

The ratio of CH_4 to CO_2 in biogas is a function of the mean oxidation state of the carbon in the feedstock to the anaerobic digester (Gujer and Zehnder 1983). Biogas upgrading is the process of increasing the biogas energy (i.e., CH_4 content) to allow biogas to be used in a far wider range of applications without requiring a CHP unit (Angelidaki et al. 2018; Pan et al. 2018; Verbeeck et al. 2018). It can allow for the direct use of biogas for energy recovery on-site (i.e., CH_4 powered vehicles) or the direct injection of CH_4 into the existing natural gas infrastructure if CH_4 content is > 96% (v/v) (Sun et al. 2015; Verbeeck et al. 2018).

Biogas upgrading may be accomplished within the digester (i.e., *in situ*) or after biogas removal from the digester (*ex situ*). One method of *in situ* biogas upgrading is biomethanation, in which exogenous hydrogen (H_2) is added to an anaerobic digester to promote hydrogenotrophic methanogenesis (Al-mashhadani et al. 2016; Luo et al. 2012; Salomoni et al. 2011). However, considerable challenges exist, and biomethanation is no longer considered suitable for use in industrial applications (Angenent et al. 2018). Methods of *ex situ* biogas upgrading include adsorption, absorption, membrane systems, cryogenics and algal biomass systems. However, these methods require expensive consumables (e.g.,

---

*Corresponding author: spyros.pavlostathis@ce.gatech.edu

adsorbent/absorbent material, membrane cartridges, etc.), large amounts of energy (i.e., cryogenics) or large areas (e.g., algae systems) (Angelidaki et al. 2018; Muñoz et al. 2015; Sun et al. 2015).

Table 1: Anaerobic digestion stages, reactions and microorganisms involved
(Deepanraj et al. 2014; Dykstra 2017)

Stage	Reactions	Microorganisms Involved
Stage I - Hydrolysis	$(C_6H_{10}O_5)_n + nH_2O = n(C_6H_{12}O_6)$	*Clostridium, Proteus, Vibrio, Bacillus, Peptococcus, Bacteriodes, Staphylococcus*
Stage II - Acidogenesis	$C_6H_{12}O_6 + 2H_2O \rightarrow 2CH_3COOH + 4H_2 + 2CO_2$ $C_6H_{12}O_6 + 2H_2 \rightarrow 2CH_3CH_2COOH + 2H_2O$ $C_6H_{12}O_6 \rightarrow CH_3CH_2CH_2COOH + 2H_2 + 2CO_2$ $C_6H_{12}O_6 \rightarrow 2CH_3CH_2OH + 2CO_2$ $C_6H_{12}O_6 \rightarrow 2CH_3CHOHCOOH$	*Lactobacillus, Escherichia, Bacillus, Staphylococcus, Pseudomonas, Sarcina, Desulfovibrio, Selenomonas, Streptococcus, Veollonella, Desulfobacter, Desulforomonas, Clostridium, Eubacterium*
Stage III - Acetogenesis	$CH_3CH_2OH + H_2O \rightarrow CH_3COOH + 2H_2$ $2CH_3CH_2OH + CO_2 \rightarrow CH_4 + 2CH_3COOH$ $CH_3CH_2COOH + 2H_2O \rightarrow CH_3COOH + 3H_2 + CO_2$ $CH_3CH_2CH_2COOH + 2H_2O \rightarrow 2CH_3COOH + 2H_2$ $CH_3CHOHCOOH + H_2O \rightarrow CH_3COOH + CO_2 + 2H_2$	*Clostridium, Syntrophomonas*
Stage IV - Methanogenesis	$CH_3COOH \rightarrow CH_4 + CO_2$ $CO_2 + 4H_2 \rightarrow CH_4 + 2H_2O$	*Methanobacterium, Methanobrevibacter, Methanoplanus* and *Methanospirillum*

Bioelectrochemical systems (BESs) may be used to enhance or intensify the production of CH_4 from anaerobic digestion, without expensive materials, or large amounts of energy. In typical anaerobic digestion, the production of an oxidized form of carbon (CO_2) along with a reduced form of carbon (CH_4) means there is an opportunity for further reduction of the CO_2 to CH_4, if enough electron equivalents and a catalyst are provided. In methanogenic BESs, an external source supplies electron equivalents to the biocathode, where microorganisms act as self-renewing catalysts for the reduction of CO_2 to CH_4. Bioanodes may be used to generate electron equivalents, reducing the requirement for external energy supply.

Many processes and parameters must be taken into account when developing a methanogenic BES. Several recent reviews have summarized materials and conditions (e.g., electrode type, voltage, methanogenic community, membrane type, temperature, etc.) used in previous methanogenic BES studies (Nelabhotla and Dinamarca 2018; Gadkari et al. 2018; Geppert et al. 2016). However, a more focused approach is needed to evaluate the practical design and development considerations for the upgrade of anaerobic digester biogas, using *in situ* and *ex situ* BESs. Thus, the objective of this chapter is to not only introduce methanogenic BES fundamentals, processes and microorganisms but to also discuss various practical aspects of methanogenic BES design, setup and operation (e.g., techniques for CO_2 delivery, electrode configurations, applied potential, catholyte recirculation, etc.), with a focus on the application of BES technology to anaerobic digester biogas upgrading.

2. Fundamentals of Methanogenesis

Methanogens are Archaea that carry out the process of methanogenesis under anaerobic conditions. They belong to five orders: *Methanobacteriales, Methanococcales, Methanomicrobiales, Methanosarcinales* and the unique hyperthermophile, *Methanopyrales* (Boone and Garrity 2001; Ferry 1993). Methanogens may also be divided according to their methanogenic pathway: acetoclastic (Equation 1, Table 2), methylotrophic (Equation 2) or hydrogenotrophic (Equation 3). Acetoclastic methanogenesis converts acetate into CO_2 and CH_4 and methylotrophic methanogenesis converts methylated compounds, such as methanol or methylated amines, into CO_2 and CH_4 (Ferry 1993). Hydrogenotrophic methanogenesis, which is of particular interest for bioelectrochemical systems, converts CO_2 and hydrogen (H_2) into CH_4. Some hydrogenotrophic methanogens are capable of directly accepting electrons from a solid surface, enabling the reduction of CO_2 to CH_4 without H_2 as an electron donor (Equation 4) (Lohner et al. 2014). As shown in Table 2, the standard redox potential of the hydrogenotrophic methanogenesis reaction (Equation 3) is higher than that of the acetoclastic (Equation 1) or methylotrophic (Equation 2) methanogenesis reactions. Thus, if H_2 and CO_2 are abundant, the hydrogenotrophic reaction becomes more favorable than acetoclastic or methylotrophic methanogenesis.

Table 2: Reactions and standard redox potentials ($E_H^{\circ\prime}$ or $\Delta E^{\circ\prime}$)[a]

No.	Reaction	$E_H^{\circ\prime}$ or $\Delta E^{\circ\prime}$ (V)
1	$CH_3COO^- + H^+ \rightarrow CH_4 + CO_2$	0.12
2	$4CH_3OH \rightarrow 3CH_4 + CO_2 + 2H_2O$	0.14
3	$CO_2 + 4H_2 \rightarrow CH_4 + 2H_2O$	0.17
4	$CO_2 + 8H^+ + 8e^- \rightarrow CH_4 + 2H_2O$	-0.24
5	$2H^+ + 2e^- \rightarrow H_2$	-0.41

[a] Redox potentials are under standard environmental conditions (i.e., 25°C and 1 atm) at pH 7.

CO_2 is the electron acceptor in the hydrogenotrophic reaction (Equation 3) and in the methanogenesis reaction by direct electron transfer from a cathode electrode surface (Equation 4). In closed systems, the equilibrium of CO_2 between liquid and gas phases is governed by Henry's Law:

$$K_H = C_{aq}/(p * P) \tag{6}$$

where K_H is the Henry's Law constant, C_{aq} is the aqueous concentration and p and P are the partial pressure (fraction) and total pressure of the gas in the headspace, respectively. In pure water at 298.15 K, the K_H value for CO_2 is 3.43 M/atm (Sander 2015). However, in biological systems, the K_H value must be corrected for temperature and the medium components. Sander (2015) outlined a method for temperature correction, with tabulated values for heat correction constants, and Weisenberger and Schumpe (1996) described a method for correcting K_H based on the quantification of dissolved anions, cations and gases.

When CO_2 dissolves into water, bicarbonate (HCO_3^-) and H_3O^+ are formed initially in what is the rate-limiting step for CO_2 hydration (Stirling and Pápai 2010). The HCO_3^- then participates in a dynamic, pH-dependent speciation between carbonate (CO_3^{2-}), HCO_3^-, carbonic acid (H_2CO_3) and the aqueous form of CO_2 ($CO_{2\,(aq)}$). Because of the difficulty of independently quantifying H_2CO_3 and $CO_{2(aq)}$, these two species are often represented together as $H_2CO_3^*$. At a neutral pH (7.0), the majority of the dissolved CO_2 is present as HCO_3^-. However, HCO_3^- must first be converted to $CO_{2(aq)}$ in order to react with unprotonated methanofuran and form carbamate in the first step of the hydrogenotrophic methanogenesis pathway (Bartoschek et al. 2000).

3. Bioelectrochemical Methane Production

BESs allow for the physical separation of an oxidation and reduction reaction that in turn allows two different sets of microbial communities to perform the two half-reactions. Thus, the conditions (e.g., pH, liquid components, gases, etc.) in both the anode and cathode influence the operation of the overall system (Krieg et al. 2018; Logan 2010).

In a typical methanogenic BES, a bioanode that oxidizes organics is separated from a biocathode that reduces CO_2 to CH_4 (Figure 1). A bioanode may oxidize acetate, releasing electrons to the anode surface

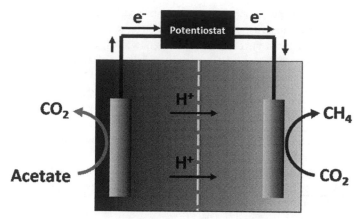

Figure 1: Schematic of a BES with a bioanode and a methanogenic biocathode. The dashed line indicates an optional proton exchange membrane that is utilized in dual chamber systems.

and protons into the surrounding anolyte. With a low applied voltage to the system, electrons from the anode travel through the circuit to the cathode where they are used by methanogens for CO_2 reduction to CH_4 (Equation 4). The modes of electron transfer from the microbial surface to the anode and from the cathode surface to methanogens are discussed further in Section 5. At the cathode surface, various reduction reactions may occur, including those described in Equations 4-5. Methanogens may receive electrons directly from a cathode electrode surface for the reduction of CO_2 to CH_4 in a process termed 'electromethanogenesis' (Cheng et al. 2009; Lohner et al. 2014). If electron equivalents are not directly used, as illustrated in Equation 4, H_2 produced from Equation 5 may be used for hydrogenotrophic methanogenesis, as shown in Equation 3. Theoretically, the electromethanogenesis reaction is more efficient than the H_2-mediated biocathode methanogenesis because energy losses occur at each electron transfer. Thus, the electromethanogenesis reaction is desired over H_2-mediated methanogenesis in methanogenic biocathodes because lower energy loss results in less required external energy input, increasing the overall energy efficiency of the BES.

Methanogenic BES efficiency may be evaluated using at least four factors: anode chemical oxygen demand (COD) removal, cathode CH_4 production, coulombic efficiency (CE) and cathode capture efficiency (CCE). Measurement of the anode COD removal can provide an estimate of how many electron equivalents were liberated during oxidation of organics at the anode. By measuring the current through the circuit over time, the charge transferred may be calculated. The ratio of total charge transferred to the theoretical charge released from COD oxidation is the calculated bioanode CE. Similarly, the CCE, measured as a ratio of total CH_4 electron equivalents produced to the total electron equivalents transferred through the circuit, can provide information on the electrocatalytic activity of the biocathode microbial community.

Electrochemical analyses may be used to evaluate the performance of bioanoades and biocathodes. Voltage measurements between the anode and cathode can indicate the cell voltage, which is proportional to cell current for a given system, as described by Ohm's law (Equation 7):

$$V = I * R \qquad (7)$$

where V = cell (anode-cathode) voltage (V), I = amperes of current (A) and R = resistance in ohms (Ω). A multimeter or electrochemical system monitor can also measure BES current and allow for resistance to be calculated. To get a better picture of the catalytic activity of an electrode, cyclic voltammetry may be used to sweep the cell or electrode voltage from one extreme to another to identify major redox peaks that indicate the voltage at which system components are oxidized and/or reduced (Harnisch and Freguia 2012; Logan 2012). Electrochemical impedance spectroscopy (EIS) is another electrochemical technique for evaluating the performance of a BES (Manohar et al. 2008; Ramasamy et al. 2008) and can be used in the development of system models.

Reported values for biocathode CH_4 production rates vary widely due to differences in system setups (e.g., BES geometry, electrode material, applied potential, etc.). In one study that compared a mixed methanogenic (MM) inoculum with an enriched hydrogenotrophic (EHM) inoculum, the biocathode mean CH_4 production rate at -0.80 V (vs. SHE), normalized to the proton exchange membrane surface area over the course of a typical feeding cycle, was 142 ± 21 and 603 ± 28 mmol/m^2-d, respectively (Dykstra and Pavlostathis 2017b). A separate study, with a biocathode poised at -0.80 V (vs. SHE), produced CH_4 at a mean rate of 200 mmol/m^2-d (Cheng et al. 2009). In comparison, a thermophilic biocathode poised at -0.8 V produced 1,103 mmol CH_4/m^2-d (Fu et al. 2015). Higher biocathode CH_4 production rates have been reported in other studies at higher poised potentials or over shorter periods of operation (e.g., -0.9 V, 5 hours) (Villano et al. 2010; Zhen et al. 2015).

Three factors affect bioelectrochemical CH_4 production: system design, microbial communities and system operation (Gadkari et al. 2018). Thus, it is important to take these items into consideration when developing a BES for biocathode CH_4 production, as discussed below.

4. System Design

BES design is important because it determines the system's ohmic losses, concentration polarization and electrode overpotentials (Krieg et al. 2018). Electrode and current collector materials will affect the ohmic resistance. Although inexpensive, carbon materials have a higher internal resistance than metals (Krieg et al. 2018). During the start-up of methanogenic biocathodes at -0.65 V (vs. SHE), biofilms on graphite foil and carbon felt electrodes produced CH_4 at a faster rate than biofilms on graphite rod electrodes although the materials' performance was the same after one year of operation (Saheb-Alam et al. 2018). When heat-treated stainless steel felt was compared with untreated stainless steel felt and graphite felt as biocathode materials poised at -0.8, -1.1 and -1.3 V (vs. SHE), the heat-treated stainless steel outperformed other cathode materials for electrocatalytic H_2 production and biocathode start-up (Liu et al. 2017a). In another study, a carbon stick with graphite felt outperformed a bare carbon stick and carbon sticks covered with various catalytic and/or conductive materials as a biocathode for CH_4 production at -0.70 V (vs. SHE) (Zhen et al. 2018). In a study comparing graphite, carbon black, carbon fiber brushes and coated graphite cathode materials, a platinum-coated carbon black cathode resulted in the highest CH_4 production. However, at more positive potentials ($\geq$ -0.55 mV vs. SHE), a plain graphite electrode performed similarly to the platinum-coated carbon black cathode (Siegert et al. 2014).

The physical properties (i.e., micro-roughness and surface area) of the electrodes can not only impact microbial adhesion but also the local diffusion profile, the diffusion layer thickness and the adhesion of redox active compounds (Champigneux et al. 2018). The surface area of the electrode is often maximized to improve reaction rates, but if the surface area is too large, biomass will become ion transfer-limited and will not colonize the entire electrode (Harrington et al. 2015). The surface area of the proton exchange membrane, which controls the rate of proton transport from anode to cathode, also affects BES performance (Geppert et al. 2016).

The ionic resistance in a BES depends on the catholyte specific ionic conductivity and the electrode distance (Krieg et al. 2018; Park et al. 2017). In a methanogenic biocathode fed with biogas, the catholyte ionic conductivity can be controlled by the catholyte composition. A large distance between the anode and cathode results in a high internal resistance that can be partially mitigated by efficient catholyte mixing (Park et al. 2017). If the distance between the anode and cathode is too small, short-circuiting of the anode and cathode species can occur in a single chamber BES (Krieg et al. 2018). Other operational parameters (e.g., pH, temperature, etc.) also affect overall BES performance (Geppert et al. 2016; Yang et al. 2018).

Efficient delivery of CO_2 to the biocathode is also required to prevent substrate limitations. Techniques for biocathode CO_2 delivery include headspace pressurization, bubbling, recirculation of the catholyte and hollow fiber membrane electrodes. Because of the relatively high solubility of CO_2, pressurizing the reactor headspace with CO_2-containing biogas is a simple and effective CO_2 delivery technique (Dykstra and Pavlostathis 2017a). When CO_2 is added by pressurizing the headspace, gaseous

CO_2 dissolves into the water toward an equilibrium concentration as described by Henry's Law. In the initial part of a batch cycle in a dual chamber system, CO_2 was transported across the membrane from cathode to anode due to the high CO_2 gradient across the two chambers and the CO_2 permeability of the proton exchange membrane (Dykstra and Pavlostathis 2017a). However, once robust methanogenic activity in the cathode was achieved, the cathode had a lower concentration of dissolved CO_2 species, due to CO_2 conversion in the cathode reaction, than in the anode where CO_2 was produced by the anode reaction. Thus, at the end of a 7-d batch cycle, the moles of CH_4 collected from the biocathode exceeded the moles of CO_2 that were initially supplied to the cathode because of the net transfer of carbon from anode to cathode by CO_2 transport across the proton exchange membrane and the subsequent conversion of CO_2 to CH_4 (Dykstra and Pavlostathis 2017a). Although effective due to the high solubility of CO_2 and fast equilibrium, headspace pressurization requires a batch operation to obtain high-purity CH_4 and relies on efficient catholyte mixing to bring CO_2 to the biocathode.

Bubbling has also been used to deliver biocathode CO_2 (Xu et al. 2014; Zeppilli et al. 2015), although a difficulty with this method is the short bubble travel paths in typical laboratory-scale cathode compartments that result in unconverted CO_2 diluting the collected CH_4 gas. Catholyte recirculation has also been used successfully for delivering CO_2 from an outside reservoir by maintaining a high catholyte CO_2 concentration while also improving biocathode CH_4 production by reducing the thickness of the diffusion layer and improving the transport of gases to and from the cathode surface (Champigneux et al. 2018; Dykstra 2017). In one methanogenic biocathode, catholyte recirculation increased the CH_4 production rate by 91% (Dykstra 2017), highlighting the importance of addressing mass transfer issues with reactor design.

Methanogenic BESs may be divided into two types: dual chamber and single chamber systems. Dual chamber systems commonly use an ion exchange membrane to separate the anode compartment from the cathode compartment. A typical H-style dual chamber system with a methanogenic biocathode is shown in Figure 2A. Other types of dual chamber systems have been developed, including plate reactors and tubular reactors (Krieg et al. 2014). The main advantage of using dual chamber systems is that the anode reactants and products can be separated from the cathode reactants and products, which prevents side reactions and can result in a higher purity cathode product. In a dual chamber methanogenic BES, acetate in the anode can be separated from the methanogens in the cathode, which reduces the possibility of acetoclastic methanogenesis in the bioanode (1 mol acetate:1 mol CO_2, 1 mol CH_4) occurring instead of electron donation to the circuit for biocathode electromethanogenesis and/or hydrogenotrophic methanogenesis (1 mol acetate: 2 mol CH_4). Another benefit of a dual chamber system is the ability to obtain high purity CH_4 in the biocathode headspace at the end of a batch feeding cycle because anode and cathode headspaces are kept separated (Logan 2010; Nelabhotla and Dinamarca 2018).

Single chamber systems consist of an anode and a cathode placed at a distance apart without a separator in between them (Figure 2B). For biogas upgrading applications, single chamber systems typically consist of an anaerobic digester with anode and cathode electrodes and an applied exterior potential (Nelabhotla and Dinamarca 2018). These types of single chamber systems do not require expensive membrane materials and can be installed on existing anaerobic digesters without the construction of a separate treatment chamber. Furthermore, without a membrane to exert additional ohmic resistance from ion transport, the internal resistance of a single chamber system is less than in a dual chamber system (Logan 2010; Nelabhotla and Dinamarca 2018). However, as mentioned previously, undesirable side reactions may occur because the anode and cathode reactants and products are comingled. Furthermore, the CH_4 purity of the final biogas is lower in a single chamber system because there is a single headspace for the anode and cathode (Krieg et al. 2014; Logan 2010).

5. Microorganisms

In a BES, microorganisms fulfill a number of important roles, including acting as inexpensive, self-renew catalysts that reduce the activation energy required for a specific reaction to occur. In a bioanode, exoelectrogenic bacteria (e.g., *Geobacter metallireducens*) oxidize organics, releasing electrons to

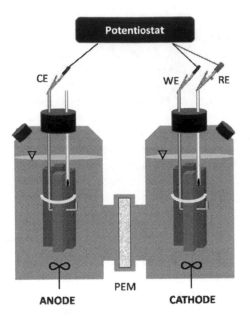 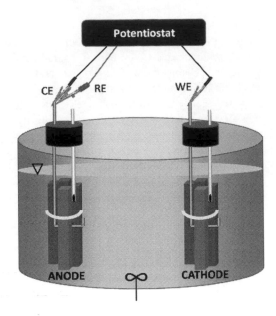

Figure 2: A dual chamber, three-electrode system (A), and a single chamber two-electrode system (B). Counter electrode, CE; working electrode, WE; reference electrode, RE; proton exchange membrane; PEM. In both electrode configurations, the working electrode is poised against the counter electrode but the reference electrode is different. In a three-electrode system, an adjacent standard reference electrode (e.g., Ag/AgCl) is the reference electrode; in a two-electrode system, the reference electrode is the same as the counter electrode. Applied voltage is measured between the working and reference electrodes.

the solid anode electrode surface and protons into the anolyte solution. A circuit then carries electron equivalents from the anode to the cathode with the assistance of a potentiostat in the case of biocathode CH_4 production. In a methanogenic biocathode, methanogens use electron equivalents from the cathode, along with protons generated at the anode, to convert CO_2 to CH_4. Both bioanode and biocathode must be maintained under oxygen-free (O_2) conditions. In the case of the bioanode, the exclusion of oxygen is required to prevent O_2 from competing with the anode surface as an electron acceptor. In a methanogenic biocathode, O_2 must be excluded to prevent toxicity to anaerobic microorganisms that are not equipped with enzymes to enable them to detoxify the free radicals produced during O_2 reduction (Fridovich 1998).

Several modes of BES electron transfer have been proposed (Figure 3A-B). In a bioanode, direct electron transfer (DET) may occur from the microorganism to the anode surface using outer membrane cytochromes or electrically-conductive pili (Kumar et al. 2018). In other cases, an electron shuttle or mediator may be used to transport electrons from the microorganism to the surface of the anode that is known as mediated electron transfer (MET). Both naturally occurring redox mediators (i.e., humic acids) and artificial mediators (e.g., neutral red, resazurin, methylene blue, etc.) may enhance bioanode performance by improving the transfer of electrons from microorganisms to the anode electrode (Martinez and Alvarez 2018; Watanabe et al. 2009).

In a methanogenic biocathode, electrons are transferred from the cathode electrode surface to methanogens for use in the reduction of CO_2 to CH_4. Some methanogens are known to be capable of receiving electrons from bacterial species through direct interspecies electron transfer (Lovley 2012). Similarly, methanogens may also receive electrons directly from a cathode electrode surface for the reduction of CO_2 to CH_4 (Cheng et al. 2009; Lohner et al. 2014). The formation of H_2 from water electrolysis theoretically occurs at a cathode potential of -0.414 V (vs. SHE), but because of the overpotentials in the system actual cathodes may require a potential of $\leq$ -0.60 V (vs. SHE) to produce H_2 from water molecules (Cheng et al. 2009; Wagner et al. 2010; Yates et al. 2014). At more negative cathode potentials (i.e., $\leq$ -0.60 V vs. SHE), H_2-mediated electron transfer likely plays a larger role in the

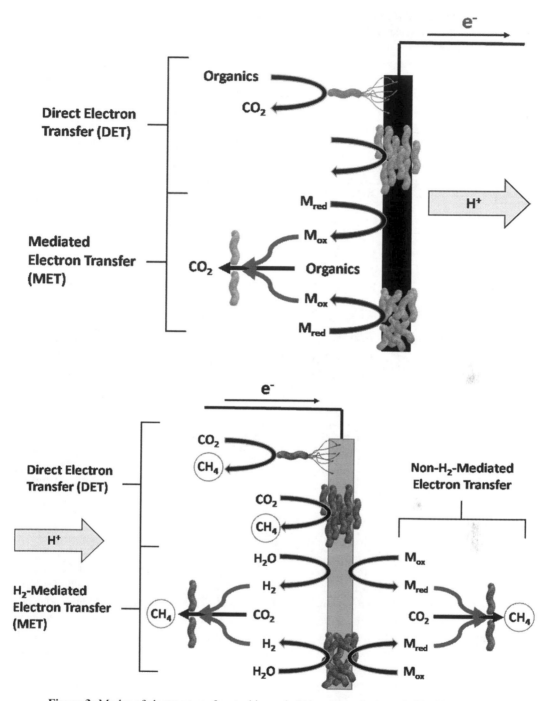

Figure 3: Modes of electron transfer at a bioanode (A) and a methanogenic biocathode (B).

production of CH_4 at a biocathode because water electrolysis, which generates H_2 at the cathode surface, becomes more favorable. Indeed, both viable and nonviable microbial cells on a biocathode can catalyze the water electrolysis reaction at -0.60 V (vs. SHE) (Yates et al. 2014). Several studies have evaluated the effect of other redox mediators on biocathode processes (Martinez and Alvarez 2018). The addition of zero valent iron (Fe^0) has been shown to improve biocathode methanogenesis and produce a redox-active precipitate that may participate in electron shuttling (Dykstra and Pavlostathis 2017c). However, few studies have specifically evaluated the effect of redox mediators on methanogenic biocathodes.

The makeup of the microbial community that colonizes a methanogenic biocathode is integral to the overall system performance. A review of recent methanogenic biocathode studies (Table 3) suggests that archaeal diversity in single cell systems is higher than in dual cell systems that separate bioanode and biocathode microbial communities. Biocathodes in dual cell systems were typically dominated by either *Methanobacterium* or *Methanobrevibacter*, while single cell systems contained these and additional genera, such as *Methanosaeta, Methanomassiliicoccus, Methanothrix, Methanomicrobiales, Methanosarcina, Methanoculleus* and *Methanothermobacter* (Table 3).

Table 3: Reported archaeal genera in single chamber and dual chamber BESs.

Biocathode Archaeal Genera	References
Single Chamber BES	
Methanobacterium	(Cerrillo et al. 2018; Dou et al. 2018; Gajaraj et al. 2017; Lee et al. 2017; Liu et al. 2016; Park et al. 2018a; Park et al. 2018b; Ren et al. 2018)
Methanobrevibacter	(Cerrillo et al. 2018)
Methanosarcina	(Gajaraj et al. 2017; Lee et al. 2017; Park et al. 2018a; Park et al. 2018b; Ren et al. 2018)
Methanosaeta	(Gajaraj et al. 2017; Ren et al. 2018; Zhao et al. 2015)
Methanoculleus	(Park et al. 2018a)
Methanomassiliicoccus	(Cerrillo et al. 2018; Park et al. 2018a)
Methanothrix	(Cerrillo et al. 2018)
Methanomicrobiales	(Gajaraj et al. 2017)
Dual Chamber BES	
Methanobacterium	(Alqahtani et al. 2018; Baek et al. 2017; Cai et al. 2018; Xu et al. 2017; Yang et al. 2018; Zhen et al. 2015)
Methanobrevibacter	(Cerrillo et al. 2017; Dykstra and Pavlostathis 2017a; Dykstra and Pavlostathis 2017b; Dykstra and Pavlostathis 2017c; Zeppilli et al. 2015)
Methanosarcina	(Xu et al. 2017; Zeppilli et al. 2015)
Methanosaeta	(Xu et al. 2017)
Methanoculleus	(Xu et al. 2017)

Although methanogens are the microorganisms involved in the conversion of CO_2 to CH_4, the bacterial community also plays a role in methanogenic biocathodes. Indeed, two biocathodes—with different bacterial communities but similar archaeal communities and amounts of microbial biomass—exhibited significantly different CH_4 production rates (Dykstra and Pavlostathis 2017b). The biocathode with a greater abundance in *Proteobacteria*, exoelectrogens and putative producers of electron shuttle mediators was capable of producing CH_4 at a rate that was nearly three-fold faster than that of the other biocathode. The cathode was only supplied with CO_2, and thus, the presence of heterotrophs indicates that organic microbial products and/or lysed cells likely were used by heterotrophs to recycle microbially-produced organic carbon (Dykstra and Pavlostathis 2017b). Thus, it is hypothesized that CH_4 in the better performing biocathode was improved through the enhanced recycling of organic cell debris into electron equivalents by *Proteobacteria* and exoelectrogens for CH_4 production as well as the possible microbial production of electron shuttle mediators that carried electron equivalents to methanogens located at a distance from the cathode surface (Dykstra and Pavlostathis 2017b). The presence of acetogens in a biocathode microbial community suggests that some electrons may be utilized for the production of acetate from CO_2, diverting electron equivalents away from CH_4 production. Indeed, *Acetobacterium* and *Treponema* have been detected at higher abundances in biocathodes with a lower CCE (Cerrillo et al. 2017; Dykstra and Pavlostathis 2017b).

Because the microbial community is so integral to the biocathode performance, initial inoculation and start-up of a biocathode are important. Successful methanogenic biocathodes have been developed using anaerobic sludge and digester mixed liquor as an inoculum (Baek et al. 2017; Dou et al. 2018; Dykstra and Pavlostathis 2017b). In one study, biocathode start-up and performance were compared between a biocathode inoculated with anaerobic digester mixed liquor and a biocathode inoculated with a hydrogenotrophic methanogenic mixed culture enriched from the anaerobic digester mixed liquor by feeding a mixture of H_2 and CO_2 (80:20; v/v). The biocathode inoculated with the enriched culture outperformed the biocathode inoculated with unenriched anaerobic digester mixed liquor (Dykstra and Pavlostathis 2017b). A rotating methanogenic biocathode was developed using inoculum from return activated sludge and discs that cycled through anode and cathode polarizations (Cheng et al. 2011). However, in a nonrotating system, the conversion of an acetate-oxidizing bioanode to a methanogenic biocathode by changing the electrode potential was reported to be a less effective start-up strategy than inoculation of a bare electrode (Saheb-Alam et al. 2018). Pretreatment of cathode chambers with antibiotics during start-up reduced the biofilm abundance of hydrogenotrophic methanogens and increased the abundance of acetoclastic methanogens. A higher CCE was also achieved that may be due to an increase in electromethanogenesis and a decrease in less efficient H_2-mediated electron transfer (Xu et al. 2017). In a single-cell system consisting of an anaerobic digester with two submerged carbon felt electrodes, better performance was achieved when biomass was allowed to develop on the electrodes under open-circuit conditions before the application of a potential (Dou et al. 2018). However, many questions still remain unanswered and further study is needed to optimize the start-up of a methanogenic biocathode.

6. System Operation

In addition to system design and microbial factors, the performance of a methanogenic BES also varies based on how the system is operated. Operational parameters that affect system performance include applied potential, temperature, mixing velocity, anolyte and catholyte buffering capacity, the concentration of H_2S in the anaerobic digester biogas that needs to be upgraded and the presence of redox mediators in the system. Other operational parameters may also affect the performance of a methanogenic BES, but further research is needed to identify influential factors and evaluate their effects.

Applied potential is an important parameter of BES operation. A typical three-electrode configuration for a methanogenic BES is shown in Figure 2A in which the working electrode cable is attached to the cathode electron collector, the reference electrode cable is attached to a reference electrode (e.g., Ag/AgCl) placed adjacent to the cathode electrode and the counter electrode is attached to the anode electron collector. In this configuration, the cathode potential may be set to a particular voltage, as measured against the adjacent reference electrode, typically -1.0 to -0.5 V (vs. SHE). In a second configuration, the cell voltage (i.e., voltage between anode and cathode) is controlled, as shown in Figure 2B, and the applied potential typically ranges between 0.2 V and 2.0 V. Thus, the voltage of both the anode and cathode is allowed to fluctuate in the second configuration as long as the overall cell voltage is constant, while only the anode voltage is allowed to fluctuate in the first configuration. Thus, the effect on methanogenic BES performance of varying the applied potential depends on how the electrodes are configured. In a dual chamber, H-style methanogenic BES configured as shown in Figure 2B, anode COD removal efficiency and CH_4 production were inhibited at cell voltages greater than 0.8 V due to an increase in cell lysis and reduction in microbial growth and activity (Ding et al. 2016). In a dual chamber system, the maximum amount of energy recovered occurs when the biocathode CH_4 production is maximized, while the energy input required to produce CH_4 (i.e., 'specific energy') is minimized. Based on tests of a biocathode at -0.80 V to -0.50 V (vs. SHE), the best performance occurred at -0.80 V (Figure 4). In a single chamber methanogenic BES, which was developed at $22\pm2°C$ with a dextrin- and peptone-fed mixed anaerobic culture and tested at a voltage range from 0.5 to 2.0 V, the fastest CH_4 production was achieved at 2.0 V.

At this applied cell voltage, CH$_4$ was produced at 105 mL/L$_{reactor}$-d and the resulting biogas was 88.5% CH$_4$ (v/v), as compared to 66.3% CH$_4$ (v/v) in the conventional anaerobic digester (Dou et al. 2018). In another study, multiple single chamber systems inoculated with anaerobic digester effluent were fed with a mixture of glucose and anaerobic digester sludge and were tested at cell voltages from 0.5 V to 1.5 V. The maximum soluble COD removal and the maximum CH$_4$ production were observed in the system with the cell voltage poised at 1.0 V (Choi et al. 2017). The difference in the optimal applied potential between systems is likely due to differences in reactor design, inoculum and operating factors.

Temperature is another factor that may affect methanogenic BES performance. Therefore, a study developed biocathodes at different temperatures (15 °C to 70 °C) while poising the cathode potential at -0.70 V (vs. SHE); the optimum temperature for CH$_4$ production was 50 °C (Yang et al. 2018). However, higher temperature also negatively affects the solubility of CO$_2$ with a nearly two-fold decrease in the solubility limit between room temperature (25 °C) and thermophilic temperatures (55 °C) (Wiebe and Gaddy 1940). Although a higher temperature may lead to an increased CH$_4$ production rate, heating a BES also requires energy input which impacts the overall energy efficiency. Thermophilic BESs for CH$_4$ production could potentially use waste heat from other processes for BES heating, although little data currently exists to determine whether thermophilic reactors are energetically feasible for methanogenic biocathode energy recovery.

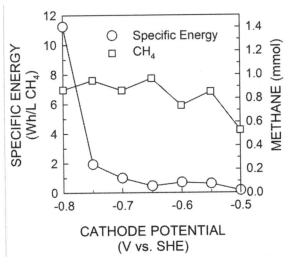

Figure 4: Specific energy (W-h/L CH$_4$) and 1-d CH$_4$ production during batch cycles of a CO$_2$-fed methanogenic biocathode operated at a range of cathode potentials (V vs. SHE).

Thermophilic cultures may also be useful in methanogenic biocathodes. In one study, a single chamber BES with a methanogenic biocathode was developed at 55°C with effluent of the anode chamber of a thermophilic microbial fuel cell and a cell potential of 0.80 V. Following biocathode biofilm development, the biocathode was transferred to a dual chamber system and poised at a cathode potential of -0.50 V vs. SHE. The developed biocathode was capable of producing CH$_4$ at -0.35 V vs. SHE, which is lower than the threshold for CH$_4$ production in most nonthermophilic biocathodes (Fu et al. 2015). A study comparing a single chamber and dual chamber thermophilic BES for CH$_4$ production determined that biogas production was accelerated in both reactors by poising a cathode at -0.80 V vs. Ag/AgCl (-0.60 V. vs. SHE) but only the dual chamber system was capable of enriching CH$_4$ in the biogas (Liu et al. 2017b).

Another important factor affecting methanogenic BES system operation is the presence of hydrogen sulfide (H$_2$S) in the biocathode influent biogas. Anaerobic digester biogas contains small amounts of H$_2$S that can be inhibitory to methanogenesis (Amha et al. 2018; Chen et al. 2008; Hilton and Oleszkiewicz 1988; Karhadkar et al. 1987). Depending on the carbon to the sulfur ratio of the feedstock entering a digester, the H_2S biogas composition may range from < 1% (v/v) for municipal wastewater sludge to $\geq$ 10% (v/v) for harvested seaweed (Peu et al. 2012). Experiments with a dual chamber methanogenic BES at various initial H_2S concentrations indicated that biocathode headspace H_2S content up to 3% (v/v) could increase biocathode CH_4 output by H_2S diffusion through the membrane and subsequent oxidation at the bioanode, increasing both current density and biocathode CH_4 production. However, at biocathode headspace H_2S content higher than 3%, the CH_4 production declined likely due to inhibition of biocathode methanogens (Dykstra 2017).

The presence of redox active substances in the cathode may also affect the operation of a methanogenic BES. In one study, zero valent iron (ZVI, Fe^0) was added, like iron filings, to a biocathode controlled at a potential of -0.80 V (vs. SHE). ZVI underwent anaerobic corrosion, producing Fe^{2+} ions and H_2, which was utilized by biocathode methanogens as an additional source of reducing power for CO_2 conversion to CH_4. However, after the addition of ZVI, the biocathode CH_4 production remained high despite the weekly replacement of catholyte and no new addition of ZVI. A redox active precipitate was identified that developed on the biocathode biofilm and likely participated in shuttling electrons to accelerate CH_4 production (Dykstra and Pavlostathis 2017c).

7. Remaining Challenges and Research Needs

BES-enhanced energy recovery from anaerobic digestion is a promising area of research. Further developments in membrane technology may also assist in economically scaling up BES technology for CH_4 production enhancement and intensification. Lower cost membranes and membranes with higher selectivity for proton and/or CO_2 transfer are also important areas for future research. Moreover, better electrode materials and designs must be developed to promote biofilm development and to enhance electron transfer between the electrode surface and microorganisms. The design of improved gas-permeable electrodes may also increase the efficiency of electron and mass transfer at the solid-liquid-gas interface. The effect of the anode electron donor on system performance must also be investigated in more depth as well as the development of new electrode materials and the utilization of membrane gas transport for achieving high CH_4 purity.

The overall design of methanogenic BESs must also be reimagined to optimize energy recovery in a full-scale system. For instance, the H-type BES is useful for laboratory testing but is not an ideal design to be scaled up for energy recovery because the distance between electrodes in the H-type BES is often larger than in other types of BESs, such as the plate-style, and thereby, increases ohmic resistance. Scalable designs must be developed that are able to be optimized for the unique solid-liquid-gas interface at the biocathode in the case of CO_2 conversion to CH_4.

The microbial communities and interactions are integral to the performance of a methanogenic biocathode and paired bioanode. However, little is currently known about how biocathode non-methanogenic bacteria and methanogens interact and influence the CH_4 output. Therefore, it is important that future research examines how both bacterial and archaeal species work together to process the flow of carbon and electrons through a methanogenic BES.

Because of the large number of operational parameters that could possibly affect methanogenic biocathode CH_4 production and energy recovery, modeling and sensitivity analyses may be useful in determining which parameters are most important for optimizing energy recovery. Future research is needed to determine how various operational parameters affect BES performance to inform future model development. Models that include bioanode and biocathode characteristics and account for dynamic biofilm development are required for accurate simulations of the start-up and steady-state performance of a methanogenic BES.

8. Conclusions

A number of questions remain to be answered before methanogenic biocathodes can be successfully scaled up to enhance CH_4 production within anaerobic digesters or to upgrade anaerobic digester biogas.

However, research interest in methanogenic BES technology is increasing and momentum currently exists for great progress to be made. The effective scaling up of methanogenic BESs will require further insights into microbial communities and interactions, use of advanced yet inexpensive electrode materials, as well as the optimization of system design and operation.

Methanogenic BESs are a promising tool for improving energy recovery from wastewater treatment by enhancing and intensifying the production of CH_4 from anaerobic digestion. By developing and deploying new solutions to improve methanogenic BES technology, fossil fuel use may be offset and WWRFs may move closer to realizing net zero energy operation.

Acknowledgment

This material is based in part upon work supported by the U.S. National Science Foundation, Graduate Research Fellowship under grant no. DGE-1148903.

References

Al-mashhadani, MKH, Wilkinson, SJ, Zimmerman, WB (2016) Carbon dioxide rich microbubble acceleration of biogas production in anaerobic digestion. Chem Eng Sci 156: 24-35.

Alqahtani, MF, Katuri, KP, Bajracharya, S, Yu, Y, Lai, Z, Saikaly, PE (2018) Porous hollow fiber nickel electrodes for effective supply and reduction of carbon dioxide to methane through microbial electrosynthesis. Adv Funct Mater 1804860.

Amha, YM, Anwar, MZ, Brower, A, Jacobsen, CS, Stadler, LB, Webster, TM, Smith, AL (2018) Inhibition of anaerobic digestion processes: Applications of molecular tools. Bioresour Technol 247: 999-1014.

Angelidaki, I, Treu, L, Tsapekos, P, Luo, G, Campanaro, S, Wenzel, H, Kougias, PG (2018) Biogas upgrading and utilization: Current status and perspectives. Biotechnol Adv 36: 452-466.

Angenent, LT, Usack, JG, Xu, J, Hafenbradl, D, Posmanik, R, Tester, JW (2018) Integrating electrochemical, biological, physical, and thermochemical process units to expand the applicability of anaerobic digestion. Bioresour Technol 247: 1085-1094.

Baek, G, Kim, J, Lee, S, Lee, C (2017) Development of biocathode during repeated cycles of bioelectrochemical conversion of carbon dioxide to methane. Bioresour Technol 241: 1201-1207.

Bartoschek, S, Vorholt, JA, Thauer, RK, Geierstanger, BH, Griesinger, C (2000) N-carboxymethanofuran (carbamate) formation from methanofuran and CO_2 in methanogenic archaea. Eur J Biochem 267: 3130-3138.

Boone, DR, Garrity, GM. 2001. Bergey's Manual of Systematic Bacteriology, Volume 1. The archaea and the deeply branching and phototrophic bacteria. Springer Science and Business Media, New York, USA, pp 1–722.

Cai, W, Liu, W, Zhang, Z, Feng, K, Ren, G, Pu, C, Sun, H, Li, J, Deng, Y, Wang, A (2018) mcrA sequencing reveals the role of basophilic methanogens in a cathodic methanogenic community. Water Res 136: 192-199.

Cerrillo, M, Viñas, M, Bonmatí, A (2018) Anaerobic digestion and electromethanogenic microbial electrolysis cell integrated system: Increased stability and recovery of ammonia and methane. Renewable Energy 120: 178-189.

Cerrillo, M, Viñas, M, Bonmatí, A (2017) Startup of electromethanogenic microbial electrolysis cells with two different biomass inocula for biogas upgrading. ACS Sustainable Chem Eng 5: 8852-8859.

Champigneux, P, Delia, M-L, Bergel, A (2018) Impact of electrode micro- and nano-scale topography on the formation and performance of microbial electrodes. Biosens Bioelectron 118: 231-246.

Chen, Y, Cheng, JJ, Creamer, KS (2008) Inhibition of anaerobic digestion process: A review. Bioresour Technol 99: 4044-4064.

Cheng, KY, Ho, G, Cord-Ruwisch, R (2011) Novel methanogenic rotatable bioelectrochemical system operated with polarity inversion. Environ Sci Technol 45: 796-802.

Cheng, S, Xing, D, Call, DF, Logan, BE (2009) Direct biological conversion of electrical current into methane by electromethanogenesis. Environ Sci Technol 43: 3953-3958.

Choi, K-S, Kondaveeti, S, Min, B (2017) Bioelectrochemical methane (CH_4) production in anaerobic digestion at different supplemental voltages. Bioresour Technol 245: 826-832.

Deepanraj, B, Sivasubramanian, V, Jayaraj, S (2014) Biogas generation through anaerobic digestion process-an overview. Res J Chem Environ 18: 80-93.

Ding, A, Yang, Y, Sun, G, Wu, D (2016) Impact of applied voltage on methane generation and microbial activities in an anaerobic microbial electrolysis cell (MEC). Chem Eng J 283: 260-265.

Dou, Z, Dykstra, CM, Pavlostathis, SG (2018) Bioelectrochemically assisted anaerobic digestion system for biogas upgrading and enhanced methane production. Sci Total Environ 633: 1012-1021.

Dykstra, CM. 2017. Bioelectrochemical Conversion of Carbon Dioxide to Methane for Biogas Upgrading. Doctoral Dissertation. Georgia Institute of Technology, Atlanta, GA, USA, pp 1-287.

Dykstra, CM, Pavlostathis, SG (2017a) Evaluation of gas and carbon transport in a methanogenic bioelectrochemical system (BES). Biotechnol Bioeng 114: 961-969.

Dykstra, CM, Pavlostathis, SG (2017b) Methanogenic biocathode microbial community development and the role of bacteria. Environ Sci Technol 51: 5306-5316.

Dykstra, CM, Pavlostathis, SG (2017c) Zero-valent iron enhances biocathodic carbon dioxide reduction to methane. Environ Sci Technol 51: 12956-12964.

Ferry, JG. 1993. Methanogenesis: Ecology, Physiology, Biochemistry and Genetics. Edition 1. Springer US, Boston, MA, USA, pp 1-536.

Fridovich, I (1998) Oxygen toxicity: A radical explanation. J Exp Biol 201: 1203.

Fu, Q, Kuramochi, Y, Fukushima, N, Maeda, H, Sato, K, Kobayashi, H (2015) Bioelectrochemical analyses of the development of a thermophilic biocathode catalyzing electromethanogenesis. Environ Sci Technol 49: 1225-1232.

Gadkari, S, Gu, S, Sadhukhan, J (2018) Towards automated design of bioelectrochemical systems: A comprehensive review of mathematical models. Chem Eng J 343: 303-316.

Gajaraj, S, Huang, Y, Zheng, P, Hu, Z (2017) Methane production improvement and associated methanogenic assemblages in bioelectrochemically assisted anaerobic digestion. Biochem Eng J 117: 105-112.

Geppert, F, Liu, D, van Eerten-Jansen, M, Weidner, E, Buisman, C, ter Heijne, A (2016) Bioelectrochemical power-to-gas: State of the art and future perspectives. Trends Biotechnol 34: 879-894.

Gujer, W, Zehnder, AJB (1983) Conversion processes in anaerobic digestion. Water Sci Technol 15: 127.

Harnisch, F, Freguia, S (2012) A basic tutorial on cyclic voltammetry for the investigation of electroactive microbial biofilms. Chem Asian J 7: 466-475.

Harrington, TD, Babauta, JT, Davenport, EK, Renslow, RS, Beyenal, H (2015) Excess surface area in bioelectrochemical systems causes ion transport limitations. Biotechnol Bioeng 112: 858-866.

Hilton, BL, Oleszkiewicz, JA (1988) Sulfide-induced inhibition of anaerobic digestion. J Environ Eng 114: 1377-1391.

Karhadkar, PP, Audic, J-M, Faup, GM, Khanna, P (1987) Sulfide and sulfate inhibition of methanogenesis. Water Res 21: 1061-1066.

Krieg, T, Madjarov, J, Rosa, LF, Enzmann, F, Harnisch, F, Holtmann, D, Rabaey, K (2018) Reactors for microbial electrobiotechnology. Adv Biochem Eng Biotechnol doi: 10.1007/10_2017_40

Krieg, T, Sydow, A, Schröder, U, Schrader, J, Holtmann, D (2014) Reactor concepts for bioelectrochemical syntheses and energy conversion. Trends Biotechnol 32: 645-655.

Kumar, M, Sundaram, S, Gnanounou, E, Larroche, C, Thakur, I (2018) Carbon dioxide capture, storage and production of biofuel and biomaterials by bacteria: A review. Biores Technol 247: 1059-1068.

Lee, B, Park, J-G, Shin, W-B, Tian, D-J, Jun, H-B (2017) Microbial communities change in an anaerobic digestion after application of microbial electrolysis cells. Biores Technol 234: 273-280.

Liu, D, Zhang, L, Chen, S, Buisman, C, ter Heijne, A (2016) Bioelectrochemical enhancement of methane production in low temperature anaerobic digestion at 10 degrees C. Water Res 99: 281-287.

Liu, D, Zheng, T, Buisman, C, Ter Heijne, A (2017a) Heat-treated stainless steel felt as a new cathode material in a methane-producing bioelectrochemical system. ACS Sustainable Chem Eng 5: 11346-11353.

Liu, SY, Charles, W, Ho, G, Cord-Ruwisch, R, Cheng, KY (2017b) Bioelectrochemical enhancement of anaerobic digestion: Comparing single- and two-chamber reactor configurations at thermophilic conditions. Bioresour Technol 245: 1168-1175.

Logan, B (2010) Scaling up microbial fuel cells and other bioelectrochemical systems. Appl Microbiol Biotechnol 85: 1665-1671.

Logan, BE (2012) Essential data and techniques for conducting microbial fuel cell and other types of bioelectrochemical system experiments. ChemSusChem 5: 988-994.

Lohner, ST, Deutzmann, JS, Logan, BE, Leigh, J, Spormann, AM (2014) Hydrogenase-independent uptake and metabolism of electrons by the archaeon *Methanococcus maripaludis*. ISME J 8: 1673-1681.

Lovley, D (2012) Electromicrobiology. Annu Rev Microbiol 66: 391-409.

Luo, G, Johansson, S, Boe, K, Xie, L, Zhou, Q, Angelidaki, I (2012) Simultaneous hydrogen utilization and in situ biogas upgrading in an anaerobic reactor. Biotechnol Bioeng 109: 1088-1094.

Manohar, AK, Bretschger, O, Nealson, KH, Mansfeld, F (2008) The use of electrochemical impedance spectroscopy (EIS) in the evaluation of the electrochemical properties of a microbial fuel cell. Bioelectrochem 72: 149-154.

Martinez, C, Alvarez, L (2018) Application of redox mediators in bioelectrochemical systems. Biotechnol Adv 36: 1412-1423.

Muñoz, R, Meier, L, Diaz, I, Jeison, D (2015) A review on the state-of-the-art of physical/chemical and biological technologies for biogas upgrading. Rev Environ Sci Biotechnol 14: 727-759.

Nelabhotla, ABT and Dinamarca, C (2018) Electrochemically mediated CO_2 reduction for bio-methane production: A review. Rev Environ Sci Biotechnol 17: 531-551.

Pan, S-Y, Chiang, P-C, Pan, W, Kim, H (2018) Advances in state-of-art valorization technologies for captured CO_2 toward sustainable carbon cycle. Crit Rev Env Sci Technol 48: 471-534.

Park, J-G, Lee, B, Shi, P, Kim, Y, Jun, H-B (2017) Effects of electrode distance and mixing velocity on current density and methane production in an anaerobic digester equipped with a microbial methanogenesis cell. Int J Hydrogen Energy 42: 27732-27740.

Park, J, Lee, B, Shi, P, Kwon, H, Jeong, S, Jun, H (2018a) Methanol metabolism and archaeal community changes in a bioelectrochemical anaerobic digestion sequencing batch reactor with copper-coated graphite cathode. Bioresour Technol 259: 398-406.

Park, J, Lee, B, Tian, D, Jun, H (2018b) Bioelectrochemical enhancement of methane production from highly concentrated food waste in a combined anaerobic digester and microbial electrolysis cell. Bioresour Technol 247: 226-233.

Petersson, A, Wellinger, A (2009) Biogas upgrading technologies - developments and innovations. IEA Bioenergy 20: 1-19.

Peu, P, Picard, S, Diara, A, Girault, R, Béline, F, Bridoux, G, Dabert, P (2012) Prediction of hydrogen sulphide production during anaerobic digestion of organic substrates. Bioresour Technol 121: 419-424.

Ramasamy, RP, Ren, Z, Mench, MM, Regan, JM (2008) Impact of initial biofilm growth on the anode impedance of microbial fuel cells. Biotechnol Bioeng 101: 101-108.

Ren, G, Hu, A, Huang, S, Ye, J, Tang, J, Zhou, S (2018) Graphite-assisted electro-fermentation methanogenesis: Spectroelectrochemical and microbial community analyses of cathode biofilms. Bioresour Technol 269: 74-80.

Saheb-Alam, S, Singh, A, Hermansson, M, Persson, F, Schnürer, A, Wilén, B-M, Modin, O (2018) Effect of start-up strategies and electrode materials on carbon dioxide reduction on biocathodes. Appl Environ Microbiol 84: e02242-17.

Salomoni, C, Caputo, A, Bonoli, M, Francioso, O, Rodriguez-Estrada, MT, Palenzona, D (2011) Enhanced methane production in a two-phase anaerobic digestion plant, after CO_2 capture and addition to organic wastes. Bioresour Technol 102: 6443-6448.

Sander, R (2015) Compilation of Henry's law constants (version 4.0) for water as solvent. Atmos Chem Phys 15: 4399-4981.

Siegert, M, Yates, MD, Call, DF, Zhu, X, Spormann, A, Logan, BE (2014) Comparison of nonprecious metal cathode materials for methane production by electromethanogenesis. ACS Sustainable Chem Eng 2: 910-917.

Stirling, As and Pápai, I (2010) H_2CO_3 forms via HCO_3^- in water. J Phys Chem B 114: 16854-16859.

Sun, Q, Li, H, Yan, J, Liu, L, Yu, Z, Yu, X (2015) Selection of appropriate biogas upgrading technology-a review of biogas cleaning, upgrading and utilisation. Renewable Sustainable Energy Rev 51: 521-532.

Verbeeck, K, Buelens, LC, Galvita, VV, Marin, GB, Van Geem, KM, Rabaey, K (2018) Upgrading the value of anaerobic digestion via chemical production from grid injected biomethane. Energy Environ Sci 11: 1788-1802.

Villano, M, Aulenta, F, Ciucci, C, Ferri, T, Giuliano, A, Majone, M (2010) Bioelectrochemical reduction of CO_2 to CH_4 via direct and indirect extracellular electron transfer by a hydrogenophilic methanogenic culture. Bioresour Technol 101: 3085-3090.

Wagner, RC, Call, DF, Logan, BE (2010) Optimal set anode potentials vary in bioelectrochemical systems. Environ Sci Technol 44: 6036-6041.

Weisenberger, S, Schumpe, A (1996) Estimation of gas solubilities in salt solutions at temperatures from 273 k to 363 k. AIChE J 42: 298-300.

Wiebe, R, Gaddy, VL (1940) The solubility of carbon dioxide in water at various temperatures from 12 to 40° and at pressures to 500 atmospheres: Critical phenomena. J Am Chem Soc 62: 815-817.

Xu, H, Giwa, AS, Wang, C, Chang, F, Yuan, Q, Wang, K, Holmes, DE (2017) Impact of antibiotics pretreatment on bioelectrochemical CH_4 production. ACS Sustainable Chem Eng 5: 8579-8586.

Xu, H, Wang, K, Holmes, DE (2014) Bioelectrochemical removal of carbon dioxide (CO_2): An innovative method for biogas upgrading. Bioresour Technol 173: 392-398.

Yang, HY, Bao, BL, Liu, J, Qin, Y, Wang, YR, Su, KZ, Han, JC, Mu, Y (2018) Temperature dependence of bioelectrochemical CO_2 conversion and methane production with a mixed-culture biocathode. Bioelectrochem 119: 180-188.

Yates, MD, Siegert, M, Logan, BE (2014) Hydrogen evolution catalyzed by viable and non-viable cells on biocathodes. Int J Hydrogen Energy 39: 16841-16851.

Zeppilli, M, Villano, M, Aulenta, F, Lampis, S, Vallini, G, Majone, M (2015) Effect of the anode feeding composition on the performance of a continuous-flow methane-producing microbial electrolysis cell. Environ Sci Pollut Res 22: 7349-7360.

Zhao, Z, Zhang, Y, Wang, L, Quan, X (2015) Potential for direct interspecies electron transfer in an electric-anaerobic system to increase methane production from sludge digestion. Sci Rep 5: 11094.

Zhen, G, Kobayashi, T, Lu, X, Xu, K (2015) Understanding methane bioelectrosynthesis from carbon dioxide in a two-chamber microbial electrolysis cells (MECs) containing a carbon biocathode. Bioresour Technol 186: 141-148.

Zhen, G, Zheng, S, Lu, X, Zhu, X, Mei, J, Kobayashi, T, Xu, K, Li, Y-Y, Zhao, Y (2018) A comprehensive comparison of five different carbon-based cathode materials in CO_2 electromethanogenesis: Long-term performance, cell-electrode contact behaviors and extracellular electron transfer pathways. Bioresour Technol 266: 382-388.

Fluidized Bed Electrodes in Microbial Electrochemistry

Sara Tejedor-Sanz[1,2,3] and Abraham Esteve-Nuñez[2,3*]

[1] Molecular Foundry, Lawrence Berkeley National Laboratory, Berkeley, CA, USA
[2] University of Alcalá, Department of Chemical Engineering, Alcalá de Henares, Madrid, Spain
[3] IMDEA Water Institute, Alcalá de Henares, Madrid, Spain

1. Introduction

To date, METs have been applied to devices in which catalysis is solely located at the electrode surface due to the need for microbial biofilm formation. Biofilms limit the performance of the system due to restricting the reactions to the electrode-biofilm interface. This presents problems like the limitations of mass transfer reducing the activity of the cells within the biofilm. The main challenge for METs to be competitive with alternative technologies is the optimization of the rate and selectivity of the desired reactions. From an engineering perspective, the kinetics can be improved by maximizing the electrode surface area, the catalyst (bacteria) concentration and the catalyst's activity. Most of the research efforts in the field have been made to increase the active area of the electrodes. To this end, the three-dimensional bed electrodes, both static and dynamic, can increase the effective surface area of the anode for bacteria adhesion. Fluidized electrodes are dynamic bed electrodes made of electrically conductive particles that are in constant motion. This represents two major advantages over flat or static electrodes: i) high mass transfer-rates and ii) large electrode surface area per unit volume of electrode. Subsequently, these benefits have allowed fluidized electrodes to be utilized in METs and improve a variety of microbial mediated processes, like volatile fatty acids oxidation, nitrogen removal and wastewater treatment.

This chapter will review the fundamentals, design concepts and applications of fluidized electrodes within the field of microbial electrochemistry. To our knowledge, this is the first review published that focuses on this specific electrode design. The different sections should provide the reader with an overview of the engineering and biological aspects that have been studied for fluidized electrodes in METs.

2. Fluidized Beds Systems: The Engineering Background

Fluidization is a process where fine particulates (solid phase) are in contact with a fluid (fluid phase) and behave in the fluid-like state for the purpose of heat or mass transfer between the phases with or without simultaneous chemical reactions. The fluid phase may be a gas or liquid. Depending on the flow rates and the properties of the fluid phase, the resulting characteristics of the fluidized bed can vary considerably. When the fluid phase is liquid, beds are generally operated in a batch mode with respect to a solid phase and in a continuous mode with respect to the fluid phase.

---

*Corresponding author: abrahamesteve@uahes

Fluidized beds are largely used in chemical and processing industries for a large variety of purposes. The most widespread application can be found within the wastewater treatment field, particularly in advanced oxidation processes and biological treatment. The latter is a well-established application employed in many large-scale plants (Heijnen et al. 1989; Bello et al. 2017). Fluidized bed reactors for treating wastewaters are biofilm-based with a suspended bed as a biomass carrier. This design can be much more efficient than other configurations used in wastewater treatment because of the following advantages that a suspended bed provides:

- A large interfacial contact between the liquid phase and the solid phase (particles) provides the biofilm with good mass and heat transport properties.
- Retention of biomass in the particles allows for the separation of the mean residence time of the liquid and the biomass.
- A high biomass concentration in the reactor allows for the treatment of wastewater with high organic loads or working at high rates.
- Fluidization of the particles eliminates clogging that can occur when working with a static solid phase (nonfluidized such as packed beds).
- The possibility of constant exchange of bed without stopping the process.

Due to these characteristics, fluidized bed reactors can operate at high loading rates with high level of biological activity. Furthermore, the reactors are easy to construct and apply at any scale, and the tubular shape of fluidized bed reactors avoids the existence of dead zones. This type of reactor design is also suitable for a continuous mode of operation and additionally provides for a compact configuration having a small footprint.

3. From Abiotic to Microbial Fluidized Bed Electrodes

Besides biological applications and processes, fluidized beds have an important role in electrochemical catalysis. By using electrically conductive particles, the bed can be converted into a 3D electroactive surface, i.e. into a fluidized electrode. This concept has attracted attention during the last decades for its application to the electrowinning of precious metals, cathodic reductions and energy storage systems (Huh 1985). In spite of a limited number of successful full-scale systems, the first studies concerning these fluidized electrodes date back to the 1960s (Trupp 1968; Goff et al. 1969; Backhurst et al. 1969). We believe that the reason this older electrode architecture has not been largely employed in real-world applications is because of the insufficient understanding of the electrical and electrochemical behavior of such electrodes. Fluidized electrodes are porous electrodes that act as a fluid in continuous motion and present complex electrochemical behaviors. One of the key factors in these complex behaviors is the potential transient events caused by the dynamic behavior and interaction of the two phases, namely solid (electrode) and fluid (electrolyte) (Hiddleston and Douglas 1970; Huh and Evans 1987). This trait affects the overall performance of the cell. Therefore, it becomes critical to properly choose the conductivity, shape, size and density of the particles as well as their manner of fluidization to increase its stability.

The main advantage of a fluidized bed electrode, as compared to conventional electrode design, is the electrode's very large effective surface area per unit volume. This greatly increases the effective current capacity beyond that of a conventional cell, improving the kinetics of the desired oxidation/reduction reactions. Additionally, the ability to vary the electrode volume within the reactor by changing the flow of the upward recirculating electrolyte allows for control of the rate of the ongoing electrochemical reactions. Figure 1 shows a direct relationship between the linear velocity (fluid flow/section of column) inside the electrochemical fluidized bed reactor with the current density produced and bed expansion by the fluidized anode for a Fe(II)/Fe(III) redox couple. As the flow of the pump is increased, the bed expands and the ratio of electrode volume to reactor volume also increases, thus providing a larger electrode-medium contact for the electrochemical reaction. As a result, the current and the oxidation of Fe(II) is enhanced in a certain window of linear velocity.

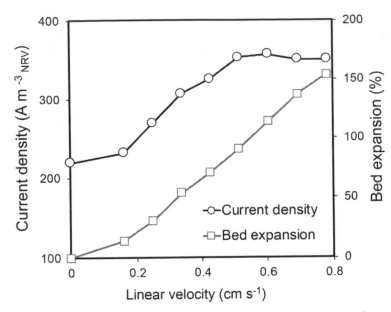

Figure 1: Electrochemical characterization of an electrochemical fluidized bed reactor. Current production and bed expansion at different linear velocities of recirculation in a ME-FBR operated with glassy carbon particles (φ=0.4-0.6 mm) with a medium containing an equimolar solution of 1 mM $[Fe(CN_6)]^{4-}/[Fe(CN_6)]^{3-}$. The fluidized electrode was polarized to 0.4 V (vs. Ag/AgCl, 3 M KCl). The bed expansion was calculated as a percentage of the quotient Δ bed height/bed height at fixed bed conditions.

Given the unique features of fluidized electrodes, researchers have used them to perform electrode-driven microbial reactions. In this case, microorganisms catalyze redox reactions by using a fluidized bed reactor either as an electron acceptor (fluidized anode) or an electron donor (fluidized cathode) in what is called a microbial electrochemical fluidized bed reactor (ME-FBR). This concept was developed and patented by the Bio*e* Group from the University of Alcalá (Spain) by merging two fields: fluidized bed reactors and MET-based systems. In this first approach, an electrochemical cell was converted into a fluidized electrode design by replacing the classic static electrode with a bed of electrically conducting particles (Figure 2). The state of fluidization was achieved by flowing electrolyte upward from the bottom of the chamber. The electrical connection with the external circuit was provided by a solid current collector immersed into the bed onto which the electrically conductive particles were continuously being discharged. The reactor had a tubular geometry that allowed the bed to expand along the column and therefore vary the fluidized electrode volume. The system was configured as a single compartment system with the counter electrode placed at the top of the column.

The potential benefits that the ME-FBR provides over other static designs within microbial electrochemistry are as follows:

- Clogging due to solids and biomass accumulation on the active area of the electrodes can be severely reduced (commonly found when using felts, mesh and static granules of carbon).
- The stimulation of the ion transport from the inner layers of the biofilm to and from the bulk liquid surrounding the anode and the cathode. This becomes important for bioelectrochemically treating wastewater since the conductivity of real wastewater is low compared to the synthetic media with high ionic contents that are used in laboratory scale systems. This is one of the main constraints of treating wastewater with electrochemical methods.
- Reduction of the internal resistance of the electrochemical cell and compensating the ohmic losses due to the low conductivity that is usually present in real wastewaters. The concentration overpotential can be minimized because of the good mixing properties in a fluidized bed system.

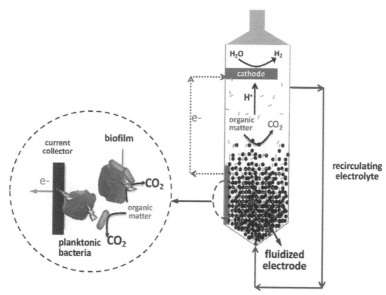

Figure 2: Microbial electrochemical fluidized bed reactor of one chamber and a tubular reactor shape. The fluidized working electrode acts as an anode (electrode potential poised by a potentiostat) and the counter electrode as a cathode.

- Reduction of pH gradients due to good mass transfer rates; the production of protons in the vicinity of the anode and OH- ions at the cathode can affect the viability of the electroactive biofilm if diffusion within the biofilm is slow (Torres et al. 2008; Scott and Yu 2015).

On top of these advantages, from an engineering and practical standpoint, the ME-FBR is attractive because it is simple. For the scaling up of METs for wastewater treatment, continuous flow with lower residence times and larger volumes of water is required. These constraints favor single compartment and membraneless systems. Other METs require complex designs with expensive components, such as an ion exchange membrane for separating anode and cathode chambers, thus complicating the possibility of up-scaling. ME-FBR is a single chamber reactor and is relatively easy to operate. Other designs are modular systems that can enhance treatment capacity by operating several modular units in parallel or serial. However, for treating large volumes of wastewater at full-scale, these kinds of configurations may not be economically viable (Zhuang et al. 2012). Because of this bottleneck, open line of research with METs being up-scaled aims to exploit the existing infrastructure (reactor vessel and complementary equipment) of wastewater treatment plants. This would eliminate a great part of the initial investment costs associated with the implementation of a bioelectrochemical wastewater treatment system. The development of hybrid MET-based systems, such as a conventional anaerobic fluidized bed reactor and a MET (resulting in a ME-FBR), a constructed wetland and a MET (METlands) (Aguirre et al. 2016) and the membrane bioreactors-MFCs (Malaeb et al. 2013) are currently gaining much attention in the field. Merging several technologies allow one to incorporate the respective advantages of each individual technology into the same treatment scheme.

4. Fluidized Bed Electrodes for Stimulating Bioelectrochemical Conversions in METs

For maximizing the connections between electroactive microbes and electrodes, many studies have focused on electrode materials and designs to provide the largest active surface area (Call et al. 2009; Santoro et al. 2014; Aguirre-Sierra et al. 2016). This is due to the location of the biocatalysis in METs at the electrode-biofilm interface where direct EET occurs. The kinetics of these reactions can be enhanced by increasing the area of the electrodes, the activity of the microorganisms within the biofilm or by optimizing the EET rates from the bacteria to the electrodes.

In this sense, METs with three-dimensional beds can increase the surface area of the anode for bacterial adhesion. These beds can remain fixed or fluidized. Fixed bed-METs provide a large ratio of electrode area to wastewater volume. In contrast, dynamic bed electrodes, such as fluidized beds, can provide good mass transport properties, better mixing, an evener temperature distribution and a large electrode surface area per unit of electrode volume. Novel scenarios have been proposed for removing the organic matter in wastewaters with dynamic electrodes, such as carbon-based capacitive mobile granules. These granules are covered by an electroactive biofilm that transfers electrons resulting from wastewater treatment in a biological reactor to the conductive granule. Subsequently, the charged granules are circulated through the anodic chamber of an external MFC in order to harvest electrons on a current collector to generate oxidized granules that are able to act as electron sinks again in the biological reactor (Deeke et al. 2015). Another dynamic bed approach used fluidized carbon granules in contact with a conductive sheet of carbon as the anode. Artificial soluble redox mediators were provided to enhance microbial electron transfer (Kong et al. 2011). However, the first fluidized electrode concept in which a bed of fluidized electrically conductive microparticles behaved as a single electrode with fluid-like properties was reported by Tejedor-Sanz et al. (Tejedor-Sanz et al. 2017b). Figure 3 shows ME-FBR design in which the bed resembles like a single and continuous fluid-like electrode although it is made of discrete electrically conductive microparticles.

4.1 Rapid Electron Discharge of Planktonic Cells on Fluidized Electrodes

The viability of using a fluidized electrode and a bacteria interacting in a synergistic fashion was first evaluated by employing electroactive bacteria from the genus *Geobacter* and a fluidized electrode as the terminal electron acceptor.

Geobacter sulfurreducens is well-known for hosting a vast network of *c*-type cytochromes that participate in the EET to electrodes as proved through spectroelectrochemical techniques (Busalmen et al. 2008). Furthermore, a severe reduction in the *c*-type cytochrome content of *G. sulfurreducens* leads to a weak electrochemical response (Estevez-Canales et al. 2014). Additionally, the network of cytochromes in planktonic *G. sulfurreducens* cells can also function as a short-term sink for electrons from acetate

Figure 3: Pictures of a ME-FBR. A: ME-FBR operated as a 3-electrode electrochemical cell with glassy carbon particles as fluidized anode and carbon felt as cathode. The numbers in the pictures stand for 1) fluidized bed anode; 2) counter electrode; 3) reference electrode; 4) current collector; 5) recirculating outlet port; 6) recirculating flow inlet port. B: Fluidized glassy carbon particles of 0.6-1 mm diameter.

metabolism when extracellular electron acceptors as iron-oxides are unavailable. This is commonly called capacitor effect that suggests the use of the cytochrome *c* network as 'iron lungs' (Esteve-Núñez et al. 2008; Lovley 2008). The same pattern has been observed in electroactive biofilms grown with anodes as the terminal electron acceptor. The electron storage capacity of *Geobacter* electroactive biofilms has been reported as comparable to that of synthetic supercapacitors with low self-discharge rates (Schrott et al. 2014; Liu et al. 2011; Malvankar et al. 2012). Furthermore, chemostat-grown planktonic cells are able to generate a rapid electrical discharge as soon as they are exposed to an electrode. These cells are so-called *plug-and-play* cells because their electroactivity allows a major reduction in the start-up period of microbial electrochemical bioreactors (Esteve-Núñez et al. 2011; Borjas et al. 2015).

When a suspension of *G. sulfurreducens* cells are grown in a chemostat under electron acceptor limiting conditions and then added to a ME-FBR, a rapid electron discharge can be observed indicating an effective EET to the fluid-like anode (Tejedor-Sanz et al. 2017b). Furthermore, the fluidized anode can be a much more efficient electron collector for *G. sulfurreducens* cells than soluble electron acceptors like AQDS. The capacity for electron recovery in a fluidized anode after cell discharge ($3 \cdot 10^{-16}$ mol e$^-$ cell^{-1}) is 18-fold higher than those estimated in previous studies using soluble electron acceptors like AQDS (Esteve-Núñez et al. 2008). Interestingly, the total charge harvested in a fluidized bed electrode is 5-fold higher than the charge collected in the same bed under static conditions. This indicates a very effective interaction between the bacteria and the conductive particles marked by rapid electron transfer from the outermost cytochromes to the fluidized anode.

4.2 Microbial Electroactivity Using Fluidized Electrodes: Planktonic vs. Biofilm Growth

Bacterial adhesion is a complex process that is affected by many factors, like the characteristics of the bacteria, the target material's surface and environmental factors (hydrodynamics or the concentration of bacteria) (Donlan 2002).

Typically, cells behave like single particles in liquid culture. The cell's association rate with the fluidized anode particles in the ME-FBR depends largely on the velocity of the recirculating liquid. Fluid motion in the bulk liquid favors bacteria adhesion because of the enhancing of cell transportation to the surface by convection (Rijnaarts et al. 1993). However, the shearing forces of the beads moving in the mixed systems can also reduce adhesion and may promote cell detachment.

As mentioned before, mass transport can be a limitation for electroactive biofilms as catalysts since they limit the reaction rates and the overall activity of the cells. These problems can be overcome with a mediated electron transfer, under which the reaction can proceed in the suspended media (Velasquez-Orta et al. 2010; Kotloski and Gralnick 2013), or by direct electron transfer not proceeding through biofilms where every single planktonic cell contributes to current production.

4.2.1 Promoting Planktonic Growth Using Fluidized Electrodes

The quantity and quality of the biomass in the reactor can be greatly affected by the construction, design, type of conductive particulate and the dynamic fluidization. Moreover, these features can create a scenario promoting either planktonic or biofilm growth. For instance, using an electrically conductive bed of glassy carbon particles, which are nonporous and smooth, allows *G. sulfurreducens* to grow in a planktonic state while performing EET from acetate metabolism (Tejedor-Sanz et al. 2017b) (Figure 4). This planktonic interaction between the cells and the fluid-like electrode has been proven viable by over two months of successful batch operation with multiple medium replacements. This means that the redox coupling between *G. sulfurreducens* planktonic cells and the fluidized anodic particles is sufficiently balanced for promoting microbial growth. *G. sulfurreducens* achieves this mode of growth through decoupling catabolism and respiration by substrate oxidation with the temporary storage of electrons in their cytochromes network and is followed by electron transfer to the fluidized anode.

The surface of the glassy carbon particles used by the authors lacked roughness, superficial imperfections and pores that reduces the ability of bacteria to anchor. Glassy carbon is a material with a

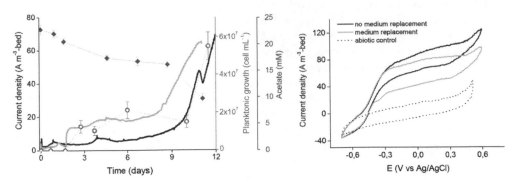

Figure 4: A: Planktonic grown of *G. sulfurreducens* (green open circles), acetate in medium (blue diamonds) and current density in a ME-FBR operated with glassy carbon particles and acetate as sole electron donor. B: Voltammograms of the fluidized anode at maximum current production in Figure 4 A under different conditions.
[Figure extracted from (Tejedor-Sanz et al. 2017b). Reproduction with permission of publishers.]

relative hydrophobicity nature and low wettability that further complicates the attachment and growth of cells. Also, this material does not possess the typical superficial functional groups key for the bacteria-electrode interactions (Fiset and Puig 2015). Furthermore, it has been reported that hydrophobic surfaces require larger periods of time to be colonized by microbes than hydrophilic surfaces in METs (Santoro et al. 2014).

4.2.2 Promoting Biofilm Growth and Bacteria Attachment Using Fluidized Electrodes

In contrast, by employing electrode particles made of porous and hydrophilic material biofilm architecture could be easily formed. In this regard, activated carbon, a material with high porosity and charged superficial functional groups bound to its surface and conferring a hydrophilic nature, can be used to operate a biofilm-based ME-FBR. Under any biofilm-based configuration, the support material determines the capacity for biomass retention that is critical for the performance of the system. Thus, the choice of the correct conductive particle is crucial for ensuring the success of the desired process. There is a large list of tested and well-studied materials used for non-electrochemical fluidized bed reactors. These include materials such as sepiolite, sand, pumice stone (Balaguer et al. 1997), biolite (Prakash and Kennedy 1996) and celite beads (Hsu and Shieh 1993). In contrast, just a few materials have been used for abiotic fluidized electrodes in reactors, such as silver (Kreysa et al. 1975), tin oxide particles coated in graphite (Lee et al. 2002) and copper (Fleischmann et al. 1971). Even fewer materials, electroconductive activated carbon (Tejedor-Sanz et al. 2017b, 2018) and glassy carbon particles (Tejedor-Sanz et al. 2017b) have been reported for colonized electroconductive fluidized particles. We expect a wider variety of materials will likely be utilized after a better understanding of the bacteria-electrode interaction has been elucidated.

In addition to the physical-chemical characteristics of the electrode, shear stress is one of the main factors that control biofilm structure, thickness and density of any surface. This shear stress among the particles is governed, in turn, by the hydraulic conditions within the reactor. This opens the possibility of tuning an electroactive biofilm thickness by varying the up-flow velocity in the ME-FBR. Efficient MET design compromises between high electrode surface area and biofilm thickness to maintain efficient mass transport and the desired metabolic activity. It has been reported that cells in the upper layer of the biofilm (beyond 30-40 μm) on flat and static electrodes are most likely inactive or not growing and therefore do not contribute to current production (Schrott et al. 2014). The superior mass transport conditions in a ME-FBR may allow one to develop thicker active biofilms as compared to using static electrodes designs. However, the biofilm thickness that has been observed in a ME-FBR treating real brewery wastewater is of *ca.* 10 μm (Tejedor-Sanz et al. 2017a, 2018). This thickness is low as compared to previously observed biofilms in either fluidized bed reactors (with nonelectrically conductive bed) or in microbial electrochemical systems based on standard electrodes (Liu and Tay 2001; Jana et al. 2014). In addition, the authors observed partial microbial coverage on the particles of the ME-FBR system after over one year of operation (Tejedor-Sanz et al. 2018). Also, it was observed that initial cell attachment occurred at the pores and cavities of the material, the regions sheltered from hydraulic shear forces and the shear stress from particle-to-particle erosion.

The operating regime can also strongly influence particle colonization in a fluidized bed. By operating under a continuous mode, one often promotes rapid biomass attachment to the carriers in order to avoid cell washout (Rijnaarts et al. 1993). The hydraulic retention time can strongly influence the rate of particle colonization. In contrast, when operating a ME-FBR in batch or discontinuous mode, the biomass may remain active in the planktonic state while respiring the electrode in motion.

Although conditions for promoting either planktonic or biofilm formation on the particles of a fluidized electrode can be selected, both bacterial living modes performing EET may coexist in a ME-FBR. These modes might exhibit distinct phenotypes or subpopulations with respect to gene transcription and growth rate, leading to competition and syntrophic interactions among them for substrate utilization and electron acceptor utilization (Cresson et al. 2008).

4.3 Fluidized Electrodes as Anodes for Organic Matter Removal

Electrochemical fluidized beds are able to effectively stimulate the degradation of organic matter by electroactive bacteria. The electrons from the microbial oxidation of organic matter are transferred to the fluidized anodic particles that are discharged when in contact with a current collector. This phenomenon has been demonstrated with a pure culture of *G. sulfurreducens*, which could efficiently couple current generation on a fluidized anode with acetate oxidation, while achieving coulombic efficiencies of up to 91%. In addition, real brewery wastewater can be treated by using a mixed culture in a ME-FBR while producing current from the process (Tejedor-Sanz et al. 2017a, 2018). This indicates that electroactive microbial communities have a widespread capacity for interacting with fluidized anodes. As it was reported, the complex composition of brewery wastewater with the conductive nature and fluid-dynamics of the fluidized anode together created an environment in which a microbially stratified biofilm was formed. These internal layers of the biofilm on the fluidized anodic particles were highly enriched with *Geobacter* species, whereas the outermost layers were colonized by other species of bacteria (Figure 5C). This suggests that *Geobacter* species in this stratified biofilm in the ME-FBR could be responsible for the direct and ultimate transfer of electrons to the anode, either from their own metabolism or from other cells capable of performing interspecies electron transfer.

During the continuous treatment of this brewery wastewater, at an organic loading rate of 1.7 kg m$^{-3}_{NRV}$ d^{-1}, removal efficiencies for organic matter as high as 95±1.4% were achieved (Figure 5A). These studies revealed that this technology was further able to operate at higher organic loading rates than the above-mentioned processes. However, this compromised the coulombic efficiency (Figure 5B). Additionally, the ME-FBR can be used to complement other already established technologies. For instance, this treatment has several advantages over anaerobic digestion such as the capacity to directly collect electricity, to treat lower substrate concentrations and to operate at a wider range of temperatures. Therefore, the ME-FBR could treat the effluent of anaerobic digesters, achieving relatively high coulombic efficiencies and recovering electricity in the form of hydrogen from water electrolysis at the counter electrode.

Overall, by optimizing parameters such as bed electrode materials, reactor design and microbial community adaptation (to maximize electron recovery) the ME-FBRs can offer a potential alternative or complementary contribution to conventional anaerobic digesters in the agroindustry field.

4.4 Fluidized Electrodes for Nutrients Removal

As occurring in anaerobic wastewater treatment bioreactors, a ME-FBR designed for organic matter removal cannot completely treat wastewater due to the remaining nutrients. Only a fraction of the nutrients that are consumed in microbial anabolism is eliminated. ME-FBR can overcome this limitation by operating with other systems in which nutrients are removed, like pre-treatment with electrocoagulation

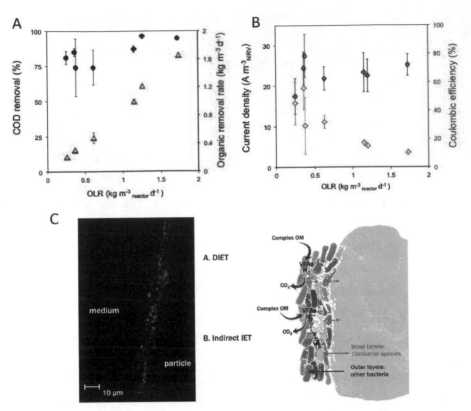

Figure 5: Organic matter removal from brewery wastewater in a ME-FBR and biofilm stratification. A: Organic matter removal in terms of COD removal and organic matter removal rate under different organic loading rates (OLR) in a ME-FBR. B: Current density production and coulombic efficiencies achieved at different ORLs. C: Microbial stratification observed using fluorescence *in situ* hybridization (FISH) in a biofilm developed on a fluidized anode composed of activated carbon particles treating a brewery wastewater. The green probe corresponds to *Geobacter* cluster, whereas the red probe to Eubacteria. [Figures extracted from (Tejedor-Sanz et al. 2018). Reproduction with permission of respective publisher.]

(Tejedor-Sanz et al. 2017a). Other designs of METs could assist with eliminating phosphorus or nitrogen as well (Virdis et al. 2010; Ichihashi and Hirooka 2012; Tejedor-Sanz et al. 2016).

4.5 Fluidized Electrodes as Cathodes

As widely reported for other microbial electrochemical systems (Nevin et al. 2010; Liu et al. 2013; Kashima and Regan 2015) bacteria can effectively accept electrons from an electrode poised at a negative potential. Thus, by setting the fluidized electrode to an adequately negative potential, it can behave as an electron donor for electroactive bacteria in order to catalyze reduction reactions. To date, this has been demonstrated for nitrate reduction and for biological hydrogen production using a mixed microbial community (Tejedor-Sanz et al. 2017). Due to the versatility of fluidized bed electrodes for culturing electroactive bacteria consortiums under either planktonic or biofilm phenotypes, the fluidized cathodes may have potential applications in bioelectrosynthesis. Here, many reactions are mediated by hydrogen or small acids such as formate or acetate and thus the conversions occur within microbial suspensions.

5. Conclusions

The work presented in this chapter supports the idea that microbial fluidized bed reactors present a potential alternative to current bioelectrochemical designs. Whether serving as an electron donor or an acceptor for electroactive bacteria, fluidized electrodes can effectively drive microbial catalysis. The interaction between the electrode and electroactive bacteria is the core of the catalytic process in METs and fluidized electrodes can highly improve the physical cell-electrode contact, a fact that can have a notable impact on the kinetics of the process.

6. Future Perspectives

To date, the research on fluidized beds in the area of microbial electrochemistry has been developed on the basis of a proof-of-concept study, and the results obtained can be taken as initial benchmarks. Nevertheless, in order to optimize these systems and obtain a better understanding of the electron transfer process occurring between fluidized anodes and bacteria a number of additional studies should be conducted.

The analysis of the parameters that affect and promote the planktonic cell growth and the biofilm development in the ME-FBR could provide insights into the mechanisms of EET in the microbial electrochemistry and in the geochemistry field as well. For instance, researching biofilm development on the fluidized electrode particles is recommended. Further studies should include factors, such as the strength of the shearing forces governing the quantity, the quality of anode colonization as well as the thickness of the electroactive and nonelectroactive biofilm as a key parameter affecting the performance of the process. An optimum biofilm thickness should be a compromise between the quantity of biomass and its metabolic state.

In order for the ME-FBR to be competitive over present technologies for wastewater treatment, one must reduce energetic losses and maximize electron recovery at the fluidized electrode. By screening key parameters that stimulate the electrogenic pathway over the methanogenic one (e.g., fluidized anode potential, bed expansion, pH, recirculating flux) the bioelectrochemical organic matter removal and the economic viability of the process should be notably enhanced.

Additionally, with further research regarding bed electrode materials and the selection or engineering of microbial strains adapted to interact with fluidized electrodes, the extracellular electron transfer rates could be highly improved for this electrode configuration. Meanwhile, the exploitation of the counter electrode reaction in the ME-FBR (and in general, in any MET) is critical for the economic sustainability of the application and should be addressed in future studies with this configuration.

Acknowledgements

We would like to thank Colin Wardman for valuable corrections and criticisms.

References

Aguirre-Sierra A, Baccheti T, Berná A, Aragón C, Salas JJ, Esteve-Núñez A, (2016) Microbial Electrochemical Systems outperform fixed-bed biofilters for cleaning-up urban wastewater. Environ Sci: Water Res Technol 2: 984–993.

Backhurst JR, Coulson JM, Goodridge F, Plimley RE, Fleischmann M (1969) A preliminary investigation of fluidized bed electrodes. J Electrochem Soc 116:1600–1607.

Balaguer MD, Vicent MT, Parfs JM (1997) A comparison of different support materials in anaerobic fluidized bed reactors for the treatment of vinasse. Environ Technol 18:539–544.

Bello MM, Abdul Raman AA, Purushothaman M (2017) Applications of fluidized bed reactors in wastewater treatment – A review of the major design operational parameters. J Cleaner Prod141:1492–1514.

Borjas Z, Ortiz JM, Aldaz A, Feliu J, Esteve-Núñez A (2015) Strategies for reducing the start-up operation of microbial electrochemical treatments of urban wastewater. Energies 8:14064–14077.

Busalmen JP, Esteve-Nuñez A, Berna A, Feliu JM (2008) C-type cytochromes wire electricity-producing bacteria to electrodes. Angew Chem Int Ed Engl 47:4874–7.

Call DF, Merrill MD, Logan BE (2009) High surface area stainless steel brushes as cathodes in microbial electrolysis cells. Environ Sci Technol 43:2179–2183.

Cresson R, Escudié R, Steyer JP, Delgenès JP, Bernet N (2008) Competition between planktonic fixed microorganisms during the start-up of methanogenic biofilm reactors. Water Res 42:792–800.

Deeke A, Sleutels THJA, Donkers TFW, Hamelers HVM, Buisman CJN, Ter Heijne A (2015) Fluidized capacitive bioanode as a novel reactor concept for the microbial fuel cell. Environ Sci Technol 49:1929–35.

Donlan RM (2002) Biofilms: microbial life on surfaces. Emerg Infec Dist. 8:881–890.

Esteve-Núñez A, Busalmen JP, Berná A, Gutiérrez-Garrán C, Feliu JM (2011) Opportunities behind the unusual ability of Geobacter sulfurreducens for exocellular respiration electricity production. Energy Environ Sci 4:2066–2069.

Esteve-Núñez A, Sosnik J, Visconti P, Lovley DR (2008) Fluorescent properties of C-type cytochromes reveal their potential role as an extracytoplasmic electron sink in Geobacter sulfurreducens. Environ Microbio 10:497–505.

Estevez-Canales M, Kuzume A, Borjas Z, Füeg M, Lovley D, Wlowsky T, Esteve-Nunez A (2014) A severe reduction in the cytochrome C content of Geobacter sulfurreducens eliminates its capacity for extracellular electron transfer. Environ Microbiol Rep 7:219–226.

Fiset E, Puig, S (2015) Modified Carbon Electrodes: A new approach for bioelectrochemical systems. J Bioremed Biodegrad 6.

Fleischmann M, Oldfield JW, Tennakoon L (1971) Fluidized bed electrodes Part IV. Electrodeposition of copper in a fluidized bed of copper-coated spheres. J Appl Electrochem 1:103–112.

Goff PL, Vergnes F, Coeuret F, Bordet J (1969) Applications of fluidized beds in electrochemistry. Ind Eng Chem 61:8–17.

Heijnen JJ, Mulder A, Enger W, Hoeks F (1989) Review on the application of anaerobic fluidized bed reactors in waste-water treatment. Chem Eng J 41:B37–B50.

Hiddleston JN, Douglas AF (1970) Current/potential relationships potential distribution in fluidized bed electrodes. Electrochim Acta 15:431–443.

Hsu Y, Shieh WK (1993) Startup of anaerobic fluidized bed reactors with acetic acid as the substrate. Biotechnol Bioeng 41:347–353.

Huh T (1985) The Performance Electrochemical Behavior of Fluidized Bed Electrodes. University of California, Berkeley.

Huh T, Evans JW (1987) Electrical electrochemical behavior of fluidized bed electrodes I. Potential transients time-averaged values. J Electrochem Soc 134:308–317.

Ichihashi O, Hirooka K (2012) Removal recovery of phosphorus as struvite from swine wastewater using microbial fuel cell. Bioresour Technol 114:303–307.

Jana PS, Katuri K, Kavanagh P, Kumar A, Leech D (2014) Charge transport in films of Geobacter sulfurreducens on graphite electrodes as a function of film thickness. Phys Chem Chem Phys 16:9039–9046.

Kashima H, Regan JM (2015) Facultative nitrate reduction by electrode-respiring Geobacter metallireducens biofilms as a competitive reaction to electrode reduction in a bioelectrochemical system. Environ Sci Technol 49:3195–3202.

Kong W, Guo Q, Wang X, Yue X (2011) Electricity generation from wastewater using an anaerobic fluidized bed microbial fuel cell. Ind Eng Chem Res 50:12225–12232.

Kotloski NJ, Gralnick JA (2013) Flavin electron shuttles dominate extracellular electron transfer by Shewanella oneidensis. MBio 4.

Kreysa G, Pionteck S, Heitz E (1975) Comparative investigations of packed fluidized bed electrodes with non-conducting conducting particles. J Appl Electrochem 5:305–312.

Lee JK, Ryu DH, Ju JB, Shul YG, Cho BW, Park D (2002) Electrochemical characteristics of graphite coated with tin-oxide copper by fluidised-bed chemical vapour deposition. J Power Sources 107:90–97.

Liu D, Lei L, Yang B, Yu Q, Li Z (2013) Direct electron transfer from electrode to electrochemically active bacteria in a bioelectrochemical dechlorination system. Bioresour Technol 148:9–14.

Liu Y, Tay JH (2001) Metabolic response of biofilm to shear stress in fixed-film culture. J Appl Microbiol 90:337–342.

Lovley DR (2008) Extracellular electron transfer: wires, capacitors, iron lungs, more. Geobiology 6:225–31.

Malaeb L, Katuri KP, Logan BE, Maab H, Nunes SP, Saikaly PE (2013) A Hybrid microbial fuel cell membrane bioreactor with a conductive ultrafiltration membrane biocathode for wastewater treatment. Environ Sci Technol 47:11821–11828.

Nevin KP, Woodard TL, Franks AE, Summers ZM, Lovley DR (2010) Microbial electrosynthesis: feeding microbes electricity to convert carbon dioxide water to multicarbon extracellular organic compounds. MBio 1:103–110.

Prakash R, Kennedy KJ (1996) Kinetics of an anaerobic fluidized bed reactor using biolite carrier. Can J Civil Eng 23:1305–1315.

Rijnaarts HHM, Norde W, Bouwer EJ, Lyklema J, Zehnder AJB (1993) Bacterial adhesion under static dynamic conditions. Appl Environ Microbiol 59:3255–3265.

Santoro C, Guilizzoni M, Correa Baena JP, Pasaogullari U, Casalegno A, Li B, Babanova S, Artyushkova K, Atanassov P (2014) The effects of carbon electrode surface properties on bacteria attachment start up time of microbial fuel cells. Carbon 67:128–139.

Schrott GD, Ordoñez MV, Robuschi L, Busalmen JP (2014) Physiological stratification in electricity-producing biofilms of Geobacter sulfurreducens. ChemSusChem 7:598–603.

Scott K, Yu EH (2015) Microbial electrochemical fuel cells: fundamentals applications. Woodhead Publishing.

Tejedor-Sanz S, Bacchetti T, Salas JJ, Pastor L, Esteve-Nuñez A (2016) Integrating a microbial electrochemical system in a classical wastewater treatment configuration for removing nitrogen from low COD effluents. Environ Sci: Water Res Technol 2:884–893.

Tejedor-Sanz S, Fernández-Labrador P, Hart S, Torres CI, Esteve-Núñez A (2018) Geobacter dominates the inner layers of a stratified biofilm on a fluidized anode during brewery wastewater treatment. Front Microbiol 9:378.

Tejedor-Sanz S, Fernández-Labrador P, Torres IC, Esteve-Núñez A (2017) Fluidized electrodes as electron acceptors or electron donors: strategies and applications. 6th General Meeting of the International Society for Microbial Electrochemistry. 3-6 October. Lisbon.

Tejedor-Sanz S, Ortiz JM, Esteve-Núñez A (2017a) Merging microbial electrochemical systems with electrocoagulation pretreatment for achieving a complete treatment of brewery wastewater. Chem Eng J 330:1068–1074.

Tejedor-Sanz S, Quejigo JR, Berná A, Esteve-Núñez A (2017b) The planktonic relationship between fluid-like electrodes and bacteria: wiring in motion. ChemSusChem 10:693–700.

Torres CI, Kato Marcus A, Rittmann BE (2008) Proton transport inside the biofilm limits electrical current generation by anode-respiring bacteria. Biotechnol Bioeng 100:872–881.

Trupp AC (1968) Dynamics of liquid fluidized beds of sphere. I Chem E Sym Ser 30:182–189.

Velasquez-Orta SB, Head IM, Curtis TP, Scott K, Lloyd JR, Canstein H von (2010) The effect of flavin electron shuttles in microbial fuel cells current production. Appl Microbiol Biotechnol 85:1373–1381.

Virdis B, Rabaey K, Rozendal RA, Yuan Z, Keller J (2010) Simultaneous nitrification, denitrification carbon removal in microbial fuel cells. Water Res 44:2970–2980.

Zhuang L, Zheng Y, Zhou S, Yuan Y, Yuan H, Chen Y (2012) Scalable microbial fuel cell (MFC) stack for continuous real wastewater treatment. Bioresour Technol 106:82–88.

Electricity Production from Carbon Monoxide and Syngas in a Thermophilic Microbial Fuel Cell

Sonia M. Tiquia-Arashiro[*]

Department of Natural Sciences, University of Michigan, Dearborn, MI 48128, USA

1. Introduction

Microbial fuel cells (MFCs) represent a novel technological solution for electricity production from biomass. In its most simple configuration, a microbial fuel cell is a device that uses microorganisms to produce an electrical current. The technology exploits the ability of microorganisms that are capable of extracellular electron transfer to an insoluble electron acceptor, such as an electrode. Logan (2008) defined microorganisms as exoelectrogens because of their capability of exocellular electron transfer. Other researchers described the microorganisms as electrochemically active bacteria (Manish and Banerjee 2008), anode-respiring bacteria (Moon et al. 2004) and electricigens (Logan 2004). The oxidation of organic chemicals by microorganisms liberates both electrons and protons. Electrons are then transferred from microorganisms to the anode and subsequently to the cathode through an electrical network. Simultaneously, protons (electron acceptor) migrating to the cathode combine with electrons and an electron acceptor, such as oxygen, to produce water. The electrical current generated is similar to that in chemical fuel cells; however, in MFCs the microbial catalysts are attached to the anode surface (Franks and Nevin 2010).

For microorganisms to produce electricity in MFCs, the cells need to transfer electrons generated along their membranes to their surfaces. While anodes and cathodes can function in microbial respiration, research has been focused on understanding microbial anodic electron transfer. Anode-respiring bacteria catalyze electron transfer in organic substrates onto the anode as a surrogate for natural extracellular electron acceptors (e.g., ferric oxides or humic substances) by a variety of mechanisms (Lovley et al. 2004; Lovley 2006; Lovely 2008; Logan 2009). Microorganisms transfer electrons to anodes either directly or via mediated mechanisms. In a direct electron transfer, microorganism requires physical contact with the electrode for the current production. The contact point between the bacteria and the anode surface requires outer membrane-bound cytochromes or putatively conductive pili called nanowires. In mediated electron transfer mechanisms, bacteria either produce or take advantage of indigenous soluble redox compounds such as quinones and flavins to shuttle electrons between the terminal respiratory enzyme and the anode surface.

---

*Corresponding author: smtiquia@umich.edu

MFCs have been successfully operated on a wide range of substrates, such as acetate, CO, H_2, glucose, galactose, butyrate, starch, marine sediments, swine wastewater, etc. (Pant et al. 2010; Hussain et al. 2011). In principle, any biodegradable material could be utilized as a fuel for electricity generation in an MFC. An ideal MFC can produce current while sustaining a steady voltage if a steady supply of substrate is maintained. MFCs do not need to use metal catalysts at the anode; instead, they use microorganisms (exoelectrogens) that biologically oxidize organic matter and transfer electrons to the electrode. Exoelectrogens, inoculated in MFCs for electricity generation, are found in marine sediment, soil, wastewater, freshwater sediment or activated sludge (Nevin et al. 2008).

The gasification of biomass at high temperatures leads to the generation of synthesis gas (syngas). Carbon monoxide (CO) and hydrogen account for 60-80% of the syngas composition with CH_4, CO_2, SO_2, H_2S and NH_3 present in smaller amounts (Sipma et al. 2006; Munasinghe and Khanal 2010). Syngas can be transformed into biofuels (such as ethanol, butanol, methane and hydrogen) that can be performed using the chemical as well as biological catalysts or by microbial transformation (Henstra et al. 2007; Tiquia 2014a; Pomaranski and Tiquia-Arashiro 2016). It can also be used to generate electricity. The microbial transformation of CO/syngas is carried out by carboxydotrophic bacteria. Carboxydotrophic bacteria use CO as their sole carbon source (Henstra et al. 2007; Oelgeschlager and Rother 2008; Nguyen et al. 2013; Tiquia-Arashiro 2014b). The ability of these microorganisms to oxidize and metabolize CO is connected to the existence of the enzyme CO-dehydrogenase (Henstra et al. 2007; Nguyen et al. 2013; Tiquia-Arashiro 2014b) that is often found in carboxydotrophic methanogens and acetogens (Sipma et al. 2006; Tiquia-Arashiro 2014c) and result predominantly in methane and acetate production, respectively. The metabolic activity of carboxydotrophic bacteria also results in the formation of H_2, ethanol, butyrate, butanol and acetate (Henstra et al. 2007; Nguyen et al. 2013; Tiquia-Arashiro 2014d; Pomaranski and Tiquia-Arashiro 2016). Some of these metabolic products can be utilized by exoelectrogens for electricity production in a mediatorless MFC in which the exoelectrogenic microorganisms transfer electrons to the anode via nanowires or self-produced mediators (Faaij et al. 1997; Steele et al. 2001; Song 2002; Liu and Logan 2004; Hussain et al. 2011; Tiquia-Arashiro 2014d). This review focuses on recent advances that have been made in electricity production from CO/syngas in a thermophilic MFC accomplished by a consortium of carboxydotrophic and CO-tolerant exoelectrogenic microorganisms.

2. Exoelectrogens

Exoelectrogens (e.g., anodophilic, electricigens or anode-respiring bacteria) are microorganisms that can transfer electrons extracellularly (Niessen et al. 2006). Different genetic groups of bacteria have shown exoelectrogenic activity in MFCs, including β-*Proteobacteria* (*Rhodoferax*) (Chaudhuri and Lovley 2003), γ-*Proteobacteria* (*Shewanella* and *Pseudomonas*) (Kim et al. 1999, 2002; Ren et al. 2007), δ-*Proteobacteria* (*Aeromonas*, *Geobacter*, *Geopsychrobacter*, *Desulfuromonas*, and *Desulfobulbus*) (Bond et al. 2002; Holmes et al. 2004a, 2004b; Pham et al. 2003), *Firmicutes* (*Clostridium*) (Park et al. 2001) and *Acidobacteria* (*Geothrix*) (Bond and Lovely 2005). It has been demonstrated that cell-bound outer membrane cytochromes and conductive pili (nanowires) play a key role in electron transfer for some *Geobacter* and *Shewanella* species (Lovely et al. 2004; Reguera et al. 2005; Gorby et al. 2006). Alternatively, some exoelectrogens, such as *Pseudomonas aeruginosa* (Rabaey et al. 2004) and *Geothrix fermentans* (Bond and Lovely 2005), excrete mediators to shuttle electrons to surfaces. Many of the exoelectrogens that produce current in an MFC are dissimilatory metal-reducing bacteria, originally isolated based on their ability to reduce insoluble metals (e.g., Fe(III) or Mn(IV) oxides) in the natural environment (Lovely 2006; Logan and Regan 2006). *Geobacter sulfurreducens* secretes riboflavin in its monolayer biofilms that interact with outer membrane *c*-type cytochromes (OM c-Cyts) (Malvankar and Lovley 2012). It also produces conductive nanowires i.e., type IV pili (made up of PilA monomer units) and OM c-Cyts (chiefly OmcZ) that mediate the DET (Jayapriya and Ramamurthy 2012; Sneider et al. 2012). *S. oneidensis*, the most versatile exoelectrogen because of its ability to reduce diverse electron acceptors (Jain et al. 2012; Leung et al. 2013), secretes two types of flavins (riboflavin and flavin mononucleotide) that help to transfer the electrons exogenously to the electrode surface (Brutinel and Gralnick 2012; Kotloski and Gralnick 2013). A single cell of *G. sulfurreducens* can generate ca. 90 fA amount of current in the MFC (Biffinger 2013), while *S. oneidensis* MR-1 can transfer the electrons to the anode surface across the cell membrane with a rate of 1.3×10^6 e$^-$ cell^{-1} s^{-1} (Mclean et al. 2010). Thermophilic exoelectrogens have been reported in several studies (Choi et al. 2004; Wrighton et al. 2008; Marshall and May 2009; Fu et al. 2013a, Fu et al. 2013b; Fu et al. 2015).

3. Carbon Monoxide/Syngas as Substrates for Electricity Generation in MFCs

The possibility of electricity production from CO and syngas in an MFC has been demonstrated by Kim and Chang (2009) in a two-stage reactor system in which CO is first microbiologically converted to fermentation products (dominantly acetate) and subsequently fed to an MFC seeded with anaerobic sludge. A maximum power output of 1.6 mW L^{-1} (normalized to the total reactor volume) and coulombic efficiency (CE) of ~ 5% were reported. Mehta et al. (2010) for the first time reported the electricity generation in an MFC directly fed with CO or syngas (a mixture of CO and H$_2$). The maximum volumetric power output and coulombic efficiency achieved in their study were 6.4 mWL^{-1} and 8.7%, respectively. Although the overall performance of the MFC directly fed with CO or syngas was marginally better than the two-stage process utilized in the study of Kim and Chang (2009), it clearly demonstrated that the microbial communities of an MFC could utilize CO or syngas as the electron donor for electricity generation. However, the adoption of an efficient gas-liquid mass transfer mechanism and reactor design optimization need to be put in place for the performance of an MFC on CO/syngas.

Based on the analysis of metabolic products, Mehta et al. (2010) concluded that the production of electricity from CO or syngas in an MFC proceeds through a multi-step biotransformation process. Several concurrent pathways are hypothesized; one is the pathway involved in CO transformation to acetate by acetogenic carboxydotrophic (CO-oxidizing) microorganisms followed by oxidation of acetate by CO-tolerant exoelectrogenic microorganisms. This pathway is hypothesized to be the foremost step responsible for electricity generation. Notably, the ability of the electricity-producing microorganisms to utilize H$_2$ as an electron donor has also been documented (Bond and Lovley 2003). Based on this observation a pathway of electricity production through H$_2$ and acetate followed by acetate conversion to electricity has been suggested. The experimental observations from the studies of Kim and Chang (2009) and Mehta et al. (2010) shows that the electricity production from syngas in an MFC poses several engineering and microbiological challenges pertaining to gas transfer limitations, selection and enrichment of microorganisms capable of efficient syngas transformation to electricity and selection of cathodic catalysts resistant to poisoning by CO and sulfur compounds. The subsequent sections of this chapter review the microbial communities and reactor designs suitable for MFC operation on CO/syngas at thermophilic temperatures.

4. Electricity Generation in Thermophilic CO/Syngas-Fed MFCs

The search for microorganisms that is capable of catalyzing the reduction of an electrode within a fuel cell has primarily been focused on bacteria that operate at mesophilic temperatures. However, anaerobic digestion studies have reported on the superiority of thermophilic operation and demonstrated a net energy gain in terms of methane yield. Microorganisms that function optimally under extreme conditions are beginning to be examined because they may serve as more effective catalysts (e.g, higher activity, greater stability, longer life, capable of utilizing a broader range of fuels) in MFCs. Considering that at the exit of the gasification process syngas temperature could be in a range of 45 to 55°C, the operation of the MFC at thermophilic temperatures might be preferable because it eliminates the need for syngas cooling and might lead to a higher biocatalytic activity (Jong et al. 2006; Mathis et al. 2008). The thermophilic conditions would also lead to a reduced oxygen solubility that is beneficial considering even trace amounts of unreacted O$_2$ diffusing through the cathode can inhibit the anaerobic carboxydotrophic microorganisms which are highly sensitive to the presence of O$_2$ (Tiquia-Arashiro 2014d).

Although the exoelectrogens studied for the generation of electricity in an MFC are predominantly mesophilic (e.g., *Geobacter sulfurreducens* or *Geobacter metallireducens*), successful MFC operation under thermophilic conditions (50 to 55°C) has been demonstrated (Choi 2004; Jong et al. 2006; Hussain et al. 2011). Hussain et al. (2011) demonstrated for the first time that electricity can be generated in a thermophilic MFC fed with syngas. The thermophilic conditions led to a higher power density and improved syngas transformation efficiency as compared to a similar MFC operated under mesophilic conditions. The Coulombic CE is also improved to 20-26% as compared to 6-9% reported for the mesophilic MFC. The supply of CO to the anodic liquid was also improved (Hussain et al. 2011). Several reasons can be cited to explain the improved MFC performance under thermophilic conditions. Firstly, thermophilic conditions affect the activation, ohmic and diffusion losses at the anode. The activation losses contribute to 5-10% of the total internal resistance in a mesophilic MFC (Logan 2008; Zhang and Liu 2010). Secondly, the electrochemical reaction rates increase with increasing temperature and thus leading to lower activation losses. Thirdly, the higher temperature affects the diffusion of the substrates in the anodic liquid, thereby influencing the concentration losses that account for 45-50% of the total internal resistance of the MFC (Logan 2008; Zhang and Liu 2010). Likely, the operation of the syngas-fed MFC at thermophilic temperatures increases the transfer rate of CO and H_2 not only to the anodic liquid but also facilitates the transport of the dissolved gasses through the stagnant liquid layer adjacent to the anode fibers, thus reducing the diffusion losses. Overall, the internal resistance of the thermophilic MFC at optimized performance is less than 50 Ω, whereas the mesophilic MFC internal resistance is above 120 Ω (Mehta et al. 2010; Hussain et al. 2011). Finally, thermophilic conditions increase the activity of the microorganisms. Up to a certain temperature, the biomass growth and substrate conversion rates increase with temperature according to the Arrhenius relationship. In general, thermophilic microorganisms feature higher growth and reaction rates as compared to the mesophilic cultures (Min et al. 2008). Therefore, a higher carboxydotrophic activity could be expected at thermophilic temperatures. In a mesophilic CO-fed MFC, the step of CO conversion to acetate appeared to limit the overall transformation rate (Mehta et al. 2010). Another advantage of the thermophilic process is the reduced O_2 solubility at elevated temperatures. While most of the O_2 diffusing through the cathode surface is consumed by the cathodic reaction, the residual O_2 diffuses to the anodic liquid and thus results in the inhibition of the anodophilic and carboxydotrophic populations (Oelgeschlager and Rother 2008) as well as competes with the anode as the final electron acceptor. The presence of trace amounts of O_2 in the anodic chamber is observed to significantly impair the power output of the mesophilic syngas-fed MFC (Hussain et al. 2011).

Mathis et al. (2008) studied thermophilic bacteria selected from sediments that colonize the anode of acetate and cellulose-fed MFCs. Cloning and sequencing of the biofilm, formed at the anode of the acetate fed MFC, showed the presence of *Deferribacters* and *Firmicutes*. Interestingly, 48 clones (out of 64) of *Firmicutes* had RFLP patterns and sequences (99%) most similar to that of *Thermincola carboxydophila*, a hydrogenogenic CO-oxidizing thermophilic microorganism (Mathis et al. 2008). These findings indicate that temperate aquatic sediments are a good source of thermophilic electrode-reducing bacteria. *Firmicutes* spp. have been also identified during thermophilic MFC operation by Wringhton et al. (2008) whose findings provided a detailed analysis of microbial community dynamics in an acetate-fed MFC inoculated with sludge collected from a thermophilic anaerobic digester. Several thermophilic metal-reducing bacteria have been studied including *Ferroglobus placidus* and *Geoglobus ahangari* that can grow at 85°C by coupling acetate oxidation to Fe (III) reduction (Tor et al. 2001). *Deferribacter thermophilus* (isolated from a petroleum reservoir in the UK) grows by the reduction of Fe (III) and Mn (IV) and nitrate in the presence of acetate, yeast extract, peptone and other carbon sources in the temperature range of 50-65°C (Greene et al. 1997). The bacterium, *Geothermobacter ehrlichii* (isolated from a hydrothermal vent), coupled acetate oxidation to Fe (III) reduction with an optimum growth temperature of 55°C. This strain is the first member of the *Geobacteraceae* family reported to be capable of thermophilic growth. Fe (III) reduction coupled to acetate oxidation has also been demonstrated by the bacterium *Thermincola ferriacetica* (Zavarzina et al. 2007). Population analysis of the exoelectrogenic microorganisms suggests a possible involvement of *Caloramator*-related bacteria in electricity generation (Fu et al. 2013b). Pure culture of *Caloramator australicus* shows electricity-generating ability, indicating that the bacterium is a new thermophilic exoelectrogen (Fu et al. 2013b). Overall, this broad range of thermophilic exoelectrogens might be capable of forming a syntrophic consortium with thermophilic carboxydotrophic microorganisms for efficient operation of a syngas-fed MFC.

5. Thermophilic Carboxydotrophs

The utilization of CO by thermophilic carboxydotrophs is catalyzed by Ni-containing CO dehydrogenases (CODHs) and acetyl-CoA synthases (ACSs). CODHs and CODH/ACS complexes are widespread among anaerobes and are found in acetogens, methanogens, sulfate reducers and iron reducers. CODH/ACS complexes catalyze both catabolic and anabolic acetyl-CoA synthesis and cleavage reactions in which CO is an intermediate that travels along a hydrophobic channel between the CODH and ACS active sites (Ragsdale 2004).

Many acetogens can grow on CO (Drake et al. 2006). CO conversion has been documented for 10 acetogens, including four moderate thermophiles e.g., *M. thermoacetica*, *Moorella thermoautotrophica*, *Thermoanaerobacter kivui* and *Moorella perchloratireducens*, described by Balk et al. (2008). *Moorella thermoautotrophica* and *M. thermoacetica* can grow on CO at high partial pressures as the sole energy source (Savage et al. 1987; Daniel et al. 1990).

The methanogenic carboxydotrophs identified by Hussain et al. (2012) in syngas-fed MFC operated at 50°C include *Methanothermobacter wolfeii*, *Methanothermobacter thermoautotrophicum* and *Methanobrevibacter arboriphilicus*. These microorganisms use H_2 and CO_2 for growth and CH_4 formation (Daniels et al. 1977; Winter et al. 1983). The ability of *M. thermoautotrophicus* and *M. arboriphilicus* to remove CO in the gas phase while growing on CO_2 and H_2 has been reported by Daniels et al. (1977), who found that *M. thermoautotrophicum* can utilize CO as the sole energy source by disproportionating CO to CO_2 and CH_4. This ability can be attributed to the presence of carbon monoxide dehydrogenase (CODH) and acetyl-CoA synthase (ACS) in the microorganism, the two metalloenzymes fundamental for growth on CO (Oelgeschlager and Rother 2008). The other uncultured archaea identified in the MFC that possess the hydrogenotrophic pathway for CH_4 formation belong to the genera *Methanobacterium* and *Methanobrevibacterium* (Wasserfallen et al. 2000).

The capacity of some sulfate-reducing bacteria to oxidize CO at low concentrations (4–20%) is long known (Yagi 1959). CO conversion by four thermophilic sulfate-reducing bacteria—*Desulfotomaculum thermoacetoxidans* CAMZ (DSM 5813) (Min and Zinder 1990), *Thermodesulfovibrio yellowstonii* ATCC 51303 (Henry et al. 1994), *Desulfotomaculum kuznetsovii* DSM 6115 (Nazina et al. 1988; Nazina et al. 1999), and *Desulfotomaculum thermobenzoicum* subsp. *thermosyntrophicum* DSM 14055 (Plugge et al. 2002)—was studied in pure cultures and co-cultures with the thermophilic hydrogenogenic carboxydotrophic bacterium *C. hydrogenoformans* (Parshina et al. 2005).

Dissimilatory Fe(III)-reducing thermophilic bacteria such as *Thermosinus carboxydovorans*, *Carboxydothermus ferreducens*, *Carboxythermus siderophilus* and *Thermicola ferriacetica* produce H_2 during the oxidation CO to CO_2. *Thermosinus carboxydivorans* grows at temperatures between 40 and 68°C (with an optimum at 60°C). This bacterium can utilize CO as its sole energy source with a doubling time of 1.15 hours leading to the formation of H_2 and CO_2 in equimolar quantities. Fe (III) is also reduced during its growth on sucrose and lactose. The dissimilatory Fe(III)-reducing, moderately thermophilic bacterium *Carboxydothermus ferrireducens* (Slobodkin et al. 2006) can grow by utilizing organic substrates or H_2 as electron donors. Apart from Fe(III) or AQDS, it can also reduce sulfite, thiosulfate, elemental sulfur, nitrate and fumarate (Slobodkin et al. 1997; Henstra and Stams 2004). *Carboxydothermus ferrireducens* can grow on CO, without hydrogen or acetate production, with ferrihydrite as the electron acceptor, forming magnetite precipitate (Slobodkin et al. 2006) or with AQDS or fumarate as electron acceptors (Henstra and Stams 2004). *Carboxydothermus siderophilus* (isolated from hot spring of Geyser Valley) produces H_2 and CO_2 along with Fe (III) and AQDS reduction during its growth on CO (Slepova et al. 2009). *Thermicola ferriacetica* (isolated from ferric deposits of a terrestrial hydrothermal spring in Kunashir Island, Russia) utilizes H_2 and acetate as energy sources with Fe (III) serving as the electron acceptor. It is also able to grow in an atmosphere of 100% CO as the sole energy source, leading to the formation of H_2 and CO_2. However, it requires 0.2 g L^{-1} of acetate as its carbon source during its growth on CO (Zavarzina et al. 2007).

Carboxydotrophic hydrogenogenic microorganisms are capable of lithotrophic metabolism based on the reaction $CO + H_2O \rightarrow CO_2 + H_2$ (Svetlitchnyi et al. 2001). All thermophilic hydrogenogenic CO oxidizers isolated so far are obligate anaerobes, including *Thermolithobacter carboxydivorans*, *Carboxydothermus hydrogenoformans*, *Thermincola carboxydophila*, *Carboxydocella thermoautotrophica*, *Themolithobacter carboxydivorans* and *Carboxydibrachium pacificum,* that produce H_2 from CO oxidation under thermophilic growth conditions (Slepova et al. 2006; Sokolova et al. 2007; Sokolova et al. 2009; Sokolova and Lebedinsky 2013; Tiquia 2014d).

Similar to the mesophilic co-culture, a co-culture of the thermophilic CO-utilizing acetogens, methanogens, sulfate reducers, iron reducers and hydrogenogens mentioned above can be co-cultured with H_2 utilizing thermophilic exoelectrogens for electricity generation in a syngas-fed MFC.

6. Design Considerations of MFCs Operating at Thermophilic Temperatures

Although the majority of MFCs have been tested at ambient or mesophilic temperatures, thermophilic systems warrant evaluation because of the potential for increased microbial activity rates on the anode. MFC studies, at elevated temperatures, have been scattered and most used designs that are already established, including the air-cathode single chambers and two-chamber designs. Previous modular MFC design by Rismani-Yazdi et al. (2007) have shown to work under mesophilic conditions (39°C) but failed at 60°C possibly because this design is not a closed system and permitted evaporation from the cathode chamber. Between 50 to 70% of the anode working volume was lost within two days. The concentrated anolyte can be detrimental to microbial metabolism and activity due to the enrichment of metabolites and cell debris. Thermophilic studies have not addressed these problems other than to note periodic anolyte and catholyte replacement (Mathis et al. 2008; Marshall and May 2009).

Jong et al. (2006) utilized a continuous flow rather than batch or fed-batch that allowed constant replacement of anolyte and catholyte in the thermophilic mediatorless MFC. The best MFC performance was achieved with 338 cm^3 h^{-1} and 11 cm^3 h^{-1} catholyte and anolyte flow rates, respectively. The catholyte required a higher flow rate likely due to the continuous evaporation of liquid from the open cathode chamber. A maximum power density of 1030 ± 340 mW/m^2 was generated continuously at 55°C with an anode retention time of 27 minutes (11 mL h^{-1}) and continuous pumping of air-saturated phosphate buffer into the cathode compartment at the retention time of 0.7 minutes (450 mL h^{-1}). While the constant replacement of anolyte and catholyte prevents the drastic liquid loss, electricity production relies on the electrochemically active biofilm alone since suspended cells are removed with the continuous flow of the anolyte. Several MFC studies have tested a range of operating temperatures and demonstrated consistently higher power densities with higher temperatures within the limits of the microbial populations (Choi 2004; Moon et al. 2006).

Carver et al. (2011) described a thermophilic MFC design that prevents evaporation that is based on the original concept elucidated by Min et al. (2008). The MFC utilized an anaerobic, glass reactor design in combination with a cathode chamber submersed in anolyte. Rather than having extensive layers of gaskets, membrane, carbon paper and polycarbonate as in the previous design (Min et al. 2008), the cathode chamber has a single rubber o-ring that prevents liquid or air crossover. The components of the cathode assembly, including the stainless screws, foil and graphite discs, have all been shown to be conducive and were securely connected. Analyses of the glucose-fed thermophilic MFC showed improved performance over 120 hours with increased maximum power of 3.3-4.5 mW m^{-2} (Carver et al. 2011). The polarization curve has three distinct sections of irreversible voltage losses: activation loss, ohmic loss and mass transfer loss. The typical initial and drastic voltage drop was not apparent, indicating lower than normal activation losses (Carver et al. 2011). This is attributed to increased reaction rates at thermophilic temperatures that lowered the activation energy and therefore the voltage necessary to maintain active anaerobic metabolism. Ohmic loss can be observed in the center of the polarization curve with the gradual decrease of voltage as current density increases (Carver et al. 2011). The slope of this overpotential section, equivalent to the voltage over current, yields an internal resistance of 9.25 ± 0.15 Ω. This value is in the general range reported for other MFCs although the experimental conditions are not comparable among the studies reviewed in the literature (He et al. 2005; Ieropoulos et al. 2010). The work by Carver et al. (2011) suggests the potential for stable, thermophilic MFC operation although optimization of biological and engineering components is necessary prior to the application of the design.

A CO or syngas-fed MFC system requires a CO-tolerant cathode. Mehta et al. (2010) used a CoTMPP cathode to generate electricity from CO with a Co load of 0.5 mg cm^{-2}. A maximum power density of 6.4 mW L^{-1} was reported. The cathode performance was tested in acetate and CO-fed MFCs. MFC operation on CO showed the best performance with the CoTMPP/FeTMPP/C cathode catalyst. Considering the high cost of Pt-based cathodes and the plausible decrease in activity with time, the use of CoTMPP/FeTMPP/C or FePc cathodes is a step forward in increasing the efficiency of CO-operated MFCs.

Membrane systems also need to be considered for improved mass transfer efficiency. Alternatives to the conventional stirred tank reactors for increased gas-liquid mass transfer include monolith packing and columnar reactors. Monolith packing consists of several narrow, straight and parallel flow channels with a large open frontal area that allows for a low flow resistance, leading to low-pressure drops and low energy losses. High volumetric mass transfer rates of ~ 1 s^{-1} and a 50–80% reduction in power consumption as compared to conventional reactors make monolith reactors economically viable option (Hickey et al. 2008; Munasinghe and Khanal 2010). Likewise, columnar reactors such as a bubble column, a trickle bed and an airlift reactor offer the advantage of a high gas-liquid mass transfer rate with low operational and maintenance costs. K_La values within the range of 18 to 860 h^{-1} have been reported for such reactors (Charpentier 1981; Bredwell et al. 1999; Munasinghe and Khanal 2010). Several reactor design improvements such as the low frequency vibration of liquid phase in the bubble column reactor, the addition of static mixers, baffles, perforated plates, jet loop and forced circulation loop in internal and external loop airlift reactors promise further increase in the gas-liquid mass transfer efficiency (Chisti et al. 1990; Vorapongsathorn et al. 2001; Krichnavaruk and Pavasant 2002; Ugwu and Ogbonna 2002; Ellenberger and Krishna 2003; Fadavi and Chisti 2005).

7. Conclusions

Electricity generation from CO/syngas predominantly takes place by a two-step process in which syngas is first converted to acetate that is then oxidized by the exoelectrogenic microorganisms to produce electricity. This pathway is accomplished by a syntrophic association of exoelectrogenic and carboxydotrophic microorganisms. With the performance of a syngas-fed MFC with mixed cultures already demonstrated, a detailed study of the CO-operated MFC might be of interest and warrant exploration. Studies can focus on (1) improving gas transfer efficiency, (2) understanding the complex transformation pathways in mixed cultures/co-cultures under thermophilic conditions and (3) determining the energetics of syntrophic cooperation between thermophilic carboxydotrophs and exoelextrogens.

The benefits of operating biochemical systems at thermophilic conditions include higher microbial activity, better substrate solubility, higher mass transfer rate and lower risk of contamination. However, one drawback is higher rates of evaporation. This review noted two answers to this problem are possible, either to run the MFC in continuous mode allowing replacement of the anolyte and catholyte or to utilize an MFC that precludes evaporation as designed by Carver et al. (2011). Further development will likely result in more efficient reactor designs and stackable MFC capable of efficient operation on gaseous substrates such as CO and H$_2$ and with power outputs suitable for commercial applications.

References

Balk M, van Gelder T, Weelink SA, Stams AJM (2008) (Per)chlorate reduction by the thermophilic bacterium *Moorella perchloratireducens* sp. nov., isolated from underground gas storage. Appl Environ Microbiol 74: 403–409.

Bond DR, Holmes DE, Tender LM, Lovley DR (2002) Electrodereducing microorganisms that harvest energy from marine sediments. Science 295: 483–485.

Bond DR, Lovley DR (2003) Electricity production by *Geobacter sulfurreducens* attached to electrodes. Appl Environ Microbiol 69:1548–1555.

Bond D R, Lovley DR (2005) Evidence for involvement of an electron shuttle in electricity generation by *Geothrix fermentans*. Appl Environ Microbiol 71: 2186–2189.

Bredwell MD, Selvaraj PT, Little MH, Worden RM (1999). Reactor design issues for synthesis-gas fermentations. Biotechnol. Prog. 15: 834–844.

Brutinel E, Gralnick J (2012) Shuttling happens: Soluble flavin mediators of extra-cellular electron transfer in *Shewanella*. Appl Microbiol Biotechnol 93: 41–48.

Carver SM, Vuorirantab P, Tuovinena OH (2011) A thermophilic microbial fuel cell design. J Power Sources 196: 3757–3760.

Chaudhuri SK, Lovley DR (2003) Electricity generation by direct oxidation of glucose in mediatorless microbial fuel cells. Nat Biotechnol 21: 1229–1232.

Choi Y (2004) Construction of microbial fuel cells using thermophilic microorganisms, *Bacillus licheniformis* and *Bacillus thermoglucosidasius*. Bull Korean Chem Soc 25: 813–818.

Chisti Y, Kasper M, Moo-Young M (1990) Mass transfer in external loop airlift bioreactors using static mixers. Can J Chem Eng 68: 45–50.

Charpentier JC (1981) Mass-transfer rates in gas–liquid absorbers and reactors. In: TB Drew, GR Cokelet, HW Hoopes Jr, T Vermeulen (eds) Advances in chemical engineering, Volume 11. Academic Publisher, New York, USA, pp 1–33.

Daniel SL, Hsu T, Dean SI, Drake HL (1990) Characterization of the hydrogen- and carbon monoxide-dependent chemolithotrophic potentials of the acetogens *Clostridium thermoaceticum* and *Acetogenium kivui*. J Bacteriol 172: 4464–4471.

Daniels L, Fuchs G, Thauer RK, Zeikus JG (1977) Carbon monoxide oxidation by methanogenic bacteria. J Bacteriol 132: 118–126.

Drake HL, Küsel K, Matthies C (2006) Acetogenic Prokaryotes. The Prokaryotes 2: 354–420.

Ellenberger J, Krishna R (2003) Shaken, not stirred, bubble column reactors: enhancement of mass transfer by vibration excitement. Chem Eng Sci 58: 705–710.

Faaij A, van Ree R, Waldheim L, Olsson E, Oudhuis A, van Wijk A, Daey-Ouwens C, Turkenburg W (1997) Gasification of biomass wastes and residues for electricity production. Biomass Bioenerg 12: 387–407.

Fadavi A, Chisti Y (2005) Gas–liquid mass transfer in a novel forced circulation loop reactor. Chem Eng J 112: 73–80.

Franks AE, Nevin K (2010) Microbial fuel cells, a current review. Energies 3: 899–919.

Fu Q, Kobayashi H, Kawaguchi H, Wakayama T, Maeda H, Sato K (2013a) A thermophilic gram-negative nitrate-reducing bacterium, *Calditerrivibrio nitroreducens*, exhibiting electricity generation capability. Environ Sci Technol 47:12583–12590.

Fu Q, Kobayashi H, Kawaguchi H, Vilcaez J, Wakayama T, Maeda H, Sato K (2013b) Electrochemical and phylogenetic analyses of current-generating microorganisms in a thermophilic microbial fuel cell. J Biosci Bioeng 115: 268–271.

Fu Q, Fukushima N, Maeda H, Sato K, Kobayashi H (2015) Bioelectrochemical analysis of a hyperthermophilic microbial fuel cell generating electricity at temperatures above 80 °C. Biosci Biotechnol Biochem. 79: 1200–1206.

Gorby YA, Yanina YS, McLean JS, Rosso KM, Moyles D, Dohnalkova A, Beveridge TJ, Chang IS, Kim BH, Kim KS, Culley DE, Reed SB, Romine MF, Saffarini DA, Hill EA, Shi L, Elias DA, Kennedy DW, Pinchuk G, Watanabe K, Ishii S, Logan B, Nealson KH, Fredrickson JK (2006) Electrically conductive bacterial nanowires produced by Shewanella oneidensis strain MR-1 and other microorganisms. Proc Natl Acad Sci USA 103: 11358–11363.

Greene AC, Patel BKC, Sheehy AJ (1997) *Deferribacter thermophilus* gen. nov., sp. nov., a novel thermophilic manganese- and iron-reducing bacterium isolated from a petroleum reservoir. Int J Syst Bacteriol 47: 505–509.

Hussain A, Guiot SR, Mehta P, Raghavan V, Tartakovsky B (2011) Electricity generation from carbon monoxide and syngas in a microbial fuel cell. Appl Microbiol Biotechnol 90: 827–36.

Hussain A, Mehta P, Raghavan V, Wang H, Guiot S, Tartakovsky B (2012) The performance of a thermophilic microbial fuel cell fed with synthesis gas. Enzyme Microb Technol 51: 163–170.

Henry EA, Devereux R, Maki JS, Gilmour CC, Woese CR, Mandelco L, Schauder R, Remsen CC, Mitchell R (1994). Characterization of a new thermophilic sulfate-reducing bacterium Thermodesulfovibrio yellowstonii, gen. nov. and sp. nov.: its phylogenetic relationship to Thermodesulfobacterium commune and their origins deep within the bacterial domain. Arch Microbiol 161: 62–69.

Henstra AM, Stams AJ (2004) Novel physiological features of *Carboxydothermus hydrogenoformans* and *Thermoterrabacterium ferrireducens*. Appl Environ Microbiol 70:7236–7240.

He Z, Minteer SD, Angenent LT (2005) Electricity generation from artificial wastewater using an upflow microbial fuel cell. Environ Sci Technol 39: 5262–5267.

Hickey R, Datta R, Tsai SP, Basu R (2008) Membrane supported bioreactor for conversion of syngas components to liquid products. US Patent.

Henstra AM, Sipma J, Rinzema A, Stams AJ. (2007) Microbiology of synthesis gas fermentation for biofuel production. Curr Opin Biotechnol 18: 200–206.

Holmes DE, Bond DE, Lovley DR (2004a) Electron transfer by *Desulfobulbus propionicus* to Fe(III) and graphite electrodes. Appl Environ Microbiol 70: 1234–1237.

Holmes DE, Nicoll JS, Bond DR, Lovley DR (2004b) Potential role of a novel psychrotolerant member of the family *Geobacteraceae, Geopsychrobacter electrodiphilus* gen. nov., sp. nov., in electricity production by a marine sediment fuel cell. Appl Environ Microbiol 70: 6023–6030.

Ieropoulos I, Winfield J, Greenman J. (2010) Effects of flow-rate, inoculum and time on the internal resistance of microbial fuel cells. Bioresour Technol 101: 3520–3525.

Jain A, Zhang X, Pastorella G, Connolly JO, Barry N, Woolley R, Marsili E (2012) Electron transfer mechanism in *Shewanella loihica* PV-4 biofilms formed at graphite electrode. Bioelectrochem 87: 28–32.

Jayapriya J, Ramamurthy V (2012) Use of non-native phenazines to improve the performance of *Pseudomonas aeruginosa* MTCC 2474 catalysed fuel cells. Biores Technol 124: 23–28.

Jiang X, Hu J, Petersen ER, Fitzgerald LA, Jackan CS, Lieber AM, Ringeisen BR, Lieber CM, Biffinger JC (2013) Probing single-to multi-cell level charge transport in *Geobacter sulfurreducens* DL-1. Nat Commun 4: 2751.

Jong BC, Kim BH, Chang IS, Liew PWY, Choo YF, Kang GS (2006) Enrichment, performance, and microbial diversity of a thermophilic mediatorless microbial fuel cell. Environ Sci Technol 40: 6449–6454.

Kim D, Chang, I (2009) Electricity generation from synthesis gas by microbial processes: CO fermentation and microbial fuel cell technology. Biores Technol 100: 4527–4530.

Kim BH, Kim HJ, Hyun MS, Park DS (1999) Direct electrode reaction of Fe(III) reducing bacterium, *Shewanella putrefaciens*. J. Microbiol. Biotechnol. 9:127–131.

Kim HJ, Park HS, M. S. Hyun, I. S. Chang, M. Kim, and B. H. Kim. 2002. A mediator-less microbial fuel cell using a metal reducing bacterium, *Shewanella putrefaciens*. Enzyme Microb. Technol. 30:145–152.

Kotloski NJ, Gralnick JA (2013) Flavin electron shuttles dominate extracellular electron transfer by *Shewanella oneidensis*. mBio 4e00553-12.

Krichnavaruk S, Pavasant P (2002) Analysis of gas–liquid mass transfer in an airlift contactor with perforated plates. Chem Eng J 89: 203–211.

Leung KM, Wanger G, El-Naggar MY, Gorby YA, Southam G, Lau WM, Yang J (2013) *Shewanella oneidensis* MR-1 bacterial nanowires exhibit p-type, tunable electronic behaviour. Nano Lett 13: 2407–2411.

Liu H, Logan BE (2004) Electricity generation using an air-cathode single chamber microbial fuel cell in the presence and absence of a proton exchange membrane. Environ Sci Technol 38: 4040–4046.

Logan BE (2004) Extracting hydrogen and electricity from renewable resources. Environ Sci Technol 38: 160A–167A.

Logan BE (2008) Microbial Fuel Cells. John Wiley and Sons Inc., Hoboken, New Jersey, USA. 216 p.

Logan BE (2009) Exoelectrogenic bacteria that power microbial fuel cells. Nat Rev Micro 7: 375–381.

Logan BE, Regan JM (2006) Electricity-producing bacterial communities in microbial fuel cells. Trends Microbiol 14: 512–518.

Lovley DR, Holmes DE, Nevin KP (2004) Dissimilatory Fe(III) and Mn(IV) reduction. Adv Microb Physiol 49: 219–286.

Lovley DR (2008) The microbe electric: conversion of organic matter to electricity. Curr Opin Biotechnol 19: 564–571.

Lovley DR, Holmes DE, Nevin KP (2004) Dissimilatory Fe(III) and Mn(IV) Reduction. Adv Microb Physiol 49: 219–286.

Lovley DR (2006) Bug juice: harvesting electricity with microorganisms. Nat Rev Micro 4: 497–508.

Mclean SJ, Wanger G, Gorby YA, Wainstein M, Mcquaid J, Ishii IS, Bretcheger O, Beyenal H, Nealson HK (2010) Quantification of electron transfer rates to a solid phase electron acceptor through the stages of biofilm formation from single cells to multi cellular communities. Environ Sci Technol 44: 2721–2727.

Malvankar NS, Lovley DR (2012) Microbial nanowires: A new paradigm for biological electron transfer and bioelectronics. Chem Sus Chem 5:1039–1046.

Manish S, Banerjee R (2008) Comparison of biohydrogen production processes. Int J Hydrogen Energy 3: 276–286.

Marshall CW, May HD (2009) Electrochemical evidence of direct electrode reduction by a thermophilic Gram-positive bacterium, *Thermincola ferriacetica*. Energy Environ Sci 2: 699–705.

Mathis B, Marshall C, Milliken C, Makkar R, Creager S, May H (2008) Electricity generation by thermophilic microorganisms from marine sediment. Appl Microbiol Biotechnol 78: 147–155.

Mehta P, Hussain A, Raghavan V, Neburchilov V, Wang H, Tartakovsky B, Guiot S (2010) Electricity generation from a carbon monoxide in a single chamber microbial fuel cell. Enzyme Microb Technol 46: 450–455.

Min H, Zinder SH (1990) Isolation and characterization of a thermophilic sulfate-reducing bacterium *Desulfotomaculum thermoacetoxidans* sp. nov. Arch Microbiol 153: 399–404.

Min B, Román Ó, Angelidaki I (2008) Importance of temperature and anodic medium composition on microbial fuel cell (MFC) performance. Biotechnol Lett 30: 1213–1218.

Moon H, Chang IS, Kang KH, Jang JK, Kim BH (2004) Improving the dynamic response of a mediator-less microbial fuel cell as a biochemical oxygen demand (BOD) sensor. Biotechnol Lett 26: 1717–1721.

Moon H, Chang IS, Kim BH (2006) Continuous electricity production from artificial wastewater using a mediator-less microbial fuel cell. Biores Technol 97: 621–627.

Munasinghe PC, Khanal SK (2010) Biomass-derived syngas fermentation into biofuels: Opportunities and challenges. Bioresour Technol 101: 5013–5022.

Nazina TN, Ivanova AE, Kanchaveli LP, Rozanova EP (1988) A new sporeforming thermophilic methylotrophic sulfate reducing bacterium, *Desulfotomaculum kuznetsovii* sp. nov. Microbiology (English Translation of Mikrobiologiya) 57: 823–827.

Nazina TN, Turova TP, Poltaraus AB, Gryadunov DA, Ivanova AE, Osipov GA, Belyaev SS (1999) Phylogenetic position and chemotaxonomic characteristics of the thermophilic sulfate reducing bacterium *Desulfotomaculum kuznetsovii*. Microbiology (English Translation of Mikrobiologiya) 68: 77–84.

Nevin KP, Richter H, Covalla SF, Johnson JP, Woodard TL (2008) Power output and Coulombic efficiencies from biofilms of *Geobacter sulfurreducens* comparable to mixed community microbial fuel cells. Environ Microbiol 10: 2505–2514.

Niessen J, Harnisch F, Rosenbaum M, Schroder U, Scholz F (2006) Heat treated soil as convenient and versatile source of bacterial communities for microbial electricity generation. Electrochem Commun 8: 869–873.

Nguyen S, Ala F, Cardwell C, Cai D, McKindles KM, Lotvola A, Hodges S, Deng Y, Tiquia-Arashiro SM (2013) Isolation and screening of carboxydotrophs isolated from composts and their potential for butanol synthesis. Environ Technol 34: 1995–2007.

Oelgeschlager E, Rother M (2008) Carbon monoxide-dependent energy metabolism in anaerobic bacteria and archaea. Arch Microbiol 190: 257–269.

Pant D, Van Bogaert G, Diels L, Vanbroekhoven K (2010) A review of the substrates used in microbial fuel cells (MFCs) for sustainable energy production. Biores Technol 101: 1533–1543.

Park HS, Kim BH, Kim HS, Kim HJ, Kim GT, Kim M, Chang IS, Park YK, Chang HI (2001) A novel electrochemically active and Fe(III)-reducing bacterium phylogenetically related to *Clostridium butyricum* isolated from a microbial fuel cell. Anaerobe 7:297–306.

Parshina SN, Kijlstra S, Henstra AM, Sipma J, Plugge CM, Stams AJM (2005) Carbon monoxide conversion by thermophilic sulfate-reducing bacteria in pure culture and in co-culture with *Carboxydothermus hydrogenoformans*. Appl Microbiol Biot 68: 390–396.

Pham CA, Jung SJ, Phung NT, Lee J, Chang IS, Kim BH, Yi H, Chun J (2003) A novel electrochemically active and Fe(III)-reducing bacterium phylogenetically related to *Aeromonas hydrophila*, isolated from a microbial fuel cell. FEMS Microbiol. Lett. 223:129–134.

Plugge C, Balk M, Stams AJM (2002) *Desulfotomaculum thermobenzoicum* subsp. *thermosyntrophicum* subsp. nov., a thermophilic syntrophic propionate-oxidizing spore-forming bacterium. Int J Syst Evol Micr 52: 391–399.

Pomaranski E, Tiquia-Arashiro SM (2016) Butanol tolerance of carboxydotrophic bacteria isolated from manure composts. Environ Technol 37: 1970–1982.

Rabaey K, Boon N, Siciliano SD, Verhaege M, Verstraete W (2004) Biofuel cells select for microbial consortia that self-mediate electron transfer. Appl Environ Microbiol 70: 5373–5382.

Ragsdale SW (2004) Life with carbon monoxide. Crit Rev Biochem Mol 39: 165–195.

Reguera G, McCarthy KD, Mehta T, Nicoll JS, Tuominen MT, Lovley DR (2005) Extracellular electron transfer via microbial nanowires. Nature 435: 1098–1101.

Ren N, Xing D, Rittmann BE, Zhao L, Xie T, Zhao X (2007) Microbial community structure of ethanol type fermentation in bio-hydrogen production. Environ Microbiol 9: 1112–1125.

Rismani-Yazdi H, Christy AD, Dehority BA, Morrison M, Yu Z, Tuovinen OH (2007) Electricity generation from cellulose by rumen microorganisms in microbial fuel cells. Biotechnol Bioeng 97: 1398–1407.

Savage MD, Wu ZG, Daniel SL, Lundie LL, Drake HL (1987) Carbon monoxide-dependent chemolithotrophic growth of *Clostridium thermoautotrophicum*. Appl Environ Microbiol 53: 1902–1906.

Sipma J, Henstra AM, Parshina SN, Lens PNL, Lettinga G, Stams AJM (2006) Microbial CO conversions with applications in synthesis gas purification and bio-desulfurization. Crit Rev Biotechnol 26: 41–65

Slepova TV, Sokolova TG, Lysenko AM, Tourova TP, Kolganova TV, Kamzolkina OV, Karpov GA, Bonch-Osmolovskaya EA (2006) *Carboxydocella sporoproducens* sp. nov., a novel anaerobic CO-utilizing/H$_2$-producing thermophilic bacterium from a Kamchatka hot spring. Int J Syst Evol Microbiol 56: 797–800.

Slepova TV, Sokolova TG, Kolganova TV, Tourova TP, Bonch-Osmolovskaya EA (2009) *Carboxydothermus siderophilus* sp. nov., a thermophilic, hydrogenogenic, carboxydotrophic, dissimilatory Fe(III)-reducing bacterium from a Kamchatka hot spring. Int J Syst Evol Microbiol 59: 213–217.

Slobodkin AI, Reysenbach A-L, Strutz N, Dreier M, Wiegel J (1997) *Thermoterrabacterium ferrireducens* gen. nov., sp. nov. a thermophilic anaerobic, dissimilatory Fe(III)-reducing bacterium from a continental hot spring. Int J Syst Bacteriol 47: 541–547.

Slobodkin AI, Sokolova TG, Lysenko AM, Wiegel J (2006) Reclassification of *Thermoterrabacterium ferrireducens* as *Carboxydothermus ferrireducens* comb. nov., and emended description of the genus *Carboxydothermus*. Int J Syst Evol Micr 56: 2349–2351.

Snider RM, Strycharz-Glaven SM, Tsoi SD, Erickson JS, Tender LM (2012) Long-range electron transport in *Geobacter sulfurreducens* biofilms is redox gradient-driven. Proc Natl Acad Sci USA. 109: 15467–15472.

Sokolova T, Hanel J, Oneynwoke RU, Reysenbach A-L, Banta A, Geyer R, Gonzalez J, Whitman WB, Wiegel J (2007) Novel chemolithotrophic, thermophilic, anaerobic bacteria *Thermolithobacter ferrireducens* gen. nov., sp. nov. and *Thermolithobacter carboxydivorans* sp. nov. Extremophiles 11: 145–157.

Sokolova TG, Henstra AM, Sipma J, Parshina SN, Stams AJ, Lebedinsky AV. (2009) Diversity and ecophysiological features of thermophilic carboxydotrophic anaerobes. FEMS Microbiol Ecol 68: 131–141.

Sokolova TG, Lebedinsky A (2013) CO-oxidizing anaerobic thermophilic prokaryotes. T Satyanarayana, J Littlechild, Y Kawarabayasi (eds) Thermophilic Microbes in the Environment and Industrial Biotechnology. Springer Publisher, New York, USA. p. 203–231.

Song C (2002) Fuel processing for low-temperature and high-temperature fuel cells: Challenges, and opportunities for sustainable development in the 21st century. Catalysis Today 77: 17–49.

Steele BCH, Heinzel A (2001) Materials for fuel-cell technologies. Nature 414: 345–352.

Svetlitchnyi V, Peschel C, Acker G, Meyer O (2001). Two membrane-associated NiFeS-carbon monoxide dehydrogenases from the anaerobic carbon-monoxide-utilizing Eubacterium *Carboxydothermus hydrogenoformans*. J. Bacteriol. 183: 5134–5144.

Tiquia-Arashiro SM (2014a) Biotechnological Applications of Thermophilic Carboxydotrophs. In: Thermophilic Carboxydotrophs and their Applications in Biotechnology. Springer International Publishing, New York, USA. p. 29–101.

Tiquia-Arashiro SM (2014b) CO-oxidizing Microorganisms. In: Thermophilic Carboxydotrophs and their Applications in Biotechnology. Springer International Publishing, New York, USA. p. 11–28.

Tiquia-Arashiro SM (2014c) Microbial CO Metabolism. In: Thermophilic Carboxydotrophs and their Applications in Biotechnology. Springer International Publishing, New York, USA. p. 5–9.

Tiquia-Arashiro SM (2014d) Thermophilic Carboxydotrophs and Their Biotechnological Applications. Springer International Publishing New York, USA. 131 p.

Tor JM, Kashefi K, Lovley DR (2001) Acetate oxidation coupled to Fe (III) reduction in hyperthermophilic microorganisms. Appl Environ Microbiol 67: 1363–1365.

Ugwu CU, Ogbonna JC (2002) Improvement of mass transfer characteristics and productivities of inclined tubular photobioreactors by installation of internal static mixers. Appl Microbiol Biotechnol 58: 600–607.

Vorapongsathorn T, Wongsuchoto P, Pavasant P (2001) Performance of airlift contactors with baffles. Chem Eng J 84: 551–556.

Wasserfallen A, Nolling J, Pfister P, Reeve J, Conway de Macario E (2000) Phylogenetic analysis of 18 thermophilic *Methanobacterium* isolates supports the proposals to create a new genus, *Methanothermobacter* gen. nov., and to reclassify several isolates in three species, *Methanothermobacter thermautotrophicus* comb. nov., *Methanothermobacter wolfeii* comb. nov., and *Methanothermobacter marburgensis* sp. nov. Int J Syst Evol Microbiol 50: 43–53.

Winter J, Lerp C, Zabel HP, Wildenauer FX, König H, Schindler F (1984) *Methanobacterium wolfei*, sp. nov., a new tungsten-requiring, thermophilic, autotrophic methanogen Syst Appl Microbiol 5: 457–466.

Wrighton KC, Agbo P, Warnecke F, Weber KA, Brodie EL, DeSantis TZ, Hugenholtz P, Andersen GL, Coates JD (2008) A novel ecological role of the *Firmicutes* identified in thermophilic microbial fuel cells. ISME J 2: 1146–1156.

Yagi, T. (1959) Enzymic oxidation of carbon monoxide. II. J. Biochem. (Tokyo) 46: 949–955.

Zavarzina DG, Sokolova TG, Tourova TP, Chernyh NA, Kostrikina NA, Bonch-Osmolovskaya EA (2007) *Thermincola ferriacetica* sp. nov., a new anaerobic, thermophilic, facultatively chemolithoautotrophic bacterium capable of dissimilatory Fe (III) reduction. Extremophiles 11: 1–7.

Zhang P, Liu Z (2010) Experimental study of the microbial fuel cell internal resistance. J Power Sources 195: 8013–8018.

Microbial Electrochemical Technology Drives Metal Recovery

Xochitl Dominguez-Benetton[1],*, Jeet Chandrakant Varia[2] and Guillermo Pozo[1]

[1] Separation and Conversion Technology, Flemish Institute for Technological Research (VITO), Boeretang 200, Mol 2400, Belgium
[2] Center for Microbial Ecology and Technology (CMET), Faculty of Bioscience Engineering, Ghent University, Coupure Links 653, 9000 Gent, Belgium

1. Introduction

A long trajectory exists on how microbes associate with metals in both natural and man-made environments (Gadd 2010). The role of microbes in the binding and mineralization of metal ions (Konhauser et al. 2008; Beveridge and Murray 1976), the interactions between microbes and a variety of metals as well as metal accumulation (White et al. 1995) and the role of microbes in the generation of acidic, metal-rich mine waters (Johnson and Hallberg 2003) have been reviewed extensively. For instance, acidophiles accelerate the dissolution of pyrite and other sulfide minerals causing substantial environmental damage due to the release of acid mine drainage (AMD) into the environment. The principles governing the generation of AMD are relatively well understood. A schematic overview of the generation of AMD is presented in Figure 1.

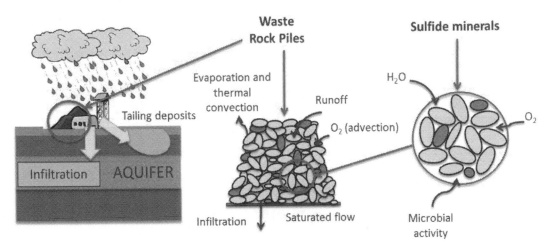

Figure 1: Microbial generation of acid mine drainage (AMD).

*Corresponding author: xoch@vito.be

From an engineering perspective, a paradigm shift has occurred over the last 35 years in the application of metal-microbial interactions. In the beginning, applications focused on the remediation of metals and radioactive elements from polluted aqueous systems (Volesky 2001; Lloyd and Lovley 2001), such as AMD. Later on, in light of the critical need for some metals and the potential of future supply risk interruptions, attention shifted to recovery, recycling and economic prospection alternatives by microbially-assisted mining (Hennebel et al. 2015; European Commission 2014). The application of microorganisms as a green methodology for the synthesis of metallic nanoparticles (NPs) has also been reported (Klaus-Joerger et al. 2001; Tiquia-Arashiro and Rodrigues 2016). The remediation of metal-containing waste streams can be combined with recovery and synthesis of novel and functional (nano) materials (Macaskie et al. 2010) under ambient conditions.

Biosorption has been used extensively to remove metals and metalloids from aqueous streams using biologically-derived materials (Volesky 2001; Anushree 2004; Bowman et al. 2018) and faradaic metal ion oxidation/reduction reactions (Veglio and Beolchini 1997; Tiquia-Arashiro 2018). Most microorganisms can create a matrix in which metals adsorb by ion-exchange or chelation at the cell wall interface, leading in some cases to the subsequent transformation of metallic states. The works of Lovley and Macaskie have much developed such microbial transformation approach and its understanding (Lovley 1995; Deplanche 2001).

Most microbes that carry out metal transformations depend on specific electron donors or acceptors and most work under the strict anaerobic conditions, thus making metal recovery slow and poorly manageable especially for metals possessing multiple valence states. The recent resurgence of the use of solid-state electrodes as electron donors or acceptors for microbial growth (Thrash and Coates 2008) opened a wide range of new possibilities, giving birth to what we know today as microbial electrochemical technologies (METs).

METs take advantage of the synergistic alliance of electrochemical and microbiological phenomena (Logan and Rabaey 2012). They have emerged as a versatile technology for applications ranging from electricity generation (Rabaey and Verstraete 2005) to the synthesis of valuable chemicals (Rabaey and Rozendal 2010). Only recently METs have been applied to remove and recover metals from aqueous matrices. Metal recovery with METs results from the interactions between microorganisms, metals, and electrodes, where the electron transfer chain associated with microbial respiration often plays a key role. This poses significant advantages with respect to systems based on the microbiology alone as the possibility to control the electric potential at which the separations or conversions take place that adds a certain degree of selectivity and rate. Compared to the more classical electrochemical counterpart, METs are less energy-intensive because of the effective electrode catalysis and microbial power generation; furthermore, the microbial metabolism can enable reactions that otherwise would not be thermodynamically or kinetically favorable (Jones and Amy 2002). METs have been classified into electricity-producing microbial fuel cells (MFCs) and electricity-consuming microbial electrolysis cell (MECs). Graphical representations of MFCs and MECs can be found in the scientific literature (Clauwaert et al. 2008).

By fine-tuning the potential at which reduction (or oxidation) occurs, it is feasible to selectively extract and separate metals. In this chapter, we outline the recent developments that make the recovery of metals possible with METs as well as deliberate on the opportunities they bring to develop sustainable and energy-efficient recovery of metal products with ample functionality.

2. Mechanisms to Transform and Recover Metals by Microbial Electrochemical Technologies

The working principle of MET for metal recovery is reasonably straightforward. Microorganisms may colonize the anode (where the oxidation occurs), the cathode (where the reduction occurs) or even both. For instance, a cathode can be driven by a power supply to directly or indirectly (typically via H_2) provide reducing power to microorganisms. The latter can use the energy gained for growth while simultaneously reducing the metallic contaminants as an electron acceptor. By fine-tuning the potential at which reduction (or oxidation) occurs, it is feasible to selectively separate metals. METs can (1) use complex solid waste or wastewaters as electron donors/acceptors, (2) reduce energy consumption compared to traditional processing and (3) recover and generate metallic commodities with lower greenhouse emissions.

Several strategies have been employed to achieve desirable metal transformations in METs. Electrochemically active microorganisms have been used either at the anode to lower the overall energy consumption of the process by oxidizing organic compounds present in, e.g., wastewater or at the cathode to catalyze the reduction or precipitation of oxidized metal species.

Based on the mechanisms used to transform and recover the metals, MET systems can be divided into four general categories. Category A includes systems with a microbial anode and an abiotic cathode. Metals are directly reduced by electrons from the cathode and the reduced form of the metal is recovered. If the metal has a high reduction potential relative to the biological anode, the system can be operated as an MFC. This is the case for e.g., Cu and Ag (Choi and Cui 2012; Ter Heijne et al. 2010). If the metal has a lower reduction potential, the system must be operated as a microbial electrolysis cell, i.e., with an external power supply. This is the case for e.g., Cd and Zn (Modin et al. 2012). In most cases, the reduced metal can be recovered as a precipitate or deposit on the cathode surface. However, other reactions are possible like for cobalt, wherein Co^{2+} can be leached from solid lithium-cobalt (Co^{3+}) oxide (Huang et al. 2013).

In Category B systems, the electrochemical reduction at the cathode is used to generate a chemical reductant that reacts with the oxidized metals in the solution. This process has been used to reduce O_2 to H_2O_2 that reacted with toxic Cr^{6+} to form less toxic Cr^{3+} and O_2 (Liu et al. 2011).

In Category C systems, the electrochemical reduction on the cathode surface is followed by a chemical or biochemical reoxidation (or further reduction) of the metal. The purpose of reducing and then reoxidizing the metal may be to enhance the power output of an MFC. By adding a redox couple with high reduction potential and fast reaction kinetics, the cathode can be operated at a higher potential compared to a conventional MFC. Examples of these include Mn^{2+}/Mn^{4+} (Rhoads et al. 2005) and Fe^{2+}/Fe^{3+} (Ter Heijne et al. 2006) that can be electrochemically reduced on the cathode surface and biologically reoxidized by bacteria in the catholyte solution.

In Category D systems, the electrochemically active microorganisms can be attached to the cathode or suspended in the catholyte. Two strategies can be summarized here for metal recovery. Firstly, microorganisms can function as bioelectrocatalysts (Hill and Higgins 1981) of electrochemical reactions at the cathode (Varia et al. 2014), facilitating the reduction of metals for which abiotic electrochemical reduction would be difficult or require a large overpotential. Secondly, microbial metabolism can be

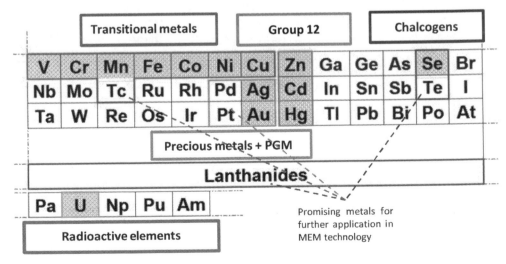

Figure 2: Reported metals recovered using MET and examined in this chapter.

stimulated by the *in situ* electrochemical production of electron donors, i.e., H_2 or direct or indirect (mediated) electron transfer to microbial cells (Thrash and Coates 2008). Microorganisms here sorb, reduce or precipitate metals within the cell wall or EPS matrix. Examples of these include reduction of U^{6+} to U^{4+} (Gregory and Lovley 2005) and Cr^{6+} to Cr^{3+} (Tandukar et al. 2009). Systems with biological cathodes may also contain biological anodes to allow operation as MFCs or to reduce the overall energy consumption of the system.

3. Key Achievements in Metal Recovery-Driven by Microbial Electrochemical Technologies

In the majority of metal recovery by MET systems, bioanodes have been implemented. Usually mixed cultures of exoelectrogenic bacteria form biofilms on carbon-based anodes. These biofilms catalyze the oxidation of organic substrates and the transfer of electrons to the anodes. Most studies reported here have used acetate as the electron donor to drive or complement the respective cathodic reaction. Acetate oxidation can be described as shown in Equation 1 (Logan et al. 2006).

$$CH_3COO^- + 4H_2O \rightarrow 2HCO_3^- + 9H^+ + 8e^- \quad -0.286 \text{ vs. SHE/V} \tag{1}$$

In classical MFC systems, oxygen has been applied as the electron acceptor at the cathode (Bard et al. 2004) (Equation 2). However, in METs metal ions often replace oxygen as the cathodic electron acceptor.

$$O_2 + 4e^- + 3H^+ \rightarrow 2H_2O \quad 1.23 \text{ vs. SHE/V} \tag{2}$$

Figure 3 summarizes metals studied to date using MET and highlights those which are highly profitable for recovery. Noteworthy reports of microbial transformation and precipitation of nanoscale crystalline transitional metals ions, include Tc^{7+} (Wildung et al. 2000), PGMs such as Pd^{4+} (Hennebel et al. 2011) and Pt^{4+} (Konishi et al. 2007), chalcogens such as Te^{4+} (Klonowska et al. 2005) and lanthanides (Deplanche et al. 2011).

Figure 4 shows some standard reduction potentials of metals. Cytochrome-*c* oxidase, an exemplarly bacterial outer membrane protein enzyme previously implicated in electron transfer to metal ions (Lovley et al. 1993), with a standard potential of 0.26 vs. SHE/V (pH 7) is also shown (Scott Mathews 1985). Cytochrome-*c* or other relational cell wall protein enzymes with similar redox potentials would be involved in the reduction of metal ions discussed. Therefore, this potential provides an approximate thermodynamic limit for metal reduction. i.e., only metals with redox potentials above 0.26 vs. SHE/V can be reduced by bacterial cells with these characteristics. Discussed subsequently, on the whole, the metal ions with standard potential below this value have found application in MET systems, while those higher in MFCs have found it predominantly with abiotic cathodes. Furthermore, metals with standard potentials higher than that of cytochrome-*c* oxidase have shown the greatest promise for application in biocathodes.

3.1 Simultaneous Metal Recovery and Energy Generation Operated as Microbial Fuel Cell Mode

Category A systems operated as MFCs have been the most investigated MET, applied for both the recovery of various metal ions and energy generation as summarized in Table 1. In some cases, metal ions are electro-deposited as zero valent metals on the cathode surface or precipitated as oxide species in solution or on the cathode. Organic oxidation by bioanodes provides electrons which drive the electrochemical reduction of metals.

3.1.1 Transitional Elements

The reduction of hexavalent chromium Cr^{6+} to the less toxic, insoluble trivalent chromium dioxide Cr_2O_3, with the simultaneous production of energy has been widely demonstrated. The formation of insoluble

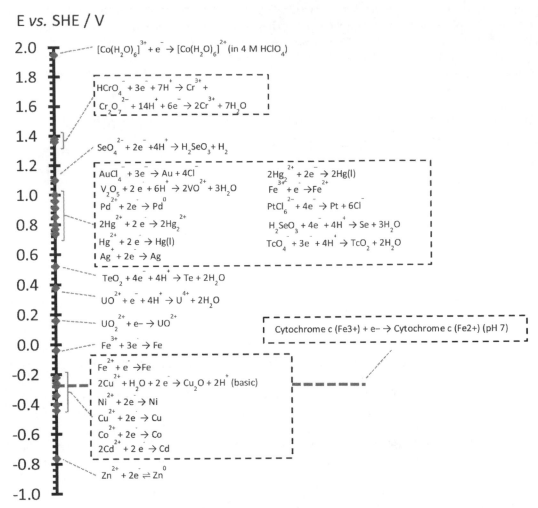

Figure 3: Redox tower of metal ion standard reduction potentials in acidic solutions and cytochrome-*c* reported vs. standard hydrogen potential.

oxides takes place via the respective electrochemical reduction and chemical precipitation (Table 1, R1-2). From the standard redox potentials of such reactions, it can be directly deduced that dichromate $Cr_2O_7^{2-}$ (1.36 vs. SHE/V) is theoretically more favorable as an electron acceptor than O_2 (1.23 vs. SHE/V) for the reduction in MFCs. Li et al. reported 99.5% removal of hexavalent chromium Cr^{6+} from real electroplating wastewater through electrochemical reduction to the less toxic trivalent chromium Cr^{3+} (Li et al. 2008). Wang et al. reported the complete reduction of Cr^{6+}, cogenerating a maximum power density of 150 mW m^{-2} (Wang et al. 2008).

Vanadium V^{5+} has been employed as an effective electron acceptor at the cathode compartment of MFCs (Zhang et al. 2010; Zhang et al. 2009).Vanadium recovery by reduction and solubilization of vanadium oxides in the cathode chamber (Table 1, R3) was proven possible. XPS analysis was used to confirm that microstructured amorphous $NaVO_4$ and V_2O_5 were the main constituents. Sulfide and total organics removal reached about 84.7% and 20.7%, respectively, whereas the V reduction efficiency was about 25.3%. Electricity generation also took place, reaching a maximum power density of approximately 572.4 mW m^{-2}.

Cu recovery was achieved in an MFC, coining the concept of 'metallurgical microbial fuel cell' that is a type of MET. Cu was recovered at the cathode in the form of pure crystals by reduction of Cu^{2+}

Table 1: Summary of MET technologies defined by category A-MFC system. Category C-MFC iron systems are also included here as reactions mechanisms occurring in the cathode are relational. The anode is biotic for all these instances.

Category	Metal	Cathode	Reaction	References
A-MFC	**Cr**	**Abiotic**	Electrochemical $Cr_2O_7^{2-} + 14H^+ + 6e^- \rightarrow 2Cr^{3+} + 7H_2O$ **R1** Chemical $2Cr^{3+} + 3H_2O \rightarrow Cr_2O_3 + 6H^+$ **R2**	(Li et al. 2008; Wang et al. 2008)
	V		Electrochemical $VO_2^+ + 2H^+ + e^- \rightarrow VO^{2+} + H_2O$ **R3**	(Zhang et al. 2010; Zhang et al. 2009)
	Cu		Electrochemical $Cu^{2+} + 2e^- \rightarrow Cu^0$ **R4**	(Ter Heijne et al. 2010; Wang et al. 2010)
			Electrochemical $2Cu^{2+} + H_2O + 2e^- \rightarrow Cu_2O + 2H^+$ **R5**	(Tao et al. 2011)
C-MFC	**Fe**	**Biotic**	Electrochemical $Fe^{3+} + e^- \rightarrow Fe^{2+}$ **R6** Microbiological $8Fe^{2+} + 8H^+ + 2O_2 \rightarrow 8Fe^{3+} + 4H_2O$ **R7**	(Ter Heijne et al. 2006)
A-MFC	**Fe**	**Abiotic**	Electrochemical $Fe^{3+} + e^- \rightarrow Fe^{2+}$ **R8** Chemical oxidation $Fe(OH)^{2+} + e^- \rightarrow Fe(OH)^+$ **R9** Chemical oxidation $Fe(OH)^+ + e^- \rightarrow Fe(OH)_2$ **R10** Chemical precipitation $Fe^{3+} + 3H_2O \rightarrow Fe(OH)_3 + 3H^+$ **R11**	(Lefebvre et al. 2012)
	Co		Electrochemical $LiCoO_2(s) + 4H^+ + e^- \rightarrow Li^+ + Co^{2+} + H_2O$ **R12**	(Huang et al. 2013; Liu et al. 2013)
	Hg		Electrochemical $2Hg^{2+} + 2e^- \rightarrow Hg_2^{2+}$ **R13** Electrochemical $Hg_2^{2+} + 2e^- \rightarrow 2Hg^0(l)$ **R14** Electrochemical $Hg^{2+} + 2e^- \rightarrow Hg^0(l)$ **R15** Chemical $2Hg^{2+} + 2Cl^- \rightarrow Hg_2Cl_2(s)$ **R16**	(Wang et al. 2011)
	Ag		Electrochemical $Ag^+ + e^- \rightarrow Ag^0$ **R17**	(Choi et al. 2012)
			Electrochemical $(AgS_2O_3)^- + e^- \rightarrow Ag^0 + S_2O_3^-$ **R18**	(Tao et al. 2012)
			Electrochemical $Ag(NH_3)_2^+ + e^- \rightarrow Ag^0 + 2NH_3$ **R19**	(Wang et al. 2013)
	Au		Electrochemical $AuCl_4^- + 3e^- \rightarrow Au^0 + 4Cl^-$ **R20**	(Choi et al. 2013)
A-MFC-- SLM	**Zn**	**Abiotic**	Chemical stripping : Zn^{2+} transport	(Fradler et al. 2014)

Source: Progress in Materials Science (2018) 94:435–461.

from solution to Cu^0 on the electrode (Table 1, R4). Removal efficiencies from the aqueous solution reached over 99% via this approach (Ter Heijne et al. 2010; Wang et al. 2010). In the presence as well as in the absence of oxygen, a similar layer of pure copper was found. In the presence of oxygen, a faster reduction (i.e., higher current at the same overpotential) was observed compared to anaerobic conditions; this being the result of combined copper and oxygen reduction and a possible catalytic effect of copper on oxygen reduction. Cu^0 can be easily released from the cathode, e.g., with an acid bath. Otherwise, Cu^0 can simply remain as a stable deposit over a supporting electrically-conductive material. Furthermore, copper deposited under aerobic conditions displays a microstructure with more octahedral features, compared to cauliflower-like features for copper precipitated under anaerobic conditions. Cu^{2+} reduction and precipitation on cathodes were also demonstrated in an MFC system at pilot-scale (Tao et al. 2011a).

Besides metallic Cu, partially-reduced Cu compounds like Cu_2O were also found on the cathode as reported (Tao et al. 2011a) and dual chamber MFC (Tao et al. 2011b). A higher initial concentration of Cu^{2+} (> 500 mg L^{-1}) resulted in the formation of $Cu_4(OH)_6SO_4$. Two electrons were required to reduce Cu^{2+} to Cu^0 completely (Table 1 R4), while only one electron was needed for the reduction of Cu^{2+} to Cu^+ (Table 1 R5). Cu removal efficiency was found dependent on the initial average Cu^{2+} concentration, reaching efficiencies from 48% to 95%, while the nature and mass of the Cu deposited was dependent on current intensity. The use of graphite as supporting electrode for cathodic Cu deposition was key for such achievements as this material minimizes the overpotential for copper reduction, whereas it maximizes the overpotential for hydrogen evolution that is paramount phenomena for industrial prospection. Other materials could provide similar features.

The dissimilative reduction of Fe^{3+} to Fe^{2+} has been extensively applied in the cathodic compartment of MFCs. Such systems do not lead to metal recovery per se but are included as they are relevant to the phenomena elaborated in this chapter. The principal objective of the academic community here is to find an alternative cathode system for O_2 reduction that would lead to higher power production. As described by the studies of Ter Heijne et al. (Ter Heijne et al. 2006, 2007, 2011), Fe^{3+} reduction at the cathode was combined with biological Fe^{2+} reoxidation to Fe^{3+} to achieve improvements in reduction reaction kinetics (Table 1 R6-7).

Fe^{3+} reduction was also studied to remove concentrated iron from AMD. In this case, reoxidation of Fe^{2+} was explained to occur through several abiotic reactions (Table 1 R8-10) in the presence of O_2 (Lefebvre et al. 2012). Reoxidation of ferrous hydroxides was followed by precipitation of Fe^{3+} in the catholyte (Table 1 R11). Part of the electrons supplied via acetate oxidation resulted in iron reduction (or ferrous oxide formation) and the other part was used to support oxygen reduction at the cathode. This caused an overall pH rise due to the transport of cations through the membrane. Indeed, ferrous iron oxidation will occur rapidly (chemically) at high pH, and it is almost impossible at low pH (2-3) unless iron-oxidizing microorganisms are present.

3.1.2 Post Transitional Elements

Huang et al. (2013) demonstrated the recovery of Co^{2+} from spent lithium-ion batteries. Co^{3+} ions were leached and reduced from lithium cobalt particles, loaded on the surface of a carbon cathode (Table 1 R12). The catholyte pH was a significant factor in cobalt leaching and power generation. Cobalt leaching efficiencies between 9% and 70% were reached, while power generation was from 33 mV m^{-3} to 258 mV m^{-3}. Co^{2+} leaching from $LiCoO_2$ with the addition of Cu^{2+} has also been demonstrated (Liu et al. 2013). A dosage of 10 mg L^{-1} Cu^{2+} improved cobalt leaching up to >300% compared to Cu^{2+}-absent controls.

Removal of Hg^{2+} in the presence of Cl^- ions was successfully demonstrated with MFCs (Wang et al. 2011) with a maximum power density of 433 mW m^{-2}. Removal efficiencies of 98-99% were reported. Products of the reduction of Hg^{2+} were verified as round microscale Hg deposits on the cathode surface and as Hg_2Cl_2 precipitates at the bottom of the cathode chamber. The electrochemical reactions concerned for Hg^{2+} reduction at the abiotic cathode are described in Table 1 (R13-15). While in the presence of Cl^-, Hg_2Cl_2 precipitation observed in the cathodic chamber would be described in Table 1 R16.

3.1.3 Precious Metals

MFC technology was applied to recover silver from organic-based wastewater while producing electrical energy. Ag^+ in the catholyte was introduced in the form of $AgNO_3$ that in aqueous solutions is fully dissociated into Ag^+ and NO_3^-. Solid metallic silver was found to be fully deposited (Table 1 R17) at the cathode surface (Choi et al. 2012). In a similar study Tao et al. (2012) reported the reduction of

Ag^+ and $(AgS_2O_3)^-$ to Ag^0 (Table 1 R18). Power densities of 109 mW m^{-3} and 35 mW m^{-3} for Ag^+ and $(AgS_2O_3)^-$ were reported, respectively. Removal efficiencies of >89% were achieved with Ag^+ solutions. For a better representation of silver wastewater, Wang et al. (2013) studied Ag recovery from ammonia-chelated silver in alkaline solutions using an MFC system. As in the above studies, bioanodes were used to generate electricity and drive the reduction of silver ions at the cathode (Table 1 R19).

The recovery of gold metal coupled with power generation using $AuCl_3^-$ has been demonstrated by Choi and Hu (Choi and Hu 2013). Power densities of up to 6.58 W m^{-2} were achieved together with recovery efficiencies as high as 99.89%. For low pH in chloride solutions Au^{3+} co-ordinates as $AuCl_4^-$ and is reduced at the cathode and deposited as elemental gold Au^0 (Table 1 R20).

The synthetic alliance of MFC integrated with a supported liquid membrane (SLM) was investigated by Fradler et al. (2014), for zinc recovery and power generation. This arrangement comprised of an anodic-chamber, a cathodic-chamber and a stripping-chamber with biocatalysts for acetate oxidation in the anode chamber. The anodic/cathodic and cathodic/stripping chambers were separated by a bipolar membrane and an SLM, respectively. Zn^{2+} concentrations were reduced from 400 mg L^{-1} to 26 mg L^{-1} and 16 mg L^{-1} in the MFC/SLM and SLM, respectively. Significant improvements in power generation were discovered with a difference of 0.233 mW and 0.094 mW for MFC/SLM and MFC systems.

Huang et al. (2015) demonstrated the recovery of Co^{2+} using an oxygen-reducing biocathode in an MFC. Biofilm biocathodes were established using mixed cultures inoculated from aerobic and anaerobic sludge, sediments and metalworking wastewaters. The *in situ* production of OH$^-$ by O_2 reduction at the biocathode led to the precipitation of amorphous microscale $Co(OH)_2$ with simultaneous energy production. Maximum power densities of 1500 mW m^{-3} were achieved under optimal conditions. SEM analysis revealed that the majority of $Co(OH)_2$ precipitates occurred on the microbial cell surface.

In another study, the reduction of Cr^{6+} was completed in strong association with electrochemically-generated H_2O_2 at an air-bubbling cathode (Liu et al. 2011). Higher concentrations of H_2O_2 were produced with activated sludge in the anode chamber and the addition of the electron mediator ADQS to the anode. The Cr^{3+} reduction could occur via direct electrochemical reduction or via electrochemically-produced H_2O_2. H_2O_2 was shown to be a superior reducer because when H_2O_2 was not formed (i.e., when the chamber was gassed with N_2), Cr^{6+} was reduced at lower rates and lower removal efficiencies (42.5% with nitrogen vs. about 100% with air).

In an original study, Li et al. (2009) demonstrated the integration of MFC and a photoelectrochemical cell (PEC), for simultaneous power generation and chromium recovery. In this system, a biotic anode was coupled to an abiotic semiconductor photovoltaic (rutile-coated) cathode. Under light irradiation, 97% of Cr^{6+} was reduced and precipitated, faster than in a non-photocatalytic process. Photoelectrons produced may also react with H_2O, OH$^-$ and O_2 to produce other reducing species, such as hydroxyl radicals ($^·$OH), superoxide radicals ($O_2^{·-}$) and H_2O_2 (Ramirez et al. 2015) that might also have contributed to the higher Cr^{6+} reduction rates.

3.2 Metal Recovery Operated in Microbial Electrolysis Cell Mode

The recovery of pure cobalt metal and H_2 production from a Co^{2+} electrolyte was reported by Jiang et al. (2014) using mixed cultures on an MEC bioanode. SEM and XPS analysis revealed initially sporadic Co metallic crystals leading to microscale to nanoscale flake blooms growing away from the cathode surface that would have potential application in electrochemical capacitors (Kong et al. 2009). Optimally applied cell potentials were 0.3-0.5 V for Co recovery and H_2 production (Table 2 R1). In a similar study, Qin et al. (2012) demonstrated the recovery of nickel in a MEC system. Ni was deposited on the cathode (Table 2, R2). The removal efficiency reached 33-99%, while the current density reached 51 to 166 A m^{-2}.

Choi et al. reported cadmium recovery using a combination of MFC and MEC (Choi et al. 2014). Two MFCs with chromium as the electron acceptor in the cathode (Cr-MFC) were connected in series to an MEC with Cd^{2+} in the cathode chamber. Here, the Cr-MFCs were used to complement the insufficient electrical potential needed to drive Cd^{2+} reduction in the MEC. High removal efficiencies of 89-93% were observed. Cadmium was recovered as metal as described in R3, Table 2.

Table 2: Summary of MEM technologies defined by category A-MEC system.

Category	Metal	Anode	Cathode	Reaction	Reference
A-MEC	Co	**Biotic**	**Abiotic**	Electrochemical $Co^{2+} + 2e^- \rightarrow Co^0$ **R1**	(Jiang et al. 2014)
	Ni			Electrochemical $Ni^{2+} + 2e^- \rightarrow Ni^0$ **R2**	(Qin et al. 2012)
A-Stacked 2MFC--MEC	Cd			Electrochemical $Cd^{2+} + 2e^- \rightarrow Cd^0$ **R3**	(Catal et al. 2009)

Source: Progress in Materials Science (2018) 94:435–461.

3.3 Biocathodes Drive Metal Recovery

In a pioneering investigation, which opened the gate to extensive research that is now dedicated to unraveling the principles and applications of biocathodic MET, Gregory and Lovley used *Geobacter sulfurreducens* for U^{6+} reduction on a biocathode. Uranium removal was effectively achieved by reducing soluble hexavalent uranium U^{6+} to the relatively insoluble tetravalent uranium U^{4+} oxide precipitate (Table 3 R1) (Gregory and Lovley 2005). The cathode served as an electron donor to *Geobacter sulfurreducens*. U^{4+} remained as a stable precipitate on the electrode in the absence of O_2. In the absence of microbes, uranium could not be reduced. With the MET, 87% of the total U^{6+} was removed, achieving a 97% current efficiency. Reports of metals recovered in biocathodes are summarized in Table 3.

Table 3: Summary of MEM technologies defined by category B system.

Category	Metal	Anode	Reaction	References
D-MFC	U	**Biotic**	Biological $UO_2^{2+} + 2e^- \rightarrow UO^{2+}$ **R1**	(Gregory and Lovley 2005)
	Se		Biological $SeO_3^{2-} + 4e^- + 6H^+ \rightarrow Se^0 + 3H_2O$ **R2**	(Catal et al. 2009)
	Cr		Biological $Cr_2O_7^{2-} + 14H^+ + 6e^- \rightarrow 2Cr^{3+} + 7H_2O$ **R3** Chemical $2Cr^{3+} + 3H_2O \rightarrow Cr_2O_3 + 6H^+$ **R4**	(Tandukar et al. 2009; Huang et al. 2011)
D-MEC	Co	**Biotic**	Microbial-electrochemical $Co^{2+} + 2e^- \rightarrow Co^0$ **R5**	(Huang et al. 2014)
	Au	**Abiotic**	Microbial-electrochemical $AuCl_4^- + 3e^- \rightarrow Au^0 + 4Cl^-$ **R6**	(Varia et al. 2014)

Source: Progress in Materials Science (2018) 94: 435–461.

In an original study, bioanodes/cathodes were applied for selenium recovery using MFCs. Simultaneous electricity generation and selenium recovery were evaluated in a single chamber air-MFC (Catal et al. 2009), leading to elemental selenite on the electrode surface and in the electrolyte. Carbon cloth was used for both the anode and cathode. High removal efficiencies of selenite were demonstrated with up to 99% selenite removal. Oxygen, and not SeO_3^{2-}, was used as the electron acceptor by the biocathode. Based on previous reports (Oremland et al. 1989), one could propose that SeO_3^{-2} was used by bacterial cultures immobilized on bioanodes, as an electron acceptor, for dissimilative respiration (Table 3 R2).

The application of biocathodes with power generation for Cr^{6+} removal as $Cr(OH)_3$ (Table 3 R3-4) was first demonstrated by Tandukar et al. (2009). The contribution of biomass decay and abiotic processes for the reduction of Cr^{6+} was minimal that would confirm that Cr^{6+} reduction was assisted by microbial activity on the cathode. Cr^{3+} precipitated as $Cr(OH)_3$ on the bacterial biomass or cathode. Following this investigation, Huang et al. (2011) studied Cr^{6+} recovery with a $Cr(OH)_3$ precipitate observed on the bacterial biomass and not the cathode.

The application of biocathodes in MECs for cobalt recovery was first reported by Huang et al. (2014) with simultaneous production of methane and acetate. 88.1% Co^{2+} was reduced to Co^0 (Table 3 R5), mainly on the bacterial surface.

Recently Varia et al. investigated the application of biocathodes for Au remediation and recovery (Varia et al. 2014; Varia et al. 2013). In the presence of bacteria, gold electro-deposition thermodynamics was influenced as revealed by positive shifts in the reverse potentials of $AuCl_4^-/Au^0$ (Table 3 R6).

4. Simultaneous Mixed Metal Recovery

All reports of MET systems studied up to 2012 focused on the recovery of a single type of metal in solution. However, in wastewaters and leachates, different metals often exist in mixtures. One example is wastewater from vanadium mining and manufacturing processes that contain a mixture of V^{5+} and Cr^{6+} ions (Zhang et al. 2012). Zhang et al. (2012) investigated the simultaneous reduction of these two ions at the cathode of an MFC. Cr^{6+} was reduced to Cr^{3+} that deposited as oxide on the cathode surface. V^{5+} was reduced to V^{4+} that remained in solution. By raising the pH from 2 to 6 at the end of the experiment, V^{4+} could be precipitated from the solution. Metal removal efficiencies of 60.9% for V and 71.4% for Cr could be obtained.

Another example of wastewater containing a mixture of several metals is leachate from solid waste incineration fly as that can contain high concentrations of Cu, Pb, Cd and Zn. Modin et al. (2012) varied the control of a microbial electrochemical system to sequentially recover these metals individually from a mixture.

Tao et al. (2014) used an MFC connected to a conventional electrolysis system to recover Cu, Pb and Zn from a real fly ash leachate. In the MFC, 97.1% of the Cu could be removed. The effluent from the MFC was fed to the electrolytic system operated at an applied cell potential of 6 V, where Pb^{2+} and Zn^{2+} were reduced simultaneously; also, 98.1% of Pb and 95.4% of Zn were removed, respectively (Tao et al. 2014).

AMD from metal mining can contain high concentrations of metals, e.g., Cu, Ni and Fe. Luo et al. (2014) investigated the use of an MEC for simultaneous reduction of these three metals in a mixed solution. Hydrogen gas was produced in combination with metal reduction. Deposits of discrete crystalline branch-like structures of Cu^0, smooth microscale thin layer of Ni^0 and flaky Fe crystals were observed on the cathode surface. The results suggested that the energy content of produced H_2 gas could offset against the energy requirements for running the reactor.

Zhang et al. (2015) reported the combination of MFC and MEC systems for recovery of Cr, Cu and Cd. MFCs using Cr^{4+} and Cu^{2+} as electron acceptors were stacked in parallel or series to drive an MEC using Cd^{2+} as the final electron acceptor. SEM and XRD confirmed the precipitation of $Cr(OH)_3$ and pure copper spherule-shaped crystals and a smooth cadmium deposit on respective cathodes.

5. Applicability and Limitations of MET in Metal Recovery

MET, as pointed out here, has the potential for metal recovery from diverse aqueous waste streams. Advantages of MET include low-cost microbial catalysts instead of expensive noble metals, potential selectivity for targeted metals, low cell potential or even simultaneous power generation, the combination of conversion and adsorption and pH correction in cathode and anode that can make a stream more treatable. However, various hurdles will have to be overcome such as instability of microbial catalysts with cold temperature, low concentration of metals limiting transport to the electrode surface, low pH metal stream rendering problematic, toxicity of metals towards catalysts, limited experience, and the need for organic electron donors or different electron sources, besides small quantities of recovered products that make market logistics difficult.

Furthermore, metals are not always present in soluble form, even in aqueous systems. In wastewater, there can be sludges or solids from which first the metals may need to be leached with pH changes to enable recovery. Therefore, pretreatment is of great importance in most industrial cases. Studies, so far, have only focused on demonstrating concepts, not optimizing reactor configuration and operational parameters. Thus, there is ample room for improvements in performance. However, it is promising that high removal/recovery efficiency can be obtained with solutions containing quite low concentrations of metals.

Several waste streams such as fly ash leachate and AMD would contain a metal mix like Cu, which can be reduced in an MFC, and metals such as Pb, Cd and Zn that cannot be spontaneously reduced under typical MEC conditions. This makes the MFC a good technology to selectively extract relatively clean Cu from complex metal mixtures. If Fe^{3+} occurs in the mix, it could also be a reduced in an MFC but it could potentially be reduced as nanoscale to macroscale precipitates of $Fe(OH)_2$ rather than deposited on the cathode.

Controlling the pH and potential of the electrodes in MET reactors are ways to selectively extract individual metals from a mix. This makes it possible to extract those metals by precipitation and recover other metals directly on the cathode by reduction.

One of the main advantages of using MET for metal removal/recovery is that less energy is required compared to conventional technologies. Therefore, the choice of electron donor at the anode is crucial. Until now, most of the studies on metal recovery have used bioanodes that oxidize acetate. This requires an organic waste stream to be available at the same location as the metal stream, which makes application of the technology limited to specific locations. On the other hand, compared to precipitation of metals with sulfides, the carbon efficiency for some MET is several times higher (Ter Heijne et al. 2010). Thus, besides the use of organics, other electron donors can be considered. Sulfur-based components may provide an interesting alternative as they are abundantly present in mining and metallurgical waste streams. Possible reactions would be the oxidation of sulfides or elemental sulfur to produce sulfates (Zhang et al. 2009; Rabaey et al. 2006).

6. Conclusions

The current state-of-the-art of MET reviewed here shows much promise for metal recovery, especially from diverse dilute metal-rich aqueous streams. However, one must emphasize that further investigation and optimization, especially for mixed metal ion systems, will be required to go beyond laboratory and pilot-scale studies. Most studies on MET technology have used bioanodes that do not actually interact directly with the metal. A number of studies have also demonstrated hybrid systems, such as the integration of SLM or photoelectrocatalyic electrodes in MET. As highlighted in this review, some studies apply microbes in the cathodic chamber or on biocathodes, which do participate and enhance metal recovery, yet this is only limited to a few individual metals. In light of numerous reports of microbial sorption and transformation phenomena scanning the width and breadth of the periodic table, further investigations of microbial interactions with metals in biocathode chambers are imperative. Novel insights in microbial electrochemical metabolism and its implications could provide new routes toward metal recovery. Furthermore, electrochemically-tunable EPS would be a promising aspect to explore for further improvement. The use of undiscovered exerogenic extremophiles as biocatalysts or nanofactories in MET could enable metal recovery from environments, such as deep marine and continental waters as well as geothermal brines.

7. Future Perspectives

Despite the wide variety of microbes that have been reported to carry out metal transformations in purely microbial systems, the same diversity has not been yet extended to MET. Only *Geobacter, Shewanella* and *Acidithiobacillus* have been studied for direct participation in metal reduction processes. Further investigation of extremophiles for the recovery and prospection of metals from extreme environments such as the deep sea is warranted. Other extreme environments such as geothermal brines and hydrothermal vents could indeed be a lucrative metal-mining direction. However, the reactor configuration of METs needs to be reengineered for these purposes. Especially, effective coupling electrobioleaching with separation techniques (e.g., membrane electrolysis), MET cathodes for metal recovery, could prove effective for such environments. Especially, MET can be applied to supplement the supply of those metals highlighted recently by the European Commission (Moss et al. 2013) as critical to foster environmental sustainability with a shift to a low carbon economy.

Acknowledgements

The contributions of XDB have been supported by the Strategic Research Fund at VITO, Project 1310236. G. Pozo acknowledges the funding from the European Union's Horizon 2020 research and innovation programme MSCA-IF-2017 under grant agreement no. 796320 (MAGDEx: Unmet magnetic properties in microparicles and nanoparticles by synthesis through gas diffusion electrocrystallisation (GDEx). XDB and GP thank for the grant received from the Flemish SIM MaRes programme for SBO project Get-A-Met. JV thanks for the support from European Union's Horizon 2020 Research and Innovation program under grant agreement no. 690088 (METGROW+ project). XDB thanks the funding received from the European Union's Horizon 2020 research and innovation programme under grant agreement no. 654100 (CHPM2030 project).

References

Anushree M (2004) Metal bioremediation through growing cells. Environ Int 30: 261–78.

Bard AJ, Stratmann M, Schäfer HJ (2004) Encyclopedia of Electrochemistry: Volume 8: Organic Electrochemistry. Wiley-VCH, Weinheim, pp. 663.

Beveridge T, Murray R (1976) Uptake and retention of metals by cell walls of Bacillus subtilis. J Bacteriol 127: 1502–1518.

Bowman N, Patel P, Sanchez S, Xu W, Alsaffar A, Tiquia-Arashiro SM (2018) Lead-resistant bacteria from Saint Clair River sediments and Pb removal in aqueous solutions. Appl Microbiol Biotechnol 102: 2391–2398.

Catal T, Bermek H, Liu H (2009) Removal of selenite from wastewater using microbial fuel cells. Biotechnol Let 31: 1211–1216.

Choi C, Cui Y (2012) Recovery of silver from wastewater coupled with power generation using a microbial fuel cell. Biores Technol 107: 522–525.

Choi C, Hu N (2013) The modeling of gold recovery from tetrachloroaurate wastewater using a microbial fuel cell. Biores Technol 133: 589–598.

Choi C, Hu N, Lim B (2014) Cadmium recovery by coupling double microbial fuel cells. Biores Technol 170: 361–369.

Clauwaert P, Aelterman P, De Schamphelaire L, Carballa M, Rabaey K, Verstraete W (2008) Minimizing losses in bio-electrochemical systems: the road to applications. App Microbiol Biotechnol 79: 9017–9013.

Deplanche K, Murray A, Mennan C, Taylor S, Macaskie L (2011) Biorecycling of precious metals and rare earth elements. Nanomatter 279-314.

European Commission (2014). Ad-hoc WG. Report on critical raw materials for the EU.

Fradler KR, Michie I, Dinsdale RM, Guwy AJ, Premier GC (2014) Augmenting Microbial Fuel Cell power by coupling with Supported Liquid Membrane permeation for zinc recovery. Water Res 55: 115–125.

Gadd GM (2010) Metals, minerals and microbes: geomicrobiology and bioremediation. Microbiol 156: 609–43.

Gregory KB, Lovley DR (2005) Remediation and recovery of uranium from contaminated subsurface environments with electrodes. Environ Sci Technol 39: 8943–8947.

Jiang L, Huang L, Sun Y (2014) Recovery of flakey cobalt from aqueous Co(II) with simultaneous hydrogen production in microbial electrolysis cells. Int J Hydrogen Ener 39: 654–663.

Hennebel T, Benner J, Clauwaert P, Vanhaecke L, Aelterman P, Callebaut R, et al. (2011) Dehalogenation of environmental pollutants in microbial electrolysis cells with biogenic palladium nanoparticles. Biotechnol Let 33: 89–95.

Hennebel T, Boon N, Maes S, Lenz M (2015) Biotechnologies for critical raw material recovery from primary and secondary sources: R&D priorities and future perspectives. New Biotechnol 32: 121–7.

Hill HAO, Higgins IJ (1981) Bioelectrocatalysis. Philos. Trans. Royal Soc. London Series A, Mathematical and Physical Sciences. 302: 267–73.

Huang L, Chai X, Cheng S, Chen G (2011a) Evaluation of carbon-based materials in tubular biocathode microbial fuel cells in terms of hexavalent chromium reduction and electricity generation. Chem Eng J 166: 652–661.

Huang L, Chai X, Chen G, Logan BE (2011b) Effect of set potential on hexavalent chromium reduction and electricity generation from biocathode microbial fuel cells. Environ Sci Technol 45: 5025–5031.

Huang L, Jiang L, Wang Q, Quan X, Yang J, Chen L (2014) Cobalt recovery with simultaneous methane and acetate production in biocathode microbial electrolysis cells. Chem Eng J 253: 281–290.

Huang L, Li T, Liu C, Quan X, Chen L, Wang A, et al. (2013) Synergetic interactions improve cobalt leaching from lithium cobalt oxide in microbial fuel cells. Biores Technol 128: 539–46.

Huang L, Liu Y, Yu L, Quan X, Chen G (2015) A new clean approach for production of cobalt dihydroxide from aqueous Co (II) using oxygen-reducing biocathode microbial fuel cells. J leaner Prod 86: 441–446.

Johnson DB, Hallberg KB (2003) The microbiology of acidic mine waters. Res Microbiol 154: 466–73.

Jones D, Amy PA (2002) Thermodynamic interpretation of microbiologically influenced corrosion. Corrosion. 58: 638–645.

Klaus-Joerger T, Joerger R, Olsson E, Granqvist C-G (2001) Bacteria as workers in the living factory: metal-accumulating bacteria and their potential for materials science. Trends Biotechnol 19: 15–20.

Klonowska A, Heulin T, Vermeglio A. (2005) Selenite and tellurite reduction by *Shewanellaoneidensis*. Appl Environ Microbiol 71: 5607–5609.

Kong LB, Lang JW, Liu M, Luo YC, Kang L (2009) Facile approach to prepare loose-packed cobalt hydroxide nanoflakes materials for electrochemical capacitors. J Power Sources 194: 1194–1201.

Konhauser K, Lalonde S, Phoenix V (2008) Bacterial biomineralization: Where to from here? Geobiol 6: 29–302.

Konishi Y, Ohno K, Saitoh N, Nomura T, Nagamine S, Hishida H, et al. (2007) Bioreductive deposition of platinum nanoparticles on the bacterium *Shewanella algae*. J Biotechnol 128: 648–653.

Lefebvre O, Neculita CM, Yue X, Ng HY (2012) Bioelectrochemical treatment of acid mine drainage dominated with iron. J Haz Mater 241–242: 411–7.

Li Y, Lu A, Ding H, Jin S, Yan Y, Wang C, et al. (2009) Cr(VI) reduction at rutile-catalyzed cathode in microbial fuel cells. Electrochem Comm 11: 1496–1149.

Li Z, Zhang X, Lei L (2008) Electricity production during the treatment of real electroplating wastewater containing Cr^{6+} using microbial fuel cell. Proc Biochem 43: 1352–1358.

Liu Y, Shen J, Huang L, Wu D (2013) Copper catalysis for enhancement of cobalt leaching and acid utilization efficiency in microbial fuel cells. J Haz Mater 262: 1–8.

Liu L, Yuan Y, Li F, Feng C (2011) In-situ Cr(VI) reduction with electrogenerated hydrogen peroxide driven by iron-reducing bacteria. Biores Technol 102: 2468–73.

Lloyd JR, Lovley DR (2001) Microbial detoxification of metals and radionuclides. Curr Opin Biotechnol 12: 248–53.

Logan BE, Hamelers B, Rozendal R, Schröder U, Keller J, Freguia S, et al. (2006) Microbial Fuel Cells: Methodology and Technology. Environ Sci Technol 40: 5181–92.

Logan BE, Rabaey K (2012) Conversion of wastes into bioelectricity and chemicals by using microbial electrochemical technologies. Science 337: 686–90.

Lovley DR (1995) Bioremediation of organic and metal contaminants with dissimilatory metal reduction. J Ind Microbiol 14: 85–93.

Lovley DR, Widman PK, Woodward JC, Phillips E (1993) Reduction of uranium by cytochrome c3 of *Desulfovibrio vulgaris*. Appl Environ Microbiol 59: 3572–3576.

Luo H, Liu G, Zhang R, Bai Y, Fu S, Hou Y (2014) Heavy metal recovery combined with H_2 production from artificial acid mine drainage using the microbial electrolysis cell. J Haz Mater 270: 153–159.

Macaskie L, Mikheenko I, Yong P, Deplanche K, Murray A, Paterson-Beedle M, et al. (2010) Today's wastes, tomorrow's materials for environmental protection. Hydrometal 104: 483–7.

Modin O, Wang X, Wu X, Rauch S, Fedje KK (2012) Bioelectrochemical recovery of Cu, Pb, Cd, and Zn from dilute solutions. J Haz Mater 235: 291–7.

Moss R, Tzimas E, Willis P, Arendorf J, Espinoza LT. (2013) Critical Metals in the Path towards the Decarbonisation of the EU Energy Sector. JRC Scientific and Policy Reports. Publications Office of the European Union.

Oremland RS, Hollibaugh JT, Maest AS, Presser TS, Miller LG, Culbertson CW (1989) Selenate reduction to elemental selenium by anaerobic bacteria in sediments and culture: biogeochemical significance of a novel, sulfate-independent respiration. Appl Environ Microbiol 55: 2333–2543.

Qin B, Luo H, Liu G, Zhang R, Chen S, Hou Y, et al. (2012) Nickel ion removal from wastewater using the microbial electrolysis cell. Biores Technol 121: 458–461.

Rabaey K, Rozendal RA (2010) Microbial electrosynthesis—revisiting the electrical route for microbial production. Nat Rev Microbiol 8: 7067–16.

Rabaey K, Verstraete W (2005) Microbial fuel cells: novel biotechnology for energy generation. Trends Biotechnol 23: 291–298.

Rabaey K, Van de Sompel K, Maignien L, Boon N, Aelterman P, Clauwaert P, et al. (2006) Microbial fuel cells for sulfide removal. Environ Sci Technol 40: 5218–5224.

Ramirez R, Arellano C, Varia J, Martinez S (2015) Visible Light-Induced Photocatalytic Elimination of Organic Pollutants by TiO_2: A Review. Curr Org Chem 19: 540–555.

Rhoads A, Beyenal H, Lewandowski Z (2005) Microbial fuel cell using anaerobic respiration as an anodic reaction and biomineralized manganese as a cathodic reactant. Environ Sci Technol 39: 4666–71.

Scott Mathews F (1985) The structure, function and evolution of cytochromes. Prog Biophys Molec Biol 45: 1–56.

Tandukar M, Huber SJ, Onodera T, Pavlostathis SG (2009) Biological chromium (VI) reduction in the cathode of a microbial fuel cell. Environ Sci Technol 43: 8159–65.

Tao HC, Gao ZY, Ding H, Xu N, Wu WM (2012) Recovery of silver from silver(I)-containing solutions in bioelectrochemical reactors. Biores Technol 111: 92–97.

Tao HC, Lei T, Shi G, Sun XN, Wei XY, Zhang LJ, et al. (2014) Removal of heavy metals from fly ash leachate using combined bioelectrochemical systems and electrolysis. J Haz Mater 264: 1–7.

Tao HC, Liang M, Li W, Zhang LJ, Ni JR, Wu WM (2011b) Removal of copper from aqueous solution by electrodeposition in cathode chamber of microbial fuel cell. J Haz Mater 189: 186–92.

Tao HC, Li W, Liang M, Xu N, Ni JR, Wu WM (2011a) A membrane-free baffled microbial fuel cell for cathodic reduction of Cu(II) with electricity generation. Biores Technol 102: 4774–4778.

Ter Heijne A, Hamelers HV, Buisman CJ (2007) Microbial fuel cell operation with continuous biological ferrous iron oxidation of the catholyte. Environ Sci Technol 41: 4130–4134.

Ter Heijne A, Hamelers HV, De Wilde V, Rozendal RA, Buisman CJ (2006) A bipolar membrane combined with ferric iron reduction as an efficient cathode system in microbial fuel cells. Environ Sci Technol 40: 5200–5205.

Ter Heijne A, Liu F, van Rijnsoever LS, Saakes M, Hamelers HVM, Buisman CJN (2011) Performance of a scaled-up Microbial Fuel Cell with iron reduction as the cathode reaction. J Power Sources 196: 7572–7577.

Ter Heijne A, Liu F, WeijdenRvd, Weijma J, Buisman CJN, Hamelers HVM (2010) Copper Recovery Combined with Electricity Production in a Microbial Fuel Cell Environ Sci Technol 44: 4376–4381.

Thrash JC, Coates JD (2008) Review: Direct and Indirect Electrical Stimulation of Microbial Metabolism. Environ Sci & Technol 42: 3921–31.

Tiquia-Arashiro SM (2018) Lead absorption mechanisms in bacteria as strategies for lead bioremediation. Appl Microbiol Biotechnol 102: 5437–5444.

Tiquia-Arashiro SM, Rodrigues D (2016) Extremophiles: Applications in Nanotechnology. Springerbriefs in Microbiology: Extremophilic Microorganisms. Springer International Publishing. 193 p.

Varia JC, Martinez SS, Velasquez-Orta S, Bull S (2014) Microbiological influence of metal ion electrodeposition: Studies using graphite electrodes, (AuCl4)−and Shewanellaputrefaciens. Electrochim Acta 115: 344–51.

Varia JC, Martínez SS, Velasquez-Orta S, Bull S, Roy S (2013) Bioelectrochemical metal remediation and recovery of Au^{3+}, Co^{2+} and Fe^{3+} metal ions. Electrochim Acta 95: 125–131.

Varia J, Zegeye A, Roy S, Yahaya S, Bull S (2014) *Shewanellaputrefaciens* for the remediation of Au^{3+}, Co^{2+} and Fe^{3+} metal ions from aqueous systems. Biochem Eng J 85: 101–109.

Veglio F, Beolchini F (1997) Removal of metals by biosorption: a review. Hydrometal 44: 301–316.

Volesky B (2001) Detoxification of metal-bearing effluents: biosorption for the next century. Hydrometal 59: 203–16.

Wang G, Huang L, Zhang Y (2008) Cathodic reduction of hexavalent chromium (Cr (VI)) coupled with electricity generation in microbial fuel cells. Biotechnol Let 30: 1959–1966.

Wang Z, Lim B, Choi C (2011) Removal of Hg^{2+} as an electron acceptor coupled with power generation using a microbial fuel cell. Biores Technol 102: 6304–6307.

Wang Z, Lim B, Lu H, Fan J, Choi C (2010) Cathodic reduction of Cu^{2+} and electric power generation using a microbial fuel cell. B Korean Chem Soc 31: 2025–30.

Wang YH, Wang BS, Pan B, Chen QY, Yan W (2013) Electricity production from a bio-electrochemical cell for silver recovery in alkaline media. Appl Ener 112: 1337–1341.

Wildung RE, Gorby YA, Krupka KM, Hess NJ, Li S, Plymale AE, et al. (2000) Effect of electron donor and solution chemistry on products of dissimilatory reduction of technetium by *Shewanellaputrefaciens*. App Environ Microbiol 66: 2451–2460.

White C, Wilkinson SC, Gadd GM (1995) The role of microorganisms in biosorption of toxic metals and radionuclides. Int Biodeterior Biodegradation 35: 17–40.

Zhang B, Feng C, Ni J, Zhang J, Huang W (2012) Simultaneous reduction of vanadium (V) and chromium (VI) with enhanced energy recovery based on microbial fuel cell technology. J Power Sources 204: 34–39.

Zhang Y, Yu L, Wu D, Huang L, Zhou P, Quan X, et al. (2015) Dependency of simultaneous Cr(VI), Cu(II) and Cd(II) reduction on the cathodes of microbial electrolysis cells self-driven by microbial fuel cells. J Power Sources 273: 1103–1113.

Zhang B, Zhao H, Shi C, Zhou S, Ni J (2009) Simultaneous removal of sulfide and organics with vanadium (V) reduction in microbial fuel cells. J Chem Technol Biotechnol 84: 1780–6.

Zhang BG, Zhou SG, Zhao HZ, Shi CH, Kong LC, Sun JJ, et al. (2010) Factors affecting the performance of microbial fuel cells for sulfide and vanadium (V) treatment. Bioproc Biosyst Eng 33: 187–194.

Plant-Microbial Fuel Cells Serve the Environment and People: Breakthroughs Leading to Emerging Applications

Lucia Zwart, Cees J.N. Buisman and David Strik[*]

Sub-Department of Environmental Technology, Wageningen University and Research, The Netherlands

1. Introduction

At the beginning of 2008, three research groups published, almost simultaneously, on the origin of the plant-microbial fuel cell (Plant-MFC). Strik et al. published the first paper on the Plant-MFC using *Glyceria maxima* (Strik et al. 2008). Others used rice plants to show that rhizodeposits were the source of electricity (De Schamphelaire et al. 2008), while a Japanese research group developed the system in a rice paddy field (Kaku et al. 2008). The technology was developed in response to the demand for new and sustainable technologies for the production of electricity. In their paper, Strik et al. (2008) described how electricity could be generated from living plants with a two-chamber setup. Since this first article, research has been conducted to better understand and improve the Plant-MFC. Researchers have tried to characterize the internal processes by looking at the microbial communities involved (Friedrich 2011; Timmers et al. 2012a) and defined the internal resistances (Timmers et al. 2012b). The design of the fuel cell has been improved by adjusting the electrolyte composition, comparing three different plant types (Helder et al. 2010), altering the physical shape from a flat plate cell to a tubular cell (Timmers et al. 2013) and integrating an oxygen reducing biocathode (Wetser et al. 2015b). Also, application of the technology in a natural environment has been investigated in different kinds of wetland (Wetser et al. 2015a). The objective of this chapter is to provide a constructive explanatory overview of the Plant-MFC and its innovative environmental technological applications. Several studies were highlighted that demonstrated new applications or steps for its technological development. As such it shows that the technology is proving its feasibility to be profitable for the environment and people.

2. Working Principle

A Plant-MFC is a type of microbial solar cell. The technology converts solar energy into electricity. Plants convert solar energy into biomass in the process of photosynthesis. Through this, 20-40% of the carbon that is converted into biomass is released again at the plant's roots in the form of rhizodeposits. Rhizodeposits is the generic term for a collection of sugars, organic acids, polymeric carbohydrates, enzymes and dead-cell plant material near plant roots. The Plant-MFC generates electricity from these rhizodeposits without the need to harvest the plant (Strik et al. 2008).

---

*Corresponding author: david.strik@wur.nl

The Plant-MFC consists of two electrodes that are connected by an electrical circuit and submerged in an electrolyte or saturated soil. Electrochemically active bacteria, present at the anode, catalyze the oxidation of rhizodeposits and other derivatives from these into CO_2, protons and electrons. At the cathode, oxygen and protons are reduced to water. The coupling of the redox reactions at both electrodes results in an electron flow that can be harvested to gain electricity.

The concept of the Plant-MFC promotes a clean, renewable, *in situ*, efficient and sustainable technology (Figure 2). There are several different characteristics of the Plant-MFC in comparison to other sustainable electricity producing technologies. The technology does not directly depend on wind intensity or competes with arable land. Wind and solar energy need storage due to the intermittent electricity generation. The Plant-MFC avoids this necessity by storing solar energy in the plant's biomass. It can be applied in many natural and man-made environments without altering the physical aspect of the land.

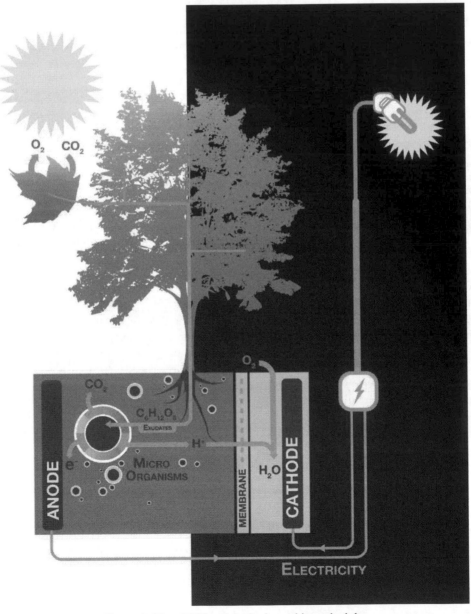

Figure 1: Plant-MFC and the basic working principles.

The greater part of scientific articles on the Plant-MFC focuses on electricity production. This *in situ* produced electricity provides many possible applications in itself. The benefit of the Plant-MFC as an electricity-producing technology in comparison to other sustainable technologies is that it can be applied in areas not suitable for agriculture, not suitable for other electricity producing technologies or areas with high aesthetic value. In the following sections, several environments where the Plant-MFC can be applied will be discussed.

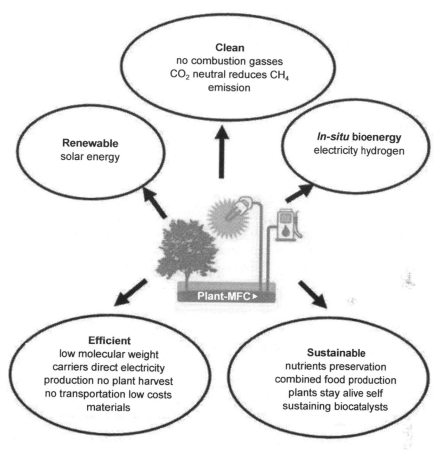

Figure 2: Concept properties of the Plant-MFC.

3. Plant-MFC Serving the Environment

3.1 Environmental Remediation

Plant-MFCs have been applied for the purpose of environmental remediation in multiple systems. Most researchers integrate the Plant-MFCs in natural or constructed wetlands. These ecosystems are able to remediate soils and water bodies because of the presence of microorganisms that are able to degrade (persistent) organic and inorganic pollutants. Plants aid in remediation by taking up nutrients, thereby reversing eutrophication. Moreover, the rhizodeposits at plant roots promote bacterial growth. These bacteria are then able to degrade organic contaminants.

3.1.2 Wastewater and Sediment Treatment

Plant-MFC can be used for the treatment of wastewater. Research on wastewater treatment with Plant-MFCs mainly focuses on the integration of the technology into constructed wetlands. Constructed wetlands are wetlands especially designed for the treatment of wastewater; the type of sediment, plant and water flow can be adjusted to a specific wastewater/pollutant. They can treat large volumes of wastewater and are nearly maintenance free. A natural potential gradient exists in constructed wetlands due to the anaerobic conditions in the deep soil vs. the more aerobic conditions in the upper sediments and water. This potential gradient makes constructed wetlands very suited environments for the application of MFCs. Integrating a Plant-MFC can enhance the treatment efficiency of a constructed wetland and generate electricity simultaneously. The combination of these two technologies is called a constructed wetland microbial fuel cell (CW-MFC).

Different kinds of pollutants have been treated with W-MFC. Several publications focused on the removal of organic carbon and inorganic nutrients from the water phase. Too many organic and (reduced) inorganic nutrients in the water phase lead to eutrophication that can disturb the ecosystem equilibrium and result in oxygen depletion. Mohan et al. (2011) researched the treatment of domestic sewage and fermented distillery wastewater by a sediment MFC with lake sediments and floating macrophytes (*Eichornia Crassipes*). The average COD removal in continuous mode was 80% for a period of 210 days. Removal efficiencies depended on the wastewater composition. A continuation of this study was made (Chiranjeevi et al. 2013) with the introduction of more plant species. The system contained a mix of emergent (tomato, rice and ferns) and submergent macrophytes. The addition of more plants resulted in higher removal efficiencies for COD and volatile fatty acids.

Chiranjeevi et al. (2013) found the power output to be wastewater specific with a higher power generation for the fermented distillery wastewater in comparison to the domestic sewage. The fermented distillery wastewater was higher in organic load. Villasenor et al. (2013) tested the functionality of a horizontal subsurface CW-MFC with *Phragmitis Australis* for synthetic wastewater at different organic loads. The CW-MFC performed well under low organic loading rates (13.9 g COD m^{-2} d^{-1}). It removed organics and nutrients and simultaneously generated electricity. However, at higher organic loading rates (61.1 g COD m^{-2} d^{-1}), organics could not be fully oxidized in the deeper parts of the constructed wetland and thereby reached water that was closer to the surface. In the upper part, the cathode was placed, and due to the organics the water became anoxic and inhibited current generation between the anode and cathode.

In order to prevent this oxygen limitation at the cathode Wu et al. (2015) designed a new type of CW-MFC. This MFC had a part of the cathode underwater and a part of it stick out above the water surface. This configuration performed well with high COD, N and P removal percentages. The continuously treated wastewater had a retention time of 10 hours.

Oon et al. (2017) researched the benefit of the plant (*Elodea Nuttallii*) and the additional aeration in an up-flow constructed wetland microbial fuel cell (UFCW-MFC). Their research showed that the plant was of significant contribution to the current generation as well as nitrification of the ammonium in the synthetic wastewater because of the oxygen loss at the plant's roots. HRT of 1 day aerating the system resulted in more dissolved oxygen around the cathode that improved the current generation as well as nitrification. As is to be expected, denitrification was slowed down by aeration. Optimum removal efficiencies can be achieved by finding a specific aerating speed.

Wang et al. (2017a) noted another importance of the plants in the CW-MFC. The plant roots created space in the substrate and prevented clogging by the biofilm. This benefitted the ion/proton transport. The presence of plants also leads to a higher concentration in electrochemically active bacteria in comparison to non-planted CW-MFCs. The researchers ascribe this to the carbohydrates and amino acids exudated by the plants that can suffice as an electron donor for the EAB.

Next to COD and nutrient removal, the removal of azo dye with CW-MFCs has been documented by several researchers (Fang et al. 2013; Yadav et al. 2012). Azo dyes are a group of synthetic dyes intensively used in the paper and textile industry. Most azo dyes are biorefractory organics and are not removed by regular sewage treatment plants (Vandevivere et al. 1998). When present in the surface water, azo dyes will block the light, thereby inhibiting photosynthetic activity. This can decrease the dissolved oxygen concentration that can have negative consequences for aquatic life. Extensive research has been performed on the biotic (J.B.A. Arends et al. 2014) or abiotic reduction of azo dyes in natural sediments and waters (Weber and Lee Wolfe 1987; Wuhrmann et al. 1980). The process of degradation is comparable to that in constructed wetlands. The reductive environment of the anaerobic soil is beneficial to the chemical or biological reduction of the azo dye bond. The presence of an anode electrode in a CW-MFC promotes the growth of reducing EAB in the soil that can aid the biodegradation process (Cabezas et al. 2015; Lu et al. 2015). Azo dye wastewater can furthermore be treated by the wetland through phytoremediation. Plant enzymes and bacteria in the rhizosphere are able to break the complex chemical structure of the dyes (Khandare and Govindwar 2015).

The degradation of azo dyes can result in the formation of aromatic amines (Balakrishnan et al. 2016; Pinheiro et al. 2004). Aromatic amines are mutagens and carcinogens. Research into biodegradation of azo dyes should, therefore, always include the determination of the derivatives. However, aromatic amines are biodegraded under aerobic conditions (Vandevivere et al. 1998). Khandare and Govindwar (2015) have found phytoremediated effluent of azo dyes to be less toxic than the untreated compounds. The different redox zones in a CW-MFC make it an ideal environment for the decolorization and degradation of azo dyes and its intermediates.

Altering the redox potential in soil by a plant or electrode also aids the growth of specific microbial communities. Bacteria can lower the activation energy of certain redox reactions when the reaction is part of their respiration. The radial oxygen loss by plant roots creates a local oxidative environment in naturally reductive waterlogged soils or soils that have become anoxic as a result of eutrophication. These theoretical considerations promote bacterial oxidation of (persistent) organic matter. Other examples of bacterially aided reactions that remove pollutant from wastewater are nitrifying and denitrifying bacteria. The bacterial reduction of sulfate to sulfide in anaerobic soils can lead to the removal of heavy metals from the aqueous phase when the metal sulfide precipitates. At the cathode, the reductive environment can be used to remove chromium in the form of $Cr(OH)_3$ (Habibul et al. 2016). Alternatively, the cathode of a Plant-MFC also used to form hydrogenperoxide that can be used for disinfection purposes (Arends et al. 2014).

3.1.3 Air Treatment (Prevention of GHG)

Alongside water and soil quality, Plant-MFCs can contribute to air quality by preventing methane emissions. Wetlands and rice paddies emit 32-47% of all methane gas (Bridgham et al. 2013; Denman et al. 2007). The global warming potential of CH_4 gas is 25 times higher than that of CO_2 and is indeed the second most important greenhouse gas (Denman et al. 2007). The aerenchyma in vascular plants is responsible for the transport of methane gas from the soil to the atmosphere. A 71-82% of wetland methane emission is plant-mediated (Liu et al. 2017). The Plant-MFC can reduce this emission greatly through the promotion of specific (electrochemically active) bacteria. These bacteria compete with methanogens archaea in the soil for electron donors and acceptors. Organic carbon from rhizodeposits makes up the primary source of electrons in many soils. The effect of Plant-MFCs on methane emissions from natural and man-made wetlands has been researched. Arends et al. (2014) created several small, rice paddy MFCs with the anode placed at the roots of the *Oryza Sativa* plants. They measured a decrease in CH_4 emission when the electric current was uninterrupted, and the concentrations of the organic substrate were low. At open cell conditions or after the supply of excess organic carbon, methanogenesis outcompeted current production.

Liu et al. (2017) measured the methane emission from a laboratory-scale CW-MFC with *Spartina Alterniflora* and found the methanogens and exoelectrogens were able to co-exist in the rhizosphere at higher organic carbon concentrations. However, when no additional glucose was added and rhizodeposits were the only carbon in the set-up, the methane emission with a closed circuit MFC was one-tenth of the emissions from the CW-MFC at open cell conditions.

Zhong et al. (2017) also demonstrated this transition in methane emission with changing organic substrate concentrations. They constructed 24 MFCs containing paddy soil with different amounts of (dried) rice straw. When the percentage of straw was lower than 1%, the CH_4 emission was significantly lower for the MFC with a closed circuit. At this low concentration of organic substrate, the electrochemically active bacteria transferring electrons to the anode for current production apparently still outcompeted the methanogens (Strik 2008). Zhong et al. (2008) provide an explanation for this increased emission with an active MFC. In the presence of an active MFC, more oxygen will be reduced in the soil near the cathode. This lower O_2 concentration might lead to a smaller number of methanotrophs bacteria. These bacteria are responsible for the oxidation of methane in the soil, preventing its release into the atmosphere.

3.2 Adding Economic Value/Protecting and Promoting Green

3.2.1 Agricultural Land

Plant-MFCs can add a secondary purpose to arable land. Rice paddies are particularly suited for the integration of Plant-MFCs because of the high water level in the parcels. Of all the cultivated crops, rice is the third most prevalent. Almost 2 million km^2 is a rice field which is 11% of the total cultivated land surface (Leff et al. 2004). Average measured Plant-MFC power output in open-air rice paddies is 6-19 mW/m^2 plant growth area (Kaku et al. 2008; Takanezawa et al. 2010; Kouzuma et al. 2014). One hectare of rice paddy could power several outdoor lights or other small electric devices. In this manner, electricity can be provided to farmers who are living in remote areas and are disconnected from the grid.

There is limited research into the long-term effects of Plant-MFCs on crop yield. However, Zhou et al. (2017) and Helder et al. (2010) looked into the effects of a Plant-MFC on plant vitality and growth rate for different wetland species over a period of several months and found no indication of negative influences. From the number of different macrophytes mentioned in Plant-MFC research (Helder et al. 2010; Zhou et al. 2017; Wang et al. 2017b), the potential to integrate the Plant-MFC with other (semi) aquatic crops is present. Examples of possible applications are hydroponics or even paddy fisheries.

3.2.2 Natural Wetlands

Plant-MFCs can also be of added value to natural wetlands. Most natural wetlands are protected areas and not automatically unsuited to agricultural practices or the construction of buildings and infrastructures. Wetlands are important natural areas for the preservation of biodiversity and provision of ecosystem services, such as flood control and coastal protection (Denman et al. 2007). Since the Plant-MFC concept is nonintrusive and barely visible, the technology can be applied in all types of wetlands without negatively affecting the ecological or aesthetic qualities. The current generated by the fuel cells can be used locally to power a light or sensor (this will be discussed in the following section on telemetry), stored in a battery or transported elsewhere for any purpose. Natural wetlands cover approximately 8.2-10.1 million km^2 worldwide (Denman et al. 2007; Lehner and Döll 2004). Wetser et al. (2017a) provided a significant potential to contribute to our total energy demand. This number would increase further when including brackish and saltwater wetlands. However, further research is needed to precisely assess to what extent it is acceptable to implement the Plant-MFC in natural areas. Wetlands are certainly of high value and are utmost important sites to protect. Possibly, wetlands restoration can be combined with Plant-MFC services to combine both nature protection as well as economic drivers. It is estimated that since the 1900s, 50% of all natural wetlands have disappeared (Denman et al. 2007). This is mainly due to human activities, such as drainage for agriculture, filling for waste disposal or infrastructure and conversion for aquaculture. Thus, indeed, providing economic value to natural wetlands through current generation can aid in their protection and restoration.

The relevant parameters for water treatment with Plant-MFCs in constructed wetlands differ from that of power production in natural wetlands. The number of papers focusing on power outputs from natural wetlands is limited. Dai et al. (2015) inserted two Plant-MFCs in a freshwater cypress-tupelo forested wetland. They studied the influence of seasonal changes (e.g., temperature, water level, organic carbon concentration, etc.) on current output. They found a strong correlation between the current and temperature. At higher temperatures, the Plant-MFCs generated more current that could possibly be a result of increased microbial activity. Extreme weather conditions such as freezing can pose problems to the technology and stop electricity production entirely. This relation to temperature was also found by Helder et al. (2013b). The average power density was 26mW/m² anode area. Wetser et al. (2015a) compared the current generation by a Plant-MFC in a salt marsh wetland with *Spartina Anglica* to that of a fen peat soil with *Phragmites Australis* in laboratory-scale setups. The salt marsh reached an average power density of 18 mW/m² plant growth area and was tenfold higher than the average power generated in the peat soil. Wetser et al. (2015b, 2017b) increased the power output by developing a biocathode. The power output of the fen peat soil with biocathode was 22 mW/m² plant growth area. The highest average power density generated in a constructed wetland was also with a biocathode Plant-MFC and reached 44 mW/m² cathode area (with 0.2 g/L glucose added) (Liu et al. 2014; Doherty et al. 2015).

3.2.3. Man-made Green Spaces

Apart from (constructed) wetlands and agricultural land, the Plant-MFC can be applied in man-made green spaces in open-air or indoors. The additional value of electricity production that the technology provides can promote the emergence of more greenery. One example is green roofs provide a building with insulation, helps to retain rainwater and also improves air quality (Helder et al. 2013b). By applying Plant-MFCs on a green roof, this list of benefits can be complemented with electricity production (Helder et al. 2013a). Other examples are green roadsides, green walls, ponds and plant pots.

4. Plant-MFC Serving the People

4.1 Telemetry

Telemetry is the process of remote data collection and transmission. The necessary components for telemetry are sensors and small monitoring systems. These components need little or occasional power input but from a reliable power source and usually sustain over a long period. The Plant-MFC meets these demands. It can provide the power *in situ* and is nearly maintenance free contrary to batteries that need replacement. Since the Plant-MFC is nonintrusive and only dependent on vegetation, temperature and soil saturation, the technology can power telemetry systems in all wetlands, estuaries and water bodies in temperate climates. This would allow for continuous monitoring without the use of an external power supply, such as a battery. There are several papers on the use of sediment MFCs or plant MFCs as a power supply for telemetry systems. They provide examples of monitoring weather conditions, water quality, pollution events and biodiversity.

The majority of papers address the technical challenges that come with using a Plant-MFC to power a telemetry system. The power output of a Plant-MFC is too low and the performance of the cell varies through time. There are different approaches to overcoming these challenges. Some researchers tried to increase the current output by electrically connecting several fuel cells, thereby effectively increasing the anode surface area (Tender et al. 2008; Liu et al. 2015; Bombelli et al. 2016; Schievano et al. 2017). Connecting several cells to one system also diminishes the change of a sensor becoming inactive due to failure or lack of current generation of a single Plant-MFC and thus increases its stability (Liu et al. 2015). Many have developed a so-called power management system (Donovan et al. 2008; Donovan et al. 2011; Donovan et al. 2013; Liu et al. 2015; Piyare et al. 2017; Rossi et al. 2017; Brunelli et al. 2017). Such a system often uses a dc-dc converter to increase the potential of the electric current coming from the Plant-MFC. Radio data transmission behavior can be adapted to the actual available electricity by adding a capacitor to the system that will store the electricity to be used intermittently. Both Piyara et al. (2017) and Brunelli et al. (2017) developed a remotely triggered wake-up receiver that allowed for minimal energy use when the system was not in use. This system could be triggered from a distance to sense or transmit data.

4.1.1 Weather

One particular field of remote monitoring that has been researched is measuring weather conditions. Plant-MFCs can power sensors to collect and transmit data on temperature, pressure, humidity, insolation and pH of the air, soil and water. Tender et al. (2008) combined several sensors in one meteorological buoy powered by a benthic-MFC that would allow weather measurements in open waters. Donovan et al. (2011) developed a monitoring device consisting of a power management system, a temperature sensor and a telemetry system that could transmit temperature data up to 10 miles. Bombelli et al. (2016) were able to power an LCD desktop weather station continuously and directly for over a week with a moss MFC system. Brunelli et al. (2017) were able to measure, store and transmit (on command) data on temperature, humidity and light intensity with their energy-neutral sensing system powered by a Plant-MFC that generated enough electricity for six months.

4.1.2 Water Quality

Monitoring watering quality can provide information on the biological and chemical condition of water bodies. This is of special importance for groundwater wells used for drinking water production and recreational areas. Several water quality variables can be measured, such as water temperature, pH, COD and the presence of aquatic life. Using MFCs for the remote monitoring of these parameters can save costs on hiring trained personnel for the execution of water quality tests. As mentioned before, Tender et al. (2008) developed a meteorological buoy that was able to measure water temperature. Donovan et al. (2013) measured and logged ultrasonic signals with a submersible ultrasonic receiver that was powered by a sediment fuel cell. The ultrasonic signals are representative of fish and other aquatic wildlife movements. Sartori and Brunelli (2016) were able to send data on groundwater level obtained with a capacitive phreatimeter over a 1 km distance powered by single sediment MFC. Zhang et al. (2011) were able to continuously measure and transmit (although at a specific frequency) temperature data with two different types of sediment fuel cells. Velasquez-Orta et al. (2017) used sediment MFC as a biosensor of organic pollution events. The system's current output was more responsive to changes in organic carbon load than to changes in temperature, pH or salinity of the water. This made it possible to detect pollution events in real-time. Schievano et al. (2017) developed a floating Plant-MFC system that was able to power a LED, buzzer or environmental sensor for over a year. The substrate was provided by rice plants that were floating on the water surface in a frame box. This system makes it possible to also monitor deeper water bodies with plant power.

4.1.3 Smart Farming

Keeping track of environmental variables can aid farmers in producing higher crop yields. The practice of systematic monitoring of agricultural land with electronic sensors is also referred to as smart farming. Farmers can avert crop losses by measuring temperature, light intensity, pH, water quality (BOD) and groundwater level either continuously or on demand. Powering biosensors with Plant-MFCs is a durable, low-cost option for farmers in comparison to the use of batteries that need frequent replacement. Another benefit of the Plant-MFC is the possibility of using the technology itself as a biosensor. Brunelli et al. (2016) created a Plant-MFC powered wireless data communication system that is able to measure light intensity and soil moisture as well as monitor plant health. The electric current output of the Plant-MFC not only powers the light and moisture sensors but is also representative of the plant's health; an actively photosynthesizing plant will release more carbon at its roots in comparison to an inactive plant. The amount of biologically degradable organics in the soil is reflected in the current output of the Plant-MFC. Since the frequency of data transition is dependent on the current, a higher frequency will express a more actively photosynthesizing plant. Brunelli et al. (2016) do emphasize the importance of finding the proper range in current output for each specific biosystem representative of an active or inactive plant. This is of importance because the current output is also related to other environmental parameters, such as temperature, pH and salinity. This combined system can give a good view of long-term plant health.

4.1.4 Security Surveillance

Another usage of Plant-MFC telemetry is security surveillance. Farmers and homeowners can protect their real estate with motion detectors. Arias-Thode et al. (2017) even powered a magnetometer to detect passing ships with an array of sediment MFCs.

4.2 Designers Application and Spin-Offs of the Plant-MFC

The plant MFC does not only exist in scientific labs. On several occasions, technology has made its entrance in the real world. Scientists and designers have used the technology in conceptual projects. These projects mainly serve demonstrative purposes and educate people on the possibilities and benefits of Plant-MFC technology. They are examples of how new scientific research can be used for demonstrative and educational purposes.

4.2.1 Nomadic Plants

Gilberto Esparza (http://plantasnomadas.com/) designed a walking robot powered by a MFC. The multi-legged robot can take up water at a riverbank and carry several plants on its back, hence its name 'Nomadic Plant'. The robot depends on power generated by electrochemically active bacteria from the degradation of organic carbon in the water and plant rhizodeposits. The robot should, therefore, be capable of living autonomously while cleaning polluted surface water for its entire robotic life. Since the 'Nomadic Plant' is an art project, not much scientific research on its performance has been published to validate the actual feasibility of its design.

4.2.2 Moss Table, Moss FM and P2P

Dr. Paolo Bombelli (https://www.bioc.cam.ac.uk/howe/members/paolo-bombello-postdoctoral-researcher-1) from Cambridge University has worked together with several scientists and designers to create a series of biophotovoltaic design pieces in the shape of everyday items. All of his collaborative designs use rhizodeposition from bryophytes as the carbon source for the MFC. The 'Moss Table is a table that encloses a multitude of small individual moss plants that provide enough electric current to power a digital clock. At present, the power generated by the table is not used, but the table serves as a demonstration of the technologies' potential. The 'Moss FM' is a serial circuit of 10 moss MFCs generating enough electricity to power a small LCD weather station continuously or a commercial radio for 80 seconds every 10 hours (Bombelli et al. 2016). The 'P2P' stands for 'plant to power' and is an installation that combines photovoltaic panels with plant MFCs. It was displayed at the botanical garden in Cambridge for a year during which time its power output was measured. The conceptual pieces that Bombelli and his partners (2016) created have been exhibited at multiple events, won several prizes and have been written about in many online design websites and blogs.

4.2.3 Floating Gardens

Schievano et al. (2017) (http://www.expo2015.org/archive/en/index.html) created floating plant MFCs with rice plants as previously described. Several floating cells were placed in ponds and lakes in Milan for the EXPO Milano 2015, a universal exposition on agriculture, nutrition and resources. Each floating cell powered a LED light to showcase the gardens in day and night. During the exposition, the performance of the cells was periodically monitored. EXPO Milano 2015 lasted for six months and attracted millions of visitors.

4.2.4 Plant-E: Living Light, Sprout 'N Spark and Larger Modules

Start-up company Plant-e (founded in 2009 as co-outcome of an EU project PlantPower (https://setis.ec.europa.eu/energy-research/content/plantpower-living-plants-microbial-fuel-cells-clean-renewable-sustainable-efficient-situ-1) is introducing the Plant-MFC technology to the market (http://www.plant-e.com/; https://www.livinglight.info/) The company sells small educational modules called "sprout and spark" for scholars and interested to construct their own Plant-MFC. Larger modules powering lights have been installed next to roadsides, at schools, companies, a museum and even in the garden of the Dutch Ministry of Economic Affairs and Climate. Plant-e is based in the Netherlands and is a spin-off company from the research performed on the Plant-MFC at the Wageningen University. However, their orders come from outside the Netherlands as well as from other continents. Designer Ermi van Oers collaborated with Plant-e to create 'living light'. This light is switched on by stroking the leaves of the plant.

4.2.5 Other Spin-Outs

Other companies like BIOO (https://www.biootech.com/) or Magical Microbes (https://www.magicalmicrobes.com) use either plants or MFC principles to develop new products to create impact using principles of the Plant-MFC. Possibly, there are more companies developing novel applications that can all contribute to making (y)our science ready for impact.

5. Conclusions

Since 2008 numerous breakthroughs on technological applications related to the Plant-MFC are leading to emerging applications (Figure 3). New companies and designers are active to implement the technology to serve people and the environment. It seems like that anywhere plants can often grow, it will bring an opportunity to integrate this emerging technology.

- Serving the environment
  - **Environmental remediation:** Wastewater and sediment treatment, Air treatment (prevention of GHG)
  - **Adding economic value; protecting and promoting green:** Agricultural land, Natural wetlands, Man-made green spaces
- Serving the people
  - **Telemetry:** Weather, Water quality, Smart farming, Security surveillance
  - **Designers applications and spin offs of the Plant-MFC**

Figure 3: Services of the Plant-MFC.

6. Future Perspectives

The Plant-MFC is an example of an ecotechnology. It is making use of 'endless' natural cycles, and its concept is clean and not harmful to the environment. Commercial application of this technology is in reach and many opportunities are to be further explored. For future economical attractive large-scale electricity generation, effective means on the installation of scaled-up systems is required. Effective use of 'cheap' or novel materials may lead to a step forward in using cheaper materials to generate plant power. Here, further understanding of the working mechanisms can help to operate Plant-MFC more effectively. Toward its application, it is important to realize that the electricity generation is always combined with, at least, the plant's functionality. This way several benefits for the environment and people are combined and will be accounted for. This can help to develop a circular economy and create a more sustainable world on a short and long term.

Acknowledgement

This research is financed by the Netherlands Organisation for Scientific Research (NWO), which is partly funded by the Ministry of Economic Affairs and Climate Policy, and co-financed by the Netherlands Ministry of Infrastructure and Water Management and partners of the Dutch Water Nexus consortium.

References

Arends JBA, Speeckaert J, Blondeel E, De Vrieze J, Boeckx P, Verstraete W, Rabaey K, Boon N (2014) Greenhouse gas emissions from rice microcosms amended with a plant microbial fuel cell. Appl Microbiol Biotechnol 98: 3205–3217.

Arends JBA, Van Denhouwe S, Verstraete W, Boon N, Rabaey K (2014) Enhanced disinfection of wastewater by combining wetland treatment with bioelectrochemical H2O2 production. Bioresour Technol 155: 352–358.

Arias-Thode YM, Hsu L, Anderson G, Babauta J, Fransham R, Obraztsova A, Tukeman G, Chadwick DB (2017) Demonstration of the SeptiStrand benthic microbial fuel cell powering a magnetometer for ship detection. J Power Sources 356: 419–429.

Balakrishnan VK, Shirin S, Aman AM, de Solla SR, Mathieu-Denoncourt J, Langlois VS (2016) Genotoxic and carcinogenic products arising from reductive transformations of the azo dye, Disperse Yellow 7. Chemosphere 146: 206–215.

Bombelli P, Dennis RJ, Felder F, Cooper MB, Madras Rajaraman Iyer D, Royles J, Harrison STL, Smith AG, Harrison CJ, Howe CJ (2016) Electrical output of bryophyte microbial fuel cell systems is sufficient to power a radio or an environmental sensor. R Soc Open Sci 3: 160249.

Bridgham SD, Cadillo-Quiroz H, Keller JK, Zhuang Q (2013) Methane emissions from wetlands: biogeochemical, microbial, and modeling perspectives from local to global scales. Glob. Chang Biol 19: 1325–1346.

Brunelli D, Rossi M, Tosato P (2017). A Radio-Triggered Wireless Sensor Platform Powered by Soil Bacteria. Proceedings 1: 568.

Brunelli D, Tosato P, Rossi M (2016) Flora Health Wireless Monitoring with Plant-Microbial Fuel Cell. Procedia Eng 168: 1646–1650.

Cabezas A, Pommerenke B, Boon N, Friedrich MW (2015). Geobacter, Anaeromyxobacter and Anaerolineae populations are enriched on anodes of root exudate-driven microbial fuel cells in rice field soil. Environ Microbiol Rep 7: 489–497.

Chiranjeevi P, Chandra R, Mohan SV (2013) Ecologically engineered submerged and emergent macrophyte based system: An integrated eco-electrogenic design for harnessing power with simultaneous wastewater treatment. Ecol Eng 51: 181–190.

Dai J, Wang J-J, Chow AT, Conner WH (2015) Electrical energy production from forest detritus in a forested wetland using microbial fuel cells. GCB Bioenergy 7: 244–252.

De Schamphelaire L, Van Den Bossche L, Hai SD, Höfte M, Boon N, Rabaey K, Verstraete W (2008) Microbial fuel cells generating electricity from rhizodeposits of rice plants. Environ Sci Technol 42: 3053–3058.

Denman KL, Denman KL, Brasseur G, Chidthaisong A, Ciais P, Cox PM, Dickinson RE, Hauglustaine D, Heinze C, Holland E, Jacob D, Lohmann U, Ramachandran S, Da PL, Dias S, Wofsy SC, Zhang X (2007). IPPC Report AR4 Climate Change 2007: The Physical Science Basis Chapter 7 Couplings Between Changes in the Climate System and Biogeochemistry, Cambridge University Press, Cambridge, United Kingdom and New York, NY, USA.

Doherty L, Zhao Y, Zhao X, Hu Y, Hao X, Xu L, Liu R (2015) A review of a recently emerged technology: Constructed wetland – Microbial fuel cells. Water Res 85: 38–45.

Donovan C, Dewan A, Heo D, Beyenal H (2008) Batteryless, Wireless Sensor Powered by a Sediment Microbial Fuel Cell. Environ Sci Technol 42: 8591–8596.

Donovan C, Dewan A, Heo D, Lewandowski Z, Beyenal H (2013) Sediment microbial fuel cell powering a submersible ultrasonic receiver: New approach to remote monitoring. J Power Sources 233: 79–85.

Donovan C, Dewan A, Peng H, Heo D, Beyenal H (2011a) Power management system for a 2.5W remote sensor powered by a sediment microbial fuel cell. J Power Sources 196: 1171–1177.

Donovan C, Dewan A, Peng H, Heo D, Beyenal H (2011b) Power management system for a 2.5 W remote sensor powered by a sediment microbial fuel cell. J Power Sources 196: 1171–1177.

Fang Z, Song H-L, Cang N, Li X-N (2013) Performance of microbial fuel cell coupled constructed wetland system for decolorization of azo dye and bioelectricity generation. Bioresour Technol 144: 165–171.

Friedrich MW (2011) The microbial ecology of electrigenic microorganisms in plant-rhizosphere based microbial fuel cells. Commun Agric Appl Biol Sci 76: 25–26.

Habibul N, Hu Y, Wang Y-K, Chen W, Yu H-Q, Sheng G-P (2016) Bioelectrochemical Chromium(VI) Removal in Plant-Microbial Fuel Cells. Environ Sci Technol 50: 3882–3889.

Helder M, Chen W-S, van der Harst EJM, Strik DPBTB, Hamelers HBVM, Buisman CJN, Potting J (2013a) Electricity production with living plants on a green roof: environmental performance of the plant-microbial fuel cell. Biofuels, Bioprod Biorefining 7: 52–64.

Helder M, Strik DPBTB, Hamelers HVM, Kuhn AJ, Blok C, Buisman CJN (2010) Concurrent bio-electricity and biomass production in three Plant-Microbial Fuel Cells using Spartina anglica, Arundinella anomala and Arundo donax. Bioresour Technol 101: 3541–3547.

Helder M, Strik DPBTB, Timmers RA, Raes SMT, Hamelers HVM, Buisman CJN (2013b) Resilience of roof-top Plant-Microbial Fuel Cells during Dutch winter. Biomass and Bioenergy 51: 1–7.

Kaku N, Yonezawa N, Kodama Y, Watanabe K (2008) Plant/microbe cooperation for electricity generation in a rice paddy field. Appl Microbiol Biotechnol 79: 43–49.

Khandare R V, Govindwar SP (2015) Phytoremediation of textile dyes and effluents: Current scenario and future prospects. Biotechnol Adv 33: 1697–1714.

Kouzuma A, Kaku N, Watanabe K (2014) Microbial electricity generation in rice paddy fields: recent advances and perspectives in rhizosphere microbial fuel cells. Appl Microbiol Biotechnol 98: 9521–9526.

Leff B, Ramankutty N, Foley JA (2004) Geographic distribution of major crops across the world. Global Biogeochem Cycles 18: n/a-n/a.

Lehner B, Döll P (2004) Development and validation of a global database of lakes, reservoirs and wetlands. J Hydrol 296: 1–22.

Liu B, Weinstein A, Kolln M, Garrett C, Wang L, Bagtzoglou A, Karra U, Li Y, Li B (2015) Distributed multiple-anodes benthic microbial fuel cell as reliable power source for subsea sensors. J Power Sources 286: 210–216.

Liu S, Feng X, Li X (2017) Bioelectrochemical approach for control of methane emission from wetlands. Bioresour Technol 241: 812–820.

Liu S, Song H, Wei S, Yang F, Li X (2014) Bio-cathode materials evaluation and configuration optimization for power output of vertical subsurface flow constructed wetland — Microbial fuel cell systems. Bioresour Technol 166: 575–583.

Lu L, Xing D, Ren ZJ (2015) Microbial community structure accompanied with electricity production in a constructed wetland plant microbial fuel cell. Bioresour Technol 195: 115–121.

Mohan SV, Mohanakrishna G, Chiranjeevi P (2011) Sustainable power generation from floating macrophytes based ecological microenvironment through embedded fuel cells along with simultaneous wastewater treatment. Bioresour Technol 102: 7036–7042.

Oon Y-L, Ong S-A, Ho L-N, Wong Y-S, Dahalan FA, Oon Y-S, Lehl HK, Thung W-E, Nordin N (2017) Role of macrophyte and effect of supplementary aeration in up-flow constructed wetland-microbial fuel cell for simultaneous wastewater treatment and energy recovery. Bioresour Technol 224: 265–275.

Pinheiro HM, Touraud E, Thomas O (2004) Aromatic amines from azo dye reduction: status review with emphasis on direct UV spectrophotometric detection in textile industry wastewaters. Dye Pigment 61: 121–139.

Piyare R, Murphy AL, Tosato P, Brunelli D (2017) Plug into a Plant: Using a Plant Microbial Fuel Cell and a Wake-Up Radio for an Energy Neutral Sensing System, in: 2017 IEEE 42nd Conference on Local Computer Networks Workshops (LCN Workshops). IEEE, pp. 18–25.

Rossi M, Tosato P, Gemma L, Torquati L, Catania C, Camalo S, Brunelli D (2017) Long range wireless sensing powered by plant-microbial fuel cell, in: Design, Automation & Test in Europe Conference & Exhibition (DATE), 2017. IEEE, pp. 1651–1654.

Sartori D, Brunelli D (2016) A smart sensor for precision agriculture powered by microbial fuel cells, in: 2016 IEEE Sensors Applications Symposium (SAS). IEEE, pp. 1–6.

Schievano A, Colombo A, Grattieri M, Trasatti SP, Liberale A, Tremolada P, Pino C, Cristiani P (2017) Floating microbial fuel cells as energy harvesters for signal transmission from natural water bodies. J Power Sources 340: 80–88.

Strik DPBTB, Hamelers (Bert) HVM, Snel JFH, Buisman CJN (2008) Green electricity production with living plants and bacteria in a fuel cell. Int J Energy Res 32: 870–876.

Takanezawa K, Nishio K, Kato S, Hashimoto K, Watanabe K (2010) Factors affecting electric output from rice-paddy microbial fuel cells. Biosci Biotechnol Biochem 74: 1271–1273.

Tender LM, Gray SA, Groveman E, Lowy DA, Kauffman P, Melhado J, Tyce RC, Flynn D, Petrecca R, Dobarro J (2008) The first demonstration of a microbial fuel cell as a viable power supply: Powering a meteorological buoy. J Power Sources 179: 571–575.

Timmers RA, Rothballer M, Strik DPBTB, Engel M, Schulz S, Schloter M, Hartmann A, Hamelers B, Buisman C (2012a) Microbial community structure elucidates performance of Glyceria maxima plant microbial fuel cell. Appl Microbiol Biotechnol 94: 537–548.

Timmers RA, Strik DPBTB, Hamelers HVM, Buisman CJN (2013) Electricity generation by a novel design tubular plant microbial fuel cell. Biomass and Bioenergy 51: 60–67.

Timmers RA, Strik DPBTB, Hamelers HVM, Buisman CJN (2012b) Characterization of the internal resistance of a plant microbial fuel cell. Electrochim Acta 72: 165–171.

Vandevivere PC, Bianchi R, Verstraete W (1998) Review Treatment and Reuse of Wastewater from the Textile Wet-Processing Industry : Review of Emerging Technologies. J Chem Technol Biotechnol 72: 289–302.

Velasquez-Orta SB, Werner D, Varia JC, Mgana S (2017) Microbial fuel cells for inexpensive continuous in-situ monitoring of groundwater quality. Water Res 117: 9–17.

Villaseñor J, Capilla P, Rodrigo MA, Cañizares P, Fernández FJ (2013) Operation of a horizontal subsurface flow constructed wetland - Microbial fuel cell treating wastewater under different organic loading rates. Water Res 47: 6731–6738.

Wang J, Song X, Wang Y, Bai J, Bai H, Yan D, Cao Y, Li Y, Yu Z, Dong G (2017a) Bioelectricity generation, contaminant removal and bacterial community distribution as affected by substrate material size and aquatic macrophyte in constructed wetland-microbial fuel cell. Bioresour Technol 245: 372–378.

Wang J, Song X, Wang Y, Bai J, Li M, Dong G, Lin F, Lv Y, Yan D (2017b) Bioenergy generation and rhizodegradation as affected by microbial community distribution in a coupled constructed wetland-microbial fuel cell system associated with three macrophytes. Sci Total Environ 607–608: 53–62.

Weber EJ, Lee Wolfe N (1987) Kinetic studies of the reduction of aromatic AZO compounds in anaerobic sediment/water systems. Environ Toxicol Chem 6: 911–919.

Wetser K (2017) Electricity from wetlands: Technology assessment of the tubular Plant Microbial Fuel Cell with an integrated biocathode.

Wetser K, Dieleman K, Buisman C, Strik D (2017) Electricity from wetlands: Tubular plant microbial fuels with silicone gas-diffusion biocathodes. Appl Energy 185: 642–649.

Wetser K, Liu J, Buisman C, Strik D (2015a) Plant microbial fuel cell applied in wetlands: Spatial, temporal and potential electricity generation of Spartina anglica salt marshes and Phragmites australis peat soils. Biomass and Bioenergy 83: 543–550.

Wetser K, Sudirjo E, Buisman CJN, Strik DPBTB (2015b) Electricity generation by a plant microbial fuel cell with an integrated oxygen reducing biocathode. Appl Energy 137: 151–157.

Wu D, Yang L, Gan L, Chen Q, Li L, Chen X, Wang X, Guo L, Miao A (2015) Potential of novel wastewater treatment system featuring microbial fuel cell to generate electricity and remove pollutants. Ecol Eng 84: 624–631.

Wuhrmann K, Mechsner K, Kappeler T (1980) Investigation on rate ? Determining factors in the microbial reduction of azo dyes. Eur J Appl Microbiol Biotechnol 9: 325–338.

Yadav AK, Dash P, Mohanty A, Abbassi R, Mishra BK (2012) Performance assessment of innovative constructed wetland-microbial fuel cell for electricity production and dye removal. Ecol Eng 47: 126–131.

Zhang F, Tian L, He Z (2011) Powering a wireless temperature sensor using sediment microbial fuel cells with vertical arrangement of electrodes. J Power Sources 196: 9568–9573.

Zhong W-H, Cai L-C, Wei Z-G, Xue H-J, Han C, Deng H (2017) The effects of closed circuit microbial fuel cells on methane emissions from paddy soil vary with straw amount. CATENA 154: 33–39.

Zhou Y, Xu D, Xiao E, Xu D, Xu P, Zhang X, Zhou Q, He F, Wu Z (2018) Relationship between electrogenic performance and physiological change of four wetland plants in constructed wetland-microbial fuel cells during non-growing seasons. J Environ Sci 70: 54-62.

PART-IV

Applications of Microbial Electrochemical Systems in Bioremediation

Degradation and Mineralization of Recalcitrant Compounds in Bioelectrochemical Systems

Yixuan Wang[1], Xinbai Jiang[2], Houyun Yang[3], Weihua Li[3], Yang Mu[1],* and Jinyou Shen[2],*

[1] CAS Key Laboratory of Urban Pollutant Conversion, University of Science and Technology of China, Hefei, China
[2] School of Environmental and Biological Engineering, Nanjing University of Science and Technology, Nanjing, China
[3] School of Environment and Energy Engineering, Anhui Jianzhu University, Hefei, China

1. Introduction

A great number of recalcitrant wastes including azo dyes, nitroaromatic compounds, halogenated compounds and heterocyclic compounds as well as furan derivatives and phenolic substances are produced by industry and agriculture. These wastes have been regarded as priority pollutants owing to their recalcitrance, mutagenicity and tendency to accumulate in the environment. Conventional physicochemical treatment technologies include carbon adsorption, chemical precipitation, filtration, photo-degradation, advanced oxidation and membrane technology (Pozo et al. 2006). Major limitations of these physicochemical methods are high-cost and generation of secondary pollutants. Conventional biological treatments including aerobic and/or anaerobic processes for organic contaminants degradation are environmentally friendly and cost-effective but have its inherent defects, e.g., high energy consumption and low degradation efficiency, corresponding to aerobic and anaerobic processes, respectively (Savant et al. 2006).

Aiming to remove recalcitrant wastes more effectively, based on the integration of a biological process and electrochemical reduction/oxidation, a newly developed and promising technology termed bioelectrochemical systems (BESs) was born. BESs are bioreactors that use microbes as catalysts to drive oxidation and/or reduction reactions at electrodes and have recently attracted much attention owing to its high efficiency, low-cost, environmental sustainability, ambient operating temperatures with biologically compatible materials and value-added by-products (Rabaey et al. 2007). In recent years researchers have realized the successful degradation of a large number of recalcitrant wastes using BESs, and there are some review papers that have focused on either the extracellular electron transfer between the microbes and electrode or the structure and configuration of BESs (Hamelers et al. 2010). However, no comprehensive review article has systematically summarized this advanced technology for the removal of recalcitrant wastes. Therefore, the objectives of this article is to review the state-of-the-art degradation and mineralization of recalcitrant compounds in BESs; moreover, discuss the scientific and technical challenges that are yet to be faced in the future.

*Corresponding authors: yangmu@ustc.edu.cn; shenjinyou@mail.njust.edu.cn

2. Azo Dyes Removal in BESs

2.1. Azo Dye Degradation at the Anode

Recent studies showed that many types of azo dyes were successfully degraded in the anode chamber of BESs (Wang et al. 2015). Azo dye degradation at the anode of the BES usually requires an electron donor (co-substrate) to create a necessary reductive condition (Figure 1A). Thus, the main degradation mechanism for azo dyes was co-metabolism in the anaerobic anode chamber or the anode side in the single chamber BESs. Electrochemically active microorganisms (EAMs) oxidize co-substrates to release electrons and protons, and these electrons are transferred to azo dyes and the electrode via various extracellular electron transfer mechanisms, such as *c*-type cytochromes, conductive pili and electron shuttles. On one hand, azo dyes are readily reduced through the splitting of azo bond with the generation of aromatic amines. On the other hand, bioelectricity can be generated from electrons transported from the anode to cathode in the external circuit. In general, the co-substrate addition typically far exceeds the stoichiometric requirements, leading to additional costs and unwanted methane production. The effects of various operational parameters, such as co-substrates (Figure 1B), chemical structures of azo dyes (Figure 1C) and enrichment procedures on the performance of azo dyes degradation at the anode of BESs have been already investigated (Cao et al. 2010; Chen et al. 2010; Hou et al. 2011a). Moreover, it was found that larger membrane pore size (Figure 1D) and surface area of the anode (Figure 1E) could improve bacterial attachment and increase azo dyes degradation and bioelectricity production (Hou et al. 2011b; Sun et al. 2012). In addition, modification of the anode with redox mediators (RMs) (Figure 1F), graphene or NiO@polyaniline (Figure 1G) could further enhance azo dye degradation at the anode and bioelectricity generation simultaneously (Sun et al. 2013; Huang et al. 2017; Guo et al. 2014; Zhong et al. 2018).

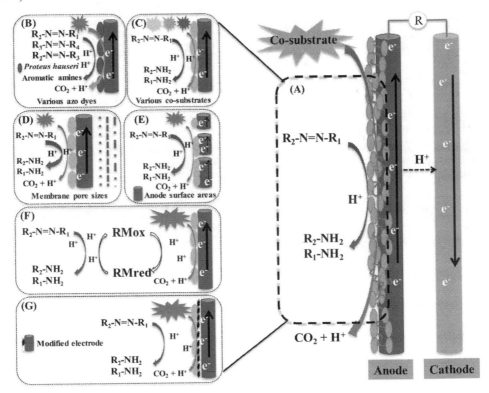

Figure 1: (A) typical azo dye degradation; (B) various azo dyes degradation using *Proteous hauseri*; azo dye degradation with (C) various co-substrates used; (D) different membranes used; (E) different surface area of anode; (F) RMs as mediators; (G) modified electrode in the anode of BESs

Many strains of EAMs have been isolated from naturally colonized anodes, such as *Geobacter* sp., *Shewanella* sp., *Klebsiella* sp., *Desulfuromonas acetoxidans*, *Rhodoferax ferrireducens*, *Pseudomonas aeruginosa*, *Ochrobactrum anthropi*, *Corynebacterium* sp., *Clostridium acetobutylicum*, *Enterobacter cloacae* and *Lactococcus lactis* (Hou et al. 2011a). Pure stains like *Proteus hauseri* ZMd44, *Pseudomonas aeruginosa*, *Geobacter sulfurreducens* and *Beta Proteobacteria* were shown to produce electricity and azo dye degradation in BESs (Chen et al. 2010; Ilamathi and Jayapriya 2018) but generally with low power densities. Therefore, the mixed cultures were widely applied in the anode of BESs owing to their stability and adaptiveness to the environmental variables and a wide range of substrates together with an enhanced electron transfer. Moreover, the mixed cultures could be obtained from various natural environments like wastewater, aquatic sediments and sewage sludge.

2.2 Azo Dye Degradation at the Cathode

Many of typical azo dyes could also be reduced in the cathode of BESs by accepting electrons from the cathode that come from the EAMs which oxidize the substrate in the anode (Figure 2A). At the abiotic cathode, it was found that the azo dyes structure (Figure 2C), catholyte pH and azo dyes concentration were the key factors affecting system performance (Liu et al. 2009). In addition, the reduction of azo dyes was enhanced by modifying the cathode electrodes with RMs (Figure 2E) (Xu et al. 2017). In recent years, reduction of azo dyes in the biocathode of BESs (Figure 2B) has also been explored, including the effect of electrode positions (Figure 2F) and pure stains *Shewanella oneidensis* (Figure 2G) (Yeruva et al. 2018; Gao et al. 2016). Moreover, various types of EAMs have been reported that possess azo dyes reduction capacities, including *Pseudomonas*, *Acinetobacter*, *Citrobacter*, *Comamonas* and *Pannonibacter*.

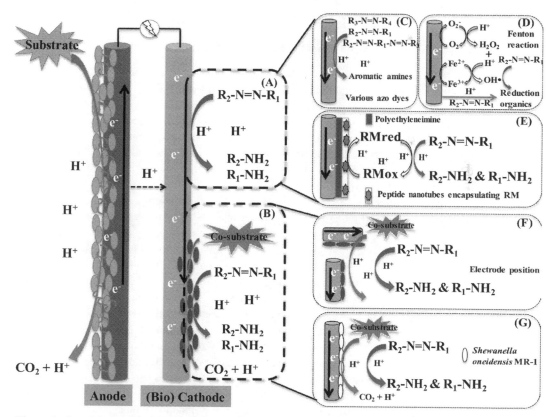

Figure 2: Azo dye degradation in (A) abiotic cathode and (B) biocathode of the BES: (C) effect of azo dye structure, (D) Fenton reaction, (E) RMs as mediators in the abiotic cathode, (F) different electrode position, (G) pure stains *Shewanella oneidensis* in biocathode.

2.3 Mineralization of Azo Dye in BESs

Azo dyes can be easily degraded under anaerobic conditions, while aromatic amine products are able to be mineralized under aerobic conditions. Liu et al. (2015) reported a self-driven degradation of azo dye in an MFC where the cathodic reduction of azo dye was effectively coupled with the anodic oxidation of its reduction intermediates. Sun et al. (2015) operated a novel BES with polarity reversion for simultaneous anaerobic/aerobic treatment of azo dye and bioelectricity generation. Azo dye first degraded in the anode, and then the resultant degradation intermediates were further mineralized after the anode was reversed to the aerobic biocathode. Oon et al. (2018) also developed a BES reactor for continuous azo dye decolorization and mineralization treatment, and the results showed that azo dye was decolorized initially in the anodic chamber and was followed by degradation of aromatic amines into less harmful products in the aerobic cathodic chamber.

2.4 Azo Dye Degradation in Integrated BESs with Other Technologies

In order to overcome the shortages of BESs, such as low degradation and mineralization efficiencies, the BESs have been integrated with other technologies to enhance the removal of azo dyes. For instance, the BESs have been successfully integrated into several traditional biological systems, including anaerobic sludge reactor, aerobic biocontact oxidation reactor, anaerobic-aerobic sequential reactor, continuous stirred tank reactor and biofilm electrode reactor, for improving degradation of the azo dyes (Pan et al. 2017; Cao et al. 2017). In addition, the BES coupled with photo-electrocatalytic and Fenton technologies have already been demonstrated to promote azo dyes degradation and mineralization (Long et al. 2017; Fu et al. 2010).

3. Nitroaromatic Compounds Removal in BESs

3.1 Nitroaromatic Compounds Degradation at the Anode

For nitroaromatic compounds degradation at the anode, co-metabolism has been demonstrated as the main removal mechanism for contaminants in the anaerobic anode chamber (Figure 3). The co-substrates could be consumed by the anode-respiration bacteria and other anaerobic bacteria to stimulate their own growth and metabolism. Then, BES anodes could provide anaerobic conditions and electrons for the reduction of oxidizing groups of nitroaromatic compounds. The existence of the anode promoted the degradation of biorefractory compounds, and electricity production consumed the co-substrates (Wang et al. 2015). The BES displayed a maximum power density of 1.778 mW m^{-2} and a maximum p-nitrophenol degradation rate of 64.69% when only 0.36 mM p-nitrophenol was used as a sole substrate (Liu et al. 2013). The electricity output of the BES increased when p-nitrophenol concentration was increased, however, p-nitrophenol degradation rate and COD removal rate decreased due to the toxicity of p-nitrophenol to microorganisms in the anode. Easily degradable organic matter such as glucose and sodium acetate could obviously improve the electricity generation and nitroaromatic compounds degradation rate. Li et al. (2010) evaluated the effect of nitrobenzene on electricity generation and simultaneous biodegradation of nitrobenzene in the BES with nitrobenzene and glucose as the anodic reactants. Nitrobenzene was degraded efficiently in the anode, but the addition of nitrobenzene, even at low concentration, significantly inhibited electricity generation primarily due to electron competition between the nitrobenzene and anode. The presence of nitrobenzene also resulted in changes in the dominant bacterial species on the anodes possibly because of nitrobenzene toxicity. The degradation of contaminants might have been enhanced through the anode biofilm acclimation to toxicity. Zhang et al. (2017) demonstrated that single chamber BES with pre-enriched bioanodes and activated carbon air-cathodes had a capability of effective nitrobenzene removal with concomitant electricity. The high nitrobenzene tolerance (7.0 mM), as well as effective nitrobenzene removal with concomitant electricity production, could be attributed to the mature electroactive biofilms on anodes pre-enriched over long-term operation under nitrobenzene-free conditions. Microbial community structure analysis indicated that the operation of the BES in the closed-

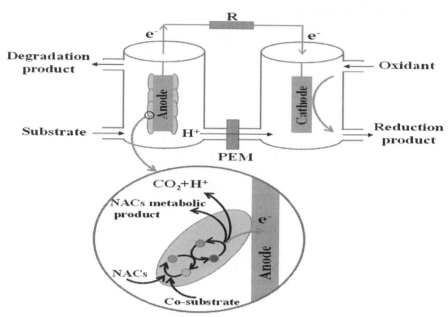

Figure 3: Schematic of nitroaromatic compounds degradation in the anode of BESs.

circuit mode promoted the growth of various functional microbial communities. The electrons generated by the anodic electrogenic bacteria in the BES were crucial for improving the efficiency of removal of nitroaromatic compounds (Zhao and Kong 2018).

3.2 Nitroaromatic Compounds Reduction at the Abiotic Cathode

It was also found that the nitro-groups of nitroaromatic compounds could be electrochemically reduced by BES abiotic cathode. At the cathode of BES, the nitro groups in the benzene were documented to electrochemical reduced orderly to nitroso, hydroxyamino and finally to amino (Guo et al. 2015). The main benefit of removing nitroaromatic compounds at the abiotic cathode is that they make it possible to optimize operational conditions on the cathode side without affecting the microbial communities on the anode (Escapa et al. 2016). Mu et al. (2009) investigated nitrobenzene removal at the abiotic cathode of BES coupled with microbial acetate oxidation at the anode. Effective reduction of nitrobenzene at a removal rate of 1.29 mol m^{-3} d^{-1} was achieved with aniline formation rate of 1.14 mol m^{-3} d^{-1} and energy recovery simultaneously. With small energy consumption at 17.06 W m^{-3}, nitrobenzene removal and aniline formation rates were significantly enhanced that reached up to 8.57 and 6.68 mol m^{-3} d^{-1}, respectively. Meanwhile, the required dosage of organic co-substrate was significantly reduced compared to conventional anaerobic biological methods. Effective removal of recalcitrant *p*-nitrophenol at rates up to 9.14 mol m^{-3} d^{-1} was achieved at an energy consumption as low as 0.010 kWh mol^{-1} *p*-nitrophenol using abiotic cathode in BESs, and the *p*-nitrophenol removal rate was enhanced with negative cathode potential, increased influent *p*-nitrophenol concentration and shortened hydraulic retention time (Shen et al. 2012). In addition, the acute toxicity of the nitrophenol effluent significantly decreased along with enhanced biodegradability after treatment in the abiotic cathode of BESs (Shen et al. 2013).

3.3 Nitroaromatic Compounds Reduction at the Biocathode

It is proved that some microorganisms are able to directly or indirectly uptake the electrons from the solid electrode and serve as the electron source for intracellular metabolism. A number of bacteria are known to selectively and completely convert nitroaromatics to their corresponding aromatic amine compounds with less toxic intermediate products generation. Wang et al. (2011) reported the conversion of nitrobenzene to aniline by using fed-batch BESs with biocathodes. When a voltage of 0.5 V was applied in the presence of glucose, 88.2% of 0.5 mM nitrobenzene was transformed to aniline within 24 hours that was 10.25 and 2.90 times higher than those with the abiotic cathode and with open circuit, respectively. Aniline was the only product detected during the bioelectrochemical reduction of nitrobenzene, whereas in abiotic conditions nitrosobenzene was observed as an intermediate of the nitrobenzene reduction to aniline. Nitroreductase is considered a key enzyme in the catalytic reduction of nitrosobenzene (Wang et al. 2011). A membraneless BES with 0.5 V power supply was developed to reduce nitrobenzene with 98% removal efficiency obtained at cathode zone, resulting in a maximum removal rate of 3.5 mol m^{-3} d^{-1} nitrobenzene (Wang et al. 2012). The main product from nitrobenzene degradation was aniline, and the production rate reached 3.06 mol m^{-3} d^{-1}. The biocathode BESs were used for *p*-nitrophenol degradation with sodium bicarbonate as the carbon source (Wang et al. 2016). The *p*-nitrophenol degradation efficiency in the biocathode BES reached 96.1% within 72 hours with an applied voltage of 0.5 V that was much higher than that obtained in the biocathode BES without applied voltage, open circuit biocathode BES or abiotic cathode BES. Liang et al. (2013) found the selective transformation of nitrobenzene to aniline maintained with biocathode communities after carbon source switchover. Continuous electrical stimulation and carbon source switchover would markedly influence on the microbial community succession. The enhanced reduction of nitroaromatic compounds in the BES could be attributed to higher diversity and the enrichment of reduction-related species (*Escherichia*, *Pseudomonas*, *Flavobacterium*, *Enterobacter* and *Desulfovibrio*), potential electroactive species (*Desulfovibrio*, *Enterobacter*, *Comamonas* and *Geobacter*) and fermentative species (*Acetobacterium*, *Kosmotoga*, *Dysgonomonas* and *Escherichia*) (Liang et al. 2013; Jiang et al. 2016). However, the enrichment of specific pollutant-reducing species was usually required prior to the startup of biocathode and the whole process is time-consuming for biocathode acclimation. To overcome this problem, a polarity inversion strategy was developed in which direct polar inversion from bioanode to biocathode was proposed (Figure 4). It was found that the acclimated electrochemically bioanode could be used as biocathode and might catalyze the reduction of different aromatic pollutants (Yun et al. 2017).

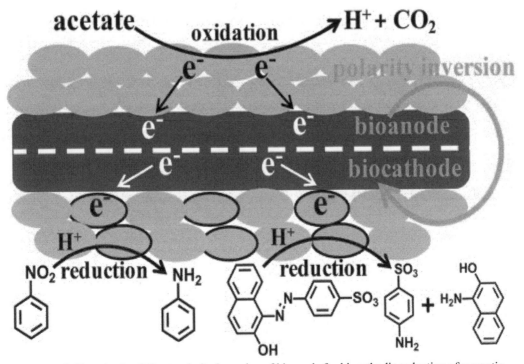

Figure 4: Hypothesis of direct polarity inversion of bioanode for biocathodic reduction of aromatic pollutants (reprinted from Yun et al. (2017), copyright 2017 with permission from Elsevier).

4. Halogenated Compounds Removal in BESs

As shown in Figure 5, various types of halogenated compounds are able to be removed through the cathodic reduction in BESs as they can serve as terminal electron acceptors under highly reducing conditions. Liang et al. (2013) realized the degradation of chloramphenicol (CAP) to amine product at the cathode of BESs with an applied voltage of 0.5 V and glucose as an intracellular electron donor. The results indicated that the degradation rate of CAP in the biocathode was accelerated compared with the abiotic cathode. Moreover, CAP was selectively converted to $AMCl_2$ with one identified intermediate, acetylated CAP (CAP-acetyl), and then was dechlorinated to another amine product (AMCl) in the biocathode BESs (Figure 5A). Wang et al. (2017) investigated the dechlorination process of hexachlorobenzene (HCB) in single chamber and membraneless soil BESs and demonstrated that a decrease in external resistance and internal resistance and an increase in phosphate buffer concentration could improve the degradation rate and removal efficiency of HCB. The HCB was degraded to pentachlorobenzene, tetrachlorobenzene and trichlorobenzene in sequence by a reductive dechlorination process in soil BESs. Mu et al. (2010) reported for the first time the reductive dehalogenation of X-ray contrast media iopromide at the cathode of BESs, driven by microbial oxidation of organics at the anode and iopromide could be completely dehalogenated in the cathode of BESs when the granular graphite cathode potential was controlled at -800 mV vs. SHE or lower. De Gusseme et al. (2011) found that biogenic palladium nanoparticles (bio-Pd) significantly enhanced diatrizoate removal by direct electrochemical reduction and by reductive catalysis using the H_2 gas produced at the cathode of BESs. The results suggested that the combination of BESs and bio-Pd in its cathode offer the potential to dehalogenate pharmaceuticals, including X-ray

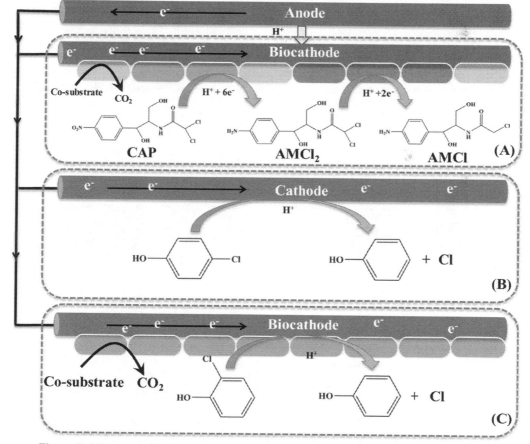

Figure 5: (A) proposed pathway of biotically cathodic CAP reduction, (B) abiotically cathodic 4-chlorophenol removal and (C) biotically cathodic 2-chlorophenol dechlorination in the BESs.

contrast media, and significantly lower the environmental burden of hospital waste streams. Aulenta et al. (2007) showed for the first time that BESs with a solid-state electrode polarized at -500 mV vs. SHE in combination with a low-potential redox mediator (methyl-viologen) could efficiently transfer electrochemical reducing equivalents to microorganisms that respire using chlorinated solvents. By this approach, the reductive transformation of TCE to harmless end-products such as ethene and ethane could be performed. Furthermore, using a methyl-viologen-modified electrode, they demonstrated that dechlorinating bacteria were able to accept reducing equivalents directly from the modified electrode surface.

In addition to studying the reduction removal of halogenated compounds at the cathode, researchers have also studied the oxidation removal of them at the anode of BESs. Pham et al. (2009) applied BESs to remove 1,2-DCA through anodic oxidation by anodophilic microbial consortia at the anode chamber for the first time, and the removal efficiency of 1,2-DCA could reach to 95% within two weeks while converting 43 ± 4% of electrons available from the removal to electricity. In addition, the production of ethylene glycol, acetate and carbon dioxide indicated that 1,2-DCA were metabolized by anodophilic bacteria through the oxidation process, probably, by means of a hydrolysis-based pathway.

5. Heterocyclic Compounds Removal in BESs

Some previous studies have indicated the effectiveness of BESs in the oxidative degradation of some representative N-heterocyclic compounds in the anode of BESs (Figure 6). Zhang et al. (2009) conducted the biodegradation of pyridine in graphite-packed BESs and found that pyridine biodegradation efficiency could reach to 95% within 24 hours. No heterocyclic intermediates were detected during pyridine biodegradation. Therefore, the metabolic pathway of pyridine under anaerobic conditions in BESs was initiated either by ring reduction or ring hydroxylation, and NH_3 was detected in the anode solution indicating that NH_3 was the final oxidation product of the pyridine biodegradation (Figure 6A). Luo et al. (2010) realized the biodegradation of indole and simultaneous power generation through continuous-fed BESs (C-BESs) and batch-fed BESs (B-BESs) (Figure 6B). When 250 mg L^{-1} indole was used as the fuel, the maximum power densities achieved from the C-BESs and B-BESs were 2.1 and

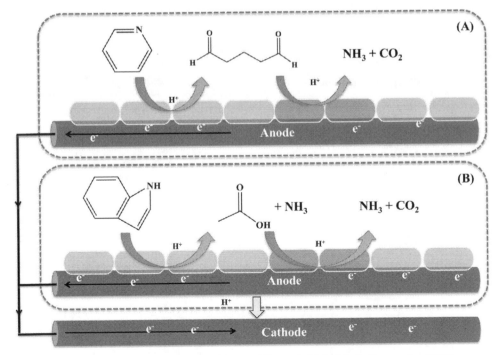

Figure 6: Oxidative degradation of (A) pyridine and (B) indole in the anode of BESs.

3.3 W m^{-3}, respectively. Indole concentrations of 250 mg L^{-1} and 500 mg L^{-1} could be removed completely in the C-BESs within 6 hours and 30 hours, respectively. Microbial community analysis showed that the presence of indole resulted in the obvious changes of the dominant bacterial species on the electrode of the C-MFC. Hu et al. (2011) investigated the possible electricity production with N-heterocyclic compounds (indole, quinoline and pyridine) degradation in the two-chamber BESs. The removal efficiencies of these N-heterocyclic compounds were 95%, 93% and 86% corresponding to indole, quinoline and pyridine, with the maximum power densities of 142.1, 203.4 and 228.8 mW m^{-2}, respectively. Additionally, some long-chain fatty acid intermediates and heterocyclic intermediates in the anode chamber were detected by using GC/MS, suggesting that pyridine was possible to be biodegraded via the anaerobic microbial metabolic pathway in BESs that was different from other published studies (Zhang et al. 2009). The reports proposed that pyridine ring was cleaved between the N and C-2 atoms, and subsequently deaminated to glutaric dialdehyde followed by successive oxidation to glutarate semialdehyde. As for indole, on the basis of products analysis, oxidation was proposed to occur at the heterocyclic double bond followed by deprotonation at nitrogen. Quinolone was recognized as the main degradation product of quinoline in numerous studies either under aerobic or anaerobic condition. Similarly, 2(1H)-quinolinone was identified in the samples taken after 12 hours, but no degradation products were observed in the samples taken near the end of the experiment that indicated that 2(1H)-quinolinone was being further transformed (Hu et al. 2011).

6. Phenolic Compounds Removal in BESs

Feng et al. (2015) accomplished the concomitant removal of phenol and nitrogen in a dual chamber BES reactor, and the bacterial analysis revealed that the phenol at the cathode chamber was removed by an oxidation process. Biotransformation of two furanic (i.e., furfural, 5-hydroxymethyl furfural) and three phenolic compounds (i.e., syringic acid, vanillic acid, 4-hydroxybenzoic acid) was explored in the anode of BESs. At an initial concentration of 1,200 mg L^{-1} of the mixture, their biotransformation rate ranged from 0.85 to 2.34 mM d^{-1} with the anode coulombic efficiency of 44-69%. The observation that the transformation rate of the five compounds increased, while current and H$_2$ production decreased indicates that the five compounds were not the direct substrates for exoelectrogens and current generation. Because there was no external electron acceptor available in the anode medium in the present study, the initial biotransformation process of the five compounds in the BESs bioanode was assumed to be fermentation, and the major identified fermentation products were catechol and phenol that were further consumed by exoelectrogens. The anode microbial community consisted of exoelectrogens, putative degraders of the five compounds and syntrophic partners of exoelectrogens (Zeng et al. 2015). Wen et al. (2013) explored the use of BESs to abiotically cathodic dechlorination of 4-chlorophenol (4-CP) where the process was driven by microbial oxidation of glucose at the anode. It was confirmed that the 4-CP reduction process was feasible in BESs and the dechlorination efficiency of 4-CP could achieve 50.3% at the cathode with energy generation. The 4-CP dechlorination efficiency reached up to 92.5% at an energy consumption of 0.549 kWh mol^{-1} 4-CP. GC-MS results showed that the major product of 4-CP dechlorination at the cathode of BESs was phenol. Strycharz et al. (2010) realized the reductive dechlorination of 2-chlorophenol by *Anaeromyxobacter dehalogenans* with an electrode serving as the electron donor. *Anaeromyxobacter dehalogenans* attached to electrodes poised at -300 mV vs. standard hydrogen electrode reductively dechlorinated 2-chlorophenol to phenol. There was no dechlorination in the absence of *A. dehalogenans*, and electrode-driven dechlorination stopped when the supply of electrons to the electrode was disrupted. The findings that microorganisms can accept electrons from electrodes for anaerobic respiration and that chlorinated aromatic compounds can be dechlorinated in this manner suggest that there may be substantial potential for treating a diversity of contaminants with microbe-electrode interactions.

7. Conclusions and Future Perspectives

Bioelectrochemical system is an emerging platform technology for enhanced and accelerated degradation and mineralization of recalcitrant compounds. In comparison with conventional treatments, the BES process including oxidation and reduction is an effective strategy for the treatment of recalcitrant wastes due to several advantages such as versatility, controllability, less sludge production, reusability of the effluent and easy operation. However, the BES technology has shown that even if there is a potential for recalcitrant compounds biodegradation and environmental bioremediation, the vast majority of studies on this technology still remain at the laboratory-scale due to a range of factors limiting its up-scaling and practical applications, such as high amplification cost, time-consuming, poor stability and the difficulty to carry out field studies. In order to realize the large-scale and engineering application of the BES-based technology, more unremitting efforts should be put into various aspects in the future. For instance, great efforts are required to better understand and manipulate the electrode-respiring microbiomes and microbe-electrode interactions and to improve integration of microbial electrorespiration-based technologies with other technologies. Moreover, more studies are necessary for new modification in electrodes and to explore low-cost membranes. In addition, there are stability issues such as the logging of electrodes and membrane fouling during long-term operation for practical recalcitrant wastes treatment. On the other hand, the design of new reactor configurations is essential to decrease internal resistance and the reactor dead zone. Bioelectrochemical degradation of recalcitrant compounds will yield even more impressive results when we overcome the limitations of the BES-based systems.

Acknowledgements

The authors wish to thank the Natural Science Foundation of China (51538012, 51922050 and 51708293), and the NSFC-EU joint program (31861133001) for financially supporting this study.

References

Aulenta F, Catervi A, Majone M, Panero S, Reale P, Rossetti S (2007) Characterization of an electro-active biocathode capable of dechlorinating trichloroethene and cis-dichloroethene to ethene. Environ Sci Technol 41(7): 2554–2559.

Cao Y, Hu Y, Sun J, Hou B (2010) Explore various co-substrates for simultaneous electricity generation and Congo red degradation in air-cathode single-chamber microbial fuel cell. Bioelectrochemistry 79(1): 71–76.

Cao X, Wang H, Li XQ, Fang Z, Li XN (2017) Enhanced degradation of azo dye by a stacked microbial fuel cell-biofilm electrode reactor coupled system. Bioresour Technol 227: 273–278.

Chen BY, Zhang MM, Chang CT, Ding Y, Lin KL, Chiou CS, Hsueh CC, Xu H (2010) Assessment upon azo dye decolorization and bioelectricity generation by *Proteus hauseri*. Bioresour Technol 101(12): 4737–4741.

De Gusseme B, Hennebel T, Vanhaecke L, Soetaert M, Desloover J, Wille K, Verbeken K, Verstraete W, Boon N (2011) Biogenic palladium enhances diatrizoate removal from hospital wastewater in a microbial electrolysis cell. Environ Sci Technol 45(13): 5737–5745.

Escapa A, Mateos R, Martínez EJ, Blanes J (2016) Microbial electrolysis cells: An emerging technology for wastewater treatment and energy recovery From laboratory to pilot plant and beyond. Renew Sust Energ Rev 55: 942–956.

Feng C, Huang L, Yu H, Yi X, Wei C (2015) Simultaneous phenol removal, nitrification and denitrification using microbial fuel cell technology. Water Res 76: 160–170.

Fu L, You SJ, Zhang GQ, Yang FL, Fang XH (2010) Degradation of azo dyes using in-situ Fenton reaction incorporated into H₂O₂-producing microbial fuel cell. Chem Eng J 160(1): 164–169.

Gao SH, Peng L, Liu Y, Zhou X, Ni BJ, Bond PL, Liang B, Wang AJ (2016) Bioelectrochemical reduction of an azo dye by a *Shewanella oneidensis* MR-1 formed biocathode. Int Biodeter Biodegr 115: 250–256.

Guo W, Cui Y, Song H, Sun J (2014) Layer-by-layer construction of graphene-based microbial fuel cell for improved power generation and methyl orange removal. Bioproc Biosyst Eng 37(9): 1749–1758.

Guo WQ, Guo S, Yin RL, Yuan Y, Ren NQ, Wang AJ, Qu DX (2015) Reduction of 4-chloronitrobenzene in a bioelectrochemical reactor with biocathode at ambient temperature for a long-term operation. J Taiwan Inst Chem E 46: 119–124.

Hamelers HV, Ter Heijne A, Sleutels TH, Jeremiasse AW, Strik DP, Buisman CJ (2010) New applications and performance of bioelectrochemical systems. Appl Microbiol Biot 85(6): 1673-1685.

Hou B, Sun J, Hu YY (2011a) Effect of enrichment procedures on performance and microbial diversity of microbial fuel cell for Congo red decolorization and electricity generation. Appl Microbiol Biot 90(4): 1563–1572.

Hou B, Sun J, Hu YY (2011b) Simultaneous Congo red decolorization and electricity generation in air-cathode single-chamber microbial fuel cell with different microfiltration, ultrafiltration and proton exchange membranes. Bioresour Technol 102(6): 4433–4438.

Hu WJ, Niu CG, Wang Y, Zeng GM, Wu Z (2011) Nitrogenous heterocyclic compounds degradation in the microbial fuel cells. Process Saf Environ 89(2): 133–140.

Huang W, Chen J, Hu Y, Chen J, Sun J, Zhang L (2017) Enhanced simultaneous decolorization of azo dye and electricity generation in microbial fuel cell (MFC) with redox mediator modified anode. Int J Hydrogen Energ 42(4): 2349–2359.

Ilamathi R, Jayapriya J (2018) Microbial fuel cells for dye decolorization. Environ Chem Lett 16: 239–250.

Jiang X, Shen J, Lou S, Mu Y, Wang N, Han W, Sun X, Li J, Wang L (2016) Comprehensive comparison of bacterial communities in a membrane-free bioelectrochemical system for removing different mononitrophenols from wastewater. Bioresour Technol 216: 645–652.

Li J, Liu G, Zhang R, Luo Y, Zhang C, Li M (2010) Electricity generation by two types of microbial fuel cells using nitrobenzene as the anodic or cathodic reactants. Bioresour Technol 101(11): 4013–4020.

Liang B, Cheng HY, Kong DY, Gao SH, Sun F, Cui D, Kong FY, Zhou AJ, Liu WZ, Ren NQ (2013) Accelerated reduction of chlorinated nitroaromatic antibiotic chloramphenicol by biocathode. Environ Sci Technol 47(10): 5353–5361.

Liang B, Cheng H, Van Nostrand JD, Ma J, Yu H, Kong D, Liu W, Ren N, Wu L, Wang A, Lee DJ, Zhou J (2014) Microbial community structure and function of Nitrobenzene reduction biocathode in response to carbon source switchover. Water Res 54: 137–148.

Liu L, Li FB, Feng CH, Li XZ (2009) Microbial fuel cell with an azo-dye-feeding cathode. Appl Microbiol Biot 85(1): 175.

Liu H, Hu TJ, Zeng GM, Yuan XZ, Wu JJ, Shen Y, Yin L (2013) Electricity generation using p-nitrophenol as substrate in microbial fuel cell. Int Biodeter Biodegr 76: 108–111.

Liu RH, Li WW, Sheng GP, Tong ZH, Lam MHW, Yu HQ (2015) Self-Driven Bioelectrochemical Mineralization of Azobenzene by Coupling Cathodic Reduction with Anodic Intermediate Oxidation. Electrochim Acta 154: 294–299.

Long XZ, Pan QR, Wang CQ, Wang H, Li H, Li XN (2017) Microbial fuel cell-photoelectrocatalytic cell combined system for the removal of azo dye wastewater. Bioresour Technol 244: 182–191.

Luo Y, Zhang R, Liu G, Li J, Li M, Zhang C (2010) Electricity generation from indole and microbial community analysis in the microbial fuel cell. J Hazard Mater 176(1): 759–764.

Mu Y, Rozendal RA, Rabaey K, Keller J (2009) Nitrobenzene removal in bioelectrochemical systems. Environ Sci Technol 43(22): 8690–8695.

Mu Y, Radjenovic J, Shen J, Rozendal RA, Rabaey K, Keller J (2010) Dehalogenation of iodinated X-ray contrast media in a bioelectrochemical system. Environ Sci Technol 45(2): 782–788.

Oon YS, Ong SA, Ho LN, Wong YS, Oon YL, Lehl HK, Thung WE, Nordin N (2018) Disclosing the synergistic mechanisms of azo dye degradation and bioelectricity generation in a microbial fuel cell. Chem Eng J 344: 236–245.

Pan Y, Wang YZ, Zhou AJ, Wang AJ, Wu ZT, Lv LT, Li XJ, Zhang K, Zhu T (2017) Removal of azo dye in an up-flow membrane-less bioelectrochemical system integrated with bio-contact oxidation reactor. Chem Eng J 326: 454–461.

Pham H, Boon N, Marzorati M, Verstraete W (2009) Enhanced removal of 1, 2-dichloroethane by anodophilic microbial consortia. Water Res 43(11): 2936–2946.

Pozo K, Harner T, Wania F, Muir DC, Jones KC, Barrie LA (2006) Toward a global network for persistent organic pollutants in air: results from the GAPS study. Environ Sci Technol 40(16): 4867–4873.

Rabaey K, Rodriguez J, Blackall LL, Keller J, Gross P, Batstone D, Verstraete W, Nealson KH (2007) Microbial ecology meets electrochemistry: electricity-driven and driving communities. ISME J 1(1): 9.

Savant DV, Abdul-Rahman R, Ranade DR (2006) Anaerobic degradation of adsorbable organic halides (AOX) from pulp and paper industry wastewater. Bioresour Technol 97(9): 1092-1104.

Shen J, Feng C, Zhang Y, Jia F, Sun X, Li J, Han W, Wang L, Mu Y (2012) Bioelectrochemical system for recalcitrant p-nitrophenol removal. J Hazard Mater 209-210: 516–519.

Shen J, Zhang Y, Xu X, Hua C, Sun X, Li J, Mu Y, Wang L (2013) Role of molecular structure on bioelectrochemical reduction of mononitrophenols from wastewater. Water Res 47(15): 5511–5519.

Strycharz SM, Gannon SM, Boles AR, Franks AE, Nevin KP, Lovley DR (2010) Reductive dechlorination of 2-chlorophenol by Anaeromyxobacter dehalogenans with an electrode serving as the electron donor. Env Microbiol Rep 2(2): 289–294.

Sun J, Li Y, Hu Y, Hou B, Xu Q, Zhang Y, Li S (2012) Enlargement of anode for enhanced simultaneous azo dye decolorization and power output in air-cathode microbial fuel cell. Biotechnol Lett 34(11): 2023–2029.

Sun J, Li W, Li Y, Hu Y, Zhang Y (2013) Redox mediator enhanced simultaneous decolorization of azo dye and bioelectricity generation in air-cathode microbial fuel cell. Bioresour Technol 142: 407–414.

Sun J, Zhang Y, Liu G, Ning X, Wang Y, Liu J (2015) Unveiling characteristics of a bioelectrochemical system with polarity reversion for simultaneous azo dye treatment and bioelectricity generation. Environ Biotechnol 99: 7295–7305.

Wang AJ, Cheng HY, Liang B, Ren NQ, Cui D, Lin N, Kim BH, Rabaey K (2011) Efficient reduction of nitrobenzene to aniline with a biocatalyzed cathode. Environ Sci Technol 45(23): 10186–10193.

Wang AJ, Cui D, Cheng HY, Guo YQ, Kong FY, Ren NQ, Wu WM (2012) A membrane-free, continuously feeding, single chamber up-flow biocatalyzed electrolysis reactor for nitrobenzene reduction. J Hazard Mater 199: 401–409.

Wang H, Luo H, Fallgren PH, Jin S, Ren ZJ (2015) Bioelectrochemical system platform for sustainable environmental remediation and energy generation. Biotechnol Adv 33(3): 317–334.

Wang X, Xing D, Ren N (2016) p-Nitrophenol degradation and microbial community structure in a biocathode bioelectrochemical system. RSC Adv 6(92): 89821–89826.

Wang H, Yi S, Cao X, Fang Z, Li X (2017) Reductive dechlorination of hexachlorobenzene subjected to several conditions in a bioelectrochemical system. Ecotox Environ Safe 139: 172–178.

Wen Q, Yang T, Wang S, Chen Y, Cong L, Qu Y (2013) Dechlorination of 4-chlorophenol to phenol in bioelectrochemical systems. J Hazard Mater 244: 743–749.

Xu H, Quan X, Xiao Z, Chen L (2017) Cathode modification with peptide nanotubes (PNTs) incorporating redox mediators for azo dyes decolorization enhancement in microbial fuel cells. Int J Hydrogen Energ 42(12): 8207–8215.

Yeruva DK, Sravan JS, Butti SK, Modestra JA, Mohan SV (2018) Spatial variation of electrode position in bioelectrochemical treatment system: Design consideration for azo dye remediation. Bioresour Technol 256: 374–383.

Yun H, Liang B, Kong DY, Cheng HY, Li ZL, Gu YB, Yin HQ, Wang AJ (2017) Polarity inversion of bioanode for biocathodic reduction of aromatic pollutants. J Hazard Mater 331: 280–288.

Zeng X, Borole AP, Pavlostathis SG (2015) Biotransformation of furanic and phenolic compounds with hydrogen gas production in a microbial electrolysis cell. Environ Sci Technol 49(22): 13667–13675.

Zhang C, Li M, Liu G, Luo H, Zhang R (2009) Pyridine degradation in the microbial fuel cells. J Hazard Mater 172(1): 465–471.

Zhang E, Wang F, Zhai W, Scott K, Wang X, Diao G (2017) Efficient removal of nitrobenzene and concomitant electricity production by single-chamber microbial fuel cells with activated carbon air-cathode. Bioresour Technol 229: 111–118.

Zhao H, Kong CH (2018) Enhanced removal of p-nitrophenol in a microbial fuel cell after long-term operation and the catabolic versatility of its microbial community. Chem Eng J 339: 424–431.

Zhong D, Liao X, Liu Y, Zhong N, Xu Y (2018) Enhanced electricity generation performance and dye wastewater degradation of microbial fuel cell by using a petaline NiO@polyaniline-carbon felt anode. Bioresour Technol 258: 125–134.

Enhanced Bioremediation of Petroleum Hydrocarbons Using Microbial Electrochemical Technology

Huan Wang, Lu Lu and Zhiyong Jason Ren[*]

Department of Civil and Environmental Engineering and the Andlinger Center for Energy and the Environment, Princeton University, Princeton, NJ 08544, USA

1. Introduction

1.1 Petroleum Hydrocarbons

Petroleum hydrocarbons are hydrocarbons derived from crude oil (or petroleum products) and popular products including oil, gasoline, diesel fuels, lubricating oil, paraffin wax and asphalt (Altgelt 2016). The major hydrocarbon components are nonpolar fractions that were generally divided into two groups: aliphatic hydrocarbons and aromatic hydrocarbons. Aliphatics mainly include alkanes, alkenes and cycloalkanes. Aromatics have one or more benzene rings as part of their structure, such as benzene, toluene, ethylbenzene, xylenes (BTEX) and polycyclic aromatic hydrocarbons (PAHs) (Speight 2014). Some products may also have polar groups containing sulfur, nitrogen and naphthenic acids.

Large quantities of petroleum hydrocarbons are released into the environment during the process of exploration, production, transportation, improper storage and refining. Hydrocarbon contamination is among the most widely spread environmental problems in soil, sediment and groundwater and range from crude oil spills to leakages of underground storage tanks (Sarkar et al. 2005). For instance, in 2010 the Deepwater Horizon oil spill discharged an estimated 4.9 million barrels of crude oil that contaminated the ocean, sediment, beaches, wetlands and estuaries (National Response Team 2011). The US EPA estimates that there are more than 3 million Underground Storage Tanks (UST) storing petroleum products in the United States and more than 439,000 USTs were confirmed leaking hydrocarbons into the ground (EPA 2004). The remediation of hydrocarbon-contaminated sites is a long-term need, and new technologies are needed to provide low-energy, low-cost and sustainable methods for hydrocarbon remediation.

1.2 The Challenges of Petroleum Hydrocarbons Remediation

Remediation technologies for petroleum hydrocarbons include physical, chemical and biological methods (Lim et al. 2016). Physical approaches include soil excavation, capping and thermal-assisted extraction. Such methods alter the contaminated matrix but do not necessarily destroy the contaminants. Another physical method is soil vapor extraction that is an *in situ* method that minimizes the problem with the

*Corresponding author: zjren@princeton.edu

landscape. However, the high demand for energy use associated with this method and the associated carbon footprint hinder its application (Ávila-Chávez and Trejo 2010). Chemical oxidation (chemOx) has been widely used for contaminated soil and groundwater treatment, but it relies on contact reactions between the injected oxidants and the contaminants. For this reason, chemOx is less or noneffective in matrices of lower permeability such as clay-rich soils where invasive methods (e.g., soil mixing) is required (Usman et al. 2012). In addition, shipping and handling chemical hazardous oxidants can be difficult and expensive (Siegrist et al. 2011). Bioremediation technologies such as biostimulation, bioaugmentation, landfarming, biopiling, bioventing, biosparging, and bioreactors have been established for *in situ* and *ex situ* treatment of petroleum hydrocarbons in soils and groundwater (Xu and Obbard 2004). Bioremediation is considered as low-cost and environmentally friendly, but current *in situ* bioremediation activities are often limited by the availability of electron acceptors and nutrients in the subsurface environment (Soares et al. 2010). Bioremediation processes often need amendments of nutrients (e.g., slow-release fertilizers, biosurfactants and biopolymers) and electron acceptors (e.g., oxygen, Fe(III), sulfate, nitrates or humic substances) (Lovley 2000). The delivery of these compounds generally faces the same soil matrix permeability problem, and the delivery of alternative electron acceptors such as NO_3^- or SO_4^{2-} are expensive and may cause secondary contamination.

1.3 Advantages of MET for Petroleum Hydrocarbon Remediation

Microbial electrochemical technology (MET) integrates microorganisms and electrochemical process for environmental and energy applications, and it was recently developed for removing hydrocarbon contaminants during the soil, sediment or groundwater remediation. Electroactive bacteria (EAB), also known as exoelectrogens, electricigens or anode-respiring bacteria (Zhang et al. 2010; Logan 2009), on or near the anode can work with other bacteria or directly oxidize biodegradable materials including petroleum hydrocarbons and respire the electrode as the electron acceptor. The electrons harvested from the anode are transferred through an external circuit to the cathode where terminal electron acceptors (e.g., oxygen) are reduced (Logan et al. 2006; Wang et al. 2015; Wang and Ren 2013). This extracellular electron transfer process was widely researched as microbial fuel cells (MFCs) to generate electricity from wastewater, but low current generation and slow treatment rate were found to be limiting factors for system scale-up. In contrast, using a similar approach for environmental remediation was considered much more feasible because compared with a high rate (a few hours of retention time) needed for wastewater treatment the rate of bioremediation is very slow and lasted many years. In this context, the facilitated electron transfer in microbial electrochemical reactors can make a big impact and enhance bioremediation, shorten remediation period and, therefore, reduce cost. Although microbial electrochemical systems (MES) present a more accurate description of the system, which is different from enzyme-based electrochemical systems (together known as bioelectrochemical systems or BES), it is more consistent with microbial electrochemical technology (MET); we follow the tradition in this chapter to use BES. The BES process should not be confused with abiotic electroremediation process such as electrokinetics as abiotic systems directly apply much higher current between distributed electrodes to charged organic and inorganic particles (Pazos et al. 2010). However, it is possible to combine these processes to improve overall remediation performance.

Microbial electrochemical remediation offers a number of advantages: 1) the electrodes provide a clean and inexhaustible source of electron acceptor without requiring chemical addition or energy input for the bioremediation; 2) The remediation process and monitoring time can be greatly shortened compared with traditional bioremediation; 3) The degradation progress can be readily monitored by self-generated nonintrusive online monitoring (current/voltage); 4) The costs of installation and maintenance are substantially lower than other active remediation methods; 5) The remediation process is energy neutral or positive, and the current generated from BES can even provide power for remote sensors.

2. Reactor Design, Material and Operational Parameters

2.1 Reactor Design for the Remediation of Hydrocarbons in Soil, Sediment and Groundwater

2.1.1 Suspended or Horizontally Embedded Electrode Design

Figure 1 shows the various designs of BES reactors. The most common design of BES configuration consists of an anode buried in the soil/sediment/groundwater and a cathode in an oxygen-rich phase (Rezaei et al. 2007; Morris and Jin 2012; Yan et al. 2012; Yu et al. 2017). One study reported 95% phenanthrene and 99% pyrene were removed during 240 days operation by combined BES and amorphous ferric hydroxide (Yan et al. 2012). For the electrode arrangement, cathodes were typically placed on the overlying water/sediment for maintaining oxygen supply, while anodes were embedded in the soil/sediment with direct contact of contaminants (Figure 1A) (Morris et al. 2009; Cao et al. 2015; Zabihallahpoor et al. 2015). In rotating cathode design, rotation of the cathode disks increases the dissolved oxygen in the overlying water. No external energy input is required when disks are rotated by natural water currents. However, the large electrode spacing causes high internal resistance and ohmic loss, so current production was low. Also, because the oxic and anoxic zones are naturally formed with spatial separation in sediment, this configuration faces challenges when used in deeper contamination sites due to high internal ohmic loss (Li and Yu 2015). Yu et.al (2017) optionally buried cathode into the soil to control the interval between the anode and cathode and increased the current.

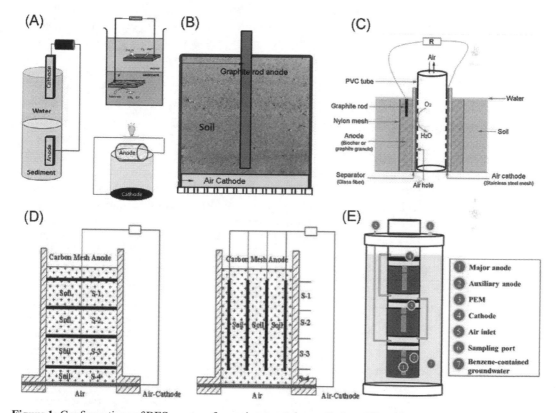

Figure 1: Configurations of BES reactors for environmental remediation: (A) suspended electrodes in sediment (Zabihallahpoor et al. 2015), (B) graphite rod insert-type snorkel (Li et al. 2016a), (C) horizontal column-type reactor (Lu et al. 2014b), (D) multi-anode system (Zhang et al. 2015) and (E) stacked multi-cathode system (Liu et al. 2018). Copyrights from (A) The Royal Society of Chemistry 2015, (B) 2016 Elsevier B.V., (C) 2014 Elsevier B.V., (D) 2014 Springer-Verlag Berlin Heidelberg and (E) 2018 Elsevier Ltd. Reproduced with permission.

2.1.2 Vertically Deployed Tubular or Column Electrode Design

Vertical designs became popular recently due to its easy installation and high performance. Typical tubular and column type reactors are constructed in sediment or soil remediation by wrapping an assembly of anode, separator and cathode layers around a tube. The cathode layer faces inside and is exposed to air, while anode faces outside and is exposed to sediment or surrounding soil matrix (Figure 1C) (Yuan et al. 2010; Lu et al. 2014a). Tubular-type air-cathode BESs with carbon cloth anode (CCA) or biochar anode (BCA) were constructed for soil remediation. Results show that biochar anode exhibited better performance for diesel degradation and current output than cloth anode due to the higher adsorption capability that provided more substrates for the anode microbes and facilitated hydrocarbon diffusion from surrounding areas toward the anode. The total petroleum hydrocarbon (TPH) removal rate almost doubled in soils close to the anode (63.5–78.7%) compared to that (37.6–43.4%) in the open-circuit controls during a period of 64 days (Lu et al. 2014a) (Table 1). Recently, a similar cylindrical configuration of BES was applied for *in situ* groundwater petroleum hydrocarbon removal. This laboratory-scale reactor, named 'the bioelectric well', consisted of graphite granules as bioanode and concentric stainless steel mesh as a cathode and was installed directly within groundwater well (Palma et al. 2018). The laboratory-scale bioelectric well was operated continuously for a period of 56 days, and phenol removal gradually increased from 12% to reported 100% after 38 days.

2.1.3 Multi-Electrode Design

The low mass transfer has been considered as a major barrier for many remediation processes. To overcome this challenge, Nielsen et al. (2007) attempted to enforce pumping or utilizing the natural advective flows of porewater. Natural hydraulic flow enhancement was also used to enhance mass transfer in pilot-scale sediment BESs (Li et al. 2017). This system can decrease internal resistance and promote current generation. However, natural hydraulic flows should be carefully controlled to prevent the oxygenation of the anode chamber. Increasing anode sizes and raising contact areas between the anode and soil/sediment were found alleviating limitation of mass transfer as well. In this respect, several BES designs such as U-type reactors with four pairs of air-cathodes and anodes and multi-anode reactors with three layers of anodes system and one air cathode were applied in soil remediation (Figure1D, E) (Wang et al. 2012; Li et al. 2014a; Zhang et al. 2015). In the U-type air-cathode BES, TPH degradation rate close to the anode (<1cm) was enhanced by 120% from 6.9% (open-circuit control) to 15.2% after 25 days, and in the meantime, a maximum power density of 0.85 mWm^{-2} was obtained. Adsorptive electrode materials such as biochar were used in connected mini-electrodes that significantly increased the influence area of bioremediation. Results showed that with adsorptive electrodes, the TPH removal rate was as high as 89.7%, a 241% increase within 120 days compared to control (Lu et al. 2014b). In addition, using carbon-fiber anode with the larger surface area could enhance BES radius of influence (ROI) as well (Li et al. 2016a). In multi-anode soil BES with the addition of glucose, the degradation rate of TPH and charge output were enhanced by 200% and 262%, respectively, with the reason explained to be boosted bacterial metabolisms according to the increase of dehydrogenase and polyphenol oxidase activities (Table1). In similar multi-anode BES, a sand amendment can also enhance soil porosity, decrease ohmic resistance (by 46%) and increase TPH degradation (up to 268%) (Table 1). Another recent design consisted vertically stacked tubular equipped with low-cost polyvinyl alcohol (PVA)-coke anode, the PVA-hydrogel elastomer (PVA-HE) separator and carbon cloth cathode for benzene removal from groundwater (Figure 1F). The BES was also used in monitoring wells or remediation wells as a permeable reactive barrier (PRB) downstream of the contaminated groundwater (Liu et al. 2018). These results indicate that stacked tubular-BES could indeed shorten the remediation duration than single electrode reactors. The maximum power density of the serially-connected stacked BES was 12.7 mW/m^2, a 3.3-fold increase over the single BES (Table 1).

Table 1: BES remediation of the total petroleum hydrocarbons and aromatic hydrocarbons in soil, sediment and groundwater.

Configuration/ Matrix	Pollutant	A_{max} (mA m^{-2})/ P_{max} mW m^{-2}	Time (day)	[a] D_{max} (%)/ Enhanced D_{max} (%)	Original concentration	Reactor volume (L)	Anode	Cathode	Reference
Embedded electrode/soil	PAHs (Anthracene, Phenanthrene, Pyrene)	65/12.1	175	54.2/160 (Anthracene)	103mg/kg soil	0.76	activated carbon fiber felt	activated carbon fiber felt	(Yu et al. 2017)
Suspended cathode/sediment	TPH	22/6.7	66	24.4/1062	16 g/kg	0.07	carbon cloth	carbon paper-Pt	(Morris and Jin 2012)
Embedded anode (amorphous ferric hydroxide)/ sediment	PAHs (phenanthrene and pyrene)	17.1mV	240	99.47/ 23.9 (phenanthrene) 94.79/21.4 (pyrene)	0.0954 mg/kg (phenanthrene) 0.0316 mg/kg (pyrene)	0.1	stainless steel	stainless steel	(Yan et al. 2012)
Embedded anode	PAHs	–/63	72	74/–	30 µg/g	350	carbon mesh	activated carbon fiber felt	(Li et al. 2017)
U-type/ soil	Petroleum hydrocarbons (TPH, PAH)	5.5/0.85	25	15.2/120	28.3g/kg soil	2.736	carbon felt	carbon mesh-Pt/C	(Wang et al. 2012)
Column-type/soil	Diesel (TPH)	70.4/8.8 (GGA) 35.2/3.4 (BCA)	120	89.7/241	12.25 g/kg soil	50 (soil)	graphite granule anode (GGA) and biocharanode (BCA).	stainless steel mesh-Pt/C	(Lu et al. 2014b)
Tubular-type/soil	Diesel (TPH)	73/– (CCA) 85.9/– (BCA)	65	73.1/237(CCA) 78.7/233(BCA)	11.46 g/kg soil	2 (soil)	carbon cloth anode (CCA) or biochar anode (BCA)	activated carbon cloth-Pt/C	(Lu et al. 2014a)
Graphite rod insert-type (mixed carbon fiber in soil)/soil	Petroleum hydrocarbons (TPH, PAH)	203/15	144	30/100	701 µg/g soil	0.72	graphite rod	activated carbon	(Li et al. 2016a)

(Contd.)

Table 1: (*Contd.*)

Configuration/ Matrix	Pollutant	A_{max} (mA m^{-2})/ P_{max} mW m^{-2}	Time (day)	$^a D_{max}$ (%)/ Enhanced D_{max} (%)	Original concentration	Reactor volume (L)	Anode	Cathode	Reference
Two single chambered reactors/sludge	petroleum (TPH)	–/53.11	17	40.1/98	413g/kg soil	0.55	graphite plates	graphite plates	(Mohan and Chandrasekhar 2011)
Multi-anode (sand amendent)/ soil	Petroleum hydrocarbons (TPH, PAH)	74/20	135	22/268	27891ng/g soil	0.324	carbon mesh	activated carbon	(Li et al. 2015)
Multi-anode (glucose addition)/soil	Petroleum hydrocarbons (TPH, PAH)	127/58	135	21/200	28322 ng/g soil	0.324	activated carbon	activated carbon	(Li et al. 2016b)
Multi-anode/soil	Petroleum hydrocarbons (TPH, PAH)	282mV	135	12.5/95	25.7 g/kg soil	0.324	carbon meshes	stainless steel mesh	(Zhang et al. 2015)
Stacked tubular-type/groundwater	benzene	57.2/12.7	12	100/100	60mg/L	3.3	graphite rod and polyvinyl alcohol (PVA)-coke	carbon cloth	(Liu et al. 2018)

$^a D_{max}$ (%) = The maximum degradation efficiency of total petroleum hydrocarbon (TPH) (%); Enhanced D_{max} (%) = (The maximum degradation efficiency of TPH in BES – degradation efficiency of TPH in natural attenuation or open circuit control)/degradation efficiency of TPH in natural attenuation or open circuit control × 100%.

2.2 BES Electrode Materials

2.2.1 Anode Materials

The BES anode materials should be low-cost, biocompatible and resistant to corrosion/biodegradation. Based on these requirements, stainless steel anode (Morris and Jin 2012; Morris et al. 2009) and carbon-based materials were commonly adopted. Stainless steel is highly conductive with good mechanic strength and is easy to install, but its relatively small specific surface area and vulnerability to corrosion in anoxic environment make it less suitable as a BES anode (Dumas et al. 2007). Carbon-based materials including graphite, carbon fiber, activated carbon and biochar are widely used. Graphite plate anodes and graphite rod anodes have been commonly adopted in early remediation studies. Carbon-fiber anode had higher TPH removal rate than carbon cloth anode in soil BES because the fiber structure provided a much higher surface area and higher conductivity (Li et al. 2016a). When adding carbon fiber in soil BES, carbon fiber was found enhancing PAH removal even if electrodes were not connected (Kronenberg et al. 2017). Reports also showed that the interspecies electron transfer (IET) might be promoted by these carbon materials and other conductors (Kouzuma et al. 2015; Cruz Viggi et al. 2014). However, such carbon fiber application by mixing with soil is difficult to deploy in real-world applications due to high cost, so optimization of electrode arrangement with carbon fiber material maybe an alternative option. Anodic materials may also affect removal efficiency. TPH removal was slightly higher in biochar anode (78.7%) compared to a carbon cloth anode (73.1%) (Lu et al. 2014a).

2.2.2 Cathode Materials

BES cathodes are also primarily carbon-based and materials used include carbon cloth, carbon felt, carbon mesh, carbon fiber, graphite rod and graphite granule, etc. In addition, a low-cost cloth (GORE-TEX) was also used as cathode material in an insertion-type soil BES (Huang et al. 2011). The pure carbon cathode alone has low oxygen reduction catalytic capability, so catalysts such as platinum (Pt) were usually loaded on the cathode surface (Scott et al. 2008). Although the loading amounts of Pt gradually decreased, it still adds significant cost and limits the lifespan of the electrode. Low-cost activated carbon cathode showed good catalytic activity and stability as well (Zhang et al. 2011). One study on saline-alkali soil remediation demonstrated similar charge output but showed longer operation with activated carbon air-cathode than Pt air-cathode (Li et al. 2014b).

2.3 Main Factors Influencing BES Remediation

2.3.1 Radius of Influence

The extension of the radius of influence (ROI) is one of the key factors that need to be considered for field BES application. The ROI directly determines the density of the BES matrix to be applied on a site that also affects the cost and deployment. The first ROI study was conducted using saline soil. The total petroleum hydrocarbon degradation rate increased only at a location close to the electrode (<1cm), while the degradation rate further from the anode (1-2 cm and 2-3 cm) was found to be similar as controls (Wang et al. 2012). Another study showed that the degradation of hydrocarbon can be enhanced both at 1 cm and 5cm (Lu et al. 2014a). A laboratory pilot reactor performed by Lu et al. (2014b) showed that ROI expanded overtime from closer to the electrode to all across the reactor within 120 days, and the predicted maximum ROI was 90 cm based on the linear correlation between the distance from the anode and the enhanced removal rate. A correlation between the ROI and the radius of BES reactor was constructed, and it was found in that condition the ROI can be 35-40 times of radius of electrode (ROE) indicating that the ROI can be extended with the increase of the radius of the electrode (Lu et al. 2014b). In addition, the ROI can be influenced by physicochemical and biological properties of the soil and contaminants. Li et al. (2015) found that increasing the porosity of soil (e.g., by sand amendment) the ROI can be increased accompanied by higher mass transfer and hydrocarbon degradation.

2.3.2 Conductivity

Food and Agriculture Organization (FAO) estimated that the total area of saline soil is about 397 million ha, and sodic soil is about 434 million ha. Salt-affected soils make up around 20% of the global irrigated land (Chhabra 2017). Petroleum hydrocarbons contaminated saline-alkali soil is a common condition because of many major oil fields located in the coastal area. While there have been few studies focused on how soil conductivity affects BES performance in hydrocarbon remediation, many studies have demonstrated that increasing conductivity in wastewater facilitates ion transfer and boosts current generation. Similarly, high ionic conductivity can facilitate ion transfer within the soil and groundwater matrix and, therefore, increases current density in BES. However, super high conductivity may also negatively affect microbial activities because salts such as sodium chloride can inhibit certain microbial growth. The soil electrical conductivity also affects electron transfer, which is a faster process than ion transfer, so it has not been considered a rate-limiting process. A recent study found that conductive materials such as carbon fiber act as electron transport promoters to promote soil electrical conductivity without inhibiting microbial mechanisms (Li et al. 2016a). In addition, clay soil and soils containing higher metal contents tend to have higher current generation due to their high electrical conductivity.

3. Application and Scale-Up of BES Remediation

3.1 Soil BES

BES for soil remediation has been demonstrated very efficient compared with natural attenuation in removing various hydrocarbon products, including phenol (Huang et al. 2011), petroleum (Wang et al. 2012), diesel (Lu et al. 2014b), lindane (Camacho-Pérez et al. 2013), dibenzothiophene (Rodrigo et al. 2014), hexachlorobenzene (Cao et al. 2015) and mixtures. Soil texture and water content (soil water

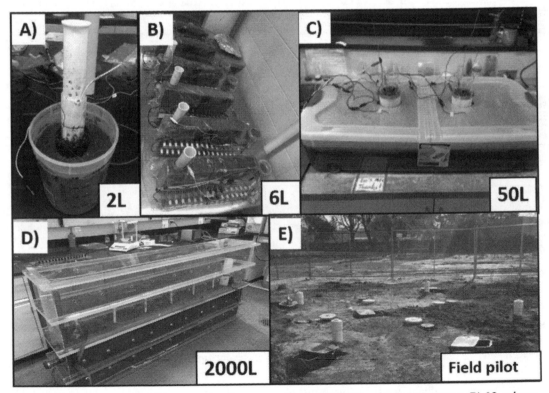

Figure 2: Soil column-type BES system development and scale-up. A) 2 L column-type reactor; B) 6 L column-type electrode reactor; C) 50 L column-type electrode reactor; D) 2,000 L continuous flow reactor and E) Field pilot systems (Courtesy: Advanced Environmental Technologies and Chevron Energy Company).

content is usually less than 60%) (Wang et al. 2012) can greatly affect BES performance, and sandy soil with high porosities and high water content generally lead to faster remediation and higher current output.

To date, most soil BES remediation research has been conducted in the laboratory with reactors smaller than 1 liter. The Ren group has worked with Advanced Environmental Technologies and Chevron Energy Company and scaled up the column type BES soil remediation system from laboratory-scale to field pilot-scale. The reactors have been scaled from 2 L to 2,000 L (Figure 2A-D) and recently a field pilot study was successfully conducted in California, USA (Figure 2E). Using these different reactors, they found that BES jumped started and enhanced diesel TPH degradation by 50-200% compared to natural attenuation that led to shortened remediation and monitoring time. Plus, the current generation can be correlated with TPH removal serving a real-time and nonintrusive monitoring method. The process produced up to 80 mA/m² electricity and can be used to power data collection sensors. They estimated ROI for one BES (1 ft) can reach 15-25 feet in a static saturated condition, and the ROI can be greatly expanded when operated in soil with active groundwater flow. They have also performed geophysical electrode measurements, which monitored the *in situ* soil resistivity, that were found to correlate well with electrical, chemical and microbiological data (Wang et al. 2015; Lu et al. 2014a; Lu et al. 2014b; Mao et al. 2016; Zuo Y. 2017).

3.2 Sediment BES

Sediment remediation was among the first BES applications used on site. One early study showed BES could increase TPH degradation by 11 folds compared with open-circuit control (24% vs. 2.1%) in 66 days (Morris and Jin 2012). Benthic microbial fuel cell (BMFC) has been widely studied in sediment although the primary goal was not for contaminant removal rather current generation. For those used in remediation, a pilot-scale BMFC with carbon fabric anode and rolling activated carbon air-cathode was used to degrade polycyclic aromatic hydrocarbons (PAHs) (Figure 3A), and results show it reduced 89% of TOC in the water phase and 70% of TOC in the sediment around the anode. The removal of PAHs was 74% that was higher than the sediment sample far away from the anode (60%) collected as control. The maximum power density was 63 ± 3 mWm⁻² on day 2, and then it was further reduced to 30 mWm⁻² on day 72 due to low substrate concentration that limited anode performance (Li et al. 2017). General oceanographic sensors and communications systems require a power of 0.6 mW in sleep mode and 20 W in active mode (Hsu et al. 2013). Large-scale BMFCs (210 cm in diameter, 150 cm in height and in 10 m depth) were designed to provide power for the temperature-depth sensors (6 V) (Figure 3B). Studies showed that the output power of the BMFC could reach 427 mW in continuous mode for 17 months (Zhou et al. 2018).

3.3 Groundwater BES

BES groundwater treatment studies are scarce but received increasing attention (Morris and Jin 2008). Groundwater generally has low TPH concentration and conductivity (less than 1 mS·cm⁻¹), so the

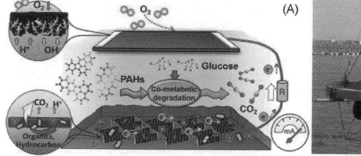

Figure 3: (A) Schematic of pilot-scale BMFC (Li et al. 2017) and (B) Structure of BMFC-sea (Zhou et al. 2018). Copyrights from 2017 Elsevier B.V. (A) and 2017 Springer-Verlag GmbH Germany (B). Reproduced with permission.

kinetics and current generation are low (Puig et al. 2012). Recently, more strategies are developed to overcome these barriers and scaled up reactors to field applications (Pous et al. 2018). A bench-scale (50 L) constructed wetland BES was established for groundwater remediation in Leuna (Sxony-Anhalt, Germany) where gasoline compounds, especially benzene and methyl-tert-butyl ether (MTBE) and NH_4^+, have been historical contaminants. The system was operated for 400 days, and results showed that benzene and MTBE were almost completely removed by days 95 and 125, while control reactors removed 80% of these compounds on day 150. Overall, the highest removal efficiencies of benzene and MTBE were observed in the upper layer (24 cm) probably due to better availability of oxygen (Wei et al. 2015). Another laboratory-scale study demonstrated that the fermentation of phenols can generate acetate that served as an electron donor which enhanced removal by 40% in close circuit BES compared to the open-circuit control. In addition, the enhanced biodegradation of acetate facilitated the removal of phenol, and power generation was up to 1.8 mW/m². Conceptual models of field-scale BES have been developed based on the laboratory-scale study (Figure 4) (Hedbavna et al. 2016). The anode is designed as a permeable reactive barrier in contact with the contaminated and oxygen-depleted groundwater, while the cathode is placed in uncontaminated groundwater where oxygen is provided through a borehole. In addition, the small tubular BES was easily scaled up by stacking units in series (Liu et al. 2018). A startup company Advance Environmental Technologies recently developed a commercial product called E-Redox (O) that has been deployed in several contaminated sites in the US and Latin American and demonstrated satisfactory results. The E-Redox is a passive groundwater remediation process that takes advantage of groundwater flow to generate redox gradients across the electrodes to accomplish remediation.

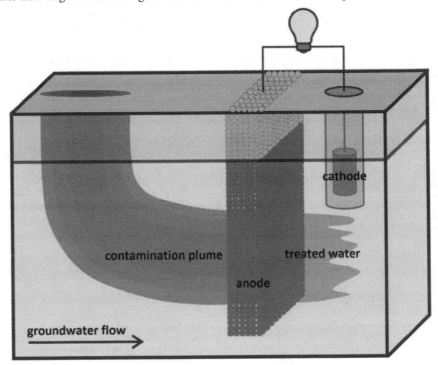

Figure 4: Conceptual model of an *in situ* BES with the anode constructed as a permeable reactive barrier (Hedbavna et al. 2016). Copyright from 2015 Elsevier Ltd. Reproduced with permission.

As more BES studies move from laboratories to fields, new and flexible BES configurations to adapt to different soil matrix, moisture level, contaminant types, site conditions and other real-world factors will need to be developed and tested. This will need to be accompanied by new and low-cost electrodes, separators and water retention layers that are anticorrosion and long-lasting. Field applications also require more understandings of geophysical conditions of the site and microbial community and degradation pathways.

4. Mechanisms and Microbial Communities in BES Remediation

Microbial electrochemical degradation of contaminants is a new type of biodegradation with a much faster rate than enhanced natural attenuation, but it largely still relies on microbial communities naturally present in soil, sediment and groundwater. Studies have investigated on interspecies electron transfer and extracellular electron transfer processes, as well as microbial community structures (Cruz Viggi et al. 2014; Scherr 2013), although a lot of questions remain to be answered.

4.1 Redox Potential

When soil, sediment or groundwater is contaminated by hydrocarbons, biodegradation occurs using the hydrocarbon contaminants as electron donors and environmental available chemicals, such as O_2, SO_4^{2-}, NO_3^- and Fe^{3+} as electron acceptors. However, because these electron acceptors are limited in the environment as compared to the a large amount of contaminant, the bioremediation process is constrained by the available electron acceptors. The different electron acceptors are used based on their respective redox potential, and Figure 5A and B show the sequence of their depletion order. O_2 has the highest redox potential and is used first by aerobic hydrocarbon-degrading bacteria. Once O_2 is depleted, anaerobic degradation occurs with the sequence of nitrate reduction, iron (III)-oxide reduction, sulfate reduction and finally methanogenic degradation (Daghio et al. 2017). The role of the electrode (anode) in this case is to provide an alternative electron acceptor for bacteria that are limited by electron acceptors, and the electrons transferred to the anode are ultimately transferred to the air-cathode that is exposed to air and catalyzes O_2 reduction. The redox potential of the anode can be regulated by external resistance in the field that can be considered as a tuning knob which influences anode redox potential for different conditions. For instance, lower resistance leads to higher anode potential where more microorganisms can gain energy by directly or indirectly respiring the anode as the electron acceptor. Such operation can lead to faster degradation and higher current and, therefore, shorten remediation period (Madigan et al. 2008; Ren et al. 2011). Daghio et al. (2016) found direct current increase when the anode potential increased from +0.2V to +0.5V for toluene degradation.

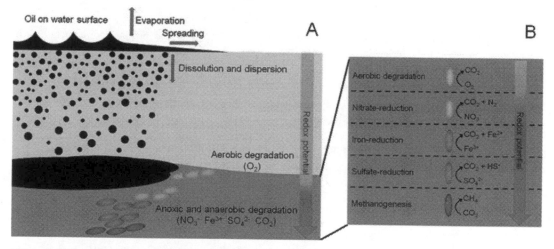

Figure 5: (A) Hydrocarbon degradation pathway after a spill in marine environments. (B) Detail of microbial degradation process in the sediment without electrode (Daghio et al. 2017). Copyright from 2017 Elsevier Ltd. Reproduced with permission.

4.2 Anodic Reactions and Microorganisms

4.2.1 Microbial Syntrophy

While BES hydrocarbon remediation is generally considered an anaerobic process, the mechanism is believed to involve consortia of microorganisms that formed syntrophic relationships to degrade recalcitrant hydrocarbons into different metabolites. Although EAB such as *Geobacter* is also known to degrade benzene and other hydrocarbon molecules, energetically and kinetically it is difficult to have one single cell to both break down complex hydrocarbons and conduct extracellular electron transfer (Lovley 2006). Therefore, it is hypothesized a syntropy exists between hydrocarbon degrading bacteria (HDB) and EAB (Figure 6). EAB firstly degrades hydrophobic hydrocarbons into hydrophilic intermediates (alcohols, acids and aldehydes, etc.) that then is transported to the anode and oxidized by EAB (Daghio et al. 2017; Wang et al. 2019) or used by methanogens to produce CH_4 (Wang et al. 2015; Li and Yu 2015). This is consistent with the typical alkane degradation pathway that starts with oxygenases in the first step to oxidize hydrocarbons into intermediates followed by adegradation by dehydrogenases. Recent studies connected microbial community structure, predictive functional gene and phylogenetic molecular ecological networks together and revealed such network to collaboratively convert petroleum hydrocarbon into electrons and carbon products.

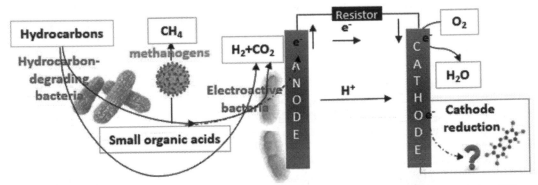

Figure 6: Hypothetical pathways of microbial electrochemical hydrocarbon degradation involving different electron donors and acceptors. The solid line represents metabolic reactions associated with oxidation and/or reduction. The dash line represents electron transfer pathway.

4.2.2 *Sulfur Cycle*

Existing hydrocarbon biodegradation in marine sediment is known to be primarily connected with sulfate reduction because sulfate is the most abundant oxidizing agent such environment (Barton and Hamilton 2007). Decades of research have established and demonstrated the process of sulfur cycling for the removal of hydrocarbons by multiple interconnecting metabolic pathways. Sulfide present in marine sediment can be oxidized at the anode to elemental sulfur or even sulfate, depending on the redox potential (Zhang et al. 2014). In the meantime, sulfide (80-90%) produced from sulfate reduction is reoxidized to sulfate through sulfur compounds of intermediate oxidation state (Fossing 2004). This process is supported by experimental findings in which significant reduction of sulfate reduction was observed in sediment with embedded electrodes than the control (Cruz Viggi et al. 2015). However, the community structure and role of microorganisms were not fully characterized.

4.3 **Cathodic Reactions and Microorganisms**

For most BES remediation systems, oxygen is served as the terminal electron acceptor via cathode catalysis (Bond and Lovley 2003). However, if oxygen is lacking, other electron acceptors may be used on the cathode. Studies have reported chlorophenols (Kong et al. 2014), nitrobenzene (Wang et al. 2011), metals (Huang et al. 2014) and other electron acceptors (Wang et al. 2015) have been used on the cathode. In some cases, the thermodynamic barrier of these reactions needs to be overcome by applying an external bias (Schamphelaire et al. 2008; Logan and Rabaey 2012). To date, studies of hydrocarbon remediation in soil and sediment are mainly focused on hydrocarbon degradation in the anode and most of them use air-cathode, while limited studies examined the biocathode microbial community. Studies showed that some EAB were also found on the cathode along with some facultative bacteria such as *Pesudomonas* and *Novosphigobium* (Clauwaert et al. 2007; Erable et al. 2010).

5. Future Perspectives

5.1 Understanding the BES Hydrocarbon Degradation Mechanisms

Hydrocarbon degradation is a complex process and affected by many different parameters, such as soil property, moisture content, geographical conditions, microbial community, nutrient condition, redox potential, conductivity, pH, temperature and more. Current BES studies still focus on reactor and engineering designs, while the fundamental degradation mechanisms are still largely unexplored. Conventional GC-FID has been widely used to measure total petroleum hydrocarbon (TPH) or carbon number fractions, but it does not provide information on the spectrum of daughter products. GC/MS analysis can provide more information, but the high complexity of petroleum hydrocarbon mixtures and the lack of specificity of the mass spectra of certain products still impede the identification. Comprehensive two-dimensional gas chromatography (GC × GC) and fourier transform ion cyclotron resonance mass spectrometry (FT-ICR MS) are very promising and hold powerful techniques that have ultrahigh resolving power for detailed product analysis. These tools have already been successfully applied to characterize the polar compounds in aerobic and anaerobic biodegradation of crude oil. However, to date, there is no study on the identification of BES hydrocarbon degradation products based on these new techniques. In addition, meta-omics tools for microbial analysis will help understand the structure of microbial communities and levels of gene expressions that are related to BES hydrocarbon degradation.

5.2 Integration with Other Processes

While BES has demonstrated good performance in enhancing natural attenuation, there are still challenges to improve efficacy. BES remediation can be combined with phytoremediation when used in wetland, and previous studies reported that electroactive bacteria on the anode can use plant root exudates as substrates for electricity production or reduction of greenhouse gas emissions (Arends et al. 2014; Lu et al. 2015). The combined degradation of pyrene and benzo[a]pyrene in contaminated sediments using BES with the treatment of the macrophyte acorus calamus has also been studied. The combination of BES and phytoremediation led to a higher degradation efficiency compared to BES or phytoremediation treatment alone (Yan et al. 2015). Another promising opportunity to utilize BES for bioremediation is to drive the production of compounds and/or reductive dehalogenation on the cathode (Wang et al. 2015; Logan and Rabaey 2012). Other technologies such as electrokinetics may also be combined with BES to improve ROI and kinetics.

5.3 Field Application

Field pilots have been tested using different configurations of BES in various conditions but field data reports have been scarce, and there exist knowledge gaps between field degradation and fundamental mechanism that are critical for regulatory approval. As aforementioned, some key factors need to be further investigated for field application. The radius of influence (ROI) needs to be expanded and efforts are needed to enable BES to be used in the unsaturated condition. The electron transfer mechanisms need to be elucidated, especially for those locations that are far away from the electrode. Bioelectrochemical systems have demonstrated great potentials for hydrocarbon remediation, and we look forward to seeing full-scale applications in the near future.

6. Conclusions

The MET platform has demonstrated good performance and scalability for enhanced remediation of petroleum hydrocarbons in soil, groundwater, and other subsurface matrix. This chapter summarizes the state-of-the-art of this application as well as system development and microbial ecology, and it provides perspectives on future development of this platform. With the rapid development of the bioelectrochemical systems and the MET platform, the commercial deployment of MET for remediation is expected soon.

References

Altgelt KH (2016) Composition and Analysis of Heavy Petroleum Fractions. CRC Press. Marcel Dekker: New York.

Arends JB, Speeckaert J, Blondeel E, De Vrieze J, Boeckx P, Verstraete W, Rabaey K, Boon N (2014) Greenhouse gas emissions from rice microcosms amended with a plant microbial fuel cell. Appl Microbiol Biotechnol 98(7): 3205-3217.

Ávila-Chávez MA, Trejo A (2010) Remediation of soils contaminated with total petroleum hydrocarbons and polycyclic aromatic hydrocarbons: Extraction with supercritical ethane. Ind Eng Chem Res 49(7): 3342-3348.

Barton LL, Hamilton WA (2007) Sulphate-reducing Bacteria: Environmental and Engineered Systems. Cambridge University Press, Cambridge.

Bond DR, Lovley DR (2003) Electricity production by *Geobacter sulfurreducens* attached to electrodes. Appl Environ Microbiol 69(3): 1548-1555.

Camacho-Pérez B, Ríos-Leal E, Solorza-Feria O, Vazquez-Landaverde PA, Barrera-Cortés J, Ponce-Noyola MT, Garcia-Mena J, Rinderknecht-Seijas N, Poggi-Varaldo, HM (2013) Preformance of an electrobiochemical slurry reactor for the treatment of a soil contaminated with lindane. J New Mat Electrochem Syst 16(3): 217-228.

Cao X, Song H, Yu C, Li X (2015) Simultaneous degradation of toxic refractory organic pesticide and bioelectricity generation using a soil microbial fuel cell. Bioresour Technol 189:87-93.

Chhabra R (2017) Soil Salinity and Water Quality. Routledge, London.

Clauwaert P, Van der Ha D, Boon N, Verbeken K, Verhaege M, Rabaey K, Verstraete, W (2007) Open air biocathode enables effective electricity generation with microbial fuel cells. Environ Sci Technol 41(21): 7564-7569.

Cruz Viggi C, Rossetti S, Fazi S, Paiano P, Majone M, Aulenta F (2014) Magnetite particles triggering a faster and more robust syntrophic pathway of methanogenic propionate degradation. Environ Sci Technol 48(13): 7536-7543.

Cruz Viggi C, Presta E, Bellagamba M, Kaciulis S, Balijepalli SK, Zanaroli G, Petrangeli Papini M, Rossetti S, Aulenta F (2015) The "Oil-Spill Snorkel": an innovative bioelectrochemical approach to accelerate hydrocarbons biodegradation in marine sediments. Front Microbiol 6: 881.

Daghio M, Vaiopoulou E, Patil SA, Suárez-Suárez A, Head IM, Franzetti A, Rabaey K (2016) Anodes stimulate anaerobic toluene degradation via sulfur cycling in marine sediments. Appl Environ Microbiol 82(1): 297-307.

Daghio M, Aulenta F, Vaiopoulou E, Franzetti A, Arends JB, Sherry A, Suárez-Suárez A, Head IM, Bestetti G, Rabaey K (2017) Electrobioremediation of oil spills. Water Res 114: 351-370.

Dumas C, Mollica A, Féron D, Basséguy R, Etcheverry L, Bergel A (2007) Marine microbial fuel cell: use of stainless steel electrodes as anode and cathode materials. Electrochim Acta 53(2): 468-473.

EPA US (2004) How to evaluate alternative clean up technologies for underground storage tank sites: a guide for corrective action plan reviewers, in: Office of Underground Storage Tanks (Ed.), Rep. No. EPA/510/R-04-002, Washington, D.C.

Erable B, Vandecandelaere I, Faimali M, Delia M-L, Etcheverry L, Vandamme P, Bergel A (2010) Marine aerobic biofilm as biocathode catalyst. Bioelectrochemistry 78(1): 51-56.

Fossing H (2004) Distribution and fate of sulfur intermediates–sulfite, tetrathionate, thiosulfate, and elemental sulfur–in marine sediments. Sulfur Biogeochemistry: Past and Present 379: 97.

Hedbavna P, Rolfe SA, Huang WE, Thornton SF (2016) Biodegradation of phenolic compounds and their metabolites in contaminated groundwater using microbial fuel cells. Bioresour Technol 200: 426-434.

Hsu L, Chadwick B, Kagan J, Thacher R, Wotawa-Bergen A, Richter K (2013) Scale up considerations for sediment microbial fuel cells. RSC Adv 3(36): 15947-15954.

Huang DY, Zhou SG, Chen Q, Zhao B, Yuan Y, Zhuang L (2011) Enhanced anaerobic degradation of organic pollutants in a soil microbial fuel cell. Chem Eng J 172(2-3): 647-653.

Huang L, Jiang L, Wang Q, Quan X, Yang J, Chen L (2014) Cobalt recovery with simultaneous methane and acetate production in biocathode microbial electrolysis cells. Chem Eng J 253: 281-290.

Kong F, Wang A, Ren HY, Huang L, Xu M,Tao H (2014) Improved dechlorination and mineralization of 4-chlorophenol in a sequential biocathode–bioanode bioelectrochemical system with mixed photosynthetic bacteria. Bioresour Technol 158: 32-38.

Kouzuma A, Kato S, Watanabe K (2015) Microbial interspecies interactions: recent findings in syntrophic consortia. Front Microbiol 6: 477.

Kronenberg M, Trably E, Bernet N, Patureau D (2017) Biodegradation of polycyclic aromatic hydrocarbons: Using microbial bioelectrochemical systems to overcome an impasse. Environ Pollut 231: 509-523.

Li H, He W, Qu Y, Li C, Tian Y, Feng Y (2017) Pilot-scale benthic microbial electrochemical system (BMES) for the bioremediation of polluted river sediment. J Power Sources 356: 430-437.

Li XJ, Wang X, Zhang YY, Ding N, Zhou QX (2014b) Opening size optimization of metal matrix in rolling-pressed activated carbon air–cathode for microbial fuel cells. Appl Energy 123: 13-18.

Li XJ, Wang X, Zhang YY, Cheng LJ, Liu J, Li F, Gao BL, Zhou QX (2014a) Extended petroleum hydrocarbon bioremediation in saline soil using Pt-free multianodes microbial fuel cells. RSC Adv 4(104): 59803-59808.

Li XJ, Wang X, Ren ZYJ, Zhang YY, Li N, Zhou QX (2015) Sand amendment enhances bioelectrochemical remediation of petroleum hydrocarbon contaminated soil. Chemosphere 141: 62-70.

Li XJ, Wang X, Zhao Q, Wan LL, Li YT, Zhou QX (2016a) Carbon fiber enhanced bioelectricity generation in soil microbial fuel cells. Biosens Bioelectron 85: 135-141.

Li XJ, Wang X, Wan LL, Zhang YY, Li N, Li DS, Zhou QX (2016b) Enhanced biodegradation of aged petroleum hydrocarbons in soils by glucose addition in microbial fuel cells. J Chem Technol Biotechnol 91(1):267-275.

Li WW, Yu HQ (2015) Stimulating sediment bioremediation with benthic microbial fuel cells. Biotechnol Adv 33(1): 1-12.

Lim MW, Von Lau E, Poh PE (2016) A comprehensive guide of remediation technologies for oil contaminated soil–present works and future directions. Marine pollution bulletin 109(1): 14-45.

Liu SH, Lai CY, Ye JW, Lin CW (2018) Increasing removal of benzene from groundwater using stacked tubular air-cathode microbial fuel cells. J Clean Prod 194: 78-84.

Logan BE, Hamelers B, Rozendal RA, Schrorder U, Keller J, Freguia S, Aelterman P, Verstraete W, Rabaey K (2006) Microbial fuel cells: Methodology and technology. Environ Sci Technol 40(17): 5181-5192.

Logan BE (2009) Exoelectrogenic bacteria that power microbial fuel cells. Nat Rev Microbiol 7(5): 375-381.

Logan BE, Rabaey K (2012) Conversion of wastes into bioelectricity and chemicals by using microbial electrochemical technologies. Science 337(6095): 686-690.

Lovley DR (2000) Anaerobic benzene degradation. Biodegradation 11(2-3):107-116.

Lovley DR (2006) Bug juice: harvesting electricity with microorganisms. Nat Rev Microbiol 4(7): 497.

Lu L, Huggins T, Jin S, Zuo Y, Ren ZJ (2014a) Microbial metabolism and community structure in response to bioelectrochemically enhanced remediation of petroleum hydrocarbon-contaminated soil. Environ Sci Technol 48(7): 4021-4029.

Lu L, Yazdi H, Jin S, Zuo Y, Fallgren PH, Ren, ZJ (2014b) Enhanced bioremediation of hydrocarbon-contaminated soil using pilot-scale bioelectrochemical systems. J Hazard Mater 274: 8-15.

Lu L, Xing D, Ren ZJ (2015) Microbial community structure accompanied with electricity production in a constructed wetland plant microbial fuel cell. Bioresour Technol 195: 115-121.

Lu L, Mao D, Wang H, Huang Z, Cui Y, Jin S, Zuo Y, Ren ZJ (2018) Microbial ecological interactions in bioelectrochemical remediation of hydrocarbon-contaminated soils with different textures. In revision review. Chem Eng J.

Madigan MT, Martinko JM, Dunlap PV, Clark DP (2008) Brock biology of microorganisms 12th edn. Int Microbiol 11: 65-73.

Mao D, Lu L, Revil A, Zuo Y, Hinton J, Ren ZJ (2016) Geophysical monitoring of hydrocarbon-contaminated soils remediated with a bioelectrochemical system. Environ Sci Technol 50(15): 8205-8213.

Mohan SV, Chandrasekhar K (2011) Self-induced bio-potential and graphite electron accepting conditions enhances petroleum sludge degradation in bio-electrochemical system with simultaneous power generation. Bioresour Technol 102(20): 9532-9541.

Morris JM, Jin S (2008) Feasibility of using microbial fuel cell technology for bioremediation of hydrocarbons in groundwater. J Environ Sci Health A - Tox Hazard Subst Environ Eng 43(1): 18-23.

Morris JM, Jin S, Crimi B, Pruden A (2009) Microbial fuel cell in enhancing anaerobic biodegradation of diesel. Chem Eng J 146(2): 161-167.

Morris JM, Jin S (2012) Enhanced biodegradation of hydrocarbon-contaminated sediments using microbial fuel cells. J Hazard Mater 213: 474-477.

National Response Team (2011) On scene coordinator report: Deepwater Horizon oil spill., https://homeport. uscg.mil/Lists/Content/Attachments/119/DeepwaterHorizonReport%20-31Aug2011%20-CD_2.pdf.

Nielsen ME, Reimers CE, Stecher HA (2007) Enhanced power from chambered benthic microbial fuel cells. Environ Sci Technol 41(22): 7895-7900.

Palma E, Daghio M, Franzetti A, Petrangeli Papini M, Aulenta F (2018) The bioelectric well: a novel approach for *in situ* treatment of hydrocarbon-contaminated groundwater. Microb Biotechnol 11(1): 112-118.

Pazos M, Rosales E, Alcántara T, Gómez J, Sanromán M (2010) Decontamination of soils containing PAHs by electroremediation: a review. J Hazard Mater 177(1-3):1-11.

Pous N, Balaguer MD, Colprim J, Puig S (2018) Opportunities for groundwater microbial electro-remediation. Microb Biotechnol 11(1): 119-135.

Puig S, Coma M, Desloover J, Boon N, Colprim Js, Balaguer MD (2012) Autotrophic denitrification in microbial fuel cells treating low ionic strength waters. Environ Sci Technol 46(4): 2309-2315.

Ren Z, Yan H, Wang W, Mench MM, Regan JM (2011) Characterization of microbial fuel cells at microbially and electrochemically meaningful time scales. Environ Sci Technol 45(6): 2435-2441.

Rezaei F, Richard TL, Brennan RA, Logan BE (2007) Substrate-enhanced microbial fuel cells for improved remote power generation from sediment-based systems. Environ Sci Technol 41(11): 4053-4058.

Rodrigo J, Boltes K, Esteve-Nuñez A (2014) Microbial-electrochemical bioremediation and detoxification of dibenzothiophene-polluted soil. Chemosphere 101: 61-65.

Sarkar D, Ferguson M, Datta R, Birnbaum S (2005) Bioremediation of petroleum hydrocarbons in contaminated soils: comparison of biosolids addition, carbon supplementation, and monitored natural attenuation. Environ Pollut 136(1): 187-195.

Schamphelaire LD, Bossche LVD, Dang HS, Höfte M, Boon N, Rabaey K, Verstraetem W (2008) Microbial fuel cells generating electricity from rhizodeposits of rice plants. Environ Sci Technol 42(8):3053-3058.

Scherr KE (2013) Hydrocarbon, InTech.

Scott K, Cotlarciuc I, Head I, Katuri K, Hall D, Lakeman J, Browning D (2008) Fuel cell power generation from marine sediments: Investigation of cathode materials. J Chem Technol Biotechnol 83(9):1244-1254.

Siegrist RL, Crimi M, Simpkin TJ (2011) *In situ* Chemical Oxidation for Groundwater Remediation. Springer Science & Business Media. New York.

Soares AA, Albergaria JT, Domingues VF, Maria da Conceição M, Delerue-Matos C (2010) Remediation of soils combining soil vapor extraction and bioremediation: Benzene. Chemosphere 80(8): 823-828.

Speight JG (2014) The Chemistry and Technology of Petroleum. CRC Press. Marcel Dekker: New York.

Usman M, Faure P, Ruby C, Hanna K (2012) Remediation of PAH-contaminated soils by magnetite catalyzed Fenton-like oxidation. Appl Catal B: Environmental 117:10-17.

Wang AJ, Cheng HY, Liang B, Ren NQ, Cui D, Lin N, Kim BH, Rabaey K (2011) Efficient reduction of nitrobenzene to aniline with a biocatalyzed cathode. Environ Sci Technol 45(23): 10186-10193.

Wang H, Ren ZJ (2013) A comprehensive review of microbial electrochemical systems as a platform technology. Biotechnol Adv 31(8): 1796-1807.

Wang H, Luo H, Fallgren PH, Jin S, Ren ZJ (2015) Bioelectrochemical system platform for sustainable environmental remediation and energy generation. Biotechnol Adv 33(3): 317-334.

Wang H, Lu L, Mao D, Huang Z, Cui Y, Jin S, Zuo Y, Ren, ZJ (2019) Dominance of electroactive microbiomes in bioelectrochemical remediation of hydrocarbon-contaminated soils with different textures. Chemosphere 235: 776-784.

Wang X, Cai Z, Zhou Q, Zhang Z, Chen C (2012) Bioelectrochemical stimulation of petroleum hydrocarbon degradation in saline soil using U-tube microbial fuel cells. Biotechnol Bioeng 109(2):426-433.

Wei M, Rakoczy J, Vogt C, Harnisch F, Schumann R, Richnow HH (2015) Enhancement and monitoring of pollutant removal in a constructed wetland by microbial electrochemical technology. Bioresour Technol 196: 490-499.

Xu R, Obbard JP (2004) Biodegradation of polycyclic aromatic hydrocarbons in oil–contaminated beach sediments treated with nutrient amendments. J Environ Qual 33(3): 861-867.

Yan Z, Song N, Cai H, Tay JH, Jiang H (2012) Enhanced degradation of phenanthrene and pyrene in freshwater sediments by combined employment of sediment microbial fuel cell and amorphous ferric hydroxide. J Hazard Mater 199: 217-225.

Yan Z, Jiang H, Cai H, Zhou Y, Krumholz LR (2015) Complex interactions between the macrophyte Acorus calamus and microbial fuel cells during pyrene and benzo [a] pyrene degradation in sediments. Sci Rep 5:10709.

Yu B, Tian J, Feng L (2017) Remediation of PAH polluted soils using a soil microbial fuel cell: influence of electrode interval and role of microbial community. J Hazard Mater 336: 110-118.

Yuan Y, Zhou SG, Zhuang L (2010) A new approach to *in situ* sediment remediation based on air-cathode microbial fuel cells. J Soils Sediments 10(7): 1427-1433.

Zabihallahpoor A, Rahimnejad M, Talebnia F (2015) Sediment microbial fuel cells as a new source of renewable and sustainable energy: present status and future prospects. RSC Adv 5(114):94171-94183.

Zhang F, Pant D, Logan BE (2011) Long-term performance of activated carbon air cathodes with different diffusion layer porosities in microbial fuel cells. Biosens Bioelectron 30(1): 49-55.

Zhang T, Gannon SM, Nevin KP, Franks AE, Lovley DR (2010) Stimulating the anaerobic degradation of aromatic hydrocarbons in contaminated sediments by providing an electrode as the electron acceptor. Environ Microbiol 12(4): 1011-1020.

Zhang T, Bain TS, Barlett MA, Dar SA, Snoeyenbos-West OL, Nevin KP, Lovley DR (2014) Sulfur oxidation to sulfate coupled with electron transfer to electrodes by *Desulfuromonas* strain TZ1. Microbiology 160(1): 123-129.

Zhang Y, Wang X, Li X, Cheng L, Wan L, Zhou Q (2015) Horizontal arrangement of anodes of microbial fuel cells enhances remediation of petroleum hydrocarbon-contaminated soil. Environ Sci Pollut Res 22(3): 2335-2341.

Zhou C, Fu Y, Zhang H, Chen W, Liu Z, Liu Z, Ying M, Zai, X (2018) Structure design and performance comparison of large-scale marine sediment microbial fuel cells in lab and real sea as power source to drive monitoring instruments for long-term work. Ionics 24(3): 797-805.

Zuo Y (2017) Bioelectrochemical Technology for Enhanced Remediation of Petroleum Hydrocarbon-contaminated soil. ISMET6 Conference, October 3-6, Lisbon, Portugal.

Electroactive Biochar: Sustainable and Scalable Environmental Applications of Microbial Electrochemical Technologies

Raul Berenguer[1,2], Stefania Marzorati[2], Laura Rago[2], Pierangela Cristiani[3], Alberto Pivato[4], Abraham Esteve Nuñez[2,6] and Andrea Schievano[*1]

[1] Instituto Universitario de Materiales, Departamento de Química Física, Universidad de Alicante UA, Apartado 99, 03080-Alicante, Spain

[2] IMDEA Water Institute, Av. Punto Com, 2, 28805 Alcalá de Henares, Madrid, Spain

[3] *e*-BioCenter, Department of Environmental Science and Policy (ESP), Università degli Studi di Milano, Via Celoria 2, 20133 Milan, Italy

[4] RSE, Ricerca Sul Sistema Energetico S.p.A., Via Rubattino 54, 20100, Milano, Italy

[5] Department of Civil, Environmental and Architectural Engineering, University of Padova, Via Marzolo No. 9, 35131 Padova, Italy

[6] University of Alcalá, Department of Chemical Engineering, Ctra. Madrid-Barcelona, km. 33,6, 28871, Alcalá de Henares, Madrid, Spain

1. Introduction

In over 15 years of laboratory-scale and pilot-scale studies, bioelectrodes for microbial electrochemical technologies (METs), which is driven by living microorganisms/biofilms, have been typically fabricated using 'technological' conductors (electrocatalysis based on, e.g., carbon-fibers, graphite and precious metals such as titanium, platinum, etc.) (Logan 2010; Schievano et al. 2016). However, when METs are thought for large-scale environmental and biogeochemical applications (e.g., wastewater treatment, soil/water bioremediation, biomass processing, etc.), the use of such materials is substantially excluded because of their high economic and environmental fabrication costs. At present, carbon materials represent the most widely used electrochemical supports mainly due to a unique combination of good conductivity, activity and chemical stability (Pant et al. 2011).

In this panorama, charcoal derived from biomass carbonization (Biochar) represents a target class of materials that might be a successful alternative to state-of-the-art because a) it has been shown to have several electroactive properties (Chacón et al. 2017; Kappler et al. 2014; Prévoteau et al. 2016); b) it can be produced at large scales (from green waste to agro-forestry residues) with a positive energy balance (recovery of syngas and bio-oils) and c) it is fully biogenic/biocompatible. Despite these apparent advantages, to date, biochar has been relatively ignored in the field of METs. After a quick search in Scopus (Article title, Abstract, Keywords Search Option), the keyword 'biochar matches with 'microbial electrochemical technology' (MET) in four documents out of 756 in this field with 'microbial electrochemical systems' (MES) in eight documents out of 1913 and with 'microbial fuel cells' (MFC)

*Corresponding author: andrea.schievano@unimi.it

in 23 documents (many related to the abiotic cathode) out of 8,730. This is a relatively low number of studies as compared to the over 11,000 documents reporting the keyword 'biochar' only in the title. In fact, biochar has been an object of intense research in fields like soil science and material science.

The relatively low conductivity of biochar might have discouraged researchers in its application in METs applications aimed at harvesting current (MFCs) or optimizing coulombic efficiencies (e.g., microbial electrolysis, electrosynthesis, etc.). Also different types of METs are recently under development for large-scale environmental applications, such as constructed wetlands and fluidized bed bioreactors (Aguirre-Sierra et al. 2016; Marzorati et al. 2018a; Prado et al. 2018; Tejedor-Sanz et al. 2017a). In such cases, the efficient current collection is of less importance over other functions, such as superficial redox reactions, local electron transfer, capacitive phenomena and interspecies electron shuttling.

In this chapter, we provide an overview of the actual knowledge of the properties of biochar that might be useful in such MET applications. A review of the literature regarding the interaction of biochar and microbial communities is also covered, considering the frameworks of both soil science and METs. Finally, we propose a new class of 'electroactive biochar' (*e*-biochar) with optimized properties to be 'tuned' by tailoring the carbonization process. The ideal *e*-biochar should simultaneously show sufficient electroconductivity and biocompatible characteristics, promote superficial interaction and electron-transfer with bacteria, act as a capacitor and at the same time, be possibly available in large amounts at relatively low-cost and impacts.

2. Biochar Among Carbon Materials

2.1 Definition

Considering the recommended terminology for carbon materials (Fitzer et al. 1995), biochar is a solid decomposition product obtained by thermochemical or hydrothermal conversion of natural organic material (molecule, polymer or biomass). These carbonization processes differ from coalification (geological process) in their much faster (many orders of magnitude) reaction rates. Hence, although they could show some common properties, biochars exclude coals (charcoals) produced by coalification that are fossil-like and non-sustainable resources. The broad term of biochar, however, also includes those biomass-derived chars subjected to modification/reaction treatments before, during or after carbonization, such as biomass-derived activated carbons (activated biochars). Nevertheless, the use of additional steps or reagents to the carbonization process increases the wastes/by-products, cost and complexity, thus reducing its sustainability for large-scale applications.

2.2 Fabrication

Biochars are mainly produced, in acceptable yields (20-80%) and times (1-24 hours), by pyrolysis or hydrothermal treatment of biomass (Figure 1) (Libra et al. 2011). These methods involve the carbonization of biomass, i.e., the thermochemical reactions leading to a progressive increase in element carbon content (Kambo and Dutta 2015).

Pyrolysis corresponds to the thermal decomposition of biomass in an inert atmosphere (Kambo and Dutta 2015). The fabrication of biochars and biomass-derived activated carbons date back to many years, but it started to gain great attention since the 70-80s of last century when the first modern scientific reviews were being published (Rodríguez-Reinoso 1986a). Next, studies on the thermal decomposition of wood and wood components greatly progressed in the 90s (Rosas et al. 2010). On the other hand, hydrothermal carbonization (HTC) of biomass is carried out in water under autogenous (steam) pressure and temperature in subcritical water. This is also called wet pyrolysis, and the product is hydrochar. During these complex processes, many reactions take place concurrently. Details on the differences of both process and products are revised elsewhere (Kambo and Dutta 2015).

The properties of biochars depend on the nature of the precursor and the experimental conditions used during carbonization. In pyrolysis, the final temperature (> 200°C), heating rate and treatment length are the key parameters (Libra et al. 2011), whereas the reaction temperature (< 350 °C) and pressure

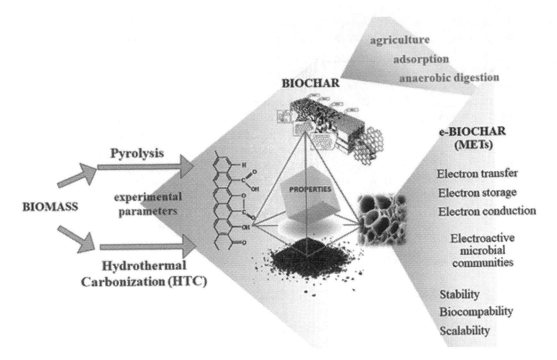

Figure 1: Biochar and *e*-biochar: key properties (structural, chemical, textural and morphological/conformational) can be tailored thanks to enable different applications.

(< 20 bar) prevail in HTC. Special attention must be paid to the effects of some metallic species, particularly alkaline metals (K, Na) and other nutrients commonly present in biomass that are well-known catalysts of the chemical activation of chars. On the other hand, several pre-, *in-situ* and/or post-modification treatments have been studied to obtain derived biochars with optimized textural and chemical features for various applications. A more detailed description of these processes can be found elsewhere (Rodríguez-Reinoso 1986b). Table 1 resumes the properties of various biochars found in literature and their relationship with the preparation conditions. The effects of experimental parameters on the final product are well understood, so their control renders biochars with substantially different properties for certain applications. However, tailoring specific properties without affecting others is not that simple.

2.3 Properties and Characterization

Biochar is a quite complex material exhibiting a wide range of different inter-correlated properties, depending on the precursor and preparation method (Figure 1). Essentially, the performance of biochar in their different applications may depend on structural, chemical and textural properties; also, the shape, size and/or conformation could be also very important.

In terms of chemical structure, biochars lie between natural organic molecules (humic acids, biopolymers, biomolecules, etc.), amorphous (non-graphitic) carbon and crystalline (graphite-like) carbons (Fitzer et al. 1995) with different levels of 2D and 3D crystalline order of polyaromatic layers. The longer the crystalline order range, the more delocalized π-electrons and, thus, the higher the electrical conductivity. Comparing biochar to organic molecules (e.g., humic acids), soil/agriculture scientists usually consider biochar as a conductive material. On the other hand, from the point of view of materials science and technology, biochars are generally amorphous/low-crystalline materials with poor electrical conductivity ($\sim 10^{-2}$-10^{-4}). Therefore, they are generally excluded from fast-response/high-currents electrochemical applications.

Biochars generally exhibit a rich variety of surface moieties with heteroatoms (mainly O but also N, P, S, etc.) and some metal species that come from the original organic precursor. Based on their

Table 1: Effect of preparation experimental parameters on the physico-chemical properties of biochars

Precursor feedstock	Pyrolysis conditions and treatments	Particle size or geometrical shape	BET* Surface area/ m² g⁻¹	Porosity range	Electrical conductivity/ mS cm⁻¹	Surface heteroatoms	Function in electrodes and observed effects	References
Pomelo peel	1000°C	n.r.*	622.2	63% micro, 36% meso	n.r.	Fe, N	Studied as coating for abiotic ORR* electrocatalysis	(Ma et al. 2016)
Dewatered sewage sludge	800°C	n.r.	265.05	Mesopores	n.r.	N, S, Fe	OER* and ORR* electrocatalysis	(Yuan et al. 2016)
Cellulose	250-500°C	0.8-2 mm	199-557	Mesopores	n.r.	N, P	Catalyst support for abiotic MeOH oxidation	(Nieva Lobos et al. 2016)
Coconut shell	800°C	Carbon paste electrode (mixed with spectro grade paraffin wax)	2536	Micropores	n.r.	n.r.	Photocatalytic hydrogen production	(Zha et al. 2016)
Cotton microfiber	700-850°C	n.r.	912.1	Mesopores	n.r.	N	ORR* abiotic electrocatalysis	(Lin et al. 2017)
Sawdust and sugarcane straw	800°C	n.r.	590	Micropores	n.r.	n.r.	Abiotic catalyst for sulfide oxidation	(Mendonça et al. 2017)

different thermal stability, the nature and concentration of these functionalities vary with the heating conditions (mainly with temperature) reached during carbonization. With increasing temperatures, the total O content is progressively reduced and, in parallel, the proportion of less-stable superficial O-groups (SOGs) (carboxylic-like and anhydride-like evolve between 200-400°C) are in favor of most stable ones (lactone-like and phenolic-like evolve between 400-700°C and ether-/carbonyl-/quinone-SOGs evolve between 700-1,100°C). Some transformations have been also reported. Above 1,200°C most O- and P-functionalities seem to be released (García-Mateos et al. 2017).

The porous texture is the last piece of this complex puzzle. A larger volume and specific surface area of micro/mesopores result in a larger number of defects in the ideal graphitic structure. This reduces conductivity but it simultaneously provides a larger number of sites with distinct reactivity and/or where functional groups (redox-active or electroactive) are attached. Without additional treatments, generally, biochars are essentially microporous materials showing more/less rugosity and define bigger pores. The microporosity (d < 2 nm) inherently develops with increasing temperature as a consequence of the loss of volatile matter up to a maximum (800-1,000°C) and then decreases due to solid reorganization resulting in pore shrinkage. Some mesoporosity is also possible. These features usually lead to micro/mesoporous surface areas ranging between 10-600 m^2/g (Table 1).

By contrast, rugosity and macroporosity (surface morphology) mainly depend on the biomass precursor. If the precursor does not pass through a fluid stage (some biopolymers/molecules are thermoplastic), the carbon-enriched char often retains (although shrunk) the characteristic architecture of the precursor. Thus, some biochars can exhibit the vascular structure of preceding plant trunks/stems with defined pores/channels of different dimensions. The analysis of pore size distribution of biochars has revealed extraordinary hierarchical pore architectures (Mendonça et al. 2017).

Finally, biochars can be prepared in multiple forms/conformations, depending on the morphology of the precursor. Generally, monolith-like biochars can be directly obtained by carbonization of plant stems, stalks, canes, etc. The carbonization of trunk wood and/or wood pieces can lead to materials with different shapes and dimensions. The use of fruit shells (almond, coconut, etc.), stones (olive, etc.), skin, seeds, etc., or smaller pieces of wood or canes enable to produce granular/particle-like biochars. The pyrolysis of fibrous biomolecules or biomass, e.g., wheat straw, sugarcane, cellulose, hemp, etc., directly leads to fiber-shaped biochars. On the other hand, powder-like biomolecules, biopolymers (lignin, etc.) or simply crushed biomass can be used to obtain powdery biochars. In addition, these powdered and/or melted (thermoplastic) biomolecules, biopolymers and biomass can be processed into a wider range of biochar conformations, including pellets, monoliths, microfibers and nanofibers, templated materials, etc. (Rosas et al. 2014).

The characterization of biochars is essential to understand their distinct performance in multiple applications. Various papers describing the techniques for the characterization of biochar and other carbon materials can be found in the literature, and they are summarized in Table 2.

2.4 Redox and Electrochemical Properties of Biochar Under Abiotic Conditions

The interfacial chemistry and electrochemistry of carbon materials have been extensively studied and reviewed in the past five/six decades (Conway 1999; Karthikeyan et al. 2015; Li Chum and Baizer 1985). In these works, however, little was focused on biochars probably because of their poor electrical conductivity and/or lower stability under demanding electrochemical conditions as compared to other carbon materials. Most known related studies are addressed to activated carbons (Biniak et al. 2001) whose properties can be tailored for specific applications. All this knowledge, however, can be also used to describe biochars.

Apart from acid-base reactivity, the most important interfacial properties of biochars affecting their behavior in electrolyte solutions can be classified into:

A. Electron-transfer (ET) reactivity (or electron exchange capacity, EEC): This property exclusively refers to the capability of biochar to donate and/or accept electrons. These reactions proceed through the different types of redox-active moieties (electron donors and electron acceptors), namely, the conjugated

Table 2: Recommended Techniques to Characterize Biochars

Property	Technique	Parameter/Property
Electrical Conductivity	Four-Point Probe Resistivity Measurements	Resistivity (Ω cm) or conductivity (S/cm)
	AC impedance	
Structural order	XRD	
	Raman	
Surface chemistry	XPS	Surface % C, N, S, O, P and metals (and qualitative)
	temperature-programmed desorption (TPD)-MS	mmol (CO_2/CO-evolving SOGs)/g e-biochar (and qualitative)
Chemical composition	Elemental analysis	Bulk % C, H, N, S, O, P, ash
Porous texture	N_2 adsorption-desorption isotherms	S_{BET}, VN_2
	CO_2 adsorption	ACO_2, VCO_2
	Hg porosimetry	VHg, porosity (%), density
Morphology	SEM	
Electrochemical	CV	Electrochemical surface area
	EIS	
Redox properties	Mediated electrochemical analysis	Electron-Accepting Capacity Electron-Donating Capacity

π-electron system in condensed aromatic structures, the surface functional groups with heteroatoms (oxygen, nitrogen, sulfur and phosphorous) (for instance, see in Figure 2A the different types of oxygen moieties available in biochars), and/or some metallic species. Most importantly, it is necessary to distinguish between: (i) redox reactions where the donor/acceptor is any compound in liquid, gas or solid phase; the electrons are directly exchanged between species and which do not necessarily imply the existence of electrical conductivity on the biochar; and (ii) electrochemical (faradic or pseudofaradic) reactions in which the ET processes involve the consumption, generation or accumulation of electricity (electric charges) on the material's structure and in which, therefore, electrical conductivity is required. Whenever the electrons involved in the interfacial reactions are externally extracted, supplied through, spread or stored across the material, a conductive structure of biochar is required. In both cases, the reactions can be irreversible or reversible.

Because of this capability in participating in redox and electrochemical ET reactions, biochars are considered as both redox-active and electroactive materials under abiotic conditions.

B. Physical adsorption of ionic and neutral species: (i) electrostatic properties are mainly due to the presence of charges on the carbon surface because of the ET and acid-base processes. These charges are usually located on surface functional groups (heteroatoms) but also distributed throughout the entire carbon surface depending on resonance effects and/or extended π bonding. In fact, the presence of π bonds and aromatic layers also contribute to the overall surface charge. All these charges must be necessarily balanced by ions and water molecules from solution that give rise to the so-called electrical double-layer. These surface solid-liquid coulombic interactions are usually described by parameters like the electrode potential and pH at point of zero charge (pHpzc); (ii) adsorption of gases (e.g., O_2, CO_2) or dissolved neutral species and compounds by van der Waals forces is also a very important property of biochars that can be profited for multiple applications.

C. Electrokinetic properties: The capability of affecting/modifying both previous properties as a consequence of external perturbations to the electrostatic equilibrium attained between the solid (biochar) and its surroundings. A known example of this type of properties is the capability of charging/discharging biochar for energy storage through the electrical double layer and/or pseudocapacitive mechanisms

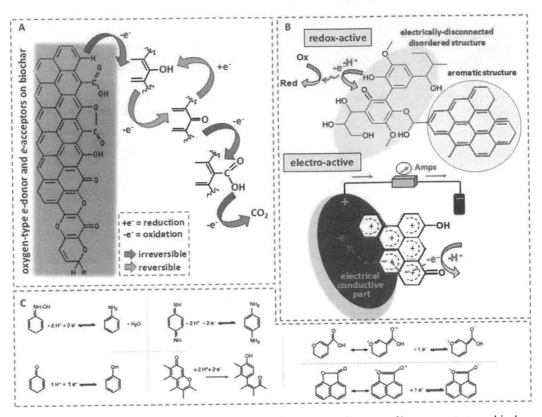

Figure 2: A) Example of O-type functionalities that can act as *e*-donors and/or *e*-acceptors on biochars, emphasizing the reversibility or irreversibility of the ET processes. B) Schematic representations of the redox-active and electro-active properties of biochar. C) Examples of N- and O-based redox couples and processes reported to show electro-activity on carbon materials. (*Source*: Béguin and Raymundo-Piñero 2013. Copyright © Batteries for Sustainability, Springer Nature. Reproduced with permission.)

(Conway 1991a). It is worthy to stress that these properties depend on not only the properties of biochar, e.g., electrical conductivity, micropore volume, type and concentration of functionalities, etc., but also on the external perturbation (e.g., scan rate, charging/discharging currents, etc.) (Conway 1991b).

The electrokinetic behavior of biochar is greatly conditioned by three features that are directly correlated: (i) the narrow microporosity where fast ion or reagent/products diffusion is hampered. This implies limited mass transfer kinetics, i.e., the surface area of biochars can be effectively accessed at very low scan rates and/or by applying low currents. For instance, a very low current is necessary to fully charge the surface of biochar; (ii) the poor conductivity of biochar, which strongly limits the kinetics of electron transfer reactions, or charge storage or effective polarization; (iii) a large amount of oxygen functional groups favors interaction with aqueous media, wettability, hydrophilic character and redox-reactions. The presence of electroactive groups can be positive for ET and charge storage, but the involved pseudofaradic reactions are time-demanding chemical reactions. These three features strongly limit the applicability of biochars in electrochemical applications where high power and current densities are required (supercapacitors, batteries and fuel cells). However, they may not be limiting factors on processes where fast kinetics and high current densities are not necessary.

2.5 Applications

Due to this extraordinary versatility, biochar has been receiving increasing attention for several applications (Ioannidou and Zabaniotou 2007). The most studied applications include the use of biochar as fuel (Waqas et al. 2018), soil amendment (Jeffery et al. 2011), substrate to store carbon over long terms, as adsorbent for CO_2 sequestration (Lehmann et al. 2006) and heavy metals or other contaminants (Lee et al. 2011) in water or air purification and in heterogeneous catalysis (Qian et al. 2015). These applications have been extensively revised elsewhere (Ahmad et al. 2014; Lehmann et al. 2011; Qian et al. 2015). More recently, the possibility of using biochar as a sustainable and cheaper electrode in different electrochemical applications and devices is receiving increasing attention (Gao et al. 2017), including supercapacitors (Berenguer et al. 2016), batteries (Banek et al. 2018) and catalyst support in fuel cells (García-Mateos et al. 2017).

Finally, the development of biochar-based bioelectrodes for METs is the object of recent studies as a platform to drive sustainable scalability of these technologies. Next sections try to give an overview of such point of view.

3. Effects of Biochar's Electrochemical Properties on Microbial Communities

3.1 In Soil and Mixed Microbial Communities

Natural organic biomass burning creates biochar that forms a considerable proportion of the soil's organic carbon in natural forests. In the Amazon-basin, numerous pre-Columbian agricultural sites are composed of variable quantities of highly stable organic black carbon in soil, and they were named the 'dark earth of the Indians' (*Terra preta de Indio*). The apparent high agronomic fertility of these sites, as compared to tropical soils in general, has always attracted interest (Atkinson et al. 2010). Over the last decades, the use of biochar to improve soil fertility and influencing biogeochemical cycles was largely studied (Komang et al. 2016). It has been shown that biochar may induce changes in soil structure and stability, nutrients and energy cycling, carbon-storage capacity, aeration, water use efficiency and disease resistance (Brussaard et al. 2007; Lehmann et al. 2011). Besides chemical and structural changes, soil amendment with biochar influences the microbial abundance and composition (Grossman et al. 2010; O'Neill et al. 2009). These microbiological shifts influence several important soil processes, such denitrification and ammonification (Yanai et al. 2007), methane oxidation (Reddy et al. 2014), biological nitrogen fixation (Rondon et al. 2007) and carbon fixation and degradation (Komang and Caroline 2016; Lehmann and Joseph 2015).

Biochar was either held responsible for important influences on biochemical pathways and gene expression dependent on intercellular signaling (Masiello et al. 2013). Those responses were often associated with physical and chemical environmental changes induced by biochar amendment or simply to biochar's sorption capability (Lehmann et al. 2011). However, it has been often impossible to fully justify such important biochar influences on soil microbial ecology. Also, in other related research, the biological activity of biofilms formed on activated carbon (AC) was not related to electroactive metabolism (Simpson 2008). In Simpson's review (2008), variations of biochemical/microbiological indicators and growth control strategies were associated mainly to modifications in nutrients or dissolved oxygen concentrations and pH, while not to the presence of an electroactive material (i.e., AC).

Only in more recent years, several studies related both AC and biochar with electroactive metabolism in different biologically-mediated processes (Table 3). In a recent review, Yuan et al. (2017) explored the more recent literature concluding that biochar could be understood as an environmentally-sustainable electron donor, acceptor or mediator in different applicative microbiologically or abiotic redox reactions. In 2014, Chen et al. demonstrated that biochar promotes direct interspecies electron transfer (DIET) in co-cultures of *Geobacter metallireducens* with *Geobacter sulfurreducens* or *Methanosarcina barkeri* (Chen et al. 2014). In this study, microbial co-cultures did not need to form aggregates, suggesting that the cells were electrically connected through the biochar which it permitted ET.

In the same year, Kappler et al. (2014) demonstrated that biochar can influence biogeochemical equilibria by direct mediation (electron-shuttling) of ET processes. The addition of biochar to soil stimulated the microbial reduction of the Fe(III) oxyhydroxide mineral ferrihydrite to siderite by the well-known electroactive heterotroph *Shewanella oneidensis*. They hypothesized that ferrihydrite reduction was improved by electron hopping among cells, biochar and ferrihydrite. An increasing of the abundance of Fe(III)-reducing bacteria as *Geobacter* spp. and *Shewanella* spp. was observed in biochar-amended paddy soils (Tong et al. 2014).

In some studies, biochar's electroactive properties improved soil-remediation strategies. Yu and his co-workers demonstrated that the presence of biochar promoted (as high as 24-fold) the electron transfer from *Geobacter sulfurreducens* to pentachlorophenol (PCP), enhancing reductive dechlorination (Yu et al. 2015). The authors affirmed that the coexistence of cells and PCP (which was adsorbed by biochar) on biochar particles was facilitating the ET between them by conductive pili or outer membrane cytochromes. In 2016, Yu et al. demonstrated that the addition of biochar in a *Geobacter sulfurreducens* anaerobic incubation induced an increase (31-fold) in biomass and in exchanged reduced equivalents (Yu et al. 2016). Around 60% of the electrons obtained from the acetate oxidation were recovered by biochar. More recently, it was demonstrated that the presence of biochar influenced arsenic speciation by the biologically-mediated process (Z. Chen et al. 2016). In other experiments, the same authors also proved that biochar influenced the sorption of both ferrous and ferric ion of $FeCl_3$ and $FeSO_4$ suspensions. Biomolecular studies showed temporal changes in bacterial genera abundances in biochar-treated samples as compared to controls. More specifically, well-known electroactive and metal-reducing genera as *Geobacter* (5-20%), *Anaeromyxobacter* (3-11%), *Desulfosporosinus* (7-21%) and *Pedobacter* (7-18%), increased their abundance.

The ability of biochar to stimulate electron transfer between Fe(II) and Fe-oxidizing bacteria was studied also for different purposes, such as metals bioleaching processes (Wang et al. 2016). In mixed-culture systems, the addiction of biochar promoted Fe-mediated bioleaching and the effective leaching of Cu. They hypothesized that biochar accelerated the oxidation of Fe(II) inducing consequent growth of Fe-oxidizing and reducing bacteria (i.e., *Sulfobacillus spp.* and *Alicyclobacillus* spp.).

Biochar's electroactive properties were also used to enhance microbial bioprocesses. In anaerobic digestion, the presence of biochar favored methanogenic species; the observed increase in methane production rate was also related to important changes in the microbial community (Lü et al. 2016). In the early stages of biofilm formation, *Methanosaeta* was the pioneer microorganism in the system when biochar pores represented the only protective environment against ammonium toxicity. Successively, when the biofilm was covering biochar pores, other mechanisms were playing important roles. The authors suggested that interspecies hydrogen transfer and/or DIET mechanisms were stimulated by the presence of biochar that facilitate the methanogenesis under stressed environment. This mechanism was also promoted by DIET between methanogens (as *Methanosaeta* and *Methanosarcina*) and other syntrophic microorganisms (as syntrophic acetogens *Syntrophomonas* and *Sporanaerobacter*). Specifically, *Methanosarcina's* ability to perform DIET in the presence of biochar was helping the resistance to high ammonium and acetate concentration as previously described (Chen et al. 2014). The study also investigated the effect of different particle size on the process, showing i.e., that the use of fine biochar promoted the production of volatile fatty acids.

3.2 Biochar in Microbial Electrochemical Technologies (METs)

In recent METs literature, a reasonable number of studies demonstrated the feasibility of biochar from biomass residues to fabricate bioelectrodes (Table 3). For anodes, many experiments used biochar obtained from a variety of biomasses, such as chestnut shells (Chen et al. 2018), crop plants (Chen et al. 2012a; Chen et al. 2012b), pinewood chips (Huggins et al. 2014) or corrugated cardboard (Chen et al. 2012c). In this specific last case, for instance, cardboard was considered an interesting starting point, being inexpensive and, more importantly, one of the most abundant packing materials in modern societies. After carbonization, the resulting material was employed as the anode in MFCs for being able to achieve satisfactory power and current densities.

Examples of cathodes modified by means of carbonized biomass were fabricated first by Yuan et al. (Yuan et al. 2013a). They were able to avoid the complex synthetic steps or specific machinery in the production of nitrogen-doped carbon materials. They converted sewage sludge into biochar and tested it as the cathode in MFCs. Very high catalytic activity for the ORR was achieved due to high surface area

Table 3: Studied in METs at laboratory-scale

Precursor feedstock	Pyrolysis conditions and treatments	Particle size or geometrical shape	BET* Surface area/ $m^2\ g^{-1}$	Porosity range	Electrical conductivity/ $mS\ cm^{-1}$	Surface heteroatoms	Function in electrodes and observed effects	References
Pine	700°C for 30 sec & 500°C for 15 min	≤0.4 mm	15	n.a	4.4	n.r.	Promotion of DIET in co-cultures	(Chen et al. 2014)
Wood chip	620°C	0.1–0.3 μm and a significant fraction of 3–30 μm	341	n.a	49.7	N, Fe	Improved microbial reduction of the Fe(III)	(Kappler et al. 2014)
Rape-straw	350°C at 20°C min⁻¹ and held constant for 4 h	0.165-mm	2.12	n.r.	n.r.	n.r.	Accelerated reductive dechlorination of PCP	(Tong et al. 2014)
Rice straw	increase at 20°C min⁻¹ up to 900°C for 1 h	0.15 mm	10.85	n.r.	2.4	n.r.	Worked as electron mediators for the dechlorination of PCP	(Yu et al. 2015)
Sieving residues	550°C for 2 h	n.r.	193.9	n.r.	n.r.	n.r.	Influenced metal speciation by increasing the relative abundance of As(V)-/Fe(III)-reducing bacteria	(Z. Chen et al. 2016)
Mature coconut shell	900°C for 1.5 h	0.3 mm	n.r.	n.r.	2.6	n.r.	Improved sediment MFC power generation and TOC removal rate	(S. Chen et al. 2016)
Activated sludge	increase at 10°C min⁻¹ up to 500°C	0.15 mm	n.r.	n.r.	n.r.	n.r.	Improved efficiency of microbial bioleaching of metals	(Wang et al. 2016)
Fruitwoods	800–900°C	2-5 mm, 0.5-1 mm and 75-150 μm	n.r.	n.r.	n.r.	n.r.	Increased tolerance to ammonium and substrate high concentrations in AD	(Lü et al. 2016)

(Contd.)

Table 3: (*Contd.*)

Precursor feedstock	Pyrolysis conditions and treatments	Particle size or geometrical shape	BET* Surface area/ m² g⁻¹	Porosity range	Electrical conductivity/ mS cm⁻¹	Surface heteroatoms	Function in electrodes and observed effects	References
Rice-straw treated with 3.2 g FeCl₃:100 g	500°C for 2 h	0.15 mm	5.48	Mesopores	n.r.	n.r.	Improved methane concentration in AD	(Qin et al. 2017)
Pine chips	800-1000°C for 8 h	0.5-1.0 mm	8.92	n.r.	n.r.	n.r.	Enhanced caproate and caprylate production via chain elongation	(Liu et al. 2017)
Rice straw	5°C min⁻¹ up to maximum T° for 2 h: 300°C, 800°C	0.15 mm	2.6 / 205	n.r.	8.4 / 20.4	n.r.	300°C: faster NO₃⁻ reduction by the enrichment of the nitrate-reducing bacteria. 800°C: promoted N₂O reduction	(Chen et al. 2017)
Pine wood lumber	1000°C	Fine frit glass filter funnel	183.0	82% Micropores	n.r.	Traces of metals	Electrocatalytic layer on a carbon cloth support for air-breathing cathode in MFCs	(Huggins et al. 2015)
Bananas	550-900°C	Ground to powder	105.2 - 172.3	n.r.	n.r.	N	Electrocatalytic layer on a carbon cloth support for air-breathing cathode in MFCs	(Yuan et al. 2014)
Chestnut shell	900°C	Natural chestnut shell shape and powder	468	71% Micropores	n.r.	N	Used in MFCs anodes	(Q. Chen et al. 2018)
Corncob	250, 350, 450, 550, 650, 750°C for 2 h	Ground to powder	n.r.	n.r.	n.r.	n.r.	Used as a layer on a carbon cloth MFC cathode for abiotic ORR*	(Li et al. 2018)

AD = anaerobic digestion

and N and Fe enrichment. In another work, bananas were employed as sources of biochar (Yuan et al. 2014) to fabricate MFCs cathodes. Activation steps with KOH followed and were found successful in an ORR catalysis enhancement. Finally, biochar found another application in the MFCs as a manganese oxide electrocatalytic support in a study by Huggins et al. (2014). The cost of the electrode's material was decreased down to 0.02 dollars, almost 5,000 times lower than a similar electrode, and was prepared from a commercially available activated carbon that supported the manganese oxide catalyst.

Biochar was used in laboratory-scale sediment MFCs, reducing the charge transfer resistance. *Firmicutes* phylum (mainly *Fusibacter* genus) was enriched in these reactors, improving power generation by up to a tenfold (S. Chen et al. 2016). In addition, organic carbon removal was enhanced because the interactions between fermentative and electroactive bacteria were boosted by the presence of biochar as confirmed by an increase in *Firmicutes*. In other words, strictly anaerobic fermentative bacteria from *Firmicutes* phylum were degrading complex organic molecules to short-chain fatty acids that were easily used by electroactive microorganisms for power production. According to the authors, the enhanced consumption of fermentation metabolites by biochar-associated electroactives boosted the growth of *Firmicutes*.

In another study, biochars obtained under different pyrolysis temperatures were characterized by distinct redox-active superficial functional groups (G. Chen et al. 2018). These redox-active moieties interacted in different ways with microbial communities, mediating different ET pathways. Two different biochars at 300 and 800°C were tested in denitrification processes and N_2O reduction to N_2 in batch reactors with NO_3^- as the only electron acceptor and nitrogen resource. The 300°C-biochar as electron donor reduced NO_3^- faster by the enrichment of the nitrate-reducing bacteria. The 800°C-biochar promoted N_2O reduction to N_2 by acting as the electron acceptor and as an electron shuttle for denitrifying bacteria. Very different microbial communities were enriched in biochar-amended reactors as compared to biochar-free controls. First, the addition of biochars induced a decrease in diversity and shifted denitrifying bacterial community; for instance, the increase of *Oceanospirillales* (74% in 800°C-biochar *vs.* 0.0031% in control) was attributed to the increase of the *Halomonas* genus. *Halomonas* genus was recently found in the cathodic biofilm of MFCs (Rago et al. 2017, 2018) and according to the authors, was probably using the 800°C-biochar as electron acceptors.

Other example was reported in another study (Qin et al. 2017) where the authors fabricated biochar from rice-straw with the addition of 3.2 g $FeCl_3$:100 g. This material added to an anaerobic digester influenced microbial community and methane production rates (an increase of around 12%). Despite the improved methane production, the composition and abundance of *Archaea* community (*Methanosarcina*, *Methanobacterium*, unclassified *Methanomicrobiales*, *Methanosaeta*, *Methanoculleus* and *Methanomassiliicoccus*) were not influenced. The main microbiological difference was represented by the enrichment of electroactive bacteria. For instance, *Pseudomonadaceae* presence increased (11.7% *vs.* 0.3% of control without biochar addition). The authors suggested that *Pseudomonadaceae* may have taken part in DIET with methanogens.

A recent study (Liu et al. 2017) showed that biochar introduced into electro-fermentation systems, stimulated the development of a more stable microorganism community structure. The absolute abundances of microorganisms and the stability of the microbial community were enhanced by the presence of biochar *vs.* control. Additionally, caproate and caprylate productions were enhanced, while products-inhibition mitigated. These results were attributed to the overall increase in electrical conductivity in the fermenter, which facilitated ET, but also a high-density structure of extracellular polymeric substances (EPS). Scanning electron microscopy images of the biofilm showed that both microbial cells and EPS were increasingly denser in the vicinity of biochar particles. Such dense biofilm structure increased ET; also, the authors hypothesized an effective defense against the toxicity of high concentrations of products and substrates. In addition, the microbial community (dominated by the bacteria *Sporanaerobacter acetigenes* and *Rummeliibacillus suwonensis* and the archaea *Methanosaeta concilii* and *Methanobacterium spp.*) was significantly different from previous similar studies where biochar was not used.

4. Microbial Electron Transfer Mechanisms on Biochar

According to the literature, the ET capability observed for biochar has been related with two different types of redox-active structures, namely, the surface moieties (the most important being the couple quinone-hydroquinone) and the conjugated π-electron system associated with condensed aromatic structures available in the biochar. These two modes/mechanisms of electron transfer have been denominated 'geobattery' (Klüpfel et al. 2014; Lovley et al. 1996) and 'geoconductor' (Sun et al. 2017), respectively, and are outlined in Figure 3.

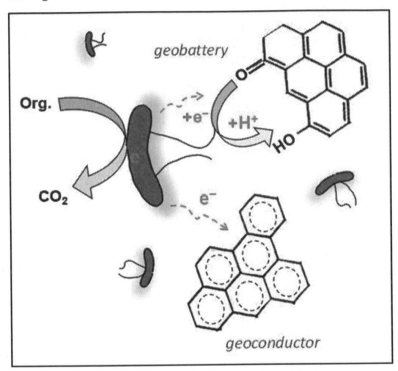

Figure 3: Scheme of the proposed 'geobattery' and 'geoconductor' electron transfer mechanisms.

4.1 ET Mediated by Surface Functional Groups ('Geobattery' Mechanism)

Although the mechanisms by which biotic and abiotic redox reactions on biochar occur are not fully understood, this process was recently compared with the electron accepting and donating capacities of electron shuttles, such as humic and fulvic acids (Klüpfel et al. 2014; Lovley et al. 1996). These reversible redox mechanisms for ET make biochars to act as rechargeable reservoirs of bioavailable electrons, i.e., the so-called geobatteries (Imhoff and Chiu 2015). The *geobattery* mechanism has been exclusively associated with the redox-active and electroactive quinoid/hydroxyl couple on carbon surfaces that can exchange electrons reversibly. This reversibility and the exchanged electrons being stored in the form of chemical bonds well fits with the concept of a battery. However, other functional groups on the surface of biochar could participate in irreversible redox reactions (see Figure 2), also counting in the overall ET with bacteria. In addition, as explained before, not only the Q/Ph couple but also other reversible electroactive species can participate (see Figure 2). Therefore, the term geobattery and the exclusive consideration of Q/Ph species do not offer a complete description of the mediated ET mechanism.

Another controversy arises when quantifying the contribution of this mechanism. The reported analytical methods to quantify the ET capacity of biochars are based on indirect electrochemical techniques, i.e., using different reagents that are chemically oxidized or reduced by the biochar, and later electrochemically regenerated (accounted). In such measurements, irreversible redox-active species on biochars, different to Q/Ph-ones, may be involved in these quantifications. On the other hand, the EAC and EDC of biochars determined by these techniques, however, are based on abiotic reactions, so they might differ from the true ET capability of biochars with bacteria. In addition, the quantified exchanged electrons could depend on the ET kinetics, and these methods require the biochar to be finely grounded and causing probably a significant modification of the material properties. In this sense, it is still unknown whether the numbers quantified by these techniques can be directly related to the biochar-bacteria ET. Therefore, other methods involving the presence of bacteria should be investigated.

4.2 Direct ET to Aromatic Rings ('geoconductor' mechanism)

Despite a generally low conductivity, biochar was demonstrated to act as a promoter of DIET for different syntrophic associations of microorganisms, thanks to its electrical conductivity (Chen et al. 2014). It was also reported that an increase in number and diversity in soil microbial community is more pronounced upon amending with high-temperature biochars vs. low-temperature ones (Khodadad et al. 2011). Later, Sun et al. (2017) reported that biochars prepared above 700°C, with a very low O/C ratio and higher conductivity, can directly transfer electrons to/from a solution redox couple i.e., through the electrode interface, three times faster than the charging and discharging cycles of surface functional groups. This faster ET was potentially proposed for new biogeochemical reactions on the 'bacteria-carbon-mineral' conductive network.

Such electroconductivity-related mechanisms for ET to/from the conductive network of biochars have been referred as to the 'geoconductor' behavior. Unlike superficial redox mechanisms in which the transferred electrons remain mainly localized in functional groups, the electroconductivity of biochar may lead to a relatively longer-distance transport of electrons through the carbon matrix. This may facilitate the access of electrons to acceptors other than biochar.

As in the case of the previous mechanism, some aspects of the geoconductor mechanism deserve a deeper discussion. First, the work of Sun et al. (2017) in which the mechanism was first proposed was carried out under abiotic conditions; thus, the significance of this mechanism for ET between biochars and bacteria is still controversial (Sun et al. 2017). Second, this work concluded that the ET kinetics involved in this mechanism is faster than in superficial ET. The kinetic analysis was based on electrochemical methods, so the response of the different materials toward interfacial charge transfer was strongly influenced also by their different intrinsic electrical conductivity. This was evident by the clear effects of scan rates on the CV profiles.

Finally, the 'geoconductor' mechanism is generally associated with long-range conductivity and/or measurable structural order. However, the presence of local-scaled conductive clusters (condensed aromatic and/or graphitic structures) might be hardly directly detected because it is disordered and/or disconnected from the rest, while being enough to promote ET among electroactive microorganisms. Although the effective dimensions of these structures are unknown, this fact would explain why, even if generally low conductive, biochars can promote ET by this mechanism. Nevertheless, larger dimensions, better arrangement or greater concentration of aromatic/graphitic clusters (i.e., implying overall higher conductivity) could contribute to promoting ET mechanism by increasing the probability of geoconductor-like ET or simply because the delocalization and movement of charges could facilitate the incorporation of new ones.

4.3 Mixed Electron Transfer Mechanism (Geobattery-Geoconductor)

The geobattery and geoconductor models represent two limit cases of ET. Both mechanisms may operate together as most biochars display surface functional groups as well as certain aromatic and/or graphitic structures. This mixed contribution of both ET mechanisms was first proposed by Sun et al. (2017) and represents a more realistic view of the phenomenon. These authors also suggested the possible prevalence of one ET mechanism, depending on the properties of the biochar. Considering the potential effects of different properties in the presence of microorganisms is, therefore, essential.

5. *E*-Biochar: A Potential Platform for Sustainable Bioelectrodes in Large-Scale METs

Merging the knowledge on the properties of biochars (section 2) and on microbial ET mechanisms (section 4), the target biochar (to be called electroactive biochar or *e*-biochar) should simultaneously gather: (i) sufficient surface are with superficial activity toward ET with microbes, (ii) sufficient conductivity to promote electron transport at desired scales and distance ranges, iii) proper physico-chemical and textural properties (porosity, surface area and roughness) to host electroactive microbial communities and to improve electron capacitance, (iv) proper biocompatibility, (v) stability and (vi) mechanical properties and conformation for specific METs applications. In addition, the whole process of fabrication, use in bioelectrodes and re-use after their life-cycle should be both environmentally sustainable and economically feasible (Figure 4). Economic studies and life cycle assessment approaches should look at the overall energetic and resources balance including biomass production, carbonization and functionalization, use in METs, METs performances in terms of carbon removal, nutrients recovery, water re-use and final agricultural application of spent biochar-based materials.

However, relatively little is known about the performance-determining properties of biochar in METs applications. Similarly, the concept of a microbial-electrochemically active material or surface area is still unclear. In fact, these definitions are traditionally related to electroactivity measured under abiotic conditions. Biochars are complex materials showing inter-related properties, affecting microbial ET in very different ways. Therefore, literature still lacks systematic studies analyzing the influence of only one property, disregarding others. In most works instead, the characterization of biochars is poor or incomplete which leads to less meaningful conclusions.

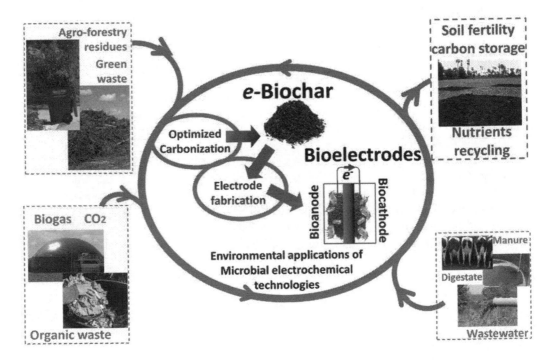

Figure 4: Electroactive biochar (*e*-biochar) circular-economy concept. Tailoring electroactive properties by specific carbonization processes would allow the use of *e*-biochar in bioelectrodes of METs applied to wastewater treatment, CO_2 reutilization, waste bioprocessing and fermentations, as well as bioremediation. End-of-life electrodes materials might be re-used in agricultural soil to improve soil fertility and ensure long-term carbon storage.

Considering this premature state-of-the-art, it is difficult to provide an unequivocal list of desired electroactive properties of biochars for METs and how to optimize carbonization processes to obtain them. Here, we try to provide an overview of what has been done so far.

5.1 Conformation and Scalability of Bioelectrodes

Optimized *e*-biochar have the potential of being more efficient and versatile (with customized fabrication design and properties), including the characteristics of both structural features of a current collector (defined conformation, mechanical rigidity, conductivity) and high electroactive surface area. Other crucial factors for the development and scaling up of METs are the performance and costs of separators as well as the structural frame (Li et al. 2011).

In a reasonable number of reports, electrodes designs were based on biochar monoliths with self-supported 3-dimensional structures (Table 4). To create an air-water interface in air-exposed cathodes, for instance, porous materials with intrinsic structural rigidity acting as separators are needed to resist water pressure under given depths. The addition of rigid separators (which does not necessarily act as membranes) between electrodes has the advantage of adding a self-structured element and due to its own presence, conveys rigidity to the overall system. Other examples are sponge-like rigid frameworks with high macroporous volumetric surface area and roughness that are ideal for 3D bioanodic surfaces.

Good examples are binder-free air-cathodes made of sintered activated carbon powders that were found to be inexpensive and easily mass manufactured (Walter et al. 2018). Other authors fabricated bioelectrodes via ligno-cellulosic biomass carbonization, while preserving the original 3D shape. Bioanodes were obtained from carbonized plant stems (kenaf and bamboo), corn cobs, marine loofah sponges, king/wild mushrooms (Chen et al. 2012a; Chen et al. 2012b; Li et al. 2014a; Yuan et al. 2013a). Rigid air-breathing biocathodes were obtained from giant cane stems that maintained their original cylindrical shape and simultaneously acted as microporous air-water separators (Marzorati et al. 2017, 2018).

Bioelectrodes or electroactive beds configurations even closer to scaled up MET applications were also proposed, even if in some cases the employed carbon materials (e.g., activated carbon granules) were from origins other than biomass (Table 4). Fluid-like bioelectrodes made of floating carbon particles are very promising configurations (Tejedor-Sanz et al. 2017a). Electroconductive granules are fluidized in a bioreactor acting as 'planktonic' electrodes supporting microbial electroactivity (Tejedor-Sanz et al. 2017b). Electrons accumulate on the material and are discharged to a collector by periodic contact. The potentials of this concept are currently under investigation at Bioe group (University of Alcalá, Spain).

Granular carbons of macroscopic size and sufficient mechanical rigidity (diameters in the range of 5-20 mm) were also the base of fixed-bed or packed-bed bioelectrodes (with either anodic or cathodic configuration) for different purposes (Ghafari et al. 2009; Liu et al. 2018; Rodrigo Quejigo et al. 2018). This design has been applied to probably the largest-scale application of MET that merges the use of electroactive material with the concept of constructed wetland. The result is the so-called 'METland' concept where the classically constructed wetland with bed biofilter made of inert material can be substituted by electro-active materials (Aguirre-Sierra et al. 2016). Based on this concept, a 20m² METland made of *e*-biochar was constructed for treating around 7 m³/day of urban wastewater. The *e*-biochar acts as an electroconductive bed for electroactive biofilm and helps in avoiding electron acceptor limitations for bacteria. The final outcome is a stimulation of the ET mechanism that resulted in a large enhancement of the biodegradation rates for organic pollutants in the wastewater with no energy cost and under extremely low growth yield (Aguirre-Sierra et al. 2016).

METland biofilters made of *e*-biochar have been also used at large-scale for enhancing anaerobic treatment in a real-scale wastewater treatment plant (serving a community of around 200 people) recently constructed by the startup company METfilter at Otos (Murcia, Spain). Interestingly, in METlands *e*-biochar was considerably more efficient for wastewater treatment than more conductive materials (coke and graphite) (Prado et al. 2018). Higher performances were observed under a wide range of operational conditions, including polarization at 0.4 and 0.6 V (vs. Ag/AgCl/Cl⁻ ref.) (Prado et al. 2018). Higher

Table 4: Self-standing 3D-shaped e-biochar bioelectrodes and potentially-scalable bioelectrodes configurations (some studies were based on non-biogenic carbons)

Precursor feedstock	Pyrolysis conditions and treatments	Particle size or geometrical shape	BET* Surface area/m² g⁻¹	Porosity range	Electrical conductivity/ mS cm⁻¹	Surface heteroatoms	Function in electrodes and observed effects	References
Pinewood sawdust pellets and chips	1000°C	26-700 mm³	0.04	Small mesopores	16-35	n.r.	Granular bioanodes in MFCs	(Huggins et al. 2014)
Giant cane (*Arundo donax L.*) stalks	900°C	Natural cylindrical shape (10 mm diameter, 10-20 cm length)	114	Micropores	11	N, P	Air-breathing biocathodes in METs, acting as self-structured air-water porous interface, with cylindrical shape	(Marzorati et al. 2017, 2018)
Kenaf (*Hibiscus cannabinus*)	1000°C	Natural cylindrical shape 4 mm/10 mm inner/outer diameters	n.r.	Macro-channels of 50-60 μm	n.r.	n.r.	Studied as bioanodes	(Chen et al. 2012a)
Bamboo	1000°C	Tubes with inner diameter: 1 mm, 1.5 mm, 2 mm and 3 mm	n.r.	n.r.	n.r.	n.r.	Bioanodes for microbial fuel cells.	(Li et al. 2014)
Pomelo peel	900°C	Sponge-like architecture	n.r.	Macroporous architecture	n.r.	n.r.	High-performance anode in microbial fuel cells	(Chen et al. 2012b)
King mushroom, wild mushroom and corn stem	1000°C	Sponge-like architecture	n.r.	Macroporous architecture	n.r.	n.r.	Anodes for microbial fuel cells.	(Karthikeyan et al. 2015)
Natural Loofah Sponge	900°C	Sponge-like architecture	445-504	Macroporous architecture	n.r.	N	Anodes for microbial fuel cells.	(Yuan et al. 2013b)

(Contd.)

BET* Surface area/m² g⁻¹ — where the asterisk notes the BET method for surface area measurement.

Material		Size					Description	Reference
Graphite granules fixed bed	-	2-3.5 mm	n.r.	n.r.	n.r.	n.r.	Measurements on single graphite granules in bed electrodes by cyclic voltammetry	(Rodrigo Queijigo et al. 2018)
Graphite particles/glassy carbon particles	-	0.42 to 0.69 mm/0.63 to 1 mm	n.r.	n.r.	n.r.	n.r.	Studies on electron transfer mechanisms in microbial electrochemical fluidized bed reactor	(Tejedor-Sanz et al. 2017a, 2017b)
unspecified	n.r.	Granules of 1–3 mm diameter	764	n.r.	n.r.	n.r.	Granular carbon biocathodes in methane-producing MET	(Liu et al. 2018)
Coke granules fixed bed	-	5-20 mm	n.r.	n.r.	n.r.	n.r.	METlands as integration of microbial electrochemical technologies with natural wastewater treatment biofilters	(Aguirre-Sierra et al. 2016)

working potentials showed higher currents for graphite but overall lower COD removal efficiency as compared to *e*-biochar. Hence, larger surface area, hierarchical pore architecture and richer electroactive surface chemistry including phenol and quinones might dominate over conductivity in some systems. Such aspects deserve deeper investigation.

6. Conclusions and Future Perspectives

Although efforts have been made to characterize the redox properties of biochars, microbial electrogenic reactions constitute quite complex processes that are difficult to characterize under abiotic conditions, so the state-of-the-art analytical methods should be reconsidered. Particularly, these methods should pay special attention to the characteristic slow kinetics of the ET processes driven by electrogenic bacteria.

The use of *e*-biochar in the METs context may represent an effective strategy to decrease costs and the environmental impact in the fabrication of electrodes process. The projections in the field towards the METs scaling up depend on the replacement of expensive components. The costs decrease is achievable, for instance, considering the local availability of biomass residues to be transformed into *e*-biochar, avoiding feedstock purchasing, extraction and transportations. The scales, characteristics and performances of carbonization facilities will play a crucial role in the sustainability and feasibility of these technological applications.

To address these issues, life cycle assessment studies should accompany basic research.

References

Aguirre-Sierra A, Bacchetti-De Gregoris T, Berná A, Salas JJ, Aragón C, Esteve-Núñez A (2016) Microbial electrochemical systems outperform fixed-bed biofilters in cleaning up urban wastewater. Environ Sci Water Res Technol 2: 984–993.

Ahmad M, Rajapaksha AU, Lim JE, Zhang M, Bolan N, Mohan D, Vithanage M, Lee SS, Ok YS (2014) Biochar as a sorbent for contaminant management in soil and water: A review. Chemosphere 99: 19–23.

Atkinson CJ, Fitzgerald JD, Hipps NA (2010) Potential mechanisms for achieving agricultural benefits from biochar application to temperate soils: a review. Plant Soil 337: 1–18.

Banek NA, Abele DT, McKenzie KR, Wagner MJ (2018) Sustainable Conversion of Lignocellulose to High-Purity, Highly Crystalline Flake Potato Graphite. ACS Sustain. Chem Eng 6: 13199–13207.

Béguin F, Raymundo-Piñero E (2013). Nanocarbons for supercapacitors. *In*: Ralph J. Brodd (Ed.). Batteries for Sustainability. Springer New York, New York, NY, pp. 393–421. doi:10.1007/978-1-4614-5791-6_12

Berenguer R, García-Mateos FJ, Ruiz-Rosas R, Cazorla-Amorós D, Morallón E, Rodríguez-Mirasol J, Cordero T (2016) Biomass-derived binderless fibrous carbon electrodes for ultrafast energy storage. Green Chem 18: 1506–1515.

Biniak S, Swiatkowski A, Pakuta M (2001) Electrochemical studies of phenomena at active carbon-electrolyte solution interfaces. *In*: Ljubisa R. Radovic (Ed.). Chemistry & Physics of Carbon. CRC Press, pp. 125–225.

Brussaard L, de Ruiter PC, Brown GG (2007) Soil biodiversity for agricultural sustainability. Agric Ecosyst Environ 121: 233–244.

Chacón FJ, Cayuela ML, Roig A, Sánchez-Monedero MA (2017) Understanding, measuring and tuning the electrochemical properties of biochar for environmental applications. Rev Environ Sci Bio/Technology 16: 695–715.

Chen G, Zhang Z, Zhang Z, Zhang R (2018) Redox-active reactions in denitrification provided by biochars pyrolyzed at different temperatures. Sci Total Environ 615: 1547–1556.

Chen Q, Pu W, Hou H, Hu J, Liu B, Li J, Cheng K, Huang L, Yuan X, Yang C, Yang J (2018) Activated microporous-mesoporous carbon derived from chestnut shell as a sustainable anode material for high performance microbial fuel cells. Bioresour. Technol. 249: 567–573.

Chen S, He G, Hu X, Xie M, Wang S, Zeng D, Hou H, Schröder U (2012a) A three-dimensionally ordered macroporous carbon derived from a natural resource as anode for microbial bioelectrochemical systems. ChemSusChem 5: 1059–1063.

Chen S, He G, Liu Q, Harnisch F, Zhou Y, Chen Y, Hanif M, Wang S, Peng X, Hou H et al. (2012c). Layered Corrugated Electrode Macrostructures Boost Microbial Bioelectrocatalysis. Energy Environ. Sci. 5(12): 9769. https://doi.org/10.1039/c2ee23344d.

Chen S, Liu Q, He G, Zhou Y, Hanif M, Peng X, Wang S, Hou H (2012b) Reticulated carbon foam derived from a sponge-like natural product as a high-performance anode in microbial fuel cells. J Mater Chem 22: 18609.

Chen S, Rotaru A-E, Shrestha PM, Malvankar NS, Liu F, Fan W, Nevin KP, Lovley DR (2014) Promoting interspecies electron transfer with biochar. Sci Rep 4: 5019.

Chen S, Tang J, Fu L, Yuan Y, Zhou S (2016) Biochar improves sediment microbial fuel cell performance in low conductivity freshwater sediment. J Soils Sediments 16: 2326–2334.

Chen Z, Wang Y, Xia D, Jiang X, Fu D, Shen L, Wang H, Li QB (2016) Enhanced bioreduction of iron and arsenic in sediment by biochar amendment influencing microbial community composition and dissolved organic matter content and composition. J Hazard Mater 311: 20–29.

Conway BE (1999). Electrochemical Supercapacitors: Scientific Fundamentals and Technological Applications. Springer US.

Conway BE (1991a) Transition from "Supercapacitor" to "Battery" Behavior in Electrochemical Energy Storage. J Electrochem Soc 138: 1539.

Conway BE (1991b) Transition from "Supercapacitor" to "Battery" Behavior in Electrochemical Energy Storage. J Electrochem Soc 138: 1539.

Fitzer E, Kochling K-H, Boehm HP, Marsh H (1995) Recommended terminology for the description of carbon as a solid (IUPAC Recommendations 1995). Pure Appl Chem 67: 473–506.

Gao Z, Zhang Y, Song N, Li X (2017) Biomass-derived renewable carbon materials for electrochemical energy storage. Mater Res Lett 5: 69–88. doi:10.1080/21663831.2016.1250834

García-Mateos FJ, Cordero-Lanzac T, Berenguer R, Morallón E, Cazorla-Amorós D, Rodríguez-Mirasol J, Cordero T (2017) Lignin-derived Pt supported carbon (submicron)fiber electrocatalysts for alcohol electro-oxidation. Appl Catal B Environ 211: 18–30.

Ghafari S, Hasan M, Aroua MK (2009) Nitrate remediation in a novel upflow bio-electrochemical reactor (UBER) using palm shell activated carbon as cathode material. Electrochim Acta 54: 4164–4171.

Grossman JM, O'Neill BE, Tsai SM, Liang B, Neves E, Lehmann J, Thies JE (2010) Amazonian anthrosols support similar microbial communities that differ distinctly from those extant in adjacent, unmodified soils of the same mineralogy. Microb Ecol 60: 192–205.

Huggins T, Wang H, Kearns J, Jenkins P, Ren ZJ (2014) Biochar as a sustainable electrode material for electricity production in microbial fuel cells. Bioresour Technol 157: 114–119.

Huggins TM, Pietron JJ, Wang, H., Ren, Z.J., Biffinger, J.C., 2015. Graphitic biochar as a cathode electrocatalyst support for microbial fuel cells. Bioresour Technol 195: 147–153. doi:10.1016/J.BIORTECH.2015.06.012

Imhoff PT, Chiu PC (2015) Biochar as a Rechargeable Geobattery to Promote Nitrogen Removal in Stormwater from Roadways. TRB annual meeting. Available at https://cait.rutgers.edu/files/CAIT-UTC-061-final.pdf

Ioannidou O, Zabaniotou A (2007) Agricultural residues as precursors for activated carbon production-A review. Renew Sustain Energy Rev 11(9): 1966-2005.

Jeffery S, Verheijen FGA, van der Velde M, Bastos AC (2011) A quantitative review of the effects of biochar application to soils on crop productivity using meta-analysis. Agric Ecosyst Environ 144: 175–187.

Kambo HS, Dutta A (2015) A comparative review of biochar and hydrochar in terms of production, physico-chemical properties and applications. Renew Sustain Energy Rev 45: 359–378.

Kappler A, Wuestner ML, Ruecker A, Harter J, Halama M, Behrens S (2014) Biochar as an Electron Shuttle between Bacteria and Fe(III) Minerals. Environ Sci Technol Lett 1: 339–344.

Karthikeyan R, Wang B, Xuan J, Wong JWC, Lee PKH, Leung MKH (2015) Interfacial electron transfer and bioelectrocatalysis of carbonized plant material as effective anode of microbial fuel cell. Electrochim Acta 157: 314–323.

Khodadad CLM, Zimmerman AR, Green SJ, Uthandi S, Foster JS (2011) Taxa-specific changes in soil microbial community composition induced by pyrogenic carbon amendments. Soil Biol Biochem 43: 385–392.

Klüpfel L, Keiluweit M, Kleber M, Sander M (2014) Redox properties of plant biomass-derived black carbon (biochar). Environ Sci Technol 48: 5601–5611.

Komang T R-S, Caroline HO (2016) Biochar Application Essential Soil Microbial Ecology. Elsevier Science. https://books.google.it/books?id=Q5bBCQAAQBAJ&dq=Biochar+Application+Essential | Soil+Microbial+Ecology&lr=&source=gbs_navlinks_s

Lee SH, Kim EY, Park H, Yun J, Kim JG (2011) In situ stabilization of arsenic and metal-contaminated agricultural soil using industrial by-products. Geoderma 161: 1–7.

Lehmann J, Gaunt J, Rondon M (2006) Bio-char sequestration in terrestrial ecosystems – A review. Mitig Adapt Strateg Glob Chang 11: 403–427.

Lehmann J, Joseph S (2015) Biochar for Environmental Management: Science, Technology and Implementation. Routledge. https://books.google.it/books?id=7V_ABgAAQBAJ&dq=Biochar+for+Environmental+Man agement:+Science,+Technology+and+Implementation&source=gbs_navlinks_s

Lehmann J, Rillig MC, Thies J, Masiello CA, Hockaday WC, Crowley D (2011) Biochar effects on soil biota – A review. Soil Biol Biochem 43: 1812–1836.

Li Chum H, Baizer MM (1985) Electrochemistry of biomass and derived materials. ACS (ACS Monogr 183), Washington, DC, USA.

Li J, Zhang J, Ye D, Zhu X, Liao Q, Zheng J (2014) Optimization of inner diameter of tubular bamboo charcoal anode for a microbial fuel cell. Int J Hydrogen Energy 39: 19242–19248.

Li M, Zhang H, Xiao T, Wang S, Zhang B, Chen D, Su M, Tang J (2018) Low-cost biochar derived from corncob as oxygen reduction catalyst in air cathode microbial fuel cells. Electrochim Acta 283: 780–788.

Li W-W, Sheng G-P, Liu X-W, Yu H-Q (2011) Recent advances in the separators for microbial fuel cells. Bioresour Technol 102: 244–252.

Libra JA, Ro KS, Kammann C, Funke A, Berge ND, Neubauer Y, Titirici M-M, Fühner C, Bens O, Kern J, Emmerich, K-H (2011) Hydrothermal carbonization of biomass residuals: a comparative review of the chemistry, processes and applications of wet and dry pyrolysis. Biofuels 2: 71–106.

Lin X, Wang X, Li L, Yan M, Tian Y (2017) Rupturing cotton microfibers into mesoporous nitrogen-doped carbon nanosheets as metal-free catalysts for efficient oxygen electroreduction. ACS Sustain Chem Eng 5: 9709–9717.

Liu D, Roca-puigros M, Geppert F, Caizán-juanarena L (2018) Granular carbon-based electrodes as cathodes in methane-producing bioelectrochemical Systems 6: 1–10.

Liu Y, He P, Shao L, Zhang H, Lü F (2017) Significant enhancement by biochar of caproate production via chain elongation. Water Res 119: 150–159.

Logan BE (2010) Scaling up microbial fuel cells and other bioelectrochemical systems. Appl Microbiol Biotechnol 85: 1665–1671.

Lovley DR, Coates JD, Blunt-Harris EL, Phillips EJP, Woodward JC (1996) Humic substances as electron acceptors for microbial respiration. Nature 382: 445–448.

Lü F, Luo C, Shao L, He P (2016) Biochar alleviates combined stress of ammonium and acids by firstly enriching Methanosaeta and then Methanosarcina. Water Res 90: 34–43.

Ma M, You S, Wang W, Liu G, Qi D, Chen X, Qu J, Ren N (2016) Biomass-derived porous Fe3C/tungsten carbide/graphitic carbon nanocomposite for efficient electrocatalysis of oxygen reduction. ACS Appl Mater Interfaces 8: 32307–32316.

Marzorati S, Goglio A, Fest-Santini S, Mombelli D, Villa F, Cristiani P, Schievano A (2018) Air-breathing bio-cathodes based on electro-active biochar from pyrolysis of Giant Cane stalks. Int J Hydrogen Energy 44: 4496–4507.

Marzorati S, Goglio A, Mombelli D, Mapelli C, Trasatti SP, Cristiani P, Schievano A (2017) Giant Cane as Low-cost Material for Microbial Fuel Cells Architectures. *In*: Proceedings of EFC2017 European Fuel Cell Technology & Applications Conference - Piero Lunghi Conference December 12-15, 2017, Naples, Italy. p 114.

Masiello CA, Chen Y, Gao, X, Liu S, Cheng HY, Bennett MR, Rudgers JA, Wagner DS, Zygourakis K, Silberg JJ (2013) Biochar and microbial signaling: Production conditions determine effects on microbial communication. Environ Sci Technol 47: 11496–11503.

Mendonça FG, Cunha IT, Soares RR, Tristão JC, Lago RM (2017) Tuning the surface properties of biochar by thermal treatment. Bioresour Technol 246: 28–33.

Nieva Lobos ML, Sieben JM, Comignani V, Duarte M, Volpe MA, Moyano EL (2016) Biochar from pyrolysis of cellulose: An alternative catalyst support for the electro-oxidation of methanol. Int J Hydrogen Energy 41: 10695–10706.

O'Neill B, Grossman J, Tsai MT, Gomes JE, Lehmann J, Peterson J, Neves E, Thies JE (2009) Bacterial community composition in Brazilian Anthrosols and adjacent soils characterized using culturing and molecular identification. Microb Ecol 58: 23–35.

Pant D, Van Bogaert G, Porto-Carrero C, Diels L, Vanbroekhoven K (2011) Anode and cathode materials characterization for a microbial fuel cell in half cell configuration. Water Sci Technol 63: 2457.

Prado A, Berenguer R, Esteve-Núñez A (2018) Electroconductive carbon biofilters for efficient and sustainable treatment of urban wastewater, *In*: CARBON 2018. The World Conference on Carbon. p. 2.

Prévoteau A, Ronsse F, Cid I, Boeckx P, Rabaey K (2016) The electron donating capacity of biochar is dramatically underestimated. Sci Rep 6: 32870.

Qian K, Kumar A, Zhang H, Bellmer D, Huhnke R (2015) Recent advances in utilization of biochar. Renew Sustain Energy Rev 42: 1055-1064.

Qin Y, Wang H, Li X, Cheng JJ, Wu W (2017) Improving methane yield from organic fraction of municipal solid waste (OFMSW) with magnetic rice-straw biochar. Bioresour Technol 245: 1058–1066.

Rago L, Cristiani P, Villa F, Zecchin S, Colombo A, Cavalca L, Schievano A (2017) Influences of dissolved oxygen concentration on biocathodic microbial communities in microbial fuel cells. Bioelectrochemistry 116: 39–51.

Rago L, Zecchin S, Marzorati S, Goglio A, Cavalca L, Cristiani P, Schievano A (2018) A study of microbial communities on terracotta separator and on biocathode of air breathing microbial fuel cells. Bioelectrochemistry 120: 18–26.

Reddy KR, Yargicoglu EN, Yue D, Yaghoubi P (2014) Enhanced Microbial Methane Oxidation in Landfill Cover Soil Amended with Biochar. J Geotech Geoenvironmental Eng 140: 0401-4047.

Rodrigo Quejigo J, Rosa LFM, Harnisch F (2018) Electrochemical characterization of bed electrodes using voltammetry of single granules. Electrochem Commun 90: 78–82.

Rodríguez-Reinoso F (1986) Preparation and characterization of activated carbons. *In*: José L. Figueiredo, Jacob A. Moulijn (Eds.). Carbon and Coal Gasification. Springer Netherlands, Dordrecht, pp. 601–642.

Rondon MA, Lehmann J, Ramírez J, Hurtado M (2007) Biological nitrogen fixation by common beans (Phaseolus vulgaris L.) increases with bio-char additions. Biol Fertil Soils 43: 699–708.

Rosas JM, Bedia J, Rodríguez-Mirasol J, Cordero T (2010) On the preparation and characterization of chars and activated carbons from orange skin. Fuel Process Technol 91: 1345–1354.

Rosas JM, Berenguer R, Valero-Romero MJ, Rodriguez-Mirasol J, Cordero T (2014) Preparation of different carbon materials by thermochemical conversion of lignin. Front Mater 1: 29.

Schievano A, Pepé Sciarria T, Vanbroekhoven K, De Wever H, Puig S, Andersen SJ, Rabaey K, Pant D (2016) Electro-fermentation – Merging electrochemistry with fermentation in industrial applications. Trends Biotechnol 34: 866–878.

Simpson DR (2008) Biofilm processes in biologically active carbon water purification. Water Res 42(12): 2839-2848.

Sun T, Levin BDA, Guzman JJL, Enders A, Muller DA, Angenent LT, Lehmann J (2017) Rapid electron transfer by the carbon matrix in natural pyrogenic carbon. Nat Commun 8: 14873.

Tejedor-Sanz S, Ortiz JM, Esteve-Núñez A (2017a) Merging microbial electrochemical systems with electrocoagulation pretreatment for achieving a complete treatment of brewery wastewater. Chem Eng J 330: 1068–1074.

Tejedor-Sanz S, Quejigo JR, Berná A, Esteve-Núñez A (2017b) The planktonic relationship between fluid-like electrodes and bacteria: Wiring in motion. ChemSusChem 10: 693–700.

Tong H, Hu M, Li FB, Liu CS, Chen MJ (2014) Biochar enhances the microbial and chemical transformation of pentachlorophenol in paddy soil. Soil Biol Biochem 70: 142–150.

Walter XA, Greenman J, Ieropoulos I (2018) Binder materials for the cathodes applied to self-stratifying membraneless microbial fuel cell. Bioelectrochemistry 123: 119–124.

Wang S, Zheng Y, Yan W, Chen L, Dummi Mahadevan G, Zhao F (2016) Enhanced bioleaching efficiency of metals from E-wastes driven by biochar. J Hazard Mater 320: 393–400.

Waqas M, Aburiazaiza AS, Miandad R, Rehan M, Barakat MA, Nizami AS (2018) Development of biochar as fuel and catalyst in energy recovery technologies. J Clean Prod 188: 477–488.

Yanai Y, Toyota K, Okazaki M (2007) Effects of charcoal addition on N_2O emissions from soil resulting from rewetting air-dried soil in short-term laboratory experiments: Original article. Soil Sci Plant Nutr 53: 181–188.

Yu L, Wang Y, Yuan Y, Tang J, Zhou S (2016) Biochar as electron acceptor for microbial extracellular respiration. Geomicrobiol J 33: 530–536.

Yu L, Yuan Y, Tang J, Wang Y, Zhou S (2015) Biochar as an electron shuttle for reductive dechlorination of pentachlorophenol by Geobacter sulfurreducens. Sci Rep 5: 16221.

Yuan H, Deng L, Qi Y, Kobayashi N, Tang J (2014) Nonactivated and activated biochar derived from bananas as alternative cathode catalyst in microbial fuel cells. Sci World 2014: 832–850.

Yuan S-J, Dai X-H, Jaroniec M, Qiao SZ, Zhang DYH, Che RC, Tang Y, Su DS, Asiri AM, Zhao DY, Lin YH (2016) An efficient sewage sludge-derived bi-functional electrocatalyst for oxygen reduction and evolution reaction. Green Chem 18: 4004–4011.

Yuan Y, Bolan N, Prévoteau A, Vithanage M, Biswas JK, Ok YS, Wang H (2017) Applications of biochar in redox-mediated reactions. Bioresour Technol 246: 271–281.

Yuan Y, Yuan T, Wang D, Tang J, Zhou S (2013a) Sewage sludge biochar as an efficient catalyst for oxygen reduction reaction in an microbial fuel cell. Bioresour Technol 144: 115–120.

Yuan Y, Zhou S, Liu Y, Tang J (2013b) Nanostructured Macroporous Bioanode based on polyaniline-modified natural loofah sponge for high-performance microbial fuel cells. Environ Sci Technol 47: 14525–14532.

Zha DW, Li LF, Pan YX, He JB (2016) Coconut shell carbon nanosheets facilitating electron transfer for highly efficient visible-light-driven photocatalytic hydrogen production from water. Int J Hydrogen Energy 41: 17370–17379.

PART-V

Materials for Microbial Electrochemical Technologies

Toward a Sustainable Biocatalyst for the Oxygen Reduction Reaction in Microbial Fuel Cells

Jean-Marie Fontmorin*, Edward M. Milner and Eileen H. Yu

School of Engineering, Newcastle University, Newcastle upon Tyne, United Kingdom

1. Introduction

In the past 10-15 years, microbial fuel cells (MFCs) have captured the attention of the scientific community for the possibility of degrading organic waste directly into electricity. Indeed, MFCs are based on the oxidation of organic matter by bacteria at the anode and on the reduction of a terminal electron acceptor at the cathode (Logan 2009). The electrons resulting from the oxidation of the organic matter flow toward the cathode through an external circuit whereas ions migrate in solution across the membrane, thus generating electrical power. The virtually inexhaustible availability of oxygen and its high value of standard equilibrium potential ($E^0 = 1.229$ V vs. Standard Hydrogen Electrode; Equation 1) make it an ideal terminal electron acceptor for redox systems.

$$O_2 + 4H^+ + 4e^- \rightarrow 2H_2O \tag{1}$$

However, the kinetics of the oxygen reduction reaction (ORR) is sluggish, associated with high overpotentials and catalysts are rare, which is why the ORR is still a bottleneck for the development of MFCs. From a more general point of view, fast kinetics for ORR would imply various spontaneous oxidation reactions, like fast corrosion of metallic materials or fast oxidative deterioration of non-metallic materials. In addition, fast kinetics and abundance of ORR catalysts would also increase the exposition of living organisms to reactive oxygen species (ROS) resulting from oxygen reduction, therefore, also increased oxidative stress responsible of aging and death of human beings in particular (Erable et al. 2012). To date, the performance of the cathode is still considered to be the main limitation to the practical development of MFCs, and one of the main challenges is to develop cheap, efficient, stable and sustainable catalysts for the ORR (Sawant et al. 2017). Catalysts for the ORR can be classified into three categories: enzymatic, chemical and microbial catalysts. With enzymatic systems, oxygen may in principle be reduced at the redox potential of the enzyme itself. For that matter, laccase from *Trametes versicolor* offers astonishing properties since its redox potential is very close to that of the O_2/H_2O redox couple. However, it presents little activity at neutral pH (Lapinsonniere et al. 2012). In addition, enzymes are sensitive to any kind of inhibition; they are expensive and their lifetime usually does not exceed a few days in operating conditions. Their immobilization on electrode surfaces requires sophisticated chemical

---

*Corresponding author: jean-marie.fontmorin@ncl.ac.uk

operations, and the electrical communication between the electrode substrate and the enzyme is also a limitation (Erable et al. 2012; Leech et al. 2012). For these reasons, it is safe saying that enzymatic catalysts are more suitable for small and disposable devices. Chemical catalysts have been extensively studied with Platinum (Pt) still being the most commonly used catalyst although it is not a sustainable option. Indeed, Pt is rare and, therefore, expensive; Pt-based catalysts are easily poisoned by carbon monoxide (CO) or contaminants present in wastewater, such as sulfides (Sopian and Daud 2006). Lately, carbon-based catalysts have been the most promising in terms of cost-effectiveness, such as activated carbon (AC) (Zhang et al. 2014). However, none has been commercially developed yet and studies show how much work there is yet to be done to develop a viable catalyst. Cost-efficiency, robustness, durability and environmental sustainability are the requirements for an ideal catalyst for ORR. In this context, what is the place for biocatalysts? Can aerobic biocathodes be reasonably considered as viable alternatives for ORR? Biocathodes offer several advantages such as being completely free, environmentally sustainable, robust and scalable without the need to be manufactured although little is currently understood regarding the exact mechanism occurring (Milner 2015). In recent years, most literature reviews have focused on the abiotic oxygen-reducing cathodes, especially on non-Pt catalysts (Sawant et al. 2017; Stacy et al. 2017; Yuan et al. 2016). This chapter is fully dedicated to the recent advances made on the understanding and the development of aerobic biocathodes. The knowledge gained on microbial communities associated with the ORR will be presented as well as the mechanisms involved, the performance that can be achieved and the impact of operational parameters.

2. The Oxygen Reduction Reaction in a Nutshell

Although oxygen is the best candidate for terminal electron acceptor in many redox systems, its slow kinetics of reduction and thus the development of low-cost, stable and sustainable catalysts remain the main challenges related to the development of MFCs.

In aqueous solutions, the ORR can occur according to two different pathways: the two-electron reaction leading to hydrogen peroxide and the four-electron reaction leading to water (Allen J. Bard 2001). In realistic conditions for MFCs (neutral pH and partial O_2 pressure of 0.2 atm), these two pathways, which are pH-dependant, are described by the following equations (Song and Zhang 2008):

$$O_2 + 2e^- + 2H^+ \rightleftharpoons H_2O_2 \tag{2}$$
$$E^{0\prime} = 0.13 \text{ V (pO}_2 = 0.2 \text{ atm, pH} = 7, [H_2O_2] = 5 \text{ mM)}$$

$$O_2 + 4e^- + 4H^+ \rightleftharpoons 2H_2O \tag{3}$$
$$E^{0\prime} = 0.60 \text{ V (pO}_2 = 0.2 \text{ atm, pH} = 7)$$

In MFCs, the four-electron pathway producing water will be preferred since double the number of electrons is transferred with the utilization of half of the amount of reactant (oxygen). However, most of MFC systems and more generally BES work around neutral pH that implies a lower concentration of H^+ and OH^- which are the main drivers for the ORR (Santoro et al. 2017). The two-electron pathway leads to the generation of H_2O_2 and free radicals that can cause degradation of the membrane and corrosion of the cathode material (Roche et al. 2010). Moreover, on carbon materials, which are the most commonly used electrode substrates in MFCs, the production of H_2O_2 predominates and justifies the need to develop catalysts for which the four-electron pathway predominates (Scott and Yu 2015). Considering that MFCs were primarily designed for wastewater treatment, ideal ORR catalysts are expected to be robust, durable, cost-effective, environmentally sustainable in addition to offering high catalytic activity.

3. Aerobic Biocathodes for the ORR

The microbial catalysis of ORR is a spontaneous phenomenon that occurs when the surface of metallic materials are exposed to air in natural environments (Erable et al. 2012). It has been studied for many years in the field of marine corrosion as it is one of the main motors of aerobic microbial corrosion. The phenomenon is called microbially influenced corrosion (MIC); for instance, when a metallic surface such as stainless steel is exposed to aerobic seawater, microorganisms can grow as a biofilm on their surface and increase the rate of the reduction of oxygen, subsequently increasing the potential of corrosion onset ('ennoblement of free potential') and inducing fast propagation of localized corrosion (Bergel et al. 2005). The first microbial oxygen reducing cathodes was developed in 1997 when it was reported that current densities of around 0.2 A m^{-2} could be produced by polarizing stainless steel electrodes at a potential of -0.2 V vs. saturated calomel electrode (SCE) in seawater (Mollica et al. 1997). The knowledge gained in the field of marine corrosion and marine biofilms was further used to design low-cost cathodes for oxygen reduction in fuel cells and later in MFCs. In 2007, Clauwaert et al. reported that it was possible to combine an acetate-oxidizing bioanode with an aerobic biocathode in an MFC. The power density of 83 ± 11 W m^{-3} MFC in the batch-fed system was achieved, demonstrating the potential of microorganisms as ORR catalysts in MFCs (Clauwaert et al. 2007). At the time, little was known about the mechanisms occurring and about the community involved, but the performance recorded was already competitive to chemical catalysts. Bacterial communities were involved in the ORR in aerobic biocathodes.

3.1 Bacterial Communities Involved in Aerobic Biocathodes

3.1.1 Aerobic Chemolithotrophic Bacteria

Similarly to the oxidation of the organic matter by bacteria at the anode, the catalysis of ORR at MFC cathodes by bacteria can occur directly or indirectly. In the first case, bacteria may directly use the electrode as the energy source. Chemolithotrophs bacteria are known for utilizing inorganic reduced compounds as a source of energy for metabolism and growth. They can be aerobic or anaerobic and most of them are also autotrophic, thus they are able to fix CO_2 through the Calvin cycle (Prescott et al. 2005). Aerobic chemolithotrophic bacteria associate substrate oxidation with oxygen reduction that they use as a terminal electron acceptor that is coupled with the generation of proton motive force and ATP. They are usually classified according to the substrate they oxidize, and we can distinguish iron oxidizing bacteria (IOB), sulfur oxidizing bacteria (SOB), ammonia oxidizing bacteria (AOB), nitrite oxidizing bacteria (NOB), manganese oxidizing bacteria (MOB) and hydrogen oxidizing bacteria (HOB) (Milner 2015).

Therefore, in environments containing manganese or iron ions, MOB and IOB are able to oxidize these compounds to MnO_2 and Fe^{3+}, respectively, that are in turn reduced back at the surface of the electrode. Oxygen is used as a terminal electron acceptor by the MOBs and IOBs, and in this case, Mn and Fe ions act as electron shuttles from the bacteria to the electrode and the catalysis of ORR is considered as 'indirect' or 'mediated'. The feasibility of this process is based on the fact that the reduction kinetics of both MnO_2 and Fe^{3+} at plain non-catalyzed electrodes are faster than that of the ORR (He and Angenent 2006; Rhoads et al. 2005). Ter Heijne et al. have studied the feasibility of an MFC using the IOB *Acidithiobacillus ferrooxidans* as a biocatalyst (ter Heijne et al. 2007; ter Heijne et al. 2006; ter Heijne et al. 2011a). By using the microorganisms in solution, power and current densities of 0.86 W m^{-2} and 4.5 A m^{-2}, respectively, were initially obtained. The further immobilization of the bacteria on biomass support particles (BSPs) improved the power density to 1.2 W m^{-2} for a similar current density. The schematic principles of this MFC are depicted in Figure 1. One of the challenges, however, is the necessity to maintain the pH of the catholyte around 2 in order to keep the ferric iron soluble. In addition, it should be noted that the constant aeration used in this study is a major limitation to the development of such a technology and would significantly reduce its economic viability. Nevertheless, the system was later scaled up to a 5-L reactor with a 0.5 m^2 cathode surface area (ter Heijne et al. 2011a). The utilization of Fe^{2+} at 6 g L^{-1} led to a maximum power density of 2 W m^{-2}. However, the cathode remained the main limiting factor as it contributed to 58% of the total internal resistance.

In the studies mentioned previously, the catalysis of ORR was mediated by the redox couple Fe^{2+}/Fe^{3+}, Fe^{2+} being biologically oxidized and Fe^{3+} being reduced back at the surface of the electrode. However, it was also reported that certain strains of IOB are able to use the electrode as the sole electron donor for energy and growth. For instance, *Acidithiobacillus ferrooxidans* was grown at pH 2 in a half-cell

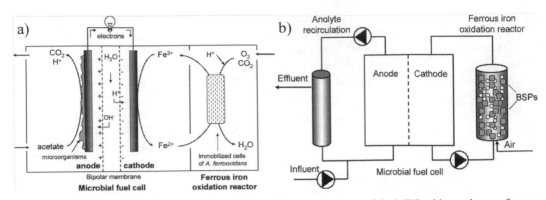

Figure 1: a) MFC with continuous ferrous oxidation, b) Process scheme of the MFC with continuous ferrous oxidation. Reprinted with permission from (ter Heijne et al. 2007). Copyright © American Chemical Society.

reactor at a potential of 0 V vs. SCE with the electrode as a sole energy source and without the addition of any mediator. Current densities up to 5 mA cm^{-2} were measured, and the onset potential for the ORR had shifted from 300 mV to a higher potential compared to a plain graphite electrode used as a control (Carbajosa et al. 2010). Similarly, *Mariprofundus ferrooxydans* PV-1, a neutraphilic obligate Fe(II)-oxidizing autotroph, was cultured using a poised electrode as the sole energy source (Summers et al. 2013). The strain was first enriched in a Fe-containing medium before being transferred to a 3-electrode cell in which an increasing cathodic current was recorded over a few weeks with an onset potential for ORR of approximately -0.2 V determined by cyclic voltammetry (CV).

The utilization of MOB to catalyze the ORR in MFCs has also been extensively studied in the past. MOBs are able to use oxygen to oxidize manganese ion into manganese oxohydroxide (MnOOH) that deposits onto the electrode surface, further leading to manganese dioxide (MnO_2). MnO_2 is then reduced electrochemically to Mn^{2+} with MnOOH as an intermediate (Erable et al. 2012). A pathway of the indirect microbial catalysis of ORR mediated by manganese compounds is presented in Figure 2.

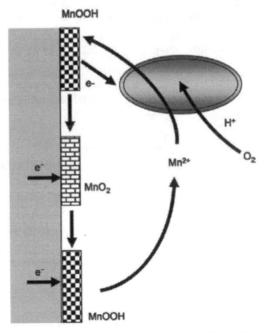

Figure 2: Detailed pathway of the indirect microbial catalysis mediated by manganese compounds. Adapted with permission from Shi et al. (2002). Copyright © Elsevier.

In 2005, Rhoads et al. had described a similar mechanism in their MFC using a glucose oxidizing anode and a manganese oxides reducing cathode with *Leptothrix discophora* as a biocatalyst (Rhoads et al. 2005). Such a cathode reached a potential of 384.5 ± 64 mV vs. SCE, and a peak power density of 127 ± 32 mW m^{-2} that was significantly higher than the power density recorded with an oxygen-reducing cathode, i.e., without manganese or biocatalyst (3.9 ± 0.7 mW m^{-2}). One advantage of this system is the possibility to work at neutral pH. Although neutrophilic IOB has also shown activity toward ORR (Summers et al. 2013), they are very difficult to isolate and can be a major limitation for the development of the technology.

One could expect that isolating bacteria that are active toward ORR could be the ideal solution to develop a high performing aerobic biocathode. However, bacterial isolation is challenging and previous studies attempting to isolate bacteria from wild biofilms reported mitigated results. In 2008, Rabaey et al. isolated autotrophic strains from an ORR-catalyzing biofilm (Rabaey et al. 2008). *Sphingobacterium sp.* and *Acinetobacter sp.*, which were dominating in the wild biofilms, exhibited electroactive properties toward ORR although current densities recorded with the isolates were much lower than those recorded with the wild biofilm. Similar behavior was observed with marine electroactive biofilms from which strains able to catalyze ORR were isolated. These strains, however, such as *Winogradskyella johsonii* and *Acinetobacter poriferorum,* achieved current densities of only a few percents of that obtained with the wild biofilms (Erable et al. 2010; Vandecandelaere et al. 2010). Although the reasons for this are still not clear yet, possible explanations are the synergetic effect coming from different species in wild biofilms, pH changes, surface modifications or underdeveloped biofilms under pure culture conditions (Little et al. 2008). Therefore, it has appeared over the years that the best strategy to develop an ORR catalyzing biofilm is to provide the right conditions (medium, pH, electrode potential, temperature, etc.) for the best bacteria to develop. This is what will be discussed in the next section.

3.1.2 Mixed Communities for the Catalysis of ORR in Aerobic Biocathodes

The utilization of environmental samples is the most popular way to develop a mixed community for the catalysis of ORR. In the first studies dealing with MFCs with aerobic biocathodes, Clauweart et al. developed a biofilm by using a sludge/sediment mixture as inoculum. After five months of operation, only α- and γ-proteobacteria were detected with two species of the genera *Pseudomonas* and *Novosphigobium* being more enriched at the air-oriented side of the cathodic felt (Clauwaert et al. 2007). In 2008, Rabaey et al. used a mixture of environmental samples from rusted metal poles, sediments and activated sludge as inoculum. After a start-up period of about 50 days, current densities of 2.2 A m^{-2} of cathode projected surface for power densities of 303 mW m^{-2}. The authors also reported that *Sphingobacterium, Acinetobacter and Acidovorax sp.*, H$_2$-oxidizing bacteria, dominated the bacterial community.

Since the first results in 2007, several studies focused on the development of aerobic biocathodes, and a wide range of bacterial species have been reported as dominant in the biofilms developed. Among them, *Alphaproteobacteria* (Clauwaert et al. 2007; Du et al. 2014; Zhang et al. 2011), *Betaproteobacteria* (Chen et al. 2008; Rabaey et al. 2008; Zhang et al. 2012a; Zhang et al. 2011), *Gammaproteobacteria* (Chen et al. 2010; Chung et al. 2011; Clauwaert et al. 2007; De Schamphelaire et al. 2010; Rabaey et al. 2008; Reimers et al. 2006; Rothballer et al. 2015; Strycharz-Glaven et al. 2013; Wang et al. 2015), Bacteroidetes (Chen et al. 2010; Chen et al. 2008; Rabaey et al. 2008; Xia et al. 2012) and other lesser well-known groups (Blanchet et al. 2014; Du et al. 2014; Rimboud et al. 2015). In order to appreciate the variety of the bacterial communities involved in these studies and their impact on the catalytic activity of the corresponding aerobic biocathodes, Milner et al. have recently ranked the biocathodes according to their onset potential for ORR (E_{ORR}) (Table 1) (Milner et al. 2016). The E_{ORR} is a key parameter to characterize any chemical or biological catalyst as it represents the activity of the catalyst toward ORR. As can be seen from Table 1, the onset potentials for ORR of the aerobic biocathodes reported range from +0.40 to -0.15 V vs. Ag/AgCl. The best performing aerobic biocathodes reported, thus, have an E_{ORR} about +0.40 V vs. Ag/AgCl and these performances were obtained with different reactor configurations. Indeed, some have been developed in half-cells (Milner et al. 2016; Rothballer et al. 2015), while some in

Table 1: Dominant bacteria recovered from aerobic biocathodes ranked according to the onset potential for ORR. The bacteria have been identified to different taxonomic levels: phylum (p), class (c), order (o), family (f) and genus (g). The method of community analysis is given for each study: pyrosequencing (PS), clone libraries (CL), denaturing gel electrophoresis (DGGE) and illumina dye sequencing (IDS). Biocathode electrode materials used in each of these studies are carbon felt (CF), carbon cloth (CC), graphite plate (GP), graphite granules (GG) and carbon brush (CB). The type of BES using in each study is given: microbial fuel cell (MFC), sediment microbial fuel cell (SMFC), microbial solar cell (MSC) and half-cell (HC) (adapted from Milner et al. 2016)

Type of BES	Elec. mat.	E_{ORR} (V)	Dominant bacteria in mixed community biocathode	Method	Reference
HC	GF	0.40	*Gammaproteobacteria (c)*	CL	(Milner et al. 2016)
HC	GP	0.40	*Gammaproteobacteria (c)*	PS	(Rothballer et al. 2015)
MFC	CC	0.40	*Xanthomonadaceae (f), Xanthomonas (g)*	CL	(Chung et al. 2011)
SMFC	GP	0.40	*Pseudomonas (g)*	CL	(Reimers et al. 2006)
SMFC	CF	0.40	*Gammaproteobacteria (c)*	DGGE	(De Schamphelaire et al. 2010)
HC	CC	0.25	*Sporosarcina (g), Brevundimonas (g)*	PS	(Rimboud et al. 2017)
MFC	CF	0.25	*Pseudomonas (g), Rhodobacteraceae (f), Sphingomonadaceae (f)*	CL	(Clauwaert et al. 2007)
MSC	GP	0.25	*Marinobacter (g)*	CL	(Strycharz-Glaven et al. 2013)
HC	GP	0.25	Chromatiaceae (f)	IDS	(Wang et al. 2015)
HC	CP	0.25	*Bacteroidetes (p)*	CL	(Xia et al. 2012)
MFC	CB	0.20	*Nitrospira (g), Nitrosomonas (g), Nitrobacter (g), Alkalilimncola (g)*	CL	(Du et al. 2014)
MFC	CC	0.20	*Rhizobiales (o), Phycisphaerales (o), Planctomycetales (o), Sphingobacteriales (o)*	PS	(Wang et al. 2013)
MFC	GG	0.15*	*Azovibrio (g), Bacteroidetes (p)*	CL	(Chen et al. 2008)
MFC	GG	0.15*	*Xanthomonas (g), Bacteroidetes (p)*	CL	(Chen et al. 2010)
HC	CC	0.00	*Deinococcus-Thermus (p), Gemmatimonadetes (p)*	PS	(Rimboud et al. 2015)
MFC	CF	-0.010	*Acinetobacter (g), Sphingobacterium (g), Acidovorax (g)*	CL	(Rabaey et al. 2008)
MFC	CC	-0.15	*Chloroflexus (g)*	PS	(Blanchet et al. 2014)
MFC	Various	-	*Comamonas (g), Sphingomonas (g), Acidovorax (g)*	CL	(Sun et al. 2012)
MFC	CB	-	*Achromobacter (g)*	PS	(Zhang et al. 2009)
MFC	CB/GG	-	*Gammaproteobacteria (c), Agrobacterium (g), Achromobacter (g)*	DGGE	(Zhang et al. 2011)

MFCs (Chung et al. 2011) and other in sediment MFCs (SMFCS) (De Schamphelaire et al. 2010; Reimers et al. 2006). In these studies, however, bacterial communities were dominated by *Gammaproteobacteria*. Many of the authors relate ability to oxidize hydrogen with ORR catalysis that stems from the study of Rabaey et al. in which all the dominant bacteria had ability to oxidize hydrogen (Rabaey et al. 2008). Indeed, *Sphingobacterium*, *Acinetobacter* and *Acidovorax* species are all known for being able to oxidize hydrogen and since then, HOB is thought to play an important role in mixed community biocathodes and HOBs were identified in biofilms catalyzing ORR in several studies (Wang et al. 2013; Zhang et al. 2012a). This also led some researchers to inoculate MFC reactors with HOB-enriched inocula in order to develop aerobic biocathodes (Zhang et al. 2011). To date, the aerobic biocathodes with the highest onset potential (> +0.35 vs. Ag/AgCl) were enriched with activated sludge in half-cell systems. Little is known about the bacteria involved in the catalysis of ORR in wild biofilms, especially about the interaction and possible synergetic effect within bacterial communities. For instance, community analysis carried out on some seawater microbial cathodes has shown no difference between the microbial composition of the biofilms that were able to catalyze ORR and those that were not (Faimali et al. 2010). Rimboud et al. have shown than the methodology followed to develop aerobic biocathodes has a great impact on the bacterial community in the biofilm although the same environmental sample is used as inoculum (Rimboud et al. 2017). In this study, biocathodes were either developed at open circuit potential poised at -0.15 vs. Ag/AgCl or by reversion of already established acetate-fed bioanodes with respective to E_{ORR} of 0.0 V, +0.25 and -0.25 V vs. Ag/AgCl. Populations from biocathodes formed under aerobic polarization and open-circuit conditions were dominated by *Sporosarcina* (bacilli). *Brevundimonas* (alphaproteobacteria) was the second dominant genus on the polarized biocathodes although it was minor on electrodes formed at OCP. Interestingly, *Sporosarcina* was already dominant in the inoculum whereas *Brevundimonas* was very minor, showing the impact of the poised potential for certain bacteria to proliferate. Communities on reversible electrodes were dominated by *Arenimonas* (Gammaproteobacteria) (Rimboud et al. 2017). The variety of bacteria involved and the wide range of onset potentials reported (Table 1) also suggest the complexity of the ORR mechanism in biofilms, showing how much work still needs to be done to understand and master these catalysts. So what do we know about the mechanism of ORR in biofilms?

3.2 Possible Mechanisms of Electron Transfer in Aerobic Biocathodes

To date, the mechanisms proposed for the ORR catalysis by bacteria include direct (DET) and indirect electron transfer mechanisms:

1. Direct ORR catalysis by extracellular enzymes such as peroxidase dismutase, catalase and peroxidase are adsorbed on the material surface. It was reported that catalase and horseradish peroxidase adsorbed on glassy carbon and pyrolytic graphite electrodes could catalyse ORR by direct electron transfer (Freguia et al. 2010; Huang and Hu 2001).
2. Direct catalysis by porphyrins and organometallic compounds entrapped in microbial biofilms. Porphyrins constitute the prosthetic group of catalase and other oxidases. Adsorbed iron porphyrins on glassy carbon or stainless steel can exhibit catalytic activity toward ORR; to give an example, they could take part in ORR catalysis in natural biofilms after enzyme degradation (Erable et al. 2012; Freguia et al. 2010; Jiang and Dong 1990; Parot et al. 2011).
3. Indirect catalysis by enzymes can reduce oxygen to hydrogen peroxide and organic acid. This mechanism only occurs in specific conditions, such as pH value around 2.9, the presence of hydrogen peroxide around 2.4 mM and can mainly apply to marine biofilms (Erable et al. 2012).
4. Indirect catalysis by metal ions complexed within extracellular polymeric substances. As described in section 3.1.1, in environments containing Fe or Mn ions, IOBs and MOBs are able to oxidize these ions to oxides that are reduced back to ions at the surface of the electrode, and oxygen is the terminal electron acceptor.
5. Direct electron transfer (DET) is from the electrode to the bacterial cell. This mechanism can be compared with the mechanism occurring in anaerobic biocathodes (Rosenbaum et al. 2011).

Freguia et al. have demonstrated the possibility that some redox active compounds excreted by the bacteria that are able to react with oxygen might be involved in the ORR mechanism (Freguia et al. 2010). They compared the catalytic activity toward ORR of two pure cultures, namely *Acinetobacter calcoaceticus* and *Shewanella putrefaciens,* to those of hemin at 1 uM concentration, 2-amino-3-dicarboxy-1,4-naphthoquinone (ΛCNQ) and catalase. These compounds catalyze the 2 e-ORR to peroxide and are used as substitutes to heme proteins, quinone and catalase enzyme. It appeared that both hemin and ACNQ could increase the E_{ORR} to about -0.1 V vs. Ag/AgCl that was similar to the E_{ORR} obtained with *Acinetobacter calcoaceticus*. In the case of *Shewanella putrefaciens*, it was concluded that membrane attached compounds were responsible for the catalytic current generated. As the bacterium is well-known to localize *c*-type cytochromes on the external surface of the outer membrane, it is likely that cytochromes or cytochrome-containing nanowires could interact directly with the cathode and exchange electrons via DET. In this study, however, catalase did not exhibit ORR activity (Freguia et al. 2010). It is reasonable to think that the catalysis of ORR in aerobic biofilms occurs via parallel pathways and that redox species secreted by bacteria, such as heme proteins or quinones, do contribute to the ORR. Nevertheless, the onsets for ORR of -0.1 V vs. Ag/AgCl recorded in this study are far from the +0.4 V vs. Ag/AgCl mentioned earlier. On the one hand, it should be kept in mind that the highest E_{ORR} were obtained with mixed communities and that no study using pure culture was able to reproduce these results. On the other hand, it shows again the complexity behind the microbial catalysis of ORR and that a synergetic combination of direct and indirect pathways is very likely.

DET mechanism forms the electrode to the bacteria and with the oxygen as a terminal electron acceptor is likely to be involved in ORR catalysis in aerobic biocathodes. This mechanism can be assimilated to that occurring in anaerobic biocathodes (Erable et al. 2012; Rosenbaum et al. 2011) but also in bioanodes with anode respiring bacteria (ARB), such as *Geobacter sulfurreducens* (Bonanni et al. 2012). *Geobacter* species have also been shown to accept electrons from poised-potential electrodes for respiration and their purified *c*-type cytochromes to be electrochemically reversible, showing that they could be involved in direct electron uptake from cathodes (Gregory et al. 2004; Rosenbaum et al. 2011). If *c*-type OMCs were coupled to an electron chain, electrons could be shuttled to cytochrome c-oxidase (an enzyme which catalyses the reduction of oxygen to water in the electron transport chain of many aerobic bacteria) located on the inner membrane and ultimately to intracellular O_2 (Milner 2015). In their study, Milner et al. have also demonstrated that the ORR catalysis is highly likely to be linked to a bacterial electron transport chain (Milner et al. 2016) after growing aerobic biocathodes for 61 days at different poised potentials; cyclic voltammetry (CV) analysis showed the apparition of several redox features (Fig. 3, features 1-3) with feature 1 being related to ORR. The subsequent addition of azide, a cytochrome c-oxidase inhibitor, led to the suppression of the redox feature 1 demonstrating the involvement of the electron transport chain in the ORR mechanism (Milner et al. 2016).

As mentioned previously, the direct ORR catalysis could also occur via mediated electron transfer (MET). It is known that bacteria such as *Shewanella oneidensis* MR1 can use both DET and MET mechanisms at anode electrodes. In the case of MET, *Shewanella oneidensis* MR1 can use both endogenous [riboflavin, flavin mononucleotide and 9,10-anthraquinone-2,7-disulphonic acid (AQDS)] and exogenous (humic acids, phenazines) redox shuttles for electron transfer (Klupfel et al. 2014; Paquete et al. 2014).

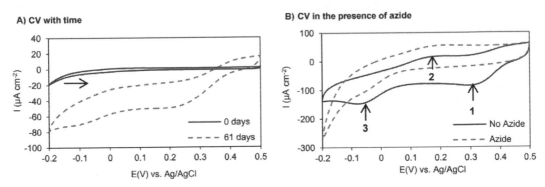

Figure 3: A) CV in the presence of oxygen at 0 and 61 days for an aerobic biocathode developed at -0.1 V vs. Ag/AgCl and (B) CV in the presence of oxygen and azide recorded at the end of the operational period for the same cell (Milner et al. 2016).

3.3 Operating Parameters Affecting the Performance of Aerobic Biocathodes

3.3.1 Effects of the Poised Potential in Half-Cell Systems

In some half-cell studies, it was demonstrated that lowering the poised potential can decrease the start-up period for the development of biocathodes and increase the catalytic current produced. Such an observation was reported by Bergel et al. who enriched marine cathode biofilms at different potentials of -0.45, -0.30 and -0.15 V vs. Ag/AgCl and noticed that the biocathodes poised at lower potentials took half the time to enrich (four days) than the ones polarised at -0.15 V vs. Ag/AgCl (Bergel et al. 2005). In another study, ter Heijne et al. inoculated half-cells with activated sludge at poised potentials of 0.05, 0.15 and 0.25 mV vs. Ag/AgCl and monitored smaller enrichment periods for cells poised at 0.05 and 0.15 mV vs. Ag/AgCl. Maximum current densities during chronoamperometry experiments were recorded for the cells poised at 0.15 mV vs. Ag/AgCl (ter Heijne et al. 2010). Xia et al., however, did not notice any impact of the poised potential on the enrichment time; the half-cells inoculated with activated sludge and containing carbon electrodes were polarized at -144, 16 and 156 mV vs. Ag/AgCl. All the cells had an identical enrichment time of 10 days and similar E_{ORR} of 0.25 V vs. Ag/AgCl. Nevertheless, current densities recorded were higher for the cells poised at -144 mV and 16 mV vs. Ag/AgCl (Xia et al. 2012). In these cells, the analysis revealed that an uncultured *Bacteroidetes* bacterium dominated the bacterial community by 80 and 75% on the electrodes poised at -144 mV and 16 mV, respectively, and only by 46% in the cells polarized at 156 mV (Xia et al. 2012). Milner et al. did not notice any impact of poised potential on the current densities recorded during chronoamperometry experiments for half-cells poised at -0.1 V and +0.2 V vs. Ag/AgCl (Milner et al. 2016). However, it is possible that the poised potential impacts on the electron transfer mechanism; CVs recorded on the biocathodes developed at different potentials showed different redox features. A common feature was related to the ORR (Figure 3B, feature 1) whereas features 2 and 3 were only observed for the biocathodes developed at -0.1 V. The authors suggested that these features could be due to a bacterially produced electron mediator, i.e., a cytochrome interacting with the electrode surface, leading to a reversible redox peak (Milner et al. 2016). Rimboud et al. have also shown that the methodology followed to develop aerobic biocathodes has a great impact on both their performance and their bacterial communities (Rimboud et al. 2017). In their study, the biocathodes developed at -0.15 vs. Ag/AgCl had an E_{ORR} of +0.25 vs. Ag/AgCl, whereas the ones grown at open circuit potential had an E_{ORR} 0.0 V. Although all populations were dominated by *Sporosarcina* (bacilli), *Brevundimonas* (alphaproteobacteria) was the second dominant genus on the polarized biocathodes, whereas it was minor on electrodes formed at OCP (Rimboud et al. 2017).

3.3.2 Effects of Electrode Materials

To date, and similarly as studies focusing on bioanodes, carbon and graphite materials have been used in most studies on aerobic biocathodes. In some studies, planar and smooth electrodes were used such as graphite plates (Renslow et al. 2011; ter Heijne et al. 2010), glassy carbon (Aldrovandi et al. 2009) or carbon paper (Freguia et al. 2010). However, most studies reported the utilization of three-dimensional materials, such as carbon and graphite felts (Carbajosa et al. 2010; Clauwaert et al. 2007; De Schamphelaire et al. 2010; Milner et al. 2016; Rabaey et al. 2008), graphite granules (Chen et al. 2008; Mao et al. 2010; Wei et al. 2011b) or graphite fiber or carbon brushes (You et al. 2009). Three-dimensional materials have been preferred because in theory bacteria can penetrate through the voids between strands, giving a higher specific surface area for the biofilm. However, the bacteria may be limited by the transfer of substrates to the inner surfaces (Wei et al. 2011b). The performance of an MFC with different cathode materials was compared: graphite felt, carbon paper and stainless steel mesh (all of the equal geometric area of 7 cm^2) were used, and peak powers of 109.5, 32.7 and 3.1 mW m^{-2} were obtained, respectively (Zhang et al. 2012c). However, 3D materials make performance more difficult to compare as surface areas that are actually active are more difficult to determine and current densities should be expressed in A m^{-3} rather than A m^{-2}. Granular materials have also been reported for aerobic biocathodes. With their high porosities (30-50%) and diameters ranging from 1.5 to 5.0 mm, they offer high surface areas for biofilms to grow. Wei et al. compared the performance of granular semi-coke (GS), granular activated carbon (GAC), graphite granules (GG) and carbon felt (CF) as biocathodes materials (Wei et al. 2011a). For these materials, diameters were ranging from 2 to 5 mm and total surface areas were 1.44×10^4 m^2, 3.43×10^4 m^2, 60.7 m^2 and 10.6 m^2, respectively. Despite the much larger surface areas for GS and GAC, the difference in MFC peak power outputs were not as significant; the values of 24.3, 20.1, 17.1 and 14.4 W m^{-3} in the order GAC > GS > CFC > GG were obtained, respectively (Wei et al. 2011a). These results confirm that higher surface areas do not necessarily translate into higher performance (linearly at least) that may be due to differences in bio-available surface areas, substrate diffusion limitation and in the case of granular materials to the quality of electrical contact between individual granules. As reported in bioanode experiments, brush electrodes have also been reported in aerobic biocathode studies as biocatalysts support (Zhang et al. 2012a; Zhang et al. 2012b; Zhang et al. 2011). It was shown that combining graphite granules with a graphite brush could improve the peak power density of an MFC (Zhang et al. 2011).

As discussed previously, cultivating aerobic biocathodes in half-cells at poised potential is a good strategy to shorten the start-up period and enrich the suitable bacterial community. However, when carbon materials are used, the possible abiotic formation of H$_2$O$_2$ should be considered. Through the generation of reactive oxygen species such as superoxide, H$_2$O$_2$ can cause oxidative stress by damaging bacterial cells (Imlay 2003) although enzymes such as catalase and superoxide dismutase can mitigate these harmful effects. Porous carbon materials such as carbon felt, carbon cloth and carbon veil or carbon brush are common electrode materials for the development of both bioanodes and biocathodes. Therefore, understanding the production of H$_2$O$_2$ on these materials is required. Milner et al. have used a novel 4-electrode system to determine at which potential H$_2$O$_2$ starts being produced on carbon felt and HNO$_3$-treated carbon felt (Milner et al. 2017). Carbon felts were used as primary electrodes whereas platinum served as secondary sensing electrode; polarizing Pt at a potential at which H$_2$O$_2$ is oxidized allowed to determine at which potential the polarized carbon felts would start producing hydrogen peroxide. Therefore, it was assumed that the highest potential at which oxidation of H$_2$O$_2$ could be detected at the second sensing electrode is the lowest potential that could be used to enrich anaerobic biocathode without risking inhibition of microbial growth and development. Authors showed that this lowest potential was -0.1 V vs. Ag/AgCl for the untreated carbon felt, and 0.0 V vs. Ag/AgCl for the HNO$_3$-treated carbon felt. This assumption was confirmed with the development of aerobic biocathodes in half-cells with untreated carbon felt in which a biocathode was successfully developed at -0.1 V but not at -0.2 V vs. Ag/AgCl (Milner et al. 2017).

In aerobic biocathodes, mass transfer of oxygen to the biofilm/electrode surface must be efficient. In sediment MFCs, this is ensured by the continuous mixing of water above the sediment, but in most of the laboratory-scale studies the electrolyte is sparged *in situ* or in an external recirculation vessel which is energy intensive. If implemented in existing facilities, such as wastewater treatment plant, the wastewater is already aerated and pumped through that would not introduce any additional energy costs. In any case, the electrode material needs to be taken into account to allow the efficient mass transfer of O$_2$. The need for energy-intensive aeration can be eliminated, or at least reduced, by using gas diffusion electrode (GDE). The GDE is composed of a 3-phase interface of solid, water and gas that overcome the problem of low O$_2$ solubility in the catholyte. At the 3-phase interface, oxygen diffuses from the air and protons are transported from the electrolyte combining at the solid catalyst surface to produce water/peroxide (Milner 2015). In the case of aerobic biocathodes, the chemical catalyst is replaced by biocatalysts that can be developed on the GDE itself. By following the protocol proposed by Cheng et al. to make home-made GDEs (Cheng et al. 2006), Xia et al. have cultivated aerobic biocathodes on carbon cloth GDEs with four diffusion layers (DLs) of PTFE and the biocathodes were grown in half-cell at a poised potential of +0.1 V vs. Ag/AgCl (Xia et al. 2013). The biocathodes developed had E$_{ORR}$ of +0.4 V vs. Ag/AgCl under passive aeration and were transferred to full single (membraneless) and dual (with membrane) chamber MFC configuration as presented in Figure 4.

The dual chamber and single chamber MFCs reached peak power densities of 554 and 199 mW m^{-2}, respectively (Xia et al. 2013). The dual chamber MFC equipped with the biocathode performed almost as

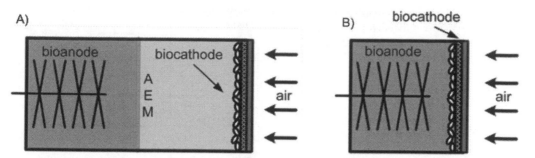

Figure 4: Schematics of a A) two-chamber, air biocathode and of a B) single chamber air biocathode configuration. Reprinted with permission from Xia et al. (2013). Copyright © American Chemical Society.

well as the MFC equipped with the Pt GDE (576 mW m^{-2}). The high difference between the performance of the single chamber and the dual chamber MFCs was attributed to the membrane limiting the transfer of O$_2$ from the cathode chamber to the anode chamber, thus increasing the flux of O$_2$ to the biocathode (Xia et al. 2013). It was also discussed that the number of DLs and PTFE content to apply on the GDE can drastically impact on the GDE performance. Cheng et al. found that 4 DLs were optimal for the carbon cloth GDE in their membraneless single chamber MFC with Pt catalyst. However, they also found fewer DLs increased the *k* value for O$_2$ mass transfer of the GDE (Cheng et al. 2006). When using biocatalyst instead of chemical catalyst, the optimal number of DLs may vary as increasing PTFE layers to the side facing solution increases hydrophobicity that would, in turn, reduce bacterial attachment. Therefore, in another study, aerobic biocathodes were cultivated at -0.3 V vs. Ag/AgCl on GDEs from carbon cloth with only 2 DLs and 10% wt PTFE instead of 30% (Wang et al. 2013). E$_{ORR}$ of +0.2 V vs. Ag/AgCl peak power output of 103 mW m^{-2} were measured and were lower values than those mentioned previously.

It seems, to date, understanding the development of aerobic biocathodes on GDEs is still insufficient. To our knowledge, only two studies focused on the topic and both used methodology initially developed for use with chemical catalysts (Wang et al. 2013; Xia et al. 2013), suggesting that optimization of engineering conditions and bacterial enrichment is needed. Moreover, although GDEs seem good alternative as a replacement to energy-intensive aeration systems, their scale up may be challenging. In addition, a similar problem encountered with GDEs with chemical catalysts, such as the development of non-electroactive biofilms, may also be problematic with aerobic biocathode GDEs.

3.3.3 Effects of Oxygen Mass Transfer

Mass transfer of oxygen is a critical parameter in aerobic biocathodes. Impact of oxygen mass transfer was studied in half-cells poised at +0.15 V vs. Ag/AgCl by decreasing dissolved oxygen (DO) concentration from 100% to 65% to 0.8% (ter Heijne et al. 2010). Such a decrease in DO concentration resulted in a decrease in current density from 241 to 194 to 15 mA m^{-2}. A similar effect was observed when decreasing the recirculation rate, showing that the aerobic biocathode performance was limited by oxygen mass transfer (ter Heijne et al. 2010). As depicted in Figure 5, electrochemical impedance spectroscopy analysis (EIS) showed that the flow rate had a direct impact on the mass transfer resistance, whereas the charge transfer resistance was unaffected (ter Heijne et al. 2011b).

In MFC configuration, it was also reported that the aeration rate impacts the cathode potential. Zhang et al. showed that by increasing the aeration rate from 0 to 300 mL min^{-1}, the cathode potential increased from -0.20 V to 0.10 V vs. Ag/AgCl and the anode potential remained constant (Zhang et al. 2008). However, increasing the aeration rate from 300 to 400 mL min^{-1} led to a sharp increase of the anode potential and a decrease of the cathode potential that were attributed to oxygen influx into the anode chamber (Zhang et al. 2008). Milner et al. confirmed these trends by following the impact of aeration rate on DO, cathode potentials and current densities (Milner 2015; Milner and Yu 2018). The authors showed that by increasing the aeration rate from 0 to 100 mL min^{-1}, the DO increased from 3 to 8 mg L^{-1} and the cathode and cell potentials rose from -400 to -100 mV vs. Ag/AgCl and from 50 to 350 mV, respectively.

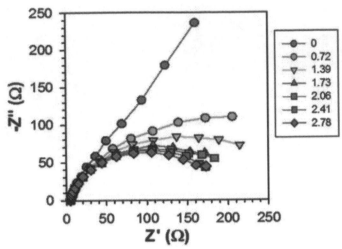

Figure 5: Nyquist plots obtained by EIS for a biocathode at different flow rates (expressed in cm s⁻¹) and constant cathode potential of 0.28 V vs. Ag/AgCl. Reprinted with permission from (ter Heijne et al. 2011b). Copyright © Royal Society of Chemistry.

When the aeration was increased to 400 mL min⁻¹, these values reached a plateau i.e., DO reached 8.5 mg L⁻¹, cathode potential was 400 mV and the cell voltage reached 650 mV (Milner 2015). A similar impact was observed on cathode current densities as it increased from 0.1 to 0.3 mA cm⁻² when the flow rate was increased from 0.4 L min⁻¹ to 1.2 L min⁻¹ (Milner and Yu 2018). It was also observed that oxygen concentration affected biofilm formation and distribution. Most bacteria were distributed at or close to the surface of the electrode where DO concentration was high, and the number of bacteria decreased from the surface toward the middle of the electrodes where the DO concentration was low (Figure 6) (Milner and Yu 2018). The impact of DO on the performance of aerobic biocathode is also dependant on the MFC

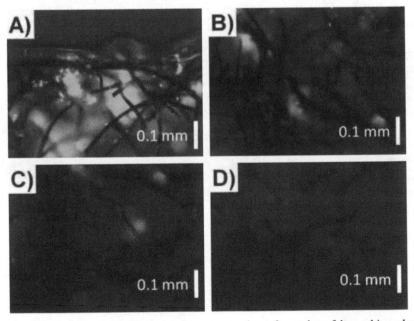

Figure 6: Epifluorescence microscope images of a cross-section of a carbon felt working electrode after the development of an aerobic biocathode. The images are ordered as consecutive frames starting from the electrode surface (A) and going into the electrode cross-section as (A), (B), (C) and then (D). Reprinted with permission from (Milner and Yu 2018). Copyright © Wiley.

configuration. Rago et al. studied the impact of DO on three MFCs of different configurations, namely air-breathing, water-submerged and assisted by photosynthetic microorganisms (Rago et al. 2017). Although the DO concentration was the highest in the photosynthetic microorganisms-assisted MFC (up to 16 mg L^{-1}), the best performance were recorded with the air-breathing cathode. The DO (2 mg L^{-1} O$_2$) was the limiting factor in the water-submerged MFC for which no aeration was provided (Rago et al. 2017). In order to better control the oxygen mass transfer to the biofilm and limit the energy-intensive aeration, the utilization of passive aeration using GDE or flow systems with recirculation should be prioritized.

4. Conclusions and Future Perspective

Developing an efficient and sustainable catalyst for the oxygen reduction reaction is a real challenge and one of the bottlenecks for the development, scale up and industrialization of MFCs. Although Pt-based catalysts are still a reference in terms of activity and performance, they are not sustainable and alternatives need to be found in order to develop the technology. A sustainable alternative is to consider biocatalysts for ORR. Aerobic biocathode biofilms have been studied in different configurations, such as half-cells and MFCs, and a wide range of bacteria was found to be able to catalyze the reduction of oxygen. However, little is still understood on the mechanism involved in ORR by microorganisms, and the results reported in the literature vary largely in terms of biocathode performance (E$_{ORR}$) and microbial communities. The best performing biocathode reported so far were developed with mixed communities often dominated by *Gammaproteobacteria* and have onset potentials of +0.4 V vs. Ag/AgCl. The utilization of pure cultures isolated from wild biofilms have always led to lower performance compared to the wild biofilms, showing that there is probably a synergetic activity within the communities of these wild biofilms. Although the mechanism(s) of electron transfer still has to be confirmed, several studies have explored the impact of different operating parameters on the performance of aerobic biocathodes. Among them the impact of the electrode materials, membrane or dissolve oxygen concentration were considered. When biocatalysts are used at both anode and cathode, these parameters should be considered carefully since substrate and oxygen cross-over, pH splitting and MFC ohmic resistance will affect the performance and stability of these catalysts and, therefore, the overall performance of the system. To date, mostly sediment MFCs have been successfully scaled up mainly as power sources for remote sensors in the environment. MFC technologies have never been considered as serious contenders for wastewater treatment, but recent studies have shown the progress made over the last few years. Pilot-scale MFCs were implemented in existing facilities, such as wastewater treatment plant, showing good performance and being energy efficient. Nevertheless, in the best performing pilot reactors, Pt-based catalysts were still used which shows that the ideal and sustainable catalyst for the oxygen reduction reaction for MFCs has not been developed yet.

Acknowledgment

The authors thank EPSRC LifesCO2R project (EP/N009746/1), EPSRC Supergen Biological Fuel Cells (EP/H019480/1) and NERC MeteoRR (NE/L014246/1) to support this work.

References

Aldrovandi A, Marsili E, Stante L, Paganin P, Tabacchioni S, Giordano A (2009) Sustainable power production in a membrane-less and mediator-less synthetic wastewater microbial fuel cell. Bioresource Technology 100(13):3252-3260

Allen J. Bard LRF (2001) Electrochemical Methods: Fundamentals and Applications, 2nd Edition. John Wiley & Sons, New York

Bergel A, Féron D, Mollica A (2005) Catalysis of oxygen reduction in PEM fuel cell by seawater biofilm. Electrochemistry Communications 7(9):900-904 doi:http://dx.doi.org/10.1016/j.elecom.2005.06.006

Blanchet E, Pecastaings S, Erable B, Roques C, Bergel A (2014) Protons accumulation during anodic phase turned to advantage for oxygen reduction during cathodic phase in reversible bioelectrodes. Bioresource Technology 173:224-230

Bonanni PS, Schrott GD, Robuschi L, Busalmen JP (2012) Charge accumulation and electron transfer kinetics in Geobacter sulfurreducens biofilms. Energy Environ Sci 5(3):6188-6195

Carbajosa S, Malki M, Caillard R, Lopez MF, Palomares FJ, Martín-Gago JA, Rodríguez N, Amils R, Fernández VM, De Lacey AL (2010) Electrochemical growth of Acidithiobacillus ferrooxidans on a graphite electrode for obtaining a biocathode for direct electrocatalytic reduction of oxygen. Biosensors and Bioelectronics 26(2):877-880 doi:http://dx.doi.org/10.1016/j.bios.2010.07.037

Chen GW, Choi SJ, Cha JH, Lee TH, Kim CW (2010) Microbial community dynamics and electron transfer of a biocathode in microbial fuel. Korean J Chem Eng 27(5):1513-1520

Chen GW, Choi SJ, Lee TH, Lee GY, Cha JH, Kim CW (2008) Application of biocathode in microbial fuel cells: cell performance and microbial community. Appl Microbiol Biot 79(3):379-388 doi:10.1007/s00253-008-1451-0

Cheng S, Liu H, Logan BE (2006) Increased performance of single-chamber microbial fuel cells using an improved cathode structure. Electrochemistry Communications 8(3):489-494

Chung K, Fujiki I, Okabe S (2011) Effect of formation of biofilms and chemical scale on the cathode electrode on the performance of a continuous two-chamber microbial fuel cell. Bioresource Technology 102(1):355-360

Clauwaert P, Van der Ha D, Boon N, Verbeken K, Verhaege M, Rabaey K, Verstraete W (2007) Open air biocathode enables effective electricity generation with microbial fuel cells. Environ Sci Technol 41(21):7564-7569

De Schamphelaire L, Boeckx P, Verstraete W (2010) Evaluation of biocathodes in freshwater and brackish sediment microbial fuel cells. Appl Microbiol Biot 87(5):1675-1687

Du Y, Feng YJ, Dong Y, Qu YP, Liu J, Zhou XT, Ren NQ (2014) Coupling interaction of cathodic reduction and microbial metabolism in aerobic biocathode of microbial fuel cell. Rsc Adv 4(65):34350-34355 doi:10.1039/c4ra03441d

Erable B, Feron D, Bergel A (2012) Microbial Catalysis of the Oxygen Reduction Reaction for Microbial Fuel Cells: A Review. Chemsuschem 5(6):975-987

Erable B, Vandecandelaere I, Faimali M, Delia ML, Etcheverry L, Vandamme P, Bergel A (2010) Marine aerobic biofilm as biocathode catalyst. Bioelectrochemistry 78(1):51-56

Faimali M, Chelossi E, Pavanello G, Benedetti A, Vandecandelaere I, De Vos P, Vandamme P, Mollica A (2010) Electrochemical activity and bacterial diversity of natural marine biofilm in laboratory closed-systems. Bioelectrochemistry 78(1):30-38

Freguia S, Tsujimura S, Kano K (2010) Electron transfer pathways in microbial oxygen biocathodes. Electrochimica Acta 55(3):813-818

Gregory KB, Bond DR, Lovley DR (2004) Graphite electrodes as electron donors for anaerobic respiration. Environ Microbiol 6(6):596-604

He Z, Angenent LT (2006) Application of bacterial biocathodes in microbial fuel cells. Electroanal 18(19-20):2009-2015

Huang R, Hu NF (2001) Direct electrochemistry and electrocatalysis with horseradish peroxidase in Eastman AQ films. Bioelectrochemistry 54(1):75-81

Imlay JA (2003) Pathways of oxidative damage. Annu Rev Microbiol 57:395-418

Jiang RZ, Dong SJ (1990) Study on the Electrocatalytic Reduction of H2o2 at Iron Protoporphyrin Modified Electrode with a Rapid Rotation-Scan Method. Electrochimica Acta 35(8):1227-1232

Klupfel L, Piepenbrock A, Kappler A, Sander M (2014) Humic substances as fully regenerable electron acceptors in recurrently anoxic environments. Nat Geosci 7(3):195-200

Lapinsonniere L, Picot M, Barriere F (2012) Enzymatic versus Microbial Bio-Catalyzed Electrodes in Bio-Electrochemical Systems. Chemsuschem 5(6):995-1005

Leech D, Kavanagh P, Schuhmann W (2012) Enzymatic fuel cells: Recent progress. Electrochimica Acta 84:223-234

Little BJ, Lee JS, Ray RI (2008) The influence of marine biofilms on corrosion: A concise review. Electrochimica Acta 54(1):2-7

Logan BE (2009) Exoelectrogenic bacteria that power microbial fuel cells. Nat Rev Micro 7(5):375-381 doi:10.1038/nrmicro2113

Mao YP, Zhang LH, Li DM, Shi HF, Liu YD, Cai LK (2010) Power generation from a biocathode microbial fuel cell biocatalyzed by ferro/manganese-oxidizing bacteria. Electrochimica Acta 55(27):7804-7808

Milner EM (2015) Development of an Aerobic Biocathode for Microbial Fuel Cells. Newcastle University

Milner EM, Popescu D, Curtis T, Head IM, Scott K, Yu EH (2016) Microbial fuel cells with highly active aerobic biocathodes. Journal of Power Sources 324:8-16 doi:http://dx.doi.org/10.1016/j.jpowsour.2016.05.055

Milner EM, Scott K, Head IM, Curtis T, Yu EH (2017) Evaluation of porous carbon felt as an aerobic biocathode support in terms of hydrogen peroxide. Journal of Power Sources 356:459-466

Milner EM, Yu EH (2018) The Effect of Oxygen Mass Transfer on Aerobic Biocathode Performance, Biofilm Growth and Distribution in Microbial Fuel Cells. Fuel Cells 18(1):4-12

Mollica A, Traverso E, Thierry D (1997) EFC Working Party report on "Aspect of Microbially Induced Corrosion". European Federation of Corrosion Series, Institute of Materials, London 22:51-63

Paquete CM, Fonseca BM, Cruz DR, Pereira TM, Pacheco I, Soares CM, Louro RO (2014) Exploring the molecular mechanisms of electron shuttling across the microbe/metal space. Front Microbiol 5

Parot S, Vandecandelaere I, Cournet A, Delia ML, Vandamme P, Berge M, Rogues C, Bergel A (2011) Catalysis of the electrochemical reduction of oxygen by bacteria isolated from electro-active biofilms formed in seawater. Bioresource Technology 102(1):304-311

Prescott LM, Harley JP, Klein DA (2005) Microbiology, McGraw-Hill Higher Education edn. McGraw-Hill Higher Education

Rabaey K, Read ST, Clauwaert P, Freguia S, Bond PL, Blackall LL, Keller J (2008) Cathodic oxygen reduction catalyzed by bacteria in microbial fuel cells. Isme J 2(5):519-527

Rago L, Cristiani P, Villa F, Zecchin S, Colombo A, Cavalca L, Schievano A (2017) Influences of dissolved oxygen concentration on biocathodic microbial communities in microbial fuel cells. Bioelectrochemistry 116:39-51

Reimers CE, Girguis P, Stecher HA, Tender LM, Ryckelynck N, Whaling P (2006) Microbial fuel cell energy from an ocean cold seep. Geobiology 4(2):123-136

Renslow R, Donovan C, Shim M, Babauta J, Nannapaneni S, Schenk J, Beyenal H (2011) Oxygen reduction kinetics on graphite cathodes in sediment microbial fuel cells. Phys Chem Chem Phys 13(48):21573-21584

Rhoads A, Beyenal H, Lewandowski Z (2005) Microbial fuel cell using anaerobic respiration as an anodic reaction and biomineralized manganese as a cathodic reactant. Environ Sci Technol 39(12):4666-4671

Rimboud M, Barakat M, Bergel A, Erable B (2017) Different methods used to form oxygen reducing biocathodes lead to different biomass quantities, bacterial communities, and electrochemical kinetics. Bioelectrochemistry 116:24-32

Rimboud M, Desmond-Le Quemener E, Erable B, Bouchez T, Bergel A (2015) The current provided by oxygen-reducing microbial cathodes is related to the composition of their bacterial community. Bioelectrochemistry 102:42-49

Roche I, Katuri K, Scott K (2010) A microbial fuel cell using manganese oxide oxygen reduction catalysts. J Appl Electrochem 40(1):13-21

Rosenbaum M, Aulenta F, Villano M, Angenent LT (2011) Cathodes as electron donors for microbial metabolism: Which extracellular electron transfer mechanisms are involved? Bioresource Technology 102(1):324-333

Rothballer M, Picot M, Sieper T, Arends JBA, Schmid M, Hartmann A, Boon N, Buisman CJN, Barriere F, Strik DPBTB (2015) Monophyletic group of unclassified gamma-Proteobacteria dominates in mixed culture biofilm of high-performing oxygen reducing biocathode. Bioelectrochemistry 106:167-176

Santoro C, Arbizzani C, Erable B, Ieropoulos I (2017) Microbial fuel cells: From fundamentals to applications. A review. Journal of Power Sources 356:225-244

Sawant SY, Han TH, Cho MH (2017) Metal-Free Carbon-Based Materials: Promising Electrocatalysts for Oxygen Reduction Reaction in Microbial Fuel Cells. Int J Mol Sci 18(1) doi:ARTN 2510.3390/ijms18010025

Scott K, Yu EH (2015) Microbial Electrochemical and Fuel Cells: Fundamentals and Applications.

Shi XM, Avci R, Lewandowski Z (2002) Electrochemistry of passive metals modified by manganese oxides deposited by Leptothrix discophora: two-step model verified by ToF-SIMS. Corros Sci 44(5):1027-1045 doi:Pii S0010-938x(01)00104-4

Song C, Zhang J (2008) Electrocatalytic Oxygen Reduction Reaction. In: Zhang J (ed) PEM Fuel Cell Electrocatalysts and Catalyst Layers: Fundamentals and Applications. Springer London, London, pp 89-134

Sopian K, Daud WRW (2006) Challenges and future developments in proton exchange membrane fuel cells. Renew Energ 31(5):719-727

Stacy J, Regmi YN, Leonard B, Fan M (2017) The recent progress and future of oxygen reduction reaction catalysis: A review. Renewable and Sustainable Energy Reviews 69:401-414 doi:https://doi.org/10.1016/j.rser.2016.09.135

Strycharz-Glaven SM, Glaven RH, Wang Z, Zhou J, Vora GJ, Tender LM (2013) Electrochemical Investigation of a Microbial Solar Cell Reveals a Nonphotosynthetic Biocathode Catalyst. Appl Environ Microbiol 79(13):3933-3942

Summers ZM, Gralnick JA, Bond DR (2013) Cultivation of an Obligate Fe(II)-Oxidizing Lithoautotrophic Bacterium Using Electrodes. Mbio 4(1)

Sun YM, Wei JC, Liang P, Huang X (2012) Microbial community analysis in biocathode microbial fuel cells packed with different materials. Amb Express 2

ter Heijne A, Hamelers HVM, Buisman CJN (2007) Microbial fuel cell operation with continuous biological ferrous iron oxidation of the catholyte. Environ Sci Technol 41(11):4130-4134

ter Heijne A, Hamelers HVM, De Wilde V, Rozendal RA, Buisman CJN (2006) A bipolar membrane combined with ferric iron reduction as an efficient cathode system in microbial fuel cells. Environ Sci Technol 40(17):5200-5205

ter Heijne A, Liu F, van Rijnsoever LS, Saakes M, Hamelers HVM, Buisman CJN (2011a) Performance of a scaled-up Microbial Fuel Cell with iron reduction as the cathode reaction. Journal of Power Sources 196(18):7572-7577

ter Heijne A, Schaetzle O, Gimenez S, Fabregat-Santiago F, Bisquert J, Strik DPBTB, Barriere F, Buisman CJN, Hamelers HVM (2011b) Identifying charge and mass transfer resistances of an oxygen reducing biocathode. Energy Environ Sci 4(12):5035-5043

ter Heijne A, Strik DPBTB, Hamelers HVM, Buisman CJN (2010) Cathode Potential and Mass Transfer Determine Performance of Oxygen Reducing Biocathodes in Microbial Fuel Cells. Environ Sci Technol 44(18):7151-7156

Vandecandelaere I, Nercessian O, Faimali M, Segaert E, Mollica A, Achouak W, De Vos P, Vandamme P (2010) Bacterial diversity of the cultivable fraction of a marine electroactive biofilm. Bioelectrochemistry 78(1):62-66

Wang Z, Leary DH, Malanoski AP, Li RW, Hervey WJ, Eddie BJ, Tender GS, Yanosky SG, Vora GJ, Tender LM, Lin B, Strycharz-Glaven SM (2015) A Previously Uncharacterized, Nonphotosynthetic Member of the Chromatiaceae Is the Primary CO_2-Fixing Constituent in a Self-Regenerating Biocathode. Appl Environ Microbiol 81(2):699-712

Wang ZJ, Zheng Y, Xiao Y, Wu S, Wu YC, Yang ZH, Zhao F (2013) Analysis of oxygen reduction and microbial community of air-diffusion biocathode in microbial fuel cells. Bioresource Technology 144:74-79

Wei FC, Liang P, Cao XX, Huang X (2011a) Use of inexpensive semicoke and activated carbon as biocathode in microbial fuel cells. Bioresource Technology 102(22):10431-10435

Wei JC, Liang P, Huang X (2011b) Recent progress in electrodes for microbial fuel cells. Bioresource Technology 102(20):9335-9344

Xia X, Sun YM, Liang P, Huang X (2012) Long-term effect of set potential on biocathodes in microbial fuel cells: Electrochemical and phylogenetic characterization. Bioresource Technology 120:26-33

Xia X, Tokash JC, Zhang F, Liang P, Huang X, Logan BE (2013) Oxygen-Reducing Biocathodes Operating with Passive Oxygen Transfer in Microbial Fuel Cells. Environ Sci Technol 47(4):2085-2091

You SJ, Ren NQ, Zhao QL, Wang JY, Yang FL (2009) Power Generation and Electrochemical Analysis of Biocathode Microbial Fuel Cell Using Graphite Fibre Brush as Cathode Material. Fuel Cells 9(5):588-596

Yuan HY, Hou Y, Abu-Reesh IM, Chen JH, He Z (2016) Oxygen reduction reaction catalysts used in microbial fuel cells for energy-efficient wastewater treatment: a review. Mater Horiz 3(5):382-401 doi:10.1039/c6mh00093b

Zhang GD, Zhao QL, Jiao Y, Wang K, Lee DJ, Ren NQ (2012a) Biocathode microbial fuel cell for efficient electricity recovery from dairy manure. Biosens Bioelectron 31(1):537-543 doi:10.1016/j.bios.2011.11.036

Zhang GD, Zhao QL, Jiao Y, Wang K, Lee DJ, Ren NQ (2012b) Efficient electricity generation from sewage sludge using biocathode microbial fuel cell. Water Res 46(1):43-52

Zhang GD, Zhao QL, Jiao Y, Zhang JN, Jiang JQ, Ren N, Kim BH (2011) Improved performance of microbial fuel cell using combination biocathode of graphite fiber brush and graphite granules. Journal of Power Sources 196(15):6036-6041 doi:10.1016/j.jpowsour.2011.03.096

Zhang JN, Zhao QL, Aelterman P, You SJ, Jiang JQ (2008) Electricity generation in a microbial fuel cell with a microbially catalyzed cathode. Biotechnol Lett 30(10):1771-1776

Zhang LX, Liu CS, Zhuang L, Li WS, Zhou SG, Zhang JT (2009) Manganese dioxide as an alternative cathodic catalyst to platinum in microbial fuel cells. Biosens Bioelectron 24(9):2825-2829

Zhang XY, Pant D, Zhang F, Liu J, He WH, Logan BE (2014) Long-Term Performance of Chemically and Physically Modified Activated Carbons in Air Cathodes of Microbial Fuel Cells. Chemelectrochem 1(11):1859-1866

Zhang YP, Sun J, Hu YY, Li SZ, Xu Q (2012c) Bio-cathode materials evaluation in microbial fuel cells: A comparison of graphite felt, carbon paper and stainless steel mesh materials. Int J Hydrogen Energ 37(22):16935-16942

Bioelectrochemical Treatment: Present Trends and Prospective

G. Velvizhi[1,2*], J. Annie Modestra[1,3], Dileep Kumar Yeruva[1,3], J. Shanthi Sravan[1,3] and
S. Venkata Mohan[1,3]

[1] Bioengineering and Environmental Sciences Lab (BEES), Department of Energy and
Environmental Engineering (E&EE), CSIR-Indian Institute of Chemical Technology (CSIR-IICT),
Hyderabad 500007, India
[2] CO$_2$ Research and Green Technology Centre Vellore Institute of Technology (VIT), Vellore, 632 014, India
[3] Academy of Scientific and Innovative Research (AcSIR), CSIR-Indian Institute of Chemical Technology
(CSIR-IICT) Campus, Hyderabad, India

1. Introduction

Treatment of complex wastewater is challenging because of their strength, variability and composition, and hence implementation of next-generation technologies is profoundly important. Although anaerobic processes are effective biological processes for the treatment of complex pollutants, requires more electron donor source for its treatment, which requires co-substrate, that subsequently increases the operational cost. Alternately, in the current research context, bioelectrochemical systems are emerging technologies toward the treatment of complex wastewaters. The functional activity of anaerobic bacteria in concomitance with solid electron acceptor exploits microbial catabolic activities to generate electrons and protons for the degradation of complex pollutants. The presence of artificially introduced electrodes induces the development of potential difference through microbial metabolism of the substrate that acts as a net driving force for pollutant removal and bioelectrogenic activity. The concept of utilizing microbes to generate electricity was initiated long back in 1789 and Luigi Galvani was the first electrochemist who discovered that the muscles of dead frog legs twitched when struck by an electrical spark (Galvani 1791). Alessandro Volta further proved that the movement was occurred due to the metal gable connected between the nerves and the muscles. In parallel, the chemical cell invention was also initiated with lead-acid battery for the storage of power and Grove developed gas battery that was termed as a fuel cell to get practical importance (Grove 1839). In 1959, Francis Bacon developed the hydrogen fuel cell with alkaline electrolyte (Carrette et al. 2001). Further advancement in fuel cell was developed and classified as polymeric membrane, ceramics, liquid electrolyte, alcohol fuel cell, operating at a high temperature of solid oxide fuel cells (SOFCs), molten carbonate fuel cells (MCFCs) and low-temperature proton exchange membrane fuel cells (PEMFCs). However, these fuel cells operated at high temperatures, incur high costs and in some cases, were highly corrosive. To overcome the aforementioned disadvantages, biological fuel cells were designed and operated under ambient and mild reaction conditions (Shukla et al. 2004). The concept of a microbial fuel cell (MFC) was intensively studied for the production

*Corresponding author: gvels@yahoo.com, velvizhi.g@vit.ac.in

of electricity with the presence of simple substrate, such as glucose, acetate, starch, etc. The simple organic substrate is catalyzed by the action of bacterial metabolism to break the chemical energy bond to electrical energy in a fuel cell setup (Venkata Mohan et al. 2008a; Kim et al. 2008; Biffinger et al. 2008). The exoelectrogens that are predominating in MFC is highly regulated and specified toward electrode as a solid electron acceptor for power generation (Venkata Mohan et al. 2008b; Venkata Mohan et al. 2010). There are also some weak electricigens that are not highly electrogenic and rely on soluble electron acceptors under variable conditions and they are now emerging in the domain different from power generation, such as degradation of specific pollutant, bioremediation, synthesis of value-based product and recovery of metal using cathode microbiome (Venkata Mohan et al. 2013; Venkata Mohan et al. 2018). Understanding the activity, dynamics and interaction of microbes in natural and simulating ecosystems of weak electricigens would broaden the scope to wider applications of microbial catalyzed electrochemical systems (Venkata Mohan et al. 2018).

The present chapter discusses the holistic aspects of bioelectrochemical treatment (BET) as bioprocesses for the treatment of wastewater. Although there are many books and papers published on MFC works exclusively on its application emphasizing bioelectrochemical treatment are very few and mechanism for the treatment of the pollutant limited. The chapter exclusively discusses on past and recent perspectives of bioelectrochemical treatment and the detailed mechanism involved at the bacteria electrode reaction for the treatment of pollutant with respect to their biotic/abiotic nature. The chapter also discusses the factors involved in the operation of BET and understanding the role of these essential parameters is crucial in up-scaling the BETs with economic viability for industrial and societal outreach.

2. Bioelectrochemical Treatment (BET)

Bioelectrochemical treatment (BET) is the process in which the reducing equivalents generated by the oxidation of organic matter by biocatalyst could be used for the treatment of complex organic and inorganic pollutants present in the wastewaters rather than the production of bioelectricity (Venkata Mohan et al. 2009, 2010; Mohan Krishna et al. 2010; Velvizhi and Venkata Mohan 2011). The weak electrigens in these processes shift towards the treatment of pollutants rather than recovering energy from the organic substrate. In BET process, the anaerobic fermentation degrades the substrate and the reducing equivalents generated in this processes get accepted by the solid electrodes and instigates the electrochemical oxidation that leads to electrolytic dissociation of the pollutant (Venkata Mohan et al. 2014a; Velvizhi 2018). The basic mechanism involved in these processes is direct and indirect anodic oxidation. In the direct oxidation, pollutants are adsorbed on the anode surface and get degraded by the anodic electron transfer reaction, and in indirect anodic oxidation, the mediators are involved as intermediates for electron shuttling between the electrode and the organic compounds (Venkata Mohan et al. 2009, 2013; Panizza and Cerisola 2009; Velvizhi and Venkata Mohan 2015). BET reported initially for the degradation of simple substrate and extended the research toward degradation of complex pollutants followed by specific pollutants and application as biosensors.

2.1 BET for the Degradation of Simple Substrate

BET utilizes various simple substrates as carbon source and studies are carried out with synthetic wastewater to understand the mechanism of degradation in the fuel cell setup. The synthetic wastewaters were designed with different simple carbon sources along with trace micro and macro minerals that are used for the biological metabolism. Acetate is been extensively used carbon source because it is easy and readily biodegradable substrate with simpler metabolism (Biffinger et al. 2008). Sun et al. (2015) reported that acetate showed lesser initiation time and higher coulombic efficiency (31%) compared to other substrates because the end product of several metabolic pathways for higher order carbon sources (Sun et al. 2015; Biffinger et al. 2008). Glucose was also a simple substrate used widely in bioelectrochemical treatment systems, since it can be oxidized easily by microbes and its degradation consist of multiple conversion pathways (Venkata Mohan ct al. 2007; Venkata Mohan et al. 2013). This substrate is prone to higher electron loss since the possibility of undergoing multiple metabolic pathways is higher hence

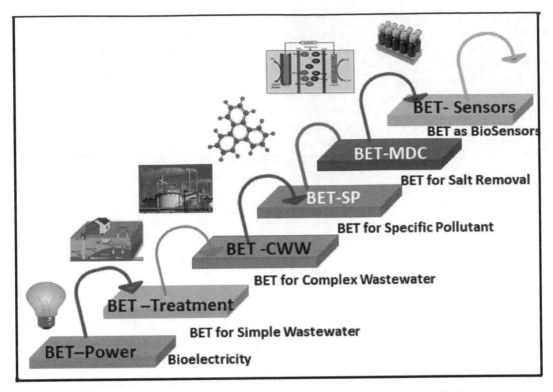

Figure 1: Prospective of microbial catalyzed electrochemical systems applications

it can be considered as electron quenching. Although, the number of electrons generated from glucose is higher than acetate, the energy conversion efficiency is less due to the electron losses (Venkata Mohan et al. 2013). Especially with mixed bacterial suspensions, this problem is significant because of the possible diverse metabolisms. Apart from glucose and acetate several other substrates, viz., sucrose, starch, butyrate, dextran, peptone, ethanol, etc., were also evaluated as anodic substrates in BET (Pant et al. 2010, 2011; Venkata Mohan et al. 2014a). The operation of BET with real field wastewaters is initiated to a certain extent but is not full-fledged and needs concern over the shift for applicability.

2.2 BET for the Degradation of Simple Wastewaters

The traditional wastewater treatment systems were developed to meet the discharge limits only by the carbonaceous and nutrient removal. BET is specifically designed to meet the discharge limits and has the potential of resource recovery and energy production from wastewater (Venkata Mohan et al. 2010). The benefits of BET, in comparison to convention treatment unit is the conversion of waste directly to electricity rather than generating biogas, less sludge production in comparison to activated sludge process and low carbon footprint using indigenous exoelectrogenic bacteria etc. In general, the domestic wastewater treatment plant with aerobic activated sludge treatment and anaerobic sludge digestion technology consumes 0.6 kWh of energy per m³ of wastewater treated, about 50% of that is for electrical energy to supply air for the aeration basins (McCarty et al. 2011). However, if more of the energy potential in wastewater were captured for use and even less were used for wastewater treatment then the wastewater treatment might become a net energy producer rather than a consumer (Logan 2004). Studies reported that MFCs consume only 10% of external energy for their operation when compared to conventional activated sludge processes (Zhang et al. 2013a, b). Hence, BET could use wastewater as a potential feedstock (Venkata Mohan et al. 2009a; Venkata Mohan et al. 2010; Lovely 2006). Bioelectrochemical systems can also use biodegradable organic compounds originating from domestic wastewaters, food, landfill leachates and many others that contain simpler compounds. Domestic wastewater is considered

Table 1: Categorization of Wastewater/Specific pollutants/Sensor through microbial catalyzed electrochemical system

	Wastewater/Specific pollutant/Sensors	Configuration	Power	Removal efficiency/ concentration (mg/l)	Reference
Wastewater	Sugar beet (SBPW)	Double chamber	14.9 mW m^{-2}	97%	Rahman et al. 2018
	Agri-food Waste	Double chamber	27 W m^{-3}	85%	Molognoni et al. 2018
	Municipal	Air-cathode MFC	103 mW/m^2	71.0%	You et al. 2006
	Municipal	Stackable horizontal MFC (SHMFC)	116 mW	79.0%	Feng et al. 2014
	Pharmaceutical Wastewater	Single chamber	1.25 mW	COD 85%	Velvizhi and Venkata Mohan 2011
Specific Pollutant	Distillery wastewater	Single chamber	0.870 mW	COD 72%	Mohanakrishna et al. 2010
	Petroleum contaminates	Single chamber	53.11 mW/m^2	Aromatic 75.54%	Chandrasekhar and Venkata Mohan 2012
	N$_2$O	Double chamber	−200 to 0 mV	0.76 and 1.83 kg N m^{-3}	Desloovert et al. 2011
	Total Nitrogen	Single	-	97%	Wu et al. 2018
	Polychlorinatedbiphenyls	Double Chamber		43%	Xu et al. 2015
	Aniline	Double Chamber	47.2 C	91%	Cheng et al. 2015
	Cr^{6+}; Pb^{2+}; Ni^{2+}	Double Chamber		30; 32.7; 8.9 mg/l	Li et al. 2015b
Salt removal	NaCl	Four MDC- Continues mode		97%	Qu et al. 2013
	Synthetic ground water	Three MDC		51.5%	Hemalatha et al. 2017
	Saline buffer	Batch MDC	29.4 W m^{-3}	80%	Ebrahimi et al. 2018
	Sea Water	SPEEK and QAPPO Membrane reactor	235 ± 7 mW m^{-2}	78.6%	Moruna et al. 2018
Sensors	BOD	Double chamber	32-1280 mg l^{-1}		Modin and Wilén 2012
	Levofloxacin	Double chamber	0·1-1000 μg l^{-1}		Zeng et al. 2017
	Cu^{2+}	Double chamber	2 mg l^{-1}		Jiang et al. 2015
	DO	Submersible	8·8 mg l^{-1}		Zhang and Angelidaki 2012
	Chromium (VI)	Double chamber	2.5-60 mg		Wu et al. 2017
	VFA		5-100 mM		Jin et al. 2017

SPEEK: Sulfonated sodium poly (ether ether ketone; QAPPO: Quaternary ammonium poly (2,6-dimethyl 1,4-phenylene oxide)

to be simple and highly biodegradable in nature with low substrate load and hence, power generation retains for only a few hours with a COD removal efficiency of 66% (Venkata Mohan et al. 2009b). Dairy wastewater is rich in nitrogenous compounds such as milk proteins that also may interfere with the power generation due to weak electricigens though reported COD removal efficiency was 95% (Venkata Mohan et al. 2010) and canteen based food waste reported removal efficiency of 65% (Goud et al. 2011).

2.3 BET for the Degradation of Complex Wastewaters

Bioelectrochemical systems are also reported to be efficient to treat complex wastewaters. BET functions as an integrated wastewater treatment system manifested by multiple unit operations such as anaerobic fermentation, electrolytic dissociation and electrochemical oxidation in a single system (Venkata Mohan et al. 2009a). Distillery wastewater with a maximum substrate loading of 16.2 kg COD/m^3 enumerated COD removal efficiency of 72%, inferring that the organic pollutants present in the spent wash wastewater are adsorbed on the anode surface and get degraded by the direct anodic electron transfer reactions and by the mediated oxidants. The oxidants produced through indirect oxidation process will have a significant influence on the complex organic removal efficiency (Mohana Krishna et al. 2010). Pharmaceutical wastewater reported COD removal efficiency of 85% in BET system for maximum organic loading of 3.96 kg COD/m^3, whereas in conventional biological processes 32% was reported (Velvizhi and Venkata Mohan 2011). However, BET retained their performances by direct anodic oxidation (DAO) mechanism that deprotonated substrate and instigated the formation of oxidation species on the anode surface due to the manifestation of bioelectrochemical reactions. The self-induced electrogenic microenvironment developed in the system showed enhanced treatment (Velvizhi and Venkata Mohan 2011). Hydrocarbon wastewater generated from petrochemicals, refinery, coking or hydrocarbon manufacturing industries that were treated for remediation of the mixture of phenanthrene and benzene to determine their applicability for *in situ* and *ex situ* treatment of hydrocarbon contaminated groundwater. This study observed >90% removal of the petroleum hydrocarbons and 79% bromate by achieving power density output of 6.75 mW/m^2 (Adelaja et al. 2017). Palm oil mill effluent with high BOD and COD of 30,000 mg/L and 50,000 mg/L reported maximum power density of 45 mW/m^2 with volumetric power density of 304 mW/m^3 and 45% COD reduction (Baranitharan et al. 2013). Mateo-Ramírez et al. (2017) reported maximum treatment efficiency of 51% when the slaughterhouse waste was used in BET system (Mateo-Ramírez et al. 2017). The complex nature of wastewater, which is of low biodegradable nature, also reported good treatment efficiency in BET system because of the flexible processes of oxidation and reduction mechanism owing to the presence of solid electrode as electron acceptor (Sreelatha et al. 2015).

2.4 BET for Specific Pollutant Removal

2.4.1 BET for Metal Removal and Recovery

Electrochemical treatment techniques could carry out redox reactions that will aid in metal removal/ recovery but the usage of solvents/chemicals and the application of potential using expensive electrode materials are the major economic constraints. The bioelectrochemical systems are cost-effective and eco-friendly processes because of the presence of bacteria as biocatalyst and have the capability to undergo redox reactions toward the metal recovery (Nancharaiah et al. 2015 and 2016). The bio-potential developed in bioelectrochemical systems by the degradation of the organic substrate acts as the driving force toward metal removal and recovery (Modestra et al. 2017). The bioelectrochemical systems usually comprise of a biotic anode and an abiotic cathode (or an abiotic anode and a biotic cathode) to accomplish the metal recovery. The metal ions can be recovered by using the *in situ* and the *ex situ* generated potential. Bacteria play a crucial role in the biotic systems by utilizing the organic substrate, thereby, liberating reducing equivalents. The reducing equivalents act as the power source for bioelectrochemical systems in reducing the metal species (Modestra et al. 2017). Various metals, viz., Ag, Au, V, Pb, Cd, Cr and Cu, have been recovered in the bioelectrochemical systems by the action of biotic anode and abiotic cathode (Wang et al. 2008; Zhang et al. 2012). Metal wastes viz., Se (IV), V (V), Ag(I), Cu(II), Mn(IV) and

Cr(VI) are removed by using a bioelectrochemical system (Nancharaiah et al. 2015; Zhang et al. 2009; Rhoads et al. 2005). The use of metals as electrode material/electron acceptor at anode chamber also has been studied that enables high electron transfer and transition in metal oxidation state. The majority of the reduction reactions take place at the cathode that aids in the removal and recovery of metal/degradation of oxidized pollutants (Nancharaiah et al. 2015). Non-precious and highly conductive hybrid electrodes can be developed to generate a higher bio-potential in bioelectrochemical systems to recover the heavy metals by the *in situ* potential rather than applying external potential. Copper was removed and recovered as deposits on the cathode when the cathode chamber was fed with a fly ash leachate (Tao et al. 2014). Bioelectrochemical systems for the removal/recovery of metals have been studied through extracellular electron transport capabilities (Harris et al. 2010) and through direct electron transfer mechanisms (Patil et al. 2012).

2.4.2 BET for Hydrocarbon

Specific hydrocarbon such as nitrobenzene, phenols, p-nitrophenol, phenanthrene and pyrene, etc., generated either from petrochemicals, refinery, coking or hydrocarbon manufacturing industries that are considered to be a potential feedstock in the bioelectrochemical systems. Studies were reported to remove nitrobenzene in a cathode compartment with a removal rate of 1.29 mol m^{-3} TCC d^{-1} (Mu et al. 2009). Luo et al. (2009) reported 95% of phenol degradation with 60 hours of retention time in BES system (Luo et al. 2009). Shen et al. (2014) reported that p-nitrophenols degradation rate was increased from 6.16 to 6.66 mol m^{-3} d^{-1} with the rise of current density from 0 to 4.71 A m^{-3} when the external current was applied in an up flow anaerobic sludge blanket reactor coupled with BES system. Phenanthrene and pyrene degradation was 99.47% and 94.79% in ferric hydroxide addition and sediment MFC conditions, respectively (Yan et al. 2012). Polyaromatic hydrocarbons (PAH) degradation was observed to be removed 82% in a dual chamber system (Morris et al. 2009). BET operation showed effective degradation of real field petroleum sludge associated driven in a single chamber system with an abiotic open-air cathode. The study infers that self-induced electrochemical oxidation developed in the system helps to degrade soluble [aliphatics, aromatics and NSO (nitrogen, sulfur and oxygen)] and insoluble (asphaltenes) fractions (Venkata Mohan and Chandrasekhar 2011). Studies also reported on the treatment of real-field petrochemical wastewater by considering BET as a post-treatment processes (Yeruva et al. 2015). SBR integrated with BET reported removal efficiency of 54.5% the presence of electrode assembly in BET system facilitates bio-potential/bioelectrogenic activity that enhances the biodegradability of complex compounds as well as the intermediates that are produced in the SBR processes (Yeruva et al. 2015).

2.4.3 BET for Nitrate and Sulfate Removal

Generally, nitrate removal through heterotrophic biological processes requires external organic matter that could result in excessive biomass production. BET systems are effective to treat nitrate through autotrophic denitrification processes in the cathode chamber due to the low redox potential in the abiotic condition by accepting the electrons generated in the anode chamber (Clauwert et al. 2009; Velvizhi and Venkata Mohan 2013). Studies also preceded using biotic cathode for simultaneous removal of organic matter and nitrate through nitrification and denitrification processes (Clauwaert et al. 2007). BET is effective to treat pharmaceutical wastewater containing nitrate with an efficiency of 47% in a single chamber system (Velvizhi and Venkata Mohan 2011). Effective biofilm formation on the electrode influences the enhanced treatment of pollutant in BET system through cell-to-cell signaling (Parsek and Greenberg 2005). Quorum sensing (QS) signaling benefited the diffusible signal factors release in the BES that enhances the extracellular polymeric substances that would improve the biofilm formation activity and result in higher current (Parsek and Greenberg 2005). Studies reported that increasing the current improves the nitrate removal performance but also results in by-product accumulation (Huang et al. 2013). Biofilm on the electrodes also shows significant acidification pH drift in the anodic biofilms and the cathodic biofilms that also influences on nitrogen removal (Cheng et al. 2012). Denitrification process also depends on the characteristics of the electron donor used and the ratio of N and COD available in the wastewaters (Pham et al. 2009; Khudzari et al. 2016). Biocathode reported simultaneous nitrification and denitrification of 99.9% nitrate removal efficiency (Li et al. 2017).

Sulfate-reducing bacteria (SRB) and sulfide oxidizing bacteria (SOB) reduces the sulfate concentration in the wastewaters however, in BET systems, the presence of electrodes induces the reduction of sulfate in the elemental form (Rabaey et al. 2006). The deposited sulfur on the electrodes acts as a mediator for the transfer of electrons to the anode (Dutta et al. 2009). They act as good redox shuttles between bacteria and insoluble electron acceptors and impart enhanced electron transfer (Straub et al. 2004). Sulfide is a well-known redox shuttle between bacteria and insoluble electron acceptors and imparts enhanced electron transfer (Straub et al. 2004). Zhao et al. (2009) reported 91% of sulfate removal using activated carbon cloth, carbon fiber veil composite anode and air-breathing dual cathodes (Zhao et al. 2009). Lee et al. (2014) reported removal of sulfate and carbon in MFC using SRB and SOB in the anodic biofilm system (Lee et al. 2014).

2.4.4 Microbial Desalination

Microbial desalination cell (MDC) is a promising approach for sustainable low-cost desalination processes that is based on the transfer of ionic species out of water in proportion to the current generated by bacteria (Cao et al. 2009; Nikhil et al. 2015; Hemalatha et al. 2017). MDC is configured through bipolar processes by addition of saline chamber between the anode and cathode chamber. The transfer of ionic species from the middle chamber results in desalination without electrical energy or water pressure (Cath et al. 2006). Cao et al. (2009) worked as a proof-of-concept for desalinating saline water in a three chambered setup reporting 90% salt removal (Cao et al. 2009). Jacobson et al. (2011) reported up flow MDC (UMDC) reactor for the reduction of salt from NaCl salt solution and artificial prepared sea water and inferred a TDS removal of 90%, while producing an energy content of 1.8 kWh. The study also compared with conventional reverse osmosis and concluded that the net energy benefit by using UMDC with RO was 4.0 kWh and UMDC has 100% water recovery, whereas RO has only 50% water recovery where the remaining was brine water (Jacobson et al. 2011). Several studies also reported on the reduction of TDS in wastewater in BET system through electrochemical salt splitting or electrolytic decomposition process where salt gets disassociated under the influence of *in situ* potential developed in BET system (Venkata Mohan et al. 2009a; Venkata Mohan and Srikanth 2011). Studies have been reported that the addition of a number of cells could significantly increase the desalination rate (Velvizhi and Venkata Mohan 2017). The multi-electrode system reported (32%) significant treatment efficiency of TDS in comparison to single electrode setup (15%) (Velvizhi and Venkata Mohan 2017).

2.5 BET as Biosensor

Bioelectrochemical systems has wider scope to be an analytical device since it contains several components viz., biological sensing elements (bacteria), substrate for the production of electrons and protons, electrodes for accepting and donating the electrons, membrane to transfer the protons and circuit connections to harvest the energy (Lei et al. 2006). Compared with conventional electrochemical sensors, biosensors have a wider applicability in remote areas for wastewater monitoring with its high sensitivity and stability (Jouanneau et al. 2014; Sara et al. 2006). The MFC-based biosensor devices could be used for testing water quality parameters, such as BOD test; generally, in conventional assays the water sample is estimated at 20°C for 5 days which is time-consuming. Physical transducers are used to measure the dissolved oxygen (DO) and the electrical/optical signals are used to estimate BOD value, hence, it also requires a power supply (Peixoto et al. 2011). However, BET-based BOD sensors are self-powered, and it readily provides voltage and current by substrate degradation and doesn't require any transducers. It responds fast, has good sensitivity, measures with a wide range and is low maintenance (Jouanneau et al. 2014). In this method, the wastewater gets treated and the current output shows a linear relationship with organic carbon strength of wastewater (Chang et al. 2004). BET can also be used for measuring biofouling on the metal surface and is a main problem in the oil and gas industry. The conventional sensor used for measuring biocorrosion is by applying an external field to detect electrical resistance changes across biofilm (George et al. 2006). The MFC-based biosensor does not require external voltage; the *in situ* potential developed in the system infers the presence of electrogenic activity that corresponds to corrosion. MFC sensors are also used to measure the toxic chemicals and conventional chromatographic methods, such as GC, HPLC, GC-MS etc., are used to measure the toxicity, however, they are expensive and are less stable. The MFC-based biosensors express the toxicity based on the inhibition rate of electrogenic rate in bacteria that correspondingly reduces the current output (Yang et al. 2015). BES also has several other potential applications to monitor an anaerobic digester such as an up-flow anaerobic fixed bed (UAFB), a gas-liquid separator and wall-jet MFCs (Liu et al. 2011; Kaur et al. 2013; Feng et al. 2013). These sensors tend to be low cost and self-powered with online monitoring capabilities and they are tested in laboratory conditions, but in real conditions, many parameters might affect the response of the BES and its biosensing capabilities and more research is still needed to test these devices under operating and real conditions. The operation of BET depends upon several factors viz., reactor configuration, electrode placement, microbe-electrode interactions, etc., for improvising the bio-electrodynamics for treatment of diverse wastewater streams.

3. Factors Influencing BET Performance

BET is a hybrid process that is partly microbial assisted electrochemical oxido-reduction and partly microbial metabolic activity for the treatment of pollutant compounds (Mohana Krishna et al. 2010; Venkata Mohan et al. 2014; Luo et al. 2009; Rozendal et al. 2008; Vargas et al. 2000). In BET, microbial energy was generated by the formation of redox equivalents via converting organic substrate into metabolic compounds by the catabolic and anabolic reactions. These redox compounds are capable enough of participating in extracellular electron transfer mechanism to transfer electrons to an external terminal acceptor i.e., electrode (Venkata Mohan et al. 2009, 2010; Yeruva et al. 2016). This creates a sink of electrons, while protons are migrated toward the cathode either through membrane or solution due to the difference in potential. Electrons pass through the circuit from the sink (anode) across the external load. The generated *in situ* potential during the transportation of redox equivalents in the BET process has the potential to treat wastewater. The *in situ* generated redox equivalents are adsorbed on the respective electrodes, where they can generate the oxidative species, by oxidizing the pollutant compounds present in the wastewater (Rozendal et al. 2008). Generally, the polluted compounds present in the wastewater itself act as an electron donor/acceptor, and its remediation gets manifested either through anodic or cathodic reduction under defined conditions. The pollutant degradation depends upon several factors, such as electrode material, substrate, biocatalyst, reactor configuration, microenvironment etc., will govern the performance of biotic anode as well as interactions between them (Modestra et al. 2016). Biocompatibility, mechanical strength, the longevity of electrode materials etc., are few of the important considerations to be taken into account, while designing bioanodes for catalyzing efficient redox reactions (Wei et al. 2011; Kumar et al. 2013). Substrate type, complexity and origin also play a prominent role in affecting the performance of bioanode as the number/concentration of reducing equivalents is dependent on the ability of bacteria to effectively metabolize the substrate (Zhao et al. 2009). Nature and origin of biocatalyst signify the metabolic and catalytic activity of substrate as well as its adherence on the electrode surface. More commonly, the mixed microbiome has been used as a biocatalyst in bioelectrochemical systems for the treatment or remediation of wastewaters. Mixed bacteria offer several advantages in comparison to a single culture in terms of ease of operation, maintenance, economic viability, robustness etc. Based on the type of wastewater/substrate/targeted pollutants, specific bacteria will be enriched from the mixed culture gradually with the course of operation that will help in enhancing the treatment (Sreelatha et al. 2015). Biofilm formation is one of the vital steps that enhance the electron transfer rate as well as substrate degradation. A well-established biofilm facilitates efficient interaction that enables rapid electron transfer reactions required for developing redox potential as well as compounds reduction. Mixed culture biofilm comprises of a diverse group of bacteria enriched in accordance with the substrate/wastewater with complex pollutants. Since the rate kinetics for electron delivery will be higher with biofilm in comparison to suspension bacteria, the aforementioned regulating parameters must be well optimized for facilitating significant compounds remediation (Harris et al. 2010). Understanding the mechanism in depth also paves an easy way for the removal of pollutants.

4. Mechanisms for the Removal of Pollutant in BET System

4.1 Role of Anodic Potential in the Treatment of Bioelectrochemical Treatment

Bioelectrochemical reactions required for treatment or reduction of pollutants is critically regulated by the interactions between bacteria and electrode (Venkata Mohan et al. 2014a, b; Patil et al. 2012). Electrode present at anode chamber acts both as electron acceptor as well as donor to catalyze the redox reactions (Figure 2). Biotic anode specifies the presence of bacteria as biocatalyst at anode that catalyzes the reactions necessitated toward treatment/reduction/recovery (Huang et al. 2013; Pant et al. 2012). Substrate metabolism by bacteria delivers electrons extracellularly into the anode chamber, and the fate of electrons is essentially dependent upon the redox potential between donor and acceptor (Venkata Mohan et al. 2008a). Besides, the interaction between bacteria and anode significantly promotes the development of bio-potential, an important reaction regulating parameter in bioelectrochemical systems (Mohana krishna et al. 2015). The anodic microorganisms play a vital role in the treatment of the organic matter in wastewater by metabolically catalyzing electrochemical oxidation. The principal mechanism involved in anodic degradation is via direct anodic oxidation (DAO) and indirect anodic oxidation (IAO) reactions. DAO, which is initiated by the microbial metabolism, will lead to the production of electrons (e^-) and protons (H^+) from simple organic compounds that migrate to the electrode surface ($E[]$) and develop the *in situ* bio-potentials (Venkata Mohan and Srikanth 2011). Due to the potential variation, reducing equivalents further react with a water molecule and lead to the formation of primary oxidizing agents such as nascent oxygen (O^*) and hydroxyl (OH^-) free radicals by electron transferring reactions

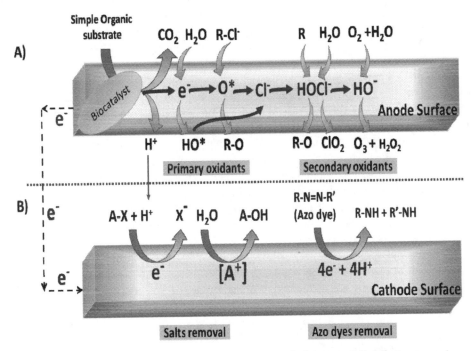

R- Pollutant compound; A-X- Salt compound; A- cationic compound; X- Anionic compound

Figure 2: Pollutant treatment mechanism by bio-electrochemical induced oxidants A) Primary and secondary oxidants formation on anode surface area, B) Salt removal and azo bond splitting on cathode surface

(Israilides et al. 1997). These primary oxidants get adsorbed on the anode at active sites and further leads to DAO either individually or in combinations with Cl⁻ ions present in the wastewater. Secondary oxidants are evolved in the system by an electrochemical reaction between the primary oxidants and pollutants in wastewater on the anode surface. These secondary oxidants of chlorine dioxide (ClO_2), hypochlorite (OCl^-), hydroxyl radicals (OH^*), ozone (O_3) and hydrogen peroxide (H_2O_2) will be mediated through the IAO process (Mohanakrishna et al. 2010; Chatzisymeon et al. 2006; Korbahti et al. 2007). The quantity of secondary oxidants formation is dependent on the increase in primary oxidants concentration. In a direct oxidation, pollutants are adsorbed on the anode surface and get degraded by the anodic electron transfer reaction, while in indirect anodic oxidation, mediators act as intermediates for electron shuttling between the electrode and the organic compounds (Venkata Mohan et al. 2009, 2013; Panizza and Cerisola 2009; Velvizhi and Venkata Mohan 2015; Guven et al. 2008). These oxidizing agents have a significant impact on wastewater treatment, especially on color removal, TDS and metal reductions. The formation of oxidative free radicals is dependent on the pollutants present in the wastewater, and the general mechanism of anodic oxidative formation is represented as below (where R is an organic substrate, X is pollutant compound and E [ ] is electrode surface area)

$$O + e^- \rightarrow O^*$$

$$R + \text{Biocatalyst} \rightarrow e^- + H^+ + CO_2$$

$$H_2O + e^- \rightarrow O^*$$

$$O^* + E [\ ] \rightarrow E [O^*]$$

$$X + E[O^*] \rightarrow X - O^* + E [\ ]$$

Formation of primary oxidants:

$$H_2O + E [\ ] + Cl^- \rightarrow E [ClOH^*] + H^+ + 2e^-$$

$$X + E [ClOH^*] \rightarrow X - O + E [\ ] + Cl^- + H^+$$

$$H_2O + e^- + E [\ ] \rightarrow E [OH^*] + O_2 + 3H^+ + 2e^-$$

$$X + E [ClOH^-] \rightarrow X - O + Cl^- + E [\ ] + 3H^+ + e^-$$

Formation of secondary oxidants:

$$H_2O + E [ClOH^-] + Cl_2 \rightarrow E [\ ] + ClO_2 + 3H^+ + 2Cl^- + e^-$$

$$O_2 + E [OH^-] \rightarrow E [\ ] + O_3 + H^+ + e^-$$

$$H_2O + E [OH^-] \rightarrow E [\ ] + H_2O_2 + H^+ + e^-$$

4.2 Role of Cathode Potential in the Treatment of Bioelectrochemical Treatment

Most of the pollutant compounds present in wastewater are nitrogenous compounds, sulfur, phosphate, etc., that are electronegative that aid in accepting the electrons more efficiently, thus, facilitating the reduction (Venkata Mohan et al. 2010; Velvizhi and Venkata Mohan 2011; Venkata Mohan and Chandrasekhar 2011). Metals that also act as electron acceptors have been widely used in bioelectrochemical systems for their reduction and recovery at the cathode (Gregory and Lovley 2005; Catal et al. 2009). Reduction reactions are very crucial for the treatment of pollutants as well as its complete degradation (Figure 3). Cathode that majorly acts as a counter electrode in bioelectrochemical systems facilitates reduction reactions by accepting the electrons and protons from anode chamber. Most of the cathode reactions are abiotic in general at which the pollutant of interest will be placed for its reduction (Venkata Mohan et al. 2008b). Xenobiotic compounds/pollutants/metals/any compounds can act as good electron acceptors at the cathode that enables reduction reactions as well as treatment (Gorby and Lovley 1992; Kondaveeti et al. 2014). Cathode potential is one of the limiting factors that determine the efficiency

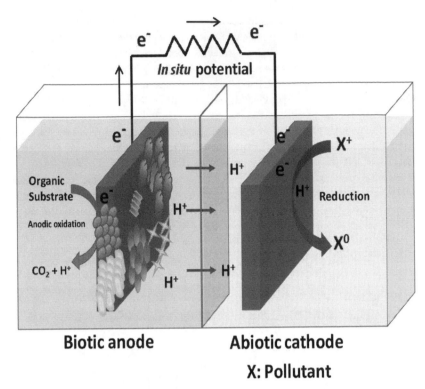

Biotic anode   Abiotic cathode

X: Pollutant

Figure 3: Electrode-bacteria interactions represented for various microenvironments at biotic anode treatment of pollutants in bio-electrochemical treatment systems. X: denotes pollutant

of the bioelectrochemical system, both in terms of treatment and power generation. In general, cathode potentials are reported to be significantly less in comparison to anode potentials in several microbial electrochemical systems. However, several strategies can be put forth to increase the half-cell potential of cathode that aid in increasing the overall efficiency of the system. The biotic cathode can be operated under various microenvironments such as aerobic, anaerobic and anoxic based on the requirement of necessary conditions for pollutant degradation (Srikanth and Venkata Mohan 2012b; He and Angenent 2006). The presence of electrochemically active or specific bacteria at cathode enhances the reduction reactions, thereby, increasing the performance of bioelectrochemical systems (Iskander et al. 2016; Venkata Mohan and Srikanth 2011).

Treatment efficiency of the cathode is influenced by various factors, such as the nature of pollutants, concentration and potential difference (Liu et al. 2009). Substrate degradation at anode lies on the efficiency of cathodic reductions and aid in the development of *in situ* potential (Rozendal et al. 2008). Biocathodes can be operated in different microenvironments via aerobic, anaerobic and microaerophilic conditions depending on the nature of pollutant and type of bacterial species. In aerobic biocathodes, favorable redox conditions support the rapid metabolic activities, thus, resulting in high substrate removal (Figure 4). Oxygen is the most common and widely reported terminal electron acceptor in various microbial catalyzed electrochemical systems. However, the pollutants or metals also act as electron acceptors at cathode based on the thermodynamic hierarchy in electronegativity. Aerobic biocathodes are rapidly emerging to treat several pollutants with simultaneous generation of bioelectricity. Energy gain associated with aerobic metabolism is higher in comparison to respective microenvironments that, in turn, are associated with the generation of reducing equivalents necessary either for treatment/reduction or power generation. In the case of certain pollutants/metals reduction, the prevalence of aerobic conditions might inhibit/retard the process and mandates the requirement of anaerobic/microaerophilic conditions.

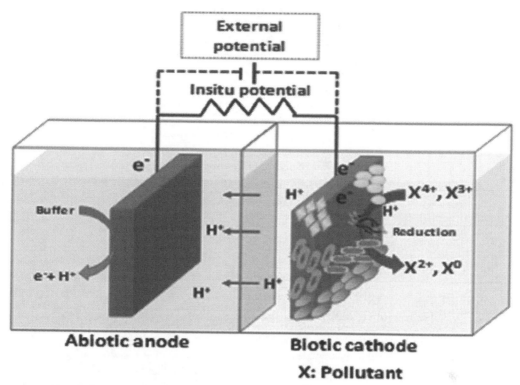

Figure 4: Electrode-bacteria interactions represented for various microenvironments at biotic cathode. X: denotes pollutant of any nature at cathode (metal/wastewater etc.)

In the case of the anaerobic cathode, the possibility of treatment of specific pollutants and toxic compounds present in the wastewater gets increased by their electron acceptance tendency as a terminal electron acceptor. Unlike anode, cathode does not exhibit higher current densities. Hence, an alternative is to use biocathodes to attain higher current densities along with pollutant removal facilitated through enhanced microbial metabolic activities. Studies are carried out in microaerophilic conditions at the cathode that proves the advantage with simultaneous oxidation and reduction behavior in some wastewaters like azo dyes that needs complete mineralization. The anaerobic condition helps in splitting the azo bond, while the aerobic condition helps in mineralization of dye metabolites (Venkata Mohan et al. 2013a, b; Sreelatha et al. 2015; Yeruva et al. 2018). Cathode facilitates treatment efficiency by transferring the electrons to electron acceptors, like azo dyes, nitrobenzene, nitrates, sulfates and metals, etc. The electron acceptance is dependent on the respective thermodynamic hierarchy of the compounds present in the wastewater. Generally, nitrates are well-known electron acceptors accounting for denitrification that reduces nitrate or nitrite to generate a proton motive force (Virdis et al. 2008; Venkata Mohan et al. 2013a). Nitrate gets converted to nitrite by consuming two electrons which further accepts one more electron forming nitric oxide (NO) and gets reduced to nitrous oxide by accepting one more electron and finally forms nitrogen by accepting another electron. On the whole, conversion of nitrate to nitrogen requires five electrons (Venkata Mohan et al. 2013a, b; Clauwaert et al. 2007; Sander 2017; Velvizhi et al. 2014), whereas few heavy metals containing groups have high redox potentials and could be used as electron acceptors while they themselves are reduced and precipitated on electrode surface prior to separation and removal. Metal removal is attributed to abiotic and biotic cathodes that undergoe the assimilation and dissimilation process. Aromatic organic compounds, such as nitrobenzene, phenols and volatile organic compounds, etc. existing in the pollutants can be treated in the cathode chamber with combinations of anaerobic reductive and aerobic oxidative pathways. These compounds can be removed by the acceptance of electrons either directly or through mediators. BET can be employed for the dechlorination of the chlorinated wastewater via direct electron transfer, redox mediator-assisted transfer and hydrogen-mediated electron transfer (Zhang et al. 2018; Yeruva et al. 2015). The oxidizing species also react with primary cationic species, viz., Na + and K +, under bio-potential leading to their removal that is especially observed as TDS removal. The dechlorination mechanism at the cathode is as follows: E is electrode surface, R is organic compound and X is halogens (Cl⁻, Br⁻). The dechlorination rate is dependent upon the imposed potential and the microbial community at the cathode. The bioremediation method by BET cathode provides the possibility to reduce the operational cost with its biological and electrochemical remediation. Along with bioremediation, it also simplifies the process and helps in the recovery of potential energy. As these free radicals and oxidation species increase the biopotential at cathode, it will aid in the increase the rate of removal of other pollutants (Torres 2014). A targeted pollutant or any complex waste stream can be degraded by enriching specific bacteria that are capable of degrading a particular compound of interest. Growth and exposure to specific pollutant/waste stream will result in gradual adaptation of specific bacteria that are able to thrive on specific waste streams toward enhancing the reduction or treatment efficiency.

$$E\,[\,]+e^-+R-X\ \rightarrow R-H+X^-$$
$$E\,[\,]+e^-+M_{ox}+R-X_2\ \rightarrow R-H+M_{red}+X^-$$
$$E\,[\,]+e^-+H^+\ \rightarrow R-H+X^-$$

4.3 Other Strategies for Enhanced Treatment

Apart from enriched biocathodes, the applied voltage can also be employed as one of the strategies to enhance treatment efficiency that intercalates with the bioelectrochemical reactions that occur on the electrode surface at which maximum interactions take place (Figure 5). This strategy also aids in the rapid development of biofilm that enriches electrochemically active microbes to participate in extracellular

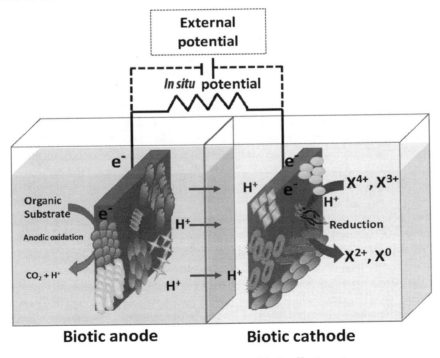

Figure 5: Electrode-bacteria interactions represented for various microenvironments at cathode and anode for treatment of pollutants in bio-electrochemical treatment systems. X: denotes pollutant of any nature at cathode (metal/wastewater etc.)

electron transfer (EET) responsible for the enhancement in reduction reactions (Kato et al. 2012). In addition, certain membrane proteins or redox active components will be expressed in accordance with the external voltage applied based on the reduction potential of targeted pollutant. The expressed redox metabolites and membrane proteins act as electron carriers that enhance the reduction reactions. The efficiency of biocathode, in turn, is dependent on the microenvironment of the anode (abiotic/biotic) as well since it provides the reducing equivalents necessary conditions for compound reduction/product recovery/bacterial growth. Operating pH is also one of the critical parameters that govern the biocathode efficiency in terms of growth and metabolic regulation of the bacterial community. Besides the treatment, biocathodes have been widely employed to recover valuable products such as H_2O_2, platform chemicals, etc., from waste substrates. Although several studies have been carried out employing biocathodes for bioremediation, the applications can be widespread by expanding the research toward optimization of process parameters for recovery of value-added products, bioelectricity, electrosynthesis, etc., that will mark the domain of biocathodes as a sustainable and eco-friendly approach. If the cathode is considered as a working electrode, mostly an anode will be chosen abiotic and an external potential will be applied on the system to drive electrons from electrolyte dissociation. On the contrary, the anode can be chosen to be biotic as well if wastewater remediation is considered to be an application with simultaneous production of reducing equivalents; the choice of the anode to be abiotic or biotic is dependent on the experimental criteria/targeted application. Electrolytic or buffer solution at anode undergoes electrolysis and provides the electron and proton a source required to facilitate reduction reactions at the cathode. However, the electrolysis reactions also take place under the application of external potential either on the whole cell or on the working electrode. Most of the biocathode studies that involve bacteria as biocatalyst at cathode may either choose abiotic anode to provide electron source or select biotic anode to provide reducing equivalents as well as to enable waste remediation.

5. Conclusions

The chapter summarizes BET as a potential process for degradation of organic/inorganic pollutants along with salts and metals contributing to the effective treatment of complex wastewaters (Yeruva et al. 2016; Nancharaiah et al. 2016; Velvizhi and Venkata Mohan 2015; Venkata Mohan et al. 2009; Luo et al. 2009). The operating conditions' mechanisms and reaction interferences involved at the anode/cathode in the presence of a solid electrode has been discussed. It has the scope of extensively changing the course of wastewater treatment by the integration of microbiological catalysis to electrochemistry. The presence of electrodes develops an *in situ* bio-potential in the microenvironment influencing the overall performance of BET for a breakdown of complex organic compounds and inorganic salts from diverse wastewaters (Vamsi Krishna et al. 2014; Cusick et al. 2011). The reduction and oxidation of multiple pollutants, in the presence of the electrodes, develop anode and cathode potentials that create an overall potential difference and increase the pollutant breakdown. The breakdown of complex pollutants would result in the formation of secondary compounds that helps in shuttling the electrons toward electrodes, thereby, achieving the treatment of wastewater as well (Venkata Mohan et al. 2014; Modestra et al. 2016). The development of potential gradient by these conductive materials offers a variety of possible oxidative and reductive mechanisms and has opened up new perspectives in BET application in the waste management domain. It can be an interdisciplinary domain involving electrochemistry, microbiology and systems engineering for wastewater treatment and achieving environmental sustainability. In comparison with the conventional physico-chemical processes, BET offsets energy, cost and chemical inputs as well as treatment time and has the capability to augment the existing biological wastewater treatment efficiency by integrating/embedding with the biological wastewater treatment systems. However, full-scale implementation of BET is not established that poses various challenges in the course. Research on BET has been conducted so far at the laboratory-scale and requires further studies on flexible, efficient and economic configurations for the translational outcome.

6. Future Perspectives

Studies on BET need to be focused majorly on improving the bioelectrocatalytic performance while developing low-cost and efficient electrode materials. BET as technology in the waste management sector needs to overcome various physical and chemical factors involved in its performance along with cost benefits and techno-economics involved. The good impact of BET in its operation also inter-depends on the electrometabolism, electrode materials, substrate and the reactor design considerations. These multiple challenges need to be addressed prior to implementation at a larger scale of operation. The cathode in BET is the major bottleneck where the oxidation-reduction kinetics majorly depends, hence, its spatial arrangement with respect to the anode needs to be specifically focused to attain higher multi-pollutant degradation. BET configurations with minimal electrode placing are a suitable option that lowers internal resistance than the high electrode spacing configurations. Conventional bioprocesses have certain limitations with respect to losses in the process and removal of salts and metal contaminants. In order to bring down these losses, alternates for utilizing the advantages of BES are utmost essential. The synergistic approach of integrating BET with the conventional processes can help to increase the efficiency of waste remediation along with the added advantage of salts and metal removal. The electrochemically-driven microbial interactions in BET could improve the catalytic rates of reaction as compared to the conventional bioprocess. These electrocatalytic reactions raise the microbial electrometabolic capabilities, thereby, influencing the overall pollutant degradation efficiencies. These electrometabolic reactions also influence on the decrease of electrochemical losses related to substrate-hydroxide-oxide binding and increasing waste utilization capabilities. The specificity of BET in wastewater treatment can extend its advantages by integration with the ecological engineered systems (EES), constructed wetland and ETPs/STPs for increasing the treatment efficiencies. The development of modular electrogenic systems for wastewater remediation is an option that can be best suitable for application in remote areas and wastewater at point sources. Stacking BET systems and integrating with other biological processes can considerably increase the treatment efficiency for different complex wastewaters. The commercialization aspect of BET can improve the sanitation, a major issue among all the developing countries and has a role to play addressing the global environmental problems.

Acknowledgment

The author wishes to thank the Director, CSIR-IICT for support and encouragement and GV wishes to thank DST-SERB, Government of India, in the form of Young Scientists Grant (SERB-Grant number YSS/2015/001438) and the management of VIT for the kind support. JAM, DKY and JSS acknowledge CSIR for providing research fellowship.

References

Adelaja O, Keshavarz T, Kyazze G (2017) Treatment of phenanthrene and benzene using microbial fuel cells operated continuously for possible in situ and ex situ applications. Int Biodeterior Biodegradation 116: 91-103.

Aelterman P (2009) Microbial fuel cells for the treatment of waste streams with energy recovery. Ph.D. Thesis, Gent University, Belgium.

Baranitharan E, Khan MR, Prasad DMR, Salihon JB (2013) Bioelectricity generation from palm oil mill effluent in microbial fuel cell using polacrylonitrile carbon felt as electrode. Water Air & Soil Pollution 224 (5): 1533.

Biffinger JC, Byrd JN, Dudley BL, Ringeisen BR (2008) Oxygen exposure promotes fuel diversity for Shewanella oneidensis microbial fuel cells. Biosens Bioelectron 23:820-826.

Butti SK, Velvizhi G, Mira LK, Sulonen Haavisto MJ, Koroglu EO, Cetinkaya AY, Singh S, Arya D, Modestra JA, Vamshi Krishna K, Verma A, Ozkaya B, Lakaniemi AM, Puhakka JA, Venkata Mohan S (2016)

Microbial electrochemical technologies with the perspective of harnessing bioenergy: Maneuvering towards up scaling. Renew Sust Energy Rev 53:462–476.

Cao X, Huang X, Liang P, Xiao K, Zhou Y, Zhang X, Logan BE (2009) A New Method for Water Desalination Using Microbial Desalination Cells. Environ Sci Technol 43 (18):7148-52.

Catal T, Bermek H, Liu H (2009) Removal of selenite from wastewater using microbial fuel cells. Biotechnol Lett 31(8):1211–1216.

Cath TY, Childress AE, Elimelech M (2006) Forward osmosis: Principles, applications, and recent developments. J Membr Sci 281:70–87.

Chang IS, Jang JK, Gil GC, Kim M, Kim HJ, Cho B.W, Kim BH (2004) Continuous determination of biochemical oxygen demand using microbial fuel cell type biosensor. Biosens Bioelectron 19:607–613.

Chatzisymeon E, Xekoukoulotakis NP, Coz A, Kalogerakis N, Mantzavinos D (2006) Electrochemical treatment of textile dyes and dye house effluents, J Hazard Mater137:998–1007.

Cheng KY, Ginige MP, Kaksonen AH (2012) Ano-Cathodophilic Biofilm Catalyzes Both Anodic Carbon Oxidation and CathodicDenitrification. Environ Sci Technol 46 (18) :10372-8.

Cusick R, Call DF, Selembo PA, Regan JM, Logan BE (2011) Anode microbial communities produced by changing from microbial fuel cell to microbial electrolysis cell operation using two different wastewaters. Bioresour Technol 102:388–394.

Dutta PK, Keller J, Yuan Z, Rozendal RA, Rabaey K (2009) Role of Sulfur during Acetate Oxidation in Biological Anodes. Environ Sci Technol 43:3839–3845.

Feng Y, Barr W, Harper WF, Jr (2013) Neural network processing of microbial fuel cell signals for the identification of chemicals present in water. J Environ Manag 120:84–92.

Galvani De viribuselectricitatis in motumusculari (1791) Lelio della Volpe, Bologna.

George RP, Muraleedharan P, Dayal RK, Khatak HS (2006) Techniques for biofilm monitoring. Corros Rev 24:123–150.

Gorby YA, Lovley DR (1992) Enzymatic uranium precipitation. Environ Sci Technol 26(1):205–207.

Goud RK, Babu PS, VenkataMohan S (2011) Canteen based composite food waste as potential anodic fuel for bioelectricity generation in single chambered microbial fuel cell (MFC): Bio-electrochemical evaluation under increasing substrate loading condition. Int. J.Hydrogen Energy 36:6210–6218.

Grattieri M, Hasan K, Minteer SD (2017) Bio-Electrochemical Systems as A Multipurpose Biosensing Tool: Present Perspective and Future Outlook Chem Electro Chem 4:834 –842.

Gregory KB, Lovley DR (2005) Remediation and recovery of uranium from contaminated subsurface environments with electrodes. Environ Sci Technol 39(22):8943–8947.

Grove WR XXIV (1839) On voltaic series and the combination of gases by platinum Philos Mag 14:127-130.

Guotao S, Thygesen A, Meyer AS (2015) Acetate is a superior substrate for microbial fuel cell initiation preceding bioethanol effluent utilization. Appl Microbiol Biotechnol 99:4905–4915.

Guven G, Perendeci A, Tanyolac A (2008) Electrochemical treatment of deproteinated whey wastewater and optimization of treatment conditions with response surface methodology. J Hazar Materi 157:69–78.

Harris HW, El-Naggar MY, Bretschger O, Ward MJ, Romine MF, Obraztsova AY, Nealson KH (2010) Electrokinesis is a microbial behavior that requires extracellular electron transport. Proc Natl Acad Sci U S A 107(1):326–331.

He Z, Angenent LT (2006) Application of bacterial biocathodes in microbial fuel cells. Electro analysis 18:2009–2015.

Heijne AT, Liu F, Weijden RVD, Weijma J, Buisman CJN, Hamelers HVM (2010) Copper recovery combined with electricity Production in a microbial fuel cell. Environ Sci Technol 44(11):4376–4381.

Hemalatha M, Butti SK, Velvizhi G, Venkata Mohan S (2017) Microbial desalination cells for ground water softening, resource recovery and power generation. Biores Technol 242:28–35.

Huang L, Chai X, Cheng S, Chen G (2011) Evaluation of carbon-based materials in tubular biocathode microbial fuel cells in terms of hexavalent chromium reduction and electricity generation. Chem Eng J 166(2):652–661.

Huang L, Wang Q, Quan X, Liu Y, Chen G (2013). Bioanodes/biocathodes formed at optimal potentials enhance subsequent pentachlorophenol degradation and power generation from microbial fuel cells. Bioelectrochemistry 94:13–22.

Huanga B, Fenga H, Dinga Y, Zhenga X, Wanga M, Li N, Shena D, Zhangc H (2013) Microbial metabolism and activity in terms of nitrate removal in bioelectrochemical systems. Electrochimica Acta 113:29–36

Iskander S, Brazil B, Novak J, He Z (2016) Resource recovery from landfill leachate using bioelectrochemical systems: opportunities, challenges, and perspectives. Bioresour Technol 201:347–354.

Israilides CJ, Vlyssides AG, Mourafeti VN, Karvouni G (1997) Olive oil waste water treatment with the use of an electrolysis system. Bioresour Technol 61:163–170.

Jacobson KS, Drew DM, He Z (2011) Efficient salt removal in a continuously operated up flow microbial desalination cell with an air cathode. Bioresour Technol 102(1):376-80.

Jouanneau S, Recoules L, Durand MJ, Boukabache A, Picot V, Primault Y, Lakel A, Sengelin M, Barillon B, Thouand G (2014) Methods for assessing biochemical oxygen demand (BOD): a review. Water Res 49:62–82.

Kato S, Hashimoto K, Watanabe K (2012) Microbial interspecies electron transfer via electric currents through conductive minerals. Proc Nat Acad Sci 109(25):10042–10046.

Katuri KP, Enright AM, O'Flaherty V, Leech D (2012) Microbial analysis of anodic biofilm in a microbial fuel cell using slaughterhouse wastewater. Bioelectrochemistry 87:164-171.

Kaur A, Kim JR, Michic I, Dinsdale RM, Guwy AJ, Premier GC (2013) Microbial fuel cell type biosensor for specific volatile fatty acids using acclimated bacterial communities. Biosens Bioelectron 47:50–55.

Kaye GI, Weber PB, Evans A, Venezia RA (1998) Efficacy of alkaline hydrolysis as an alternative method for treatment and disposal of infectious animal waste. Journal of the American Association for Laboratory Animal Science 37(3): 43-4

Kondaveeti, S., Lee, S.H., Park, H.D., Min, B., 2014. Bacterial communities in a bioelectrochemical denitrification system: the effects of supplemental electron acceptors. Water Res 51(0):25–36.

Korbahti BK, Aktas N, Tanyolac A (2007) Response surface optimization of electrochemical treatment of textile dye wastewater, J Hazard Mater 145:277–286.

Kumar GG, Sarathi VGS, Nahm KS (2013) Recent advances and challenges in the anode architecture and their modifications for the applications of microbial fuel cells. Biosens Bioelectron 43:461–475.

Carrette L, Friedrich KA, Stimming U (2001) Fuel Cells – Fundamentals and Applications of Fuel Cells 1:5-39

Lee DJ, Liu X, Weng HL (2014) Sulfate and organic carbon removal by microbial fuel cell with sulfate-reducing bacteria and sulfide-oxidising bacteria anodic biofilm. Bioresour Technol 156:14–19.

Lei Y, Chen W, Mulchandani, A (2006) Microbial biosensors. Anal Chim Acta 568:200–210

Li H, Zuo W, Tian Y, Zhang J, Di S, Li L, Su X (2017) Simultaneous nitrification and denitrification in a novel membrane bioelectrochemical reactor with low membrane fouling tendency. Environ Sci Pollut Res. 24:5106–5117.

Liu L, Li FB, Feng CH, Li XZ (2009) Microbial fuel cell with an azo-dye-feeding cathode. Appl Microbiol Biotechnol 85:175–183.

Liu Z, Liu J, Zhang S, Xing XH, Su Z (2011) Microbial fuel cell based biosensor for in situ monitoring of anaerobic digestion process. Bioresour Technol 102:10221–10229

Logan BE (2004) Peer reviewed: extracting hydrogen and electricity from renewable resources. Environ Sci Technol 38 (9):160A–167A.

Lovley DR (2006) Bug juice: harvesting electricity with microorganisms. Nat Rev Microbiol 4 (7): 497–508.

Luo H, Liu G, Zhang R., Jin S (2009) Phenol degradation in microbial fuel cells, Chem Engg J 147:259-264.

Mahmoud M, Parameswaran P, Torres CI, Rittmann BE (2014) Fermentation pre-treatment of landfill leachate for enhanced electron recovery in a microbial electrolysis cell. Bioresour Technol 151:151–158.

Mateo-Ramírez F, Addi H, Jose F, Hernandez-Fernandez Godínez C, De Los Ríos AP, Lotfi M, El Mahi M, Blanco LJL (2017) Air breathing cathode-microbial fuel cell with separator based on ionic liquid applied to slaughterhouse wastewater treatment and bio-energy production. J Chem Tech and Biotech 92:642–648.

McCarty PL, Bae J, Kim J (2011) Domestic wastewater treatment as a net energy producer can this be achieved? Environ Sci Technol 45:7100–7106.

Modestra JA, Chiranjeevi P, Venkata Mohan S (2016) Cathodic material effect on electron acceptance towards bioelectricity generation and wastewater treatment. Renew Energ 98:178–187.

Mohammed TA, Birima AH, Noor MJMM, Muyibi SA, Idris A (2008) Evaluation of using membrane bioreactor for treating municipal wastewater at different operating conditions. Desalination 221:502–510.

Mohana krishna G, Venkata Mohan S, Sarma PN (2010) Bio-electrochemical treatment of distillery wastewater in microbial fuel cell facilitating decolorization and desalination along with power generation, J Hazard Mater 177:487- 494.

Mohana krishna G, Srikanth S, Pant D (2015) Bioelectrochemical Systems (BES) for Microbial Electro remediation:Garima Kaushik (eds) An Advanced Wastewater Treatment Technology, In Book: Applied Environmental Biotechnology: Present Scenario and Future Trends. Springer, London, UK. pp 130-145.

Mohana krishna G, Venkata Mohan S, Sarma PN (2010) Bio-electrochemical treatment of distillery wastewater in microbial fuel cell facilitating decolorization and desalination along with power generation. J Hazard Mater 177:487–494.

Morris JM, Jin S, Crimi B, Pruden A (2009) Microbial fuel cell in enhancing anaerobic biodegradation of diesel. Chem Engg J 146(2):161-167.

Mu Y, Rabaey K, Rozendal RA, Yuan Z, Keller J (2009) Decolorization of azo dyes in bioelectrochemical systems. Environ sci Technol 43(13): 5137-5143

Nagendranatha Reddy C, Naresh Kumar A, Annie Modestra J, Venkata Mohan S, (2014) Induction of anoxic microenvironment in multi-phase metabolic shift strategy during periodic discontinuous batch mode operation enhances treatment of azo dye wastewater. Bioresour Technol 165: 241–249.

Nagendranatha Reddy C, Naresh Kumar A, Venkata Mohan S, (2018) Metabolic phasing of anoxic-PDBR for high rate treatment of azo dye wastewater. J Hazard Mat 343:49–58.

Nancharaiah YV, Venkata Mohan S, Lens PN (2016) Recent advances in nutrient removal and recovery in biological and bioelectrochemical systems. Bioresour Technol 215:173–185.

Nancharaiah YV, Venkata Mohan S, Lens PNL (2015) Removal and recovery of metal ions in microbial fuel cells: a review. Bioresour Technol 195:102–114.

Nikhil GN, Venkata Subhash G, Yeruva DR, Venkata Mohan S (2015) Synergistic yield of dual energy forms through biocatalyzed electrofermentation of waste: stoichiometric analysis of electron and carbon distribution Energy 88:281–291.

Panizza M, Cerisola G (2009) Direct And Mediated Anodic Oxidation of Organic Pollutants. Chem Rev 109: 6541–6569.

Pant D, Singh A, Bogaert VG, Olsen SI, Nigam PS, Diels L, Vanbroekhoven K (2012) Bioelectrochemical systems (BES) for sustainable energy production and product recovery from organic wastes and industrial wastewaters. RSC Adv 2:1248–1263.

Pant D, Singh A, Van Bogaert G, Gallego YA, Diels L, Vanbroekhoven K (2011) An introduction to the life cycle assessment (LCA) of bioelectrochemical systems (BES) for sustainable energy and product generation: relevance and key aspects. Renew Sustain Energy Rev 15 (2):1305-1313.

Pant D, Van Bogaert G, Diels L, Vanbroekhoven K (2010) A review of the substrates used in microbial fuel cells (MFCs) for sustainable energy production. Bioresour Technol 101 (6): 1533-1543

Parsek MR, Greenberg EP (2005) Ociomicrobiology: the connections between quorum sensing and biofilms, Trends Microbiol. 13: 27.

Patil SA, Hägerhäll C, Gorton L (2012) Electron transfer mechanisms between microorganisms and electrodes in bioelectrochemical systems. Bioanal Rev 4:159–192.

Peixoto L, Min B, Martins G, Brito AG, Kroff P, Parpot P, Angelidaki I, Nogueira R (2011) In situ microbial fuel cellbased biosensor for organic carbon. Bioelectrochemistry 81:99–103

Rabaey K, Rozendal RA (2010) Microbial electro synthesise revisiting the electrical route for microbial production, Nat Rev Microbiol 8:706-716

Rabaey K, Van de Sompel K, Maignien L, Boon N, Aelterman P, Clauwaert P, De Schamphelaire L, Pham H, Vermeulen J, Verhaege M, Lens P, Verstraete W (2006) Microbial Fuel Cells for Sulfide Removal. Environ Sci Technol 40 (17):5218–5224.

Rhoads A, Beyenal H, Lewandowski Z (2005) Microbial fuel cell using anaerobic respiration as an anodic reaction and biomineralized manganese as a cathodic reactant. Environ Sci Technol 39(12): 4666–4671.

Rozendal R, Hubertus VM, Rabaey K, Keller J, Buisman JN (2008) Towards practical implementation of bioelectrochemical wastewater treatment. Trends in biotechnology 26:450-459

Sadhukhan J, Lloyd JR, Scott K, Premier KC, Yu EH, Curtis T, Head IM (2016) A critical review of integration analysis of microbial electrosynthesis (MES) systems with waste biorefineries for the production of biofuel and chemical from reuse of CO2, Renew. Sustain Energy Rev. 56:116-132

Sander EM, Virdis B, Freguia S (2017) Bioelectrochemical nitrogen removal as a polishing mechanism for domestic wastewater treated effluents. Water Sci Technol.76:3150-3159

Sara RM, Maria JLDA, Damia B (2006) Biosensors as useful tools for environmental analysis and monitoring. Anal BioanalChem 386:1025–1041.

Shena J, Xua X, Jiang X, Hua C, Zhang L, Sun X, Li J, Wang Y (2014) Coupling of a bioelectrochemical system for p-nitrophenol removal in an upflow anaerobic sludge blanket reactor. Water Res 67:11–18.

Shukla AK, Suresh P, Berchmans S, Rajendran A (2004) Biological fuel cells and their applications Curr. Sci. 87: 455-468.

Sreelatha S, Velvizhi G, Naresh Kumar A, Venkat Mohan S (2015) Functional behavior of bio-electrochemical treatment system with increasing azo dye concentrations: Synergistic interactions of biocatalyst and electrode assembly. Bioresour. Technol. 213:11–20.

Srikanth S, Venkata Mohan S (2012a) Influence of terminal electron acceptor availability to the anodic oxidation on the electrogenic activity of microbial fuel cell (MFC). BioresourTechnol 123:480–487.

Srikanth S, Venkata Mohan S (2012b). Change in electrogenic activity of the microbial fuel cell (MFC) with the function of biocathode microenvironment as terminal electron accepting condition: Influence on over potentials and bio-electro kinetics. Bioresour. Technol. 119: 242-251.

Straub KL, Schink B (2004) Ferrihydrite-Dependent Growth of Sulfuro spirillum deleyianum through Electron Transfer via Sulfur Cycling. Appl Environ Microbiol 70:5744–5749.

Tao HC, Lei T, Shi G, Sun XN, Wei XY, Zhang LJ, Wu WM (2014) Removal of heavy metals from fly ash leachate using combined bioelectrochemical systems and electrolysis. J Hazard Mater 264:1–7.

Tao HC, Liang M, Li W, Zhang LJ, Ni JR, Wu WM (2011) Removal of copper from aqueous solution by electrodeposition in cathode chamber of microbial fuel cell. J Hazard Mater 189: 186–192.

Torres CI (2014) On the importance of identifying, characterizing, and predicting fundamental phenomena towards microbial electrochemistry applications. Cur Opn Biotechnol 27:107–114.

Vamsi Krishna K, Omprakash S, Venkata Mohan S (2014) Bioelectrochemical treatment of paper and pulp wastewater in comparison with anaerobic process: integrating chemical coagulation with simultaneous power production. Bioresour Technol 174:142–151.

Vargas A, Soto S, Moreno J, Buitro G (2000). Observer-based time-optimal control of an aerobic SBR for chemical and petrochemical wastewater treatment, Water Sci Tech 42:163-170.

Velvizhi G (2018) Overview of Bioelectrochemical Treatment (BET) Systems for Wastewater Remediation Accepted 'Microbial Electro-chemical Technology: Fuels, Chemicals and Remediation' Volume I of a new series on 'Biomass, Biofuels and Biochemicals' Edited S.Venkata Mohan, SunitaVarjani and Ashok Pandey. Elsevier

Velvizhi G, Goud RK, Venkata Mohan S (2014) Anoxic bio-electrochemical system for treatment of complex chemical wastewater with simultaneous bioelectricity generation, Bioresour Technol 151:214–22.

Velvizhi G, Venkata Mohan S (2011) Biocatalyst behavior under self-induced electrogenic microenvironment in comparison with anaerobic treatment: evaluation with pharmaceutical wastewater for multi-pollutant removal. Bioresour Technol 102:10784–10793.

Velvizhi G, Venkata Mohan S (2013) In Situ System Buffering Capacity dynamics on bioelectrogenic activity during the remediation of wastewater in microbial fuel cell. Environ Prog Sus Energy 33:453-462

Velvizhi G, Venkata Mohan S (2015) Bioelectrogenic role of anoxic microbial anode in the treatment of chemical wastewater: Microbial dynamics with bioelectro-characterization. Water Res 70:52–63.

Venkata Mohan S, Shanthi Sravan J, Butti,SK, Vamshi Krishna K, Annie Modestra J, Velvizhi G, Naresh Kumar A, Varjani S, Pandey A (2018) Microbial Electro-chemical Technology: Fuels, Chemicals and Remediation' Volume I of a new series on 'Biomass, Biofuels and Biochemicals' Edited S.Venkata Mohan, Sunita Varjani and Ashok Pandey.Elsevier

Venkata Mohan S, Raghuvulu SV, Srikanth S, Sarma PN (2007) Bioelectricity production by meditorless microbial fuel cell (MFC) under acidophilic condition using wastewater as substrate: influence of substrate loading rate. Current Science 92(12):1720-1726.

.Venkata Mohan S, Raghavulu SV, Peri D, Sarma PN (2009) Integrated function of microbial fuel cell (MFC) as bio-electrochemical treatment system associated with bioelectricity generation under higher substrate load. Biosens Bioelectron 24:2021–2027

Venkata Mohan S, Chandrasekhar SK (2011) Self-induced bio-potential and graphite electron accepting conditions enhances petroleum sludge degradation in bio-electrochemical system with simultaneous power generation. Bioresour Technol 102(20):9532–9541.

Venkata Mohan S, Mohanakrishna G, Reddy BP, Saravanan R, Sarma PN (2008a) Bioelectricity generation from chemical wastewater treatment in mediatorless (anode) microbial fuel cell (MFC) using selectively enriched hydrogen producing mixed culture under acidophilic microenvironment. Biochem Eng J 39:121–130.

Venkata Mohan S, Saravanan R, Raghavulu SV, Mohanakrishna G, Sarma PN (2008b). Bioelectricity production from wastewater treatment in dual chambered microbial fuel cell (MFC) using selectively enriched mixed microflora: effect of catholyte. Bioresour Technol 99:596–603. doi:10.1016/j.biortech.2006.12.026

Venkata Mohan S, Mohanakrishna G, Velvizhi G, Babu VL, Sarma PN (2010) Bio-catalyzed electrochemical treatment of real field dairy wastewater with simultaneous power generation. Biochem Eng J 51:32–39.

.Venkata Mohan S, Saravanan R, Raghavulu SV, Mohanakrishna G, Sarma PN (2008b) Bioelectricity production from wastewater treatment in dual chambered microbial fuel cell (MFC) using selectively enriched mixed microflora: effect of catholyte. BioresourTechnol 99:596–603.

Venkata Mohan S, Srikanth S (2011) Enhanced wastewater treatment efficiency through microbially catalyzed oxidation and reduction: synergistic effect of biocathode microenvironment. Bioresour Technol 102(22):10210–10220.

Venkata Mohan S, Srikanth S, Sarma PN (2009) Non-catalyzed microbial fuel cell (MFC) with open air cathode for bioelectricity generation during acidogenic wastewater treatment. Bioelectrochemistry 75:130–135.

Venkata Mohan S, Srikanth S, Velvizhi G, Lenin Babu M (2013) Microbial Fuel Cells for Sustainable Bioenergy Generation: Principles and Perspective Applications, in: Biofuel Technologies. Springer Berlin Heidelberg, Berlin, Heidelberg, pp 335–368.

Venkata Mohan S, Velvizhi G, Krishna KV, Lenin Babu M (2014a) Microbial catalyzed electrochemical systems: a bio-factory with multi-face applications. Bioresour Technol 165: 355–364.

Venkata Mohan S, Velvizhi G, Modestra JA, Srikanth S (2014b) Microbial fuel cell: critical factors regulating bio-catalyzed electrochemical process and recent advancements. Renew Sust Energ Rev 40:779–797.

Wang G, Huang L, Zhang Y (2008) Cathodic reduction of hexavalent chromium [Cr(VI)] coupled with electricity generation in microbial fuel cells. Biotechnol Lett 30(11):1959–1966.

Wei J, Liang P, Huang X (2011) Recent progress in electrodes for microbial fuel cells. Bioresour Technol 102 (20):9335–9344.

Xiao Y, XuXie MH, Yang HZ, Zeng GM (2015) Comparison of the treatment for isopropyl alcohol wastewater from silicon solar cell industry using SBR and SBBR, Int J Environ Sci Technol 12:2381-2388.

Yan Z, Song N, CaiaJoo-HwaTay H, Jianga H (2012) Enhanced degradation of phenanthrene and pyrene in freshwater sediments by combined employment of sediment microbial fuel cell and amorphous ferric hydroxide. J Hazard Mat 199-200:217–225.

Yang H, Zhou M, Liu M, Yang W, Gu, T (2015) Microbial fuel cells for biosensor applications Biotechnol Lett 37:2357–2364

Yeruva DK, Srinivas J, Velvizhi G, Naresh Kumar A, Swamy YV, Venkata Mohan S (2015) Integrating sequencing batch reactor with bio-electrochemical treatment or augmenting remediation efficiency of complex petrochemical wastewater. Bioresour Technol 188:33–42.

Yeruva DK, Velvizhi G, Venkata Mohan S (2016) Coupling of aerobic/anoxic and bioelectrogenic processes for treatment of pharmaceutical wastewater associated with bioelectricity generation. Renew Ene 98:171–177.

Yeruva YD, ShanthiSravan J, Butti SK, Annie Modestra J, Venkata Mohan S (2018) Spatial variation of electrode position in bioelectrochemical treatment system: Design consideration for azo dye remediation; Bioresour Technol 256:374–383

Zhang B, Zhao H, Shi C, Zhou S, Ni J (2009) Simultaneous removal of sulfide and organics with vanadium (V) reduction in microbial fuel cells. J Chem Technol Biotechnol 84(12): 1780–1786.

Zhang LJ, Tao HC, Wei XY, Lei T, Li JB, Wang AJ, Wuc WJ (2012) Bioelectrochemical recovery of ammonia–copper (II) complexes from wastewater using a dual chamber microbial fuel cell. Chemosphere 89:1177–1182.

Zhang F, Ge Z, Grimaud J, Hurst J, He Z (2013a) Long-term performance of liter scale microbial fuel cells treating primary effluent installed in a municipal wastewater treatment facility. Environ Sci Technol 47:4941–4948.

Zhang F, Ge Z, Grimaud J, Hurst J, He Z (2013b) In situ investigation of tubular microbial fuel cells deployed in an aeration tank at a municipal wastewater treatment plant. Bioresour Technol 136:316–321.

Zhang S, You J, Kennes C, Cheng Z, Ye J, Chen D, Chen J, Wang L (2018) Current advances of VOCs degradation by bioelectrochemical systems: A Review. Chemical Engineering, 334: 2625-2637

Zhao F, Rahunen N, Varcoe JR, Roberts AJ, Rossa CA, Thumser AE, Robert CTS (2009) Factors affecting the performance of microbial fuel cells for sulfur pollutants removal. Bioscns. Bioelectron. 24:1931–1936.

Graphene Electrodes in Bioelectrochemical Systems

Pier-Luc Tremblay[1,2]**, Yuan Li**[1,2]**, Mengying Xu**[1,2]**, Xidong Yang**[1,2] **and Tian Zhang**[1,2*]

[1] State Key Laboratory of Silicate Materials for Architectures, Wuhan University of Technology, Wuhan P.R. China
[2] School of Chemistry, Chemical Engineering and Life Science, Wuhan University of Technology, Wuhan, P.R. China

1. Introduction

In recent years, a large number of scientists have worked intensively on the development of novel, more efficient and less polluting energy technologies. One avenue of research that has been explored the most is the development of bioelectrochemical systems (BES), where electrochemical reactors harness the transformative potential of biological material for different energy-related purposes (Javed et al. 2018; Rabaey and Rozendal 2010; Yu et al. 2018). For instance, BES has been developed for the microbial conversion of organic carbon molecules into electricity, H_2 or methane as well as for the microbial conversion of carbon dioxide and electricity into valuable organic carbon molecules (ElMekawy et al. 2017; Jiang and Jianxiong Zeng 2018).

BESs also include biosensors that are devices usually combining a biological element with electrochemical hardware to sense target molecules of interest (Justino et al. 2013; Justino et al. 2017). Well-designed biosensors can detect a low concentration of target compounds over a large linear detection range. Biosensors have been developed for sensing a wide range of molecules, including sugars, biomarkers, heavy metals, antibodies, antigens, pesticides, metabolites, antibiotics and so on (Justino et al. 2017).

Electrodes are critical components of BES, and they have been the object of relentless optimization efforts over the years (Justino et al. 2017; Aryal et al. 2017a; ElMekawy et al. 2017). Multiple electrode materials have been tested for BES with the objective of improving productivity and efficiency while keeping the cost to a minimum. This includes noble metals, earth-abundant metals, metal alloys, metal oxides, carbonaceous materials, polymers and ionic liquids. Some of these materials have been used alone or as composites. Additionally, the same material can be employed for BES electrode fabrication in different physical conformation, such as nanoparticles, nanoribbons, nanowires, mesh, foam, felt, cloth, three-dimensional (3D) porous structure, aerogel, hydrogel, flake, nanosheets, etc.

Among carbonaceous materials, different types of graphene (Gr) have been incorporated in the fabrication of BES electrodes (Justino et al. 2017; Aryal et al. 2017a; ElMekawy et al. 2017). A single-layer Gr was produced for the first time in 2004 (Novoselov et al. 2004). Since then, it has been considered as a revolutionary material because of its extraordinary physico-chemical properties and its

---

*Corresponding author: tzhang@whut.edu.cn

low fabrication cost. This chapter will focus on the ongoing research effort to develop and optimize high-performance Gr-based electrodes for BES.

2. Graphene Structure and Properties

Graphene is a two-dimensional nanomaterial composed of a dense conjugated sp^2 carbon network forming a hexagonal honeycomb lattice (Figure 1) (Geim and Novoselov 2007). It has excellent optical, electrical and mechanical properties and has important application prospects in materials science, micro/nano processing, energy-related domains, biomedicine and drug delivery. Gr also has outstanding electrochemical properties, such as wide electrochemical window, good electrochemical stability, small charge transfer resistance, high electrocatalytic activity and fast electron transfer (ET) rate.

It has been reported that the charge carrier mobility of suspended single layer Gr can be in excess of 200,000 cm^2 V^{-1} s^{-1} (Bolotin et al. 2008). Consequently, electrons travel in Gr at a Fermi velocity of 106 m s^{-1} without scattering which is a process known as ballistic conduction. It is because of its ability to quickly transport electrons, which is 200 times greater than silicon, that Gr has a high electrical conductivity (X. Wang et al. 2008; Liu et al. 2010). These unique electrical properties are tunable toward specific applications via different synthesis procedures that are attractive for BES.

Graphene also exhibits promising physical properties, including high mechanical strength, high thermal conductivity and good elasticity (Lee et al. 2008; Kim et al. 2009). Its maximal specific surface area is estimated to be 2,630 m^2 g^{-1} and it is several times higher than that of other carbon-based materials (Stoller et al. 2008).

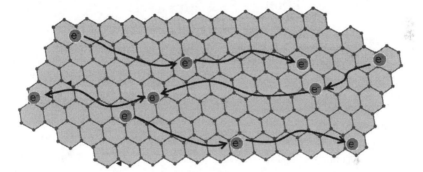

Figure 1: Hexagonal honeycomb lattice structure of graphene and high electron mobility.

3. Graphene-Modified Anode in Microbial Fuel Cell

Microbial fuel cell (MFC) is a major bioelectrochemical technology developed more intensively in the last few decades (Hasany et al. 2016). In an MFC, a microbial catalyst oxidizes organic carbon molecules and transfers electrons to an anode for the production of electricity (Li et al. 2018) (Figure 2). The MFC concept was demonstrated by Potter in 1910 with a system where microbial cultures were combined with platinum electrodes for the production of electricity (Potter 1911). The anode is a key part of MFC and properties of anode material can significantly influence ET between microorganisms and electrodes as well as power density and energy conversion efficiency (Li et al. 2017). In the quest for an ideal anode, Gr has been widely employed to enhance the electrochemical catalytic activity and the performance of MFC (Z. Liu et al. 2017). This section introduces Gr-containing anodes developed for MFC in recent years.

3.1 MFC Principle

A standard dual chambered MFC comprises an anodic chamber and a cathodic chamber separated by an ion exchange membrane (IEM) (Figure 2). The microbial catalyst grows on the anode and oxidizes organic substrate while transferring electrons to the anode and releasing protons into the electrolytes.

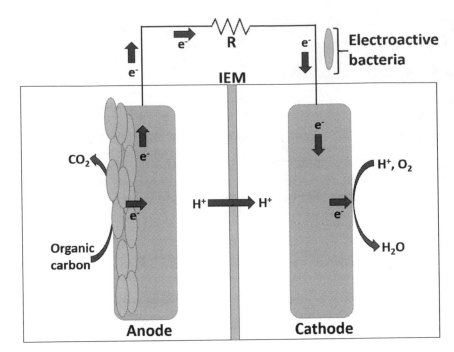

Microbial Fuel Cell

Figure 2: Representation of a two chamber MFC reactor.

Multiple organic substrates can be used to drive MFC, including volatile fatty acids, wastewaters and lignocellulosic biomass hydrolysates (Z. Liu et al. 2017; Pant et al. 2010; Rabaey et al. 2005). In the cathodic chamber, oxygen is reduced into water with electrons from the cathode and protons coming from the anodic chamber that passed through the IEM. In this system, the anode and the cathode are connected with wires and form an electrical circuit.

Various microbes such as proteobacteria, phototrophic bacteria as well as yeast have been shown to be electrochemically active, thus capable of transferring electrons to a solid electrode (Kumar et al. 2016). Bacteria can transfer electron via four extracellular electron transfer (EET) mechanisms: 1) *c*-type cytochromes bound to the cell membrane, 2) conductive pili or nanowires, 3) soluble electron mediators and 4) the electrochemical oxidation of reduced primary metabolites (Saratale et al. 2017). Microbes capable of direct electron transfer (DET) via *c*-type cytochromes or nanowires to a solid acceptor such as a MFC anode are called exoelectrogens. Examples of exoelectrogens used as biocatalyst for MFC include *Geobacter metallireducens*, *Geobacter sulfurreducens* and *Shewanella putrefaciens* (Kumar et al. 2016; Zhou et al. 2013). Alternatively, microorganisms can also serve as a biocatalyst in MFCs by indirectly transferring electrons to an anode via artificially-amended soluble mediators or secreted natural mediators (e.g., flavins and pyocyanin) (Kotloski and Gralnick 2013; Brutinel and Gralnick 2012). Examples of biocatalyst relying on this EET strategy include *Shewanella oneidensis* and *Pseudomonas spp.* (Zhou et al. 2013).

One of the main technical challenges for the industrial application of MFC is the low ET efficiency between microbes and the anode. There are many factors influencing ET rates in MFC reactors, such as the type of substrate, the biocatalyst employed, the reactor design, the kind of membrane or other component separating the anodic and cathodic chambers and finally, the electrode material and spatial arrangement. Among all of them, the anode electrode plays a pivotal role since its surface is the heart of the MFC system where the microbial catalyst respires and grows (ElMekawy et al. 2017). The properties of anode material, such as conductivity, roughness, surface area per volume ratio, porosity, biocompatibility, flexibility, robustness and manufacturing cost can have a major impact on MFC efficiency, productivity and scalability (Li et al. 2017).

3.2 Graphene Usage for MFC Anode

Electrodes in electrochemical systems can be made of noble metal materials, such as platinum (Pt), gold (Au) and silver (Ag). For commercial application, cost reduction is paramount and may be achieved by replacing noble metal electrodes with less expensive materials, such as iron (Fe), copper (Cu), nickel (Ni), aluminum (Al) stainless steel (SS) or corrosion-resistant carbonaceous materials. Commercially-available carbonaceous porous materials, such as graphite felt (GF), carbon cloth (CC), carbon paper (CP) and carbon felt (CF) have been widely used as electrodes for MFC because of their good electrical conductivity and physicochemical stability (Li et al. 2017; Picot et al. 2011; Song et al. 2012; Chaudhuri and Lovley 2003; Liu et al. 2004; Logan et al. 2007; Wang et al. 2009). However, power density, energy conversion efficiency and ET rate between microbes and these types of electrodes are often restricted by important limitations, including a low surface area to volume ratio and high internal resistance (Sonawane et al. 2017; Zou et al. 2008).

In recent years, Gr has been used in the fabrication of novel anodes with the objective of improving the performance of MFC systems (Table 1) (Z. Liu et al. 2017; ElMekawy et al. 2017). Gr has been shown to promote better power density and energy conversion efficiency partly because of its high specific surface area and biocompatibility that enables more substantial bacteria growth on the electrode surface resulting in higher ET fluxes (Zhang et al. 2011; Liu et al. 2012; Jain et al. 2012). This is important since low efficiency of MFC is mainly due to inefficient EET, activation losses, ohmic losses and mass transport losses (Rismani-Yazdi et al. 2008; Hutchinson et al. 2011; Ahn and Logan 2013). Employing Gr-containing electrode may help overcome some of these problems because it has the potential to increase the number of active sites for ET from microbes to the anode.

One controversial aspect of Gr is that several studies have suggested that it has antimicrobial properties (Liu et al. 2011). For instance, reduced graphene oxide (rGO) may induce cell wall disruption via rGO edges and oxidative stress (Liu et al. 2011; Hu et al. 2010). On the contrary, results from other research groups indicate that Gr has no intrinsic antibacterial activity and even enhances bacterial growth (ElMekawy et al. 2017; Ruiz et al. 2011). It has been suggested that the detrimental effect on bacteria attributed to Gr in the past may be due to contaminants from the Gr synthesis processes (Ruiz et al. 2011).

Another concern with the usage of rGO for the fabrication of bioelectrochemical anodes is that both this material and bacteria are negatively-charged that means there is electrostatic repulsion. This may lead to a reduction of the amount of bacteria attached on the Gr surface that translates into lower ET rate (Y. Wang et al. 2013). This is one of the reasons why Gr has been combined with conducting polymer, metal oxides or nanomaterials for the fabrication of optimal anodes capable of enhancing MFC output via the augmentation of bacterial attachment and ET rates. Multiple graphene oxide (GO) and rGO anodes as well as derived composite electrodes have been fabricated for MFC by different methods, such as direct deposition, electrochemical reduction, layer-by-layer self-assembly, chemical doping metal functionalization and bioreduction (ElMekawy et al. 2017).

3.2.1 Anode Surface Modified with Graphene Only

Extensive research has been conducted to develop high-performance anodes with surface modified with pure Gr. For instance, a MFC system equipped with a Gr-SS mesh (GSM) anode had an outstanding maximum power density of 2.668 W m^{-2} that was 18 times higher than that of bare SS mesh anode (Table 1). The enhanced performance of the GSM anode was due to the increased specific surface area of the electrode, better bacterial adhesion to rGO and faster EET between biofilm and electrode (Zhang et al. 2011).

The Gr fabrication method can also significantly influence MFC performance. In one case, an electroactive rGO-hybridized 3D macroporous biofilm was designed by self-assembling of GO with *S. oneidensis* MR-1. This hybrid electroactive biofilm highly enhanced EET rate because bacteria were encapsulated within an rGO scaffold during the self-assembly process, resulting in a 25-fold increase of the outward current in MFC compared to that of an *S. oneidensis* biofilm devoid of rGO (Yong et al. 2014).

Table 1: Examples of graphene-modified anode for MFC

Graphene and composite	Anode base material	Maximum power density	Microbial catalyst	Substrate	Reference
Gr	SSM[a]	2.67 Wm^{-2}	*E. coli*	Glucose and yeast extract	Zhang et al. 2011
Gr	SSM	3.22 Wm^{-2}	Mixed	Acetate	Zheng et al. 2015
GO-agarose foam	SSM	0.786 Wm^{-2}	*S. putrefaciens*	Lactate	Yang et al. 2015
Gr-Au	CP[b]	0.51 Wm^{-2}	*S. oneidensis*	Lactate	Zhao et al. 2015
Crumpled Gr particles	CC[c]	3.60 Wm^{-3}	Mixed	Acetate	Luo et al. 2011
N-doped graphene nanosheets	CC	1.008 Wm^{-2}	*E. coli*	glucose	Kirubaharan et al. 2015
Gr aerogel	-	2.381 Wm^{-3}	Anaerobic sludge	Acetate	F. Yu et al. 2018
GO nanoribbons	CP	0.0342 Wm^{-2}	*S. oneidensis*	Lactate	Huang et al. 2011
Biofilm-rGO	CC	0.84 Wm^{-2}	*S. oneidensis*	Lactate	Yong et al. 2014
Biofilm-rGO	CC	1.905 Wm^{-2}	anaerobic sludge	Acetate	Y. Yuan et al. 2012
3D rGO-Ni foam	Ti wire	661 Wm^{-3}	*S. oneidensis*	Trypticase soy broth	H. Wang et al. 2013
Gr-PANI[d]	CC	1.390 Wm^{-2}	anaerobic sludge	Acetate	Hou et al. 2013
Ionic liquid[e]-Gr	CP	0.601 Wm^{-2}	*S. oneidensis*	Lactate	Zhao et al. 2013
PPy[f]-rGO	CC	1.07 Wm^{-2}	*E. coli*	LB medium	Gnana kumar et al. 2014a
Gr-PTFE[g]	SSM	2.668 Wm^{-2}	*E. coli*	Glucose	Zhang et al. 2011
Gr-PU[h]	SSM	1.57 Wm^{-2}	Mixed	Glucose	Xie et al. 2012
Nafion-Gr	Stainless steel fiber felts	2.142 Wm^{-2}	Mixed	Acetate	J. Hou et al. 2014
PANI on Gr nanoribbons	CP	0.17 Wm^{-2}	*S. oneidensis*	Lactate	Zhao et al. 2013b
3D-Gr-PANI foam	NA[i]	0.77 Wm^{-2}	*S. oneidensis*	Lactate	Yong et al. 2012
Gr-SnO$_2$	CC	1.624 Wm^{-2}	*E. coli*	Glucose	Mehdinia et al. 2014
Gr-TiO$_2$	CP	1.060 Wm^{-2}	*S. oneidensis*	Lactate	Zhao et al. 2014
Gr-MWCNTs-Fe$_3$O$_4$ Foam	SSM	882 Wm^{-3}	*S. oneidensis*	Lactate	R.-B. Song et al. 2016
GO -CNTs Poly$_N$-isopropylacrylamide hydrogel composite	NA	0.44 Wm^{-2}	*E. coli*	Glucose	Kumar et al. 2014

[a]SSM: Stainless steel mesh, [b]CP: Carbon paper, [c]CC: Carbon cloth, [d]PANI: Polyaniline, [e]Ionic liquid: 1-(3-aminopropyl)-3-methylimidazolium bromide, [f]PPy: Polypyrrole, [g]PTFE: polytetrafluoroethylene, [h]PU: polyurethane, [i]NA: not applicable.

Another MFC study doped Gr with nitrogen to improve ET rate. An anode made of Gr sheets doped with nitrogen was fabricated by plasma enhanced chemical vapor deposition (Kirubaharan et al. 2015). Gr-based anode was enhanced by the N-doping because it increased pore density leading to an augmentation of the specific surface area available for EET by bacteria. It also generated defects on the Gr sheets and augmented the edge plane exposure that improved catalytic activity (Kirubaharan et al. 2015).

Graphene conformation can be altered to augment the electrode specific surface area available for the adhesion of the bacterial catalyst in MFC and improve ET rate. For instance, rGO sheets that were crumpled by capillary compression to increase surface area to volume ratio and resistance to aggregation were used for the fabrication of an MFC anode (Luo et al. 2011). An MFC system equipped with an anode modified with crumpled rGO sheets had a higher maximum power density of 3.6 W m^{-3} compared to a reactor equipped with an anode modified with flat standard rGO sheets that had a power density of 2.7 W m^{-3} (Table 1). In a second example, Huang and his coworkers achieved a significant enhancement of ET rate by modifying an MFC anode with GO nanoribbons that have a high length to diameter ratio and act as conductive nanowires (Huang et al. 2011).

A different strategy to exploit the full potential of Gr is to fabricate an intricated 3D network where multiple faces of a bacterial cell inserted into a pore can interact with the anode surface. For MFC, optimum performance of 3D-Gr anode is achieved with macropore larger than 50 nm compared to micropore (<2 nm) or mesopore (2–50 nm) because the bacterial diameter is usually 0.25 to 2 μm (J. Hou et al. 2014). An example of this approach is the fabrication of a nickel foam anode coated with 3D-rGO (rGO-Ni) that was used for the MFC and had a maximal volumetric power density of 661 W m^{-3} when normalized with the anode material volume (H. Wang et al. 2013). Compared to a conventional electrode, the 3D rGO-Ni anode provided a large accessible surface area for microbial colonization and electron mediators as well as a uniform macroporous scaffold for effective mass diffusion of the electrolyte. Recently, Yu and his coworkers constructed a porous 3D-Gr aerogel as bioanode using a hydrothermal reduction method. This anode had a maximum power density of 2.38 W m^{-3} and high specific capacity (3,670 F m^{-2}). The long-term electricity generation stability, which can be defined as the time span to stabilize in the 20% amplitude range of the maximum voltage in a cycle period operation, reached approximately 100 hours that was 36 times higher than that of a CP anode (F. Yu et al. 2018).

3.2.2 Anode Surface Modified with Graphene and Conductive Polymer Composite

Among all the conductive polymers, polyaniline (PANI) is commonly used for electrochemical applications because it is inexpensive, stable and easy to fabricate. Combining Gr with PANI has been shown to significantly enhance MFC performance. A power density of 1.390 W m^{-2} was obtained in an MFC reactor equipped with a Gr-PANI modified CC anode that was three times higher than the unmodified CC electrode (Table 1) (Hou et al. 2013). In this context, Gr not only functions as an extraordinary conductive material but also provides a large substrate surface for PANI. Moreover, this hybrid electrode promotes biofilm formation because of the strong electrostatic interaction between negatively-charged bacteria and positively-charged PANI backbone. Other conducting polymers also used in combination with Gr for the fabrication of modified anodes in MFC include: polyaniline (PANI), ionic liquid:1-(3-aminopropyl)-3-methylimidazolium bromide, polypyrrole (PPy), polytetrafluoroethylene PTFE, polyurethane (PU) and Nafion (Zhang et al. 2011; J. Hou et al. 2014; Gnana kumar et al. 2014a; Xie et al. 2012; Zhao et al. 2013a). Among these, the maximum power density of 2.142 W m^{-2} was obtained in an MFC reactor equipped with a Nafion-Gr modified SS fiber felts anode (Hou et al. 2014).

3.2.3 Anode Surface Modified with Graphene and Metal Oxide Composite

Metal oxides have been used to modify the anode of MFC systems in order to enhance their performance (Mehdinia et al. 2014; Benetton et al. 2010). Nanostructured metal oxide semiconductors have a high specific surface area, good biocompatibility and chemical stability. Among the metal oxide semiconductors, high consideration has been given to SnO$_2$ and TiO$_2$ because of their beneficial properties for bioelectrochemical applications, such as high electrical conductivity and low-cost. Both SnO_2 and TiO_2 have been combined with Gr to modify anodes of MFC systems. SnO_2 nanoparticles are more conductive than TiO_2 which is an advantage for MFC (Mehdinia et al. 2014; Zhao et al. 2014). SnO_2 was intertwined with rGO on the surface of a CC substrate, and this composite anode achieved a power density of 1.624 W m^{-2} that was almost 5 times higher than a control anode made of unmodified CC (Table 1) (Mehdinia et al. 2014). For this rGO-SnO_2-CC anode, SnO_2 and Gr show synergistic benefits combining good biocompatibility with higher charge transfer efficiency.

3.2.4 Anode Surface Modified with Graphene and Nanomaterial Composite

Nanomaterials such as carbon nanotube (CNT) have been combined with Gr to improve MFC performance. CNT has a high mechanical strength, a large specific surface area and a good electrical conductivity. Song and his coworkers fabricated a 3D macroporous anode made of Gr sheets, Fe_3O_4 nanospheres and multiwalled (MW)-CNT to enhance the long-term performance of an MFC system. With this anode, they obtained a maximum power density of 882 W m^{-3} (Table 1) (R.-B. Song et al. 2016). Generally, during the Gr fabrication process, 3D Gr sheets irreversibly aggregate via π–π stacking, thus dramatically reducing the accessible surface area for bacterial adhesion. To solve this problem, this novel anode comprised MWCNT that are thought to prevent GS from restacking, leading to an increase of microbes-electrode contacts and ET rate. Fe_3O_4 nanospheres act as bioaffinity anchors for the biocatalyst and help to maintain the metabolic activity of dissimilatory iron-reducing bacteria *S. oneidensis* MR-1 and support their long-term growth. In another example of the synergistic benefit of combining Gr with CNT for MFC anode, Kumar et al. fabricated via a suspension polymerization method a GO-CNT composite electrode that utilized hydrophilic GO to disperse CNT through non-covalent π–π stable interaction and form a composite hydrogel (Kumar et al. 2014). An MFC system equipped with this GO-CNT hydrogel anode had a maximum power density of 434 mW m^{-2}.

4. Graphene-Modified Cathode in Bioelectrochemical Systems

The cathode of BES can also be modified with Gr. MFC, microbial electrolysis cell (MEC), as well as microbial electrosynthesis (MES) systems have been developed with pristine Gr or composite Gr cathodes. MEC is related to MFC but protons and electrons from biological oxidation reactions at the anode are used by the cathode to produce H_2 or other chemicals (Figure 3A) (Kadier et al. 2016). In MES, the microbial catalyst is growing in the cathodic chamber where it acquires electrons from the cathode for the reduction of CO_2 into chemicals of interest (Figure 3B) (Rabaey and Rozendal 2010).

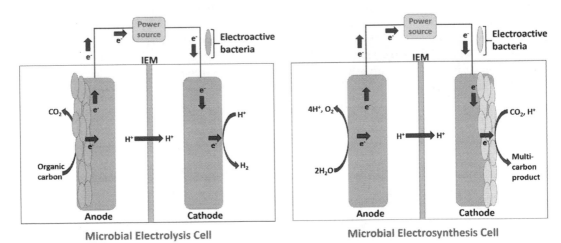

Figure 3: Representation of two-chamber reactors for A) H_2-producing MEC and B) MES.

4.1 Microbial Electrolysis Cell Principle

In 2004, Logan et al. developed a proof-of-concept MEC reactor producing H_2 from wastewater containing organic carbon substrates oxidized by microorganisms (Liu et al. 2005; Liu and Logan 2004). In MEC, electrons and protons derived from biological oxidation reactions at the anode are used for electrochemical reduction reactions at the cathode resulting in the generation of valuable products, such as H_2, methane or H_2O_2. Many different reactors have been developed for MEC, such as dual chambered MEC, single chambered MEC, continuous-flow MEC and integrated MEC systems (Zhou et al. 2013). Standard dual chambered MEC reactor works in the same way as a dual chambered MFC reactor with the exception that the cathodic reaction is not oxygen reduction into water but usually protons reduction into H_2. The other major difference compared to MFC is that MEC needs an external voltage supplier because the coupled redox reactions are thermodynamically unfavorable. Still, the external electrical power needed is less than for water electrolysis because the microbial oxidation of organic carbon substrates provides a substantial part of the energy required for the cathodic reduction reactions.

4.2 Microbial Electrosynthesis Cell Principle

The principle of MES is that autotrophic microbes use reducing equivalents derived from the cathode of a BES to reduce CO_2 into biofuels or other valuable compounds. (Rabaey and Rozendal 2010; Zhang and Tremblay 2017). Standard MES reactors include an anode and a cathode separated by an IEM and connected by an electrical circuit (Tremblay et al. 2017). Biological CO_2 reduction by MES required external electric energy. Therefore, one of the main applications of MES is the storage of electricity into the chemical bonds of ready-to-use compounds. During MES, an oxidation process occurs at the anode. This can either be an abiotic reaction such as water splitting or a biological oxidation reaction such as the ones driving electricity generation in MFC or H_2 evolution in MEC. The benefit of employing biological oxidation reactions is that less electricity is required to operate the MES reactor. Various products have been synthesized from CO_2 and electricity by MES, including formate, methane, acetate, 2-oxobutyrate, propionate, butyrate, wax esters, ethanol, 1-butanol, isobutanol and 3-methyl-1-butanol (Batlle-Vilanova et al. 2017; Lehtinen et al. 2017; Ammam et al. 2016; Zaybak et al. 2013; Marshall et al. 2013; Nevin et al. 2011; Ganigué et al. 2015).

4.3 Graphene-Modified Cathodes for MFC

Although biodegradation of organic substrate occurs at the anode in MFC reactors, cathode performance where the electron acceptors are reduced also contributes to the whole bioelectrochemical cell productivity (Lu and Li 2012). Oxygen is widely used as an electron acceptor in MFC because of its availability and high reduction potential. However, oxygen reduction reaction (ORR) is sluggish and thus requires the presence of accelerating catalysts. Precious metal catalysts such as platinum (Feng et al. 2012), palladium (Koenigsmann et al. 2012) and gold (Oh and Nazar 2012) exhibit fast ORR catalytic activities, but their high cost, limited availability and high susceptibility to biological and chemical fouling often prevent large-scale applications. To make MFC more scalable, great efforts have been deployed to develop alternative to cathodes made of precious metals, such as carbonaceous materials (Qiao et al. 2010; Zhou et al. 2011), earth-abundant metal alloys and metal oxides (Li et al. 2016; Wu et al. 2014; Gnana Kumar et al. 2014b). Among these, Gr has been of great interest because of its intrinsic quality and its low-cost of fabrication (Novoselov et al. 2004).

Examples of MFC cathodes modified with Gr reported in the literature are numerous (Table 2). For instance, Zhuang et al. found that the usage of a 3D-Gr-biofilm composite coated on a CC cathode enhanced electrocatalytic activity toward ORR and improved MFC maximum power density by 103% (Zhuang et al. 2012). Xiao et al. modified a CC cathode with crumpled Gr particles (Xiao et al. 2012). The maximum power density with this modified cathode was 3.3 W m^{-3} which was 11 times higher than the unmodified CC. Zhang et al. synthesized a cathode modified with Gr functionalized with iron tetrasulfophthalocyanine (FeTsPc) (Zhang et al. 2012). In this study, FeTsPc-Gr was mixed with 0.1%

Nafion and deposited on glassy carbon (GC) electrode. A maximum power density of 817 mW m^{-2} was reached with the FeTsPc-Gr-GC cathode that was 1.56 times higher than that of a FeTsPc-GC cathode and comparable to that of platinum on carbon (Pt/C)-GC cathode. In another study, a cathode made of nitrogen-doped Gr brushed on CP was developed (Liu et al. 2013). With this cathode, MFC had a maximum voltage output of 650 mV and a maximum power density of 776 mW m^{-2}. Both performance indicators were higher than what was observed with an MFC equipped with a Pt/C-modified cathode that had a maximum voltage output of 610 mV and a maximum power density of 750 mW m^{-2}.

Several studies have developed composite cathodes for MFC made of Gr and MnO$_2$, which is a catalyst widely used for ORR (Table 2). For instance, Wen et al. developed a cathode modified with a MnO$_2$-Gr nanosheet catalyst generating a maximum power density of 2,083 mW m^{-2}, and it was 1.4 times higher than that of MFC equipped with a cathode modified with pure MnO$_2$ catalyst (Wen et al. 2012). Khilari et al. synthesized α-MnO$_2$ nanotubes (MnO$_2$NT)-Gr composite and coated it on a CP substrate (Khilari et al. 2013). When this MnO$_2$NT-Gr-CP cathode was used in a single chambered MFC, a maximum volumetric power density of 4.68 W m^{-3} was achieved. This performance was higher than that of a MnO$_2$NT-MWCNT (3.94 W m^{-3}) and that of a MnO$_2$NT-Vulcan XC (2.2 W m^{-3}) composite cathodes. Gnana Kumar and his coworkers reported that MFC equipped with a CC cathode modified with nanotubular-shaped α-MnO$_2$-GO composite exhibited a maximum power density of 3.359 W m^{-2}, and it was 7.8-fold higher than that of the unmodified electrode and was comparable to a CC cathode coated with Pt/C (Gnana Kumar et al. 2014b).

Table 2: Examples of graphene-modified cathode for MFC

Graphene and composite	Cathode base material	Maximum power density	Microbial catalyst/ substrate	Reference
3D-Gr[a]-biofilm	CC[b]	323.2 mW m^{-2}	Anaerobic sludge/Acetate	Zhuang et al. 2012
Crumpled Gr	CC	3.3 W m^{-3}	Anaerobic sludge/Acetate	Xiao et al. 2012
FeTsPc[c]-Gr	CP[d]	817 mW m^{-2}	*Escherichia coli*/Glucose	Zhang et al. 2012
Nitrogen-doped Gr	CP	776 mW m^{-2}	Activated sludge/Acetate	Liu et al. 2013
MnO$_2$-Gr nanosheet	-PTFE[e] -Activated carbon powder -Stainless steel net	2083 mW m^{-2}	Anaerobic sludge/Acetate	Wen et al. 2012
MnO$_2$NT[f]-Gr	CP	4.68 W m^{-3}	Anaerobic consortia/ Acetate	Khilari et al. 2013
α-MnO$_2$-GO	CC	3.359 W m^{-2}	Anaerobic sludge/Acetate	Gnana Kumar et al. 2014b
PB[g]-Gr	Activated carbon particles	15.88 W m^{-3}	Anaerobic sludge/ Glucose	Xu et al. 2016
Nitrogen- doped Gr-CoNi-N-BCNT[h]	CC	2.0 W m^{-2}	MFC microbial enrichment/Acetate	Hou et al. 2016
Gr-biofilm	GF[i]	163.8 mW m^{-2}	Anaerobic sludge/ Glucose/Cr(VI)	T. Song et al. 2016
rGO-PEDOT[j]-Fe$_3$O$_4$	CC	3525 mW m^{-2}	Anaerobic sludge/ Glucose	Gnana kumar et al. 2016

[a]Gr: graphene. [b]CC: carbon cloth. [c]FeTsPC: tetrasulfophthalocyanine. [d]CP: carbon paper. [e]PTFE: polytetra-fluoroethylene. [f]MnO$_2$NT: MnO$_2$ nanotube. [g]PB: Prussian blue. [h]BCNT: bamboo-like carbon nanotube. [i]GF: graphite felt. [j]PEDOT: poly(3,4-ethylenedioxythiophene)

In another study, prussian blue (PB) alone or PB-Gr nanocomposite were deposited on the surface of active carbon particles, and the resulting particles were employed in MFC systems as 3D cathodes for ORR. After long-term operation, the cell voltage and maximum power density of MFC with PB particles cathode decreased sharply from 530 mV and 15.63 W m^{-3} to 395 mV and 7.5 W m^{-3}, respectively. For MFC with PB-Gr particles cathode, voltage and maximum power density only decreased slightly from 530 mV and 16.26 W m^{-3} to 470 mV and 15.88 W m^{-3}. Additionally, the PB-Gr nanoparticles cathode had higher kinetic activity and faster reactions than PB-MFC cathode (Xu et al. 2016).

In 2016, a novel Gr catalyst prepared by a one-step synthesis strategy was reported by Hou et al. (Hou et al. 2016). The catalyst featured nitrogen-doped bamboo-like carbon nanotube (BCNT) with inner cavities in which CoNi alloy particles were encapsulated with multiple nitrogen-doped Gr layers (N-Gr-CoNi-N-BCNT). An MFC reactor equipped with a CC cathode coated with N-Gr-CoNi-N-BCNT yielded an average current density of 6.7 A m^{-2}, which was slightly lower than that of Pt/C, but with less mass transfer potential loss. The maximum power density of the N-Gr-CoNi-N-BCNT-CC cathode was 2.0 ± 0.1 W m^{-2}. The cost of N-Gr-CoNi-N-BCNT for constructing one m^2 cathode electrode is 200 times lower than that of Pt/C.

Other examples of Gr-containing cathodes for MFC include a GF cathode covered with a biofilm mixed with Gr (T. Song et al. 2016). Besides current generation, this MFC system was concomitantly reducing hexavalent chromium Cr(VI)) in the cathodic chamber. Cr(VI) is a highly mobile carcinogenic compound released in the environment by various industries (Park et al. 2007). The MFC technology can be used to reduce Cr(VI) to Cr(III) that is significantly less mobile and toxic (G. Wang et al. 2008). The maximum power density of this system was 5.7 times higher than that of an MFC equipped with an unmodified GF biocathode. Additionally, the Gr-modified biocathode-driven MFC removed 100% of 40 mg ml^{-1} Cr(VI) within 48 hours compared to 58.3% for the MFC with a GF biocathode.

Recently, Gnana Kumar et al. modified a CC cathode with a ternary composite made of rGO, poly(3,4-ethylenedioxythiophene) (PEDOT) and Fe$_3$O$_4$ (Gnana kumar et al. 2016). The maximum power density and current density of an air-cathode MFC system equipped with a rGO-PEDOT-Fe$_3$O$_4$-CC cathode were 3,525 mW m^{-2} and 9,153 mA m^{-2}, respectively. The maximum power density of this novel cathode was 1.1 times higher than that of a Pt/C catalyst. The rGO-PEDOT-Fe$_3$O$_4$-CC cathode also had superior durability of 600 hours comparable to commercially available Pt/C.

4.4 Graphene-Modified Cathodes for MEC

In MEC, cathode material has an important role to ensure fast and sustained H$_2$ evolution (Brown et al. 2014; Logan 2010). As with MFC, one of the key challenges for MEC commercialization is how to reduce the cost of the cathode (Rozendal et al. 2008). Many non-precious metals such as steel mesh, nickel and cobalt-molybdenum alloys as well as carbonaceous materials have been developed as cost-effective alternative cathodes for MEC (Call et al. 2009; Kadier et al. 2015; Jeremiasse et al. 2011). However, these cathodes have a low catalytic property and often reach insufficient H$_2$ evolution rate (Y. Hou et al. 2014; Wang et al. 2012).

Recently, Gr-modified cathodes achieved good catalytic performance in MEC (Table 3). For instance, Dai et al. synthesized a nano-Mg(OH)$_2$-Gr composite and used it to coat a CP cathode. At an external bias of 0.7V, the H$_2$ evolution rate of 0.63 m^3 m^{-3} d^{-1} of an MEC equipped with this novel composite cathode was slightly higher than the value obtained with a Pt/C cathode. The nano-Mg(OH)$_2$-Gr-CP cathode was comparable with the Pt/C cathode in terms of current density and energy efficiency (Dai et al. 2016). Cai et al. developed a nickel foam-Gr cathode for MEC and showed a significant enhancement of H$_2$ evolution rate compared to bare nickel foam. The nickel foam-Gr cathode had an average H$_2$ evolution rate of 1.31 mL H$_2$ mL^{-1} reactor d^{-1} at a 0.8 V external bias that was comparable to a Pt/C cathode (Cai et al. 2016). Hou et al. developed a third type of Gr-containing MEC cathode made of a 3D hybrid of layered MoS$_2$-nitrogen-doped Gr nanosheet aerogels (3D MoS$_2$-N-Gr). A high output current density of 0.36 mA cm^{-2} with a H$_2$ evolution rate of 0.19 m^3 H$_2$ m^{-3} d^{-1} was obtained with the 3D MoS$_2$-N-Gr hybrid at a 0.8 V bias that was significantly higher than that of MoS$_2$ nanosheets and N-Gr cathodes alone and comparable to that of a Pt/C catalyst (Y. Hou et al. 2014). In a slightly different strategy, Su et al. constructed a Gr-modified CC biocathode (Su et al. 2015). At 1.1 V bias, the H_2 evolution rate of this MEC system was 2.49 m^3 per m^3 per day with 89.12 % electron recovery in H_2 at a current density of 14.07 A m^{-2} that were 2.83 times, 1.38 times and 2.06 times that of the unmodified CC biocathode, respectively.

Table 3: Examples of graphene-modified cathode for MEC

Graphene and composite	Cathode base material	Current density	Microbial catalyst/ substrate	Ext. bias	HER[a]	CE[b]/ERE[c] (%)	Reference
nMg(OH)$_2$-Gr[d]	CP[e]	9.21 A m^{-2}	Pre-acclimated bacteria/ Acetate	0.7 V	0.63 m^3 m^{-3} d^{-1}	83.0 (CE)	Dai et al. 2016
nickel foam-Gr	Nickel foam	1.10 mA cm^{-2}	Activated sludge/ Acetate	0.8 V	1.31 ml ml^{-1} d^{-1}	86.17 (CE)	Cai et al. 2016
3D-MoS$_2$-N-Gr[f] aerogel	-	0.36 mA cm^{-2}	?/Acetate	0.8 V	0.19 m^3 m^{-3} d^{-1}	24.3 (CE)	Y. Hou et al. 2014
Gr-biofilm	CC[g]	14.07 A m^{-2}	?/Sucrose	1.1 V	2.49 m^3 m^{-3} d^{-1}	89.12 (ERE)	Su et al. 2015

[a]HER: H_2 evolution reaction. [b]CE: coulombic efficiency. [c]ERE: electron recovery efficiency. [d]Gr: graphene. [e]CP: carbon paper. [f]N-Gr: nitrogen-doped graphene. [g]CC: carbon cloth.

4.5 Graphene-Modified Cathodes for MES

One of the first studies on MES was done in 2010 and reported approximately 123 mM day^{-1} m^{-2} of acetate produced from CO_2 with a graphite stick cathode and steady consumption of current by the bacterial catalyst *Sporomusa ovata* (Nevin et al. 2010). Since then, different types of Gr-modified cathodes have been developed to improve MES productivity (Table 4). For instance, a novel 3D Gr-nickel foam cathode was fabricated by Song et al. for MES with a mixed community as microbial catalyst (Song et al. 2018). The Gr coating considerably increased the available surface area for microbial adhesion as well as the ET rate. A 1.8 times increase of the volumetric acetate production rate was reported with the 3D Gr-nickel foam cathode set at a potential of -1.05 V vs. Ag/AgCl compared with untreated nickel foam. Also, 70% of the electrons consumed were used for acetate production.

Table 4: Examples of graphene-modified cathode for MES

Graphene and composite	Cathode base material	Current density	Microbial catalyst	Cathode potential (vs. SHE[a])	Acetate production rate	CE[b] (%)	Reference
Gr[c]-nickel foam	Nickel foam	-10.2 A m^{-2}	Anaerobic sludge	-0.85 V	0.187 g l^{-1} d^{-1}	70.0	Song et al. 2018
3D-rGO[d]	CF[e]	-2450 mA m^{-2}	*Sporomusa ovata*	-0.69 V	925.5 mM m^{-2} d^{-1}	86.5	Aryal et al. 2016
rGO-TEPA[f]	CC[g]	-2358 mA m^{-2}	*S. ovata*	-0.69 V	1052 mM m^{-2} d^{-1}	83.3	Chen et al. 2016
rGO-biofilm	CF	-4.9 A m^{-2}	Anaerobic sludge	-0.85 V	0.17 g l^{-1} d^{-1}	77.0	Song et al. 2017
Free-standing rGO paper	-	-2580 mA m^{-2}	*S. ovata*	-0.69 V	168.5mmol m^{-2} d^{-1}	90.7	Aryal et al. 2017

[a]SHE: standard hydrogen electrode. [b]CE: coulombic efficiency. [c]Gr:graphene. [d]rGO:reduced graphene oxide. [e]CF; carbon felt. [f]TEPA: tetraethyl pentamine. [g]CC: carbon cloth.

A different approach consists of the fabrication of cathode made of Gr coated on a carbonaceous substrate. Aryal and his coworkers reported the development of a 3D-rGO functionalized CF cathode enabling faster ET to the microbial catalyst *S. ovata* in an MES reactor (Aryal et al. 2016). Modification of the CF cathode with the 3D-rGO network increased the electrosynthesis rate of acetate from CO_2 by a 6.8-fold to 925.5 mM m^{-2} d^{-1} at a potential of -690 mV vs. standard hydrogen electrode (SHE). The 3D-rGO-CF cathode also improved biofilm density and current consumption to 2,450 mA m^{-2}. Coulombic efficiency (CE) of electrons consumed to acetate was 86.5%. In a second study, Chen et al. coated rGO functionalized with tetraethylene pentamine (rGO-TEPA) on a CC cathode with a simple and inexpensive method (Chen et al. 2016). The presence of rGO-TEPA led to the formation of a substantial biofilm with a unique spatial arrangement on the cathode surface. Contrary to rGO, TEPA is a positively-charged molecule that can form stronger electrostatic interactions with negatively-charged bacteria, such as *S. ovata*. MES with rGO-TEPA-CC cathode set at a potential of -690 mV vs. SHE, and a methanol-adapted *S. ovata* strain generated 11.8 times more acetate with a rate of 1,052 mM m^{-2} d^{-1}. The current consumption was 2,358 mA m^{-2} and CE was 83%. A characteristic shared by both studies is that modification of the cathode with rGO or rGO-TEPA significantly increases the surface area available for microbial interactions that is probably one of the reasons why MES was more performant. In another study, a CF cathode covered with an rGO-biofilm self-assembled *in situ* was tested for MES by Song et al. (2017). In this system, a volumetric acetate production rate of 0.17 g L^{-1} d^{-1} was achieved by a mixed community at a cathode potential of -850 mV vs. SHE with a CE of 77% and a final acetate titer of 7.1 g L^{-1} after 40 days.

Cathode made of pure Gr is also an interesting alternative for MES. In 2017, Aryal et al. fabricated and tested a freestanding rGO paper cathode set at a potential of -690 mV vs. SHE in a MES reactor driven by *S. ovata* (Aryal et al. 2017b). The acetate production rate was 8-fold faster at 168.5 mmol $m^{-2}d^{-1}$ compared to a MES system equipped with a CP cathode. The current density was 2,580 mA m^{-2} and CE was 90.7%.

5. Graphene in Biosensors

Biosensors are a special type of chemical sensor that uses biologically-active elements, such as enzymes, microorganisms, animal or plant tissue sections, antigens, antibodies and nucleic acids as sensing components for the highly selective detection of target molecules (Peña-Bahamonde et al. 2018). Biosensors detect reaction between target molecules and biological sensing elements and translate it into an electric signal through different types of the converter to establish the concentration of the measured compounds. Often, the core part of biosensors is a three-electrode system with a reference electrode (RE), a counter-electrode (CE) and a working electrode (WE) that comprises the biological sensing element (Figure 4). Biosensors can be a relatively inexpensive option for the design of portable instruments enabling refined analytical measurements at disseminated locations. Due to their simple operation and high sensitivity, they may be used in a wide range of applications and in complex media, including blood and interstitial fluid. These uses include cell typing, the detection of large proteins, viruses, antibodies, DNA, electrolytes, drugs, pesticides and other compounds (Cornell et al. 1997).

Because of its planar shape and chemical structure, Gr has several advantageous properties for the fabrication of performant biosensor compared to other materials (Table 5). For instance, the atomic thickness of Gr sheets and their high surface to volume ratio make this material highly sensitive to changes in localized environmental conditions that are an important advantage in the sensing field because all carbon atoms interact directly with the analytes. As a result, Gr can be used to fabricate biosensor with high sensitivity (Justino et al. 2017). Incorporating Gr in electrochemical biosensors design has resulted in multiple sensing strategies with applications in diverse areas such as clinical diagnosis and food safety (Y. Song et al. 2016).

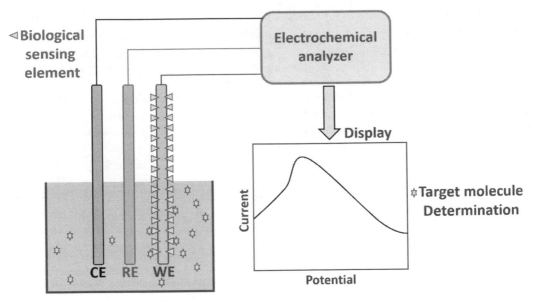

Figure 4: Representation of an electrochemical biosensor. CE: counter electrode. RE: reference electrode. WE: working electrode.

5.1 Graphene-Based Biosensor for Glucose

The development of reliable, simple and rapid glucose detection methods is of great importance in many fields, including clinical diagnosis, medicine, food industry and environmental monitoring. Sensors for detecting glucose are mainly classified into glucose oxidase (GOx)-based sensors and non-enzymatic sensors. GOx is widely used in the construction of various glucose biosensors due to its high sensitivity and selectivity. Recently, Pu et al. proposed a continuous blood glucose monitoring microsystem harnessing the outstanding physico-chemical propertied of Gr (Pu et al. 2016). Their biosensor system comprises a three-electrode GOx-based bioelectrochemical device integrated into a polydimethylsiloxane microfluidic chip that is used to transdermally extract interstitial fluid for glucose measurement in a noninvasive manner. A single Gr layer was affixed with gold nanoparticles (AuNP) to the WE of the biosensor to construct a composite nanostructure. Gr presence on the WE surface improved electrical activity and enabled the detection of low glucose concentration. Direct electrodeposition of AuNP onto the Gr layer increased ET rate from the active center of GOx to the electrode and enhanced sensitivity. Finally, GOx was immobilized on the surface of the composite nanostructure by electrochemical polymerization. This Gr-containing biosensor could precisely measure glucose in a linear range from 0 to 162 mg dl^{-1} with a detection limit of 1.44 mg dl^{-1}. In comparison, commercial glucometers can measure concentration ranging from 20 to 500 mg dl^{-1} that is not sensitive enough for diluted interstitial fluid samples where glucose concentration can be as low as 2 mg dl^{-1}.

5.2 Graphene-Based Biosensor for Hydrogen Peroxide

The development of rapid, selective, sensitive, inexpensive, stable and reliable analytical tools for the quantitative determination of hydrogen peroxide (H_2O_2) has attracted much attention in multiple fields, including pharmaceutical, environmental, clinical, biological, chemical and food industries (Yang et al. 2013; Jiang et al. 2013; Jia et al. 2014). Compared with spectrofluorometry, spectrophotometry and chromatography, bioelectrochemical methods could be more effective for *in situ* and real-time analysis of H_2O_2 due to their fast response, good selectivity, high sensitivity, excellent reproducibility and facile operation (L. Yuan et al. 2012; Xu et al. 2014). Liu et al. fabricated a Gr-based biosensor for H_2O_2 detection (Y. Liu et al. 2017). They described a simple method for preparing a porous Gr network

Table 5: Examples of graphene-modified working electrode for electrochemical biosensor

Target molecule	Working electrode	Performance	Reference
Glucose	-PDSM[a] microfluidic chip -Glass substrate -AuNP[b] on Gr[c] layer -Glucose oxidase	-Detection limit: 1.44 mg dl^{-1} -Linear range: 0 to 162 mg dl^{-1}	Pu et al. 2016
Hydrogen peroxide	-GC[d] substrate -Porous Gr network etched with AgNP[e] -Horseradish peroxidase	-Detection limit: 26.7 pM -Linear range: 80 pM to 835 μM	Y. Liu et al. 2017
Ethanol	-GC substrate -rGO[f] film with BSA[g] -Luminophore Ru(bpy)$_3^{2+}$ -Alcohol dehydrogenase	-Detection limit: 0.1 μM -Linear range: 1 to 2000 μM	Gao et al. 2013
ssDNA[h]	-GC substrate -Gr layer -ssDNA -Sandwich assay with target ssDNA and AuNP-marked ssDNA probe -Silver staining	-Detection limit: 72 pM -Linear range: 200 pM to 500 nM	Lin et al. 2011
ssDNA	- ssDNA-labeled GOQD[i] -GO[j] as a quencher	-Detection limit: 0.008 nM -Linear range: 0.05 nM to 50 nM	He and Fan 2018
CEA[k]	-GC substrate -Nanocomposite made of rGO, AuNP and PICA[l] -Ionic liquid modification -Anti-CEA antibody	-Detection limit: 0.02 ng ml^{-1} -Linear range: 0.02 to 90 ng ml^{-1}	Zhao et al. 2016
Cholesterol	-GC substrate -Cerium oxide-Gr matrix -Luminol -H$_2$O$_2$-generating cholesterol oxidase	-Detection limit: 4.0 μM -Linear range: 12 μM to 7.2 mM	Zhang et al. 2013
OP[m]	-GC - CS[n]-TiO$_2$-rGO multilayered matrix -Acetylcholinesterase	-Detection limit: 29 nM -Linear range: 0.036 μM to 22.6 μM	Cui et al. 2018
OP	-Gr electrode fabricated by IML[o] -PtNP[p]	-Detection limit: 3 nM -Linear range: 0.1 to 1 μM -Response time of 5 seconds	Hondred et al. 2018

[a]PDSM: polydimethylsiloxane. [b]AuNP: gold nanoparticle. [c]Gr: graphene. [d]GC: glassy carbon. [e]AgNP: silver nanoparticle. [f]rGO: reduced graphene oxide. [g]BSA: bovine serum albumin. [h]ssDNA: single-stranded DNA. [i]GOQD: graphene oxide quantum dot. [j]GO: graphene oxide. [k]CEA: carcinoembryonic antigen. [m]OP: organophosphorus pesticide. [n]CS: chitosan. [o]IML: inkjet mask lithography. [p]PtNP: platinum nanoparticle.

(PGN) by etching silver nanoparticles (AgNPs) and combining it with horseradish peroxidase (HRP) on a GC electrode (Table 5). Because of the structure and large surface area of PGN, this biosensor could load a large amount of HRP that increased its sensitivity. The H$_2$O$_2$ detection limit was 26.7 pM with a very wide linear detection range covering 7 orders of magnitude.

5.3 Graphene-Based Biosensor for Ethanol

Ethanol is a basic raw material and solvent, mainly used in food, chemical, military and pharmaceutical industries. Therefore, on-site monitoring and ethanol detection under multiple circumstances is an important sensing process. Guo et al. developed a novel Gr-based electrochemiluminescence (ECL) biosensor for the detection of ethanol (Gao et al. 2013). ECL biosensors measure luminescent signals emitted by chemical species excited by ET reactions at the surface of an electrode. In this system, an rGO film was directly formed on the surface of a GC electrode. During this process, the luminophore Ru(bpy)$_3^{2+}$ was immobilized on the WE and GO was reduced *in situ* to rGO by bovine serum albumin (BSA). The other important biological component of the WE besides BSA was an alcohol dehydrogenase (ADH) that is the enzyme interacting with ethanol. With this Gr-based ECL biosensor, the ethanol detection limit was 0.1 μM with a linear range from 1 to 2,000 μM.

5.4 Graphene-Based Biosensor for DNA

Sensitive and selective detection of DNA is an important technological goal due to the importance of this molecule in human health and in other fields of life science. Many disorders and cancers are linked to DNA sequence variations. In the case of cancer, detection of tumor DNA is widely done in hospitals by real-time PCR or via sequencing techniques, but these methods can be complicated and costly. Thus, biosensors represent an interesting alternative and vast research efforts have been deployed in that direction (Table 5). For instance, Lin et al. developed a Gr-based electrochemical biosensor for DNA detection (Lin et al. 2011). In their system, single-stranded DNA (ssDNA) molecules have been attached to a Gr-modified WE via π-π stacking. Then, target DNA sequences and oligonucleotide probes labeled with AuNPs were hybridized in a sandwich assay format. Subsequently, silver was deposited on the AuNPs and was detected by differential pulse voltammetry. Due to the high DNA loading capacity of Gr and the unique signal amplification of silver-stained AuNPs, this biosensor exhibited good analytical performance, wide linear detection range between 200 pM to 500 nM and a low detection limit of 72 pM. He et al. designed a different Gr-based biosensor with ssDNA-labeled GO quantum dots (GOQD) as a fluorescent probe and GO as a quencher (He and Fan 2018). The detection principle is that when ssDNA-labeled GOQD hybridizes with complementary target DNA to form double-strand(ds) DNA-labeled GOQD, fluorescence emission becomes higher. In the absence of complementary target DNA, GO quenches the fluorescence of ssDNA-labeled GOQD. This biosensor was developed for the ultrasensitive detection of NOS terminator gene sequences that are associated with the genetically-modified organism (GMO). The detection limit of this biosensor was 0.008 nM, and the linear detection range was 0.05 nM to 50 nM.

5.5 Graphene-Based Biosensor for Antigen

Antigens detection can be used to diagnose diseases and to establish a plan for medical treatment. For instance, carcinoembryonic antigen (CEA) is a type of glycoprotein produced in the gastrointestinal tract during embryonic development. Certain forms of cancer such as colon and rectum cancer can cause higher blood levels of CEA in an adult. Thus, CEA is employed as a tumor marker via blood test to predict the prospect of cancer. Zhao et al. fabricated a Gr-containing label-free electrochemical immunosensor for CEA (Zhao et al. 2016). It is based on a nanocomposite consisting of rGO, AuNP and poly(indole-6-carboxylic acid) (PICA) deposited on a GC electrode and modified with ionic liquid. Anti-CEA antibodies interacting with CEA have then affixed on this composite WE. With this biosensor, CEA detection limit was 0.02 ng ml^{-1} and the linear detection range was between 0.02 to 90 ng ml^{-1}.

5.6 Graphene-Based Biosensor for Cholesterol

Cholesterol is an essential compound of mammalian metabolism involved in the synthesis of the cell membrane, hormones, bile acids and vitamins. However, cholesterol must be routinely monitored because accumulation in artery walls can result in cardiovascular diseases, such as atherosclerosis, hypertension coronary heart disease, myocardial and cerebral infarction (Ruecha et al. 2011). Zhang et al. prepared a simple and sensitive ECL cholesterol biosensor with a cerium oxide-Gr composite deposited on a GC electrode (Zhang et al. 2013). This system relies on the oxidation of luminol by H_2O_2 that is generated by a cholesterol oxidase (ChOx) from cholesterol and O_2. Upon oxidation, luminol emits light at 425 nm. Besides ChOx immobilization, the cerium oxide-Gr matrix amplified the luminol ECL signal because of the outstanding electrocatalytic activity of CeO_2 toward H_2O_2 and the high electrical conductivity of Gr. This biosensor had a detection limit of 4.0 µM and a linear detection range for cholesterol going from 12 µM to 7.2 mM. In addition, the biosensor has excellent reproducibility, long-term stability and selectivity.

5.7 Graphene-Based Biosensor for Pesticides

Pesticides are widely used in agriculture to protect seeds and crops. Because of their toxicity and persistence in the biosphere, their presence in soil and water must be continuously monitored. The current standard procedures for the determination of pesticides in soil and water samples that include high-performance liquid chromatography and gas chromatography-mass spectrometry are accurate but time-consuming (Bäumner and Schmid, 1998). Therefore, it would be advantageous to develop a simpler, quicker and potentially less expensive method to detect pesticides in various samples such as biosensors (Table 5). Cui et al. have developed a novel Gr-based organophosphorus pesticide (OP) biosensor (Cui et al. 2018). In this system, a multi-layered matrix made of chitosan (CS), TiO_2 sol-gel and rGO (CS-TiO_2-rGO) was deposited on a GC electrode. The enzyme acetylcholinesterase (AChE) was then immobilized in the CS-TiO_2-rGO matrix. AChE hydrolyzes acetylcholine into thiocholine that can be electrooxidized by an electrode. When OP is present, AChE activity is inhibited and ET from thiocholine to the electrode diminishes proportionally to the concentration of OP. This biosensor had a detection limit of 29 nM for the model OP dichlorvos and a linear detection range of 0.036 µM to 22.6 µM. The CS-TiO_2-rGO matrix had a mesoporous structure and a large specific surface providing a biocompatible environment for AChE. The specific role of rGO was to improve the sensitivity of the biosensor.

Hondred et al. develop a different Gr-containing biosensor to detect OP (Hondred et al. 2018). Firstly, a Gr WE was fabricated with a newly developed technology called inkjet mask lithography that is a thin-film manufacturing technology (Hondred et al. 2017). Then, the printed Gr WE was engraved with a laser and subjected to the electrochemical deposition of platinum nanoparticles (PtNP) to form a nano/microstructure with high specific surface area and electrical conductivity. The enzyme phosphotriesterase (PTE) was then conjugated to the surface of the PtNP-modified Gr WE. PTE hydrolyzes the OP paraxon into *p*-nitrophenol that is an electroactive molecule that can be sensed by an electrode. This biosensor had a 3 nM detection limit for paraxon and a linear detection range of 0.1 to 1 µM with a response time of 5 seconds.

6. Conclusions and Future Perspectives

The aim of this chapter was to describe how Gr has been used in recent years in the fabrication of high-performance BES electrodes. Literature shows that using Gr alone or as part of composite electrodes usually resulted in MFC, MEC, MES or biosensor systems that are more performant when compared with similar reactors equipped with electrodes made of other metallic or carbonaceous materials. In certain cases, Gr-based electrodes had BES performance comparable to Pt/C electrodes that are significantly more expensive. When both higher performance and lower fabrication cost are considered, Gr could facilitate the transposition of BES technologies from the laboratory to the industry. The full spectrum and potential of Gr applications in the fabrication of BES electrodes are not known yet. Many more Gr-containing composite electrodes with different physical conformation can be developed and tested into MFC, MEC, MES or biosensor systems. At the current rhythm of research published on this topic, the future of Gr materials in BES looks exciting and full of promises.

References

Ahn Y, Logan BE (2013) Altering anode thickness to improve power production in microbial fuel cells with different electrode distances. Energy Fuels 27: 271–276.

Ammam F, Tremblay P-L, Lizak DM, Zhang T (2016) Effect of tungstate on acetate and ethanol production by the electrosynthetic bacterium *Sporomusa ovata*. Biotechnol Biofuels 9: 163.

Aryal N, Ammam F, Patil SA, Pant D (2017a) An overview of cathode materials for microbial electrosynthesis of chemicals from carbon dioxide. Green Chem 19: 5748–5760.

Aryal N, Halder A, Tremblay P-L, Chi Q, Zhang T (2016) Enhanced microbial electrosynthesis with three-dimensional graphene functionalized cathodes fabricated via solvothermal synthesis. Electrochim Acta 217: 117–122.

Aryal N, Halder A, Zhang M, Whelan PR, Tremblay P-L, Chi Q, Zhang T (2017b) Freestanding and flexible graphene papers as bioelectrochemical cathode for selective and efficient CO_2 conversion. Sci Rep 7: 9107.

Batlle-Vilanova P, Ganigué R, Ramió-Pujol S, Bañeras L, Jiménez G, Hidalgo M, Balaguer MD, Colprim J, Puig S (2017) Microbial electrosynthesis of butyrate from carbon dioxide: Production and extraction. Bioelectrochemistry 117: 57–64.

Bäumner AJ, Schmid RD (1998) Development of a new immunosensor for pesticide detection: a disposable system with liposome-enhancement and amperometric detection. Biosens Bioelectron 13: 519–529.

Benetton XD, Navarro-Ávila SG, Carrera-Figueiras C (2010) Electrochemical evaluation of Ti/TiO$_2$-polyaniline Anodes for Microbial Fuel Cells using Hypersaline Microbial Consortia for Synthetic-wastewater Treatment. J New Mater Electrochem Syst 13: 1–6.

Bolotin KI, Sikes KJ, Jiang Z, Klima M, Fudenberg G, Hone J, Kim P, Stormer HL (2008) Ultrahigh electron mobility in suspended graphene. Solid State Commun 146: 351–355.

Brown RK, Harnisch F, Wirth S, Wahlandt H, Dockhorn T, Dichtl N, Schröder U (2014) Evaluating the effects of scaling up on the performance of bioelectrochemical systems using a technical scale microbial electrolysis cell. Bioresour Technol 163: 206–213.

Brutinel ED, Gralnick JA (2012) Shuttling happens: soluble flavin mediators of extracellular electron transfer in *Shewanella*. Appl Microbiol Biotechnol 93: 41–48.

Cai W, Liu W, Han J, Wang A (2016) Enhanced hydrogen production in microbial electrolysis cell with 3D self-assembly nickel foam-graphene cathode. Biosens Bioelectron 80: 118–122.

Call DF, Merrill MD, Logan BE (2009) High surface area stainless steel brushes as cathodes in microbial electrolysis cells. Environ Sci Technol 43: 2179–2183.

Chaudhuri SK, Lovley DR (2003) Electricity generation by direct oxidation of glucose in mediatorless microbial fuel cells. Nat Biotechnol 21: 1229–1232.

Chen L, Tremblay P-L, Mohanty S, Xu K, Zhang T (2016) Electrosynthesis of acetate from CO_2 by a highly structured biofilm assembled with reduced graphene oxide–tetraethylene pentamine. J Mater Chem A 4: 8395–8401.

Cornell BA, Braach-Maksvytis VL, King LG, Osman PD, Raguse B, Wieczorek L, Pace RJ (1997) A biosensor that uses ion-channel switches. Nature 387: 580–583.

Cui H-F, Wu W-W, Li M-M, Song X, Lv Y, Zhang T-T (2018) A highly stable acetylcholinesterase biosensor based on chitosan-TiO$_2$-graphene nanocomposites for detection of organophosphate pesticides. Biosens Bioelectron 99: 223–229.

Dai H, Yang H, Liu X, Jian X, Liang Z (2016) Electrochemical evaluation of nano-Mg(OH)$_2$/graphene as a catalyst for hydrogen evolution in microbial electrolysis cell. Fuel 174: 251–256.

ElMekawy A, Hegab HM, Losic D, Saint CP, Pant D (2017) Applications of graphene in microbial fuel cells: The gap between promise and reality. Renew Sust Energ Rev 72: 1389–1403.

Feng Y, Shi X, Wang X, Lee H, Liu J, Qu Y, He W, Kumar SMS, Kim BH, Ren N (2012) Effects of sulfide on microbial fuel cells with platinum and nitrogen-doped carbon powder cathodes. Biosens Bioelectron 35: 413–415.

Ganigué R, Puig S, Batlle-Vilanova P, Balaguer MD, Colprim J (2015) Microbial electrosynthesis of butyrate from carbon dioxide. Chem Commun 51: 3235–3238.

Gao W, Chen Yunsheng, Xi J, Lin S, Chen Yaowen, Lin Y, Chen Z (2013) A novel electrochemiluminescence ethanol biosensor based on tris(2,2'-bipyridine) ruthenium (II) and alcohol dehydrogenase immobilized in graphene/bovine serum albumin composite film. Biosens Bioelectron 41: 776–782.

Geim AK, Novoselov KS (2007) The rise of graphene. Nat Mater 6: 183–191.

Gnana kumar G, Kirubaharan CJ, Udhayakumar S, Ramachandran K, Karthikeyan C, Renganathan R, Nahm KS (2014a) Synthesis, Structural, and Morphological Characterizations of Reduced Graphene Oxide-Supported Polypyrrole Anode Catalysts for Improved Microbial Fuel Cell Performances. ACS Sustain Chem Eng 2: 2283–2290.

Gnana Kumar G, Awan Z, Suk Nahm K, Xavier JS (2014b) Nanotubular MnO_2/graphene oxide composites for the application of open air-breathing cathode microbial fuel cells. Biosens Bioelectron 53: 528–534.

Gnana kumar G, Joseph Kirubaharan C, Yoo DJ, Kim AR (2016) Graphene/poly(3,4-ethylenedioxythiophene)/Fe_3O_4 nanocomposite – An efficient oxygen reduction catalyst for the continuous electricity production from wastewater treatment microbial fuel cells. Int J Hydrog Energy 41: 13208–13219.

Hasany M, Mardanpour MM, Yaghmaei S (2016) Biocatalysts in microbial electrolysis cells: A review. Int J Hydrog Energy 41: 1477–1493.

He Y, Fan Z (2018) A novel biosensor based on DNA hybridization for ultrasensitive detection of NOS terminator gene sequences. Sens Actuators B Chem 257: 538–544.

Hondred JA, Breger JC, Alves NJ, Trammell SA, Walper SA, Medintz IL, Claussen JC (2018) Printed Graphene Electrochemical Biosensors Fabricated by Inkjet Maskless Lithography for Rapid and Sensitive Detection of Organophosphates. ACS Appl Mater Interfaces 10: 11125–11134.

Hondred JA, Stromberg LR, Mosher CL, Claussen JC (2017) High-Resolution Graphene Films for Electrochemical Sensing via Inkjet Maskless Lithography. ACS Nano 11: 9836–9845.

Hou J, Liu Z, Yang S, Zhou Y (2014) Three-dimensional macroporous anodes based on stainless steel fiber felt for high-performance microbial fuel cells. J Power Sources 258: 204–209.

Hou J, Liu Z, Zhang P (2013) A new method for fabrication of graphene/polyaniline nanocomplex modified microbial fuel cell anodes. J Power Sources 224: 139–144.

Hou Y, Yuan H, Wen Z, Cui S, Guo X, He Z, Chen J (2016) Nitrogen-doped graphene/CoNi alloy encased within bamboo-like carbon nanotube hybrids as cathode catalysts in microbial fuel cells. J Power Sources 307: 561–568.

Hou Y, Zhang B, Wen Z, Cui S, Guo X, He Z, Chen J (2014) A 3D hybrid of layered MoS_2/nitrogen-doped graphene nanosheet aerogels: an effective catalyst for hydrogen evolution in microbial electrolysis cells. J Mater Chem A 2: 13795–13800.

Hu W, Peng C, Luo W, Lv M, Li X, Li D, Huang Q, Fan C (2010) Graphene-Based Antibacterial Paper. ACS Nano 4: 4317–4323.

Huang Y-X, Liu X-W, Xie J-F, Sheng G-P, Wang G-Y, Zhang Y-Y, Xu A-W, Yu H-Q (2011) Graphene oxide nanoribbons greatly enhance extracellular electron transfer in bio-electrochemical systems. Chem Commun 47: 5795–5797.

Hutchinson AJ, Tokash JC, Logan BE (2011) Analysis of carbon fiber brush loading in anodes on startup and performance of microbial fuel cells. J Power Sources 196: 9213–9219.

Jain A, Zhang X, Pastorella G, Connolly JO, Barry N, Woolley R, Krishnamurthy S, Marsili E (2012) Electron transfer mechanism in *Shewanella loihica* PV-4 biofilms formed at graphite electrode. Bioelectrochemistry 87: 28–32.

Javed MM, Nisar MA, Ahmad MU, Yasmeen N, Zahoor S (2018) Microbial fuel cells as an alternative energy source: current status. Biotechnol Genet Eng Rev 34: 216–242.

Jeremiasse AW, Bergsma J, Kleijn JM, Saakes M, Buisman CJN, Cohen Stuart M, Hamelers HVM (2011) Performance of metal alloys as hydrogen evolution reaction catalysts in a microbial electrolysis cell. Int J Hydrog Energy, 36: 10482–10489.

Jia N, Huang B, Chen L, Tan L, Yao S (2014) A simple non-enzymatic hydrogen peroxide sensor using gold nanoparticles-graphene-chitosan modified electrode. Sens Actuators B Chem 195: 165–170.

Jiang F, Yue R, Du Y, Xu J, Yang P (2013) A one-pot 'green' synthesis of Pd-decorated PEDOT nanospheres for nonenzymatic hydrogen peroxide sensing. Biosens Bioelectron 44: 127–131.

Jiang Y, Jianxiong Zeng R (2018) Expanding the product spectrum of value added chemicals in microbial electrosynthesis through integrated process design-A review. Bioresour Technol pii: S0960-8524(18)31205-7.

Justino CIL, Gomes AR, Freitas AC, Duarte AC, Rocha-Santos TAP (2017) Graphene based sensors and biosensors. Trends Anal Chem 91: 53–66.

Justino CIL, Rocha-Santos TAP, Duarte AC, Rocha-Santos TAP (2013) Advances in point-of-care technologies with biosensors based on carbon nanotubes. Trends Anal Chem 45: 24–36.

Kadier A, Kalil MS, Abdeshahian P, Chandrasekhar K, Mohamed A, Azman NF, Logroño W, Simayi Y, Hamid AA (2016) Recent advances and emerging challenges in microbial electrolysis cells (MECs) for microbial production of hydrogen and value-added chemicals. Renew Sust Energ Rev 61: 501–525.

Kadier A, Simayi Y, Chandrasekhar K, Ismail M, Kalil MS (2015) Hydrogen gas production with an electroformed Ni mesh cathode catalysts in a single-chamber microbial electrolysis cell (MEC). Int J Hydrog Energy 40: 14095–14103.

Khilari S, Pandit S, Ghangrekar MM, Das D, Pradhan D (2013) Graphene supported α-MnO_2 nanotubes as a cathode catalyst for improved power generation and wastewater treatment in single-chambered microbial fuel cells. RSC Adv 3: 7902–7911.

Kim Keun Soo, Zhao Y, Jang H, Lee SY, Kim JM, Kim Kwang S, Ahn J-H, Kim P, Choi J-Y, Hong BH (2009) Large-scale pattern growth of graphene films for stretchable transparent electrodes. Nature 457: 706–710.

Kirubaharan CJ, Santhakumar K, Gnana kumar G, Senthilkumar N, Jang J-H (2015) Nitrogen doped graphene sheets as metal free anode catalysts for the high performance microbial fuel cells. Int J Hydrog Energy 40: 13061–13070.

Koenigsmann C, Sutter E, Chiesa TA, Adzic RR, Wong SS (2012) Highly Enhanced Electrocatalytic Oxygen Reduction Performance Observed in Bimetallic Palladium-Based Nanowires Prepared under Ambient, Surfactantless Conditions. Nano Lett 12: 2013–2020.

Kotloski NJ, Gralnick JA (2013) Flavin electron shuttles dominate extracellular electron transfer by *Shewanella oneidensis*. MBio 4: e00553-12.

Kumar GG, Hashmi S, Karthikeyan C, GhavamiNejad A, Vatankhah-Varnoosfaderani M, Stadler FJ (2014) Graphene Oxide/Carbon Nanotube Composite Hydrogels—Versatile Materials for Microbial Fuel Cell Applications. Macromol Rapid Commun 35: 1861–1865. https://doi.org/10.1002/marc.201400332

Kumar R, Singh L, Zularisam AW (2016) Exoelectrogens: Recent advances in molecular drivers involved in extracellular electron transfer and strategies used to improve it for microbial fuel cell applications. Renew Sust Energ Rev 56: 1322–1336.

Lee C, Wei X, Kysar JW, Hone J (2008) Measurement of the elastic properties and intrinsic strength of monolayer graphene. Science 321: 385–388.

Lehtinen T, Efimova E, Tremblay P-L, Santala S, Zhang T, Santala V (2017) Production of long chain alkyl esters from carbon dioxide and electricity by a two-stage bacterial process. Bioresour Technol 243: 30–36.

Li S, Cheng C, Thomas A (2017) Carbon-Based Microbial-Fuel-Cell Electrodes: From Conductive Supports to Active Catalysts. Adv Mater 29: 1602547.

Li Y, Tremblay P-L, Zhang T (2018) Anode Catalysts and Biocatalysts for Microbial Fuel Cells. In: PP Kundu, K Dutta (eds) Progress and Recent Trends in Microbial Fuel Cells. Elsevier, pp. 143–165.

Li Z, Shao M, Zhou L, Yang Q, Zhang C, Wei M, Evans DG, Duan X (2016) Carbon-based electrocatalyst derived from bimetallic metal-organic framework arrays for high performance oxygen reduction. Nano Energy 25: 100–109.

Lin L, Liu Y, Tang L, Li J (2011) Electrochemical DNA sensor by the assembly of graphene and DNA-conjugated gold nanoparticles with silver enhancement strategy. Analyst 136: 4732–4737.

Liu C, Alwarappan S, Chen Z, Kong X, Li C-Z (2010) Membraneless enzymatic biofuel cells based on graphene nanosheets. Biosens Bioelectron 25: 1829–1833.

Liu H, Cheng S, Logan BE (2005) Power Generation in Fed-Batch Microbial Fuel Cells as a Function of Ionic Strength, Temperature, and Reactor Configuration. Environ Sci Technol 39: 5488–5493.

Liu H, Logan BE (2004) Electricity Generation Using an Air-Cathode Single Chamber Microbial Fuel Cell in the Presence and Absence of a Proton Exchange Membrane. Environ Sci Technol 38: 4040–4046.

Liu H, Ramnarayanan R, Logan BE (2004) Production of Electricity during Wastewater Treatment Using a Single Chamber Microbial Fuel Cell. Environ Sci Technol 38: 2281–2285.

Liu J, Qiao Y, Guo CX, Lim S, Song H, Li CM (2012) Graphene/carbon cloth anode for high-performance mediatorless microbial fuel cells. Bioresour Technol 114: 275–280.

Liu S, Zeng TH, Hofmann M, Burcombe E, Wei J, Jiang R, Kong J, Chen Y (2011) Antibacterial Activity of Graphite, Graphite Oxide, Graphene Oxide, and Reduced Graphene Oxide: Membrane and Oxidative Stress. ACS Nano 5: 6971–6980.

Liu Y, Liu H, Wang C, Hou S-X, Yang N (2013) Sustainable energy recovery in wastewater treatment by microbial fuel cells: stable power generation with nitrogen-doped graphene cathode. Environ Sci Technol 47: 13889–13895.

Liu Y, Liu X, Guo Z, Hu Z, Xue Z, Lu X (2017) Horseradish peroxidase supported on porous graphene as a novel sensing platform for detection of hydrogen peroxide in living cells sensitively. Biosens Bioelectron 87: 101–107.

Liu Z, Zhou L, Chen Q, Zhou W, Liu Y (2017) Advances in Graphene/Graphene Composite Based Microbial Fuel/Electrolysis Cells. Electroanalysis 29: 652–661.

Logan B, Cheng S, Watson V, Estadt G (2007) Graphite Fiber Brush Anodes for Increased Power Production in Air-Cathode Microbial Fuel Cells. Environ Sci Technol 41: 3341–3346.

Logan BE (2010) Scaling up microbial fuel cells and other bioelectrochemical systems. Appl Microbiol Biotechnol 85: 1665–1671.

Lu M, Li SFY (2012) Cathode Reactions and Applications in Microbial Fuel Cells: A Review. Crit Rev Environ Sci Technol 42: 2504–2525.

Luo J, Jang HD, Sun T, Xiao L, He Z, Katsoulidis AP, Kanatzidis MG, Gibson JM, Huang J (2011) Compression and Aggregation-Resistant Particles of Crumpled Soft Sheets. ACS Nano 5: 8943–8949.

Marshall CW, Ross DE, Fichot EB, Norman RS, May HD (2013) Long-term operation of microbial electrosynthesis systems improves acetate production by autotrophic microbiomes. Environ Sci Technol 47: 6023–6029.

Mehdinia A, Ziaei E, Jabbari A (2014) Facile microwave-assisted synthesized reduced graphene oxide/tin oxide nanocomposite and using as anode material of microbial fuel cell to improve power generation. Int J Hydrog Energy 39: 10724–10730.

Nevin KP, Hensley SA, Franks AE, Summers ZM, Ou J, Woodard TL, Snoeyenbos-West OL, Lovley DR (2011) Electrosynthesis of organic compounds from carbon dioxide is catalyzed by a diversity of acetogenic microorganisms. Appl Environ Microbiol 77: 2882–2886.

Nevin KP, Woodard TL, Franks AE, Summers ZM, Lovley DR (2010) Microbial electrosynthesis: feeding microbes electricity to convert carbon dioxide and water to multicarbon extracellular organic compounds. MBio 1: e00103-10.

Novoselov KS, Geim AK, Morozov SV, Jiang D, Zhang Y, Dubonos SV, Grigorieva IV, Firsov AA (2004) Electric field effect in atomically thin carbon films. Science 306: 666–669.

Oh SH, Nazar LF (2012) Oxide Catalysts for Rechargeable High-Capacity Li–O_2 Batteries. Adv Energy Mater 2: 903–910.

Pant D, Van Bogaert G, Diels L, Vanbroekhoven K (2010) A review of the substrates used in microbial fuel cells (MFCs) for sustainable energy production. Bioresour Technol 101: 1533–1543.

Park D, Lim S-R, Yun Y-S, Park JM (2007) Reliable evidences that the removal mechanism of hexavalent chromium by natural biomaterials is adsorption-coupled reduction. Chemosphere 70: 298–305.

Peña-Bahamonde J, Nguyen HN, Fanourakis SK, Rodrigues DF (2018) Recent advances in graphene-based biosensor technology with applications in life sciences. J Nanobiotechnology 16: 75.

Picot M, Lapinsonnière L, Rothballer M, Barrière F (2011) Graphite anode surface modification with controlled reduction of specific aryl diazonium salts for improved microbial fuel cells power output. Biosens Bioelectron 28: 181–188.

Potter MC 1911. Electrical effects accompanying the decomposition of organic compounds. Proc R Soc Lond B 84: 260–276.

Pu Z, Zou C, Wang R, Lai X, Yu H, Xu K, Li D (2016) A continuous glucose monitoring device by graphene modified electrochemical sensor in microfluidic system. Biomicrofluidics 10: 011910.

Qiao Y, Bao S-J, Li CM (2010) Electrocatalysis in microbial fuel cells—from electrode material to direct electrochemistry. Energy Environ Sci 3: 544–553.

Rabaey K, Boon N, Höfte M, Verstraete W (2005) Microbial phenazine production enhances electron transfer in biofuel cells. Environ Sci Technol 39: 3401–3408.

Rabaey K, Rozendal RA (2010) Microbial electrosynthesis - revisiting the electrical route for microbial production. Nat Rev Microbiol 8: 706–716.

Rismani-Yazdi H, Carver SM, Christy AD, Tuovinen OH (2008) Cathodic limitations in microbial fuel cells: An overview. J Power Sources 180: 683–694.

Rozendal RA, Hamelers HVM, Rabaey K, Keller J, Buisman CJN (2008) Towards practical implementation of bioelectrochemical wastewater treatment. Trends Biotechnol 26: 450–459.

Ruecha N, Siangproh W, Chailapakul O (2011) A fast and highly sensitive detection of cholesterol using polymer microfluidic devices and amperometric system. Talanta, 84: 1323–1328.

Ruiz ON, Fernando KAS, Wang B, Brown NA, Luo PG, McNamara ND, Vangsness M, Sun Y-P, Bunker CE (2011) Graphene Oxide: A Nonspecific Enhancer of Cellular Growth. ACS Nano 5: 8100–8107.

Saratale GD, Saratale RG, Shahid MK, Zhen G, Kumar G, Shin H-S, Choi Y-G, Kim S-H (2017) A comprehensive overview on electro-active biofilms, role of exo-electrogens and their microbial niches in microbial fuel cells (MFCs). Chemosphere 178: 534–547.

Sonawane JM, Yadav A, Ghosh PC, Adeloju SB, (2017) Recent advances in the development and utilization of modern anode materials for high performance microbial fuel cells. Biosens Bioelectron 90: 558–576.

Song R-B, Zhao C-E, Jiang L-P, Abdel-Halim ES, Zhang J-R, Zhu J-J (2016) Bacteria-Affinity 3D Macroporous Graphene/MWCNTs/Fe$_3$O$_4$ Foams for High-Performance Microbial Fuel Cells. ACS Appl Mater Interfaces 8: 16170–16177.

Song T, Fei K, Zhang H, Yuan H, Yang Y, Ouyang P, Xie J (2018) High efficiency microbial electrosynthesis of acetate from carbon dioxide using a novel graphene–nickel foam as cathode. J Chem Technol Biotechnol 93: 457–466.

Song T, Jin Y, Bao J, Kang D, Xie J (2016) Graphene/biofilm composites for enhancement of hexavalent chromium reduction and electricity production in a biocathode microbial fuel cell. J Hazard Mater 317: 73–80.

Song T, Tan W, Wu X, Zhou CC (2012) Effect of graphite felt and activated carbon fiber felt on performance of freshwater sediment microbial fuel cell. J Chem Technol Biotechnol 87: 1436–1440.

Song T-S, Zhang H, Liu H, Zhang D, Wang H, Yang Y, Yuan H, Xie J (2017) High efficiency microbial electrosynthesis of acetate from carbon dioxide by a self-assembled electroactive biofilm. Bioresour Technol 243: 573–582.

Song Y, Luo Y, Zhu C, Li H, Du D, Lin Y (2016) Recent advances in electrochemical biosensors based on graphene two-dimensional nanomaterials. Biosens Bioelectron 76: 195–212.

Stoller MD, Park S, Zhu Y, An J, Ruoff RS (2008) Graphene-Based Ultracapacitors. Nano Lett 8: 3498–3502.

Su M, Wei L, Qiu Z, Jia Q, Shen J (2015) A graphene modified biocathode for enhancing hydrogen production. RSC Adv 5: 32609–32614.

Tremblay P-L, Angenent LT, Zhang T (2017) Extracellular Electron Uptake: Among Autotrophs and Mediated by Surfaces. Trends Biotechnol 35: 360–371.

Wang G, Huang L, Zhang Y (2008) Cathodic reduction of hexavalent chromium [Cr(VI)] coupled with electricity generation in microbial fuel cells. Biotechnol Lett 30: 1959–1966

Wang H, Wang G, Ling Y, Qian F, Song Y, Lu X, Chen S, Tong Y, Li Y (2013) High power density microbial fuel cell with flexible 3D graphene–nickel foam as anode. Nanoscale 5: 10283–10290.

Wang L, Chen Y, Huang Q, Feng Y, Zhu S, Shen S (2012) Hydrogen production with carbon nanotubes based cathode catalysts in microbial electrolysis cells. J Chemi Technol Biotechnol 87: 1150–1156.

Wang X, Cheng S, Feng Y, Merrill MD, Saito T, Logan BE (2009) Use of carbon mesh anodes and the effect of different pretreatment methods on power production in microbial fuel cells. Environ Sci Technol 43: 6870–6874.

Wang X, Zhi L, Müllen K (2008) Transparent, Conductive Graphene Electrodes for Dye-Sensitized Solar Cells. Nano Lett 8: 323–327.

Wang Y, Zhao C, Sun D, Zhang J-R, Zhu J-J (2013) A Graphene/Poly(3,4-ethylenedioxythiophene) Hybrid as an Anode for High-Performance Microbial Fuel Cells. ChemPlusChem 78: 823–829.

Wen Q, Wang S, Yan J, Cong L, Pan Z, Ren Y, Fan Z (2012) MnO$_2$–graphene hybrid as an alternative cathodic catalyst to platinum in microbial fuel cells. J Power Sources 216: 187–191.

Wu Z-Y, Chen P, Wu Q-S, Yang L-F, Pan Z, Wang Q (2014) Co/Co$_3$O$_4$/C–N, a novel nanostructure and excellent catalytic system for the oxygen reduction reaction. Nano Energy 8: 118–125.

Xiao L, Damien J, Luo J, Jang HD, Huang J, He Z (2012) Crumpled graphene particles for microbial fuel cell electrodes. J Power Sources 208: 187–192.

Xie X, Yu G, Liu N, Bao Z, Criddle CS, Cui Y (2012) Graphene–sponges as high-performance low-cost anodes for microbial fuel cells. Energy Environ Sci 5: 6862–6866.

Xu J, Li Q, Yue Y, Guo Y, Shao S (2014) A water-soluble BODIPY derivative as a highly selective "Turn-On" fluorescent sensor for H$_2$O$_2$ sensing *in vivo*. Biosens Bioelectron 56: 58–63.

Xu L, Zhang G, Chen J, Yuan G, Fu L, Yang F (2016) Prussian blue/graphene-modified electrode used as a novel oxygen reduction cathode in microbial fuel cell. J Taiwan Inst Chem Eng 58: 374–380.

Yang Y, Fu R, Wang H, Wang C (2013) Carbon nanofibers decorated with platinum nanoparticles: a novel three-dimensional platform for non-enzymatic sensing of hydrogen peroxide. Microchim Acta 180: 1249–1255.

Yang L, Wang S, Peng S, Jiang H, Zhang Y, Deng W, Tan Y, Ma M, Xie Q (2015) Facile Fabrication of Graphene-Containing Foam as a High-Performance Anode for Microbial Fuel Cells. Chemistry 21: 10634–10638.

Yong Y-C, Dong, X-C, Chan-Park MB, Song H, Chen P (2012) Macroporous and monolithic anode based on polyaniline hybridized three-dimensional graphene for high-performance microbial fuel cells. ACS Nano 6: 2394–2400.

Yong Y-C, Yu Y-Y, Zhang X, Song H (2014) Highly Active Bidirectional Electron Transfer by a Self-Assembled Electroactive Reduced-Graphene-Oxide-Hybridized Biofilm. Angew Chem Int Ed 53: 4480–4483.

Yu F, Wang C, Ma J (2018) Capacitance-enhanced 3D graphene anode for microbial fuel cell with long-time electricity generation stability. Electrochim Acta: 259: 1059–1067.

Yu Z, Leng X, Zhao S, Ji J, Zhou T, Khan A, Kakde A, Liu P, Li X (2018) A review on the applications of microbial electrolysis cells in anaerobic digestion. Bioresour Technol 255: 340–348.

Yuan L, Lin W, Xie Y, Chen B, Zhu S (2012) Single Fluorescent Probe Responds to H_2O_2, NO, and H_2O_2/NO with Three Different Sets of Fluorescence Signals. J Am Chem Soc 134: 1305–1315.

Yuan Y, Zhou S, Zhao B, Zhuang L, Wang Y (2012) Microbially-reduced graphene scaffolds to facilitate extracellular electron transfer in microbial fuel cells. Bioresour Technol 116: 453–458.

Zaybak Z, Pisciotta JM, Tokash JC, Logan BE (2013) Enhanced start-up of anaerobic facultatively autotrophic biocathodes in bioelectrochemical systems. J Biotechnol 168: 478–485.

Zhang M, Yuan R, Chai Y, Wang C, Wu X (2013) Cerium oxide–graphene as the matrix for cholesterol sensor. Anal Biochem 436: 69–74.

Zhang T, Tremblay P-L (2017) Hybrid photosynthesis-powering biocatalysts with solar energy captured by inorganic devices. Biotechnol Biofuels 10: 249.

Zhang Y, Mo G, Li X, Ye J (2012) Iron tetrasulfophthalocyanine functionalized graphene as a platinum-free cathodic catalyst for efficient oxygen reduction in microbial fuel cells. J Power Sources 197: 93–96.

Zhang Y, Mo G, Li X, Zhang W, Zhang J, Ye J, Huang X, Yu C (2011) A graphene modified anode to improve the performance of microbial fuel cells. J Power Sources 196: 5402–5407.

Zhao C, Wang W-J, Sun D, Wang X, Zhang J-R, Zhu J-J (2014) Nanostructured Graphene/TiO_2 Hybrids as High-Performance Anodes for Microbial Fuel Cells. Chemistry 20: 7091–7097.

Zhao C, Wang Y, Shi F, Zhang J, Zhu J-J (2013a) High biocurrent generation in *Shewanella*-inoculated microbial fuel cells using ionic liquid functionalized graphene nanosheets as an anode. Chem Commun 49: 6668–6670.

Zhao C, Gai P, Liu C, Wang X, Xu H, Zhang J, Zhu J-J (2013b) Polyaniline networks grown on graphene nanoribbons-coated carbon paper with a synergistic effect for high-performance microbial fuel cells. J Mater Chem A 1: 12587–12594.

Zhao C, Gai P, Song R, Zhang J, Zhu J-J (2015) Graphene/Au composites as an anode modifier for improving electricity generation in *Shewanella*-inoculated microbial fuel cells. Anal Methods 7: 4640–4644.

Zhao D, Wang Y, Nie G (2016) Electrochemical immunosensor for the carcinoembryonic antigen based on a nanocomposite consisting of reduced graphene oxide, gold nanoparticles and poly(indole-6-carboxylic acid). Microchim Acta 183: 2925–2932.

Zheng S, Yang F, Chen S, Liu L, Xiong Q, Yu T, Zhao F, Schröder U, Hou H (2015) Binder-free carbon black/stainless steel mesh composite electrode for high-performance anode in microbial fuel cells. J Power Sources 284: 252–257.

Zhou M, Chi M, Luo J, He H, Jin T (2011) An overview of electrode materials in microbial fuel cells. J Power Sources 196: 4427–4435.

Zhou M, Wang H, Hassett DJ, Gu T (2013) Recent advances in microbial fuel cells (MFCs) and microbial electrolysis cells (MECs) for wastewater treatment, bioenergy and bioproducts. J Chem Technol Biotechnol 88: 508–518.

Zhuang L, Yuan Y, Yang G, Zhou S (2012) *In situ* formation of graphene/biofilm composites for enhanced oxygen reduction in biocathode microbial fuel cells. Electrochem Commun 21: 69–72.

Zou Y, Xiang C, Yang L, Sun L-X, Xu F, Cao Z (2008) A mediatorless microbial fuel cell using polypyrrole coated carbon nanotubes composite as anode material. Int J Hydrog Energy 33: 4856–4862.

PART-VI

Design of Microbial Electrochemical Systems: Toward Scale-up, Modeling and Optimization

Scale-Up of Bioelectrochemical Systems for Energy Valorization of Waste Streams

Raúl M. Alonso[1], M. Isabel San-Martín[1], Raúl Mateos[1], Antonio Morán[1] and Adrián Escapa[1,2*]

[1] Chemical and Environmental Bioprocess Engineering Group Natural Resources Institute (IRENA) University of León Avda de Portugal 41 Leon 24009 Spain
[2] Department of Electrical Engineering and Automatic Systems, University of León Campus de Vegazana s/n 24071 León Spain

1. Introduction

The use of microbial fuel cells (MFC) for electricity production from organic wastes was the first practical application envisaged for bioelectrochemical systems (BES). With the discovery that BES can also be operated in electrolytic mode (microbial electrolysis cells-MEC) the rage of energy products expanded to fuel gases (H_2 and CH_4), thus bringing new opportunities for energy recovery from waste streams. Despite the list of potential applications of BES has widened dramatically during the past decade (biosensors, desalination, bioremediation, etc.), energy uses still enjoy a great interest among the scientific community (Beegle and Borole 2018; Do et al. 2018; Puig et al. 2017). This is evidenced by the fact that most of the scale-up experiences of BES carried out to date involve energy valorization of waste streams either as electricity (Dong et al. 2015) or fuel production (Cusick et al. 2011). Although commercial application seems to be within reach, practical use of BES for energy valorization of wastes still presents important challenges that are mainly economical and technical in nature. Thus, when scaling up BES, researchers and engineers run mainly into one or several of these issues: (i) keeping low electrode overpotential (Heidrich et al. 2014), (ii) developing economically competitive designs (Cusick et al. 2011), (iii) managing gas feeding/production (Cotterill et al. 2017), (iv) managing electrical power (Ge et al. 2015) or (v) developing stable biofilms (Do et al. 2018).

Recent reviews have mainly focused on the fundamentals of BES (Santoro et al. 2017) and exploring the perspectives of using BES for resources recovery from wastes (Seelam et al. 2018), bioremediation (Sevda et al. 2018) or wastewater treatment (Choudhury et al. 2017). This chapter addresses many of the issues listed in the previous paragraph, reviewing the answers and solutions provided by researchers and focusing on reactor configurations, flow regimes and stacking approaches. It also outlines some of the most significant scale-up experiences for both MFC and MEC, describing benchmarks and current performance levels. In addition, as BES produces/requires energy at low voltages, a section is devoted to discussing energy management systems and strategies. Finally, since this chapter is mainly devoted to practical aspects of BES, it ends by reviewing some market niches and alternatives that can provide BES an easy entrance to the market of environmental technologies.

*Corresponding author: adrianescapa@unileon.es

2. Keeping over Potentials under Check: The Importance of Geometry, Flow Regimes and Architecture

Ideally, the aim of the scaling process is the selection of a set of design conditions that guarantee that the effect of the operational variables over the process keeps constant regardless of the size of the units. Unfortunately, the absence of linearity and the rapid increase of internal losses with scale (Premier et al. 2012) make it not that easy for BES. Regarding internal losses, they are attributed to overpotentials (the difference between the thermodynamic and the observed reduction potential of a half reaction) that which can be classified into three categories: activation, ohmic and concentration overpotentials (Logan et al. 2006). Activation overpotentials have to do with electrode reactions and the electron transfer rates and as such, they are not largely affected by scale (improved electrode materials, surface modifications or catalysts are relevant strategies to ameliorate activation overpotentials). In contrast, concentration and ohmic overpotentials greatly depend on the size of the reactor since they are caused by factors, such as mass and charge transport limitations, inadequate hydrodynamic regimes or deficient electrode configuration. These are typical concerns for scientists and engineers as they largely determine the decision-making process during the scale-up of BES and play a significant role in the selection of reactor geometry and configuration and flow regimes.

2.1 Reactor Geometry

Based on reactor geometry, electrochemical reactor designs can be broadly classified as 2-dimensional and 3-dimensional electrode configurations (Walsh and Reade 1994). In addition, regarding the motion of the electrodes, each of these categories can be further subdivided into static electrodes and moving electrodes. During the past 15 years, researchers have come up with a whole set of BES configurations that can be included within those categories, many of which have been tested on large size reactors (Table 1). As with conventional electrochemical systems, parallel plate reactors with all its variations (filter press, cassettes, plate-in-tank) have become the most popular cell design as they are relatively easy to manufacture and handle. Moreover, parallel plate configurations have the advantages of encouraging uniform potential distribution (Walsh and Reade 1994), thus promoting efficient use of all the electrode surface area. Filter presses represent a simple and compact design that favors stacking (Escapa et al. 2015; Gil-Carrera et al. 2013b; Liang et al. 2018). Still, when stacked in series by means of bipolar plates (the cathode of one cell functions as the anode of the next cell) there exists the risk of voltage reversal, a phenomenon that adversely influences the performance of the affected BES and arises as a result of substrate starvation and/or differences in the internal resistance of one or more cells within a stack (Logan et al. 2010). Cassette configuration also provides a compact design (Baeza et al. 2017; Heidrich et al. 2014) although maintenance works may be difficult as both the anode and the cathode are enclosed within a single frame/case and individual access to any of the two electrodes may get hampered. In an attempt to alleviate this issue, (He et al. 2015) developed a simplified cassette design in which the anodes and cathodes can be removed individually from the tank. Interestingly, this configuration enabled relatively high power generation (11 W m^{-3}) from low strength wastewater (WW).

Concentric cylinders represent another common reactor typology in BES, and the advantages of these designs have been acknowledged from the very beginning (Liu et al. 2004; Rabaey et al. 2005). Tubular arrangements have been proposed as an advantageous configuration in BES as they provide a means for increasing the size of the reactors while maintaining relative spatial distribution, and their prismatic geometry enhances their manufacturability (Premier et al. 2012). Still in tubular configurations, the anode and cathode are usually wrapped one around the other (with a separator that electrically isolates them) in a cylinder that can make maintenance works of the electrodes difficult. The very first large-scale experience of a BES was precisely a tubular MFC (Keller and Rabaey 2008) that consisted of 12 cylindrical units with the anode on the inside and the cathode on the outside to favor contact with air with a total volume of 1 m^3. Low conductivity and low buffer capacity of the wastewater (brewery wastewater) and biomass proliferation on the cathode side were the critical factors that limited the reactor

Table 1: Classification of the different geometries of reactors used in the scaling-up of BES

Electrode configuration	Motion	Type	Advantages	Drawbacks	References
2-dimensional	Static	Filter press	Easy to manufacture	Favors stacking	Escapa et al. 2015; Gil-Carrera et al. 2013b; Liang et al. 2018
		Filter press with bipolar plates	Maximize compacity	Risk of voltage reversal	Logan 2010
		Cassettes	Promote stacking	Maintenance difficults	Baeza et al. 2017; Heidrich et al. 2014; He et al. 2015
		Concentric cylinders	Favours scale-up while maintaining spatial distribution	Inconvenient separator location	Liu et al. 2004; Rabaey et al. 2005; Keller and Rabaey 2008; Gil-Carrera et al. 2013c
	Rotating	Rotating impeller-electrode	Improves mass transfer in an easy to control manner	Ensuring gas tightness	Park et al. 2018
3-dimensional	Static	Packed bed electrode	Increase electrodic area with a low cost approach	Electrode clogging	Wu et al. 2016; Vilajeliu-Pons et al. 2017
		Brush electrode	High specific area while avoiding excessive pressure drop	Low compacity	Logan et al. 2007; Dong et al. 2015; Cusick et al. 2011
		Porous electrode	High specific area	Indeseable collateral reactions	Sleutels et al. 2011; Rader and Logan 2010; Hussain et al. 2017
	Moving bed	Fluidized granular electrode	High specific area and enhanced mass transfer	Challenging operating conditions tuning	Tejedor-Sanz et al. 2017

performance. Tubular designs have also been investigated in MECs. Gil-Carrera et al. (2013c) built an 8 L modular tubular MEC for low strength wastewater treatment. Despite the energy, the consumption was relatively low and comparable to conventional wastewater treatment technologies (activated sludge). The energy produced as hydrogen was not enough to offset the energy usage, probably because of the very low organic matter concentration in the feeding.

In an effort to improve mixing and ameliorate concentration overpotentials, Park et al. (2018) developed a rotating electrode (anode) for a BES immersed in an AD reactor. The rotating anode was made of stainless steel and served as the impeller. The cathode consisted of a cylinder attached to the inside of the reactor walls of the digester and was made of stainless steel as well. This design was tested in a 1 L reactor, and it helped to avoid volatile fatty acid accumulation and decreases in pH at high organic loading rates where AD usually fails.

3D configurations have also aroused a considerable interest (and they are becoming increasingly popular in large-scale BES) as they provide a means to minimize mass transport as well as ohmic losses (Wu et al. 2016). It was precise with the aim of attenuating them that Wu et al. (2016) built a stacked 72

L MFC with 3D electrodes (anodes and cathodes) made of granulated activated packed bed electrodes to enhance kinetics (via facilitating biomass enrichment) and to improve mass transport. In addition, the authors paid special attention to providing the MFC with narrow and flat chambers and a large ion exchange membrane area per unit of reactor volume to facilitate charge transport and thus reduce ohmic losses. Despite these advantages, 3D packed bed electrodes are not without problems due to electrode crushing and the clogging of granular graphite that lead to a significant performance decline in a stacked 65 L MFC packed bed electrode (graphite granules) for swine manure treatment (Vilajeliu-Pons et al. 2017). Another popular 3D electrode used in BES, and one that can help to overcome the limitations of packed bed electrodes, is the brush electrode that displays large specific surface area (18,200 m^2 m^{-3}) and high porosity (95%) (Logan et al. 2007). Carbon brushes woven onto titanium wires have been successfully used in the anodes of large-scale BES for real wastewater treatment as in a 90 L baffled MFC that produced enough energy to power the ancillary equipment (Dong et al. 2015) or in a 1,000 L MEC that also used stainless steel brushes cathodes (instead of a Pt-based cathode) to reduce the costs (Cusick et al. 2011). It is important to note that 3D electrodes usually result in greater electrode spacing compared to flat and tubular reactors that undoubtedly can negatively affect the performance.

2.2 Flow Configurations

The hydrodynamics of the electrolytes have an unquestionably large influence on mass and charge transport phenomena. Thus, selecting adequate flow configurations is an important design choice to keep under control electrochemical losses. With the aim of increasing volumetric current densities while avoiding the severe mass transport limitations that usually accompany them, Sleutels et al. (2011) forced the anolyte to flow through a porous anode as would occur in a conventional fuel cell (Stuve 2014) that allowed increasing current density by almost 100%. A variant of this flow-through configuration, one that results in simplified reactor designs as no separator must exist between anode and cathode, is that in which the electrolyte flows first through the anode and then through a cathode (Rader and Logan 2010). A significant feature of this flow configuration that can only be used in MECs is that it makes it hard to obtain pure hydrogen as it promotes its conversion into methane (Rader and Logan 2010). Although this may seem a disadvantage at first sight, it really is an opportunity for promoting the practical implementation of BES as it will be discussed in section 2 in this chapter. Hussain et al. (2017) demonstrated the applicability of this configuration in a 1 L MEC. The reactor allowed to simultaneously remove organic matter and nitrogen (under microaerobic conditions) improving up to 75% methane production compared to an anaerobic control.

A more usual flow configuration, one that has been repeatedly used in large-scale BES, is the flow along with single and parallel plates (Baeza et al. 2017; Feng et al. 2014; Heidrich et al. 2014; Liang et al. 2018; Tartakovsky et al. 2017). In this configuration, it is usual to use an external recirculation loop to provide mixing (Escapa et al. 2015; Gil-Carrera et al. 2013b) or baffles to promote turbulence (Heidrich et al. 2014) inside the anodic and/or cathodic chambers and attenuate mass transfer limitations. Alongside these flow configurations, typical of conventional electrochemical systems, some other more innovative configurations that incorporate the characteristics of bioreactors have been tested. For instance, Katuri et al. (2014) developed a BES for methane production that incorporated a porous, nickel-based hollow fiber membrane that functioned as both the cathode and also as a membrane for filtering the effluent. However, this design has not yet been tested at a larger scale.

2.3 The Problem of the Ionic Exchange Membrane

A general distinction is usually made into separated and non-separated BES reactors (Krieg et al. 2014) that refer to the existence or absence of a membrane interposed between the two electrodes. These two basic configurations are also referred to as membrane and membraneless or single compartment and two compartment (Hussain et al. 2018). This distinction is relevant in the field of BES since membranes do not necessarily need to be used in bioelectrochemical reactors as they do in conventional electrochemical systems (Logan et al. 2015). Membraneless BES simplify reactor configuration and maintenance, reduces the cost of the materials and limits the need of extra ancillary equipment and the operational costs required to manage two independent electrolytes (Escapa et al. 2016). Another interesting feature of membraneless BES is that the absence of a membrane facilitates the circulation of ions between the electrodes that translate into lower internal resistance. Despite these advantages, membraneless configurations also present severe drawbacks. Lacking a physical barrier that separates the two electrodic environments, it is difficult to avoid interferences between them. For instance, in membraneless MFCs there exists the risk that cathodic oxygen diffuses to the anode, thus promoting the proliferation of aerobic microorganisms that would affect the anode performance. Similarly, in membraneless MECs, cathodic hydrogen can diffuse back to the anode where it can get reoxidized, giving place to the hydrogen recycling phenomenon that artificially increases the electrical energy usage.

2.4 Stacking BES

Building large-scale BES can be achieved through two basic strategies. An obvious approach—as small size units tend to be more efficient (Dewan et al. 2008)—would consist of stacking a certain number of smaller BES units up to the required reactor volume. This of course represents an impractical and costly approach, especially for those applications that would demand to handle large amounts of (residual) effluents. The opposite strategy, i.e., building a large size reactor made up of one single unit, does not provide a practical solution as manufacturing large units might be a challenging issue and internal losses increase dramatically with the reactor size (Dewan et al. 2008). Thus, the only economically and technically feasible strategy seems to be a combination of the two preceding ones. Realizing this brings out the issue of how to connect hydraulically and electrically several units that comprise the whole stack and the two possible strategies at hand are series and parallel connection. Here, again combining both approaches seem to be a more suitable solution as they involve pros and cons. From an electrical point of view, connecting several units in series has the advantage of allowing increasing the voltage output and limiting the amount of circulating current. So, the BES system becomes less reliant on the need of power management systems as it will be shown in section 3 [handling big currents at low voltages is always problematic and usually entails the use of large energy management equipment (Escapa et al. 2012)]. Still, this configuration is only possible when the electrodes do not share the same electrolyte (or even the same feedline) (Santoro et al. 2018) as this might give rise to ionic shortcut currents. Another issue behind this configuration is the risk of appearance of voltage reversal as mentioned earlier.

From a hydraulic point of view, series connection (sequential treatment) has been used in scaled-up BES to improve the quality of the treated effluent (Gil-Carrera et al. 2013a) although the low COD concentration in the downstream units limits their performance (Asensio et al. 2017; He et al. 2016). Another issue with this sequential treatment is that microbial biofilms might differ significantly from one unit to another as a result of different substrates becoming available to microorganisms through the treatment. This would translate in different bioelectrochemical performances (Tartakovsky et al. 2017) and thus, hydraulic retention times might need to be adjusted to optimize performance (Janicek et al. 2014).

3. Scaling-up BES for Energy Valorization of Wastewater

It is perhaps the field of wastewater treatment where the greatest potential for practical application of BES lies as almost all BES typologies that incorporate a bioanode (MFC, MEC, MDC, etc.) can potentially fit within it. In fact, and to date, most of the scale-up experiences have been carried out for wastewater treatment for MFCs and MECs.

3.1 Upscaling MFCs

It has been estimated that maximum theoretical attainable power from an MFC could reach as much as 17-19 W m^{-2} (Logan et al. 2015) although it seems unlikely that large-scale MFC developments or even laboratory-scale prototypes would get even near these figures (Pandey et al. 2016). For instance, the first lab-scale prototypes using real wastewater as substrate produced power in the range of several tens of mW m⁻² (Liu et al. 2004; Rodrigo et al. 2007). Despite more recent advances in MFC configuration and architectures and electrode materials, power production levels have not improved significantly (Hindatu et al. 2017). The first ever large-scale MFC (Keller and Rabaey 2008) brought to light the difficulties of using real wastewaters in scaled up systems (e.g., low buffer capacity, low conductivity and the presence of complex organic matter in actual wastewaters are the main factors limiting power production in large-scale MFCs). But not only wastewater itself imposes great challenges in up scaling these technologies. The sluggishness of the cathodic oxygen reduction reaction (Dekker et al. 2009) and the need to maintain a tradeoff between short electrode spacing and oxygen diffusion, which can result in a loss of performance and even in the inhibition of anode bacteria, are also recognized as main difficulties in developing large-scale BES (Logan et al. 2015). Still, recent scale-up experiences leave some room for optimism. In Dong et al. (2015), the authors built and operated a 90 L MFC made of five easily stackable, easily maintainable modules for brewery wastewater treatment. The plant was energy self-sufficient, producing enough energy to power the pumping system, achieving a COD removal efficiency of about 85%. Another interesting experience is the modular MFC (432 units, 300 L working volume) designed by Ieropoulos et al. (2016) to remove organic carbon from urine. Field tests showed how the ~300 mW produced by the MFC were enough to power the light system (1.8 W LED light) in the restroom (using a bank of supercapacitors to store energy) although COD removal was relatively low (30%).

A significant obstacle in using MFCs for energy valorization of waste streams is that energy production and organic contamination removal can be competing objectives, and so, when the focus is organic matter removal rather than power generation, MFCs may be far from becoming energy-neutral systems. This is because maximum power production usually occurs at currents below peak current, while COD removal achieves its maximum at peak current. For instance, in Lu et al. (2017) the authors built and operated a 20 L MFC that achieved almost 100% COD removal from brewery WW, but power production was three orders of magnitude lower than power consumption. Still, it was recognized that the MFC design and operation were not optimal, and the energy balance could be improved by reducing the hydraulic retention time, using cathodic catalysts or improving the energy harvesting strategy.

3.2 Upscaling MECs

Hydrogen-producing MECs represent, together with MFCs, the two typologies where most large-scale developments for wastewater treatment have been carried out to date. The 5.4 $L_{H2} L^{-1}_{MEC} d^{-1}$ reported by (Hrapovic et al. 2010) using a gas diffusion cathode with electrodeposited nickel or the impressive 50 $L_{H2} L^{-1}_{MEC} d^{-1}$ achieved by (Jeremiasse et al. 2010) using a nickel foam cathode (associated energy usage of about 2.5 kWh m⁻³$_{H2}$ in both cases) can be set as benchmarks for this technology. The referred studies were carried out on laboratory-scale reactors, using synthetic effluents and under controlled conditions. As with MFCs, it seems unlikely that larger-scale systems fed with real WW can keep up these performance figures, the reason being that similar challenges as those described above for MFCs can be found when scaling MECs. In addition to that, MECs need to face the challenge of recovering and managing the gas produced on the cathode, an issue that imposes no minor difficulties. In fact, dealing with gases, rather than current densities or internal resistances, has been seen as the main limiting factor in the scale-up of MECs (Cotterill et al. 2017). To begin with, the presence of hydrogen in a membraneless MEC can give rise to the hydrogen recycling phenomenon that can cause severe performance losses (Escapa et al. 2015). In addition, the diffusivity of hydrogen makes it difficult to confine the gas within a closed space and may cause the embrittlement of steel, all of which has prompted researchers to find alternatives to hydrogen as the energy end product. In this sense, methane occupies undoubtedly a preeminent position as its occurrence often happens naturally in MECs; it is easier to handle, and there already exists an infrastructure for methane commercialization (Moreno et al. 2016). In a recent article, (Tartakovsky et al. 2017) explored this possibility in a 20 L membraneless MEC conceived to treat urban wastewater in arctic zones. The MEC was capable of achieving almost 100% BOD removal efficiencies, and despite the low temperatures (as low as 5°C), the energy balance in the MEC was positive thanks to the relatively low energy consumption and the enhanced methane production (which was fairly constant in the range 5-23°C). Moreover, the reactor design was relatively simple (flow-through bioanode-biocathode setup incorporated into a septic tank divided by two pairs of electrodes into three compartments), all of which improves its commercial perspectives. Hydrogen-producing MECs usually require more complex designs to avoid the issues discussed above. However, they allow producing hydrogen with high levels of purity as in the 130 L MEC developed by Baeza et al. (2017) that achieved hydrogen production rates of about 0.03 $L_{H2}L^{-1}_{MEC}$ d^{-1} (95% purity) using urban wastewater as substrate. Energy balance was positive although COD removal was relatively low (25%). Cotterill et al. (2017) tested a similar MEC design (although a little larger i.e., 175 L) fed with urban wastewater as well reporting again high purity hydrogen production (93%). Despite operating at a lower temperature (~11°C), this did not seem to have an effect on COD removal; in fact it improved significantly compared to the work of Baeza et al. (2017), i.e., 63%. Nevertheless, low temperatures seemed to have a quite negative impact on hydrogen production rates that were an order of magnitude below that reported in Baeza et al. (2017).

4. Energy Harvesting and Power Control in BES

Energy exploitation of waste streams by MFCs has to cope with the low voltages and power densities displayed, usually below 0.8 V and 2 W m^{-2}, respectively, (Wang et al. 2015) that are not enough to power conventional electric loads. Increasing power and voltage levels simply by building large units or connecting smaller units in series is not always easy due to the inherent difficulties described in section 1.4. Thus, it was early acknowledged the need to develop power management interfaces to adapt the MFC to the load in terms of power and voltage levels. One of the simplest attempts consists of using a capacitor to store the energy produced by the MFC, so it can be later discharged through the load at a higher power level. This intermittent mode of operation has the added advantages that it increases the efficiency of power generation (Dewan et al. 2009), increases coulombic efficiencies and prevents the occurrence of voltage reversal (Kim et al. 2011). A more elaborated approach involves the use of charge pumps that are DC-DC electronic converters that incorporate a set of capacitors that can be connected in different combinations to increase voltage levels. Charge pumps require a minimum input voltage of 0.3 V to produce an output of up to 2.4 V, consuming a significant amount of current during operation and thus limiting the energy efficiency (Wang et al. 2015). Moreover, the output voltage of charge pumps (up to 2.4 V) is insufficient for most practical applications and so, boost converters are usually required to adapt to the power requirements of the load (Wang et al. 2015).

Despite these electronic devices helping to make the energy produced by an MFC usable, they do not guarantee its optimization. It is well-known that MFCs deliver maximum power when the resistance of the load matches the internal resistance of the MFC. The problem is that internal resistance in an MFC largely depends on operating conditions (temperature, substrate concentration, biofilm coverage of the electrodes, etc.) that will be likely subjected to frequent variations in real life applications, demanding suitable control strategies and control systems to optimize the electrical settings. Maximum power point tracking (MPPT) is a technique to maximize power extraction from an MFC under all conditions that can be implemented using different algorithms and techniques (Ge et al. 2015; Recio-Garrido et al. 2016). It relies on continuously determining the internal resistance of the MFC and matching the load impedance of the power management system. In addition to that, the use of an MPPT can help to avoid the power overshoot phenomenon (a phenomenon that deteriorates MFC performance when high current densities are demanded) (Alaraj et al. 2017). Power management systems and MPPT have been successfully used to power small electronic devices (Meehan et al. 2011) or wireless sensors (Erbay et al. 2014) that demand relative low currents (milliamp level) but also large equipment like pumps that require relative large currents to operate (amp level). For instance, in (Dong et al. 2015) the energy produced by a 90 L MFC comprising five MFC modules provided with five independent capacitor-based power management systems was enough to power its pumping system. The capacitors were charged independently for four minutes and then discharged in series for one minute to raise the voltage. A similar approach was followed by Ge et al. (2015) in a 200 L MFC treating actual wastewater where the authors used a commercially available energy harvesting device to raise voltage to 5 V and running a DC motor (pumps). Process control of MECs has not received as much attention as with MFCs. With MECs, control strategies are directed to adjusting the applied potential to maximize hydrogen production while avoiding excessive power consumption (Tartakovsky et al. 2011) or to avoid voltage reversal in electrically series connected MECs (Andersen et al. 2013).

5. Future Perspectives of Using BES for Energy Valorization of Wastes: Opportunities for Fast Commercial Deployment

Initial phases of commercial development are critical for any novel technology, and identifying those niches where they can find a fast and easy entrance in the market is of no minor importance. For BES, integration with well established technologies, to ameliorate some of their weaknesses or to exploit untapped synergies, can provide a suitable answer to this question (Figure 1).

Anaerobic digestion and BES are usually seen as competing technologies in the field of energy valorization of organic wastes. However, under some circumstances they can be also regarded as collaborating technologies (Escapa et al. 2016). For instance, integrating a MEC within a digester can promote biogas production and methane richness (Geppert et al. 2016) and stabilize the digestion process by delaying volatile fatty acid (VFA) build-up (Moreno et al. 2018). A further positive aspect is that BES can be relatively easily merged within already operating anaerobic digesters (Bo et al. 2014) and multiple alternatives for integration have been devised in this regard (Beegle and Borole 2017).

Septic tanks are inefficient systems for organic matter degradation. Incorporating BESs can make them more effective in terms of BOD removal (Logan and Regan 2006), and some authors have seen this as a feasible option for protecting public health in remote or developing communities (Yazdi et al. 2015). Moreover, this concept has already been tested on a 20 L reactor, yielding very promising results i.e., BOD removal close to 100% with a positive energy balance even at relatively low temperatures (Tartakovsky et al. 2017).

Integration of BES in the water line of a wastewater treatment plant perhaps represents the most straightforward application although maybe not the easiest. This is mainly because the techno-economic feasibility of BES as a standalone technology is still not clear. Preliminary economic assessments reveal that to make feasible the implementation of BES within the water line of a WWTP, the capital costs for

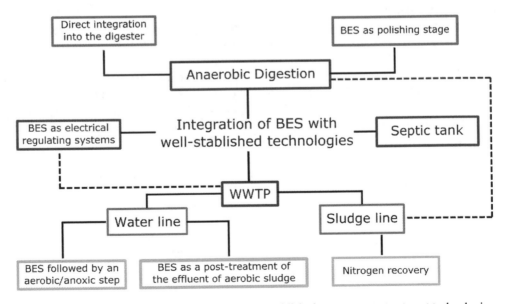

Figure 1: Niches for integration of BES into well-established waste stream treatment technologies

BES should not be higher than 1.5-2.5 € A^{-1} (Escapa et al. 2012; Modin and Gustavsson 2014). This does not seem realistic, even more having in mind the relatively low current densities developed by these systems operating with real WW. Still, the operational flexibility of BES allows for their implementation within other areas of WWTPs, e.g., sludge line (Escapa et al. 2014; Rosenbaum et al. 2010). Here, BES can benefit from a higher COD concentration that can make their use more competitive. In addition, high nitrogen concentration could enable BES to remove/recover this nutrient more easily than in the water line (Puig et al. 2017; San-Martín et al. 2018), thus further improving the commercial perspectives.

BES can also provide a link between the chemical and electrical power industries, opening interesting novel opportunities for both sectors. The increasing share of renewable generation in the electric mix and their intermittent nature can be a source of grid stability issues resulting from a mismatch between load and source (e.g., fluctuations in the renewable energy generation). Here, MECs can become an alternative to more conventional energy storage technologies such as batteries and supercapacitors. For the particular case of WWTP, this would open the way for using these facilities as electrical regulating systems, providing a means for stocking any power surplus as combustible gas (H$_2$ or CH$_4$) (Moreno et al. 2016).

Finally, another commercial niche can be found in the integration with other electrochemical technologies. Redox flow batteries (RFB) are electrochemical energy storage devices that provide interesting energy management capabilities as they allow decoupling energy output from power output (Wang and Sprenkle 2016). RFB have already been regarded as a potential strategic partner in BES deployment as they may provide an alternative for storing reductive power obtained in MECs when treating waste streams (Santos et al. 2017).

6. Conclusions

Perhaps their ability to convert residual organic matter into chemical or electrical energy is the most interesting feature of BES. The number of scale-up developments aimed at exploring this potentiality has proliferated during the past decade, giving the impression that commercial application might be within reach. Recent advances in the scale of ~100 L allow for some optimism in this regard. MFCs operating on real wastewater have proved to be capable of producing enough energy to power ancillary equipment (pumps, control systems, etc.), and MECs can produce enough gas fuel (either hydrogen or methane) to produce a positive energy balance. Researchers and engineers have come up with a plethora of novel reactor architectures and flow configurations, always with the aim of minimizing internal losses (overpotentials) and developing technically and economically and competitive designs. Other relevant practical issues such as obtaining usable power or developing suitable control and optimization strategies have also been addressed, all of which is helping to pave the way toward market development.

Acknowledgment

The authors acknowledge the "Ente Regional de la Energía de Castilla y León (EREN)" for the support of project ref: EREN_2019_L3_ULE. The authors are also greateful to the INTERREG VA España Portugal (POCTEP) territorial cooperation programme (project ref: 0688_BIOVINO_6_E)

References

Alaraj M, Radenkovic M, Park J-D (2017) Intelligent energy harvesting scheme for microbial fuel cells: Maximum power point tracking and voltage overshoot avoidance. Journal of Power Sources Elsevier 342: 726–732.

Andersen SJ, Pikaar I, Freguia S, Lovell BC, Rabaey K, Rozendal R A (2013) Dynamically Adaptive Control System for Bioanodes in Serially Stacked Bioelectrochemical Systems. Environmental Science & Technology American Chemical Society 47(10): 5488–5494.

Asensio Y, Mansilla E, Fernandez-Marchante CM, Lobato J, Cañizares P, Rodrigo MA (2017) Towards the scale-up of bioelectrogenic technology: stacking microbial fuel cells to produce larger amounts of electricity. Journal of Applied Electrochemistry 47(10): 1115–1125.

Baeza JA, Martínez-Miró À, Guerrero J, Ruiz Y, Guisasola A (2017) Bioelectrochemical hydrogen production from urban wastewater on a pilot scale. Journal of Power Sources 356: 500–509.

Beegle JR, Borole AP (2017) An integrated microbial electrolysis-anaerobic digestion process combined with pretreatment of wastewater solids to improve hydrogen production. Environmental Science: Water Research & Technology 3(6): 1073-1085.

Beegle JR, Borole A P (2018) Energy production from waste: Evaluation of anaerobic digestion and bioelectrochemical systems based on energy efficiency and economic factors. Renewable and Sustainable Energy Reviews Pergamon 96: 343–351.

Bo T, Zhu X, Zhang L, Tao Y, He X, Li D, Yan Z (2014) A new upgraded biogas production process: Coupling microbial electrolysis cell and anaerobic digestion in single-chamber barrel-shape stainless steel reactor. Electrochemistry Communications 45: 67–70.

Choudhury P, Uday USP, Mahata N, Nath Tiwari O, Narayan Ray R, Kanti Bandyopadhyay T, Bhunia B (2017) Performance improvement of microbial fuel cells for waste water treatment along with value addition: A review on past achievements and recent perspectives. Renewable and Sustainable Energy Reviews Pergamon 79: 372–389.

Cotterill SE, Dolfing J, Jones C, Curtis TP, Heidrich E S (2017) Low Temperature Domestic Wastewater Treatment in a Microbial Electrolysis Cell with 1 m ^{2} Anodes: Towards System Scale-Up. Fuel Cells 17(5): 584-592.

Cusick RD, Bryan B, Parker D, Merrill MD, Mehanna M, Kiely PD, Liu G, Logan BE (2011) Performance of a pilot-scale continuous flow microbial electrolysis cell fed winery wastewater. Applied Microbiology and Biotechnology 89(6): 2053–2063.

Dekker A, Ter Heijne A, Saakes M, Hamelers HVM, Buisman CJN (2009) Analysis and Improvement of a Scaled-Up and Stacked Microbial Fuel Cell. Environmental Science & Technology 43(23): 9038–9042.

Dewan A, Beyenal H, Lewandowski Z (2008) Scaling up microbial fuel cells. Environmental science & technology 42(20): 7643–7648.

Dewan A, Beyenal H, Lewandowski Z (2009) Intermittent Energy Harvesting Improves the Performance of Microbial Fuel Cells. Environmental Science & Technology 43(12): 4600–4605.

Do MH, Ngo HH, Guo WS, Liu Y, Chang SW, Nguyen DD, Nghiem LD, Ni B J (2018) Challenges in the application of microbial fuel cells to wastewater treatment and energy production: A mini review. Science of The Total Environment Elsevier 639: 910–920.

Dong Y, Qu Y, He W, Du Y, Liu J, Han X, Feng Y (2015) A 90-liter stackable baffled microbial fuel cell for brewery wastewater treatment based on energy self-sufficient mode. Bioresource Technology 195: 66–72.

Erbay C, Carreon-Bautista S, Sanchez-Sinencio E, Han A (2014) High Performance Monolithic Power Management System with Dynamic Maximum Power Point Tracking for Microbial Fuel Cells. Environmental Science & Technology 48(23): 13992–13999.

Escapa A, Gómez X, Tartakovsky B, Morán A (2012) Estimating microbial electrolysis cell (MEC) investment costs in wastewater treatment plants: Case study. International Journal of Hydrogen Energy 37(24): 18641–18653.

Escapa A, Mateos R, Martínez EJ, Blanes J (2016) Microbial electrolysis cells: An emerging technology for wastewater treatment and energy recovery: From laboratory to pilot plant and beyond. Renewable and Sustainable Energy Reviews 55: 942–956.

Escapa A, San-Martín MI, Mateos R, Morán A (2015) Scaling-up of membraneless microbial electrolysis cells (MECs) for domestic wastewater treatment: Bottlenecks and limitations. Bioresource Technology 180: 72–78.

Escapa A, San Martin MI, Moran A (2014) Potential use of Microbial Electrolysis Cells (MECs) in domestic wastewater treatment plants for energy recovery. Frontiers in Energy Research 2(19).

Feng Y, He W, Liu J, Wang X, Qu Y, Ren N (2014) A horizontal plug flow and stackable pilot microbial fuel cell for municipal wastewater treatment. Bioresource technology 156(0): 132–138.

Ge Z, Wu L, Zhang F, He Z (2015) Energy extraction from a large-scale microbial fuel cell system treating municipal wastewater. Journal of Power Sources 297: 260-264.

Geppert F, Liu Van D, Eerten-Jansen M,Weidner E, Buisman C, Ter Heijne A (2016) Bioelectrochemical Power-to-Gas: State of the Art and Future Perspectives. Trends in Biotechnology 34(11): 879-894.

Gil-Carrera L, Escapa A, Carracedo B, Morán A, Gómez X (2013) Performance of a semi-pilot tubular microbial electrolysis cell (MEC) under several hydraulic retention times and applied voltages. Bioresource Technology 146: 63–69.

Gil-Carrera L, Escapa A, Mehta P, Santoyo G, Guiot SR, Morán A, Tartakovsky B (2013) Microbial electrolysis cell scale-up for combined wastewater treatment and hydrogen production. Bioresource Technology 130: 584–591.

Gil-Carrera L, Escapa A, Moreno R, Morán A (2013) Reduced energy consumption during low strength domestic wastewater treatment in a semi-pilot tubular microbial electrolysis cell Journal of Environmental Management 122: 1–7.

He W, Wallack MJ, Kim KY, Zhang X, Yang W, Zhu X, Feng Y, Logan BE (2016) The effect of flow modes and electrode combinations on the performance of a multiple module microbial fuel cell installed at wastewater treatment plant. Water Research 105: 351-360.

He W, Zhang X, Liu J, Zhu X, Feng Y, Logan B (2015) Microbial Fuel Cells with an Integrated Spacer and Separate Anode and Cathode Modules. Environ Sci: Water Res Technol 2(1): 186–195.

Heidrich ES, Edwards SR, Dolfing J, Cotterill SE, Curtis TP (2014) Performance of a pilot scale microbial electrolysis cell fed on domestic wastewater at ambient temperatures for a 12 month period. Bioresource technology 173(0): 87–95.

Hindatu Y, Annuar MSM, Gumel AM (2017) Mini-review: Anode modification for improved performance of microbial fuel cell. Renewable and Sustainable Energy Reviews 73: 236–248.

Hrapovic S, Manuel MF, Luong JHT, Guiot SR, Tartakovsky B (2010) Electrodeposition of nickel particles on a gas diffusion cathode for hydrogen production in a microbial electrolysis cell. International Journal of Hydrogen Energy 35(14): 7313–7320.

Hussain A, Lebrun FM, Tartakovsky B (2017) Removal of organic carbon and nitrogen in a membraneless flow-through microbial electrolysis cell. Enzyme and Microbial Technology 102: 41-48.

Hussain SA, Perrier M, Tartakovsky B (2018) Real-time monitoring of a microbial electrolysis cell using an electrical equivalent circuit model. Bioprocess and biosystems engineering Springer 41(4): 543–553.

Ieropoulos IA, Stinchcombe A, Gajda I, Forbes S, Merino-Jimenez I, Pasternak G, Sanchez-Herranz D, Greenman J (2016) Pee power urinal microbial fuel cell technology field trials in the context of sanitation. Environ Sci: Water Res 2(2): 336–343.

Janicek A, Fan Y, Liu H (2014) Design of microbial fuel cells for practical application: a review and analysis of scale-up studies. Biofuels 5(1): 79–92.

Jeremiasse AW, Hamelers HVM, Saakes M, Buisman CJN (2010) Ni foam cathode enables high volumetric H2 production in a microbial electrolysis cell. Asian Hydrogen Energy Conference 2009 35(23): 12716–12723.

Katuri K, Werner C, Jimenez Sandoval R, Chen W, Logan B, Lai Z, Amy GL, Saikaly P E (2014) A Novel Anaerobic Electrochemical Membrane Bioreactor (AnEMBR) with Conductive Hollow-fiber Membrane for Treatment of Low-organic Strength Solutions. Environmental science & technology 48: 12833–12841.

Keller J, Rabaey K (2008) Experiences from MFC pilot plant operation. First international symposium on microbial fuel cells.

Kim Y, Hatzell MC, Hutchinson AJ, Logan BE (2011) Capturing power at higher voltages from arrays of microbial fuel cells without voltage reversal. Energy & Environmental Science 4(11): 4662–4667.

Krieg T, Sydow A, SchrÄder U, Schrader J, Holtmann D (2014) Reactor concepts for bioelectrochemical syntheses and energy conversion. Trends in biotechnology 32(12): 645-65.

Liang P, Duan R, Jiang Y, Zhang X, Qiu Y, Huang X (2018) One-year operation of 1000-L modularized microbial fuel cell for municipal wastewater treatment. Water Research 141: 1-8.

Liu H, Ramnarayanan R, Logan BE (2004) Production of Electricity during Wastewater Treatment Using a Single Chamber Microbial Fuel Cell. Environmental science & technology 38(7): 2281–2285.

Logan B, Call D, Merrill M, Cheng S (2010) Cathodes for microbial electrolysis cells and microbial fuel cells. Patent US 20100119920 A1.

Logan BE, Cheng S, Watson V, Estadt G (2007) Graphite Fiber Brush Anodes for Increased Power Production in Air-Cathode Microbial Fuel Cells. Environmental science & technology 41(9): 3341–3346.

Logan BE, Hamelers B, Rozendal R, Schröder U, Keller J, Freguia S, Aelterman P, Verstraete W, Rabaey K (2006) Microbial Fuel Cells: Methodology and Technology. Environmental Science & Technology 40(17): 5181–5192.

Logan BE, Regan JM (2006) Microbial fuel cells—challenges and applications. ACS Publications: 5172-5180.

Logan BE, Wallack MJ, Kim KY, He W, Feng Y, Saikaly PE (2015) Assessment of Microbial Fuel Cell Configurations and Power Densities. Environmental Science and Technology Letters 2(8): 206–214.

Lu M, Chen S, Babanova S, Phadke S, Salvacion M, Mirhosseini A, Chan S, Carpenter K, Cortese R, Bretschger O (2017) Long-term performance of a 20-L continuous flow microbial fuel cell for treatment of brewery wastewater. Journal of Power Sources Elsevier 356: 274–287.

Meehan A, Gao H, Lewandowski Z (2011) Energy Harvesting With Microbial Fuel Cell and Power Management System. IEEE Transactions on Power Electronics 26(1): 176–181.

Modin O, Gustavsson DJI (2014) Opportunities for microbial electrochemistry in municipal wastewater treatment–an overview. Water Science and Technology 69(7): 1359–1372.

Moreno R, Martínez E, Escapa A, Martínez O, Díez-Antolínez R, Gómez X (2018) Mitigation of Volatile Fatty Acid Build-Up by the Use of Soft Carbon Felt Electrodes: Evaluation of Anaerobic Digestion in Acidic Conditions. Fermentation Multidisciplinary Digital Publishing Institute 4(1), 2: 1-13.

Moreno R, San-Martín MI, Escapa A, Morán A (2016) Domestic wastewater treatment in parallel with methane production in a microbial electrolysis cell. Renewable Energy 93: 442–448.

Pandey P, Shinde VN, Deopurkar RL, Kale SP, Patil SA, Pant D (2016) Recent advances in the use of different substrates in microbial fuel cells toward wastewater treatment and simultaneous energy recovery. Applied Energy 168: 706–723.

Park J, Lee B, Shin W, Jo S, Jun H (2018) Application of a rotating impeller anode in a bioelectrochemical anaerobic digestion reactor for methane production from high-strength food waste. Bioresource Technology 259: 423-432.

Premier GC, Kim JR, Michie I, Popov A, Boghani H, Fradler K, Dinsdale RM, Guwy A J (2012) Issues of scale in microbial fuel cells and bioelectrochemical systems. World Renew Energy Forum WREF 2012. Renew Energy Soc Annu Conf: 4918–4925.

Puig S, Baeza JA, Colprim J, Cotterill S, Guisasola A, He Z, Heidrich E, Pous N (2017) Niches for bioelectrochemical systems in sewage treatment plants. Innovative Wastewater Treatment & Resource Recovery Technologies: Impacts on Energy Economy and Environment IWA Publishing: 96–107.

Rabaey K, Clauwaert P, Aelterman P, Verstraete W (2005) Tubular Microbial Fuel Cells for Efficient Electricity Generation. Environmental science & technology 39(20): 8077–8082.

Rader GK, Logan BE (2010) Multi-electrode continuous flow microbial electrolysis cell for biogas production from acetate. The 1st Iranian Conference On Hydrogen & Fuel Cell 35(17): 8848–8854.

Recio-Garrido D, Perrier M, Tartakovsky B (2016) Modeling optimization and control of bioelectrochemical systems. Chemical Engineering Journal 289: 180–190.

Rodrigo MA, Cañizares P, Lobato J, Paz R, Sáez C, Linares JJ (2007) Production of electricity from the treatment of urban waste water using a microbial fuel cell. Journal of Power Sources 169(1): 198–204.

Rosenbaum M, Agler MT, Fornero JJ, Venkataraman A, Angenent L T (2010) Integrating BES in the wastewater and sludge treatment line (eds) Bioelectrochemical Systems: from extracellular electron transfer to biotechnological application.IWA Publishing, London, pp 393–421.

San-Martín MI, Mateos R, Carracedo B, Escapa A, Morán A (2018) Pilot-scale bioelectrochemical system for simultaneous nitrogen and carbon removal in urban wastewater treatment plants. Journal of Bioscience and Bioengineering.

Santoro C, Arbizzani C, Erable B, Ieropoulos I (2017) Microbial fuel cells: From fundamentals to applications: A review. Journal of Power Sources 356: 225–244.

Santoro C, Flores-Cadengo C, Soavi F, Kodali M, Merino-Jimenez I, Gajda I, Greenman J, Ieropoulos I, Atanassov P (2018) Ceramic Microbial Fuel Cells Stack: power generation in standard and supercapacitive mode. Scientific Reports 8(1): 3281.

Santos MSS, Peixoto L, Mendes A, Alves MM (2017) Microbially charged redox flow battery : coupling a bioelectrochemical cell with a redox flow battery. ISMET 6 - General Meeting of the International Society for Microbial Electrochemistry and Technology: 95-95.

Seelam JS, Maesara SA, Mohanakrishna G, Patil SA, ter Heijne A,Pant D (2018) Resource Recovery From Wastes and Wastewaters Using Bioelectrochemical Systems. Waste Biorefinery: 535–570.

Sevda S, Sreekishnan TR, Pous N, Puig S, Pant D (2018) Bioelectroremediation of perchlorate and nitrate contaminated water: A review. Bioresource Technology 255: 331–339.

Sleutels THJA, Hamelers HVM, Buisman CJN (2011) Effect of mass and charge transport speed and direction in porous anodes on microbial electrolysis cell performance, Special Issue: Biofuels - II: Algal Biofuels and Microbial Fuel Cells 102(1): 399–403.

Stuve EM (2014) Electrochemical Reactor Design and Configurations. Encyclopedia of Applied Electrochemistry: 568–578.

Tartakovsky B, Kleiner Y, Manuel M (2017) Bioelectrochemical anaerobic sewage treatment technology for Arctic communities. Environmental Science and Pollution Research Environmental Science and Pollution Research: 1–7.

Tartakovsky B, Mehta P, Santoyo G, Guiot SR (2011) Maximizing hydrogen production in a microbial electrolysis cell by real-time optimization of applied voltage. International Conference on Hydrogen Production 36(17): 10557–10564.

Vilajeliu-Pons A, Puig S, Salcedo-Davila I, Balaguer MD, Colprim J (2017) Long-term assessment of six-stacked scaled-up MFCs treating swine manure with different electrode materials. Environmental Science: Water Research & Technology 3(5): 947–959.

Walsh F, Reade G (1994) Design and performance of electrochemical reactors for efficient synthesis and environmental treatment. Part 1. Electrode geometry and figures of merit. Analyst 119(5): 791–796.

Wang H, Park J, and Ren ZJ (2015) Practical Energy Harvesting for Microbial Fuel Cells: A Review. Environmental science & technology 49: 3267–3277.

Wang W, Sprenkle V (2016) Energy storage: Redox flow batteries go organic. Nature Chemistry 8 (3): 204-206.

Wu S, Li H, Zhou X, Liang P, Zhang X, Jiang Y, Huang X (2016) A novel pilot-scale stacked microbial fuel cell for efficient electricity generation and wastewater treatment .Water Research Pergamon 98: 396–403.

Yazdi H, Alzate-Gaviria L, Ren Z J (2015) Pluggable microbial fuel cell stacks for septic wastewater treatment and electricity production. Bioresource Technology 180: 258–263.

Innovative Bioelectrochemical-Anaerobic Digestion Coupled System for Process Monitoring Optimization and Product Purification of Wastewater

Yifeng Zhang*, Xiangdan Jin, Mingyi Xu and Rusen Zou

Department of Environmental Engineering, Technical University of Denmark,
DK-2800 Kongens Lyngby, Denmark

1. Introduction

On a global scale, the rapid growths in population, industrialization and urbanization have been increasing fossil fuels based energy consumption (Abas et al. 2015). Increasing depletion on fossil fuel reserves, and concern on greenhouse gas emissions during the combustion of fossil fuels has increased the demand for renewable energy source (Bauer et al. 2016). In the last decade, many governments and countries have published energy policies to mandate an incentive for renewable energy. According to the U.S. Energy Information Administration, about 10% of total U.S. energy consumption was from renewable energy sources in 2016. This number will double around 2050. Similarly, the EU has set an ambitious goal to increase the share of renewable energies in overall EU energy consumption to 20% by 2020 (Böhringer et al. 2009). As a long-term goal, 80-95% of greenhouse gas emissions will be reduced, compared to levels in 1990, by 2050 (Sovacool 2013).

1.1 Anaerobic Digestion (AD): Promising Technology but Challenges Remain

AD process, a complex biological process in which microorganisms convert diverse wastes into biogas in the absence of oxygen, has been widely applied for simultaneous biogas production and wastewater treatment. Energy from waste biomass is regarded as one of the most dominant future renewable energy sources (Appels et al. 2011). Therefore, the application of AD has received increasing attention in recent decades due to its beneficial properties including: (i) various types of biomass and waste are suitable as substrates, and co-digestion brings superior efficiencies in most cases; (ii) the digestate is nutrient rich and can be utilized in agriculture or organic amendment; (iii) carbon-neutral process without greenhouse gas emissions; (iv) it is feasible in both large-scale industrial facilities and small-scale ones which are easy to be installed in rural areas.

In general, four different steps are involved in the AD process, including hydrolysis, acidogenesis, acetogenesis and methanogenesis. These steps are catalyzed by several groups of microorganisms

*Corresponding author: yifz@env.dtu.dk

(Angelidaki et al. 2011) (Figure 1). During hydrolysis with the help of fermentative bacteria, polymeric compounds are converted into smaller units, such as glucose, xylose, amino acids and long-chain fatty acids (Xue et al. 2015). In the second stage, hydrolysis products are further transformed into smaller compounds, such as hydrogen, carbon dioxide, alcohols and short-chain volatile fatty acids (VFAs) (Karthikeyan et al. 2016). In the third stage, VFAs, lactate and alcohols are transformed to acetate by acetogenic bacteria (Seitz et al. 2016). In the methanogenesis stage, biogas composed of CH_4 and non-combustible CO_2 is produced (Demirel and Scherer 2008). About 70% of CH_4 is generated from acetate by acetoclastic methanogens, while another 30% of methane is produced from H_2/CO_2 by hydrogenotrophic methanogens.

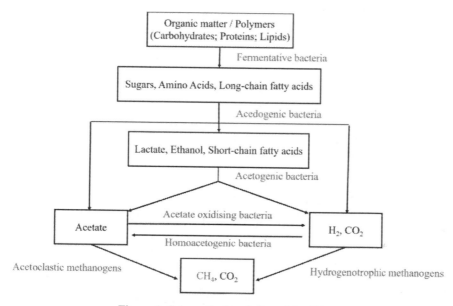

Figure 1: Schematic description of the AD process.

For a healthy AD process, it is very important to keep a balance between products obtained from the previous step and substrates for the next step. However, it is not always easy to keep a stable AD process and constant biogas yield due to the diverse needs and sensitivities of the involved microorganisms (Chen et al. 2008). For instance, the presence of less biodegradable lignocellulose such as grass, wood and pulping waste could deteriorate the AD process. Thus, physicochemical pre-treatments may be needed. Methanogens are strict anaerobes and likely more sensitive to toxicity and interruptions. Once the AD process is out of balance, VFAs will accumulate and lead to undesirable low pH. Moreover, CH_4 which is the valuable product in biogas usually accounts for 60-70% of the final gas volume. The presence of a large amount of CO_2 and trace amounts of non-inflammable components in biogas can reduce its calorific value and can cause the corrosive problem to engines as well (Andriani et al. 2014). These challenges, if not addressed, will lead to poor biogas yield and quality, process failures and serious economic losses.

1.2 Bioelectrochemical Systems (BESs): A Versatile Technology Reconstructing Nexus of Water and Energy

BESs are a promising technology that can use bacterial as a catalyst to oxidize organic waste and produce electricity or valuable products (Berk and Canfield 1964). Most of BES shares one common principle that oxidation occurs in anode, while reduction occurs in the cathode. In general, electron donors in the anode (e.g., organic waste streams) are oxidized by a specific group of microbial consortia (i.e., exoelectrogens) and then electrons are released (Wang et al. 2015). Electroactive microorganism will use a certain amount of electrons for metabolism. Subsequently, the residual electrons are transferred to the cathode through an external circuit. In cathode chamber, the electrons derived from the anode are then captured by the abiotic or biotic acceptors. The electrons can be utilized for electricity generation (microbial fuel cells or MFCs) or used for chemicals production (microbial electrolysis cells or MECs; microbial electrosynthesis or MES) or used for water desalination (microbial desalination cells, MDCs) (Logan and Rabaey 2012; Zhang and Angelidaki 2014; Cao et al. 2009).

1.3 Integration of BESs with AD Process Driving Breakthrough Innovations

Based on the principal feature, BESs can take advantage of the control or detection of electrons exchange during a biochemical or abiotic process to realize different goals, which can potentially integrate with AD process for varied applications, such as for pre-treatment of feedstocks, VFAs monitoring, recovery of VFAs and ammonia, biogas upgrading and production of methane from CO_2, etc. The schematic illustration of the principle for the integration of the two systems is shown in Figure 2. A few excellent reviews indeed have discussed the combination of BESs-AD systems (Zabranska and Pokorna 2018; Kuntke et al. 2018). However, most of the review articles focused mainly on a specific system or application, a comprehensive overview of the key BESs-AD-coupled systems and applications is still missing, especially in a context that more and more new systems and applications with better performance have been recently reported. Thus, to provide a timely update in this important research field, this review has discussed various integrated BES-AD systems in terms of innovation, working principle, application and system performance. We hope with such knowledge, more and more innovations and breakthrough in the large-scale application could be achieved in the years to come.

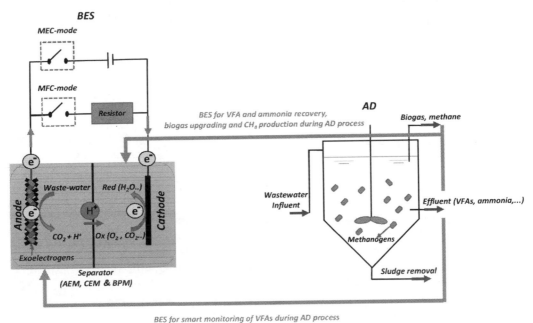

Figure 2: The schematic illustration of the principle for the integration of BESs and AD.

2. BES as Pre-treatment to Enhance the Wastewater Degradability

Various substrates, nearly all kinds of organic matter, can be used for biogas production through AD technology. However, the organic compound removal efficiency and biogas production rate are not always adequate, especially for the decomposition of complex organic substrates with poor degradability, such as waste activated sludge (WAS) and lignocellulosic materials. The main limiting factor is associated with a slow rate of hydrolysis (Abudi et al. 2016). To enhance the AD performance, reduce the hydraulic retention time (HRT) and diminish the digester volume; different pre-treatment methods have been proposed including mechanical, chemical, physical, thermal, biological and combinations of all these (Ariunbaatar et al. 2014). Although these conventional methods can enhance the solubility of the complex organic matter and therefore improve the AD performance (i.e., the high capital cost and energy consumption) a need of chemicals and generation of by-products often prohibit their application in full-scale applications (Ariunbaatar et al. 2014).

Recently, a new concept of integrating BES with AD to accelerate the substrate degradation and biogas yield has received great attention. Table 1 summarizes the recent studies in which the digestibility and biogas potential were enhanced in integrated BES-AD systems. In most of the studies, electrodes were inserted into the AD reactors directly and electricity was supplied. A lack of membranes between electrodes could get rid of the problems of fouling and dramatic pH changes, and it can also reduce the capital cost to some extent. The reactor served as the anaerobic digester and the bioelectrochemical enhancement was conducted *in situ*. For instance, Asztalos and Kim (2015) established an electrically-assisted digester (EAD; it is an anaerobic digester equipped with an MEC bioanode and cathode) treating WAS under ambient temperature conditions. An 1.2 V external voltage was supplied and three solids retention time (SRT, 7, 10 and 14 days) were chosen. Along with the SRT, the chemical oxygen demand (COD) removal and volatile suspended solids (VSS) removal in EAD were improved by 5-10% compared to those in the control digester in which the electrodes were disconnected. Correspondingly, the CH_4 production in EAD changed from 25.6 to 14.0 mL/d with the increasing SRT, while the value decreased from 17.4 to 11.0 mL/d in the control digester. Apart from the enhanced performance contributed by bioelectrochemical reactions, the energy recovery was higher than 326% that indicated that EAD can be operated as a net energy producer (Asztalos and Kim 2015). The anode electrode could not only facilitate microbes to attach but also serve as an electron donor. Feng et al. (2015) operated a cylindrical anaerobic reactor equipped with a pair of Fe tube anode and graphite pillar cathode under different supplied voltage. After a batch run, with 0.3 V the VSS removal was 11% higher than that in the control reactor without electrodes. The residual solid sludge was less with applied voltage suggesting that electrolysis improved the hydrolysis and acidification of sludge. The CH_4 production reached 170.2 L/kg-VSS at 0.3 V that was 22.4% higher than that in the control. The better performance of the MEC-anaerobic reactor was contributed by the bioaugmentation effect of archaea and bacteria. The energy recovery was 624 and 765 kJ for control and with an external voltage of 0.3 V, respectively. The energy input was only 4.9 kJ for 0.3 V that indicated that the MEC-anaerobic digester could raise the energy recovery (Feng et al. 2015). A proper electric current may stimulate the microbes in two ways: i) direct stimulation of microbial metabolism to accelerate the cell growth and ii) influence on cultivation ambient (such as pH and alkalinity) (Chen et al. 2016). In the work reported by Feng et al. (2016), the effect of applied voltage on a coupled bioelectrochemical AD treating sewage sludge was studied at ambient temperature. The species of exoelectrogens which could secrete some endogenous electron shuttles to enhance the methanogenic activity of anaerobic bacteria enriched on the anode surface. Exoelectrogens also grew on the cathode surface to produce CH_4 via CO_2 reduction (Equation 1). The process performance was improved in terms of organic matter removal and CH_4 production with 0.3-0.5 V external voltage (Feng et al. 2016).

$$\text{Cathode: } CO_2 + 8H^+ + 8e^- \rightarrow CH_4 + 2H_2O \quad E^0 = -0.24 \text{ V vs. SHE} \tag{1}$$

Due to the bioelectrochemical reaction in the combined BES and AD systems, the microbial community can be altered in diversity and population. The syntrophic interaction between exoelectrogens and anaerobic fermentation bacteria has the potential to accelerate complex substrate degradation (Zhao et al. 2016). Microbial communities such as hydrogenotrophic methanogens (Xiao et al. 2018; Feng et al. 2016) and acetoclastic methanogens (Lee et al. 2017) favorable for CH_4 production prefer to be enriched and activated. Higher energy recovery and shorter HRTs can also be obtained in the BES-AD systems compared to the conventional AD technology (Song et al. 2016). Gratifyingly, immense success has been achieved in laboratory-scale BES-AD reactors. However, more efforts should be made toward the alteration of the microbial community in BES-AD and the relations among the enhancement of complex substrate decomposition and CH_4 production. Up-scaling of BES-AD for waste treatment and bioenergy

Table 1: Overview of the wastes treated in integrated BES-AD systems

Substrates for BES-AD	Microbial inoculum	Reactor working volume	Anode	Cathode	HRT (d)	Temperature (°C)	Main results	References
WAS from secondary clarifiers in a local WWTP (7.89±1.88 gCOD/L, 5.2±1.2 gVSS/L)	Digested sludge from the lab-scale reactors	180 mL	Carbon fiber brushes	Stainless steel mesh	7, 10, 14	22.5±0.5	The electrically assisted digesters achieved enhanced VSS and COD removal (5-10%)	Asztalos and Kim 2015
Raw sludge from a WWTP (106.5±8.1 g/L TOCD, 62.7±3.8 g/L VSS)	Digested sludge from a lab-scale UASB	2.0 L	Fe tube electrode	Graphic pillar electrode	Batch mode (22 days for a batch)	35	At 0.3 V, VSS removal and methane production were increased by 11% and 22.4%, respectively	Feng et al. 2015
Raw sludge from the secondary sedimentation of a WWTP (685-820 mg/L SOCD, 29.65-29.73 g/L VS)	Digested sludge from a long-term lab-scale CSTR	700 mL	Activated carbon fiber textile	Same as the anode	Batch mode (29 days for a batch)	35±2	At 0.6 V, VS removal and methane production were enhanced by 26.6% and 76.2%, respectively	Chen et al. 2016
Sewage sludge from a WWTP (620-2213 mg/L SCOD)	Active sludge from a mesophilic e-AD	12 L	Graphite fiber fabric loaded with multiwall carbon nanotube and Ni	Same as the anode with a separator (non-woven polypropylene)	20	25±2	The process stability and performance in terms of organic matter removal and methane production were better at 0.3 and 0.5 of applied voltage.	Feng et al. 2016
Pretreated WAS from the secondary sedimentation of a WWTP (17.5±0.2 g/L TOCD, 14.0±0.7 g/L VSS)	-	500 mL	Graphite brush	Carbon cloth covered with Pt	Batch mode (Over 40 days for a batch)	20-25	The carbon degradation of VFAs, polysaccharides and proteins was accelerated by 22%, 43% and 48%, respectively	Liu et al. 2016

(Contd.)

Table 1: *(Contd.)*

Substrates for BES-AD	Microbial inoculum	Reactor working volume	Anode	Cathode	HRT (d)	Temperature (°C)	Main results	References
WAS from a sewage treatment plant (36.6±0.5 g/L TOCD, 28.8±5.2 g/L VS)	Sewage sludge from a anaerobic digester	12 L	Graphite fiber fabric loaded with multiwall carbon nanotube and Ni	Same as the anode with a separator (non-woven polypropylene)	5, 10, 15, 20	35	With a small voltage (0.3 V) to the MEC-AD system, the energy efficiency and VS reduction can be increased without any inhibitory effects	Song et al. 2015
WAS from secondary sedimentation in a local WWTP (52.3±1.1 g/L TOCD, 40.7±0.9 g/L VSS)	Refluxed liquid of anaerobic sludge of a waste sludge treatment plant	500 mL	Graphite brush	Graphite rod	Batch test (51 days for one batch)	22±2	The decomposition of complex substrates were accelerated, as well as the methane production	Zhao et al. 2016
Food waste in a sewage treatment plant (63 g/L TOCD, 37 g/L VS)	Effluent from a anaerobic digester	15 L	Graphite carbon coated with Ni	Graphite carbon coated with Fe and Cu	20	35	The conversion of organic matter to VFAs was improved along with a 40% increase in bacterial population	Lee et al. 2017
Raw sludge from a aeration tank of a WWTP (15.8±0.08 g/L TOCD, 6.02±0.22 g/L VSS)	Sludge from an anaerobic digester	180 mL	Ti/Ru alloy mesh plate	Ti/Ru alloy mesh plate	Batch test (21 days for one batch)	37±1	The decreases in SCOD and VSS were enhanced by the applied voltages	Xiao et al. 2018

production is still a big challenge. Future research could focus on the improvement of stability and reduction of costs for the system.

3. BES for Smart Monitoring of VFAs During AD Process

Effective and rapid methods to monitor the AD process is essential since process instability triggered by organic overload, toxic inhibition, ammonia inhibition and clogging can result in failures and serious economic losses. Process imbalance causes a series of phenomena like VFAs accumulation, pH decrease and low biogas yield. Parameters such as pH, alkalinity, methane concentration and biogas outflow rate, VFAs concentrations are common indicators for the AD process monitoring (Liu et al. 2011). VFAs concentrations are the favorite indicators as they can reflect the metabolic state of the biochemical process (Falk et al. 2014). Currently, conventional VFAs analytical methods such as titration, GC, HPLC and mid-infrared spectroscopy have been established due to the low cost, simplicity or accurate. However, responses from those methods cannot be real-time, while failures in a biogas plant may be just a matter of a few hours. Recently, BES-based biosensors have gained increasing interest due to their sustainability, real-time response, cost-effectiveness and portability (Yu et al. 2018). BES-based biosensors have been demonstrated their feasibility for determination of biochemical oxygen demand (BOD), COD, dissolved oxygen (DO), microbial activity and toxic components (Zhang and Angelidaki 2011; Sun et al. 2015a).

In recent years, BES has been employed to detect VFAs concentrations during AD processes. Kaur et al. (2013) used typical H-type MFCs to discriminate and measure specific VFAs (i.e., acetate, propionate and butyrate). The biofilm was acclimated by each VFA as sole carbon source. Cross-sensitivity tests of MFC reactors proved the selectivity since acetate-enriched and propionate-enriched reactors only responded to acetate and propionate, respectively. Two electrochemical methods i.e., coulombic efficiency (CE) and cyclic voltammetry (CV) were chosen to evaluate the MFC technology. The correlation between VFAs concentrations and recovered charge exhibited good linearity. However, the response of CE was longer than 20 hours when the VFA concentration was higher than 20 mg/L. By using CV with a consistent scan rate, distinctive oxidation peaks of VFAs were obtained. The response time was only 1-2 minutes. Linear correlations between oxidation peak current and VFA concentrations were obtained. However, the detection range (5-40 mg/L) was too low for practical applications (Kaur et al. 2013). Afterward, Kaur et al. (2014) further modified the anode by using natural and conductive polymers to immobilize the microbial community. Compared to an unmodified anode, the voltage output, stability, sensitivity and repeatability with the modified anode was greatly improved. The VFA detection limit was increased to 60 mg/L, but further improvement is still required (Kaur et al. 2014). Based on the previous achievement, Kaur et al. (2015) established a single chamber MFC and set the anode potential to a fixed voltage corresponding to the oxidation peak from CVs. The degradation of VFAs was enhanced under the poised potentials and the detection range of the sensor was improved to 200 mg/L. Nevertheless, measurement of current responses with high deviation errors (10-50%) exhibited quite a low accuracy and poor repeatability. Calibration was recommended to be conducted frequently (Kaur et al. 2015).

Recently, the application of a BES sensor for the AD process control under real conditions has been reported (Schievano et al. 2018). Air-cathode membraneless MFCs were established and also served as AD reactors. Digestate and four kinds of feedstock were dosed to start the AD reactors while the electrical signal was recorded. It was found that the MFC signal increased along with VFAs concentrations below 1,000 mg_{AC}/L. With different substrates, a negative correlation between the potential signal and peak of VFAs concentrations (2,500-4,500 mg_{AC}/L) was obtained. Obvious inhibition on bioanode as well as system performance under high VFAs concentrations suggested that the MFC be applied as early-warning and on-line monitoring system in full-scale AD plants (Schievano et al. 2018). However, further investigation on the sensor stability and the recovery time after shock-inhibition are required. Furthermore, the decreased signal caused by shock-inhibition or VFAs consumption should be judged carefully. In another work, the implementation of a microbial electrochemical sensor (MESe) in an AD process has been reported (Kretzschmar et al. 2018). The MESe setup that was mainly composed of an anode with pre-acclimated biofilm, count electrode and reference electrode was integrated into a 10 L laboratory-scale AD reactor directly. Chronoamperometry (CA) measurement at 0.2 V was conducted and the amperometric signal of the MESe setup changed consistently with the acetate concentrations. The system could serve as a real-time sensor of the VFA concentrations in the AD process. A linear correlation was found between the current signal and the acetate concentration from 100 to 500 mg/L. However, after several days of monitoring, it was found that major parts of the biofilm disappeared, and the residual was only loosely attached to the anode due to the mechanic abrasion of the biofilm. Even with a full-surrounding flow shield over the anode, a constant decline of the sensor signal within 1-8 days was observed indicating inhibition of the current production occurred (Kretzschmar et al. 2018). Therefore, when BES sensors are integrated with real AD reactors, the impact of the AD process parameters or toxic compounds that impairs the sensor's functionality should be avoided. Consequently, some investigations (Jin et al. 2016; Jin et al. 2017a) reported MDC/MEC-based biosensors for VFAs measurement in the middle or cathode chamber. Unlike the previous work where anode biofilm was exposed to the digestate directly, biofilm was separated from the digestate samples by an anion exchange membrane (AEM). Influencing parameters on the system performance such as microorganisms, various organic matter and toxicants in the digestate could be eliminated. Only ionized VFAs transported through AEM to the anode where they were oxidized by electroactive bacteria. Linear relationships between current signals and VFAs concentrations were obtained. Good accuracy, reproducibility and selectivity with a short response time of the biosensors have been observed. The detection range was widened significantly up to hundred mM/L, therefore, dilution of sample was not required (Jin et al. 2016; Jin et al. 2017).

System upscaling is always a challenge when integrating BES with AD. Meanwhile, the downscale or even miniaturization of a BES biosensor for the AD process monitoring should also be pursued. A smaller size reactor can enhance the mass transfer, shorten response time and accelerate the system performance. Therefore, the technology will be less costly and more practical.

4. BESs for VFA Recovery from Digestate

VFAs as important intermediates of AD processes always present at a high concentration level in digestate that need to be further removed or recovered for wastewater purification (Nam et al. 2010; Wu and Modin 2013). However, the separation or recovery of VFA from digestate, which has a complicated composition, is still a major challenge (Cerrillo et al. 2016). Compared to conventional membrane-based processes, such as nanofiltration or electrodialysis, BESs hold economic and environmental-friendly potentials for recovery of VFA from AD process (Zhang and Angelidaki 2015a; Kelly and He 2014). Recently, an innovative BES termed microbial bipolar electrodialysis cell (MBEDC), which integrates bipolar membrane (BPM) electrodialysis and microbial electrosynthesis, has been developed as sustainable downstream technologies for simultaneous waste-derived VFA recovery, hydrogen and alkali production and wastewater treatment (Kuntke et al. 2012). In this process, with a small amount of external voltage applied, organic matters in the wastewater are consumed by exoelectrogenic bacteria in the anode, while protons are reduced to hydrogen gas at the cathode. Between the anode and cathode, the chambers are separated by several different membranes. The pH of electrolyte is balanced by BPM, while the ionic equilibrium is re-established as results of the combined effect of anion exchange membrane (AEM) and cation exchange membrane (CEM). Thus, the negatively charged carboxylate ions can be transferred to the certain chamber and were then they are further accumulated and recovered. The system achieved a VFAs recovery efficiency of 98.3% at an applied voltage of 1.2 V using synthetic fermentation broth. After three consecutive batches, the VFAs concentration was concentrated by 2.96 times. According to energy balance analysis, net energy (5.20-6.86 kWh/kg-VFA recovered) can be produced in such a recovery system at all the applied voltages (0.8-1.4 V). Beside digestate and fermentation broth, the reported system could also be applied to other waste streams that are rich in VFAs. Although promising, several challenges still need to be addressed. Firstly, the system selectivity should be improved by using more specific membranes that could only allow the transportation of VFAs. Secondly, more cost-effective and *in situ* applicable reactor design should be pursued to simplify the recovery process and reduce the operating costs. Thirdly, long-term operation in continuous mode using real waste streams should be investigated to identify potential limitations.

5. BESs for Ammonia Recovery during AD Processes

Ammonia is a prevalent compound in various waste streams, and it can be produced when proteins, urea and nucleic acids are digested during anaerobic digestion (González-Fernández and García-Encina 2009). However, it has been reported that a high concentration of ammonia ($\geq$1.5 g-N/L) may inhibit the AD process and cause disturbance in the anaerobic reactor (Hejnfelt and Angelidaki 2009). The phenomenon of ammonia inhibition has been considered as one of the key challenges in the AD process (Nielsen and Angelidaki 2008). Recently, BESs have been demonstrated as promising approach to recover ammonia from wastewater and thereby, preventing the ammonia inhibition during AD process (Kuntke et al. 2012; Ieropoulos et al. 2012; Kuntke et al. 2011). Compared to conventional technologies on ammonia recovery, BESs such as MFC and MEC can realize direct ammonia recovery via the interactions among biological catalyst, membranes and electrodes that is less energy-intensive and more environmentally and economically sustainable (Wu and Modin 2013; Kuntke et al. 2012). Electricity assisted MFC-based system termed as R^2-BES has been developed for the recovery of ammonium and phosphorus (Zhang et al. 2014). This system takes advantage of bioelectricity generation from the oxidation of organic compounds to drive ammonium migration out of wastewater and at the same time, uses hydroxide ions produced from the cathode reaction as a medium to exchange phosphate ions from wastewater. Under an applied voltage of 0.8 V, the R^2-BES removed 83.4 $\pm$ 1.3% of ammonium nitrogen and 52.4 $\pm$ 9.8% of phosphate. In addition, MFC has even been employed to remove ammonia as a pre-treatment method before anaerobic digestion (Inglesby and Fisher 2012). An advanced flow-through anaerobic digester with an integrated recirculation loop MFC (ADMFC) was investigated for the production of methane using *Arthrospira maxima* as the sole feedstock. The maximum methane yield and energy efficiency significantly increased to 173 $\pm$ 38 mL CH_4 per g VS and 29.7 $\pm$ 6.8%, respectively. In order to cope with higher ammonia load in influent, abiotic operation of MEC has also been attempted for ammonia recovery with the synthetic wastewater (Desloover et al. 2012; Christiaens et al. 2017). With an energy consumption of 5 kWh kg^{-1} N removed, NH_4^+ flux of 120 g N m^{-2} d^{-1} was obtained along with NH_4^+ charge transfer efficiency of 96%. As a result, the ammonium concentration in the digestate decreased from 2.1 to 0.8 $\sim$ 1.2 g N L^{-1}. When fed with real digested pig slurry, continuous assays using an MEC were also performed by applying punctual pulses of VFAs in order to evaluate how the system responds to malfunction periods (Cerrillo et al. 2016). The results showed the anode compartment of the MEC successfully re-stabilized the AD process, and the ammonium diffusion was enhanced with a removal efficiency achieved up to 60%.

Although BESs such as MFC and MEC are promising, several limitations still need to be addressed before practical application. For instance, feeding ammonia-rich streams directly into anode may inhibit the exoelectrogenic activity of anodic biofilm (Nam et al. 2010). In addition, the characteristics such as pH of electrolyte may be significantly changed that will lead to a difficulty in reuse or for further treatment. Therefore, a novel submersible MDC (SMDC) was developed for *in situ* ammonia recovery and electricity production by submerging SMDC into a continuous stirred tank reactor (CSTR) (Zhang and Angelidaki 2015a). In a batch mode, the ammonia level in the CSTR was reduced from 6 to 0.7 g-N/L during 30 days operation. The average recovery rate was 80 g-N/m²/d. The ammonia recovery was mainly driven by NH_4^+ migration and free NH_3 diffusion. In a continuous mode, this hybrid system achieved an ammonia recovery rate of 86 g-N/m²/day and 112% extra biogas production (Zhang and Angelidaki 2015b). In an MDC system, the high energy consumption of extensive aeration in the cathode might be one of the key challenges. Therefore, a novel bipolar bioelectrodialysis system was developed on the basis of MDC in realizing simultaneous recovery of ammonia and sulfate from waste streams and hydrogen production that could offset the energy costs (Zhang and Angelidaki 2015c). In this system, NH_4^+ and SO_4^{2-} could be captured in the form of NH_3 (from NH_4^+ and OH^-) and H_2SO_4. The nitrogen and sulfate fluxes of 5.1 g NH_4^+-N/m²/d and 18.9 g SO_4^{2-}/m²/d were obtained at an applied voltage of 1.2 V.

6. BESs for Biogas Upgrading

Biogas is renewable energy generated from biomass. Its wide utilization contributes to sustainable bioenergy ecosystems (Sun et al. 2015b). Biogas mainly contains CH_4, CO_2 and trace compounds such as H_2S, H_2, NH_3, N_2 and O_2. Since the large share of CO_2 with 30-40% of gas volume, the low heating value impedes the industrial application of raw biogas. CO_2 removal is necessary in order to inject CH_4 enriched biogas into the natural gas grids or compressed for vehicles directly (Sun et al. 2015b). Commercial technologies which are extensively being used today are water scrubbing, pressure swing adsorption, chemical adsorption, organic solvent scrubbing, membrane separation, etc. Despite being applied in full-scale, issues such as extensive energy consumption, solvent disposal, corrosion and difficulty in CO_2 recovery should be addressed. In this context, BESs can be a possible alternative for biogas upgrading by physicochemically separate and capture CO_2.

In MECs with a small voltage input (approximately 0.2-1.2 V), alkaline can be produced through water splitting in the cathode chamber that can be used to absorb CO_2. Zeppilli et al. (2016) developed an MEC with biocathode to investigate the CO_2 removal mechanisms. Two types of the membrane (PEM and AME) were used. The anode was poised at +0.2 V (vs. standard hydrogen electrode, SHE) and alkalinity was generated in the cathode. Synthetic biogas (30% CO_2 and 70% N_2) was bubbled into the cathode compartment. According to the mass balance analysis of inorganic carbon, the authors found that the overall CO_2 removal was mainly attributed to alkalinity adsorption that accounted for 73% and 94% in the MEC using PEM and AEM, respectively. Meanwhile, CO_2 reduction to CH_4 by hydrogenotrophic methanogens in the biocathode accounted for 4% and 15% of CO_2 removal in the MEC using PEM and AEM, respectively. The energy required for CO_2 removal was 0.78 kWh/Nm³ CO_2 removed (anode poised at +0.2 V) in the AEM-MEC that was lower than that in conventional technologies such as water scrubbing (Zeppilli et al. 2016). Later, Zeppilli et al. (2017) studied the CO_2 removal in MEC using real effluents from two-phase AD reactor. The bioanode was fed with real digestate and the biocathode was fed by a CO_2-rich synthetic gas. The MEC performance, in terms of the electric current generation, methane production and CO_2 removal was improved by using the digestate with more biodegradable COD taken from the first stage. Coulombic efficiency of 119±28% was observed in this system. The electrons were transferred to cathode biofilm for CO_2 reduction that contributed to CO_2 removal of 3.5±0.9 mmol C/d. On the other hand, net alkalinity generation from electrolysis strongly absorbed CO_2 that caused CO_2 removal of 25±3 mmol C/d (Zeppilli et al. 2017). Therefore, using BESs as post-treatment for AD processes can realize both liquid effluent treatment and biogas upgrading that has many merits such as low-strength COD removal, low microbial growth, CO_2 removal and comparable energy consumption.

By employing BPM in BESs, alkali and acid can be produced with externally applied voltage that can be used for CO_2 capture and regeneration, respectively. Chen et al. (2012) established microbial electrolysis desalination and chemical production cell (MEDCC) with four chambers using a BPM. When the 1.0 V voltage was applied, the pH values of 12.9 and 0.68 were achieved in the cathode chamber and acid-production chamber after 18 hours, respectively. The energy input for such a system was around 7.46×10-5 kWh (Chen et al. 2012). Based on the previous work, Chen et al. (2013) developed a bipolar membrane electrodialysis (BPMED) that can produce alkaline for on-site biogas upgrading. With the electricity supplied by an MFC, alkaline was produced and pH in the alkaline generation chamber reached 9.8. When the MFC was replaced by an external power supply at 0.5 V, pH of 11.6 can be obtained in the alkaline generation chamber. The produced alkali solution was suitable for biogas upgrading where biomethane with the purity of 100% was achieved (Chen et al. 2013). More recently, Jin et al. (2017b) built a microbial electrolytic capture, separation and regeneration cell (MESC) for *in situ* biogas upgrading. With external voltage supply (1.0, 1.2 and 1.4 V), CO_2 from synthetic biogas (40% CO_2 and 60% CH_4) was absorbed by alkalinity solution in the cathode and turned into bicarbonate that was then transported to the acid generation chamber where CO_2 was regenerated and collected. The maximum methane content was up to 97.0±0.2%. The process cost around 0.17 kWh of energy to treat every cubic meter of raw biogas. Moreover, H_2 was also generated in the cathode chamber during the process and could potentially compensate part (approximately 23.4%) of the energy consumption (Jin et al. 2017b). The advantages of such biogas upgrading system include low investment and operation cost, a lack of chemical supply, mild operation condition and are environmentally friendly. Therefore, integration of BES as a post-treatment unit for biogas upgrading is very promising.

7. BES for CH_4 Production Using CO_2 as a Sole Carbon Source

In BESs, methane could be produced directly from CO_2. Many studies have demonstrated that hydrogenotrophic methanogens could be enriched in biocathode of BESs for the biological reduction of CO_2 into CH_4. When sufficient H_2 was generated at the cathode, the reduction of CO_2 could be realized according to Equation 2. Villano et al. (2010) developed an MEC with a biocathode in which CO_2 generated from organics oxidation in the anode was directed to the biocathode enriched by hydrogenotrophic methanogens. In BESs for biogas upgrading, a process called 'electromethanogenesis' has been proposed that claimed this process could achieve CO_2 reduction by receiving electrons directly from cathode electrode (Blasco-Gómez et al. 2017). In another process, CO_2 could be reduced by hydrogenotrophic methanogens using H_2 generated from electrolysis in the cathode. Both of these mechanisms were responsible for CO_2 removal. When cathode potential was poised more negative than -650 mV (vs. SHE), H_2 was produced at the cathode for CO_2 reduction. The authors also reported that methanogens could accept electron directly from the cathode electrode instead of using H_2 as an electron source (Equation 1). The contribution of these two mechanisms to methane production was highly determined by the set cathode potential. When the cathode potential was poised more negative, the CH_4 production rate increased (Villano et al. 2010). It was also found that the digestate can be further polished in the anode of BESs for organic removal and raw biogas can be sent into the cathode for biogas upgrading.

$$\text{Cathode: } 4H_2 + CO_2 \rightarrow CH_4 + 2HO_2 \qquad \Delta G^{0\prime} = -130.4 \text{ KJ/mol } CH_4 \qquad (2)$$

The concept of integration of BESs and AD for bioproduction of CH_4 has been proved by Villano et al. (2013). The anode potential of BES reactor was set at +0.2 V (vs. SHE). Low-strength AD effluent, which was prepared by a diluted stream containing acetate, was used as a substrate in the anode. A gaseous stream (30% CO_2 and 70% N_2) simulating the biogas was introduced into the biocathode. In the anode, the effluent COD concentration was around 38±6 mg COD/L as results of high removal efficiency of 94%. At the cathode, pH was adjusted by the gaseous stream to promote the methanogenesis activity. CO_2 reduction to CH_4 was mainly driven by direct extracellular electron transfer since 79% of the electric current was recovered as CH_4 (Villano et al. 2013). Apart from the excellent potential to refine digestate and biogas, other advantages such as low sludge production and limited ammonium transportation to the cathode make the technology very promising. The long-term stability should be further demonstrated before field application.

Xu et al. (2014) proposed a new approach to remove CO_2 through the integration of H-type MEC with AD. In the *ex situ* system, biogas was first generated from a digester bottle and then it was injected into the biocathode of MEC. In the *in situ* system, simultaneous biogas production and upgrading were achieved in the cathode which served as a digester. The inlet CO_2 content kept around 30% and the outlet value was below 10% in the *ex situ* system while the *in situ* system performed better due to the faster gas-liquid transfer rate. A hydrogenotrophic methanogen (*Methanobacterium petrolearium*) was detected as the dominant species responsible for CO_2 reduction at the cathode. In addition to biological reduction, CO_2 adsorption by alkali produced at the cathode also contributed to biogas upgrading (Xu et al. 2014). Bo et al. (2014) proposed a process which integrated MEC and AD for *in situ* methane production. A barrel-shaped reactor made of stainless steel which served as a cathode was employed as an AD reactor as well. When anode was inserted into the AD reactor and 1.0 V external voltage was supplied, the CH_4 content from the integrated system was upgraded to 98% and 2.3 times increase in CH_4 yield was obtained as well compared to the single AD process without MEC. In the reactor, hydrogenotrophic electromethanogens (e.g. *Methanospirillum*) were found as dominant species and their electrochemical activity was demonstrated via cyclic voltammetry (CV) technology (Bo et al. 2014). Overall, integrating BESs in AD systems to improve CH_4 concentration is an appealing concept. Upscaling the system for large-scale operations is necessary before practical application.

8. Conclusions and Future Perspectives

In conclusion, the novel concepts of using BES for recovery of valuable resources, monitoring of VFAs concentrations or biogas upgrading has been extensively studied. Through promising, further improvement and optimization are still needed in the future study. In the BES-based VFA biosensor, energy consumption for aeration could be removed by employing an air-cathode. Besides, a single chamber biosensor that can be submerged into an AD reactor should be developed to further reduce the construction costs. Correspondingly, the performance of the future systems in terms of response time, reliability and detection range should be investigated and compared with existing studies. Currently, the reported biosensor reactors are with a volume of several hundred milliliters. Miniaturization of the reactors could be one of the goals in the future in order to accelerate the mass transfer, shorten the response time and reduce the capital cost. For biogas upgrading, treatment capacity needs to be further increased to meet the industrial demand. Stack operation of BESs could be an alternative solution to achieve high process efficiency with respect to biogas purification. For most the integrated BES-AD systems, new architecture design should be developed and investigated in order to reduce the construction costs. Among others, long-term operation and system upscaling are the key issues to be explored in the future. Last but not least, a detailed environmental and economic impact assessment should be performed to accelerate the commercialization of BES-AD systems.

Acknowledgment

The authors would like to acknowledge financial support from the China Scholarship Council. This research was supported financially by The Danish Council for Independent Research (DFF-1335-00142) and Novo Nordisk Foundation (NNF16OC0021568).

References

Abas N, Kalair A, Khan N (2015) Review of fossil fuels and future energy technologies. Futures 69: 31-49.

Abudi ZN, Hu Z, Xiao B, Abood AR, Rajaa N, Laghari M (2016) Effects of pretreatments on thickened waste activated sludge and rice straw co-digestion: Experimental and modeling study. J Environ Manage 177: 213-222.

Andriani D, Wresta A, Atmaja TD, Saepudin A (2014) A review on optimization production and upgrading biogas through CO_2 removal using various techniques. Appl Biochem Biotechnol 172: 1909-1928.

Angelidaki I, Karakashev D, Batstone DJ, Plugge CM, Stams AJ (2011) Biomethanation and its potential. Methods Enzymol 494: 327-351.

Appels L, Lauwers J, Degrève J, Helsen L, Lievens B, Willems K, Van Impe J, Dewil R (2011) Anaerobic digestion in global bio-energy production: potential and research challenges. Renew Sust Energ Rev15: 4295-4301.

Ariunbaatar J, Panico A, Esposito G, Pirozzi F, Lens PN (2014) Pretreatment methods to enhance anaerobic digestion of organic solid waste. Appl Energy 123: 143-156.

Asztalos JR, Kim Y (2015) Enhanced digestion of waste activated sludge using microbial electrolysis cells at ambient temperature. Water Res 87: 503-512.

Böhringer C, Rutherford TF, Tol RS (2009) The EU 20/20/2020 targets: An overview of the EMF22 assessment. Energy Econ 31: S268-S273.

Bauer N, Mouratiadou I, Luderer G, Baumstark L, Brecha RJ, Edenhofer O, Kriegler E (2016) Global fossil energy markets and climate change mitigation–an analysis with REMIND. Clim Change136: 69-82.

Berk RS, Canfield JH (1964) Bioelectrochemical energy conversion. Appl Microbiol 12: 10–12.

Blasco-Gómez R, Batlle-Vilanova P, Villano M, Balaguer MD, Colprim J, Puig S (2017) On the edge of research and technological application: a critical review of electromethanogenesisInt J Mol Sci 18: 874.

Bo T, Zhu X, Zhang L, Tao Y, He X, Li D, Yan Z (2014) A new upgraded biogas production process: coupling microbial electrolysis cell and anaerobic digestion in single-chamber, barrel-shape stainless steel reactor. Electrochem commun 45: 67-70.

Cao X, Huang X, Liang P, Xiao K, Zhou Y, Zhang X, Logan BE (2009) A new method for water desalination using microbial desalination cells. Environ Sci Technol 43: 7148-7152.

Cerrillo M, Viñas M, Bonmatí A (2016) Removal of volatile fatty acids and ammonia recovery from unstable anaerobic digesters with a microbial electrolysis cell. Bioresour Technol 219: 348-356.

Chen M, Zhang F, Zhang Y, Zeng RJ (2013) Alkali production from bipolar membrane electrodialysis powered by microbial fuel cell and application for biogas upgrading. Appl Energy 103: 428-434.

Chen S, Liu G, Zhang R, Qin B, Luo Y (2012) Development of the microbial electrolysis desalination and chemical-production cell for desalination as well as acid and alkali productions. Environ Sci Technol 46: 2467-2472.

Chen Y, Cheng JJ, Creamer KS (2008) Inhibition of anaerobic digestion process: a review. Bioresour Technol 99: 4044-64.

Chen Y, Yu B, Yin C, Zhang C, Dai X, Yuan H, Zhu N (2016) Biostimulation by direct voltage to enhance anaerobic digestion of waste activated sludge. RSC Adv 6(2): 1581-1588.

Christiaens MER, Gildemyn S, Matassa S, Ysebaert T, De Vrieze J, Rabaey K (2017) Electrochemical ammonia recovery from source-separated urine for microbial protein production. Environ Sci Technol 51(22):13143-13150

Demirel B, Scherer P (2008) The roles of acetotrophic and hydrogenotrophic methanogens during anaerobic conversion of biomass to methane: a review. Rev Environ Sci Bio7(2): 173-190.

Desloover J, Woldeyohannis AA, Verstraete W, Boon N, Rabaey K (2012) Electrochemical resource recovery from digestate to prevent ammonia toxicity during anaerobic digestion. Environ Sci Technol 46(21): 12209-12216.

Falk HM, Reichling P, Andersen C, Benz R (2014) Online monitoring of concentration and dynamics of volatile fatty acids in anaerobic digestion processes with mid-infrared spectroscopy. Bioprocess Biosyst Eng 38: 237-249.

Feng Q, Song YC, Bae BU (2016) Influence of applied voltage on the performance of bioelectrochemical anaerobic digestion of sewage sludge and planktonic microbial communities at ambient temperature. Bioresour Technol 220: 500-508.

Feng Y, Zhang Y, Chen S, Quan X (2015) Enhanced production of methane from waste activated sludge by the combination of high-solid anaerobic digestion and microbial electrolysis cell with iron–graphite electrode. Chem Eng J 259: 787-794.

González-Fernández C, García-Encina PA (2009) Impact of substrate to inoculum ratio in anaerobic digestion of swine slurry. Biomass and Bioenerg 33(8): 1065-1069.

Hejnfelt A, Angelidaki I (2009) Anaerobic digestion of slaughterhouse by-products. Biomass Bioenerg 33: 1046-1054.

Ieropoulos I, Greenman J, Melhuish C (2012) Urine utilisation by microbial fuel cells; energy fuel for the future. Phys Chem Chem Phys 14: 94-98.

Inglesby AE, Fisher AC (2012) Enhanced methane yields from anaerobic digestion of Arthrospira maxima biomass in an advanced flow-through reactor with an integrated recirculation loop microbial fuel cell. Energ Environ Sci 5: 7996-8006.

Jin X, Angelidaki I, Zhang Y (2016) Microbial electrochemical monitoring of volatile fatty acids during anaerobic digestion. Environ Sci Technol 50: 4422-4429.

Jin X, Li X, Zhao N, Angelidaki I, Zhang Y (2017a) Bio-electrolytic sensor for rapid monitoring of volatile fatty acids in anaerobic digestion process. Water Res 111: 74-80.

Jin X, Zhang Y, Li X, Zhao N, Angelidaki I (2017b) Microbial electrolytic capture, separation and regeneration of CO_2 for biogas upgrading. Environ Sci Technol 51: 9371-9378.

Karthikeyan O.P, Selvam A, Wong JW (2016) Hydrolysis–acidogenesis of food waste in solid–liquid-separating continuous stirred tank reactor (SLS-CSTR) for volatile organic acid production. Bioresour Technol 200: 366-373.

Kaur A, Dinsdale R, Guwy A, Premier G (2015) Improved dynamic response and range in microbial fuel cell-based volatile fatty acid sensor by using poised potential. In: A Sayigh (eds) Renewable Energy in the Service of Mankind Vol I, Springer, Cham, pp 183-192.

Kaur A, Ibrahim S, Pickett CJ, Michie IS, Dinsdale RM, Guwy AJ, Premier GC (2014) Anode modification to improve the performance of a microbial fuel cell volatile fatty acid biosensor. Sensor Actuat B-Chem 201: 266-273.

Kaur A, Kim JR, Michie I, Dinsdale RM, Guwy AJ, Premier GC (2013) Microbial fuel cell type biosensor for specific volatile fatty acids using acclimated bacterial communities. Biosens Bioelectron 47: 50-55.

Kelly PT, He Z (2014) Nutrients removal and recovery in bioelectrochemical systems: A review. Bioresour Technol 153: 351-360.

Kretzschmar J, Böhme P, Liebetrau J, Mertig M, Harnisch F (2018) Microbial electrochemical sensors for anaerobic digestion process control–performance of electroactive biofilms under real conditions. Chem Eng Technol 41: 687-695.

Kuntke P, Geleji M, Bruning H, Zeeman G, Hamelers HVM, Buisman CJN (2011) Effects of ammonium concentration and charge exchange on ammonium recovery from high strength wastewater using a microbial fuel cell. Bioresour Technol 102: 4376-4382.

Kuntke P, Śmiech KM, Bruning H, Zeeman G, Saakes M, Sleutels THJA, Hamelers HVM, Buisman CJN (2012) Ammonium recovery and energy production from urine by a microbial fuel cell. Water Res 46: 2627-2636.

Kuntke P, Sleutels T, Arredondo MR, Georg S, Barbosa SG, ter Heijne A, Hamelers HVM, Buisman CJN (2018) (Bio)electrochemical ammonia recovery: progress and perspectives. Appl Microbiol Biotechnol 102: 3865-3878.

Lee B, Park JG, Shin WB, Tian DJ, Jun HB (2017) Microbial communities change in an anaerobic digestion after application of microbial electrolysis cells. Bioresour Technol 234: 273-280.

Liu Z, Liu J, Zhang S, Xing X, Su Z (2011) Microbial fuel cell based biosensor for in situ monitoring of anaerobic digestion process. Bioresour Technol 102: 10221-10229.

Logan BE, Rabaey K (2012) Conversion of wastes into bioelectricity and chemicals by using microbial electrochemical technologies. Science 337: 686-690.

Nam JY, Kim HW, Shin HS (2010) Ammonia inhibition of electricity generation in single-chambered microbial fuel cells. J Power Sources 195: 6428-6433.

Nielsen HB, Angelidaki I (2008) Strategies for optimizing recovery of the biogas process following ammonia inhibition. Bioresour Technol 99: 7995-8001.

Schievano A, Colombo A, Cossettini A, Goglio A, D'Ardes V, Trasatti S, Cristiani P (2018) Single-chamber microbial fuel cells as on-line shock-sensors for volatile fatty acids in anaerobic digesters. Waste Manage 71: 785-791.

Seitz KW, Lazar CS, Hinrichs KU, Teske AP, Baker BJ (2016) Genomic reconstruction of a novel, deeply branched sediment archaeal phylum with pathways for acetogenesis and sulfur reduction. ISME J 10: 1696-1705.

Song YC, Feng Q, Ahn Y (2016) Performance of the bio-electrochemical anaerobic digestion of sewage sludge at different hydraulic retention times. Energy Fuel 30: 352-359.

Sovacool BK (2013) Energy policymaking in Denmark: implications for global energy security and sustainability. Energy Policy 61: 829-839.

Sun JZ, Kingori GP, Si RW, Zhai DD, Liao ZH, Sun DZ, Zheng T, Yong YC (2015a). Microbial fuel cell-based biosensors for environmental monitoring: a review. Wat Sci Tech 71: 801-809.

Sun Q, Li H, Yan J, Liu L, Yu Z, Yu X (2015b) Selection of appropriate biogas upgrading technology-a review of biogas cleaning, upgrading and utilisation. Renew Sust Energ Re 51: 521-532.

Villano M, Aulenta F, Ciucci C, Ferri T, Giuliano A, Majone M (2010) Bioelectrochemical reduction of CO_2 to CH_4 via direct and indirect extracellular electron transfer by a hydrogenophilic methanogenic culture. Bioresour Technol 101: 3085-90.

Villano M, Scardala S, Aulenta F, Majone M (2013) Carbon and nitrogen removal and enhanced methane production in a microbial electrolysis cell. Bioresour Technol 130: 366-371.

Wu X, Modin O (2013) Ammonium recovery from reject water combined with hydrogen production in a bioelectrochemical reactor. Bioresour Technol 146: 530-536.

Xiao B, Chen X, Han Y, Liu J, Guo X (2018) Bioelectrochemical enhancement of the anaerobic digestion of thermal-alkaline pretreated sludge in microbial electrolysis cells. Renew Energ 115: 1177-1183.

Xu H, Wang K, Holmes DE (2014) Bioelectrochemical removal of carbon dioxide (CO_2): An innovative method for biogas upgrading. Bioresour Technol 173: 392-398.

Xue Y, Liu H, Chen S, Dichtl N, Dai X, Li N (2015) Effects of thermal hydrolysis on organic matter solubilization and anaerobic digestion of high solid sludge. Chem Eng J 264: 174-180.

Yu Z, Leng X, Zhao S, Ji J, Zhou T, Khan A, Kakde A, Liu P, Li X (2018) A review on the applications of microbial electrolysis cells in anaerobic digestion. Bioresour Technol 255: 340-348.

Zabranska J, Pokorna D (2018) Bioconversion of carbon dioxide to methane using hydrogen and hydrogenotrophic methanogens. Biotechnol Adv 36: 707-720.

Zhang F, Li J, He Z (2014) A new method for nutrients removal and recovery from wastewater using a bioelectrochemical system. Bioresour Technol 166: 630-634.

Zeppilli M, Lai A, Villano M, Majone M (2016) Anion vs cation exchange membrane strongly affect mechanisms and yield of CO_2 fixation in a microbial electrolysis cell. Chem Eng J 304: 10-19.

Zeppilli M, Pavesi D, Gottardo M, Micolucci F, Villano M, Majone M (2017) Using effluents from two-phase anaerobic digestion to feed a methane-producing microbial electrolysis. Chem Eng J 328: 428-433.

Zhang Y, Angelidaki I (2011) Submersible microbial fuel cell sensor for monitoring microbial activity and BOD in groundwater: focusing on impact of anodic biofilm on sensor applicability. Biotechnol Bioeng 108: 2339-2347.

Zhang Y, Angelidaki I (2014) Microbial electrolysis cells turning to be versatile technology: recent advances and future challenges. Water Res 56: 11-25.

Zhang Y, Angelidaki I (2015a) Submersible microbial desalination cell for simultaneous ammonia recovery and electricity production from anaerobic reactors containing high levels of ammonia. Bioresour Technol 177: 233-239.

Zhang Y, Angelidaki I (2015b) Counteracting ammonia inhibition during anaerobic digestion by recovery using submersible microbial desalination cell. Biotechnol Bioeng112: 1478-1482.

Zhang Y, Angelidaki I (2015c) Recovery of ammonia and sulfate from waste streams and bioenergy production via bipolar bioelectrodialysis. Water Res 85: 177-184.

Zhao Z, Zhang Y, Ma W, Sun J, Sun S, Quan X (2016) Enriching functional microbes with electrode to accelerate the decomposition of complex substrates during anaerobic digestion of municipal sludge. Biochem Eng J 111: 1-9.

Modeling of Wastewater Treatment and Methane Production in a Flow-through Microbial Electrolysis Cell

Javier Dario Coronado[1,2], M. Perrier[2], B. Srinivasan[1] and B. Tartakovsky[1,2*]

[1] Département de Génie Chimique, École Polytechnique de Montréal, C.P.6079 Succ.,
 Centre-Ville Montréal, QC, Canada H3C 3A7b
[2] National Research Council of Canada, 6100 Royalmount Ave., Montréal, QC, Canada H4P 2R2

1. Introduction

Bioelectrochemical systems (BESs) have considerable potential for sustainable energy production from organic wastes (Logan et al. 2008; Logan and Regan 2006). Two types of BESs can be distinguished: microbial fuel cell (MFC) and microbial electrolysis cell (MEC). While MFC features direct electricity production from organic materials (Lovley 2008), MEC produces energy carriers such as hydrogen or methane when low voltage (e.g., 0.8-1.4 V; it is below the practical threshold for water electrolysis) is applied to the electrodes (Liu et al. 2005; Rozendal et al. 2006; Villano et al. 2011).

Over time, MEC research evolved from a hydrogen-producing technology (Hu et al. 2008; Selembo et al. 2009) to a bioprocess with the capacity to produce several value-added resources (Zou and He 2018). With an increasing interest in the development of bioelectrochemical power-to-gas (BEP2G) technology (Geppert et al. 2016), biomethane production in MECs attracts research efforts focused on better understanding of the underlying bioelectrochemical principles as well as on optimizing the operational conditions of the biocathode where methane is produced (Dykstra and Pavlostathis 2017; Hou et al. 2015; Villano et al. 2011).

Commonly, MECs are built with electrode compartments separated by means of a proton-exchange membrane so that different anolyte and catholyte solutions can be used (Lovley 2008). In the case of a membraneless flow-through MEC, the compartments are separated by a porous non-conductive divider so that liquid flows freely from the first electrode compartment (i.e., anode) to the second compartment (i.e., cathode). As the bioelectrochemical half-reaction takes place in the first compartment, the water-soluble products obtained from this reaction are transported by liquid flow to the second compartment where the second bioelectrode catalyzes the second half-reaction. This flow-through MEC concept was recently introduced to achieve energy efficient wastewater treatment (Tartakovsky et al. 2017). It was shown to efficiently treat high-strength wastewater in a broad range of temperatures (5-25°C) and produced biogas containing more than 60% methane.

Optimization and scale-up of the flow-through MEC for treating municipal, agricultural and industrial waste streams involve both engineering and scientific efforts that can be facilitated by using

*Corresponding author: Boris.Tartakovsky@cnrc-nrc.gc.ca

modeling techniques. Furthermore, MEC capacity to produce energy carriers is considered as a promising approach for converting pollutants, such as CO_2, into biomethane or high-value chemicals (Lu and Ren 2016; Sadhukhan et al. 2016; Zhen et al. 2017). An adequate dynamic model of an MEC is instrumental in bridging the BES fundamentals with the development of a fully operational industrial-scale MEC.

Dynamic models have been previously used to perform bioprocess analysis and to optimize MFC (Boghani et al. 2013; Gatti and Milocco 2017; Hamelers et al. 2011; Marcus et al. 2007; Park et al. 2017; Pinto et al. 2010; Recio-Garrido et al. 2016; Zhang and Halme 1995). Pinto et al. (2011a) developed one of the first bioelectrochemical dynamic models of an MEC. The main objective of that study was to simulate hydrogen production in the MEC with a gas diffusion Ni-based cathode. The model, developed for process control and optimization purposes, considers a multi-population anodic microbial community but lacks a detailed description of the cathodic reactions.

Hussain et al. (2018) proposed a model that uses a simple equivalent electrical circuit for internal resistance and capacitance estimation. While this model was successfully used for monitoring a flow-through MEC, it does not predict biomass growth and carbon source consumption, and therefore, cannot be used for process design and optimization. Overall, there is a lack of detailed MEC models suitable for dynamic modeling of the complex microbial communities and bioelectrochemical transformations within an MEC.

This study describes a bioelectrochemical dynamic model of a membraneless flow-through MEC. The model is capable of simulating the dynamics of microbial growth and carbon source consumption in the anode and cathode compartments of the MEC. The model is calibrated using results of MEC operation on acetate as well as on protein-based synthetic wastewater. Furthermore, the model is used to propose an improved MEC design targeting high purity methane production.

2. Materials and Methods

2.1 Wastewater Composition and Analytical Methods

The stock solution of wastewater was either acetate or protein-based (synthetic wastewater). The acetate stock solution was composed of (per L) anhydrous sodium acetate (40 g), yeast extract (0.83 g), ammonium chloride (18.7 g), potassium chloride (74.1 g), potassium phosphate dibasic (32.0 g) and potassium phosphate monobasic (20.4 g). The stock solution was diluted via an influent stream of deionized water to obtain the desired acetate concentration. The solution of trace metals was added to the influent stream (1 mL per 1L) to provide essential microelements. The composition of the trace metal solution can be found elsewhere (Pinto et al. 2011). The protein-based wastewater had an average total COD concentration of 750 mg L^{-1}. The stock solution of protein wastewater contained (per L) pepticase (7 g), beef extract (2.8 g), ammonium bicarbonate (1 g), urea (0.8 g) and microelements. A detailed composition can be found elsewhere (Tartakovsky et al. 2017).

Biogas production was measured using a miniature flip-flop gas meter (milligascounter, Ritter Apparatus, Bochum, Germany). Gas composition was analyzed by using a gas chromatograph as described by Tartakovsky et al. (2017). Acetate concentration was analyzed using an Agilent 6890 gas chromatograph (Agilent, Wilmington, DE, USA). Soluble and total COD concentrations were measured according to Standard Methods (APHA 1995).

2.2 MEC Design and Operation

A horizontal flow MEC was constructed using Plexiglas plates. The MEC had an equally sized anode and cathode compartments separated by a 2 mm thick piece of geotextile (nonconductive porous cloth). Granular activated carbon (GAC) was used both as anode and cathode material, occupying the entire volume of each electrode compartment. The total liquid MEC volume was 1.7 L with a headspace volume of 0.3 L. The MEC was operated as a flow-through reactor with the influent stream entering the anode compartment and the effluent stream exiting after the cathode compartment as shown in Figure 1. To ensure adequate liquid mixing, an external recirculation line returned a fraction of the effluent stream

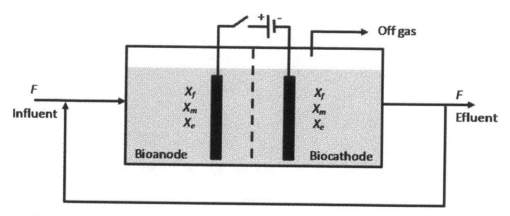

Figure 1: Schematic diagram of the horizontal flow-through MEC with porous electrodes.

to the influent as shown in Figure 1 at a rate of 13.8 L d⁻¹. Biogas produced in the anode and cathode compartments exited through a common off-gas line located at the top of the reactor. Also, MEC was maintained at room temperature (22-24°C).

The inoculation procedure consisted of adding 100 mL of homogenized anaerobic sludge (Lassonde Inc., Rougement, QC, Canada) that had a volatile suspended solids content of 22-25 g L⁻¹ to each electrode compartment. Initially, the MEC was fed with a synthetic acetate wastewater solution (1,000 mg L⁻¹ of acetate or 1,070 mg L⁻¹ as COD) at a flow rate of 2 L d⁻¹ corresponding to a hydraulic retention time (HRT) of 1.17 days. To test MEC response to increased carbon source concentration, between days 12 and 18 the influent acetate concentration had doubled. Following the acetate-fed phase, the electrode compartments were reloaded with virgin GAC that were re-inoculated with the same amount of anaerobic sludge as used in the previous test and fed protein-based wastewater instead. During this phase, the influent COD concentration was still maintained at 1,000 – 1,200 mg L⁻¹, and the flow rate was maintained at 55 L d⁻¹. While biogas flow and composition measurements were attempted throughout both tests, a gas leak prevented accurate flow rate measurements during the acetate test. The problem was only fixed during the protein wastewater test with gas production measurements available starting from day 45 of the test.

2.3 Electrical Measurements and Numerical Methods

The MEC was operated at an applied voltage of 1.5 V using a PW18-1.8AQ power supply (Kenwood Corp, Tokyo, Japan) interfaced with a computer. The current was measured with a 15 Ohm shunt resistance (R_{ext}) according to the diagram shown in Figure 1.

First order differential equations of the process model were solved in Matlab R2017a (Mathworks, Natick, MA, USA) using the *ode45s* function. Parameter estimation was carried out by minimizing the difference between experimental data and model outputs. The following objective function (mean square error, MSE) was used:

$$\text{MSF} = \frac{1}{n} \sum_{k=0}^{n} \left(\frac{y_i^{exp} - y_i^{mod}}{\underline{Y}} \right)^2 \tag{1}$$

In this, n is the number of measurements, y_i^{exp} is the ith measured value, y_i^{mod} is the ith model output and $\underline{Y}$ is the average of all measured values.

3. Model Formulation

To account for microbial growth and biotransformations in the anode and cathode compartments of the flow-through MEC, the proposed dynamic model considers three key microbial trophic groups:

(i) hydrolyzing and fermentative microorganisms (X_f), (ii) methanogenic microorganisms (X_m) and (iii) electroactive microorganisms (X_e). Accordingly, the model describes the transformation of solid (particulate) organic materials (S_p) into easily degradable dissolved organic matter, e.g., acetate (S_s), that can be metabolized by both the methanogenic and electroactive populations. Also, hydrogen (S_{H2}) production and consumption at the cathode are considered. This sequence of microbially catalyzed pathways and the corresponding microbial populations leading to S_s formation and then to the production of methane (CH_4) and carbon dioxide (CO_2) is rather simplified. In particular, hydrolysis of S_p and the multi-step transformation of the hydrolysis products to S_s are described as a single step reaction performed by X_f. In another simplification, CH_4 formation both from S_s and H_2 is assumed to be carried out by the same microbial population (X_m).

The model describes two pathways of CH_4 production. In one pathway, CH_4 is produced from S_s by methanogenic populations (X_m) through the acetoclastic methanogenesis, i.e., S_s is equated to acetate. In this case, the following stoichiometric equation can be written:

$$CH_3COOH \ X_m \rightarrow CH_4 + CO_2. \tag{2}$$

The second pathway accounts for CH_4 formation through bioreactions involving both electroactive and methanogenic microorganisms. In the anode compartment, electroactive (anodophilic) microorganisms consume S_s and transfer electrons to the anode. Simultaneously, CO_2 and protons are released:

$$CH_3COOH + 2H_2O \ Xe \rightarrow 2CO_2 + 8H^+ + 8e \tag{3}$$

In the cathode compartment, H_2 is produced by electroactive (cathodotrophic) microorganisms, and subsequently utilized by the hydrogenotrophic methanogens to produce CH_4 according to the following equations:

$$2H^+ + 2e^- \ X_e \rightarrow H_2 \tag{4}$$

$$4H_2 + CO_2 \ X_m \rightarrow CH_4 + 2H_2O \tag{5}$$

While direct electromethanogenesis was demonstrated (Cheng et al. 2009; Villano et al. 2010), we assume a more general pathway of CH_4 formation at the cathode through an intermediate step of H_2 formation. Notably, both the methanogenic (Equation 3) and electroactive-hydrogenotrophic (Equations. 4-5) pathways result in the same theoretical methane yield from acetate of 1 mol/mol or 0.37 L g^{-1}.

Kinetic equations describing the hydrolysis of particulate organic materials and the growth and metabolism of X_f, X_m, and X_e microbial populations on the soluble carbon source were largely adapted from works of Recio-Garrido et al. (2016) and Recio-Garrido et al. (2018). In particular, the specific growth rate of the hydrolysing and fermentative population (μ_f) is assumed to be dependent on particulates (S_p) concentration and described as:

$$\mu_f = \mu_{maxf} \left(\frac{S_p}{K_{sf} + S_p} \right) \tag{6}$$

where μ_{maxf} is the maximum growth rate and K_{sf} is the half saturation constant.

The methanogens (X_m) can grow both on S_s and S_{H2}, hence, the specific growth rate of this population (μ_m) is described as a summation of two Monod expressions:

$$\mu_m = \mu_{mS} + \mu_{mH_2} = \mu_{max,mS} \left(\frac{S_s}{K_{sm} + S_s} \right) + \mu_{max,mH_2} \left(\frac{S_{H_2}}{K_{H_2m} + S_{H_2}} \right), \tag{7}$$

where $\mu_{max,mS}$ and μ_{max,mH_2} are the maximum growth rate on S_s and S_{H_2}, respectively, K_{sm} and K_{H_2m} are the corresponding half saturation constants.

To describe the growth of the electroactive (X_e) population in the anode and cathode compartments, two kinetic dependences are developed. In the anode compartment the specific growth rate (μ_e) is assumed to depend on S_s and applied a voltage (V_{app}) according to the multiplicative Monod equation as follows:

$$\mu_e = \mu_{maxe}\left(\frac{S_s}{K_{se}+S_s}\right)\left(\frac{V_{app}}{K_v+V_{app}}\right), \tag{8}$$

where μ_{maxe} is the maximum growth rate K_{se} and K_v are the half-saturation constants. Importantly, the Monod-like dependence of the specific growth rate on applied voltage is introduced to describe experimentally observed dependence of electroactive anodophilic microorganisms on applied voltage (Ditzig et al. 2007; Tartakovsky et al. 2011).

In the cathode compartment, the specific growth rate (μ_e) depends on S_s and electrical current:

$$\mu_e = \mu_{maxe}\left(\frac{S_s}{K_{se}+S_s}\right)\left(\frac{I_{MEC}}{K_i+I_{MEC}}\right), \tag{9}$$

where K_i is the half saturation constant describing the dependence of electroactive growth at the cathode on MEC current.

Finally, the population of hydrolyzing and fermentative microorganisms (X_f) is assumed to hydrolyze S_p (Morgenroth et al. 2002):

$$b_h = k_{hf}\left(\frac{S_p}{K_p+S_p}\right) \tag{10}$$

where k_{hf} is the maximum hydrolysis rate and K_p is the half saturation constant.

It should be noted that in the anode compartment CH_4 is only produced through acetoclastic methanogenic activity, while at the cathode CH_4 is produced through both acetoclastic and hydrogeno-trophic methanogenesis (from bioelectrochemically produced H_2).

In a bioelectrode-based MEC, electrical current depends on multiple factors, including the activity of electroactive microorganisms in the electrode compartments, carbon source availability and applied voltage. An empirical approach to current modeling was chosen, where a multiplicative Monod expression is used to describe current (I_{MEC}) dependence on these factors according to the following expression:

$$I_{MEC} = \delta\left(\frac{S_s^a}{K_{se}+S_s^a}\right)\left(\frac{S_s^c}{K_{se}+S_s^c}\right)\left(\frac{V_{app}}{K_v+V_{app}}\right)X_e^a X_e^c \tag{11}$$

where δ is the current conversion factor, S_s^a and S_s^c are the dissolved organic carbon concentrations in the anode and cathode compartments, respectively, and X_e^a and X_e^c are the anode and cathode electroactive populations.

The material balance equations of the model are based on the assumption of ideal mixing in each electrode compartment, i.e., the flow-through MEC is modeled as two continuously stirred tank reactors (CSTRs) in series. A detailed description of the material balance equations is provided in Appendix A.

Overall, the model describes the dynamics of the MEC state variables (biomass and carbon source concentrations) in response to variations in wastewater flow (F), composition (concentrations of S_s and S_p) and applied voltage.

4. Model Calibration

The initial choice of model parameters (growth and consumption rates, yields and physical constants) was based on the MFC bioelectrochemical model proposed by Recio-Garrido et al. (2016). These parameters are described in Table 1. Considering a relatively small number of available experimental results, only maximum growth rates of the methanogenic and electroactive populations ($\mu_{max,mS}$, $\mu_{max,mH}$ and μ_{maxe}) in Equations 7-9 and the current conversion factor (δ) in Equation 11 were estimated. The choice of these parameters was based on the sensitivity analysis of the model (results not shown). Parameter values were manually adjusted to minimize MSE in Equation 1. An attempt to use a zero-order numerical parameter

Table 1: Model parameters

Parameter	Description	Carbon source		Units
		Acetate	COD	
μ_{maxf}	Maximum specific growth rate of X_f	-	0.10	d^{-1}
K_{sf}	Half-saturation coefficient of S_p for X_f	30	30	mg–S_p L^{-1}
$\mu_{max,ms}$	Maximum specific growth rate of X_m on S_s	0.10	0.05	d^{-1}
K_{sm}	Half-saturation coefficient of S_s for X_m	100	100	mg–S_s L^{-1}
μ_{max,mH_2}	Maximum specific growth rate of X_m on S_{H_2}	0.10	0.10	d^{-1}
K_{H_2m}	Half-saturation coefficient of S_{H_2} for X_m	1	1	mg–S_{H_2} L^{-1}
μ_{maxs}	Maximum specific growth rate of X_e	0.15	0.12	d^{-1}
K_{Se}	Half-saturation coefficient of S_s for X_e	40	40	mg–S_s L^{-1}
Kv	Half-saturation coefficient of V_{app} for X_e at the bioanode	0.30	0.30	V
K_i	Half-saturation coefficient of I_{MEC} for X_e at the biocathode	10	10	mA
k_{hf}	Maximum specific hydrolysis rate for X_f	-	0.50	
K_p	Half-saturation coefficient of S_p for X_f	-	100	mg-S_p L^{-1}
X_{max}	Total biomass maximum attainable density	2000	2000	mg L^{-1}
$X_{max,e}$	Electroactive biomass maximum attainable density	500	500	mg L^{-1}
b_f	Decay rate of X_f	0.01	0.01	d^{-1}
D	Dilution rate ($D = F/V$)			mL h^{-1} L^{-1}
b_m	Decay rate of X_m	0.01	0.01	d^{-1}
b_e	Decay rate of X_e	0.01	0.01	d^{-1}
f_p	Fraction of dead X_m yielding inert particulates	80	80	%
$Y_{\frac{X_f}{S_p}}$	Yield factor for X_f growth on S_{H_2}	-	0.07	mg–X_f mg–S_s^{-1}
$Y_{\frac{X_m}{S_s}}$	Yield factor for X_m growth on S_s	0.20	0.12	mg–X_m mg–S_s^{-1}
$Y_{\frac{X_e}{S_s}}$	Yield factor for X_e growth on S_e	0.20	0.12	mg–X_e mg–S_s^{-1}
$Y_{\frac{X_m}{S_{H2}}}$	Yield factor for X_m growth on S_{H_2}	0.10	0.10[a]	mg–X_m mg–S_{H2}^{-1}
$Y_{\frac{X_e}{S_{H2}}}$	Yield factor for X_e growth on S_{H_2}	0.10	0.10[a]	mg–X_e mg–S_{H2}^{-1}
$Y_{\frac{X_{mH2}}{CH_4}}$	Yield factor (CH$_4$ from X_m growth on S_{H_2})	-	2.50	mg–X_m mg–CH_4^{-1}
$Y_{\frac{X_{mS}}{CH_4}}$	Yield factor (CH$_4$ from X_m growth on S_s)	-	0.32	mg–S_S mg–CH_4^{-1}
δ	Current conversion factor	8.5e-4	3.6e-4	mA L mg–X_e^{-1}

$Y_{\frac{X_{mS}}{CO_2}}$	Yield factor (CO_2 from X_m growth on S_s)	–		0.32	mg–X_m mL–CO_2^{-1}
$Y_{\frac{X_{mH2}}{CO_2}}$	Yield factor (CO_2 from X_m growth on S_{H_2})	–		2.50	mg–X_m mL–CO_2^{-1}
$Y_{\frac{X_e}{CO_2}}$	Yield factor (CO_2 from X_e growth on S_s)	–		11.25	mg–X_e mL–CO_2^{-1}

[a] 0.01 was used in steady-state simulations to avoid H_2 release at the cathode.

estimation algorithm to further minimize the objective function in Equation 1 did not provide considerable improvement, likely due to the small number of experimental results.

4.1 MEC Operation on Acetate

MEC operation on acetate was intended for estimated model parameters related to the activity of X_e and X_m populations. Since biogas flow and composition during this experiment was not measured, model parameters related to CH_4 production were not estimated. As mentioned above, only four model parameters ($\mu_{max,ms}$, $\mu_{max,mH}$, μ_{maxe}, δ) were estimated. The resulting values are given in Table 1.

A comparison of model outputs with the measured acetate (S_s) concentrations show that the model adequately describes acetate degradation in the MEC, including the period of operation at an increased influent acetate concentration between days 12-18 (Figure 2A). Figure 2B compared experimentally measured and calculated values of MEC current (I). The observed current increase during, approximately, the first 17 days of MEC operation agrees with the predicted dynamics of X_e growth shown in Figure 3. Both experimental results and model predictions show that maximum current is achieved when both the anode and cathode electroactive populations are at their highest attainable value ($X_{max,e}$ parameter of the model).

Although the concentration of microbial populations X_e and X_m were not measured, model simulations can still be used to predict the dynamics of these populations. Figure 3A shows model predictions for X_m and X_e growth in the anode compartment, while the growth of these populations in the cathode compartment is shown in Figure 3B. As can be seen from these simulations, the model predicts fast growth of the electroactive population at the anode so that a steady-state X_e concentration corresponding to the maximum attainable cell density ($X_{max,e}$) is reached after eight days of MEC operation. Accordingly, the X_m concentration at the anode declined during this period. The steady-state concentration of the electroactive population (X_e) at the cathode is reached after 17 days of MEC operation, i.e., MEC performance is limited by the cathodic reaction during this time.

Model simulations allow for the analysis of S_s concentrations in the anode and cathode compartments. As can be expected for a flow-through system, acetate concentration is greater at the anode than at the cathode (Figure 2A). Higher acetate concentration in the anode compartment promotes the growth of both anodophilic and methanogenic microorganisms. Because of the slower specific growth rate of X_m ($\mu_{max,mS} < \mu_{max,e}$; Table 1), the electroactive population outcompetes the methanogens and approaches its maximum attainable density, $X_{max,e}$. At the cathode, X_e growth is limited by low S_s concentration resulting in a longer time to reach steady-state.

4.2 MEC Operation on Protein-Based Wastewater

The model was also calibrated using results of MEC operation on protein-based wastewater. In this case, three microbial populations were accounted for in the model: X_e and X_m and the newly added (X_f) fermentative population. Also, in this test biogas measurements were carried out between days 45-70, thus, allowing for CH_4 yield estimation.

As shown in Figure 4A, once the model parameters (μ_{maxm}, μ_{maxe}, δ) were re-estimated, the measured and predicted concentrations of soluble CODs in the MEC effluent (cathode compartment) agreed well.

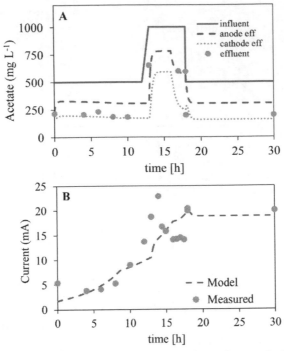

Figure 2: (A) Measured and simulated concentrations of acetate in the anode and cathode compartments of flow-through MEC during operation on acetate; (B) Measured and simulated MEC current (I).

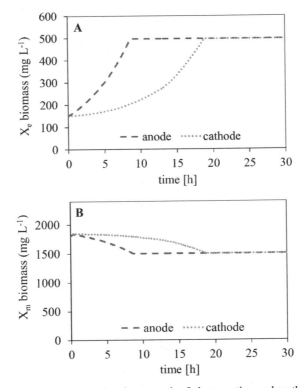

Figure 3: Model outputs showing growth of electroactive and methanogenic microbial populations in the (A) anode and (B) cathode compartments.

Throughout the experiment, cathodic S_s concentration fluctuated between 200-250 mg L^{-1}, while the predicted S_s concentration in the anode compartment, fluctuated at around 450 mg L^{-1}. This implies that the anode compartment removed up to 60% of the carbon source fed to the MEC, while the cathode compartment removed approximately 20% for combined removal efficiency of 80%. Progressive increase of MEC current shown in Figure 4B was indicative of the proliferation of electroactive microorganisms in the anode and cathode compartments, i.e., over time the degradation pathway shifted from the predominantly anaerobic degradation described by Equation 2 (due to a relatively large amount of X_m introduced with the initial anaerobic inoculum) toward a larger contribution of the electroactive population (Equations 3-5). Notably, both degradation pathways resulted in the same methane yield, thus there was no observable change in the amount of methane produced (Figure 4C).

While MEC design did not allow for individual measurements of biogas production and composition in each electrode compartment, model simulations were used for a more detailed analysis. As shown in Figure 4C, a similar amount of CH_4 is produced in each compartment, but the model predicts a higher percentage of CH_4 in the off-gas of the cathode compartment as compared to the anode compartment (Figure 4D). Indeed, the acetoclastic methanogenic (X_m) activity in the anode compartment results in S_s conversion to CH_4 according to Equation 2, i.e., both CH_4 and CO_2 are produced. Furthermore, electroactive activity at the anode also results in CO_2 production (Equation 3). As a result, the anode off-gas is predicted to contain more than 50% CO_2. In the cathode compartment, some CH_4 is produced through acetoclastic methanogenesis (Equation 2) with additional CH_4 produced from the bioelectrochemical pathway (Equations 4 and 5), thus increasing the CH_4 percentage. Essentially, the cathode compartment provides an *in situ* biogas upgrade through CO_2 conversion to CH_4.

Also, the model was used to simulate the distribution of X_f, X_m and X_e populations in the two electrode compartments (Figure 5). Similar to the simulation results obtained for MEC operation on acetate, the methanogenic population is predicted to decrease from the high concentration provided by inoculation with anaerobic sludge to lower values, while the electroactive populations in both compartments

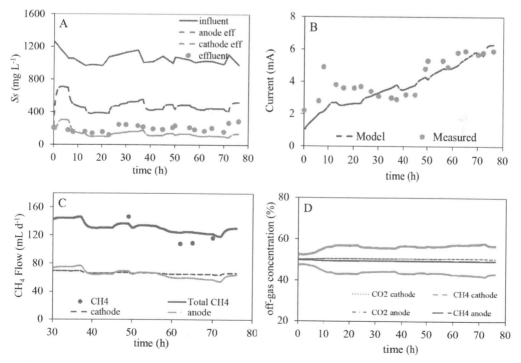

Figure 4: MEC performance during operation on protein-based wastewater. (A) Influent and effluent COD concentrations, (B) current, (C) simulated and measured methane production, (D) simulated CH_4 and CO_2 concentrations in the anode and cathode compartments.

proliferate until reaching the maximum attainable concentration (Figure 5A-B). Notably, slower X_e growth as compared to MEC operation on acetate is predicted, especially in the cathode compartment i.e., steady-state concentration is only reached after 65 days. As expected, an increase in X_f population at the anode is predicted as this compartment is fed with wastewater containing solids. Simultaneously, this population declines to a near-zero value at the cathode as solids are mostly hydrolyzed in the anode compartment thus resulting in low S_s in the cathode compartment (Figure 5C).

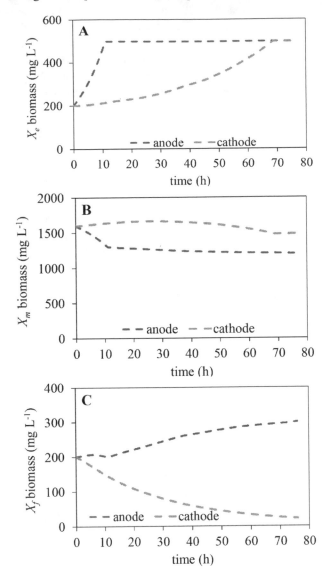

Figure 5: Simulated biomass concentrations during MEC operation on protein-based wastewater: (A) electroactive (X_e) microorganisms, (B) methanogenic (X_m) microorganisms, and (C) fermentative and hydrolysing (X_f) microorganisms.

5. Analysis of Flow-through MEC Configuration

In addition to such design parameters as electrode compartment geometry and size, the flow-through MEC also provides a choice related to electrode configuration. Indeed, wastewater can be fed to either anode or cathode compartment, i.e., the anode-cathode or cathode-anode configurations can be used. This section demonstrates the application of the model for comparing these two configurations by predicting biogas production and composition in each electrode compartment. This analysis is carried out using model parameters estimated during MEC operation on protein-based wastewater. In all simulations the influent wastewater flow rate (F) is set to 0.5 L d^{-1} (F), the wastewater is assumed to contain 1,350 mg L^{-1} of soluble organics (S_s) and 150 mg L^{-1} of particulates (S_p) and a constant compartment volume (V) of 425 mL is assumed. Also, biogas flow rate and composition are compared under steady-state conditions. Each mode of operation is calculated by integrating the model for 150 days.

5.1 Anode-Cathode Configuration

The flow-through MEC used in the experiments had a single headspace compartment (Figure 1), i.e., biogas produced in the anode and cathode compartments was mixed before gas composition measurements. Model simulations enable prediction of CH_4 production in each electrode compartment. Furthermore, by varying MEC current by changing the applied voltage or changing $X_{max,e}$ parameter of the model (e.g., changing the percentage of electroactive microorganisms), CH_4 production trends in each electrode compartment as a function of MEC current can be predicted. As shown in the previous section, the model predicts that in the cathode compartment the biogas contains a significantly higher percentage of CH_4 as opposed to similar CH_4 and CO_2 concentrations in the anode off-gas (Figure 4D). High CH_4 content in the cathode off-gas can be explained by a combination of acetoclastic and hydrogenotrophic pathways of CH_4 production in this compartment. CH_4 formation from H_2 consumes CO_2 (Equation 5), thus this reaction results in an increased percentage of CH_4 in the cathode off-gas. It can be hypothesized that high purity (e.g., 100%) CH_4 can be produced in the cathode compartment of a flow-through MEC essentially resulting in *in situ* biogas upgrade. However, biogas produced at the anode will contain elevated levels of CO_2.

It should be noted that CH_4 production in the cathode compartment strongly depends on the amount of carbon source consumed by X_e population at the anode. The dependence of biogas production and composition on the electroactive activity can be evaluated by varying the $X_{max,e}$ (maximum attainable X_e density) parameter of the model. Here, $X_{max,e}$ is correlated with MEC current. Figure 6A shows the predicted steady-state values of CH_4 flow in the anode and cathode compartments at different current values. As can be seen from this graph, CH_4 production at the cathode increases at higher current values, while it decreases at the anode. Accordingly, CO_2 production increases at the anode and also decreases at the cathode. Figure 6B compares the anode and cathode biogas composition at different currents. It predicts that pure CH_4 can be produced at the cathode starting from a current of 6 mA. At the same time, CO_2 concentration in the anode compartment increases with increasing current however never reaching 100%. This implies that the anode-cathode MEC configuration cannot achieve complete separation of CH_4 and CO_2 due to CH_4 production in the anode compartment by acetoclastic methanogens from a soluble carbon source (acetate). Also, Figure 6C shows that COD consumption in the anode compartment increases with increasing current resulting in increased removal efficiency.

Overall, steady-state analysis of the anode-cathode configuration suggests that production of high purity CH_4 at the cathode requires improved electroactive activity both in the anode (anodophilic S_s consumption) and cathode (bioelectrochemical H_2 production) compartments. Several approaches to increasing electroactive populations can be suggested, including electrode materials with lower internal resistance and increased electrode surface area for electroactive biofilm formation.

5.2 Cathode-Anode Configuration

The cathode-anode MEC configuration in which wastewater is first fed to the cathode compartment and then flows to the anode compartment can be used as a way to increase the amount of methane produced at the cathode. It can be proposed that if carbon source is fed to the cathode instead of the anode, more carbon source becomes available for acetoclastic methanogenesis. The remaining carbon source is consumed in the cathode compartment by the electroactive anodophilic population. Notably, previous research demonstrates a higher affinity of anodophilic microorganisms for acetate (Gil et al. 2003; Martin

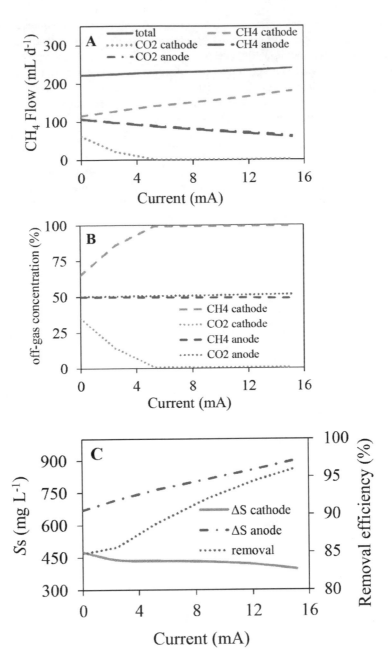

Figure 6: Performance analysis of the anode-cathode MEC configuration. (A) Production of CH_4 and CO_2 as a function of current, (B) CH_4 and CO_2 concentrations as a function of current, (C) S_s consumption in the anode and cathode compartments and acetate removal efficiency.

et al. 2010). Accordingly, the model assumes a smaller K_s (half-saturation coefficient) value for the X_e population as compared to the X_m population, 40 vs. 100 mg L^{-1}, respectively (Table 1). A smaller K_s value implies that at low S_s concentrations in the anode compartment the electroactive population is expected to outcompete the methanogenic population (Pinto et al. 2011a). In fact, coulombic efficiency of over 90% was observed in MFCs operated at low acetate levels (Pinto et al. 2011b; Recio-Garrido et al. 2017).

Figure 7 shows steady-state CH_4 and CO_2 flow rates in the anode and cathode compartments of the cathode-anode MFC configuration at different currents. As in the anode-cathode MEC configuration, the total CH_4 production remains nearly constant (somewhat lower CH_4 production at high current values is due to incomplete H_2 consumption). At greater current values (obtained by increasing the $X_{max,e}/X_{max}$ ratio) the CH_4 flow in the cathode compartment increases, while it decreases in the anode compartment. The increase in cathodic CH_4 is attributed to increased production of electrolytic H_2 at the cathode that is then used for CH_4 production according to Equations 4-5. A comparison of predicted CH_4 and CO_2 concentrations suggests that starting from a current of 10 mA, pure CH_4 is produced. At the same time, CH_4 concentration in the anode compartment decreases with increasing current, i.e., complete CH_4 and CO_2 separation can be achieved in this MEC configuration. In terms of COD removal (wastewater treatment), the cathode-anode configuration results in similar COD removal rates and efficiency as shown in Figure 7C.

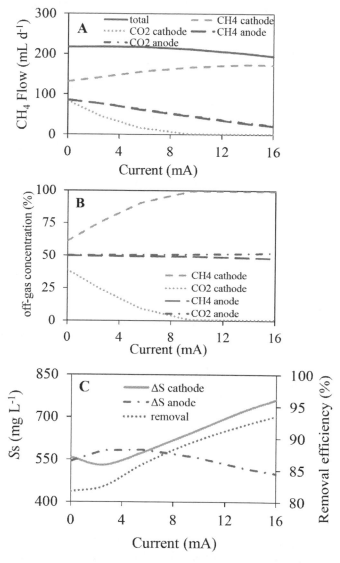

Figure 7: Performance analysis of cathode-anode flow-through MEC: (A) Production of CH_4 and CO_2 as a function of current, (B) CH_4 and CO_2 concentrations as a function of current, (C) S_s consumption in the anode and cathode compartments and acetate removal efficiency.

Overall, by comparing the anode-cathode and cathode-anode MEC configurations, it can be concluded that while the anode-cathode configuration provides a higher CH_4 purity in the cathode compartment at a lower current, this configuration also results in CH_4 and CO_2 production in the anode compartment due to acetoclastic methanogenic activity, i.e., CH_4 is always produced in the first (anode) compartment. In the cathode-anode configuration, acetoclastic CH_4 production in the second (anode) compartment is lower due to carbon source-limited conditions. Near complete CH_4/CO_2 separation can be achieved providing operating conditions (organic load and applied voltage) are optimized.

6. Future Perspectives

The work describes a dynamic model of a membraneless flow-through microbial electrolysis cell (MEC) with a bioanode and a biocathode. The model was calibrated using experimental results of MEC operation on acetate and protein-based synthetic wastewaters and then used to compare different electrode compartment arrangements. It was demonstrated that a near-complete separation of CH_4 and CO_2 can be achieved in the cathode-anode configuration under optimal operating conditions (e.g., wastewater flow and/or applied voltage optimization). While additional experiments are needed to refine the model parameters, validate the model and confirm the model predictions, modeling results thus far predict that the flow-through MEC configuration can be used to achieve *in situ* biogas upgrade to high purity CH_4. Here, CO_2 and CH_4 separation between the compartments enable bioelectrochemical conversion of remaining cathodic CO_2 to CH_4 (CH_4 bioelectrosynthesis).

Acknowledgment

JDC acknowledges the support of the Fonds de recherche du Quebec - Nature et technologies (FRQNT).

References

APHA, AWWA and WEF (1995) Standard methods for the examination of water and wastewater, 19 ed. American Public Health Association, Washington, DC.

Boghani HC, Kim JR, Dinsdale RM, Guwy AJ, Premier GC (2013) Analysis of the dynamic performance of a microbial fuel cell using a system identification approach. J Power Sources 238: 218-226.

Cheng S, Xing D, Logan B (2009) Direct biological conversion of electrical current into methane by electromethanogenesis. Env Sci Technol 43: 3953-3958.

Ditzig J, Liu H, Logan BE (2007) Production of hydrogen from domestic wastewater using a bioelectrochemically assisted microbial reactor (BEAMR). Int J Hydrogen Energy 32: 2296-2304.

Dykstra CM, Pavlostathis SG (2017) Evaluation of gas and carbon transport in a methanogenic bioelectrochemical system (BES). Biotechnology and bioengineering 114: 961-969.

Gatti MN, Milocco RH (2017). A biofilm model of microbial fuel cells for engineering applications. Int. J. Energy Environ. Eng. 8: 303-315.

Geppert F, Liu D, van Eerten-Jansen M, Weidner E, Buisman C, ter Heijne A (2016) Bioelectrochemical power-to-gas: State of the art and future perspectives. Trends in biotechnology 34: 879-894.

Gil GG, Chang IS, Kim BH, Kim M, Jang JK, Park HS, Kim HJ (2003) Operational parameters affecting the performance of a mediator-less microbial fuel cell. Biosensors and Bioelectronics 18: 327-334.

Hamelers HVM, Ter Heijne A, Stein N, Rozendal RA, Buisman CJ (2011) Butler-Volmer-Monod model for describing bio-anode polarization curves. Bioresource Technol 102: 381-387.

Hou Y, Zhang R, Luo H, Liu G, Kim Y, Yu S, Zeng J (2015) Microbial electrolysis cell with spiral wound electrode for wastewater treatment and methane production. Process Biochem 50: 1103-1109.

Hu H, Fan Y, Liu H (2008) Hydrogen production using single-chamber membrane-free microbial electrolysis cells. Water research 42: 4172-4178.

Hussain SA, Perrier M, Tartakovsky B (2018) Real-time monitoring of a microbial electrolysis cell using an electrical equivalent circuit model. Bioprocess Biosys. Eng. 41: 543-553.

Liu H, Grot S, Logan BE (2005) Electrochemically assisted microbial production of hydrogen from acetate. Env Sci Technol 39: 4317-4320.

Logan BE, Call D, Cheng S, Hamelers HV, Sleutels TH, Jeremiasse AW, Rozendal RA (2008) Microbial electrolysis cells for high yield hydrogen gas production from organic matter. Environmental science & technology 42: 8630-8640.

Logan BE, Regan JM (2006) Electricity-producing bacterial communities in microbial fuel cells. TRENDS in Microbiology 14: 512-518.

Lovley DR (2008) The microbe electric: conversion of organic matter to electricity. Current Opinion Biotechnol 19: 564–571.

Lu LL, Ren ZJ (2016) Microbial electrolysis cells for waste biorefinery: A state of the art review. Bioresource technology 215: 254-264.

Marcus AK, Torres CI, Rittmann BE (2007) Conduction-based modeling of the biofilm anode of a microbial fuel cell. Biotechnol Bioeng 98: 1171-1182.

Martin E, Savadogo O, Guiot SR, Tartakovsky B (2010) The influence of operational conditions on the performance of a microbial fuel cell seeded with mesophilic sludge. Biochem Eng J 51: 132-139.

Morgenroth E, Kommedal R, Harremoës P (2002) Processes and modeling of hydrolysis of particulate organic matter in aerobic wastewater treatment–a review. Water Science and Technology 45: 25-40.

Park JD, Roane TM, Ren ZJ, Alaraj M (2017) Dynamic modeling of a microbial fuel cell considering anodic electron flow and electrical charge storage. Appl Energ 193: 507-514.

Pinto RP, Srinivasan B, Escapa A, Tartakovsky B (2011a) Multi-Population Model of a Microbial Electrolysis Cell. Env Sci Technol 45: 5039-5046.

Pinto RP, Srinivasan B, Guiot SR, Tartakovsky B (2011b) The effect of real-time external resistance optimization on microbial fuel cell performance. Wat Res 45: 1571-1578.

Pinto RP, Srinivasan B, Manuel MF, Tartakovsky B (2010) A two-population bio-electrochemical model of a microbial fuel cell. Bioresource Technol 101: 5256-5265.

Recio-Garrido D, Adekunle A, Perrier M, Raghavan V, Tartakovsky B (2017) Wastewater treatment and online chemical oxygen demand estimation in a cascade of microbial fuel cells. Ind Eng Chem Res 56: 12471-12478.

Recio-Garrido D, Kleiner Y, Colombo A, Tartakovsky B (2018) Dynamic model of a municipal wastewater stabilization pond in the arctic. Wat Res 144: 444-453.

Recio-Garrido D, Perrier M, Tartakovsky B (2016) Combined bioelectrochemical-electrical model of a microbial fuel cell. Bioprocess Biosyst Eng 289: 180-190.

Rozendal RA, Hamelers HVM, Euverink GJW, Metz SJ, Buisman CJN (2006) Principle and perspectives of hydrogen production through biocatalyzed electrolysis. Int J Hydrogen Energ 31: 1632-1640.

Sadhukhan J, Lloyd JR, Scott K, Premier GC, Yu EH, Curtis T, Head IM (2016) A critical review of integration analysis of microbial electrosynthesis (MES) systems with waste biorefineries for the production of biofuel and chemical from reuse of CO2. Renewable and Sustainable Energy Reviews 56: 116-132.

Selembo PA, Perez JM, Lloyd WA, Logan BE (2009) High hydrogen production from glycerol or glucose by electrohydrogenesis using microbial electrolysis cells. International journal of hydrogen energy 34: 5373-5381.

Tartakovsky B, Kleiner Y, Manuel, MF (2017) Bioelectrochemical anaerobic sewage treatment technology for Arctic communities. Environ Sci Pollut Res https://doi.org/10.1007/s11356-017-8390-1.

Tartakovsky B, Mehta P, Santoyo G, Guiot SR (2011) Maximizing hydrogen production in a microbial electrolysis cell by real-time optimization of applied voltage. Int J Hydrogen Energy 36: 10557-10564.

Villano M, Aulenta F, Ciucci C, Ferri T, Giuliano A, Majone M (2010) Bioelectrochemical reduction of CO2 to CH4 via direct and indirect extracellular electron transfer by a hydrogenophilic methanogenic culture. Bioresource Technol 101: 3085-3090.

Villano M, Monaco G, Aulenta F, Majone M (2011) Electrochemically assisted methane production in a biofilm reactor. J Power Sources 196: 9467-9472.

Zhang XC, Halme A (1995) Modelling of a microbial fuel cell process. Biotechnology Letters 17: 809-814.

Zhen G, Lu X, Kumar G, Bakonyi P, Xu K, Zhao Y (2017) Microbial electrolysis cell platform for simultaneous waste biorefinery and clean electrofuels generation: Current situation, challenges and future perspectives. Progress in Energy and Combustion Science 63: 119-145.

Zou S, He Z (2018) Efficiently "pumping out" value-added resources from wastewater by bioelectrochemical systems: A review from energy perspectives. Water Res 131: 62-73.

Material Balances of the Model

Material balance equations are provided for hydrolyzing and fermentative (X_f), methanogenic (X_m) and electroactive (X_e) microbial populations as well as for particulate (S_p), soluble (S_s) carbon sources and hydrogen (S_{H_2}). The flow-through MEC using reactor-in-series approach with ideal mixing in each electrode compartment. For each electrode compartment the following X_f, X_m, and X_e balance equations are used:

$$\frac{dX_f}{dt} = (\mu_f - b_f - D\alpha)X_f + D\beta X_f^{in} \tag{A1}$$

$$\frac{dX_m}{dt} = (\mu_{mS} + \mu_{mH_2} - b_m - D\alpha)X_m + D\beta X_m^{in} \tag{A2}$$

$$\frac{dX_e}{dt} = (\mu_e - b_e - D\alpha)X_e + D\beta X_e^{in} \tag{A3}$$

Biomass transport between electrode compartments is accounted for using parameter β (transport factor). This parameter is equal to 1 for the first compartment and 0 for the second compartment. The biomass detachment process is described by assuming the maximum attainable biomass concentration (X_{max}). Total biomass concentration in each electrode compartment is calculated as:

$$X_{tot} = X_f + X_m + X_e \tag{A4}$$

This is compared with the value of X_{max}. The washout factor (α) is calculated as:

$$\alpha = \frac{1}{1 + e^{(X_{max} - X_{tot})}}. \tag{A5}$$

Since the electroactive microorganisms grow at the electrode surface, a different (smaller) maximum attainable biomass concentration is assumed for this trophic group ($X_{max,e}$) and the corresponding washout factor is calculated as:

$$\alpha_e = \frac{1}{1 + e^{(X_{maxe} - X_e)}} \tag{A6}$$

S_p, S_s and S_{H_2} material balances account for substrate hydrolysis, biotransformations and transport. For each electrode compartment the S_p balance is described as:

$$\frac{dS_p}{dt} = -\frac{\mu_f}{Y_{\frac{X_f}{S_p}}}X_f + (1 - f_p)(b_f X_f + b_m X_m + b_e X_e) - b_h X_f + D(S_p^{in} - S_p^{in}) \tag{A7}$$

Material balances for S_s account for the growth of electroactive and acetoclastic methanogenic microorganisms:

$$\frac{dS_s}{dt} = -\frac{\mu_{mS}}{Y_{\frac{X_m}{S_s}}}X_m - \frac{\mu_e}{Y_{\frac{X_e}{S_s}}}X_e + b_h X_f + D(S_s^{in} - S_s^{out}) \tag{A8}$$

S_{H_2} material balances are defined differently for the anode and the cathode. In both electrode compartments, the material balance accounts for the growth of hydrogenotrophic methanogens. In the cathode, the production of hydrogen by cathodotrophes is also considered.

$$\frac{dS_{H_2}^a}{dt} = -\frac{\mu_{mH_2}}{Y_{\frac{X_m}{S_{H_2}}}} X_m^a + D\,(S_{H_2}^{in} - S_{H_2}^{out}) \tag{A9}$$

$$\frac{dS_{H_2}^c}{dt} = -\frac{\mu_{mH_2}}{Y_{\frac{X_m}{S_{H_2}}}} X_m^c + \frac{\mu_e}{Y_{\frac{X_e}{S_{H_2}}}} X_e^a + D\,(S_{H_2}^{in} - S_{H_2}^{out}) \tag{A10}$$

CH_4 and CO_2 are considered as the model outputs. In each electrode compartment, CH_4 is produced due to methanogenic activity. Hence, methane production is calculated by combining the acetoclastic and hydrogenotrophic activity. Total CH_4 production is calculated as the sum of produced methane in both electrode compartments.

$$Q_{CH_4} = \left(\frac{\mu_{mH_2}}{Y_{\frac{X_{mH_2}}{CH_4}}} + \frac{\mu_{mS}}{Y_{\frac{X_{mS}}{CH_4}}} \right) V X_m \tag{A11}$$

At the cathode, the calculation of produced CO_2 accounts for its production from S_s and to its consumption by hydrogenotrophic methanogens. At the anode, the calculation of CO_2 also accounts for its production from S_s consumption. Total methane production is calculated as the sum of produced methane in both electrode compartments.

$$Q_{CO_2}^a = \left(\frac{\mu_{mS}}{Y_{\frac{X_{mS}}{CO_2}}} - \frac{\mu_{mH_2}}{Y_{\frac{X_{mH_2}}{CO_2}}} \right) V X_m + \frac{\mu_e}{Y_{\frac{X_e}{CO_2}}} X_e \tag{A12}$$

$$Q_{CO_2}^c = \left(\frac{\mu_{mS}}{Y_{\frac{X_{mS}}{CO_2}}} - \frac{\mu_{mH_2}}{Y_{\frac{X_{mH_2}}{CO_2}}} \right) V X_m \tag{A13}$$

Understanding the Significance of Current Density in Microbial Electrochemical Cells

Hyung-Sool Lee* and **Abid Hussain**

[1] Department of Civil and Environmental Engineering, University of Waterloo 200 University Ave W, Waterloo ON N2L 3G1, Canada

1. Introduction

Microbial electrochemical cells (MxCs) enable us to capture electrons from reduced compounds (e.g., organic compounds) because of the unique metabolism of exoelectrogens. These microorganisms, which include *Geobacter, Shewanella, Escherichia, Aeromonas, Pseudomonas* and so on (Bond and Lovley 2003; Logan and Rabaey 2012; Marsili et al. 2008; Sharma et al. 2016; Velasquez-Orta et al. 2010), are capable of transferring electrons from donor substrate to the anode as the terminal electron sink through extracellular electron transfer (EET). This unique biotechnology can separate and store substrate electrons in circuited coulombs that allow recovery of value-added products from organic waste and wastewater or synthesis of carbon-neutral biochemicals (e.g., acetate, ethanol, etc.) (Logan and Rabaey 2012; Lee et al. 2010). MxCs have been researched for 20 years approximately and have great potentials to energy-efficient waste and wastewater treatment or resource recovery biorefinery. However, full-scale MxCs have not been applied in the field yet. Electric power from microbial fuel cells (MFCs) is too small to be used as renewable energy. Hydrogen gas generated from microbial electrolysis cells (MECs) is a good energy carrier but the use of the hydrogen is limited by market demands (price, production rate, purity, etc.) and infrastructure related to H_2 utilization. Proof-of-concept experiments demonstrated biochemical synthesis using biocathode, but its fundamental understanding and application are still at a nascent stage. How could we engineer MxCs for helping establish a sustainable society? Although value-added products recovered from MxCs are not significant, there is an important phenomenon in MxCs, i.e., anaerobic respiration using electrodes. This feature allows energy-efficient oxidation of reduced contaminants (e.g., biochemical oxygen demand (BOD) in organic waste and wastewater. Considering ~50% of annual operating and maintenance costs originated from aeration in domestic wastewater treatment plants, anaerobic removal of BOD without causing noxious by-products is still an interesting concept. Many works of literature have tested MxCs for organic waste and wastewater treatment (Hussain et al. 2016; 2017; Heidrich et al. 2013; 2014; Nam et al. 2014; Modestra et al. 2015; Lalaurette et al. 2009; Lu et al. 2012a), but there is limited information on fundamental understanding of MxC challenges to organic waste and wastewater treatment and resource recovery. In this review, we discuss MxCs as anaerobic waste and wastewater treatment technologies focusing on current density closely related to BOD removal rate. First, we will discuss the inherent challenges of MFCs to organic waste and wastewater treatment.

*Corresponding author: hyung.will.lee@gmail.com

Second, the significance of the current density of MxCs as an anaerobic treatment technology will be thoroughly reviewed, focusing on biological parameters. Third, we will mechanistically evaluate the implications of biological parameters to improve current density. Fourth, MxC limitations to resource recovery (mainly H_2) and anaerobic treatment will be discussed in line with biofilm kinetics. Finally, we will summarize the perspectives of MxCs to waste and wastewater treatment.

2. Challenges of Electric Power Generation in Microbial Fuel Cells (MFCs)

Recovery of electric power from organic waste and wastewater using MFCs gained tremendous attention after mediatorless MFCs was first proved in late 1990 (Kim et al. 1999a, b). Although many studies have been conducted to improve the powder density of MFCs for ~20 years, electric power is still too small to provide significant benefits to society as renewable electrical energy. Large energy losses (microbial growth, ohmic loss in separator and wastewater and energy losses in electrodes, including activation and concentration energy losses) account for such small electric power, but MFCs have unique features related to small energy different from chemical fuel cells. First, electron donor to MFCs is typically complex organics and can be dissolved in water (i.e., organic waste and wastewater); in comparison, electron donor to chemical fuel cells is a gaseous compound, like H_2 gas. Hence, MFCs face slow mass transport as compared to chemical fuel cells. This also limits the design of stacked MFCs in series that is essential for improving power density (voltage boost-up). Second, heterogeneous composite catalysts that consist of exoelectrogens, non-exoelectrogens and extracellular polymeric substances (EPS) oxidize the organics and transfer electrons to the anode. Third, wastewater that has relatively low conductivity (2-5 mS/cm) is the main electrolyte for MFCs, while selective ion-transfer membranes that have a high ionic conductivity act as electrolytes in chemical fuel cells. Due to these characteristics, MFCs have inherent limitations of kinetics unlike chemical fuel cells.

The pH gradient on electrodes is another inherent challenge of MFCs due to the heterogeneous liquid electrolyte (wastewater) where a variety of ions (Na^+, K^+, Mg^{2+}, Ca^{2+}, Cl^- SO_4^{2-}, etc.) in mM range are present. Metabolic activity of exoelectrogens is well maintained at neutral pH, but it is readily inhibited at acidic pH. Proton concentration close to neutral pH is by a factor of 10,000-20,000 less than other cations; as a result, protons are accumulated in a biofilm anode or an anode chamber of dual chamber MFCs (Chae et al. 2008; Leong et al. 2013; Rozendal et al. 2006; Torres et al. 2008c). To keep neutral pH in the anode, anion exchange membranes transferring hydroxyl ions for charge neutrality have been used for dual chamber MFCs where alkaline conditions (~ pH 11) is maintained in a cathode chamber. This approach efficiently allows neutral pH in the anode, but the pH-driven energy loss on the cathode is significant over 0.2 V (0.056 V energy loss/unit pH). As an alternative, single chamber MFCs equipped with air-diffusion cathodes were examined, but the substantial pH gradient on the cathode (~ 3-4 pH units) decreased voltage as large as 0.3 V (Popat et al. 2012; Torres et al. 2008a).

Generating electric power from organic waste and wastewater is very attractive, but engineering MFCs in large-scale have significant technical challenges that originate from inherent limitations of MFCs. For these reasons, MFCs have been developed as a biosensor to BOD, toxic chemicals or dissolved methane (Chen and Smith 2018; Di-Lorenzo 2016; Gao et al. 2017; Jiang et al. 2018).

3. Significance of Current Density in Microbial Electrochemical Cells (MxCs)

There is a technical challenge of recovering large electric power from organic waste and wastewater with MFCs as discussed above. However, MFCs can be modified to produce other value-added products such as hydrogen peroxide, hydrogen gas, methane gas and biochemicals (acetate, ethanol, etc.) on a cathode that are generally called microbial electrochemical cells (MxCs). We should supply external energy for producing such value-added products that means focusing more on energy-demanding processes against energy-producing MFCs. However, the economic profits of the value-added products from MxCs can be more than small electric power recovered from MFCs on one hand. On the other hand, input electric power given to MxCs would be smaller than that used for aeration in wastewater treatment. To apply MxCs into resource recovery biorefinery or energy-efficient wastewater treatment, we should address at least two requirements. First, minimizing energy losses is essential to improve energy efficiency in MxCs. Second, high current density should be produced in MxCs because current density is proportional to the production rate of value-added products. The profit of value-added products against input electric power to MxCs is uncertain because of immature MxC technology for the recovery and small market for the recovered products. In comparison, the application of MxCs for wastewater treatment is highly possible if high current density proportional to fast COD removal rate can be kept in MxCs.

Equation 1 describes the removal rate of chemical oxygen demand (COD) in bulk liquid by exoelectrogens, typically expressed with dual-limitation kinetics.

$$-\frac{dS_d}{dt} = f_e^o q_{max,app} X_a \frac{S_d}{S_d + K_{sd,app}} \frac{S_{a,OMP}}{S_{a,OMP} + K_{sa,app}} \tag{1}$$

where, S_d is donor substrate (g COD/m³), t is reaction time (d), f_e^o is the fraction of electrons used for catabolism, $q_{max,app}$ is the apparent maximum specific substrate utilization rate (g COD/g VS-d), X_a is the concentration of active exoelectrogens (g VS/m³), $K_{sd,app}$ is the apparent half-saturation concentration of electron donor (g COD/m³), $S_{a,OMP}$ is concentration of an intracellular terminal electron acceptor (i.e., an outer membrane protein) (g COD/m³) and $K_{sa,app}$ is the apparent half-saturation concentration of an intracellular terminal electron acceptor (g COD/m³). We used apparent kinetic parameters here because bioreactors are run with mixed culture in wastewater treatment.

Equation 1 can be rearranged for steady-state biofilm systems where (-dS/dt) term is expressed as flux (g COD/m²-d). This flux is equivalent to current density (A/m² of anode geometric surface area) in biofilm anodes since current (A) is conceptually same to COD removal rate (g COD/d). Hence, Equation 1 becomes:

$$j = 0.14 f_e^o q_{max,app} X_f L_f \frac{S_d}{S_d + K_{sd,app}} \frac{S_{a,OMP}}{S_{a,OMP} + K_{sa,app}} \tag{2}$$

where j is current density (A/m²), X_f is the density of active exoelectrogens in a biofilm anode (g VS/m³), L_f is biofilm thickness (m) and 0.14 is the conversion factor (0.14 A= 1 g COD/d).

The maximum current density is close to ~10 A/m² of anode surface area in several literatures (Lee et al. 2009a; Parameswaran et al. 2013; Torres et al. 2007; Torres et al. 2008b) that is equivalent to 0.072 kg COD/m²; some works reported higher current density (An and Lee 2013; Hu et al. 2009; Ichihashi et al. 2014; Sleutels et al. 2012) but not much different from 10 A/m². Therefore, here, we use 10 A/m² for discussing engineering applications of MxCs in wastewater treatment. Organic loading rate typically ranges from 0.8 to 1.2 kg COD/m³-d in activated sludge processes treating domestic wastewater. Multiple MxCs will be required to have organic loading rates similar to activated sludge, but constructing numerous MxCs will need substantial investment costs and may complicate operation and maintenance of MxCs, such as electrode connection and water pipelines among multiple MxCs (inflow, outflow and circulation). Alternatively, anode modulation that uses multiple anodes in a given anode chamber can economically treat domestic wastewater without significant increase of MxC's scale (An and Lee 2013; Lanas et al. 2014; Dhar et al. 2016). Anode packing density (m² of anode surface area per m³ of anode chamber) required for the organic loading rates (0.8 to 1.2 kg COD/m³-d) is computed at 11-17 m²/m³ in anode modulated MxCs, given that high current density (~10 A/m²) is produced equivalently throughout multiple anodes. It is not challenging to design anodes meeting the required packing density. For instance, the carbon brush developed by Dr. Logan's group can provide ~3,500 m²/m³ (Lanas et al. 2014). Lee and Rittmann (2010) used a bundle of carbon fibers having 2,530 m²/m³. Several literatures modulated anodes improving the anode packing density significantly without causing the volume change (An and Lee 2013;

Cui et al. 2016; Dhar et al. 2015; He et al. 2016; Kim et al. 2010; Lanas et al. 2014; Sim et al. 2015). However, Dhar et al. (2016) proved that increase of anode surface area (or anode packing density) does not simply increase current density and, accordingly, COD removal rate in MxCs because current density is not equally produced in individual anodes of a multiple-anode MxC. They showed that ohmic energy loss changed the electric potential of each anode module and affected biofilm community in each anode. As a result, different biofilm communities did not produce current density equally in anode modules, and the improvement of current density was limited in the multiple-anode MxC. This literature indicates the limitation of engineering design and configuration of MxCs for improving current density on one hand. On the other hand, Dhar et al. (2016) support the significance of biological parameters related to current density and COD removal rate in biofilm anodes.

4. Limiting Parameters to Current Density

Literatures have reported that many factors can affect current density in MxCs, directly or indirectly, that include substrate type, substrate concentration, biofilm community, biofilm thickness, pH (or buffer concentration), anode potential, ohmic resistance, configuration of MxCs, electrode type and so on (Chae et al. 2009; Dhar and Lee 2014; Hussain et al. 2017; Kadier et al. 2014; Kiely et al. 2011; Lee et al. 2016; Lu et al. 2012c; Nam et al. 2011; Sleutels et al. 2009). Literature seems to indicate that most environmental and operating parameters could influence current density, complicating comprehension of limiting factors to current density. Here, we mechanistically analyze influential parameters to current density, mainly focusing on anode kinetics, based on intracellular electron transfer (IET) and extracellular electron transfer (EET) kinetics. IET means electron transfer from an electron donor to an intracellular terminal electron acceptor, such as outer membrane proteins (Lee et al. 2016; Torres et al. 2010). EET indicates electron transfer from the outer membrane proteins to the anode. Figure 1 shows a schematic of IET and EET from the donor substrate to the anode for exoelectrogen.

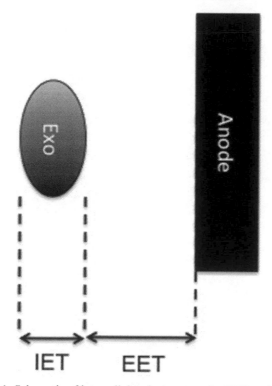

Figure 1: Schematic of intracellular electron transfer (IET) and extracellular electron transfer (EET) for exoelectrogen. Exo: exoelectrogens.

Hence, the current density is the function of two-step kinetics of IET and EET (Lee et al. 2009a, 2010, 2016; Torres et al. 2010). The complex electron transfer can be simplified to IET at anode-polarized conditions (Lee et al. 2009a, 2016; Torres et al. 2008b, 2010) where EET kinetics is saturated to current density (see Equation 2). In addition, $S_{a,OMP}$ term can be also saturated to form unity at anode-polarized conditions, and IET kinetics in Equation 2 becomes Equation 3.

$$j = 0.14 f_e^o q_{max,app} X_f L_f \frac{S_d}{S_d + K_{sd,app}} \tag{3}$$

Equation 3 shows that six parameters of f_e^o, $q_{max,app}$, X_f, L_f, S_d and $K_{sd,app}$ are involved in current density. The parameters of f_e^o, $q_{max,app}$, X_f, L_f and $K_{sd,app}$ are related to exoelectrogen biofilms. S_d term typically represents BOD concentration in MxC effluent, so S_d is almost fixed term. The term f_e^o indicates fraction of electrons used for catabolism directly related to coulombic efficiency in MxCs. Hence, the type of exoelectrogens and their population in a biofilm anode (i.e. X_f) determine f_e^o.

Table 1 summarizes coulombic efficiency of MxCs closely associated with f_e^o term. The current from substrate oxidation is much larger than decay current, and thus, f_e^o would be a range of 0.45-0.95, almost doubling current density at the maximum. This explains that high f_e^o does not only mean more energy recovery but also higher current density in MxCs (see Equation 3). The f_e^o term is closely related to X_f because non-exoelectrogens divert substrate electrons to electron sinks other than the anode. For instance, current density and energy efficiency decrease in an MxC fed with fermentable substrate as compared to non-fermentable substrate due to X_f decrease (e.g., outgrowth of methanogens and fermenters) (Call et al. 2009; Catal et al. 2015; Gil-Carrera et al. 2013b; Heidrich et al. 2014; Hussain et al. 2017; Lee et al. 2008; Lu et al. 2012b, c; Miceli et al. 2014). To maintain high X_f and f_e^o in biofilm anodes, we should operate MxCs at favorable conditions for exoelectrons to outcompete fermenters and methanogens i.e., simple forms of substrate (non-fermentable substrate), negative anode potential, neutral pH, high alkalinity and so on (Dhar and Lee 2014; Lee et al. 2008; Lee and Rittmann 2010; Lee et al. 2009a, b, 2010).

Table 1: Coulombic efficiency (CE) and current density in pure culture MxCs

Inoculum	Electron donor	Current density	CE	Reference
Geobacter Sulfurreducens	Acetate	160 A/m^3	82±8%	(Call et al. 2009)
Geobacter Metallireducens	Acetate	110 A/m^3	78±5%	(Call et al. 2009)
Thermincola Ferriacetica	Acetate	7-8 A/m^2	93%	(Parameswaran et al. 2013)
Desulfovibrio Strain - G11	Lactate	0.17-0.76 A/m^2	45%	(Croese et al. 2011)
Geoalkalibacter spp.	Acetate	8.3 A/m^2	95%	(Badalamenti et al. 2013)
Rhodoferax ferrireducens	Glucose	31 mA/m^2	83%	(Chaudhuri and Lovley 2003)
Rhodopseudomonas palustris DX-1	Acetate	1 A/m^2	~60%	(Xing et al. 2008)

At steady-state condition for S_d, high $q_{max,app}$ and low $K_{sd,app}$ values can increase current density as described in Equation 3. Keeping high $q_{max,app}$ and low $K_{sd,app}$ can be different from maintaining high X_f and f_e^o terms in a biofilm anode. In other words, the enrichment of exoelectrons in a biofilm anode cannot simply result in high $q_{max,app}$ and low $K_{sd,app}$ since the kinetic parameters could be deviated in exoelectrogens. Table 2 summarizes q_{max} and K_s for *Geobacter sulfurreducens*, *Geobacter enriched cultures and Shewanella oneidensis* MR-1; there is limited information on Monod kinetic parameters for exoelectrogens, so we discuss the implication of kinetic parameters only with Table 2. $K_{s,app}$ values range from 0.64 to 820 mg COD/L, changing current density because of $S_d/(S_d + K_{s,app})$ term in Equation 3. The $q_{max,app}$ term seems consistent at 20.9-25.6 mg COD/mg VS-d. In comparison, $q_{max} X_f$ is largely deviated from 9.2×10^4 to 2.10×10^6 although *Geobacter* genus was dominant in biofilm anodes, potentially leading to ~23 times higher or lower current density in MxCs.

Table 2: Apparent Monod kinetic parameters of exoelectrogens in biofilm anodes

Culture	$K_{s,app}$ (mgCOD/L)	$q_{max}X_f$ (mgCOD/L-d)	$q_{max,app}$ (mgCOD/ mgVS-d)	Reference
Geobacter sulfurreducens	0.64[a]		20.9 to 22.7[a]	(Esteve-Núñez et al. 2005)
Shenwanella oneidensis MR-1	820[a]		25.6[a]	(Tang et al. 2007)
Geobacter enrichment culture	119[a]	1.12×10^6	22.3[a]	(Lee et al. 2009a)
Thermincola ferriacetica	-	2.10×10^6	-	(Parameswaran et al. 2013)
Geobacter enrichment culture	168	1.26×10^5	-	(Dhar et al. 2016)
Geobacter enrichment culture	156	6.4×10^5	-	(Dhar et al. 2016)
Geobacter enrichment culture	274	9.2×10^4	-	(Lee et al. 2016)

[a] True values. Acetate was commonly used for electron donor in these literatures.

Due to limited information on kinetic parameters of exoelectrogens, we cannot conclude *Geobacter sulfurreducens* is the most kinetically efficient exoelectrogens. However, many literatures have commonly reported predominance of *Geobacter* genus in biofilm anodes generating high current density (Call et al. 2009; Commault et al. 2013; Dhar et al. 2016; Rotaru et al. 2015; Lee et al. 2009a, 2010; Torres et al. 2008b, 2010) and *Geobacter* enrichment shows the best kinetic feature in biofilm anodes (Lee et al. 2009a; Torres et al. 2008b, 2010). Hence, selecting and enriching *Geobacter* genus in a biofilm anode is one means to keep high q_{max} and low K_s for improving current density; the enrichment of exoelectrogens having poor kinetics will end up with low to moderate current density in MxCs despite having high coulombic efficiency.

Procedures for enriching *Geobacter spp.* in a biofilm anode from mixed culture inocula are not well established, but many literatures have commonly reported the significance of anode potential for proliferation of *Geobacter* genus in the biofilm. Negative anode potential close to -0.05 to -0.15 V (vs. standard hydrogen electrode) well proliferates *Geobacter spp.* in biofilm anodes (Commault et al. 2013; Torres et al. 2008b, 2009). In comparison, biofilm community becomes very diverse at positive anode potential (Kumar et al. 2013; Torres et al. 2009; Dhar et al. 2016). Anode potential is a significant, controllable factor for enriching *Geobacter* in biofilm anodes. Hence, controlling anode potential with a potentiostat or equivalent devices will be important for keeping high $q_{max,app}$ and low $K_{sd,app}$ in biofilm anodes and consequently generating high current density in MxCs.

Equation 3 shows that biofilm thickness (L_f) also influences current density. Increasing L_f does not increase current density proportionally since large L_f can decrease X_f and the density of active exoelectrogens in a biofilm anode. In a thick biofilm anode proton accumulation can readily acidify inner biofilms. Several literatures have reported proton build-up in inner biofilms where exoelectrogens are dead or metabolically inactive (Babauta et al. 2011; Dhar et al. 2017; Marcus et al. 2011; Torres et al. 2008c). For this reason, L_f should be optimized for maximizing X_f to given wastewater having relatively constant buffer concentration. In addition, substrate gradient in a thick biofilm anode may stimulate diversification of exoelectrogens (i.e., oligotrophs and non-oligotrophs).

The type of substrate is another factor influencing biofilm community. As discussed above, the non-fermentable substrate (e.g., acetate) can stimulate the outgrowth of exoelectrogens affecting f_e^o and finally kinetic parameters in biofilm anodes. In addition to the microbial competition, particulates in waste and wastewater can compete with exoelectrogens for space in a biofilm anode or cause anode clogging (Dhar and Lee 2014), hence, the wastewater containing a high concentration of suspended solids (SS) and solid waste are not ideal for MxCs. Building the syntrophic interaction between fermenters and exoelectrogens is essential for treating such waste and wastewater with hybrid MxCs or MxCs in combination with other bioreactors (Cui et al. 2017; Cui et al. 2014; Dhar et al. 2015; Gao et al. 2014). Alternatively, MxCs need pre-treatment of the waste and wastewater instead of direct application of MxCs as reported in literatures (Choi and Ahn 2014; Jiang et al. 2010; Zhang et al. 2017)

5. MxC Application to Organic Waste and Wastewater Treatment

The hydrogen production rate in MECs is sluggish as compared to dark fermentative biohydrogen. The volumetric hydrogen production rate is as high as ~8 m^3 H_2/m^3-d in dark fermentation (Lee et al. 2010), while it is mostly less than 1 m^3 H_2/m^3-d in MECs (see Table 3). Configuration of reactors or electrodes, operating conditions, inocula and type of substrate do not improve the hydrogen production rate much as shown in Table 3. The highest volumetric current density is ~200 A/m^3 equivalent to 2.2 m^3 H_2/m^3-d assuming full recovery of coulombs as hydrogen gas, but the observed hydrogen production rate is less than 1 m^3 H_2/m^3-d probably due to H_2 recycle, H_2 loss or both (Lee and Rittmann 2010; Lee et al. 2009a, b). Volumetric current density is useful for the design of MxCs but not ideal for understanding fundamental kinetics of anode respiration described in Equation 3. Given that the maximum current density ~10 A/m^2 of anode surface area is generated in multiple-anode MxCs (i.e., anode modulation), the anode packing density required for producing 8 m^3 H_2/m^3-d is only 74 m^2/m^3. We can easily achieve this anode packing density with commercial carbon electrodes as discussed above that supports the significance of biological kinetic parameters in biofilm anodes for MxC commercialization as a renewable biohydrogen technology.

It seems that MxCs do not work well as anaerobic wastewater treatment, mainly due to small current density as shown in Table 3. For dilute wastewater treatment (domestic wastewater), relatively long hydraulic retention time (HRT) is needed despite moderate COD removal. COD removal of ~90% was observed but HRT is as long as 108 hours. Hence, MxCs cannot compete with existing activated sludge processes or other anaerobic technologies (e.g., anaerobic membrane bioreactors) for municipal wastewater treatment. Similar patterns are consistently observed for MxC application to high strength organic wastewater treatment, such as food wastewater, byproduct streams from fermentation or other bioenergy processes. The poor performance of MxCs was also found in direct treatment of waste activated sludge that means that current MxCs are not standalone treatment processes to dilute high-strength wastewater and even organic solids. Together with the engineering approach to electrodes and reactors, we should revisit biofilm anode kinetics and optimize them to given waste or wastewater in order to increase their current density.

6. Future Perspectives

Considering costs, infrastructures and market demands for recovered products, (i.e., H_2), the realization of MxC biorefinery from organic waste and wastewater would need a long period. It seems realistic to drive MxC commercialization as an energy-efficient waste and wastewater treatment technology by maximizing MxC benefits of (1) no aeration, (2) less sludge production and (3) reuse of recovered products. For that, we should engineer biofilm anodes to improve exoelectrogen's kinetics. Using a potentiostat will be simple but important to enrich kinetically efficient exoelectrogens in biofilm anodes that can have high f_e^o, high $q_{max,app}$, high X_f and low $K_{sd\,app}$. There will be large markets for MxC-based organic waste and wastewater treatment if MxCs can generate high current density from organic waste and wastewater.

For dilute wastewater treatment, MxCs should improve effluent quality and addition of micro-filtration or ultra-filtration to MxCs would be one option to improve final effluent quality; however, the removal of nutrients is still limited. Nitrogen control will be very challenging, while coagulants can readily control phosphorus. Nitritation and anammox processes can be supplemented as post-treatment to MxCs for nitrogen removal. As an alternative to anammox, membrane biofilm reactors

Table 3: Performance of MxCs on wastewater and organic waste

Waste type	Influent COD concentration (g/L)	Current density A/m³	Coulombic efficiency (%)	HRT (h)	Applied voltage (V)	Temperature (°C)	COD removal (%)	Hydrogen production rate LH₂/L-d	Reference
Low Strength Wastewater (WW)									
Domestic WW	0.204-0.481	0.04$^{\delta}$	26	108	0.2-0.6	30	90	0.4§	(Ditzig et al. 2007)
Domestic WW	0.486	60	38	6	1	30	51	0.3	(Escapa et al. 2012)
Domestic WW	0.27	18.6	23	10	0.9	25	76	0.05	(Gil-Carrera et al. 2013a)
Domestic WW	0.160	14.3	64	24	0.9	30	58	0.28	(Cusick et al. 2010)
Domestic WW	0.4-0.5	9	22	14	0.6	25	74	0.06§§	(Hussain et al. 2017)
Domestic WW	0.12-0.14	3	10	24	0.6	25	20	0.02§§	(Hussain et al. 2016)
High Strength WW									
Industrial WW	3.81	2.1$^{\delta\delta}$	12	46	0.7	30	90	0.58	(Tenca et al. 2013)
Potato Processing WW	1.9-2.5	13-14	80	48	0.9	30	79	0.74	(Kiely et al. 2011)
Food Processing WW	1.83	2.5$^{\delta\delta}$	35	150	0.7	30	67	0.35	(Tenca et al. 2013)
Dairy WW	1.1	24	66$^{\#}$		0.95	25	NA	0.03§§§	(Hou et al. 2015)
Winery WW	0.58	18 - 20	50	72	0.9	30	47	0.17	(Cusick et al. 2010)
Swine WW	2	106	70	184	0.5	30	75	1	(Wagner et al. 2009)
Pilot Scale – MEC									
Domestic WW	0.1-0.45	0.3$^{\delta\delta}$	55	24	0.6-1.1	13.5-21	34	0.015	(Heidrich et al. 2013)
Domestic WW	0.1-1.9	NA	41	24	0.6-1.1	1-22	66	0.007	(Heidrich et al. 2014)
Domestic WW	0.1-0.3	36$^{\delta\delta\delta}$	11	30	0.2€	25	67	NA	(Brown et al. 2014)
Winery WW	8	7.4	NA	24	0.9	31	62	0.19	(Cusick et al. 2011)

(Contd.)

Table 3: *(Contd.)*

Waste type	Influent COD concentration (g/L)	Current density A/m³	Coulombic efficiency (%)	HRT (h)	Applied voltage (V)	Temperature (°C)	COD Removal (%)	Hydrogen production rate LH$_2$/L-d	Reference
Fermentation Effluent									
Cellulose Fermentation	4-6	52	57		0.058	25	15-20	0.43	(Wang et al. 2011)
Spent Wash Fermentation	8§	NA	NA	48	0.6	28	68	0.015	(Modestra et al. 2015)
Corn Strover Fermentation	1.5-2.5	1.15**	NA	NA	0.5	NA	65	1	(Lalaurette et al. 2009)
Cellubiose Fermentation	1.5-2.5	1.15**	NA	NA	0.5	NA	65	0.2	(Lalaurette et al. 2009)
Sugar Beet Juice	4-6	3.6$^{\delta\delta}$	70	480	0.4€	25	12-20	0.96	(Dhar et al. 2015)
Molasses fermentation WW	6.5	158	87	20	0.6	25	NA	1.41	(Lu et al. 2009)
Crude Glycerol Fermentation	3.8-4	34.8	24	NA	1	RT	40.6	0.05	(Chookaew et al. 2014)
Lignocellulosic Biorefinery Byproducts									
De-oiled Refinery WW	0.4 - 1	2.1$^{\delta\delta}$	NA	120	0.7	30	79	NA	(Ren et al. 2013)
Cellulose Fermentation WW	0.70-0.9	34-36$^{C\delta\delta}$	65-70	24	0.6-1.2	30	76	0.49	(Nam et al. 2014)
Hydrolysates of Wheat Straw	1.8-4	100 -200	600-80	NA	0.7	25	62	0.61-1	(Thygesen et al. 2011)
Waste Activates Sludge (WAS)									
Raw WAS	0.396	8	28	210	0.8	19	22	0.056	(Lu et al. 2012a)
Alkaline Pre-treated WAS	2.47	129	30	210	0.8	19	28	0.91	(Lu et al. 2012a)
Fermented WAS Effluent	2-6	8-9$^{\delta\delta}$	73	48	0.8	22	50	1.7	(Liu et al. 2012)

Expressed in $^\delta$mA/cm²; $^{\delta\delta}$A/m2; $^{\delta\delta\delta}$μA/cm²; $^\epsilon$Voltage expressed Vs Ag/AgCl; § expressed in LH$_2$/gCOD; §§CH$_4$ yield – L/gCOD; §§§CH$_4$ yield – L/L-d; **Maximum current density on synthetic feed; §VFA concentration; # energy efficiency using hydrogen gas produced from MxCs can reduce nitrite to dinitrogen (i.e., MxCs-nitritation-MBfRs). MxCs can be also used as BOD scavengers as polishing processes to existing treatment systems for high strength organic wastewater, like industrial wastewater, to meet surcharge regulations (e.g., <BOD 300 mg/L). MxC application to organic solid treatment [e.g., anaerobic digestion (AD)] had been tested that consistently showed small improvement of VS reduction. However, recent works have reported an increase of methane yield or VS reduction in AD engineering direct interspecies electron transfer (DIET) (Zhao et al. 2018; Peng et al. 2018), after DIET between *Geobacter* and *Methanosaeta* was reported by Dr. Lovely's group (Rotaru et al. 2013). Extensive studies are required to prove and comprehend DIET in AD. In microbiological viewpoint, it will be very interesting to understand complicated syntrophy among fermenters, acetogens, exoelectrogens and methanogens involving DIET pathways. In an engineering point of view, improving the performance of AD simply by adding conductive materials or immersing MxCs into existing AD systems is very attractive that can open new avenues for MxCs. It is expected that a variety of engineered MxCs will be designed and applied for organic waste and wastewater treatment in the near future, but we should be reminded that production of high current density (i.e., anode biofilm kinetics) is key for the success of MxC-based organic waste and wastewater treatment.

7. Conclusions

This study reviewed MxC application to organic waste and wastewater treatment, focusing on current density and related kinetic parameters. Given that EET would not limit current density in biofilm anodes, six parameters of f_e^o, $q_{max,app}$, X_f, L_f, S_d and $K_{sd,app}$ mainly affect current density. Enrichment of exoelectrogens can increase f_e^o and X_f, but it does not simply improve current density because of other limitations in $q_{max,app}$, $K_{sd,app}$ and L_f. It seems that *Geobacter* spp. is one of the most kinetically efficient exoelectrogens and the control of anode potential is one means to enrich *Geobacter* genus in biofilm anodes. Thick L_f mathematically increases current density, but X_f term can decrease at thick L_f due to acidification of inner biofilms, hence, L_f should be optimized. The current MxCs cannot meet requirements as a standalone wastewater treatment technology mainly due to poor treatability and resource recovery rate related to small current density. To achieve current density applicable to field application, the biological parameters of f_e^o, $q_{max,app}$, X_f and L_f should be optimized in a biofilm anode, along with engineering approach.

Acknowledgement

This review is dedicated to Moungjin, Hyoseo and Will.

References

An J, Lee HS (2013) Implication of endogenous decay current and quantification of soluble microbial products (SMP) in microbial electrolysis cells. RSC Adv 3: 14021-14028.

Babauta JT, Nguyen HD, Beyenal H (2011) Redox and pH microenvironments within *Shewanella oneidensis* MR-1 biofilms reveal an electron transfer mechanism. Environ Sci Technol 45: 6654-6660.

Badalamenti JP, Krajmalnik-Brown R, Torres CI (2013) Generation of high current densities by pure cultures of anode-respiring *Geoalkalibacter* spp. under alkaline and saline conditions in microbial electrochemical cells. mBio 4:e00144-13.

Bond DR, Lovley DR (2003) Electricity production by *Geobacter sulfurreducens* attached to electrodes. Appl Environ Microbiol 69: 1548-1555.

Brown RK, Harnisch F, Wirth S, Wahlandt H, Dockhorn T, Dichtl N, Schröder U (2014) Evaluating the effects of scaling up on the performance of bioelectrochemical systems using a technical scale microbial electrolysis cell. Biores Technol 163: 206-213.

Call DF, Wagner RC, Logan BE (2009) Hydrogen production by Geobacter species and a mixed consortium in a microbial electrolysis cell. Appl Environ Microbiol 75: 7579-7587.

Catal T, Lesnik KL, Liu H (2015) Suppression of methanogenesis for hydrogen production in single-chamber microbial electrolysis cells using various antibiotics. Biores Technol 187:77-83.

Chae KJ, Choi MJ, Lee JW, Kim KY, Kim IS (2009) Effect of different substrates on the performance, bacterial diversity and bacterial viability in microbial fuel cells. Biores Technol 100: 3518-3525.

Chae KJ, Choi M, Ajayi FF, Park W, Chang, IS, Kim IS (2008) Mass transport through a proton exchange membrane (nafion) in microbial fuel cells. Energy & Fuels 22: 169-176.

Chaudhuri SK, Lovley DR (2003) Electricity generation by direct oxidation of glucose in mediatorless microbial fuel cells. Nature Biotech 21: 1229.

Chen S,Smith AL (2018) Methane-driven microbial fuel cells recover energy and mitigate dissolved methane emissions from anaerobic effluents. Environ Sci: Water Res Technol 4: 67-79.

Choi J, Ahn Y (2014) Increased power generation from primary sludge in microbial fuel cells coupled with prefermentation. Bioprocess Biosyst Eng 37: 2549-2557.

Chookaew T, Prasertsan P, Ren ZJ (2014) Two-stage conversion of crude glycerol to energy using dark fermentation linked with microbial fuel cell or microbial electrolysis cell. New Biotechnol 31: 179-184.

Commault AS, Lear G, Packer MA,Weld RJ (2013) Influence of anode potentials on selection of *Geobacter* strains in microbial electrolysis cells. Biores Technol 139: 226-234.

Croese E, Pereira MA, Euverink GJW, Stams AJM, Geelhoed JS (2011) Analysis of the microbial community of the biocathode of a hydrogen-producing microbial electrolysis cell. Appl Microbiol Biotechnol 92: 1083-1093.

Cui D, Cui MH, Lee HS, Liang B, Wang HC, Cai WW, Cheng HY, Zhuang XL, Wang AJ (2017) Comprehensive study on hybrid anaerobic reactor built-in with sleeve type bioelectrocatalyzed modules. Chemical Eng J 330: 1306-1315.

Cui D, Guo YQ, Lee HS, Wu WM, Liang B, Wang AJ, Cheng, HY (2014) Enhanced decolorization of azo dye in a small pilot-scale anaerobic baffled reactor coupled with biocatalyzed electrolysis system (ABR–BES): A design suitable for scaling-up. Biores Technol 163:254-261.

Cui MH, Cui D, Lee HS, Liang B, Wang AJ, Cheng HY (2016) Effect of electrode position on azo dye removal in an up-flow hybrid anaerobic digestion reactor with built-in bioelectrochemical system. Sci Reports 6: 25223.

Cusick RD, Bryan B, Parker DS, Merrill MD, Mehanna M, Kiely PD, Liu G, Logan BE (2011) Performance of a pilot-scale continuous flow microbial electrolysis cell fed winery wastewater. Appl Microbiol Biotechnol 89: 2053-2063.

Cusick RD, Kiely PD, Logan BE (2010) A monetary comparison of energy recovered from microbial fuel cells and microbial electrolysis cells fed winery or domestic wastewaters. Int J Hydrogen Energy 35: 8855-8861.

Dhar BR (2016) Extracellular electron transfer in microbial electrochemical cells, University of Waterloo, Waterloo, ON, Canada.

Dhar BR, Elbeshbishy E, Hafez H, Lee HS (2015) Hydrogen production from sugar beet juice using an integrated biohydrogen process of dark fermentation and microbial electrolysis cell. Biores Technol 198: 223-230.

Dhar BR, Lee HS (2014) Evaluation of limiting factors for current density in microbial electrochemical cells (MXCs) treating domestic wastewater. Biotechnol Reports 4: 80-85.

Dhar BR, Ryu H, Santo Domingo JW, Lee HS (2016) Ohmic resistance affects microbial community and electrochemical kinetics in a multi-anode microbial electrochemical cell. J Power Sources 331: 315-321.

Dhar BR, Sim J, Ryu H, Ren H, Santo Domingo JW, Chae J, Lee HS (2017) Microbial activity influences electrical conductivity of biofilm anode. Water Res 127: 230-238.

Di-Lorenzo M (2016) Use of microbial fuel cells in sensors. In: K Scott, EH Yu (eds) Microbial electrochemical and fuel cells, Chapter 11. Woodhead Publishing, Boston, USA, pp 341-356.

Ditzig J, Liu H, Logan BE (2007) Production of hydrogen from domestic wastewater using a bioelectrochemically assisted microbial reactor (BEAMR). Int J Hydrogen Energy 32: 2296-2304.

Escapa A, Gil-Carrera L, García V, Morán A (2012) Performance of a continuous flow microbial electrolysis cell (MEC) fed with domestic wastewater. Biores Technol 117: 55-62.

Esteve-Núñez A, Rothermich M, Sharma M, Lovley D (2005) Growth of *Geobacter sulfurreducens* under nutrient-limiting conditions in continuous culture. Environ Microbiol 7: 641-648.

Gao Y, Ryu H, Rittmann BE, Hussain A, Lee HS (2017) Quantification of the methane concentration using anaerobic oxidation of methane coupled to extracellular electron transfer. Biores Technol 241: 979-984.

Gao Y, Ryu H, Santo Domingo JW, Lee HS (2014) Syntrophic interactions between H_2-scavenging and anode-respiring bacteria can improve current density in microbial electrochemical cells. Biores Technol 153: 245-253.

Gil-Carrera L, Escapa A, Mehta P, Santoyo G, Guiot SR, Morán A, Tartakovsky B (2013a) Microbial electrolysis cell scale-up for combined wastewater treatment and hydrogen production. Biores Technol 130: 584-591.

Gil-Carrera L, Escapa A, Moreno R, Morán A (2013b) Reduced energy consumption during low strength domestic wastewater treatment in a semi-pilot tubular microbial electrolysis cell. JEnviron Management 122: 1-7.

He W, Wallack MJ, Kim K, Zhang, X, Yang W, Zhu X, Feng Y, Logan BE (2016) The effect of flow modes and electrode combinations on the performance of a multiple module microbial fuel cell installed at wastewater treatment plant. Water Res 105: 351-360.

Heidrich ES, Dolfing J, Scott K, Edwards SR, Jones C, Curtis TP (2013) Production of hydrogen from domestic wastewater in a pilot-scale microbial electrolysis cell. Appl Microbiol Biotechnol 97: 6979-6989.

Heidrich ES, Edwards SR, Dolfing J, Cotterill SE, Curtis TP (2014) Performance of a pilot scale microbial electrolysis cell fed on domestic wastewater at ambient temperatures for a 12 month period. Biores Technol 173: 87-95.

Hou Y, Zhang R, Luo H, Liu G, Kim Y, Yu S, Zeng J (2015) Microbial electrolysis cell with spiral wound electrode for wastewater treatment and methane production. Process Biochem 50: 1103-1109.

Hu H, Fan Y, Liu H (2009) Hydrogen production in single-chamber tubular microbial electrolysis cells using non-precious-metal catalysts. Int J Hydrogen Energy 34: 8535-8542.

Hussain A, Lebrun FM, Tartakovsky B(2017) Removal of organic carbon and nitrogen in a membraneless flow-through microbial electrolysis cell. Enzyme Microbial Technol 102: 41-48.

Hussain A, Manuel M, Tartakovsky B (2016) A comparison of simultaneous organic carbon and nitrogen removal in microbial fuel cells and microbial electrolysis cells. J Environ Management 173: 23-33.

Ichihashi O, Vishnivetskaya TA, Borole AP (2014) High performance bioanode development for fermentable substrates via controlled electroactive biofilm growth. ChemElectroChem 1: 1940-1947.

Jiang JQ, Zhao QL, Wang K, Wei LL, Zhang GD, Zhang JN (2010) Effect of ultrasonic and alkaline pretreatment on sludge degradation and electricity generation by microbial fuel cell. Water Sci Technol 61: 2915-2921.

Jiang Y, Yang X, Liang P, Liu P, Huang X (2018) Microbial fuel cell sensors for water quality early warning systems: Fundamentals, signal resolution, optimization and future challenges. Renew Sustain Energy Rev 81: 292-305.

Kadier A, Simayi Y, Kalil MS, Abdeshahian P, Hamid AA (2014) A review of the substrates used in microbial electrolysis cells (MECs) for producing sustainable and clean hydrogen gas. Renew Energy 71: 466-472.

Kiely PD, Cusick R, Call DF, Selembo PA, Regan JM, Logan BE (2011) Anode microbial communities produced by changing from microbial fuel cell to microbial electrolysis cell operation using two different wastewaters. Biores Technol 102: 388-394.

Kim BH, Ikeda T, Park HS, Kim HJ, Hyun MS, Kano K, Takagi K Tatsumi H (1999a) Electrochemical activity of an Fe(III)-reducing bacterium, *Shewanella putrefaciens* IR-1, in the presence of alternative electron acceptors. Biotechnol Techniques 13: 475-478.

Kim BH, Kim HJ, Hyun MS, Park DH (1999b) Direct electrode reaction of Fe(III)-reducing bacterium, *Shewanella putrefaciens*. J Microbiol Biotechnol 9: 127-131.

Kim JR, Premier GC, Hawkes FR, Rodríguez J, Dinsdale RM, Guwy AJ (2010) Modular tubular microbial fuel cells for energy recovery during sucrose wastewater treatment at low organic loading rate. Biores Technol 101: 1190-1198.

Kumar A, Siggins A, Katuri K, Mahony T, O'Flaherty V, Lens P, Leech D (2013) Catalytic response of microbial biofilms grown under fixed anode potentials depends on electrochemical cell configuration. Chemical Eng J 230: 532-536.

Lalaurette E, Thammannagowda S, Mohagheghi A, Maness PC, Logan BE (2009) Hydrogen production from cellulose in a two-stage process combining fermentation and electrohydrogenesis. Int J Hydrogen Energy 34: 6201-6210.

Lanas V, Ahn Y, Logan BE (2014) Effects of carbon brush anode size and loading on microbial fuel cell performance in batch and continuous mode. J Power Sources 247: 228-234.

Lee HS, Dhar BR, An J, Rittmann BE, Ryu H, Santo Domingo, JW, Ren H, Chae J (2016) The roles of biofilm conductivity and donor substrate kinetics in a mixed-culture biofilm anode. Environ Sci Technol 50: 12799-12807.

Lee HS, Parameswaran P, Kato-Marcus A, Torres CI, Rittmann BE (2008) Evaluation of energy-conversion efficiencies in microbial fuel cells (MFCs) utilizing fermentable and non-fermentable substrates. Water Res 42: 1501-1510.

Lee HS, Rittmann BE (2010) Significance of biological hydrogen oxidation in a continuous single-chamber microbial electrolysis cell. Environ Sci Technol 44: 948-954.

Lee HS, Torres CI, Rittmann BE (2009a) Effects of substrate diffusion and anode potential on kinetic parameters for anode-respiring bacteria. Environ Sci Technology 43: 7571-7577.

Lee HS, Torres CI, Parameswaran P, Rittmann BE (2009b) Fate of H_2 in an upflow single-chamber microbial electrolysis cell using a metal-catalyst-free cathode. Environ Sci Technol 43: 7971–7976

Lee HS, Vermaas WFJ, Rittmann BE (2010) Biological hydrogen production: prospects and challenges. Trends Biotechnol 28: 262-271.

Leong JX, Daud WRW, Ghasemi M, Liew KB, Ismail M (2013) Ion exchange membranes as separators in microbial fuel cells for bioenergy conversion: A comprehensive review. Renew Sust Energy Rev 28: 575-587.

Liu W, Huang S, Zhou A, Zhou G, Ren N, Wang A, Zhuang G (2012) Hydrogen generation in microbial electrolysis cell feeding with fermentation liquid of waste activated sludge. Int J Hydrogen Energy 37: 13859-13864.

Logan BE, Rabaey K (2012) Conversion of wastes into bioelectricity and chemicals by using microbial electrochemical technologies. Science 337: 686-690.

Lu L, Ren N, Xing D, Logan BE (2009) Hydrogen production with effluent from an ethanol–H_2-coproducing fermentation reactor using a single-chamber microbial electrolysis cell. Biosensors Bioelectronics 24: 3055-3060.

Lu L, Xing D, Liu B, Ren N (2012a) Enhanced hydrogen production from waste activated sludge by cascade utilization of organic matter in microbial electrolysis cells. Water Res 46: 1015-1026.

Lu L, Xing D, Ren N (2012b) Bioreactor performance and quantitative analysis of methanogenic and bacterial community dynamics in microbial electrolysis cells during large temperature fluctuations. Environ Sci Technol 46: 6874-6881.

Lu L, Xing D, Ren N (2012c) Pyrosequencing reveals highly diverse microbial communities in microbial electrolysis cells involved in enhanced H2 production from waste activated sludge. Water Res 46: 2425-2434.

Marcus AK, Torres CI, Rittmann BE (2011) Analysis of a microbial electrochemical cell using the proton condition in biofilm (PCBIOFILM) model. Biores Technol 102:253-262.

Marsili E, Baron DB, Shikhare ID, Coursolle D, Gralnick JA, Bond DR (2008) *Shewanella* secretes flavins that mediate extracellular electron transfer. Proc Nat Acad Sci 105: 3968-3973.

Miceli JF, Garcia-Peña I, Parameswaran P, Torres CI, Krajmalnik-Brown R (2014) Combining microbial cultures for efficient production of electricity from butyrate in a microbial electrochemical cell. Biores Technol 169: 169-174.

Modestra JA, Babu ML, Mohan SV (2015) Electro-fermentation of real-field acidogenic spent wash effluents for additional biohydrogen production with simultaneous treatment in a microbial electrolysis cell. Separat Purificat Technol 150: 308-315.

Nam JY, Tokash JC, Logan BE (2011) Comparison of microbial electrolysis cells operated with added voltage or by setting the anode potential. Int J Hydrogen Energy 36: 10550-10556.

Nam JY, Yates MD, Zaybak Z, Logan BE (2014) Examination of protein degradation in continuous flow, microbial electrolysis cells treating fermentation wastewater. Biores Technol 171: 182-186.

Parameswaran P, Bry T, Popat SC, Lusk BG, Rittmann BE, Torres CI (2013) Kinetic, electrochemical, and microscopic characterization of the thermophilic, anode-respiring bacterium *Thermincola ferriacetica*. Environ Sci Technol 47:4934-4940.

Peng H, Zhang Y, Tan D, Zhao Z, Zhao H, Quan X (2018). Roles of magnetite and granular activated carbon in improvement of anaerobic sludge digestion. Biores Technol 249: 666-672

Popat SC, Ki D, Rittmann BE, Torres CI (2012) Importance of OH− transport from cathodes in microbial fuel cells. ChemSusChem 5: 1071-1079.

Ren L, Siegert M, Ivanov I, Pisciotta JM, Logan BE (2013) Treatability studies on different refinery wastewater samples using high-throughput microbial electrolysis cells (MECs). Biores Technol 136: 322-328.

Rotaru AE, Woodard TL, Nevin KP, Lovley DR (2015) Link between capacity for current production and syntrophic growth in *Geobacter* species. Front Microbiol 6: 744.

Rotaru AE, Shrestha PM, Liu F, Shrestha M, Shrestha D, Embree M, Zengler K, Wardman C, Nevn KP, Lovley DR (2013) A new model for electron flow during anaerobic digestion: direct interspecies electron transfer to Methanosaeta for the reduction of carbon dioxide to methane. Energ Environ Sci 15: 51-55.

Rozendal RA, Hamelers HVM, Buisman CJN (2006) Effects of membrane cation transport on ph and microbial fuel cell performance. Environ Sci Technol 40: 5206-5211.

Sharma SCD, Feng C, Li J, Hu A, Wang H, Qin D, Yu CP (2016) Electrochemical characterization of a novel exoelectrogenic bacterium strain SCS5, isolated from a mediator-less microbial fuel cell and phylogenetically related to *Aeromonas jandaei*. Microbes Environ 31:, 213-225.

Sim J, An J, Elbeshbishy E, Ryu H, Lee HS (2015) Characterization and optimization of cathodic conditions for H2O2 synthesis in microbial electrochemical cells. Biores Technol 195: 31-36.

Sleutels THJA, Heijne AT, Buisman CJN, Hamelers HVM (2012) Bioelectrochemical systems: An outlook for practical applications. ChemSusChem 5: 1012-1019.

Sleutels THJA, Lodder R, Hamelers HVM, Buisman CJN (2009) Improved performance of porous bio-anodes in microbial electrolysis cells by enhancing mass and charge transport. Int J Hydrogen Energy 34: 9655-9661.

Tang J, Meadows AL, Keasling JD (2007) A kinetic model describing *Shewanella oneidensis* MR-1 growth, substrate consumption, and product secretion. Biotechnol Bioeng 96: 125-133.

Tenca A, Cusick RD, Schievano A, Oberti R, Logan BE (2013) Evaluation of low cost cathode materials for treatment of industrial and food processing wastewater using microbial electrolysis cells. Int J Hydrogen Energy 38: 1859-1865.

Thygesen A, Marzorati M, Boon N, Thomsen AB, Verstraete W (2011) Upgrading of straw hydrolysate for production of hydrogen and phenols in a microbial electrolysis cell (MEC). Appl Microbiol Biotechnol 89: 855-865.

Torres CI, Kato Marcus A, Rittmann BE (2007) Kinetics of consumption of fermentation products by anode-respiring bacteria. Appl Microbiol Biotechnol 77: 689-697.

Torres CI, Krajmalnik-Brown R, Parameswaran P, Marcus AK, Wanger G, Gorby YA, Rittmann BE (2009) Selecting anode-respiring bacteria based on anode potential: phylogenetic, electrochemical, and microscopic characterization. Environ Sci Technol 43: 9519-9524.

Torres CI, Lee HS, Rittmann BE (2008a) Carbonate species as OH− carriers for decreasing the pH gradient between cathode and anode in biological fuel cells. Environ Sci Technol 42: 8773-8777.

Torres CI, Marcus AK, Lee HS, Parameswaran P, Krajmalnik-Brown R, Rittmann BE (2010) A kinetic perspective on extracellular electron transfer by anode-respiring bacteria. FEMS Microbiol Rev 34: 3-17.

Torres CI, Marcus AK, Parameswaran P, Rittmann BE (2008b) Kinetic experiments for evaluating the nernst− monod model for anode-respiring bacteria (ARB) in a biofilm anode. Environ Sci Technol 42:6593-6597.

Torres CI, Marcus AK, Rittmann BE (2008c) Proton transport inside the biofilm limits electrical current generation by anode-respiring bacteria. Biotechnol Bioeng 100: 872-881.

Velasquez-Orta SB, Head IM, Curtis TP, Scott K, Lloyd JR, von Canstein H (2010) The effect of flavin electron shuttles in microbial fuel cells current production. Appl Microbiol Biotechnol 85:1373-1381.

Wagner RC, Regan JM, Oh SE, Zuo Y, Logan BE (2009) Hydrogen and methane production from swine wastewater using microbial electrolysis cells. Water Res 43:1480-1488.

Wang A, Sun D, Cao G, Wang H, Ren N, Wu WM, Logan BE (2011) Integrated hydrogen production process from cellulose by combining dark fermentation, microbial fuel cells, and a microbial electrolysis cell. Biores Technol 102: 4137-4143.

Xing D, Zuo Y, Cheng S, Regan JM, Logan BE (2008) Electricity generation by *Rhodopseudomonas palustris* DX-1. Environ Sci Technol 42: 4146-4151.

Zhang Y, Zhao YG, Guo L, Gao M (2017) Two-stage pretreatment of excess sludge for electricity generation in microbial fuel cell. Environ Technol: 1-10.

Zhao Z, Li Y, Yu Q, Zhang Y (2018) Ferroferric oxide triggered possible direct interspecies electron transfer between *Syntrophomonas* and *Methanosaeta* to enhance waste activated sludge anaerobic digestion. Bioresource Technology 250: 79-85. Ferroferric oxide triggered possible direct interspecies electron transfer between *Syntrophomonas* and *Methanosaeta* to enhance waste activated sludge anaerobic digestion.

Index